Introduction to Food Engineering

Fourth Edition

Food Science and Technology
International Series

A complete list of books in this series appears at the end of this volume.

Introduction to Food Engineering

Fourth Edition

R. Paul Singh

Department of Biological and Agricultural Engineering and
Department of Food Science and Technology
University of California
Davis, California

Dennis R. Heldman

Heldman Associates
Mason, Ohio

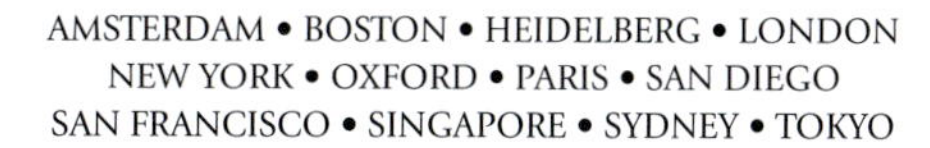
AMSTERDAM • BOSTON • HEIDELBERG • LONDON
NEW YORK • OXFORD • PARIS • SAN DIEGO
SAN FRANCISCO • SINGAPORE • SYDNEY • TOKYO

Academic Press is an imprint of Elsevier

Academic Press is an imprint of Elsevier
30 Corporate Drive, Suite 400, Burlington, MA 01803, USA
525 B Street, Suite 1900, San Diego, California 92101-4495, USA
84 Theobald's Road, London WC1X 8RR, UK

Library of Congress Cataloging-in-Publication Data

APPLICATION SUBMITTED

British Library Cataloguing-in-Publication Data
A catalogue record for this book is available from the British Library.

ISBN: 978-0-12-370900-4

Printed in China

08 09 10 9 8 7 6 5 4 3 2 1

About the Authors

R. Paul Singh and Dennis R. Heldman have teamed up here once again, to produce the fourth edition of Introduction to Food Engineering; a book that has had continuing success since its first publication in 1984. Together, Drs. Singh and Heldman have many years of experience in teaching food engineering courses to students, both undergraduates and graduates; along with Dr. Heldman's experience in the food processing industry, is once again apparent in their approach within this book. The authors' criteria for the careful selection of topics, and the way in which this material is presented, will enable students and faculty to reap the full benefits of this combined wealth of knowledge.

Singh is a distinguished professor of food engineering at the University of California, Davis, where he has been teaching courses on topics in food engineering since 1975. The American Society of Agricultural Engineers (ASAE) awarded him the Young Educator Award in 1986. The Institute of Food Technologists (IFT) awarded him the Samuel Cate Prescott Award for Research in 1982. In 1988, he received the International Award from the IFT, reserved for a member of the Institute who "has made outstanding efforts to promote the international exchange of ideas in the field of food technology." In 1997, he received the Distinguished Food Engineering Award from the Dairy and Food Industry Suppliers Association and ASAE, with a citation recognizing him as a "world class scientist and educator with outstanding scholarly distinction and international service in food engineering." In 2007, ASAE awarded him the Kishida International Award for his worldwide contributions in food engineering education. He was elected a fellow of both IFT and ASAE in 2000 and the International Academy of Food Science and Technology in 2001. He has helped establish food engineering programs in Portugal, Indonesia, Argentina, and India and has lectured extensively on food engineering topics in 40 different nations around the world. Singh has authored, or co-authored, fourteen books and published more than two hundred technical papers. His research program at Davis addresses study of heat and mass transfer in foods during processing using mathematical simulations and seeking sustainability in the food supply chain. In 2008, Singh was elected to the National Academy of Engineers "for innovation and leadership in food engineering research and education." The honor is one of the highest professional distinctions for engineers in the United States.

Currently, Heldman is the Principal of Heldman Associates, a consulting business dedicated to applications of engineering concepts to food processing for educational institutions, industry and government. He is an Adjunct Professor at the University of California-Davis and Professor Emeritus at the University of Missouri. His research interests focus on use of models to predict thermophysical properties of foods and the development of simulation models for processes used in food manufacturing. He has been author or co-author of over 150 research papers and is Co-Editor of the

Handbook of Food Engineering, and Editor of the **Encyclopedia of Agricultural, Food and Biological Engineering** and an **Encyclopedia of Biotechnology in Agriculture and Food** to be published in 2009. Heldman has taught undergraduate and graduate food engineering courses at Michigan State University, University of Missouri and Rutgers, The State University of New Jersey. He has held technical administration positions at the Campbell Soup Company, the National Food Processors Association, and the Weinberg Consulting Group, Inc. He has been recognized for contributions, such as the DFISA-ASAE Food Engineering Award in 1981, the Distinguished Alumni Award from The Ohio State University in 1978, the Young Researcher Award from ASAE in 1974, and served as President of the Institute of food Technologists (IFT) in 2006–07. In addition, Heldman is Fellow in the IFT (1981), the American Society of Agricultural Engineers (1984), and the International Academy of Food Science & Technology (2006).

Foreword

Nine out of ten Food Science students would probably claim the Food Engineering course as the most difficult one in their undergraduate curriculum. Although part of the difficulty may be related to how food engineering is taught, much of the difficulty with food engineering stems from the nature of the material. It's not necessarily that food engineering concepts are more difficult than other food science concepts, but food engineering is based on derivations of equations, and the quantitative manipulation of those equations to solve problems.

From word problems to integral calculus, the skills required to master food engineering concepts are difficult for many Food Science students. However, these concepts are integral to the required competencies for an IFT-approved Food Science program, and are the cornerstone for all of food processing and manufacturing. It is critical that Food Science graduates have a good understanding of engineering principles, both because they are likely to need the concepts during the course of their career but also because they will most certainly need to interact with engineers in an educated manner. Food Science graduates who can use quantitative engineering approaches will stand out from their co-workers in the field.

Fortunately, two of the leading food engineers, Paul Singh and Dennis Heldman, have teamed up to write a textbook that clearly and simply presents the complex engineering material that Food Scientists need to know to be successful. In this fourth edition of a classic Food Engineering textbook, Singh and Heldman have once again improved the book even further. New chapters on process control, food packaging, and process operations like filtration, centrifugation and mixing now supplement the classic chapters on mass and energy balances, thermodynamics, heat transfer and fluid flow. Furthermore, numerous problems have now been solved with MATLAB, an engineering mathematical problem solver, to enhance student's math skills.

A good textbook should clearly and concisely present material needed by the students and at a level appropriate to their backgrounds. With chapters that are broken down into short, manageable sections that promote learning, the easy-to-follow explanations in the 4th Edition of Singh and Heldman are aimed at the perfect level for Food Scientists. Numerous example problems, followed by practice problems, help students test their understanding of the concepts. With fifteen chapters that cover the fundamental aspects of engineering and their practical application to foods, this book is an ideal text for courses in both food engineering and food processing. It will also serve as a useful reference for Food Science graduates throughout their career.

Richard W. Hartel
Professor of Food Engineering
University of Wisconsin-Madison

Preface

The typical curriculum for an undergraduate food science major in the United States and Canada requires an understanding of food engineering concepts. The stated content of this portion of the curriculum is "Engineering principles including mass and energy balances, thermodynamics, fluid flow, and heat and mass transfer". The expectations include an application of these principles to several areas of food processing. Presenting these concepts to students with limited background in mathematics and engineering science presents a significant challenge. Our goal, in this text book, is to provide students, planning to become food science professionals, with sufficient background in engineering concepts to be comfortable when communicating with engineering professionals.

This text book has been developed specifically for use in undergraduate food engineering courses taken by students pursuing a four-year degree program in food science. The topics presented have been selected to illustrate applications of engineering during the handling, processing, storage, packaging and distribution of food products. Most of the topics include some descriptive background about a process, fundamental engineering concepts and example problems. The approach is intended to assist the student in appreciating the applications of the concepts, while gaining an understanding of problem-solving approaches as well as gaining confidence with the concepts.

The scope of the book ranges from basic engineering principles, based on fundamental physics, to several applications in food processing. Within the first four chapters, the concepts of mass and energy balance, thermodynamics, fluid flow and heat transfer are introduced. A significant addition to this section of the fourth edition is an introduction to the concepts of process control. The next four chapters include applications of thermodynamics and heat transfer to preservation processes, refrigeration, freezing processes and evaporation processes used in concentration of liquid foods. Following the chapters devoted to the concepts of psychrometrics and mass transfer, several chapters are used to present applications of these concepts to membrane separation processes, dehydration processes, extrusion processes and packaging. Finally, a new chapter in this edition is devoted to supplemental processes, including filtration, centrifugation and mixing.

Most features of the first three editions of this book are included in this fourth edition. Chapters include modest amounts of descriptive material to assist the student in appreciating the process applications. Although equations are developed from fundamental concepts, the equations are used to illustrate the solution to practical problems. Most chapters contain many example problems to illustrate various concepts and applications, and several examples are presented in spreadsheet program format. At the end of most chapters, lists of problems are provided for the student to use in gaining confidence with problem-solving skills, and the more difficult problems are identified.

The focus of additions to the fourth edition has been on evolving processes and related information. Chapter 2 has been expanded to include information on properties of dry food powders and applications during handling of these products. The new material on process controls in Chapter 3 will assist students in understanding the systems used to operate and control food manufacturing operations. Numerous revisions and additions in the preservation process chapter provide information on applications of evolving technologies for food preservation. Completely new chapters have been included on the subjects of supplemental processes (filtration, centrifugation, mixing) and extrusion processes. Finally, a separate chapter has been devoted to food packaging, to emphasize applications of engineering concepts in selection of packaging materials and prediction of product shelf-life.

The primary users of this book are the faculty involved in teaching students pursuing an undergraduate degree in Food Science. The approaches used to present the concepts and applications are based on our own combined teaching experiences. Faculty members are encouraged to select chapters and associated materials to meet the specific objectives of the course being taught. The descriptive information, concepts and problems have been organized to provide maximum flexibility in teaching. The organization of the information in the book does serve as a study guide for students. Some students may be able to solve the problems at the end of chapters after independent study of the concepts presented within a given chapter. For the purposes to enhance learning, many illustrations in the book are available in animated form at www.rpaulsingh.com. This website also contains most of the solved examples in an electronic form that allow "what-if" analysis.

The topics presented in this book can be easily organized into a two-course sequence. The focus of the first course would be on engineering concepts and include information from Chapters 1 through 4, and the second course would focus on applications using Chapters 5 to 8. Alternatively, Chapters 9 and 10 could be added to the course on fundamentals, and the applications from Chapters 11 through 15 would be used in the second course. The chapters on applications provide an ideal basis for a process-based capstone course.

A new feature in this edition is the inclusion of several problems that require the use of MATLAB®. We are indebted to Professor Thomas R. Rumsey for generously sharing several of these problems that he has used in his own teaching. We thank Ms. Barbara Meierhenry for her valuable assistance in editing the original manuscript.

We appreciate the many recommendations from colleagues, and the encouragement from students, as received over a period of nearly 25 years. All of these comments and suggestions have been valuable, and have made the continuous development of this book a rewarding experience. We will continue to respond to communications from faculty members and students as the concepts and applications of food engineering continue to evolve.

R. Paul Singh
Dennis R. Heldman

Contents

Chapter 1

Introduction

Physics, chemistry, and mathematics are essential in gaining an understanding of the principles that govern most of the unit operations commonly found in the food industry. For example, if a food engineer is asked to design a food process that involves heating and cooling, then he or she must be well aware of the physical principles that govern heat transfer. The engineer's work is often expected to be quantitative, and therefore the ability to use mathematics is essential. Foods undergo changes as a result of processing; such changes may be physical, chemical, enzymatic, or microbiological. It is often necessary to know the kinetics of chemical changes that occur during processing. Such quantitative knowledge is a prerequisite to the design and analysis of food processes. It is expected that prior to studying food engineering principles, the student will have taken basic courses in mathematics, chemistry, and physics. In this chapter, we review some selected physical and chemical concepts that are important in food engineering.

1.1 DIMENSIONS

A physical entity, which can be observed and/or measured, is defined qualitatively by a dimension. For example, time, length, area, volume, mass, force, temperature, and energy are all considered dimensions. The quantitative magnitude of a dimension is expressed by a unit; a unit of length may be measured as a meter, centimeter, or millimeter.

Primary dimensions, such as length, time, temperature, and mass, express a physical entity. Secondary dimensions involve a combination of primary dimensions (e.g., volume is length cubed; velocity is distance divided by time).

All icons in this chapter refer to the author's web site, which is independently owned and operated. Academic Press is not responsible for the content or operation of the author's web site. Please direct your web site comments and questions to the author: Professor R. Paul Singh, Department of Biological and Agricultural Engineering, University of California, Davis, CA 95616, USA. Email: rps@rpaulsingh.com.

Equations must be dimensionally consistent. Thus, if the dimension of the left-hand side of an equation is "length," the dimension of the right-hand side must also be "length"; otherwise, the equation is incorrect. This is a good method to check the accuracy of equations. In solving numerical problems, it is also useful to write the units of each dimensional quantity within the equations. This practice is helpful to avoid mistakes in calculations.

1.2 ENGINEERING UNITS

Physical quantities are measured using a wide variety of unit systems. The most common systems include the Imperial (English) system; the centimeter, gram, second (cgs) system; and the meter, kilogram, second (mks) system. However, use of these systems, entailing myriad symbols to designate units, has often caused considerable confusion. International organizations have attempted to standardize unit systems, symbols, and their quantities. As a result of international agreements, the *Système International d'Unités*, or the SI units, have emerged. The SI units consist of seven base units, two supplementary units, and a series of derived units, as described next.

1.2.1 Base Units

The SI system is based on a choice of seven well-defined units, which by convention are regarded as dimensionally independent. The definitions of these seven base units are as follows:

1. Unit of length (meter): The *meter* (m) is the length equal to 1,650,763.73 wavelengths in vacuum of the radiation corresponding to the transition between the levels $2p_{10}$ and $5d_5$ of the krypton-86 atom.

2. Unit of mass (kilogram): The *kilogram* (kg) is equal to the mass of the international prototype of the kilogram. (The international prototype of the kilogram is a particular cylinder of platinum-iridium alloy, which is preserved in a vault at Sèvres, France, by the International Bureau of Weights and Measures.)

3. Unit of time (second): The *second* (s) is the duration of 9,192,631,770 periods of radiation corresponding to the transition between the two hyperfine levels of the ground state of the cesium-133 atom.

4. Unit of electric current (ampere): The *ampere* (A) is the constant current that, if maintained in two straight parallel conductors

Table 1.1 SI Base Units

Measurable attribute of phenomena or matter	Name	Symbol
Length	meter	m
Mass	kilogram	kg
Time	second	s
Electric current	ampere	A
Thermodynamic temperature	kelvin	K
Amount of substance	mole	mol
Luminous intensity	candela	cd

of infinite length, of negligible circular cross-section, and placed 1 m apart in vacuum, would produce between those conductors a force equal to 2×10^{-7} newton per meter length.

5. Unit of thermodynamic temperature (Kelvin): The *Kelvin* (K) is the fraction 1/273.16 of the thermodynamic temperature of the triple point of water.

6. Unit of amount of substance (mole): The *mole* (mol) is the amount of substance of a system that contains as many elementary entities as there are atoms in 0.012 kg of carbon 12.

7. Unit of luminous intensity (candela): The *candela* (cd) is the luminous intensity, in the perpendicular direction, of a surface of 1/600,000 m^2 of a blackbody at the temperature of freezing platinum under a pressure of 101,325 newton/m^2.

These base units, along with their symbols, are summarized in Table 1.1.

1.2.2 Derived Units

Derived units are algebraic combinations of base units expressed by means of multiplication and division. For simplicity, derived units often carry special names and symbols that may be used to obtain other derived units. Definitions of some commonly used derived units are as follows:

1. Newton (N): The *newton* is the force that gives to a mass of 1 kg an acceleration of 1 m/s^2.

2. Joule (J): The *joule* is the work done when due to force of 1 N the point of application is displaced by a distance of 1 m in the direction of the force.
3. Watt (W): The *watt* is the power that gives rise to the production of energy at the rate of 1 J/s.
4. Volt (V): The *volt* is the difference of electric potential between two points of a conducting wire carrying a constant current of 1 A, when the power dissipated between these points is equal to 1 W.
5. Ohm (Ω): The *ohm* is the electric resistance between two points of a conductor when a constant difference of potential of 1 V, applied between theses two points, produces in this conductor a current of 1 A, when this conductor is not being the source of any electromotive force.
6. Coulomb (C): The *coulomb* is the quantity of electricity transported in 1 s by a current of 1 A.

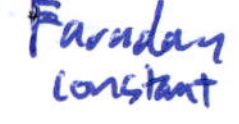

7. Farad (F): The *farad* is the capacitance of a capacitor, between the plates of which there appears a difference of potential of 1 V when it is charged by a quantity of electricity equal to 1 C.
8. Henry (H): The *henry* is the inductance of a closed circuit in which an electromotive force of 1 V is produced when the electric current in the circuit varies uniformly at a rate of 1 A/s.
9. Weber (Wb): The *weber* is the magnetic flux that, linking a circuit of one turn, produces in it an electromotive force of 1 V as it is reduced to zero at a uniform rate in 1 s.
10. Lumen (lm): The *lumen* is the luminous flux emitted in a point solid angle of 1 steradian by a uniform point source having an intensity of 1 cd.

Examples of SI-derived units expressed in terms of base units, SI-derived units with special names, and SI-derived units expressed by means of special names are given in Tables 1.2, 1.3, and 1.4, respectively.

1.2.3 Supplementary Units

This class of units contains two purely geometric units, which may be regarded either as base units or as derived units.

1. Unit of plane angle (radian): The *radian* (rad) is the plane angle between two radii of a circle that cut off on the circumference an arc equal in length to the radius.

Table 1.2 Examples of SI-Derived Units Expressed in Terms of Base Units

Quantity	SI Unit	
	Name	**Symbol**
Area	square meter	m^2
Volume	cubic meter	m^3
Speed, velocity	meter per second	m/s
Acceleration	meter per second squared	m/s^2
Density, mass density	kilogram per cubic meter	kg/m^3
Current density	ampere per square meter	A/m^2
Magnetic field strength	ampere per meter	A/m
Concentration (of amount of substance)	mole per cubic meter	mol/m^3
Specific volume	cubic meter per kilogram	m^3/kg
Luminance	candela per square meter	cd/m^2

Table 1.3 Examples of SI-Derived Units with Special Names

Quantity	SI Unit			
	Name	**Symbol**	**Expression in terms of other units**	**Expression in terms of SI base units**
Frequency	hertz	Hz		s^{-1}
Force	newton	N		$m\ kg\ s^{-2}$
Pressure, stress	pascal	Pa	N/m^2	$m^{-1}\ kg\ s^{-2}$
Energy, work, quantity of heat	joule	J	N m	$m^2\ kg\ s^{-2}$
Power, radiant flux	watt	W	J/s	$m^2\ kg\ s^{-3}$
Quantity of electricity, electric charge	coulomb	C		s A
Electric potential, potential difference, electromotive force	volt	V	W/A	$m^2\ kg\ s^{-3}\ A^{-1}$
Capacitance	farad	F	C/V	$m^{-2}\ kg^{-1}\ s^4\ A^2$
Electric resistance	ohm	Ω	V/A	$m^2\ kg\ s^{-3}\ A^{-2}$
Conductance	siemens	S	A/V	$m^{-2}\ kg^{-1}\ s^3\ A^2$
Celsius temperature	degree Celsius	°C		K
Luminous flux	lumen	lm		cd sr
Illuminance	lux	lx	lm/m^2	m^{-2} cd sr

Table 1.4 Examples of SI-Derived Units Expressed by Means of Special Names

Quantity	SI Unit		Expression in terms of SI base units
	Name	Symbol	
Dynamic viscosity	pascal second	Pa s	m^{-1} kg s^{-1}
Moment of force	newton meter	N m	m^2 kg s^{-2}
Surface tension	newton per meter	N/m	kg s^{-2}
Power density, heat flux density, irradiance	watt per square meter	W/m^2	kg s^{-3}
Heat capacity, entropy	joule per kelvin	J/K	m^2 kg s^{-2} K^{-1}
Specific heat capacity	joule per kilogram kelvin	J/(kg K)	m^2 s^{-2} K^{-1}
Specific energy	joule per kilogram	J/kg	m^2 s^{-2}
Thermal conductivity	watt per meter kelvin	W/(m K)	m kg s^{-3} K^{-1}
Energy density	joule per cubic meter	J/m^3	m^{-1} kg s^{-2}
Electric field strength	volt per meter	V/m	m kg s^{-3} A^{-1}
Electric charge density	coulomb per cubic meter	C/m^3	m^{-3} s A
Electric flux density	coulomb per square meter	C/m^2	m^{-2} s A

Table 1.5 SI Supplementary Units

Quantity	SI Unit	
	Name	Symbol
Plane angle	radian	rad
Solid angle	steradian	sr

2. Unit of solid angle (steradian): The *steradian* (sr) is the solid angle that, having its vertex in the center of a sphere, cuts off an area of the surface of the sphere equal to that of a square with sides of length equal to the radius of the sphere.

The supplementary units are summarized in Table 1.5.

Example 1.1

Determine the following unit conversions to SI units:

a. a density value of 60 lb_m/ft^3 to kg/m^3
b. an energy value of 1.7×10^3 Btu to kJ
c. an enthalpy value of 2475 Btu/lb_m to kJ/kg
d. a pressure value of 14.69 psig to kPa
e. a viscosity value of 20 cp to Pa s

Solution

We will use conversion factors for each unit separately from Table A.1.2.

a. *Although a composite conversion factor for density,* $1\ lb_m/ft^3 = 16.0185\ kg/m^3$, *is available in Table A.1.2, we will first convert units of each dimension separately. Since*

$$1\ lb_m = 0.45359\ kg$$

$$1\ ft = 0.3048\ m$$

Thus,

$$(60\ lb_m/ft^3)(0.45359\ kg/lb_m)\left(\frac{1}{0.3048}\ m/ft\right)^3$$
$$= 961.1\ kg/m^3$$

An alternative solution involves the direct use of the conversion factor for density,

$$\frac{(60\ lb_m/ft^3)(16.0185\ kg/m^3)}{(1\ lb_m/ft^3)} = 961.1\ kg/m^3$$

b. *For energy*

$$1\ Btu = 1.055\ kJ$$

Thus,

$$\frac{(1.7 \times 10^3\ Btu)(1.055\ kJ)}{(1\ Btu)} = 1.8 \times 10^3\ kJ$$

c. *For enthalpy, the conversion units for each dimension are*

$$1\ Btu = 1.055\ kJ$$

$$1\ lb_m = 0.45359\ kg$$

Thus,

$$(2475\ Btu/lb_m)(1.055\ kJ/Btu)\left(\frac{1}{0.45359\ kg/lb_m}\right)$$
$$= 5757\ kJ/kg$$

Alternately, using the composite conversion factor for enthalpy of

$$1\,Btu/lb_m = 2.3258\,kJ/kg$$

$$\frac{(2475\,Btu/lb_m)(2.3258\,kJ/kg)}{(1\,Btu/lb_m)} = 5756\,kJ/kg$$

d. *For pressure*

$$psia = psig + 14.69$$

The gauge pressure, 14.69 psig, is first converted to the absolute pressure, psia (see Section 1.9 for more discussion on gauge and absolute pressures).

$$14.69\ psig + 14.69 = 29.38\ psia$$

The unit conversions for each dimension are

$$1\,lb = 4.4483\,N$$

$$1\,in = 2.54 \times 10^{-2}\,m$$

$$1\,Pa = 1\,N/m^2$$

Thus,

$$(29.28\,lb/in^2)(4.4482\,N/lb)\left(\frac{1}{2.54 \times 10^{-2}\,m/in}\right)^2\left(\frac{1\,Pa}{1\,N/m^2}\right)$$

$$= 201877\,Pa$$

$$= 201.88\,kPa$$

Alternatively, since

$$1\,psia = 6.895\,kPa$$

$$\frac{(29.28\,psia)(6.895\,kPa)}{(1\,Psia)} = 201.88\,kPa$$

e. *For viscosity*

$$1\,cp = 10^{-3}\,Pa\,s$$

Thus,

$$\frac{(20\,cp)(10^{-3}\,Pa\,s)}{(1\,cp)} = 2 \times 10^{-2}\,Pa\,s$$

Example 1.2

Starting with Newton's second law of motion, determine units of force and weight in SI and English units.

Solution

a. *Force*

Newton's second law of motion states that force is directly proportional to mass and acceleration. Thus,

$$F \propto ma$$

Using a constant of proportionality ***k****,*

$$F = kma$$

where in ***SI units****,*

$$k = 1 \frac{N}{kg\ m/s^2}$$

Thus,

$$F = 1\left(\frac{N}{kg\ m/s^2}\right)(kg)(m/s^2)$$

$$F = 1N$$

In ***English units*** *the constant* ***k*** *is defined as*

$$k = \frac{1}{32.17}\frac{lb_f}{lb_m\ ft/s^2}$$

More commonly, another constant g_c *is used where*

$$g_c = 1/k = 32.17\left(\frac{lb_m}{lb_f}\right)\left(\frac{ft}{s^2}\right)$$

Thus

$$F = \frac{ma}{g_c}$$

or

$$F = \frac{1}{32.17}\left(\frac{lb_f}{lb_m\ ft/s^2}\right)(lb_m)(ft/s^2)$$

$$F = \frac{1}{32.17} lb_f$$

b. *Weight*

*Weight **W'** is the force exerted by the earth's gravitational force on an object. Weight of 1 kg mass can be calculated as*

*In **SI units**,*

$$\begin{aligned} W' &= kmg \\ &= \left(1\frac{N}{kg\ m/s^2}\right)(1\ kg)\left(9.81\frac{m}{s^2}\right) \\ &= 9.81 N \end{aligned}$$

*In **English units**,*

$$\begin{aligned} W' &= kmg \\ &= \frac{1}{32.17}\left(\frac{lb_f}{lb_m\ ft/s^2}\right)(1\ lb_m)(32.17\ ft/s^2) \\ &= 1\ lb_f \end{aligned}$$

1.3 SYSTEM

A **system** is any region prescribed in space or a finite quantity of matter enclosed by a **boundary**, real or imaginary. The boundary of a system can be real, such as the walls of a tank, or it can be an imaginary surface that encloses the system. Furthermore, the boundary may be stationary or moveable. For example, in Figure 1.1, the system boundary encloses a tank, piping, and a valve. If our analysis had concerned only the valve, we could have drawn the system boundary just around the valve.

The composition of a system is described by the components present inside the system boundary. Once we choose the boundaries of a system, then everything outside the boundary becomes the **surroundings**. The analysis of a given problem is often simplified by how we select a system and its boundaries; therefore, proper care must be exercised in so doing.

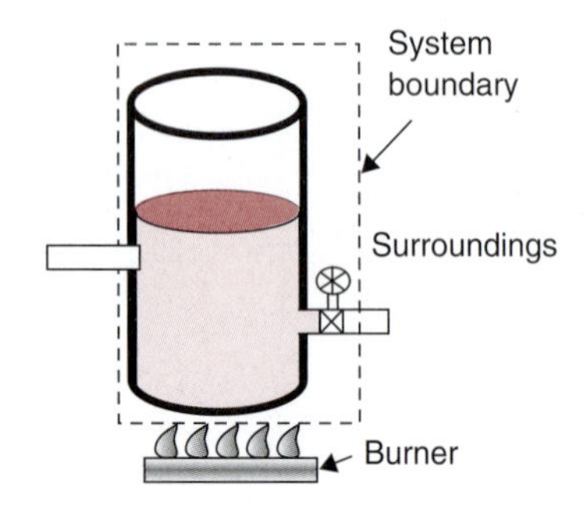

■ Figure 1.1 A system containing a tank with a discharge pipe and valve.

A system can be either **open** or **closed**. In a closed system, the boundary of the system is impervious to flow of mass. In other words, a closed system does not exchange mass with its surroundings. A closed system may exchange heat and work with its surroundings, which may result in a change in energy, volume, or other properties of the system, but its mass remains constant. For example, a system boundary that contains a section of the wall of a tank (Fig. 1.2) is impervious to the flow of matter, and thus in this case we are dealing with a closed system. In an open system (also called a control volume), both heat and mass can flow into or out of a system boundary (also called control surface). As shown in Figure 1.1, heat and water flow across the system boundary.

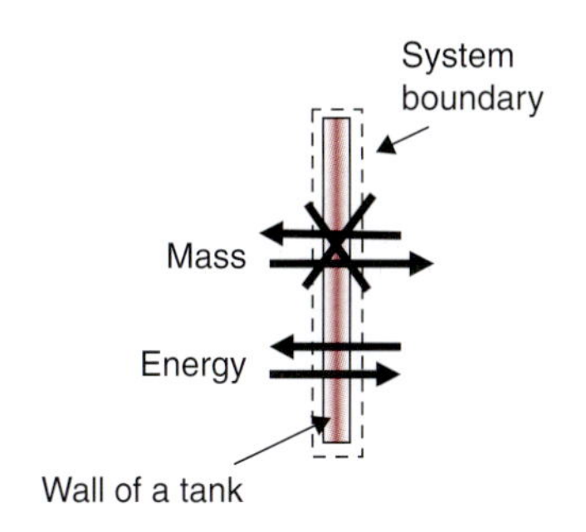

Figure 1.2 A closed system containing the wall.

Depending on the problem at hand, the system selected may be as simple as just the wall of a tank, or several parts, such as a tank, valve, and piping as we considered in Figure 1.1. As we will see later in Section 1.14, a system boundary may even enclose an entire food processing plant.

When a system does not exchange mass, heat, or work with its surroundings, it is called an **isolated** system. An isolated system has no effect on its surroundings. For example, if we carry out a chemical reaction in an insulated vessel such that no exchange of heat takes place with the surroundings, and if its volume remains constant, then we may consider that process to be occurring in an isolated system.

If either in a closed or an open system, no exchange of heat takes place with the surroundings, it is called an **adiabatic** system. Although we are unlikely to achieve perfect insulation, we may be able to approach near adiabatic conditions in certain situations. When a process occurs at a constant temperature, often with an exchange of heat with the surroundings, then we have an **isothermal** system.

Note that the system boundaries do not have to be rigid; in fact, they can be flexible and expand or contract during a process. An example of a piston and a cylinder illustrates the moving boundaries of a system. As shown in Figure 1.3, consider a system boundary that encloses only the gas. The piston and the cylinder therefore are surrounding the system. The system boundary in this case is flexible. When the cylinder moves to the right, the system boundary expands; when it moves to the left, it contracts. This is an example of a closed system, because no transfer of mass (gas) takes place across the system boundary. As an extension of this example, we can also locate a heater under the piston; because of heat transfer across the boundary, the gas will expand and the piston will move to the right.

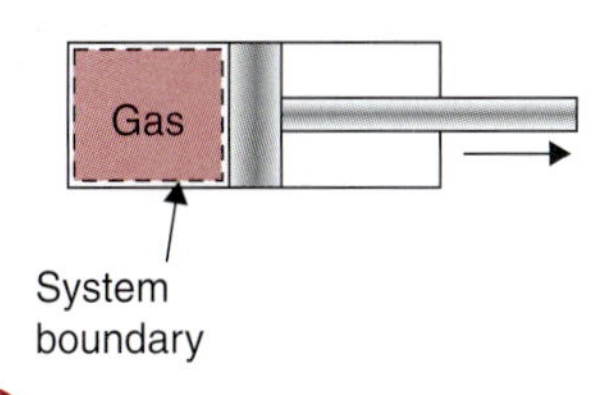

W

Figure 1.3 A system with a flexible boundary.

1.4 STATE OF A SYSTEM

Next, let us consider the **state** of a system, which refers to the equilibrium condition of the system. When a system is at equilibrium, we

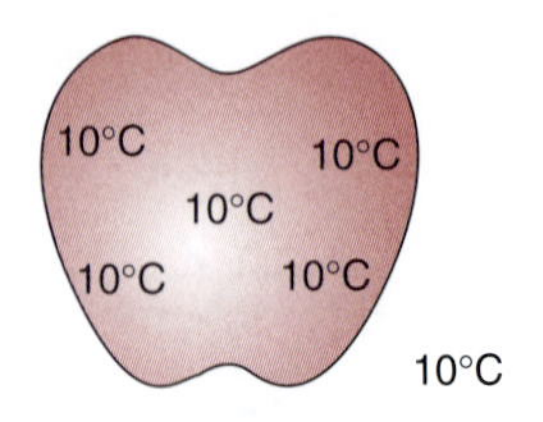

■ **Figure 1.4** An apple in a thermal equilibrium with a uniform internal temperature of 10°C.

can either measure its properties or calculate them to obtain a complete description of the state of the system. At equilibrium, all properties of a system will have fixed values. If any property value changes, then the state of the system will change. Consider an apple with a uniform internal temperature of 10°C (Fig. 1.4); it is in **thermal equilibrium**. Similarly, if the pressure in an object is the same throughout, it is in **mechanical equilibrium**. Although the pressure may vary due to a gravity-induced elevation within the system, this variation in pressure is often ignored in thermodynamic systems. When we have two phases, such as with solid crystals in a saturated liquid, and their mass remains constant, we have **phase equilibrium**. Furthermore, in situations when the chemical composition of a material remains constant with time, we have **chemical equilibrium**. This implies that there is no chemical reaction taking place. For a system to be considered in equilibrium, we must have all preceding conditions of equilibrium satisfied.

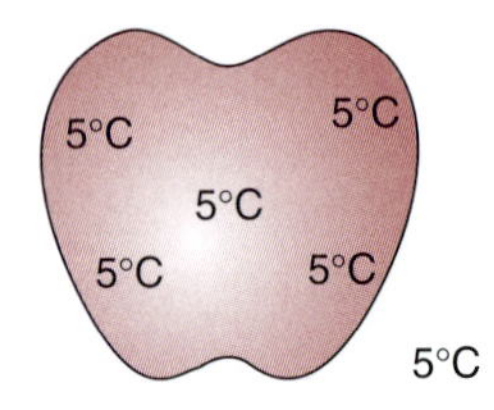

■ **Figure 1.5** The final state of an apple when placed in a 5°C environment.

When a system undergoes a change of state, then a **process** is said to have taken place. The **path** of the process may involve many different states. A complete description of a process involves initial, intermediate, and final states along with any interactions with the surroundings. For example, when the apple shown in Figure 1.4 is placed in a 5°C environment, it will subsequently attain a final state at a uniform internal temperature of 5°C (Fig. 1.5). The apple in this example went through a cooling process that caused a change in state. In this case, its temperature was initially uniform at 10°C but was changed to a final uniform temperature of 5°C. The path of the process is shown in Figure 1.6.

The previous example of the apple illustrates that we can always describe the state of any system by its **properties**. To fix the state of a system, we specify the values of its properties.

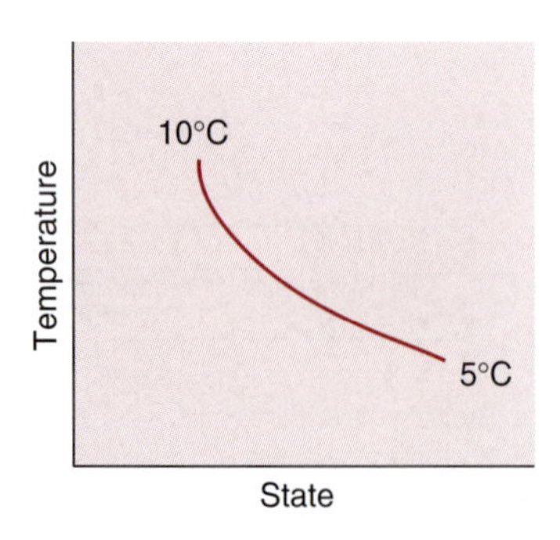

■ **Figure 1.6** A path of a process to cool an apple from 10°C to 5°C.

Properties are those observable characteristics, such as pressure, temperature, or volume, that define the equilibrium state of a thermodynamic system. Properties do not depend on how the state of a system is attained; they are only functions of the state of a system. Therefore, properties are independent of the path by which a system reaches a certain state. We can categorize properties as extensive and intensive.

1.4.1 Extensive Properties

The value of an extensive property depends on the extent or the size of a system. For example, mass, length, volume, and energy depend on the size of a given system. These properties are additive; therefore, an extensive property of a system is the sum of respective partial property values of the system components. We can determine if a property is extensive by simply doubling the size of the system; if the property value doubles, then it is an extensive property.

1.4.2 Intensive Properties

Intensive properties do not depend on the size of a system. Examples include temperature, pressure, and density. For a homogeneous system, we can often obtain an intensive property by dividing two extensive properties. For example, mass divided by volume, both extensive properties, gives us density, which is an intensive property.

There are also specific properties of a system. Specific properties are expressed per unit mass. Thus, specific volume is volume/mass, and specific energy is energy/mass.

1.5 DENSITY

Density is defined as mass per unit volume, with dimensions (mass)/(length)^3. The SI unit for density is kg/m^3. Density is an indication of how matter is composed in a body. Materials with more compact molecular arrangements have higher densities. The values of density for various metals and nonmetals are given in Appendix A.3. Density of a given substance may be divided by density of water at the same temperature to obtain specific gravity.

There are three types of densities for foods: solid density, particle density, and bulk density. The values of these different types of densities depend on how the pore spaces present in a food material are considered.

If the pore spaces are disregarded, the solid density of most food particles (Table 1.6) is 1400–1600 kg/m^3, except for high-fat or high-salt foods (Peleg, 1983).

Particle density accounts for the presence of internal pores in the food particles. This density is defined as a ratio of the actual mass of a particle to its actual volume.

Bulk density is defined as the mass of particles occupied by a unit volume of bed. Typical values of bulk densities for food materials are given in Table 1.7. This measurement accounts for the void space between the particles. The void space in food materials can be

Table 1.6 Solid Densities of Major Ingredients of Foods

Ingredient	kg/m^3	Ingredient	kg/m^3
Glucose	1560	Fat	900–950
Sucrose	1590	Salt	2160
Starch	1500	Citric acid	1540
Cellulose	1270–1610	Water	1000
Protein (globular)	~1400		

Source: Peleg (1983)

Table 1.7 Bulk Density of Selected Food Materials

Material	Bulk density (kg/m^3)
Beans, cocoa	1073
Beans, soy, whole	800
Coconut, shredded	320–352
Coffee beans, green	673
Coffee, ground	400
Coffee, roasted beans	368
Corn, ear	448
Corn, shelled	720
Milk, whole dried	320
Mustard seed	720
Peanuts, hulled	480–720
Peas, dried	800
Rapeseed	770
Rice, clean	770
Rice, hulled	320
Sugar, granulated	800
Wheat	770

described by determining the **porosity**, which is expressed as the volume not occupied by the solid material.

Thus,

$$\text{Porosity} = 1 - \frac{\text{Bulk density}}{\text{Solid density}} \tag{1.1}$$

The interparticle porosity may be defined as follows:

$$\text{Interparticle porosity} = 1 - \frac{\text{Bulk density}}{\text{Particle density}} \tag{1.2}$$

Relationships have been developed to determine density based on experimental data. For example, for skim milk

$$\rho = 1036.6 - 0.146T + 0.0023T^2 - 0.00016T^3 \tag{1.3}$$

where T is temperature in degrees Celsius.

1.6 CONCENTRATION

Concentration is a measure of the amount of substance contained in a unit volume. It may be expressed as weight per unit weight, or weight per unit volume. Normally, concentration is given in percentage when weight per unit weight measurement is used. Thus, a food containing 20% fat will contain 20 g of fat in every 100 g of food. Concentration values are also expressed as mass per unit volume—for example, mass of a solute dissolved in a unit volume of the solution.

Another term used to express concentration is **molarity**, or molar concentration. Molarity is the concentration of solution in grams per liter divided by the molecular weight of the solute. To express these units in a dimensionless form, **mole fraction** may be used; this is the ratio of the number of moles of a substance divided by the total number of moles in the system.

Thus, for a solution containing two components, A and B, with number of moles n_A and n_B, respectively, the mole fraction of A, X_A, is

$$X_A = \frac{n_A}{n_A + n_B} \tag{1.4}$$

Concentration is sometimes expressed by **molality**. The molality of a component A in a solution is defined as the amount of a component per unit mass of some other component chosen as the solvent. The SI unit for molality is mole per kilogram.

A relationship between molality, M'_A, and mole fraction, X_A, for a solution of two components, in which the molecular weight of solvent B is M_B, is

$$X_A = \frac{M'_A}{M'_A + \frac{1000}{M_B}} \tag{1.5}$$

Both molality and mole fraction are independent of temperature.

Example 1.3

Develop a spreadsheet on a computer to calculate concentration units for a sugar solution. The sugar solution is prepared by dissolving 10 kg of sucrose in 90 kg of water. The density of the solution is 1040 kg/m^3. Determine

- **a.** concentration, weight per unit weight
- **b.** concentration, weight per unit volume
- **c.** °Brix
- **d.** molarity
- **e.** mole fraction
- **f.** molality
- **g.** Using the spreadsheet, recalculate (a) to (f) if (1) the sucrose solution contains 20 kg of sucrose in 80 kg of water, and density of the solution is 1083 kg/m^3; (2) the sucrose solution contains 30 kg of sucrose in 70 kg of water, and density of the solution is 1129 kg/m^3.

Solution

1. *The spreadsheet is written using Excel™, as shown in Figure E1.1.*
2. *The results from the spreadsheet calculation are shown in Figure E1.2.*
3. *Once the spreadsheet is prepared according to step (1), the given values are easily changed to calculate all other unknowns.*

	A	B
1	**Given**	
2	Amount of sucrose	10
3	Amount of water	90
4	Density of solution	1040
5		
6	Volume of solution	=(B2+B3)/B4
7	Concentration w/w	=B2/(B2+B3)
8	Concentration w/v	=B2/B6
9	Brix	=B2/(B2+B3)*100
10	Molarity	=B8/342
11	Mole fraction	=(B2/342)/(B3/18+B2/342)
12	Molality	=(B2*1000)/(B3*342)

■ **Figure E1.1** Spreadsheet for calculation of sugar solution concentration in Example 1.3.

	A	B	C	D	E
1	**Given**				**Units**
2	Amount of sucrose	10	20	30	kg
3	Amount of water	90	80	70	kg
4	Density of solution	1040	1083	1129	kg/m^3
5					
6	Volume of solution	0.0962	0.0923	0.0886	m^3
7	Concentration w/w	0.1	0.2	0.3	kg solute/kg solution
8	Concentration w/v	104	216.6	338.7	kg solute/m^3 solution
9	Brix	10	20	30	(kg solute/kg solution)*100
10	Molarity	0.30	0.63	0.99	mole solute/liter of solution
11	Mole fraction	0.0058	0.0130	0.0221	
12	Molality	0.325	0.731	1.253	mole solute/liter of solution

■ **Figure E1.2** Results of the spreadsheet calculation in Example 1.3.

1.7 MOISTURE CONTENT

Moisture content expresses the amount of water present in a moist sample. Two bases are widely used to express moisture content; namely, moisture content wet basis and moisture content dry basis.

Moisture content wet basis (MC_{wb}) is the amount of water per unit mass of moist (or wet) sample.

Thus,

$$MC_{wb} = \frac{\text{mass of water}}{\text{mass of moist sample}} \tag{1.6}$$

Moisture content dry basis (MC_{db}) is the amount of water per unit mass of dry solids (bone dry) present in the sample.

Thus,

$$MC_{db} = \frac{\text{mass of water}}{\text{mass of dry solids}} \tag{1.7}$$

A relationship between MC_{wb} and MC_{db} may be developed as follows:

$$MC_{wb} = \frac{\text{mass of water}}{\text{mass of moist sample}} \tag{1.8}$$

$$MC_{wb} = \frac{\text{mass of water}}{\text{mass of water} + \text{mass of dry solids}} \tag{1.9}$$

Divide both numerator and denominator of Equation (1.9) with mass of dry solids:

$$MC_{wb} = \frac{\text{mass of water/mass of dry solids}}{\dfrac{\text{mass of water}}{\text{mass of dry solids}} + 1} \tag{1.10}$$

$$MC_{wb} = \frac{MC_{db}}{MC_{db} + 1} \tag{1.11}$$

This relationship is useful to calculate MC_{wb} when MC_{db} is known. Similarly, if MC_{wb} is known, then MC_{db} may be calculated from the following equation:

$$MC_{db} = \frac{MC_{wb}}{1 - MC_{wb}} \tag{1.12}$$

The moisture content values in the preceding equations are expressed in fractions. Note that moisture content dry basis may have values greater than 100%, since the amount of water present in a sample may be greater than the amount of dry solids present.

Example 1.4 Convert a moisture content of 85% wet basis to moisture content dry basis.

Solution

a. *$MC_{wb} = 85\%$*

b. *In fractional notation, $MC_{wb} = 0.85$*

c. *From equation,*

$$MC_{db} = \frac{MC_{wb}}{1 - MC_{wb}}$$

$$= \frac{0.85}{1 - 0.85}$$

$$= 5.67$$

or

$$MC_{db} = 567\%$$

Example 1.5

Develop a table of conversions from moisture content wet basis to moisture content dry basis between 0% MC_{wb} to 90% MC_{wb} in steps of 10%.

Solution

a. *Since repetitive computations are involved, a spreadsheet is prepared as follows.*

b. *In an Excel spreadsheet (Fig. E1.3), enter 0 to 90 in steps of 10 in column A.*

c. *Enter the formula given in Equation (1.12) in cell B2, modified to account for percent values and following spreadsheet notation,*

$$MC_{db} = A2/(100 - A2) * 100$$

d. *Copy cell B2 into cells B3 to B11.*

e. *The output is obtained as shown on the spreadsheet in Figure E1.4.*

f. *A plot of the values in columns A and B may be obtained using the chart command of Excel. This plot (Fig. E1.5) is useful in converting moisture content values from one basis to another.*

	A	B
1	Moisture content (wb)	Moisture content (db)
2	0	=A2/(100−A2)*100
3	10	=A3/(100−A3)*100
4	20	=A4/(100−A4)*100
5	30	=A5/(100−A5)*100
6	40	=A6/(100−A6)*100
7	50	=A7/(100−A7)*100
8	60	=A8/(100−A8)*100
9	70	=A9/(100−A9)*100
10	80	=A10/(100−A10)*100
11	90	=A11/(100−A11)*100

Figure E1.3 Spreadsheet for converting moisture content wet basis to moisture content dry basis in Example 1.5.

■ **Figure E1.4** Results of the spreadsheet calculation in Example 1.5.

	A	B
1	Moisture content (wb)	Moisture content (db)
2	0	0.00
3	10	11.11
4	20	25.00
5	30	42.86
6	40	66.67
7	50	100.00
8	60	150.00
9	70	233.33
10	80	400.00
11	90	900.00

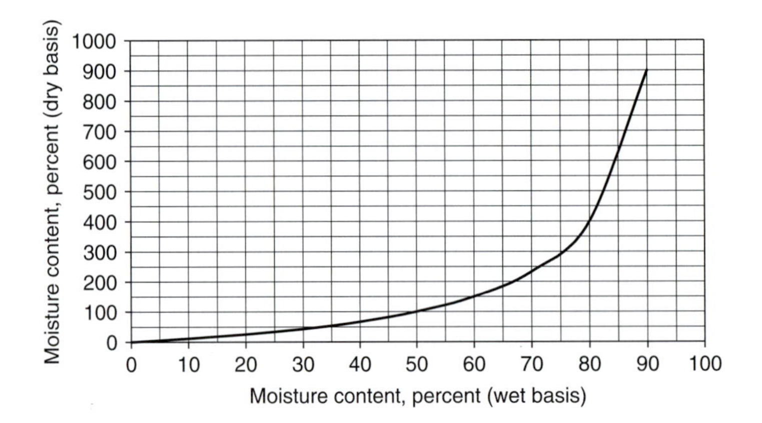

■ **Figure E1.5** Plot of moisture content wet basis versus moisture content dry basis.

1.8 TEMPERATURE

Temperature is one of those properties that defies a precise scientific definition. We generally perceive temperature as a measure of our physiological response to "hotness" or "coldness." However, physiological response is subjective, and it does not provide us with an objective measure. For example, holding a block of steel at 40°C gives a much colder sensation than holding a block of wood also at 40°C. An accurate measure of temperature is possible because of the way the properties of many materials change due to heat or cold. Furthermore, these changes are both reliable and predictable—a necessary prerequisite to accurate measurement of temperature.

A thermometer is a commonly used instrument to measure temperature; simply, it gives us a numerical measure of the degree of hotness. Typically, in a glass thermometer, a material such as mercury or alcohol is present inside a glass capillary. This material expands in response to heat. Its coefficient of expansion is much higher than that of glass. The movement of this material in the glass capillary, on a preselected scale, gives us the measure of temperature. Other instruments used in measuring temperature include thermocouple, resistance temperature detector, thermistor, and pyrometers (to be discussed later in Chapter 3).

The thermodynamic basis for the thermometer is the Zeroth Law of Thermodynamics, first described by R. H. Fowler in 1931. According to this law, "if two bodies are in thermal equilibrium with a third body, they are also in thermal equilibrium with each other." This implies that if the third body is selected as a thermometer, and the temperature of the two bodies is the same, then the two bodies are in thermal equilibrium with each other, even when they may not be in contact with each other.

The statement of the Zeroth Law of Thermodynamics appears rather trivial; however, it cannot be deduced from the other two laws of thermodynamics.

The temperature scale according to the SI units is the Celsius scale, named after a Swedish astronomer, Celsius. In the English system of units, we use the Fahrenheit scale, named after the German instrument maker G. Fahrenheit. Both these scales use two reference points. The ice point is a temperature of ice and water mixture in equilibrium with saturated air at one atmospheric pressure. The ice point for the Celsius scale is 0°C and 32°F in the Fahrenheit scale. The boiling point, when a mixture of liquid and water vapor are in equilibrium at one atmospheric pressure, is 100°C in the Celsius scale and 212°F in the Fahrenheit scale.

In addition to the temperature scales, there is a thermodynamic temperature scale that does not depend on the properties of any material. In SI units, the scale is the Kelvin scale, with a temperature unit of Kelvin (K not °K, according to convention). On the Kelvin scale, the lowest temperature is 0 K, although this temperature has not actually been measured. A corresponding scale in English units is the Rankine scale, with the temperature unit expressed as R.

The Kelvin and Celsius scales are related by the following function:

$$T(\mathrm{K}) = T(^{\circ}\mathrm{C}) + 273.15 \tag{1.13}$$

In most engineering calculations, the number in this equation is rounded off to 273.

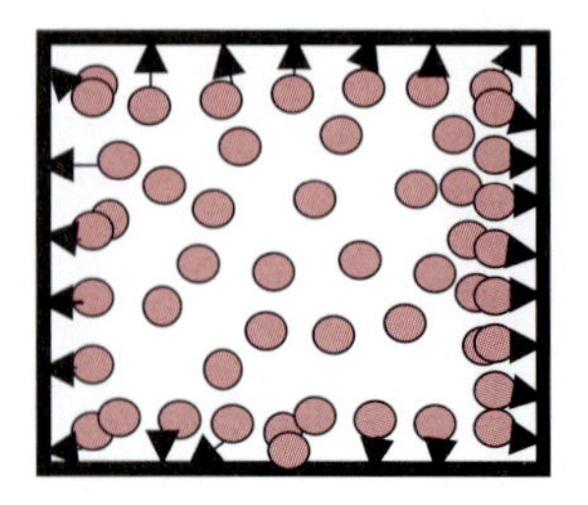

■ **Figure 1.7** Gas molecules exerting force on the inside of a chamber.

It is also important to recognize that the actual scale division in the Kelvin and Celsius scales is exactly the same. Therefore, if we are concerned with difference in temperature, then either the Celsius or Kelvin scales may be used. Thus,

$$\Delta T(\mathrm{K}) = \Delta T(^\circ\mathrm{C}) \tag{1.14}$$

For example, consider a liquid food whose specific heat value is reported as 3.5 kJ/(kg °C). The units of specific heat, kJ/(kg °C), suggest that 3.5 kJ of heat are required per kilogram of the liquid food to raise its temperature by 1°C. Therefore, whenever we have temperature in the denominator, we are actually considering a unit difference in temperature, since 1° change in the Celsius scale is the same as a unit change in the Kelvin scale. Therefore, the specific heat of the given liquid food may also be reported as 3.5 kJ/(kg K).

1.9 PRESSURE

Figure 1.7 illustrates a gas contained in a chamber. The gas molecules strike the inside surface of the chamber and exert a force normal to the surface. When the fluid is at equilibrium, the force exerted by the fluid per unit area of the inside chamber surface is called pressure. If we take a differential section of the chamber surface area, d*A*, and consider that the force acting normal to it is d*F*, then pressure is

$$P = \frac{\mathrm{d}F}{\mathrm{d}A} \tag{1.15}$$

Pressure is an intensive property of a system. The pressure of a fluid contained in a chamber increases with depth, because the weight of the fluid increases with depth.

Pressure may be expressed as force per unit area. The dimensions of pressure are $(\text{mass})(\text{time})^{-2}(\text{length})^{-1}$. In the SI system, the units are $\mathrm{N/m^2}$. This unit is also called a pascal (named after Blaise Pascal[1]). Since the pascal unit is small in magnitude, another unit, bar, is used, where

$$1\ \text{bar} = 10^5\ \text{Pa} = 0.1\ \text{MPa} = 100\ \text{kPa}$$

[1] Blaise Pascal (1623–1662), a French philosopher and mathematician, was the founder of the modern theory of probabilities. He studied hydrostatic and atmospheric pressure and derived Pascal's law of pressure. He is credited with inventing the first digital computer and the syringe. In addition to studying physical sciences, he became a scholar of religion, and in 1655 he wrote *Les Provinciales*, a defense of Jansenism against the Jesuits.

The standard atmospheric pressure is defined as the pressure produced by a column of mercury 760 mm high. The standard atmospheric pressure can be expressed, using different systems of units, as

$$1\,\text{atm} = 14.696\,\text{lb/in}^2 = 1.01325\,\text{bar} = 101.325\,\text{kPa}$$

Zero pressure characterizes absolute vacuum. When pressure is measured relative to absolute vacuum, it is called **absolute pressure**. However, when we use a pressure measuring device such as a pressure gauge, it is usually calibrated to read zero at one atmospheric pressure. Therefore, these devices are actually reading the difference between the absolute pressure and the local atmospheric pressure. A pressure measured by a gauge is often called **gauge pressure**, and it is related to the atmospheric pressure based on the following expression:

$$P_{\text{absolute}} = P_{\text{gauge}} + P_{\text{atmosphere}} \quad \text{(for pressures greater than } P_{\text{atmosphere}}\text{)} \tag{1.16}$$

$$P_{\text{vacuum}} = P_{\text{atmosphere}} - P_{\text{absolute}} \quad \text{(for pressures below } P_{\text{atmosphere}}\text{)} \tag{1.17}$$

A visual description of the relationships between the various terms used to define pressure is given in Figure 1.8.

In expressing units for **vacuum** in the English system, the atmospheric pressure is referred to as 0 inches of mercury. Perfect vacuum is 29.92 inches of mercury. Thus, 15 inches of mercury has a higher pressure

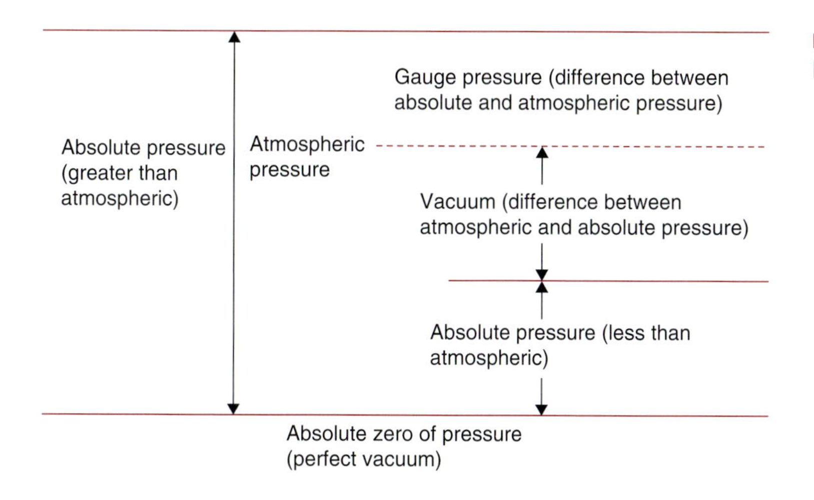

Figure 1.8 Illustration of the relationships between the terms used to define pressure.

than 20 inches of mercury. In the SI system, the convention to express vacuum is opposite from that of the English system, and the units are given in pascals. At perfect vacuum, the absolute pressure is 0 Pa (recall that one atmospheric pressure is 101.325 kPa). The relationship between the English system and the SI system for expressing vacuum can be written as

$$P_{\text{atmosphere}} = 3.38638 \times 10^3(29.92 - I) \tag{1.18}$$

where $P_{\text{atmosphere}}$ is in Pa, I is inches of mercury.

Pressure is the term used to express this property for liquids and gases. For solids, we use the term **normal stress** instead of pressure. In situations involving fluid flow, pressure is often expressed in terms of height or **head** of a fluid. The height of a fluid that can be supported by the pressure acting on it can be written mathematically as

$$P = \rho g h \tag{1.19}$$

where P is absolute pressure (Pa), ρ is fluid density (kg/m^3), g is acceleration due to gravity (9.81 m/s^2), and h is height of fluid (m).

Thus, two atmospheric pressures will support

$$\frac{2 \times (101.325 \times 10^3 \text{ N/m}^2)}{(13,546 \text{ kg/m}^3)(9.81 \text{ m/s})} = 1525 \text{ mm of mercury}$$

Consider a tank filled with cold water to a height of 7 m, as shown in Figure 1.9. The pressure exerted by the water at any point on the bottom of the tank is independent of the diameter of the tank but depends on the height of the water in the tank. This height or elevation of water in the tank is called the **static head**. As shown in the figure, a pressure gauge located at the bottom of the tank indicates a pressure of 0.69 bar (10 psig, i.e., a gauge pressure of 10 lb/in^2) exerted by a column of water of height 7 m. Thus, the static head at location 1 is 7 m of water. If there is liquid other than water in the tank, then the indicated pressure will be different because of the different specific gravity of the liquid. Thus, if the tank contained gasoline (specific gravity = 0.75), the same pressure of 0.69 bar will be exerted by a column of gasoline 9.38 m high, and the static head will be 9.38 m of gasoline. If the tank contained mercury (specific gravity = 13.6),

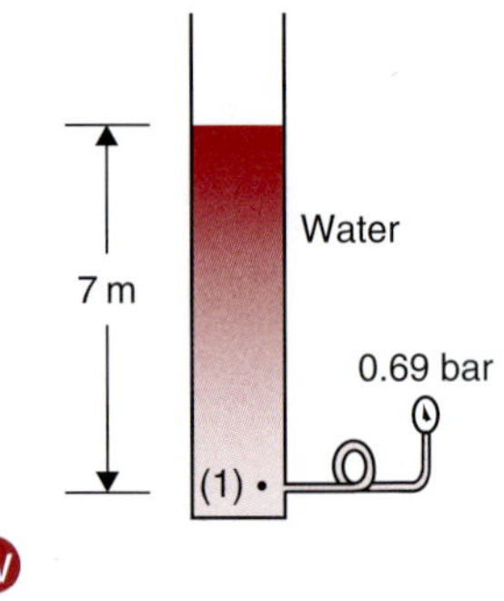

W

Figure 1.9 Pressure head of a column of water.

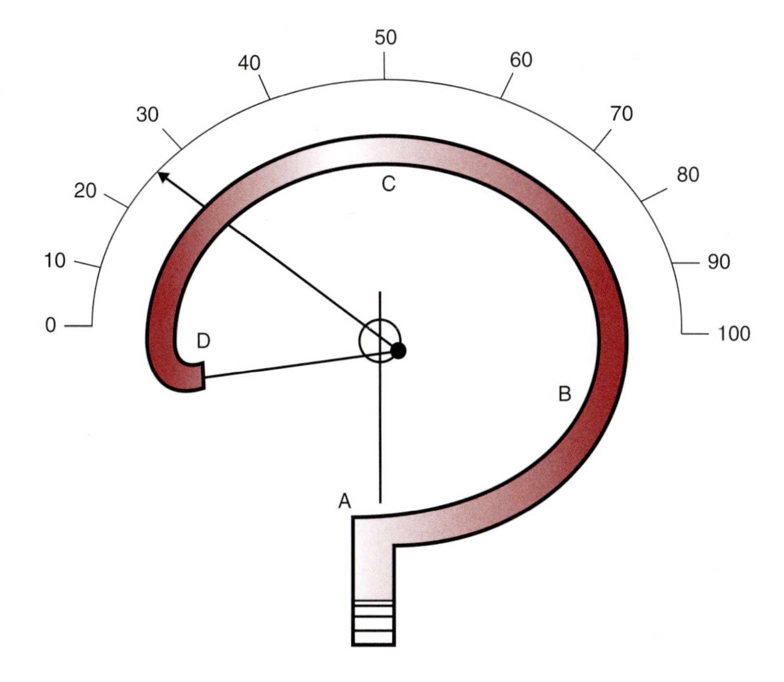

■ **Figure 1.10** A Bourdon tube pressure sensor.

the same pressure at location 1 will be exerted by a column 0.518 m high and the static head designated as 0.518 m of mercury.

The static head may be converted to pressure using the following formula:

$$\text{Pressure (bar)} = \frac{\text{Static head (m)}}{10.2} \times \text{Specific gravity} \quad (1.20)$$

In fluid flow problems, two additional terms are often encountered, namely, static pressure and impact pressure. **Static pressure** is the pressure measured by a device if it is moving with the same velocity as the fluid velocity. **Impact pressure** is force per unit area perpendicular to the direction of flow when the fluid is brought reversibly to rest.

The pressure of a fluid is measured using a variety of instruments, including a Bourdon tube, a manometer, and a pressure transducer. A Bourdon tube is shown in Figure 1.10. It consists of an oval-shaped arm ABCD. An increase in the internal pressure extends the arm, and the movement of the pointer on the dial is calibrated to indicate the pressure.

1.10 ENTHALPY

Enthalpy is an extensive property, expressed as the sum of internal energy and the product of pressure and volume:

$$H = E_i + PV \tag{1.21}$$

where H is enthalpy (kJ), E_i is internal energy (kJ), P is pressure (kPa), and V is volume (m^3).

Enthalpy may also be expressed per unit mass as follows:

$$H' = E_i' + PV' \tag{1.22}$$

where H' is enthalpy per unit mass (kJ/kg), E_i' is internal energy per unit mass (kJ/kg), and V' is specific volume (m^3/kg).

Note that enthalpy is an energy quantity only in special cases. For example, the enthalpy of air in a room is not an energy quantity, because the product of pressure and specific volume in this case is not an energy quantity. The only energy of the air in the room is its internal energy. When a fluid enters or leaves an open system, the product of pressure and volume represents flow energy. In this case, enthalpy of the fluid represents the sum of internal energy and flow energy.

Enthalpy value is always given relative to a reference state in which the value of enthalpy is arbitrarily selected, usually zero for convenience. For example, steam tables give enthalpy of steam, assuming that at 0°C the enthalpy of saturated liquid (i.e., water) is zero.

1.11 EQUATION OF STATE AND PERFECT GAS LAW

The thermodynamic properties of a simple system are established by any two independent properties. A functional relationship between the properties of a system is called an equation of state. Values of two properties of a system help establish the value of the third property.

For a perfect gas, an equation of state is a relationship between pressure, volume, and temperature. The equation may be written as

$$PV' = RT_A \tag{1.23}$$

or

$$P = \rho RT_A \tag{1.24}$$

where P is absolute pressure (Pa), V' is specific volume (m^3/kg), R is the gas constant (m^3 Pa/[kg K]), T_A is absolute temperature (K), and ρ is density (kg/m^3).

At room temperature, real gases such as hydrogen, nitrogen, helium, and oxygen follow the perfect gas law (very closely).

The equation of state for a perfect gas may also be written on a mole basis as

$$PV = nR_0T_A \tag{1.25}$$

where V is the volume (of m kg or n mol), m^3; $R_0 = M \times R$ is the universal gas constant, independent of the nature of a gas, 8314.41 m^3 Pa/(kg mol K); and M is the molecular weight of the substance.

1.12 PHASE DIAGRAM OF WATER

Water is considered to be a pure substance. It has a homogeneous and invariable chemical composition, even though it may undergo a change in phase. Therefore, liquid water or a mixture of ice and liquid water, and steam, a mixture of liquid water and water vapor, are pure substances.

When a substance occurs as a vapor at the saturation temperature and pressure, it is called *saturated vapor*. Saturation temperature is the temperature at which vaporization takes place at a given pressure. This pressure is called the *saturation pressure*. Thus, water at 100°C has a saturation pressure of 101.3 kPa. When the temperature of the vapor is greater than the saturation temperature at the existing pressure, it is called *superheated vapor*.

When a substance occurs in a liquid state at its saturation temperature and pressure, it is called a *saturated liquid*. If at the saturation pressure, the temperature of the liquid is lowered below the saturation temperature, it is called a *subcooled liquid*. In the case where, at the saturation temperature, a substance exists partly as liquid and partly as vapor, the ratio of the mass of water vapor to the total mass of the substance is expressed as the quality of the vapors. For example, if steam has 0.1 kg water and 0.9 kg vapor, the quality of steam is 0.9 divided by 1.0 (which represents the total mass of steam); thus steam quality equals 0.9 or 90%.

A phase diagram of water, shown in Figure 1.11, is useful to study pressure–temperature relationships between various phases. This diagram gives the limiting conditions for solid, liquid, and gas (or vapor) phases. At any location within the areas separated by the curves, the pressure and temperature combination permits only one phase (solid, liquid, or vapor) to exist. Any change in temperature and pressure up to the points on the curves will not change the phase. As shown in Figure 1.11, the sublimation line separates the solid phase from the vapor phase, the fusion line separates the solid phase from the liquid phase, and the vaporization line separates the liquid phase from the vapor phase. The three lines meet at a triple point. The triple point identifies a state in which the three phases—solid, liquid, and vapor—may all be present in equilibrium. The triple point for water is at 0.01°C.

The phase diagram shown in Figure 1.11 is useful in examining processes conducted at constant pressure with change of phase. For example, line AA′ is a constant-pressure process conducted at a low temperature, where ice sublimates into the vapor phase. There is no liquid phase in this case. Line BB′ represents a heating process at or above atmospheric pressure, where initially solid ice melts into the liquid state, followed by vaporization of water at a higher temperature.

Phase diagrams are important in studying processes such as extraction, crystallization, distillation, precipitation, and freeze concentration.

Figure 1.11 Phase diagram for water.

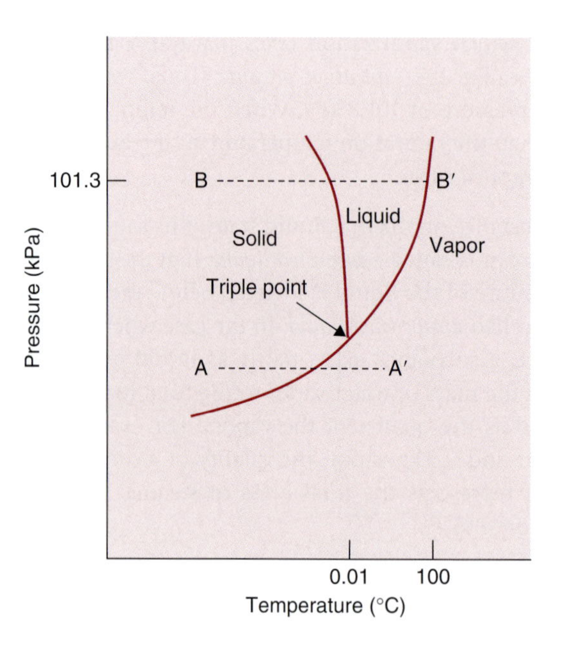

1.13 CONSERVATION OF MASS

The principle of conservation of mass states that:

> *Mass can be neither created nor destroyed. However, its composition can be altered from one form to another.*

Even in the case of a chemical reaction, the composition of mass of a reactant and the product before and after the reaction may be different, but the mass of the total system remains unaltered. When the chemical reactions are absent, the composition of a system as well as its mass remains the same for a closed system. We can express the conservation of mass principle as an equation written in words as

$$\begin{array}{c}\text{Rate of mass entering}\\\text{through the boundary}\\\text{of a system}\end{array} - \begin{array}{c}\text{Rate of mass exiting}\\\text{through the boundary}\\\text{of a system}\end{array} = \begin{array}{c}\text{Rate of mass}\\\text{accumulation within}\\\text{the systerm}\end{array} \quad (1.26)$$

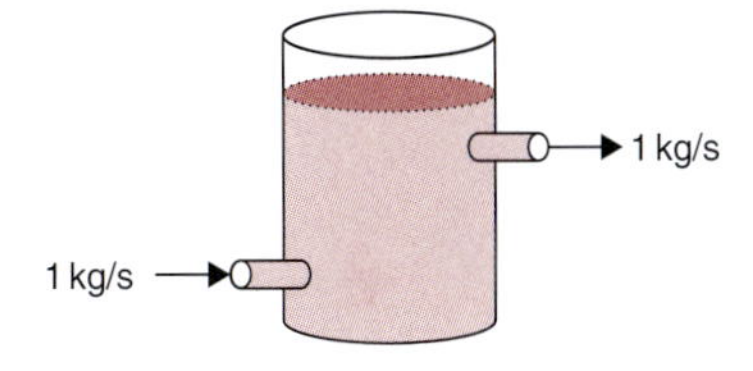

■ **Figure 1.12** Liquid flow in and out of a tank.

If the rate of mass accumulation within a system is zero, then the rate of mass entering must equal rate of mass leaving the system. For example, as shown in Figure 1.12, if the level of milk in a tank remains constant, and the milk flow rate at the inlet is 1 kg/s, then the flow rate of milk at the exit must also be 1 kg/s.

Next, let us convert the previous word equation into a mathematical form. To do this, we will refer to Figure 1.13, which shows a system with mass inlet and exit streams. Although only one inlet and one exit stream is shown, there may be more than one stream entering and exiting a control volume. Therefore, for a general case, the rate of mass flow entering the system is

$$\dot{m}_{\text{inlet}} = \sum_{i=1}^{n} \dot{m}_i \quad (1.27)$$

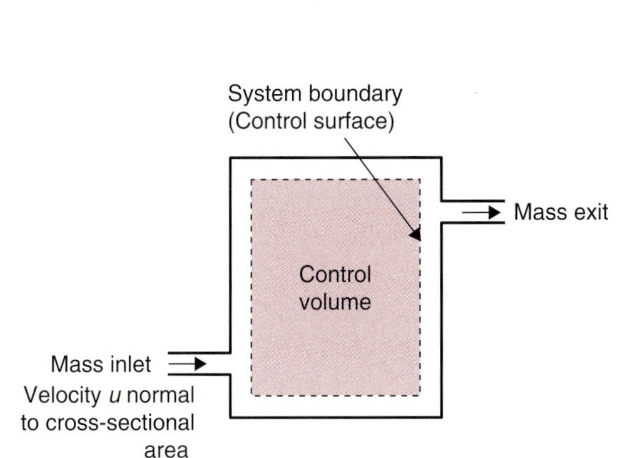

■ **Figure 1.13** A control volume.

where subscript i denotes the inlet, and n is the number of inlets to a system.

The rate of mass flow exiting the system is

$$\dot{m}_{\text{exit}} = \sum_{e=1}^{p} \dot{m}_e \tag{1.28}$$

where subscript e denotes the exit, and p is the number of exits from a system.

The rate of mass accumulation within the system boundary, expressed as a function of time, is

$$\dot{m}_{\text{accumulation}} = \frac{\mathrm{d}m_{\text{system}}}{\mathrm{d}t} \tag{1.29}$$

Then substituting into the word Equation (1.26) we obtain

$$\dot{m}_{\text{inlet}} - \dot{m}_{\text{exit}} = \frac{\mathrm{d}m_{\text{system}}}{\mathrm{d}t} \tag{1.30}$$

Typically, mass flow rate is easier to measure than other flow properties such as velocity. When instead of mass flow rate, we may measure velocity of the flow along with the density of the fluid, the mathematical analysis involves integral expressions, as shown in the following section.

1.13.1 Conservation of Mass for an Open System

Consider a section of a pipe used in transporting a fluid. For a control volume shown, for this open system a fluid with a velocity, u, is entering the system across a differential area dA. Recall that velocity is a vector quantity, possessing both magnitude and direction. As seen in Figure 1.13, only the component of velocity vector normal to the area dA will cross the system boundary. The other component, u_{tan}, (tangent to the area) has no influence on our derivation. Thus, if the fluid particle crossing the boundary has a velocity u_n, then the rate of mass flow into the system may be expressed as

$$\mathrm{d}\dot{m} = \rho u_n \mathrm{d}A \tag{1.31}$$

Integrating over a finite area

$$\dot{m} = \int_A \rho u_n \mathrm{d}A \tag{1.32}$$

The previous equation for mass flow rate will apply for inlet and exit cases.

The total mass of the system may be expressed as a product of its volume and density, or

$$m = \int_V \rho \mathrm{d}V \tag{1.33}$$

Substituting this quantity in the word Equation (1.26), we obtain

$$\int_{A_{\text{inlet}}} \rho u_n \mathrm{d}A - \int_{A_{\text{exit}}} \rho u_n \mathrm{d}A = \frac{\mathrm{d}}{\mathrm{d}t} \int_V \rho \mathrm{d}V \tag{1.34}$$

The previous equation is somewhat complicated because of the integral and differential operators. However, this expression may be simplified for two common situations encountered in engineering systems. First, if the flow is uniform, then all measurable properties of the fluid are uniform throughout the cross-sectional area. These properties may vary from one cross-sectional area to another, but at the same cross-section they are the same in the radial direction. For example, fruit juice flowing in a pipe has the same value of its properties at the center and the inside surface of the pipe. These properties may be density, pressure, or temperature. For a uniform flow, we can replace the integral signs with simple summations, or

$$\sum_{\text{inlet}} \rho u_n \mathrm{d}A - \sum_{\text{outlet}} \rho u_n \mathrm{d}A = \frac{\mathrm{d}}{\mathrm{d}t} \int_V \rho \frac{\mathrm{d}V}{\mathrm{d}t} \tag{1.35}$$

The second assumption we will make is that of *steady state*—that is, the flow rate does not change with time, although it may be different from one location to another. If there is no change with time, then the right-hand term must drop out. Thus, we have

$$\sum_{\text{inlet}} \rho u_n \mathrm{d}A = \sum_{\text{outlet}} \rho u_n \mathrm{d}A \tag{1.36}$$

Furthermore, if we are dealing with an incompressible fluid—a good assumption for most liquids—there is no change in density. Thus

$$\sum_{\text{inlet}} u_n \mathrm{d}A = \sum_{\text{outlet}} u_n \mathrm{d}A \tag{1.37}$$

The product of velocity and area is the volumetric flow rate. Thus, according to the conservation of mass principle, for a steady, uniform, and incompressible flow, the volumetric flow remains unchanged. For compressible fluids such as steam and gases, the inlet mass flow rate will be the same as the exit mass flow rate.

1.13.2 Conservation of Mass for a Closed System

Recall that in a closed system, mass cannot cross system boundaries. Therefore, there is no time rate of change of mass in the system, or

$$\frac{dm_{\text{system}}}{dt} = 0 \tag{1.38}$$

or

$$m_{\text{system}} = \text{constant} \tag{1.39}$$

1.14 MATERIAL BALANCES

Material balances are useful in evaluating individual pieces of equipment, such as a pump or a homogenizer, as well as overall plant operations consisting of several processing units—for example, a tomato paste manufacturing line, as shown in Figure 1.14. Compositions of raw materials, product streams, and by-product streams can be evaluated by using material balances.

The following steps should be useful in conducting a material balance in an organized manner.

1. Collect all known data on mass and composition of all inlet and exit streams from the statement of the problem.
2. Draw a block diagram, indicating the process, with inlet and exit streams properly identified. Draw the system boundary.
3. Write all available data on the block diagram.
4. Select a suitable basis (such as mass or time) for calculations. The selection of basis depends on the convenience of computations.
5. Using Equation (1.30), write material balances in terms of the selected basis for calculating unknowns. For each unknown, an independent material balance is required.
6. Solve material balances to determine the unknowns.

The use of material balances is illustrated in the following examples.

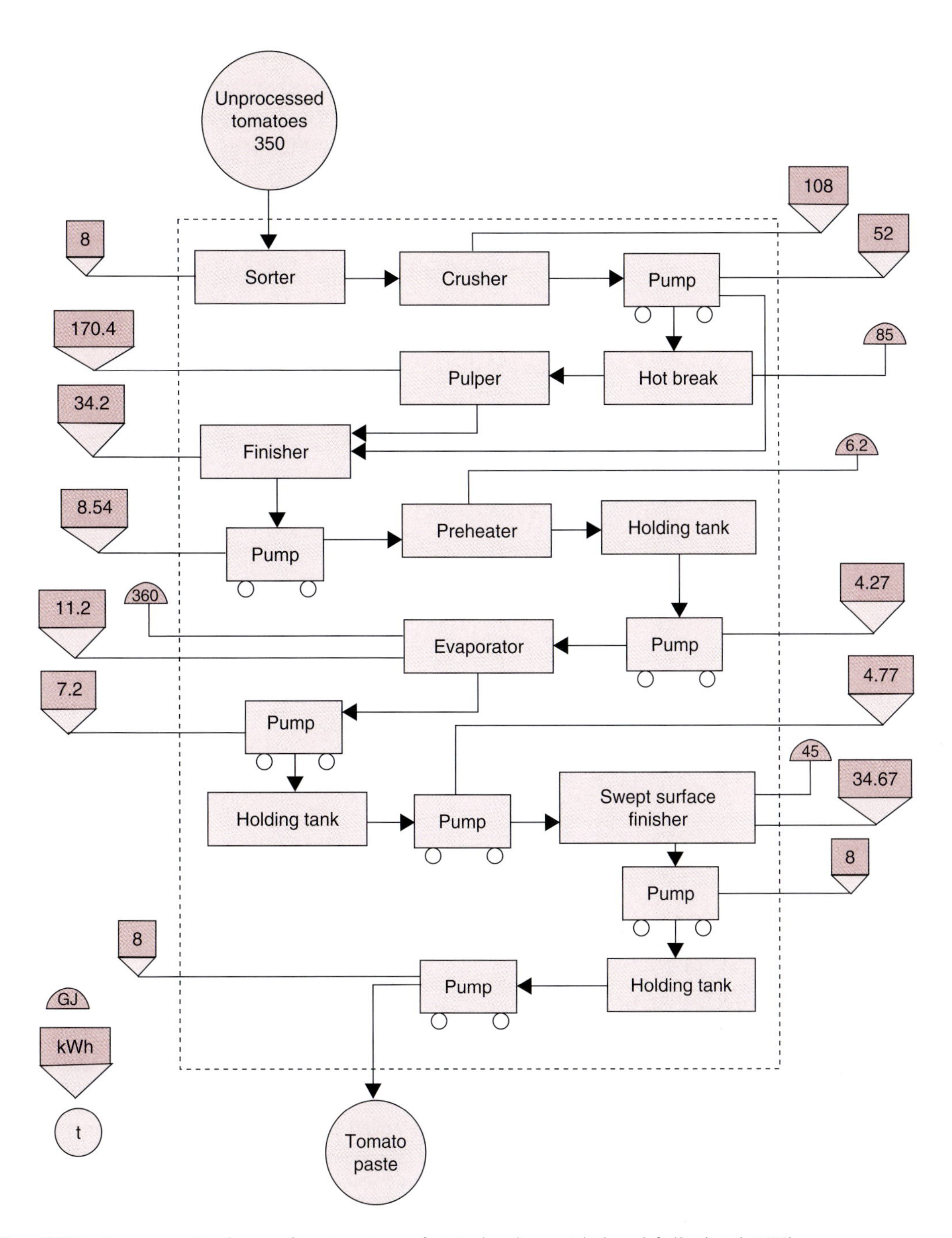

■ **Figure 1.14** Energy accounting diagram of tomato paste manufacturing based on an eight-hour shift. (Singh et al., 1980)

Example 1.6

In a furnace, 95% of carbon is converted to carbon dioxide and the remainder to carbon monoxide. By material balance, predict the quantities of gases appearing in the flue gases leaving the furnace.

Given

Carbon converted to $CO_2 = 95\%$
Carbon converted to $CO = 5\%$

Solution

1. *Basis is 1 kg of carbon*
2. *The combustion equations are*

$$C + O_2 = CO_2$$

$$C + \tfrac{1}{2}O_2 = CO$$

3. *From these equations, 44 kg carbon dioxide is formed by combustion of 12 kg carbon, and 28 kg carbon monoxide is formed by combustion of 12 kg carbon.*
4. *Then, the amount of* CO_2 *produced,*

$$\frac{(44\,\text{kg}\,CO_2)(0.95\,\text{kg C burned})}{12\,\text{kg C burned}} = 3.48\,\text{kg}\,CO_2$$

5. *Similarly, the amount of CO produced,*

$$\frac{(28\,\text{kg CO})(0.05\,\text{kg C burned})}{12\,\text{kg C burned}} = 0.12\,\text{kg CO}$$

6. *Thus, the flue gases contain 3.48 kg* CO_2 *and 0.12 kg CO for every kilogram of carbon burned.*

Example 1.7

A wet food product contains 70% water. After drying, it is found that 80% of original water has been removed. Determine (a) mass of water removed per kilogram of wet food and (b) composition of dried food.

Given

Initial water content = 70%
Water removed = 80% of original water content

Solution

1. *Select basis = 1 kg wet food product*
2. *Mass of water in inlet stream = 0.7 kg*
3. *Water removed in drying = 0.8(0.7) = 0.56 kg/kg of wet food material*
4. *Write material balance on water,*

$$\text{Water in dried food} = 0.7(1) - 0.56 = 0.14\,\text{kg}$$

5. *Write balance on solids,*

$$0.3(1) = \text{solids in exit stream}$$

$$\text{Solids} = 0.3\,\text{kg}$$

6. *Thus, the dried food contains 0.14 kg water and 0.3 kg solids.*

Example 1.8

A membrane separation system is used to concentrate total solids (TS) in a liquid food from 10% to 30%. The concentration is accomplished in two stages with the first stage resulting in release of a low-total-solids liquid stream. The second stage separates the final concentration product from a low-total-solids stream, which is returned to the first stage. Determine the magnitude of the recycle stream when the recycle contains 2% TS, the waste stream contains 0.5% TS, and the stream between stages 1 and 2 contains 25% TS. The process should produce 100 kg/min of 30% TS.

Given

(Fig. E1.6)
Concentration of inlet stream = 10%
Concentration of exit stream = 30%

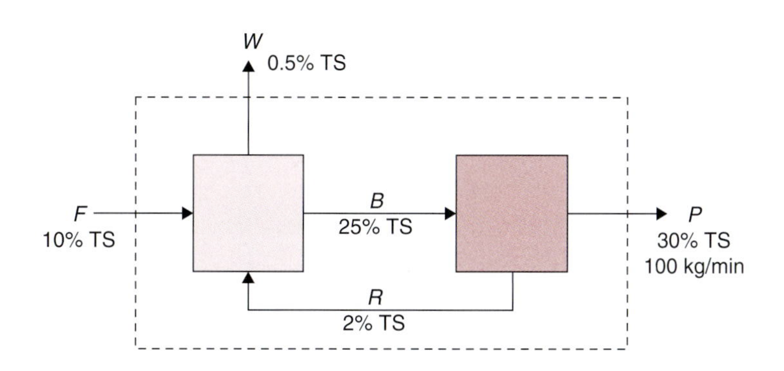

Figure E1.6 A schematic arrangement of equipment described in Example 1.8.

Concentration of recycle stream = 2%
Concentration of waste stream = 0.5%
Concentration of stream between two stages = 25%
Mass flow rate of exit stream = 100 kg/min

Solution

1. *Select 1 min as a basis.*
2. *For the total system*

$$F = P + W$$
$$Fx_F = Px_P + Wx_w$$
$$F = 100 + W$$
$$F(0.1) = 100(0.3) + W(0.005)$$

where x is the solids fraction.

3. *For the first stage*

$$F + R = W + B$$
$$Fx_F + Rx_R = Wx_W + Bx_B$$
$$F(0.1) + R(0.02) = W(0.005) + B(0.25)$$

4. *From step (2)*

$$(100 + W)(0.1) = 30 + 0.005W$$
$$0.1W - 0.005W = 30 - 10$$
$$0.095W = 20$$
$$W = 210.5\ \text{kg/min}$$
$$F = 310.5\ \text{kg/min}$$

5. *From step (3)*

$$310.5 + R = 210.5 + B$$
$$B = 100 + R$$
$$310.5(0.1) + 0.02R = 210.5(0.005) + 0.25B$$
$$31.05 + 0.02R = 1.0525 + 25 + 0.25R$$
$$4.9975 = 0.23R$$
$$R = 21.73\ \text{kg/min}$$

6. *The results indicate that the recycle stream will be flowing at a rate of 21.73 kg/min.*

Example 1.9

Potato flakes (moisture content = 75% wet basis) are being dried in a concurrent flow drier. The moisture content of the air entering the drier is 0.08 kg of water per 1 kg dry air. The moisture content of air leaving the drier is 0.18 kg water per 1 kg of dry air. The air flow rate in the drier is 100 kg dry air per hour. As shown in Figure E1.7, 50 kg of wet potato flakes enter the drier per hour. At steady state, calculate the following:

a. What is the mass flow rate of "dried potatoes"?
b. What is the moisture content, dry basis, of "dried potatoes" exiting the drier?

Given

Weight of potato flakes entering the drier F = 50 kg
Time = 1 h

Solution

1. *Basis = 1 h*
2. *Mass of air entering the drier = mass of dry air + mass of water*

$$I = 100 + 100 \times 0.08$$

$$I = 108 \text{ kg}$$

3. *Mass of air leaving the drier = mass of dry air + mass of water*

$$E = 100 + 100 \times 0.18$$

$$E = 118 \text{ kg}$$

4. *Total balance on the drier*

$$I + F = E + P$$

$$108 + 50 = 118 + P$$

$$P = 40 \text{ kg}$$

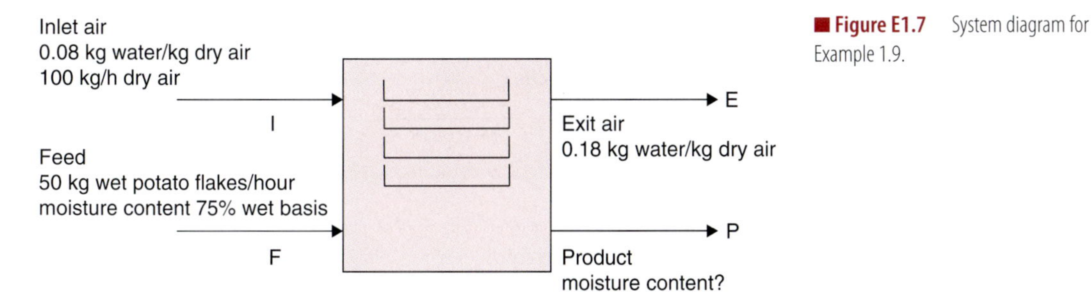

■ **Figure E1.7** System diagram for Example 1.9.

5. *Solid balance on the drier*

Solid content in feed is calculated from the definition of the wet basis moisture content (Eq. (1.6)), rewriting Equation (1.6) as

$$1 - MC_{wb} = 1 - \frac{\text{mass of water}}{\text{mass of moist sample}}$$

or

$$1 - MC_{wb} = \frac{\text{mass of dry solids}}{\text{mass of moist sample}}$$

or

$$\text{Mass of dry solids} = \text{Mass of moist sample}\ (1 - MC_{wb})$$

Therefore,

$$\text{Mass of solid content in feed} = F(1 - 0.75)$$

If y is the solid fraction in the product stream P, then solid balance on the drier gives

$$0.25\ F = y \times P$$

$$y = \frac{0.25 \times 50}{40}$$

$$= 0.3125$$

Thus,

$$\frac{\text{mass of dry solids}}{\text{mass of moist sample}} = 0.3125$$

or

$$1 - \frac{\text{mass of dry solids}}{\text{mass of moist sample}} = 1 - 0.3125$$

Therefore, moisture content (wet basis) in the exit potato stream is

$$1 - 0.3125 = 0.6875$$

6. *The wet basis moisture content is converted to dry basis moisture content*

$$MC_{db} = \frac{0.6875}{1 - 0.6875}$$

$$MC_{db} = 2.2\ \text{kg water per 1 kg dry solids}$$

7. *The mass flow rate of potatoes exiting the drier is 40 kg at a moisture content of 2.2 kg per 1 kg dry solids.*

Example 1.10

An experimental engineered food is being manufactured using five stages, as shown in Figure E1.8. The feed is 1000 kg/h. Various streams have been labeled along with the known composition values on the diagram. Note that the composition of each stream is in terms of solids and water only. Stream *C* is divided equally into streams *E* and *G*. Product *P*, with 80% solids, is the desired final product. Stream *K* produces a by-product at the rate of 450 kg/h with 20% solids. Calculate the following:

a. Calculate the mass flow rate of product *P*.
b. Calculate the mass flow rate of recycle stream *A*.
c. Calculate the mass flow rate of recycle stream *R*.

Given

Feed = 1000 kg/h
Solid content of P = 80%
Mass rate of stream K = 450 kg/h
Solids in stream K = 20%

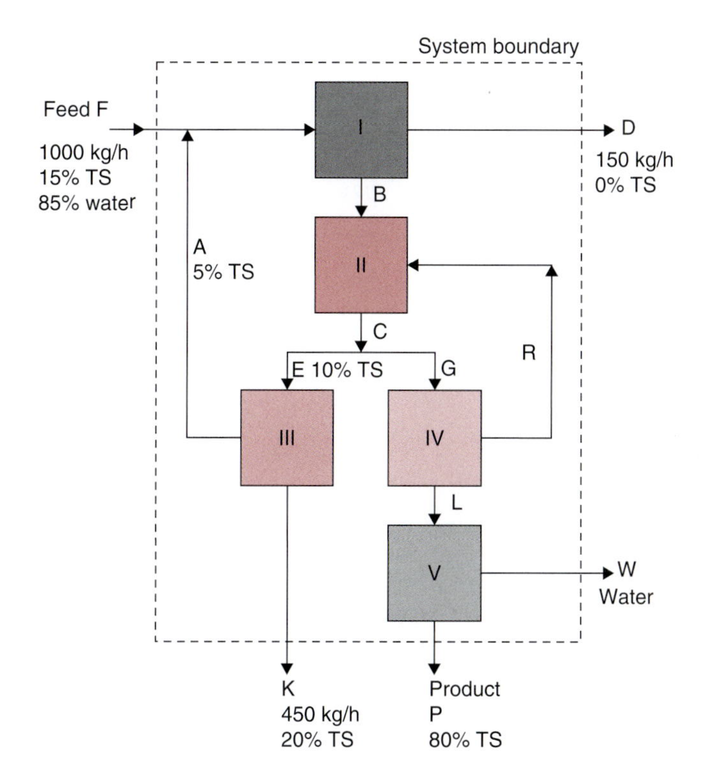

Figure E1.8 A flow sheet of an experimental food manufacturing system.

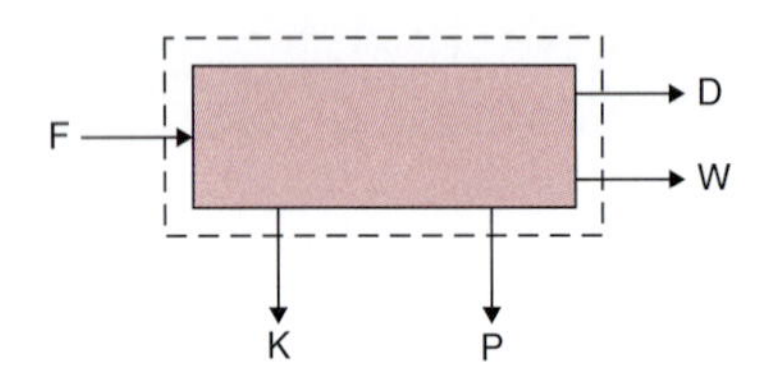

■ **Figure E1.9** Total system for Example 1.10.

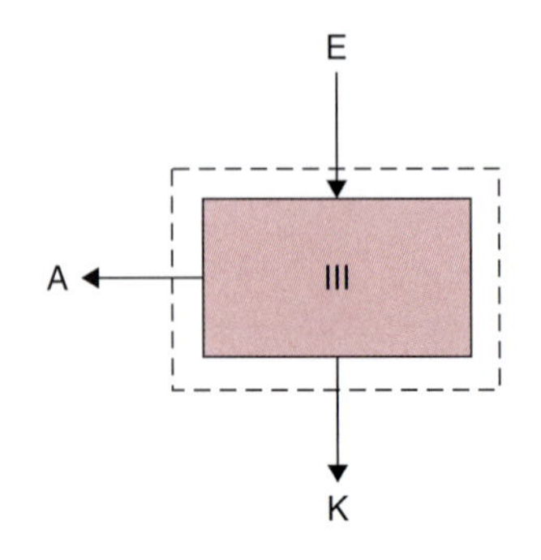

■ **Figure E1.10** Illustration of stage III of the system in Example 1.10.

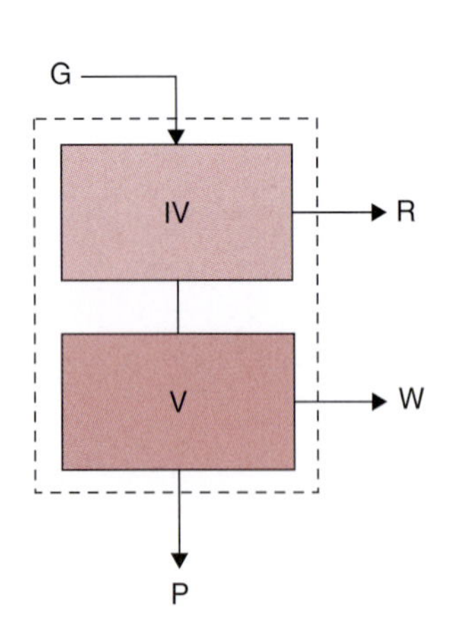

■ **Figure E1.11** Illustration of stages IV and V of the system in Example 1.10.

Solution

1. *Basis = 1 h*
2. *Consider total system, solid balance (Fig. E1.9).*

$$0.15 \times F = 0.2 \times K + 0.8 \times P$$

$$0.15 \times 1000 = 0.2 \times 450 + 0.8 \times P$$

$$150 = 90 + 0.8 \times P$$

$$P = \frac{60}{0.8} = 75\,kg$$

$$P = 75\,kg$$

3. *Consider stage III (Fig. E1.10).*
   Total balance

$$E = A + K; \quad E = A + 450 \tag{1}$$

   Solid balance

$$0.1E = 0.05A + 0.2K$$

$$0.1E = 0.05A + 0.2 \times 450$$

$$0.1E = 0.05A + 90 \tag{2}$$

   Solve preceding Equations (1) and (2) simultaneously.

$$E = 1350\,kg$$

$$A = 900\,kg$$

4. *Since C is divided equally into E and G,*

$$G = 1350\,kg \text{ with } 10\% \text{ solid}$$

5. *For total system, conduct total balance to find W.*

$$F = K + P + D + W$$

$$1000 = 450 + 75 + 150 + W$$

$$W = 325\,kg$$

6. *Consider stages IV and V together (Fig. E1.11).*

$$G = R + W + P$$

$$1350 = R + 325 + 75$$

$$R = 950\,kg$$

7. *The mass flow rates of stream P, A, and R are 75 kg, 900 kg, and 950 kg, respectively.*

1.15 THERMODYNAMICS

The science of thermodynamics gives us the foundation to study commonly occurring phenomena during processing of foods. A typical approach to studying any food process may be to first observe a phenomenon, make experimental measurements to confirm the validity of the observation, develop a mathematical basis, and then apply the knowledge gained to an engineering process at hand. This observational scheme is very similar to what is often the thermodynamic approach to examining physical systems.

In examining engineering processes, we are concerned mostly with a macroscopic view. The branch of thermodynamics that deals with this macroscopic approach is called *classical thermodynamics*. Another branch of thermodynamics, called *statistical thermodynamics*, is concerned with what happens at a molecular level, and the average behavior of a group of molecules is considered.

In food engineering, many of the processes of concern to a food engineer are applications of thermodynamics. For example, we may need to calculate heat and work effects associated with a given process. In other instances, the maximum work obtainable from a process may be the key calculation, or we may need to determine how to carry out a process with minimum work. Furthermore, we will encounter the need to determine relationships that exist between various variables of a system when it is at equilibrium.

When we conduct experiments and then want to know the behavior of a given system, the laws of thermodynamics are most useful in the analysis. Based on an experimental foundation, classical thermodynamics is concerned with the macroscopic properties of a system. These properties may be directly measured or they may be calculated from other properties that are directly measurable. For example, we can measure pressure of a gas enclosed in a chamber by attaching a pressure gauge to the chamber.

Thermodynamics also helps us in determining the potential that defines and determines the equilibrium. By knowing potential, we can determine the direction a process will undertake. Although thermodynamics may not tell us how long a process will take to arrive at its final state, it does help in determining what the final state will be. Thus, time is not a thermodynamic variable, and we will rely on other procedures to determine the rate of a process—another important area of calculations for food engineers.

1.16 LAWS OF THERMODYNAMICS

1.16.1 First Law of Thermodynamics

The first law of thermodynamics is a statement of the conservation of energy. The law states:

The energy of an isolated system remains constant.

Stated in other words,

Energy can be neither created nor destroyed but can be transformed from one form to another.

Energy can be either stored within an object or transferred to another one, such as in the form of thermal or mechanical energy. If we increase the elevation of an object, its potential energy will increase. The increased potential energy will remain stored in the object until we move it again. Similarly, we can increase the thermal energy of an object by transferring heat into it and observing an increase in temperature.

Energy can also be transformed from one form to another. For example, in a hydroelectric plant, as water falls from a high elevation onto the blades of a turbine, the potential energy of water is converted into mechanical energy in the turbine, and the generator then converts the mechanical energy into electrical energy. The electrical energy is transmitted to homes or factories where it is further converted to other useful forms such as thermal energy in electric heaters.

During the energy conversion or transmission processes, there is also generation of heat, often misstated as "loss" of energy, when it is actually conversion of energy to other forms that may not be directly useful for the intended purpose. For example, when electrical energy is converted into mechanical energy in an electric motor, the energy "loss" may be 10 to 15%. The "loss" in this case is the conversion of part of the electrical energy into heat due to friction. Although we can convert all mechanical energy into heat because all processes are assumed reversible, we cannot convert all heat into work, as will be evident when we consider the second law of thermodynamics.

1.16.2 Second Law of Thermodynamics

The second law of thermodynamics is useful in examining the direction of energy transfer or conversion. The following two statements of the second law are by Rudolf Clausius[2] and Lord Kelvin,[3] respectively.

> *No process is possible whose sole result is the removal of heat from a reservoir (system) at one temperature and the absorption of an equal quantity of heat by a reservoir at a higher temperature.*

> *No process is possible whose sole result is the abstraction of heat from a single reservoir and the performance of an equivalent amount of work.*

The second law of thermodynamics helps explain why heat always flows from a hot object to a cold object; why two gases placed in a chamber will mix throughout the chamber, but will not spontaneously separate once mixed; and why it is impossible to construct a machine that will operate continuously while receiving heat from a single reservoir and producing an equivalent amount of work.

The second law of thermodynamics assigns both quantity and quality to energy. The importance of this law is evident in any process—the path of a process is always toward that of decreasing quality. For example, a hot bowl of soup left to itself on a table cools down. In this case, the quality of energy degrades. Energy of a higher quality (at a higher temperature) transfers from the soup to the surroundings and converts into less useful forms of energy.

1.17 ENERGY

Energy is a scalar quantity. It was first hypothesized by Newton to express kinetic and potential energies. We cannot observe energy directly, but we can measure it using indirect methods and analyze its value. Energy may be in different forms, such as potential, kinetic, chemical, magnetic, or electrical.

---

[2] Rudolf Clausius (1822–1888), a German mathematical physicist, is credited with making thermodynamics a science. In 1850, he presented a paper that stated the second law of thermodynamics. He developed the theory of the steam engine, and his work on electrolysis formed the basis of the theory of electrolytic dissociation.

[3] Lord Kelvin (1824–1907) was a Scottish mathematician, physicist, and engineer. At the age of 22, he was awarded the chair of natural philosophy at the University of Glasgow. He contributed to the development of the law of conservation of energy, the absolute temperature scale (which was named after him), electromagnetic theory of light, and mathematical analysis of electricity and magnetism. A prolific writer, he published over 600 scientific papers.

Potential energy of a system is by virtue of its location with respect to the gravitational field. If an object has a mass m, located at elevation h, and acceleration due to gravity is g, then the potential energy is

$$E_{PE} = mgh \tag{1.40}$$

Kinetic energy of an object is due to its velocity. If an object is moving with a velocity u, and it has mass m, then its kinetic energy is

$$E_{KE} = \frac{1}{2}mu^2 \tag{1.41}$$

Both kinetic and potential energies are *macroscopic;* that is, they represent energy of a system due to its entire being. This is in contrast to internal energy, which is due to the *microscopic* nature of a system. At the molecular scale, the atoms of a substance are continuously in motion. They move in random direction, collide with each other, vibrate, and rotate. Energies related to all these movements, including energy of attraction between the atoms, is combined into one lump sum and is called the internal energy.

Internal energy is an extensive property, and it is independent of the path of a process. Although we cannot measure an absolute value of internal energy, we can relate changes in internal energy to other properties such as temperature and pressure.

In many engineering systems, one or two forms of energy may dominate whereas others can be neglected. For example, when a sugar beet is dropped from a conveyor into a bin, the potential and kinetic energy of the sugar beet changes, but other energy forms such as chemical, magnetic, and electrical do not change and may be neglected in the analysis. Similarly, when tomato juice is heated in a hot-break heater, the potential or kinetic energy of the juice does not change, but the internal energy will change as temperature increases.

The total energy of a system can be written in the form of an equation as

$$E_{TOTAL} = E_{KE} + E_{PE} + E_{ELECTRICAL} + E_{MAGNETIC} + E_{CHEMICAL} + \cdots + E_i \tag{1.42}$$

where E_i is the internal energy, kJ.

If the magnitudes of all other energy forms are small in comparison with the kinetic, potential, and internal energies, then

$$E_{TOTAL} = E_{KE} + E_{PE} + E_i \tag{1.43}$$

1.18 ENERGY BALANCE

The first law of thermodynamics states that energy can be neither created nor destroyed. We may express this in the form of a word equation as

$$\text{Total energy entering the system} - \text{Total energy leaving the system} = \text{Change in the total energy of the system} \quad (1.44)$$

Therefore, when a system is undergoing any process, the energy entering the system minus that leaving the system must equal any change in the energy of the system, or

$$E_{in} - E_{out} = \Delta E_{system} \quad (1.45)$$

We can also write the energy balance per unit time as a rate expression:

$$\dot{E}_{in} - \dot{E}_{out} = \Delta \dot{E}_{system} \quad (1.46)$$

We use the dot above the E to note that the units are per unit time. Therefore, $\dot{E}_{in}$ is the rate of energy at the inlet, J/s.

When applying the first law of thermodynamics to engineering problems, we need to account for all forms of energy that are important for the given system. We will consider each major form of energy that is important in the analysis of food engineering problems and discuss each based on the type of a system at hand—closed or open.

1.19 ENERGY BALANCE FOR A CLOSED SYSTEM

Recall that for a closed system, energy may transfer across the system boundaries, but the boundaries are impervious to transfer of mass. The key interactions between a system and its surroundings are due to heat transfer and different forms of work. We will first consider each of these interactions and then combine them into an energy balance, as suggested in the First Law of Thermodynamics.

1.19.1 Heat

Heat transfer between a system and its surroundings is probably the most prevalent form of energy that we observe in many food engineering systems. Heat plays a major role in cooking, preservation, and creating new food products with unique properties.

Heat is an energy form that is easy to sense because of its association with temperature. We know that heat transfers from a hot object to a cold one because of the temperature difference. Heat transfer plays a significant role in food engineering systems, and thus we will devote a separate chapter (Chapter 4) to examining heat transfer in more detail. It is sufficient to note here that heat exchange between a system and its surroundings is temperature driven.

We will denote heat with a symbol Q, with units of joule (J). A **sign convention** is used in thermodynamics regarding transfer of heat across a system boundary. If heat transfer is from a system to its surroundings, then Q is **negative**. On the other hand, if heat is transferring into a system from its surroundings (such as in heating of a potato), then heat transfer Q is **positive**.

If we consider heat transfer per unit time, then we express it as rate of heat transfer, q, with the units J/s or watts (W).

Thermal energy, Q, can be determined if the heat capacity c is known. Thus,

$$Q = m \int_{T_1}^{T_2} c\mathrm{d}T \tag{1.47}$$

If the path for energy transfer is under constant pressure, then

$$Q = m \int_{T_1}^{T_2} c_\mathrm{p}\mathrm{d}T \tag{1.48}$$

where c_p is the specific heat capacity at constant pressure, J/(kg K).

Under constant volume conditions,

$$Q = m \int_{T_1}^{T_2} c_\mathrm{v}\mathrm{d}T \tag{1.49}$$

where c_v is the specific heat capacity at constant volume, J/(kg K).

The numeric values of c_p and c_v are similar for solids and liquids; however, they may be considerably different for gases.

1.19.2 Work

Work encompasses all interactions between a system and its surroundings that are not a result of temperature difference. There are many such interactions. Examples include the movement of a piston in an engine, an electric wire conducting electric current across a system boundary, and a shaft transmitting energy from a motor to another piece of equipment that is enclosed in the system boundary.

The symbol used for work is W, and its units are joules (J). The **sign convention** used for work (W) is that any time work is done *by* a system, W is positive; when work is done *on* a system, W is negative. This is opposite to that of heat transfer.

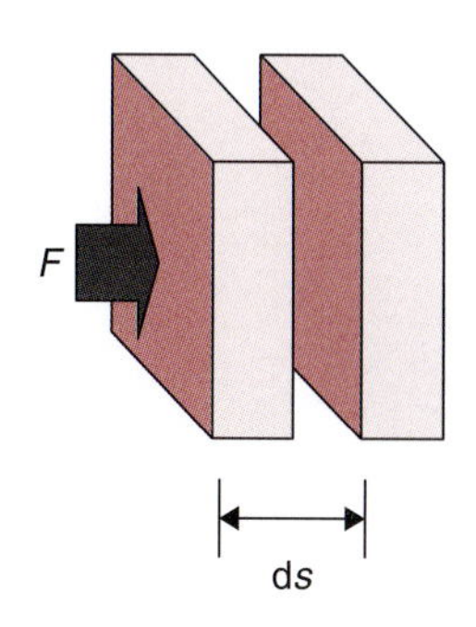

Figure 1.15 Work associated with movement of an object.

A general mathematical formulation for any work interaction is developed as follows. As shown in Figure 1.15, consider a case where an object moves a small distance ds due to the application of force F. The work done on the system may then be calculated as the product of force and distance, or

$$dW = -F ds \tag{1.50}$$

The negative sign reflects the sign convention stated earlier. We can calculate the total work done in moving the object from location 1 to 2 as

$$W_{1-2} = -\int_1^2 F ds = F(s_1 - s_2) \tag{1.51}$$

The work interaction between a system and its surroundings can be attributed to several mechanisms, such as a moving boundary, gravitational forces, acceleration, and shaft rotation. We will develop mathematical expressions for each of these in the following sections.

1.19.2.1 *Work Due to a Moving Boundary*

A common example of an energy delivery system is an engine where gas enclosed in a cylinder moves a piston. Another case is that of compressing a gas, such as in a bicycle pump, where a piston is moved to compress a gas enclosed in the cylinder. In these examples, the system boundary moves due to application of force, and work is transferred across the system boundary.

Let us consider the case of a piston and a cylinder, as shown in Figure 1.16; the system boundary is drawn around the gas. Note that the piston and cylinder are not a part of the system, but they belong to the surroundings of the system. Next, we place the cylinder on a heater, and apply a constant pressure to the piston. As the gas heats, it expands and causes the piston to move from location 1 to 2. Since the system boundary is flexible, it expands as the piston moves outward. In this case, work is being done by the expanding gas, or by the system.

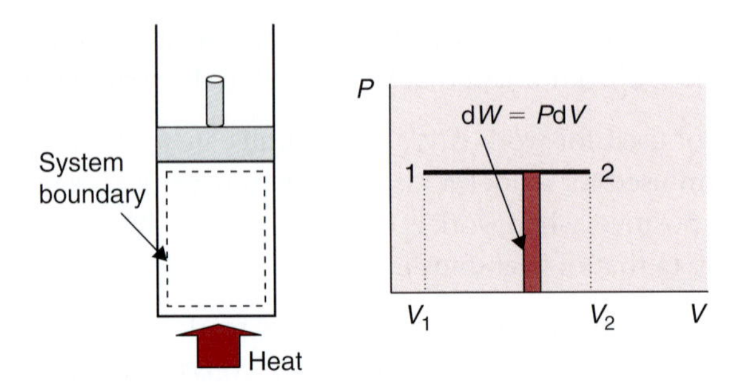

■ **Figure 1.16** Work due to moving boundary.

When the piston moves by a small distance, d*s*, then the differential amount of work done by the system is force, *F*, times the distance, *ds*:

$$dW = Fds \tag{1.52}$$

But, from Equation (1.15), Force/Area = Pressure. Therefore, if the cross-sectional area of the piston is *A*, then

$$dW = PAds = PdV \tag{1.53}$$

If the piston moves from location 1 to 2, then

$$W_{1-2} = \int_1^2 PdV \tag{1.54}$$

According to gas laws, the relationship between pressure and volume is inversely related; when pressure increases, volume decreases (or the gas is compressed), and when pressure decreases, the volume increases (or the gas expands). Therefore, in our example, as the pressure is kept constant and the gas expands due to application of heat, then the volume V_2 will be larger than V_1, and the work W_{1-2} will be positive. This agrees with our sign convention that the work is being done by the system; in moving the piston from location 1 to 2, the gas expands. On the other hand, if there is no heat supplied to the cylinder, and the gas is compressed by the movement of the piston with a downward stroke, then the final volume V_2 will be smaller than initial volume V_1, and the work calculated using Equation (1.54) would be negative, indicating that work is being done *on* the system.

It is important to understand that the work interaction and heat transfer are mechanisms of energy transfer across a system boundary. They are not properties; therefore, they depend on the path taken during a process. In case of the work associated with a moving boundary, as presented in this section, we need to know the pressure volume path. A typical path, shown in Figure 1.16, depicts a process from state 1 to 2. As pressure remains constant, the volume changes from V_1 to V_2. The area under the curve is the work done.

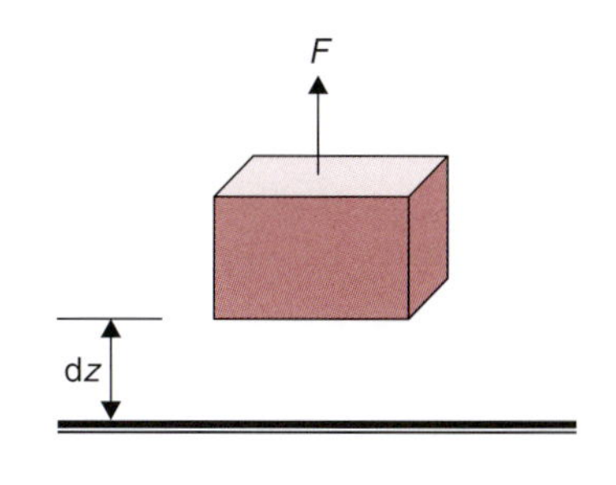

■ **Figure 1.17** Work due to raising an object.

1.19.2.2 *Work Due to Gravitational Forces*

The work done by or against the force of gravity may be calculated by using the definition of force according to Newton's[4] second law of motion:

$$F = mg \tag{1.55}$$

If an object of mass m is lifted a small distance dz, as shown in Figure 1.17, the required work is

$$\mathrm{d}W = F\mathrm{d}z \tag{1.56}$$

or, substituting Equation (1.55)

$$\mathrm{d}W = mg\mathrm{d}z \tag{1.57}$$

To lift an object from location 1 to 2:

$$\int_1^2 \mathrm{d}W = \int_1^2 mg\mathrm{d}z \tag{1.58}$$

or

$$W = mg(z_2 - z_1) \tag{1.59}$$

From Equation (1.59), we can see that the work done due to gravitational forces is equivalent to the change in potential energy of the system.

1.19.2.3 *Work Due to Change in Velocity*

If an object is moving at a velocity u_1 and we want to determine the work required to change the velocity to u_2, we will again use Newton's second law of motion. Accordingly, force is

$$F = ma \tag{1.60}$$

[4] Sir Isaac Newton (1643–1727), an English physicist and mathematician, laid the foundation of calculus, discovered the composition of white light, studied mechanics of planetary motion, derived the inverse square law, and in 1687, authored the *Principia*.

but acceleration is expressed as

$$a = \frac{du}{dt} \tag{1.61}$$

If the object moves a small distance ds during time dt, then the velocity is

$$u = \frac{ds}{dt} \tag{1.62}$$

The definition of work is

$$W = F ds \tag{1.63}$$

Substituting Equations (1.61) and (1.62) in Equation (1.63)

$$W = m\frac{du}{dt}u dt \tag{1.64}$$

Simplifying and setting up integrals,

$$W = m\int_1^2 u du \tag{1.65}$$

we obtain

$$W = \frac{m}{2}(u_2^2 - u_1^2) \tag{1.66}$$

Thus, the work done in changing velocity is equal to the change in kinetic energy of the system.

1.19.2.4 *Work Due to Shaft Rotation*

Transmitting energy by a rotating shaft is common in many energy delivery systems. For example, in an electric motor, a rotating shaft provides a source of mechanical energy to the connected equipment. Similarly, in an automobile, the energy from an engine is transmitted to the wheels by a rotating shaft. As shown in Figure 1.18, the torque Ω applied to a shaft is determined if the force and radius are known. Thus,

$$\Omega = Fr \tag{1.67}$$

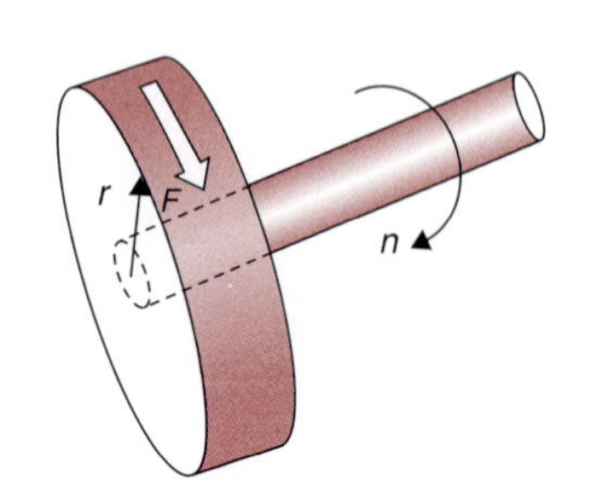

Figure 1.18 Work due to a rotating shaft.

If the movement along the circumference is distance s, and the number of revolutions is n, then the distance traveled along the circumference during n revolutions is

$$s = (2\pi r)n \tag{1.68}$$

Since the product of force and distance is work, from Equations (1.67) and (1.68), we get

$$W = \frac{\Omega}{r}(2\pi r)n \tag{1.69}$$

$$W = (2\pi n)\Omega \tag{1.70}$$

1.19.2.5 *Work Due to Frictional Forces*

If friction is present in the system, work must be done to overcome it. If we denote frictional energy as E_f, in units of J, then work due to frictional forces is

$$W = -E_f \tag{1.71}$$

1.19.2.6 *Energy Balance*

According to the first law of thermodynamics, the total change in energy of a closed system is equal to the heat added to the system minus the work done by the system. Using appropriate signs based on conventions, this can be written mathematically as

$$\Delta E = Q - W \tag{1.72}$$

The total change in energy ΔE of a system is composed of internal thermal energy E_i, kinetic energy E_{KE}, and potential energy E_{PE}. Thus,

$$\Delta E_i + \Delta E_{KE} + \Delta E_{PE} = Q - W \tag{1.73}$$

Based on discussions in Section 1.19.2, for the expansion of a gas in a cylinder (Fig. 1.16), a complete equation for work may be written as

$$W = \int P dV - \Delta E_{KE} - \Delta E_{PE} - E_f \tag{1.74}$$

Equation (1.74) may be rearranged as

$$W + E_f = \int P dV - \Delta E_{KE} - \Delta E_{PE} \tag{1.75}$$

Eliminating W, ΔE_{KE}, and ΔE_{PE} between Equation (1.73) and Equation (1.75),

$$\Delta E_i = Q + E_f - \int P dV \tag{1.76}$$

From an elementary theorem of calculus, we know that

$$d(PV) = P dV + V dP \tag{1.77}$$

Or, integrating,

$$\Delta PV = \int PdV + \int VdP \tag{1.78}$$

Thus, we can write

$$\int PdV = \Delta PV - \int VdP \tag{1.79}$$

Or substituting Equation (1.79) into Equation (1.76),

$$\Delta E_i + \Delta PV = Q + E_f + \int VdP \tag{1.80}$$

Writing Equation (1.80) in expanded form, noting that Δ for internal energy means final minus initial energy in time, and Δ for other terms means out minus in,

$$E_{i,2} - E_{i,1} + P_2V_2 - P_1V_1 = Q + E_f + \int VdP \tag{1.81}$$

$$(E_{i,2} + P_2V_2) - (E_{i,1} + P_1V_1) = Q + E_f + \int VdP \tag{1.82}$$

In Section 1.10, $E_i + PV$ was defined as enthalpy, H. Thus,

$$H_2 - H_1 = Q + E_f + \int VdP \tag{1.83}$$

Enthalpy H is used widely in process calculations. Tabulated values of enthalpy are available for many substances, such as steam, ammonia, and food products (see Tables A.4.2, A.6.2, and A.2.7).

For a heating process under constant pressure, friction is absent, and the third term on the right-hand side in Equation (1.83) is zero; thus,

$$H_2 - H_1 = Q \tag{1.84}$$

or

$$\Delta H = Q \tag{1.85}$$

Constant pressure processes are encountered most commonly in food processing applications. Thus, from Equation (1.85), the change in enthalpy is simply called the **heat content**.

The change in enthalpy, ΔH, of a system can be determined by actually measuring the change in heat content, Q, for a batch heating process, provided the process occurs at a constant pressure. A calculation procedure may be used to determine the change in enthalpy by using either measured or tabulated properties. We will consider two cases: a process involving sensible heating/cooling, and another process where heating/cooling involves a phase change.

Sensible heating at constant pressure

If the heating process involves an increase in temperature from T_1 to T_2, then

$$\Delta H = H_2 - H_1 = Q = m \int_{T_1}^{T_2} c_p \mathrm{d}T \qquad (1.86)$$

$$\Delta H = mc_p(T_2 - T_1) \qquad (1.87)$$

where c_p is heat capacity (J/[kg °C]), m is mass, T is temperature, and 1 and 2 are initial and final values.

Heating at constant pressure involving phase change

Heating/cooling processes involving latent heat may occur where the temperature remains constant while the latent heat is added or removed. For example, when ice is melted, latent heat of fusion is required. Similarly, latent heat of vaporization must be added to water to vaporize it into steam. The latent heat of fusion for water at 0°C is 333.2 kJ/kg. The latent heat of vaporization of water varies with temperature and pressure. At 100°C, the latent heat of vaporization of water is 2257.06 kJ/kg.

Example 1.11

Five kilograms of ice at −10°C is heated to melt it into water at 0°C; then additional heat is added to vaporize the water into steam. The saturated vapors exit at 100°C. Calculate the different enthalpy values involved in the process. Specific heat of ice is 2.05 kJ/(kg K). Specific heat of water is 4.182 kJ/(kg K), latent heat of fusion is 333.2 kJ/kg, and latent heat of vaporization at 100°C is 2257.06 kJ/kg.

Given

Figure E1.12 shows a plot of temperature versus enthalpy. Note that temperature remains constant in regions that involve latent heat.

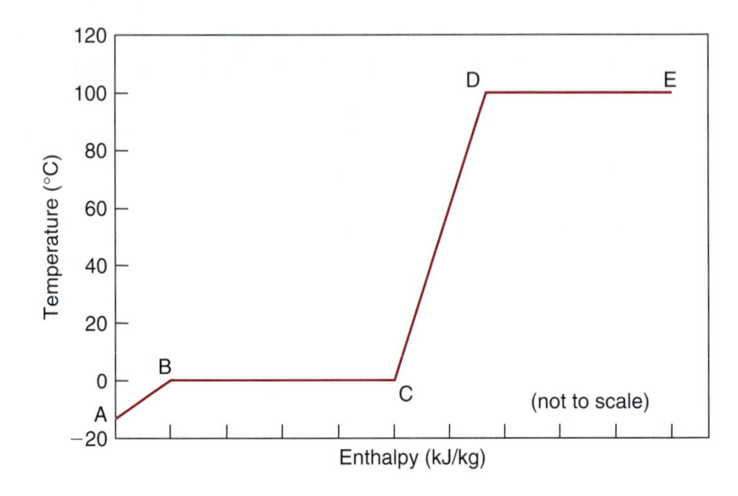

Figure E1.12 A plot of temperature versus enthalpy for melting of ice and vaporization of water.

Solution

The enthalpy calculations are made separately for each zone in Figure E1.12.

1. *Zone A–B*

$$\Delta H_{AB} = \Delta Q = m\int_{-10}^{0} c_p dT$$

$$= 5(kg) \times 2.05\left(\frac{kJ}{kg°C}\right)(0+10)°C$$

$$= 102.5\frac{kJ}{kg}$$

2. *Zone B–C*

$$\Delta H_{BC} = mH_{latent}$$

$$= 5(kg) \times 333.2\left(\frac{kJ}{kg}\right)$$

$$= 1666\ kJ$$

3. *Zone C–D*

$$\Delta H_{CD} = \Delta Q = m\int_{0}^{100} c_p dT$$

$$= 5(kg) \times 4.182\left(\frac{kJ}{kg\ °C}\right) \times (100-0)(°C)$$

$$= 2091 kJ$$

4. *Zone D–E*

$$\begin{aligned}\Delta H_{DE} &= mH_{latent} \\ &= 5(kg)\times 2257.06\left(\frac{kJ}{kg}\right) \\ &= 11,285.3\,kJ\end{aligned}$$

5. *Total change in enthalpy*

$$\begin{aligned}\Delta H &= \Delta H_{AB} + \Delta H_{BC} + \Delta H_{CD} + \Delta H_{DE} \\ &= 102.5 + 1666 + 2091 + 11,285.3 \\ &= 15,144.8\,kJ\end{aligned}$$

It is evident that almost 70% of the enthalpy is associated with the vaporization process.

1.20 ENERGY BALANCE FOR AN OPEN SYSTEM

The transfer of mass across a system boundary, in addition to work and energy, characterizes open systems. Any mass that enters or exits the system carries a certain amount of energy into or out of the system, respectively. Therefore, we need to account for the change in the energy of the system due to mass flow. The work associated with flow sometimes is referred to as **flow work**. We can calculate flow work by determining the work involved in pushing a certain mass through the system boundary. Consider a differential element of fluid of uniform properties entering an open system (Fig. 1.19). If the cross-sectional area of the element is A and the pressure of the fluid is P, then the force required to push this element through the system boundary is

$$F = PA \tag{1.88}$$

If the element is pushed a distance L, then the work done on the fluid element is

$$W_{\text{mass flow}} = FL = PAL = PV \tag{1.89}$$

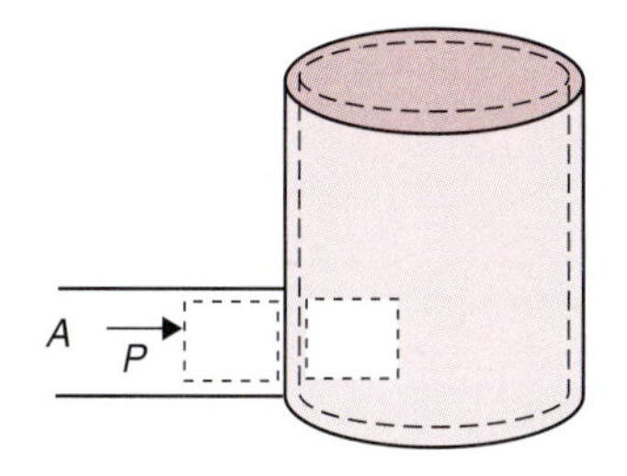

Figure 1.19 Movement of a liquid volume.

According to Equation (1.43), the total energy of the fluid element shown in Figure 1.19 includes the kinetic, potential, and internal energies. In addition, we must account for the energy associated with the flowing fluid or flow work. Therefore,

$$E = E_{\text{i}} + E_{\text{KE}} + E_{\text{PE}} + PV \tag{1.90}$$

or, substituting individual terms for the energy components,

$$E = E_{\mathrm{i}} + \frac{mu^2}{2} + mgz + PV \tag{1.91}$$

1.20.1 Energy Balance for Steady Flow Systems

When a system is in a steady state, its properties do not change with time. They may be different from one location to another. This is a very common situation encountered in numerous engineering systems. For a steady state system, there is no change in the energy of the system with time. Using this condition, we can modify the rate form of the energy balance as follows:

$$\dot{E}_{\mathrm{in}} = \dot{E}_{\mathrm{out}} = \Delta \dot{E}_{\mathrm{system}} = 0 \tag{1.92}$$

Therefore,

$$\dot{E}_{\mathrm{in}} = \dot{E}_{\mathrm{out}} \tag{1.93}$$

1.21 A TOTAL ENERGY BALANCE

Substituting all the individual terms presented in the preceding sections into Equation (1.93), we get

$$\begin{aligned} Q_{\mathrm{in}} + W_{\mathrm{in}} + \sum_{j=1}^{p} m_{\mathrm{i}} \left(E'_{i,j} + \frac{u_j^2}{2} + gz_j + P_j V'_j \right) \\ = Q_{\mathrm{out}} + W_{\mathrm{out}} + \sum_{e=1}^{q} m_e \left(E'_{i,e} + \frac{u_e^2}{2} + gz_e + P_e V'_e \right) \end{aligned} \tag{1.94}$$

In Equation (1.94) we have used E' for internal energy per unit mass and V' for specific volume. This general equation is for a system that may have p inlet streams or q exit streams for a fluid to flow in or out, respectively. If there is only one inlet (location 1) and exit (location 2) for a system, then

$$Q_{\mathrm{m}} = \left(\frac{u_2^2}{2} + gz_2 + \frac{P_2}{\rho_2} \right) - \left(\frac{u_1^2}{2} + gz_1 + \frac{P_1}{\rho_1} \right) + (E'_{i,2} - E'_{i,1}) + W_{\mathrm{m}} \tag{1.95}$$

where Q_{m} and W_{m} are the transfer of heat and work per unit mass, respectively. The specific volume V' is replaced with $1/\rho$ in Equation (1.95). A more detailed discussion of energy balances in steady flow systems will be presented in Chapter 2. The following examples illustrate the use of energy balances in food processing applications.

Example 1.12

A tubular water blancher is being used to process lima beams (Fig. E1.13). The product mass flow rate is 860 kg/h. It is found that the theoretical energy consumed for the blanching process amounts to 1.19 GJ/h. The energy lost due to lack of insulation around the blancher is estimated to be 0.24 GJ/h. If the total energy input to the blancher is 2.71 GJ/h,

a. Calculate the energy required to reheat water.
b. Determine the percent energy associated with each stream.

Given

Product mass flow rate = 860 kg/h
Theoretical energy required by product = 1.19 GJ/h
Energy lost due to lack of insulation = 0.24 GJ/h
Energy input to the blancher = 2.71 GJ/h

Approach

We will first write an energy balance and then solve for the unknowns.

Solution

1. *Select 1 h as a basis.*
2. *Energy balance may be written as follows:*
   Energy input to blancher = energy out with product + energy loss due to lack of insulation + energy out with water
3. *Substituting appropriate values in the energy balance,*

$$2.71 = 1.19 + 0.24 + E_W$$

we get

$$E_W = 1.28 \text{ GJ/h}$$

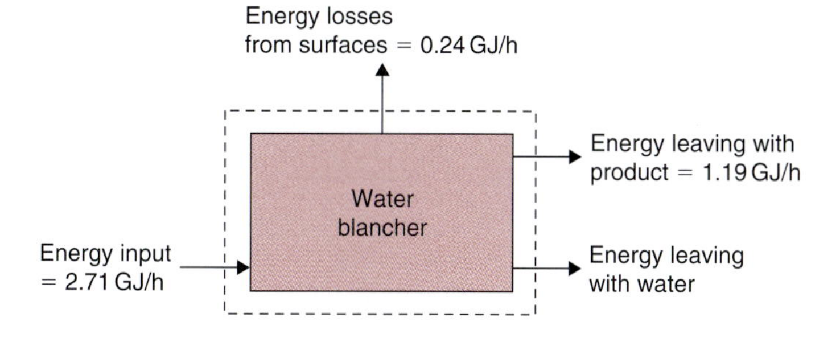

Figure E1.13 A schematic of a water blancher.

Thus, the blancher requires 2.71 − 1.28 = 1.43 GJ/h to both reheat water and maintain it at conditions necessary to accomplish the blanching process.

4. *These values can be converted to percentage of total thermal input as follows:*

$$\text{Energy out with product} = \frac{(1.19)}{2.71}(100) = 43.91\%$$

$$\text{Energy loss due to lack of insulation} = \frac{(0.24)(100)}{2.71} = 8.86\%$$

$$\text{Energy out with water} = \frac{(1.28)(100)}{2.71} = 47.23\%$$

5. *The results indicate that this water blancher operates at about 44% thermal energy efficiency.*

Example 1.13

Steam is used for peeling potatoes in a semicontinuous operation. Steam is supplied at the rate of 4 kg per 100 kg of unpeeled potatoes. The unpeeled potatoes enter the system with a temperature of 17°C, and the peeled potatoes leave at 35°C. A waste stream from the system leaves at 60°C (Fig. E1.14). The specific heats of unpeeled potatoes, waste stream, and peeled potatoes are 3.7, 4.2, and 3.5 kJ/(kg K), respectively. If the heat content (assuming 0°C reference temperature) of the steam is 2750 kJ/kg, determine the quantities of the waste stream and the peeled potatoes from the process.

Given

Mass flow of steam = 4 kg per 100 kg of unpeeled potatoes
Temperature of unpeeled potatoes = 17°C
Temperature of peeled potatoes = 35°C
Temperature of waste stream = 60°C

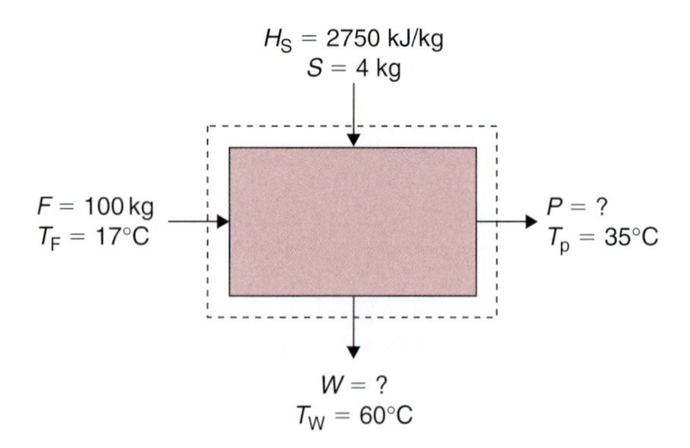

■ Figure E1.14 A block diagram showing various streams described in Example 1.13.

Specific heat of unpeeled potatoes = 3.7 kJ/(kg K)
Specific heat of peeled potatoes = 3.5 kJ/(kg K)
Specific heat of waste stream = 4.2 kJ/(kg K)
Heat content of steam = 2750 kJ/kg

Solution

1. *Select 100 kg of unpeeled potatoes as a basis.*
2. *From mass balance*

$$F + S = W + P$$

$$100 + 4 = W + P$$

$$W = 104 - P$$

3. *From energy balance*

$$Fc_p(T_F - 0) + SH_s = Wc_p(T_w - 0) + Pc_p(T_p - 0)$$

$$100(3.7)(17) + 4(2750) = W(4.2)(60) + P(3.5)(35)$$

$$6290 + 11{,}000 = 252W + 122.5P$$

4. *From step (3)*

$$17{,}290 = 252(104 - P) + 122.5P$$

$$252P - 122.5P = 26{,}208 - 17{,}290 = 8918$$

$$P = 68.87\ \text{kg}$$

$$W = 35.14\ \text{kg}$$

1.22 POWER

Power is defined as the rate of doing work, using the dimensions (mass)(length)^2(time)$^{-3}$ and the SI unit of watts (W). In the English system of units, the commonly used unit for power is horsepower (hp), where 1 hp = 0.7457 kW.

1.23 AREA

Area is a quantitative measure of a plane or a curved surface. It is defined as the product of two lengths. In the SI system, the unit is square meters (m^2).

The surface area of a food product is required in a number of process calculations. For example, when calculating heat and mass transfer across the surface of a food material, the surface area of the food must

Table 1.8 Surface Area of Foods

	Mean surface area (cm^2)
Apple, Delicious	140.13
Pear, Bartlett	145.42
Plum, Monarch	35.03
Egg (60 g)	70.5

Source: Mohsenin (1978)

be known. Certain physical processes increase the surface area; for example, prior to spray drying, a liquid stream is converted to spray droplets, increasing the surface area of the liquid and enhancing the drying process. The surface areas of some foods are given in Table 1.8. Certain food processing applications require that the surface-area-to-volume ratio be known. For example, in the canning of foods, a higher surface-area-to-volume ratio will result in faster heating of the geometrical center of the container, thus minimizing overheating of the product. For this reason, retortable pouches are often considered better than cylindrical cans. Because of their slab shape, the pouches have high surface area per unit volume, thus resulting in a more rapid heating of the slowest heating point when compared with cylindrical cans. Among various geometric shapes, a sphere has the largest surface-area-to-volume ratio.

PROBLEMS

1.1 The following unit conversions are illustrations of the SI system of units. Convert:

a. a thermal conductivity value of 0.3 Btu/(h ft °F) to W/(m °C)

b. a surface heat transfer coefficient value of 105 Btu/(h ft^2 °F) to W/(m^2 °C)

c. a latent heat of fusion value of 121 Btu/lb_m to J/kg

1.2 An empty metal can is heated to 90°C and sealed. It is then placed in a room to cool to 20°C. What is the pressure inside the can upon cooling? Assume that can contains air under ideal conditions.

1.3 A food is initially at a moisture content of 90% dry basis. Calculate the moisture content in wet basis.

1.4 10 kg of food at a moisture content of 320% dry basis is dried to 50% wet basis. Calculate the amount of water removed.

***1.5** A batch of 5 kg of food product has a moisture content of 150% dry basis. Calculate how much water must be removed from this product to reduce its moisture content to 20% wet basis.

1.6 A liquid product with 10% product solids is blended with sugar before being concentrated (removal of water) to obtain a final product with 15% product solids and 15% sugar solids. Determine the quantity of final product obtained from 200 kg of liquid product. How much sugar is required? Compute mass of water removed during concentration.

1.7 A food product is being frozen in a system capable of removing 6000 kJ of thermal energy. The product has a specific heat of 4 kJ/(kg °C) above the freezing temperature of −2°C, the latent heat of fusion equals 275 kJ/kg, and the frozen product has a specific heat of 2.5 kJ/(kg°C) below −2°C. If 10 kg of product enters the system at 10°C, determine the exit temperature of the product.

1.8 A liquid food product is being cooled from 80°C to 30°C in an indirect heat exchanger using cold water as a cooling medium. If the product mass flow rate is 1800 kg/h, determine the water flow rate required to accomplish product cooling if the water is allowed to increase from 10°C to 20°C in the heat exchanger. The specific heat of the product is 3.8 kJ/(kg K) and the value for water is 4.1 kJ/(kg K).

1.9 Milk is flowing through a heat exchanger at a rate of 2000 kg/h. The heat exchanger supplies 111,600 kJ/h. The outlet temperature of the product is 95°C. Determine the inlet temperature of the milk. The product specific heat is 3.9 kJ/(kg °C).

1.10 A steel bucket contains 4 liters of water at 12°C. An electric immersion heater rated at 1400 Watts is placed in the bucket. Determine how long it will take for water to heat to 70°C. Assume that the empty bucket weighs 1.1 kg. The specific heat of steel is 0.46 kJ/(kg°C). Use an average specific heat of water of 4.18 kJ/(kg°C). Disregard any heat loss to the surroundings.

---

*Indicates an advanced level in solving.

***1.11** Use MATLAB® to plot the enthalpy versus temperature for water over the range of temperature from –40°C to 40°C using the following data taken from Pham et al. (1994). Find enthalpy data for water from a standard source (e.g., Green and Perry, 2008) and plot it along with the following data from Pham. Discuss possible reasons for differences in the results.

T (°C)	Enthalpy (kJ/kg)
−44.8	−7.91
−36.0	6.46
−27.6	23.2
−19.5	40.0
−11.4	56.8
−3.35	73.1
0.04	123
0.07	236
0.08	311
0.18	385
3.95	433
10.5	460
17.0	487
23.4	514
29.8	541
36.2	568
42.5	595

LIST OF SYMBOLS

A	area (m^2)
c	specific heat (kJ/[kg°C])
c_p	specific heat at constant pressure (kJ/[kg°C])
c_v	specific heat at constant volume (kJ/[kg°C])
E	energy (J/kg)
E_i	internal energy (kJ)
E'_i	specific internal energy (kJ/kg)
E_{KE}	kinetic energy (kJ/kg)
E_{PE}	potential energy (kJ/kg)
F	force (N)

*Indicates an advanced level in solving.

g	acceleration due to gravity (m/s^2)
h	height of fluid (m)
H	enthalpy (kJ)
H'	enthalpy per unit mass (kJ/kg)
I	height of mercury column (in)
m	mass (kg)
$\dot{m}$	mass flow rate (kg/s)
MC_{db}	moisture content, dry basis (kg water/kg dry product)
MC_{wb}	moisture content, wet basis (kg water/kg wet product)
M'	molality (mol solute/kg solvent)
M	molecular weight
n	number of moles
P	pressure (Pa)
Q	heat content (kJ/kg)
ρ	density (kg/m^3)
R	gas constant (m^3 Pa/[kg mol K])
R_0	universal gas constant, 8314.41 (m^3 Pa/[kg mol K])
τ	time constant (s)
T	temperature (°C)
U	internal energy (kJ/kg)
u	velocity (m/s)
V'	specific volume (m^3/kg)
V	volume (m^3)
W	work (kJ)
x	mass fraction (dimensionless)
X_A	mole fraction of A
z	distance (m)
Ω	torque

BIBLIOGRAPHY

Cengel, Y. A. and Boles, M. A. (2006). *Thermodynamics. An Engineering Approach*, 5th ed. McGraw Hill, Boston.

Chandra, P. K. and Singh, R. P. (1994). *Applied Numerical Methods for Agricultural Engineers*. CRC Press, Inc., Boca Raton, Florida.

Earle, R. L. (1983). *Unit Operations in Food Processing*, 2nd ed. Pergamon Press, Oxford.

Green, D. W. and Perry, R. H. (2008). *Perry's Chemical Engineer's Handbook*, 8th ed. McGraw-Hill Book Co., New York.

Himmelblau, D. M. (1967). *Basic Principles and Calculations in Chemical Engineering*, 2nd ed. Prentice-Hall, Englewood Cliffs, New Jersey.

Mohsenin, N. N. (1978). *Physical Properties of Plant and Animal Materials: Structure, Physical Characteristics and Mechanical Properties*, 2nd ed. Gordon and Breach Science Publishers, New York.

Peleg, M. (1983). Physical characteristics of food powders. In *Physical Properties of Foods*, M. Peleg and E. B. Bagley, eds. AVI Publ. Co, Westport, Connecticut.

Pham, Q. T., Wee, H. K., Kemp, R. M., and Lindsay, D. T. (1994). Determination of the enthalpy of foods by an adiabatic calorimeter. *J. Food Engr.* **21**: 137–156.

Singh, R. P. (1996). *Computer Applications in Food Technology*. Academic Press, San Diego.

Singh, R. P. (1996). Food processing. In *The New Encyclopaedia Britannica*, Vol. 19, 339–346, 405.

Singh, R. P. and Oliveira, F. A. R. (1994). *Minimal Processing of Foods and Process Optimization – An Interface*. CRC Press, Inc., Boca Raton, Florida.

Singh, R. P. and Wirakartakusumah, M. A. (1992). *Advances in Food Engineering*. CRC Press, Inc., Boca Raton, Florida.

Singh, R. P., Carroad, P. A., Chinnan, M. S., Rose, W. W., and Jacob, N. L. (1980). Energy accounting in canning tomato products. *J. Food Sci.* **45**: 735–739.

Smith, P. G. (2003). *Introduction to Food Process Engineering*. Kluwer Academic/Plenum Publishers, New York.

Toledo, R. T. (2007). *Fundamentals of Food Process Engineering*, 3rd ed. Springer Science+Business Media, New York.

Watson, E. L. and Harper, J. C. (1988). *Elements of Food Engineering*, 2nd ed. Van Nostrand Reinhold, New York.

Chapter 2

Fluid Flow in Food Processing

In any commercial food processing plant, the movement of liquid foods from one location to another becomes an essential operation. Various types of systems are used for moving raw or unprocessed liquid foods as well as processed liquid products before packaging. The range of liquid foods encountered in a processing plant is extremely wide, encompassing foods with distinctly different flow properties, from milk to tomato paste. The design of these systems in food processing is significantly different from most other applications because of the essential need for sanitation to maintain product quality. The transport system must be designed to allow for ease and efficiency in cleaning.

In this chapter, we will concern ourselves mostly with the flow of fluids. Fluid is a general term used for either gases or liquids. Most of our discussion will deal with liquid foods. A fluid begins to move when a force acts upon it. At any location and time within a liquid transport system, several types of forces may be acting on a fluid, such as pressure, gravity, friction, thermal effects, electrical charges, magnetic fields, and Coriolis forces. Both the magnitude and direction of the force acting on a fluid are important. Therefore, a force balance on a fluid element is essential to determine which forces contribute to or oppose the flow.

From our daily experience with handling different kinds of fluids, we know that if pressure at one location within a fluid system is higher than another, the fluid moves toward the region of lower pressure. Gravity causes the flow of fluid from higher to lower elevations. A fluid moving to a lower elevation undergoes a decrease in its potential energy, while its kinetic energy increases. With the presence of thermal gradients, heated fluids experience a decrease in density, causing lighter fluid to rise while denser fluid takes its place.

Conceptually we may visualize that inside a fluid in motion one imaginary layer of fluid is sliding over another. The viscous forces act tangentially on the area between these imaginary layers, and they tend to

All icons in this chapter refer to the author's web site, which is independently owned and operated. Academic Press is not responsible for the content or operation of the author's web site. Please direct your web site comments and questions to the author: Professor R. Paul Singh, Department of Biological and Agricultural Engineering, University of California, Davis, CA 95616, USA. Email: rps@rpaulsingh.com.

oppose the flow. This is why if you spill honey—a highly viscous food—it moves much more slowly than milk, which has a substantially lower viscosity. All fluids exhibit some type of viscous behavior distinguished by a flow property called *viscosity*. We will examine these factors and their role in the design of equipment for transporting different types of liquid foods and ingredients to different locations within a processing plant.

2.1 LIQUID TRANSPORT SYSTEMS

A typical transport system consists of four basic components, namely, tanks, pipeline, pump, and fittings. Figure 2.1 illustrates a simple milk pasteurization line. Raw milk enters the balance tank prior to the

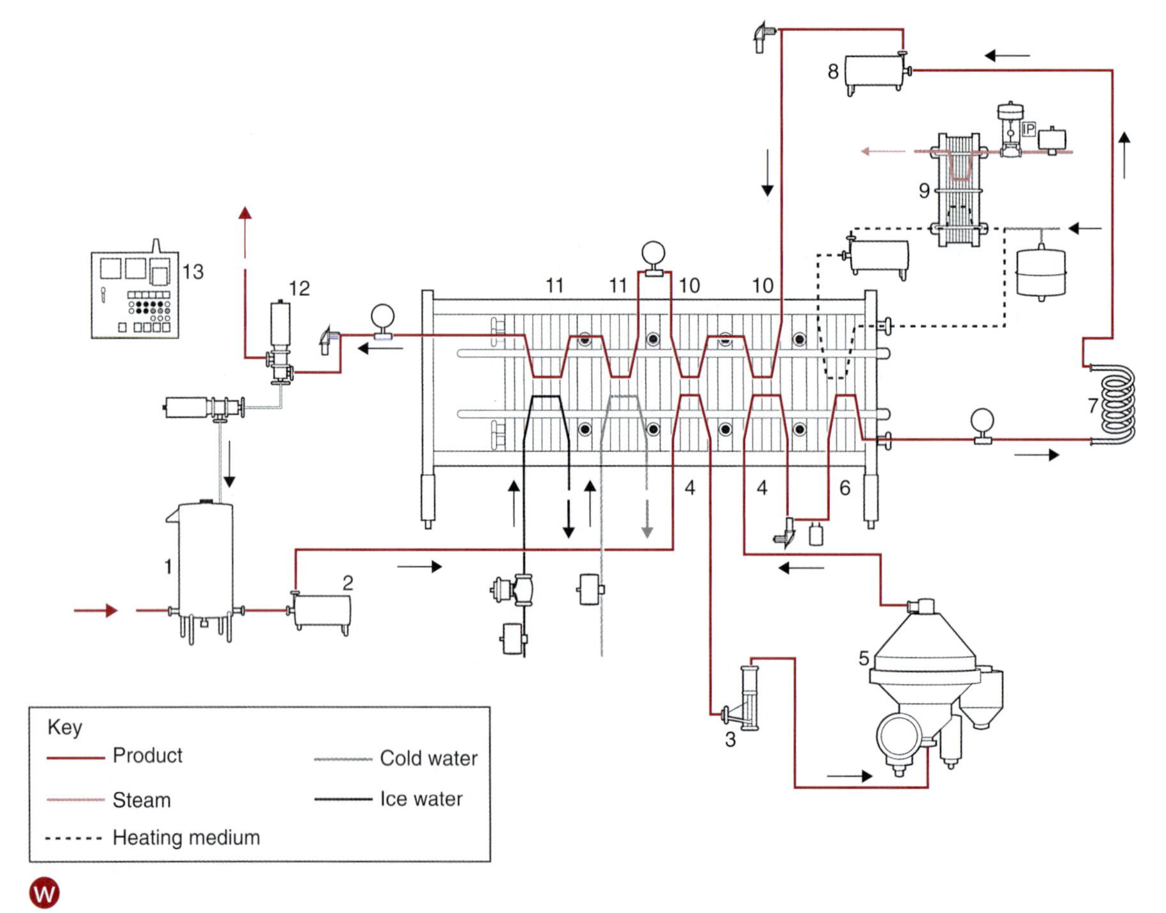

Figure 2.1 Production line for milk processing. (1) balance tank, (2) feed pump, (3) flow controller, (4) regenerative preheating section, (5) centrifugal clarifier, (6) heating section, (7) holding tube, (8) booster pump, (9) hot water heating system, (10) regenerative cooling sections, (11) cooling sections, (12) flow diversion valve, (13) control panel. (Courtesy of Tetra Pak Processing Systems AB)

pasteurization process and finally exits from the flow diversion valve. Between the tank and the valve is the conduit, or pipeline, for milk flow. Unless flow can be achieved by gravity, the third primary component is the pump, where mechanical energy is used for product transport. The fourth component of the system consists of fittings such as valves and elbows, used to control and direct flow. The tanks used in these types of systems may be of any size and configuration. In addition to the basic components of a transport system, there may be additional processing equipment as part of the system, such as a heat exchanger to pasteurize milk, as shown in Figure 2.1.

2.1.1 Pipes for Processing Plants

Fluids (liquids and gases) in food processing plants are transported mostly in closed conduits—commonly called pipes if they are round, or ducts if they are not round. Although sometimes used in processing plants, open drains generally are avoided, for sanitation reasons. The pipelines used for liquid foods and their components have numerous unique features. Probably the most evident feature is the use of stainless steel for construction. This metal provides smoothness, cleanability, and corrosion resistance. The corrosion resistance of stainless steel is attributed to "passivity"—the formation of a surface film on the metal surface when exposed to air. In practice, this surface film must reform each time after the surface is cleaned. If the protective surface film is impaired, which could occur from failure to establish passivity or any action resulting in film removal, the site is susceptible to corrosion. Therefore, stainless-steel surfaces require care to maintain corrosion resistance, especially after cleaning. A detailed description of corrosion mechanisms is given by Heldman and Seiberling (1976).

A typical pipeline system for liquid food transport contains several essential components. In addition to the straight lengths of the pipe, which may vary in diameter from 2 to 10 cm, elbows and tees are essential for changing the direction of product movement. As shown in Figure 2.2, these components are welded into the pipeline system and may be used in several different configurations. Another component is the valve used to control the flow rate; an air-actuated valve is illustrated in Figure 2.3. This valve may be operated remotely, often based on some type of preset signal.

It is essential that all components of the pipeline system contribute to sanitary handling of the product. The stainless-steel surfaces ensure smoothness needed for cleaning and sanitizing. In addition, proper use

■ **Figure 2.2** A typical liquid food processing system, illustrating pipelines and pipeline components. (Courtesy of CREPACO, Inc.)

of the system provides the desired corrosion prevention. Since cleaning of these systems most often is accomplished by cleaning-in-place (CIP), we must account for this factor in the initial design of the system.

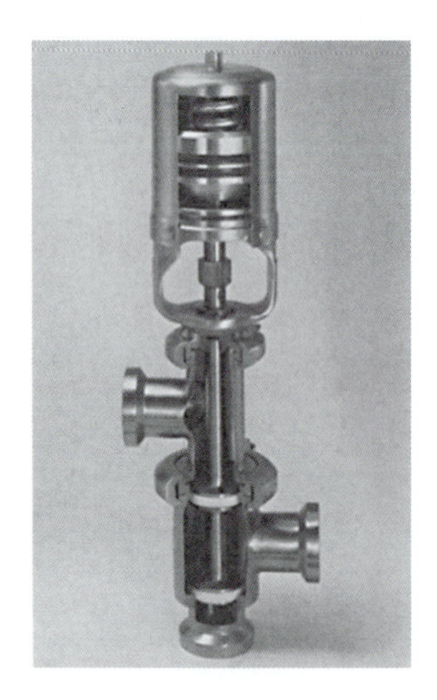

■ **Figure 2.3** An air-actuated valve for liquid foods. (Courtesy of Cherry-Burrell Corporation)

2.1.2 Types of Pumps

Except for situations where gravity can be used to move liquid products, some type of mechanical energy must be introduced to overcome the forces opposing transport of the liquid. The mechanical energy is provided by the pumps. There are numerous types of pumps used in the industry. As shown in Figure 2.4, pumps may be classified as **centrifugal** or **positive displacement**. There are variations within each of these types, as shown in the figure.

2.1.2.1 *Centrifugal Pumps*

The use of centrifugal force to increase liquid pressure is the basic concept associated with operation of a centrifugal pump. As illustrated in Figure 2.5, the pump consists of a motor-driven impeller enclosed in a case. The product enters the pump at the center of impeller rotation and, due to centrifugal force, moves to the impeller periphery. At this point, the liquid experiences maximum pressure and moves through the exit to the pipeline.

Most sanitary centrifugal pumps used in the food industry use two-vane impellers (Fig. 2.5). Impellers with three and four vanes are

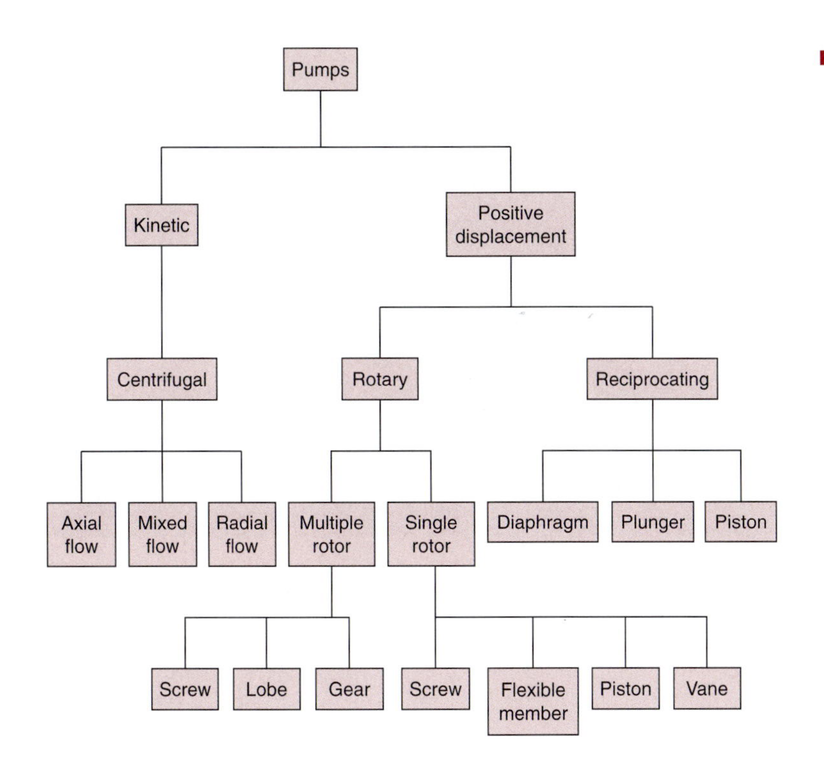

Figure 2.4 Classification of pumps.

available and may be used in some applications. Centrifugal pumps are most efficient with low-viscosity liquids such as milk and fruit juices, where flow rates are high and pressure requirements are moderate. The discharge flow from centrifugal pumps is steady. These pumps are suitable for either clean and clear or dirty and abrasive liquids. They are also used for pumping liquids containing solid particles (such as peas in water). Liquids with high viscosities, such as honey, are difficult to transport with centrifugal pumps, because they are prevented from attaining the required velocities by high viscous forces within the product.

Flow rates through a centrifugal pump are controlled by a valve installed in the pipe and connected to the discharge end of the pump. This approach provides an inexpensive means to regulate flow rate, including complete closure of the discharge valve to stop flow. Since this step will not damage the pump, it is used frequently in liquid food processing operations. However, blocking flow from a centrifugal pump for long periods of time is not recommended, because of the possibility of damage to the pump. The simple design of the centrifugal pump makes it easily adaptable to cleaning-in-place functions.

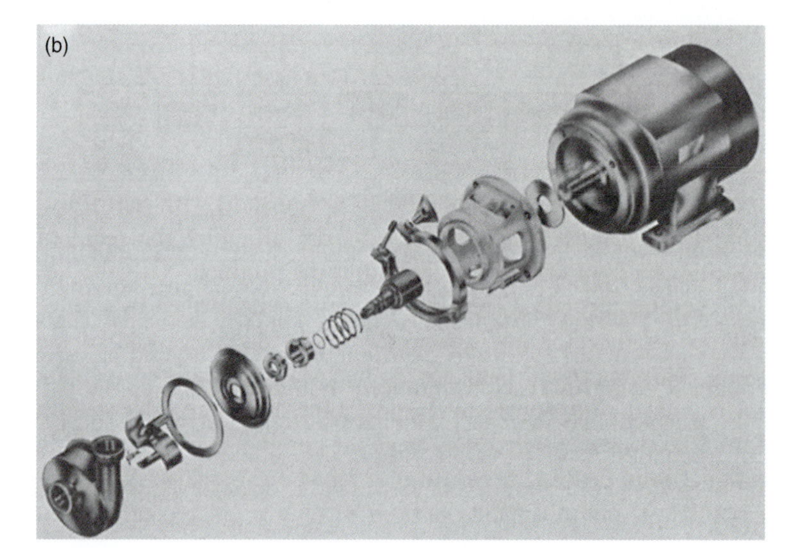

■ **Figure 2.5** (a) Exterior view of a centrifugal pump. (Courtesy of Cherry-Burrell Corporation) (b) A centrifugal pump with components. (Courtesy of CREPACO, Inc.)

2.1.2.2 *Positive Displacement Pumps*

By application of direct force to a confined liquid, a positive displacement pump produces the pressure required to move the liquid product. Product movement is related directly to the speed of the moving parts within the pump. Thus, flow rates are accurately controlled by the drive speed to the pump. The mechanism of operation also allows a positive displacement pump to transport liquids with high viscosities.

■ **Figure 2.6** A positive displacement pump with illustration of internal components. (Courtesy of Tri-Canada, Inc.)

f0060

A **rotary pump**, illustrated in Figure 2.6, is one type of positive displacement pump. Although there are several types of rotary pumps, the general operating concept involves enclosure of a pocket of liquid between the rotating portion of the pump and the pump housing. The pump delivers a set volume of liquid from the inlet to the pump outlet. Rotary pumps include sliding vane, lobe type, internal gear, and gear type pumps. In most cases, at least one moving part of the rotary pump must be made of a material that will withstand rubbing action occurring within the pump. This is an important feature of the pump design that ensures tight seals. The rotary pump has the capability to reverse flow direction by reversing the direction of rotor rotation. Rotary pumps deliver a steady discharge flow.

The second type of positive displacement pump is the **reciprocating pump**. As suggested by the name, pumping action is achieved by application of force by a piston to a liquid within a cylinder. The liquid moves out of the cylinder through an outlet valve during forward piston movement. Reciprocating pumps usually consist of several cylinder–piston arrangements operating at different cycle positions to ensure more uniform outlet pressures. Most applications are for low-viscosity liquids requiring low flow rates and high pressures. The reciprocating pumps deliver a pulsating discharge flow.

2.2 PROPERTIES OF LIQUIDS

The transport of a liquid food by one of the systems described in the previous section is directly related to liquid properties, primarily viscosity and density. These properties will influence the power requirements for liquid transport as well as the flow characteristics within the pipeline. An understanding of the physical meaning associated with these properties is necessary in order to design an optimal transport system.

Later in this chapter, we will examine different approaches used for measurement of these properties.

2.2.1 Terminology Used in Material Response to Stress

Fluid flow takes place when force is applied on a fluid. Force per unit area is defined as **stress**. When force acting on a surface is perpendicular to it, the stress is called **normal stress**. More commonly, normal stress is referred to as **pressure**. When the force acts parallel to the surface, the stress is called **shear stress**, σ. When shear stress is applied to a fluid, the fluid cannot support the shear stress; instead the fluid deforms, or simply stated, it flows.

The influence of shear stress on solids and liquids leads to a broad classification of such materials as plastic, elastic, and fluid.

In the case of an **elastic** solid, when shear stress is applied, there is a proportional finite deformation, and there is no flow of the material. On removal of the applied stress, the solid returns to its original shape.

A **plastic** material, on the other hand, deforms continuously on application of shear stress; the rate of deformation is proportional to the shear stress. When the shear stress is removed, the object shows some recovery. Examples include Jell-O® and some types of soft cheese.

A **fluid** deforms continuously on application of shear stress. The rate of deformation is proportional to the applied shear stress. There is no recovery; that is, the fluid does not retain or attempt to retain its original shape when the stress is withdrawn.

When normal stress or pressure is applied on a liquid, there is no observed appreciable effect. Thus, liquids are called **incompressible** fluids, whereas gases are **compressible** fluids, since increased pressure results in considerable reduction in volume occupied by a gas.

2.2.2 Density

The density of a liquid is defined as its mass per unit volume and is expressed as kg/m^3 in the SI unit system. In a physical sense, the magnitude of the density is the mass of a quantity of a given liquid occupying a defined unit volume. The most evident factor is that the magnitude of density is influenced by temperature. For example, the density of water is maximum at 4°C and decreases consistently with increasing temperature (Fig. 2.7).

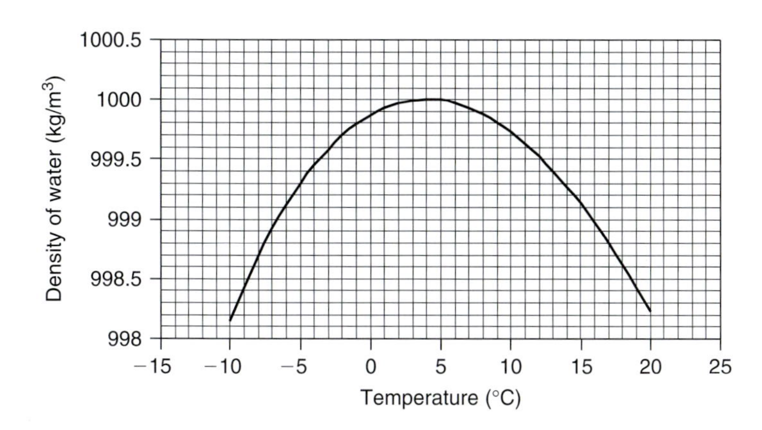

■ **Figure 2.7** Density of water as a function of temperature.

Densities of liquids are most often measured by a hand hydrometer. This instrument measures specific gravity, which is the ratio of the density of the given liquid to the density of water at the same temperature. The measuring instrument is a weighted float attached to a small-diameter stem containing a scale of specific-gravity values. The float will sink into the unknown liquid by an amount proportional to the specific gravity, and the resulting liquid level is read on the stem scale. However, when converting specific gravity values to density, care must be taken to ensure that the value for density of water at the measurement temperature is used.

2.2.3 Viscosity

A fluid may be visualized as matter composed of different layers. The fluid begins to move as soon as a force acts on it. The relative movement of one layer of fluid over another is due to the force, commonly called shearing force, which is applied in a direction parallel to the surface over which it acts. From Newton's second law of motion, a resistance force is offered by the fluid to movement, in the opposite direction to the shearing force, and it must also act in a direction parallel to the surface between the layers. This resistance force is a measure of an important property of fluids called **viscosity**.

With different types of fluids, we commonly observe a wide range of resistance to movement. For example, honey is much more difficult to pour out of a jar or to stir than water or milk. Honey is considerably more viscous than milk. With this conceptual framework, we will consider a hypothetical experiment.

Consider two parallel plates that are infinitely long and wide separated by a distance dy, as shown in Figure 2.8a. First, we place a solid block

Figure 2.8 (a) A steel block enclosed between two plates. (b) A fluid enclosed between two plates.

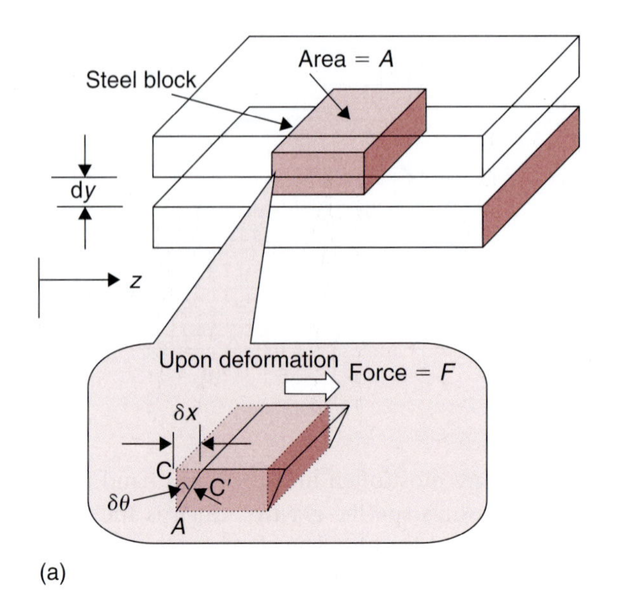

of steel between the two plates and firmly attach the steel block to the plates so that if we move a plate, the attached surface of steel block must also move with it. The bottom plate is then anchored so that it remains fixed during the entire experiment. Next, we apply a force, F, to the top plate so that it moves by a small distance, δx, to the right. Due to this displacement of the plate, an imaginary line in the steel block, AC, will rotate to AC′, and the angle of deflection will be $\delta\theta$. The force in the steel resisting the movement will be acting at the steel–plate interface, in the direction opposite to the applied force F. This opposing force acts on area A, the contact area between the plate and the steel block. The opposing force equals σA, where σ is the shear stress (force per unit area). Experimental evidence indicates that for solid materials such as steel, the angular deflection $\delta\theta$ is proportional to the shearing stress σ. When the force is removed, the steel block returns to its original shape. Therefore, steel is called an elastic material.

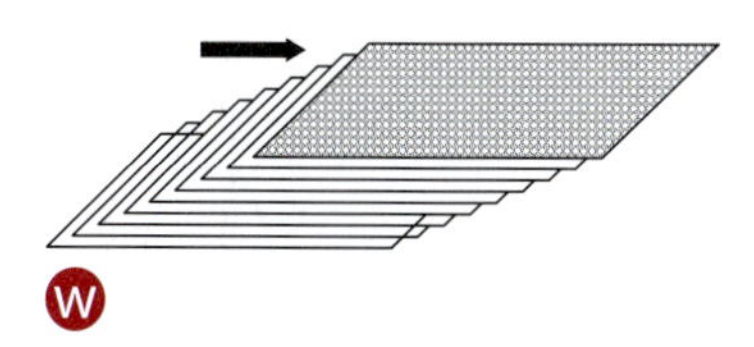

W

Figure 2.9 Illustration of drag generated on underlying cards as the top card in a deck is moved. This is analogous to the movement of the top layer of a fluid.

If we carry out the same experiment with a fluid inserted between the two plates instead of steel (Fig. 2.8b), our observations will be remarkably different. The bottom plate is anchored and remains fixed throughout the experiment. We apply force F to the top plate. After a short, transient interval, the top plate will continue to move with a velocity du as long as the force keeps acting on the top plate. The fluid layer immediately below the top plate actually "sticks" to it and moves to the right with a velocity du, whereas the bottom-most layer sticks to the bottom fixed plate and remains stationary. Between these two extreme layers, the remaining layers will also move to the right, with each successive layer dragged along by the layer immediately above it. A velocity profile, as shown in Figure 2.8b, develops between the top and bottom plate. This situation is analogous to the deck of cards shown in Figure 2.9. If the top card is moved to the right, it drags the card immediately below it, and that card drags the one below it, and so on. The drag force depends upon the frictional resistance offered by the surface in contact between the cards.

Referring to Figure 2.8b, if, in a small increment of time δt, the line AC′ deflects from AC by an angle $\delta\theta$, then

$$\tan \delta\theta = \frac{\delta x}{\mathrm{d}y} \tag{2.1}$$

For a small angular deflection,

$$\tan \delta\theta \approx \delta\theta \tag{2.2}$$

Therefore,

$$\delta\theta = \frac{\delta x}{dy} \tag{2.3}$$

But, linear displacement, δx, is equal to the product of velocity and time increment, or

$$\delta x = du\,\delta t \tag{2.4}$$

Therefore,

$$\delta\theta = \frac{du\,\delta t}{dy} \tag{2.5}$$

Equation (2.5) implies that the angular displacement depends not only on the velocity and separation between the plates, but also on time. Therefore, in the case of fluids, the shearing stress must be correlated to the rate of shear rather than shear alone, as was done for solid materials. The rate of shear, $\dot{\gamma}$, is

$$\dot{\gamma} = \lim \frac{\delta\theta}{\delta t} \tag{2.6}$$

or

$$\dot{\gamma} = \frac{du}{dy} \tag{2.7}$$

Thus, shear rate is the relative change in velocity divided by the distance between the plates. Newton observed that if the shearing stress σ is increased (by increasing force, F), then the rate of shear, $\dot{\gamma}$, will also increase in direct proportion.

$$\sigma \propto \dot{\gamma} \tag{2.8}$$

or

$$\sigma \propto \frac{du}{dy} \tag{2.9}$$

Or, removing the proportionality by introducing a constant, μ,

$$\sigma = \mu \frac{du}{dy} \tag{2.10}$$

where μ is the coefficient of viscosity, or simply viscosity, of the fluid. It is also called "absolute" or "dynamic" viscosity.

Liquids that follow Equation (2.10), exhibiting a direct proportionality between shear rate and shear stress, are called Newtonian liquids. When

Table 2.1 The Viscosity of Some Common Materials at Room Temperature

Liquid	Viscosity, approximate (Pa s)
Air	10^{-5}
Water	10^{-3}
Olive oil	10^{-1}
Glycerol	10^{0}
Liquid honey	10^{1}
Golden syrup	10^{2}
Glass	10^{40}

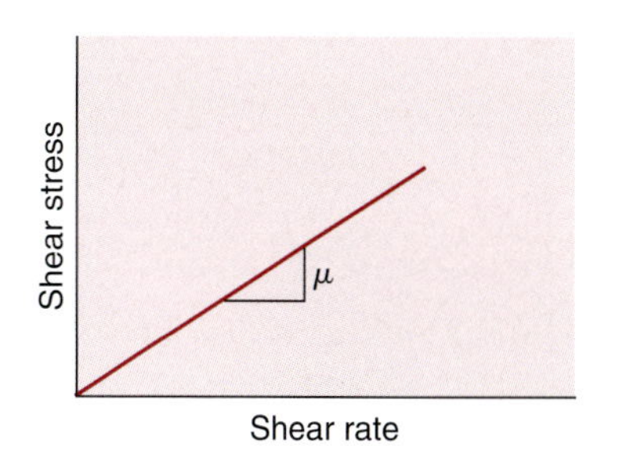

Figure 2.10 Shear stress vs shear rate behavior for a Newtonian fluid.

shear stress is plotted against shear rate, a straight line is obtained passing through the origin (Fig. 2.10). The slope of the line gives the value for viscosity, μ. Water is a Newtonian liquid; other foods that exhibit Newtonian characteristics include honey, fluid milk, and fruit juices. Table 2.1 gives some examples of coefficients of viscosity. Viscosity is a physical property of the fluid and it describes the resistance of the material to shear-induced flow. Furthermore, it depends on the physicochemical nature of the material and the temperature.

Liquids that do not follow Equation (2.10) are called non-Newtonian liquids, and their properties will be discussed later in Section 2.9.

Shear stress σ is obtained using Equation (2.10). Since force is expressed by N (newtons) and area by m^2 (meters squared), shear stress is expressed in units of Pa (Pascal) as follows,

$$\sigma \equiv \frac{\mathrm{N}}{\mathrm{m}^2} \equiv \mathrm{Pa}$$

Note that it is common to see τ used as a symbol for shear stress in literature. However, because the Society of Rheology recommends the use of σ for shear stress, we will use this symbol consistently in this book. In the cgs units, shear stress is expressed as dyne/cm^2, where,

$$1\,\mathrm{Pa} \equiv 10\,\mathrm{dyne/cm^2}$$

The term $\dot{\gamma}$, or du/dy in Equation (2.10), is called rate of shear or shear rate. It is the velocity gradient set up in the liquid due to the applied shear stress as seen in Equation (2.7). Its units are s^{-1}, determined by dividing the change in velocity (m/s) by distance (m). Therefore, the unit of viscosity, μ, in the SI system of units is pascal-second (Pa s) obtained as follows.

$$\mu \equiv \frac{\sigma}{\dot{\gamma}} \equiv \frac{\text{Pa}}{\text{s}^{-1}} \equiv \text{Pa s}$$

Frequently, the viscosity of liquids is expressed as millipascal-second or mPa s, where

$$1000 \text{ mPa s} = 1 \text{ Pa s}$$

The unit Pa s may also be expressed as:

$$\mu \equiv \text{Pa s} \equiv \left(\frac{\text{N}}{\text{m}^2}\right)\text{s} \equiv \left(\frac{\text{kg m}}{\text{s}^2\text{m}^2}\right)\text{s} \equiv \frac{\text{kg}}{\text{m s}}$$

In cgs units, where shear stress in dyne/cm^2 and shear rate is s^{-1}, viscosity is expressed as Poise (named after Poiseulle), or

$$\mu \equiv \frac{\text{dyne s}}{\text{cm}^2} \equiv \text{poise}$$

In literature, viscosity of liquids is often expressed in centipoise (0.01 poise). Note the following conversion factors:

$$1 \text{ poise} \equiv 0.1 \text{ Pa s}$$

and

$$1 \text{ cP} = 1 \text{ mPa s}$$

The viscosity of water at ambient temperatures is about 1 centipoise (or 1 mPa s), whereas the viscosity of honey is 8880 centipoise. According to Van Wazer (1963), the human eye can distinguish between differences in viscosity of fluids in the range of 100 to 10,000 cP. Above 10,000 cP the material appears to be like a solid. Thus, a liquid with a viscosity of 600 cP will appear twice as "thick" as a liquid with a viscosity of 300 cP.

Although dynamic viscosity, μ, is commonly used, an alternative term used to express viscosity is kinematic viscosity, ν. When Newtonian liquids are tested by capillary viscometers such as Ubbeholde or Cannon-Fenske (to be discussed later in Section 2.8.1), the force of gravity is used to move a liquid sample through a capillary. Therefore, the density of the liquid plays an important role in calculations. Kinematic viscosity is commonly used to express viscosity of nonfood materials such as lubricating oils. It is related to dynamic viscosity as follows:

$$\text{Kinematic voscosity} = \frac{\text{dynamic viscosity}}{\text{density}}$$

or

$$\nu = \frac{\mu}{\rho} \tag{2.11}$$

The units of kinematic viscosity are

$$\nu \equiv \frac{\text{m}^2}{\text{s}}$$

In cgs units, kinematic viscosity is expressed in units of *stokes* (abbreviated as S) or centistokes (cS). This unit is named after Sir George Stokes (1819–1903), a Cambridge physicist who made major contributions to the theory of viscous fluids. Where

$$1\,\text{S} = 100\,\text{cS}$$

or

$$1\,\text{cS} = \frac{1\,\text{mm}^2}{\text{s}}$$

water has a kinematic viscosity of 1 mm^2/s at 20.2°C.

Example 2.1

In a quality control test, viscosity of a liquid food is being measured with a viscometer. A shear stress of 4 dyne/cm^2 at a shear rate of 100 s^{-1} was recorded. Calculate the viscosity and express it as Pa s, cP, P, kg/m s and mPa s.

Given

Shear stress $= 4$ *dyne/cm*^{2}
Shear rate $= 100\,s^{-1}$

Approach

We will use the definition of viscosity given in Equation (2.10) to calculate the viscosity. For unit conversions, note that

$$1\,\text{dyne/cm}^2 = 1\,\text{g/(cm s}^2) = 0.1\,\text{kg/(m s}^2) = 0.1\,\text{N/m}^2 = 0.1\,\text{Pa}$$

Solution

1. *Shear stress in SI units is*

$$\sigma = \frac{4[\text{dyne/cm}^2] \times 0.1[\text{kg/(m s}^2)]}{1[\text{dyne/cm}^2]}$$

$$\sigma = 0.4 \text{ kg/(m s}^2)$$

$$\sigma = 0.4 \text{ Pa}$$

2. *Viscosity in Pa s*

$$\mu = \frac{0.4[\text{Pa}]}{100[\text{s}^{-1}]} = 0.004 \text{ Pa s}$$

3. *Viscosity in P*

$$\mu = \frac{4[\text{dyne/cm}^2]}{100[\text{s}^{-1}]}$$

$$= \frac{0.04[\text{dyne s)/cm}^2]}{1[\text{dyne s/cm}^2]/1[\text{P}]} = 0.04 \text{ P}$$

4. *Viscosity in cP*

$$\mu = \frac{0.04[\text{P}]}{1[\text{P}]/100[\text{cP}]}$$

$$\mu = 4 \text{ cP}$$

5. *Viscosity in kg/(m s)*

Since $1 \text{ Pa} = 1 \text{ kg/(m s}^2)$,

$$\mu = 0.004 \text{ kg/(m s)}$$

6. *Viscosity in mPa s*

Since $1 \text{ mPa s} = 1 \text{ cP}$

$$\mu = 4 \text{ mPa s}$$

Example 2.2

Determine the dynamic and kinematic viscosity of air and water at 20°C and 60°C.

Given

Temperature of water = 20°C

Approach

We will obtain values of absolute and kinematic viscosity from Tables A.4.1 and A.4.2.

Solution

1. *From Table A.4.1 for water at 20°C*
   a. *Dynamic viscosity* $= 993.414 \times 10^{-6}$ *Pa s*
   b. *Kinematic viscosity* $= 1.006 \times 10^{-6}\ m^2/s$
2. *From Table A.4.1 for water at 60°C*
   a. *Dynamic viscosity* $= 471.650 \times 10^{-6}$ *Pa s*
   b. *Kinematic viscosity* $= 0.478 \times 10^{-6}\ m^2/s$
3. *From Table A.4.2 for air at 20°C*
   a. *Dynamic viscosity* $= 18.240 \times 10^{-6}$ *Pa s*
   b. *Kinematic viscosity* $= 15.7 \times 10^{-6}\ m^2/s$
4. *From Table A.4.2 for air at 60°C*
   a. *Dynamic viscosity* $= 19.907 \times 10^{-6}$ *Pa s*
   b. *Kinematic viscosity* $= 19.4 \times 10^{-6}\ m^2/s$

As seen from these results, with increasing temperature the dynamic viscosity of water decreases, whereas for gases it increases. The influence of temperature on viscosity of liquids is more pronounced than for gases. The dynamic viscosity of air is much less than that of water.

2.3 HANDLING SYSTEMS FOR NEWTONIAN LIQUIDS

In a food processing plant, liquid foods are processed in a variety of ways, such as by heating, cooling, concentrating, or mixing. The transport of liquid foods from one processing equipment to another is achieved mostly by using pumps, although gravity systems are used when feasible. Depending on the velocity of the liquid and the internal viscous and inertial forces, different types of flow characteristics are obtained. The energy required to pump a liquid will be different under different flow conditions. In this section, we will look at quantitative methods to describe flow characteristics of liquid foods.

In subsequent sections, we will refer to fluid flow along a stream line. At any instant of time, we may consider an imaginary curve in the fluid, called a **stream line**, along which fluid moves (Fig. 2.11). No fluid movement occurs across this curve. The velocity of fluid at any point along the stream line is in a tangential direction along the line. When bunched together into a stream tube, stream lines provide a good indication of the instantaneous flow of a fluid.

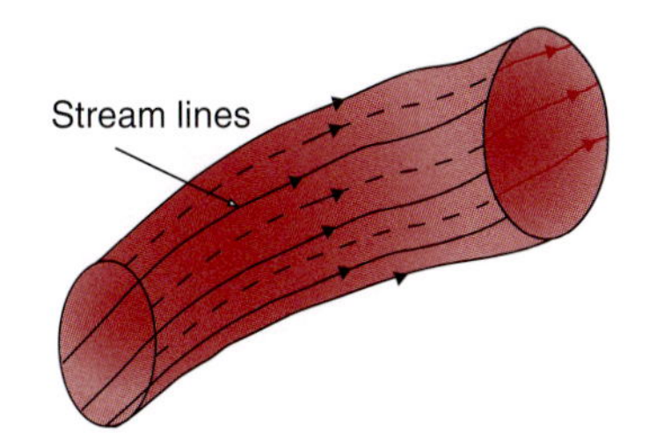

Figure 2.11 Stream lines forming a stream tube. Flow occurs only along stream lines, not across them.

2.3.1 The Continuity Equation

The principle of conservation of matter is frequently used to solve problems related to fluid flow. To understand this important principle,

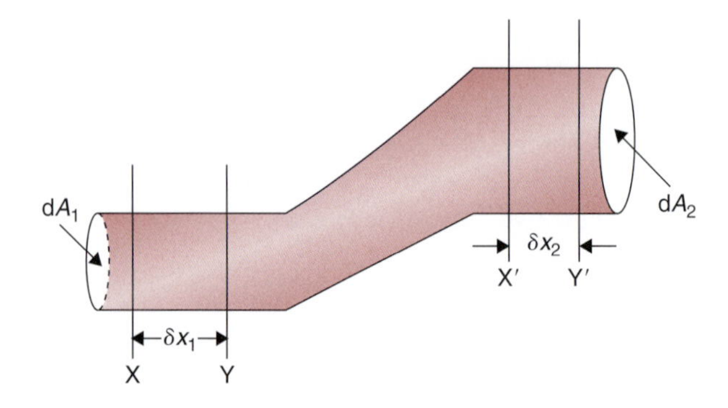

Figure 2.12 Fluid flow through a pipe of varying cross-section.

consider a fluid flowing in a pipeline, as shown in Figure 2.12. Since the fluid is moving, suppose that in time step δt, the fluid occupying space XX′ moves to space YY′. The distance between X and Y is δx_1 and between X′ and Y′ is δx_2. The cross-sectional area at X is dA_1, and at X′ it is dA_2. We have purposely selected different cross-sectional areas at the two ends to show that our derivation will be applicable to such variations. For the matter to be conserved, the mass contained in space XX′ must be equal to that in YY′. We also note that the fluid contained in space YX′ is common to both initial and final space. Therefore, the mass of fluid in space XY must equal that in space X′Y′. Therefore,

$$\rho_1 A_1 \delta x_1 = \rho_2 A_2 \delta x_2 \tag{2.12}$$

Dividing by the time step, δt,

$$\rho_1 A_1 \frac{\delta x_1}{\delta t} = \rho_2 A_2 \frac{\delta x_2}{\delta t} \tag{2.13}$$

or,

$$\rho_1 A_1 \bar{u}_1 = \rho_2 A_2 \bar{u}_2 \tag{2.14}$$

where $\bar{u}$ is the average velocity.

Equation (2.14) is the Equation of Continuity. We can express this equation on the basis of either mass flow or volumetric flow rate. In Equation (2.14)

$$\rho A \bar{u} = \dot{m} \tag{2.15}$$

where $\dot{m}$ is the mass flow rate, kg/s. The mass flow rate is a function of density ρ, the cross-sectional area A of the pipe or tube, and the mean velocity $\bar{u}$ of the fluid. Equation (2.14) shows that the mass flow rate remains constant under steady state conditions.

For an incompressible fluid, such as for liquids, the density remains constant. Then, from Equation (2.14)

$$A_1\bar{u}_1 = A_2\bar{u}_2 \tag{2.16}$$

where

$$A\bar{u} = \dot{V} \tag{2.17}$$

The volumetric flow rate $\dot{V}$ is a product of cross-sectional area A of the pipe and the mean fluid velocity, $\bar{u}$. According to Equation (2.17), under steady state conditions, the volumetric flow rate remains constant.

The preceding mathematical development will be valid only if we use mean velocity, $\bar{u}$, for the given cross-section. The use of the bar on symbol $\bar{u}$ indicates that it represents a mean value for velocity. We will observe later in Section 2.3.4 that the velocity distribution of a fully developed flow in a pipe is, in fact, parabolic in shape. At this stage, we need to ensure that only mean velocity is selected whenever Equation (2.14) is used.

Example 2.3

The volumetric flow rate of beer flowing in a pipe is 1.8 L/s. The inside diameter of the pipe is 3 cm. The density of beer is 1100 kg/m^3. Calculate the average velocity of beer and its mass flow rate in kg/s. What is the mass flow rate? If another pipe with a diameter of 1.5 cm is used, what will be the velocity for the same volumetric flow rate?

Given

Pipe diameter = 3 cm = 0.03 m
Volumetric flow rate = 1.8 L/s = 0.0018 m^3/s
Density = 1100 kg/m^3

Approach

First, we will calculate velocity, $\bar{u}$, from the given volumetric flow rate using Equation (2.17). Then we will use Equation (2.15) to obtain mass flow rate.

Solution

1. *From Equation (2.17),*

$$\text{Velocity, } \bar{u} = \frac{0.0018[m^3/s]}{\dfrac{\pi \times 0.03^2}{4}[m^2]} = 2.55 \text{ m/s}$$

2. *From Equation (2.15),*

$$\text{Mass flow rate} = \dot{m} = 1100[kg/m^3] \times \frac{\pi \times 0.03^2}{4}[m^2] \times 2.55[m/s]$$

$$\dot{m} = 1.98\,kg/s$$

3. *New velocity if the diameter of the pipe is halved and the volumetric flow rate is kept the same:*

$$\bar{u} = \frac{0.0018[m^3/s]}{\frac{\pi \times 0.015^2}{4}[m^2]} = 10.19\,m/s$$

4. *Note that by halving the diameter, the velocity is increased fourfold.*

2.3.2 Reynolds Number

We can conduct a simple experiment to visualize the flow characteristics of a liquid by carefully injecting a dye into the liquid flowing in a pipe. At low flow rates, the dye moves in a straight-line manner in the axial direction, as shown in Figure 2.13. As the flow rate increases to some intermediate level, the dye begins to blur at some distance away from the injection point. The blurring of the dye is caused by movement of some of the dye in the radial direction. At high flow rates, the dye becomes blurred immediately upon injection. At these high flow rates, the dye spreads in a random manner along both the radial and axial direction. The straight-line flow observed at low flow rates is called **laminar** flow; at intermediate flow rates, the flow is called **transitional** flow; and the erratic flow obtained at higher flow rates is called **turbulent** flow.

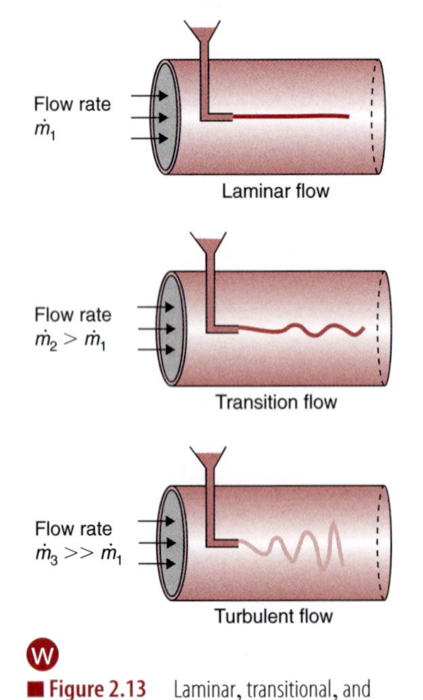

Figure 2.13 Laminar, transitional, and turbulent flow in a pipe.

The flow characteristics for laminar flow are influenced by liquid properties, flow rate, and the dimensions of liquid–solid interfaces. As the mass flow rate is increased, the forces of momentum or inertia increase; but these forces are resisted by friction or viscous forces within the flowing liquid. As these opposing forces reach a certain balance, changes in the flow characteristics occur. Based on experiments conducted by Reynolds (1874),[1] the inertial forces are a function

[1] Osborne Reynolds (1842–1912) was a British physicist, engineer, and educator. He was appointed the first professor of engineering at Owens College, Manchester, where he retired in the same position in 1905. His major work was in the study of hydrodynamics. He developed a theory of lubrication, studied the condensation process, and provided a mathematical foundation (in 1883) to the turbulence phenomenon in fluid flow. His work resulted in important redesign of boilers and condensers, and development of turbines.

of liquid density, ρ, tube diameter, D, and average fluid velocity, $\bar{u}$. The viscous forces, on the other hand, are a function of liquid viscosity. A dimensionless number, called a Reynolds number, is defined as the ratio of the inertial to the viscous forces:

$$N_{Re} = \frac{\text{inertial forces}}{\text{viscous forces}} \tag{2.18}$$

or

$$N_{Re} = \frac{\rho \bar{u} D}{\mu} \tag{2.19}$$

If instead of average velocity, the mass flow rate, $\dot{m}$, is measured or given, then substituting Equation (2.15) in Equation (2.19) and rearranging terms, we obtain

$$N_{Re} = \frac{4\dot{m}}{\mu \pi D} \tag{2.20}$$

A Reynolds number is most useful in quantitatively describing the flow characteristics of a fluid flowing either in a pipe or on the surfaces of objects of different shapes. We no longer need to limit ourselves to qualitative descriptions of flow such as low, intermediate, or high. Instead, we can use a Reynolds number to specifically identify how a given liquid would behave under selected flow conditions.

The Reynolds number provides an insight into energy dissipation caused by viscous effects. From Equation (2.18), when the viscous forces have a dominant effect on energy dissipation, the Reynolds number is small, or flow is in a laminar region. As long as the Reynolds number is 2100 or less, the flow characteristics are laminar or stream line. A Reynolds number between 2100 and 4000 signifies a transitional flow. A Reynolds number greater than 4000 indicates turbulent flow denoting small influence of viscous forces on energy dissipation.

Example 2.4

A 3 cm inside diameter pipe is being used to pump liquid food into a buffer tank. The tank is 1.5 m diameter and 3 m high. The density of the liquid is 1040 kg/m^3 and viscosity is 1600×10^{-6} Pa s.

a. What is the minimum time to fill the tank with this liquid food if it is flowing under laminar conditions in the pipe?

b. What will be the maximum time to fill the tank if the flow in the pipe is turbulent?

Given

Pipe diameter = 3 cm = 0.03 m
Tank height = 3 m
Tank diameter = 1.5 m
Density of liquid = 1040 kg/m^3
Viscosity of liquid = 1600 × 10^{-6} Pa s = 1600 × 10^{-6} kg/m s

Approach

For (a), we will use the maximum Reynolds number in the laminar range of 2100 and calculate the flow rate. For (b), we will use a minimum Reynolds number in the turbulent region of 4000 and calculate the flow rate. The time to fill the tank will be obtained from the volume of the tank and the volumetric flow rate.

Solution

Part (a)

1. *From Equation (2.19), maximum velocity under laminar conditions is*

$$\bar{u} = \frac{2100\,\mu}{\rho D} = \frac{2100 \times 1600 \times 10^{-6}\,[\text{kg/ms}]}{1040\,[\text{kg/m}^3] \times 0.03\,[\text{m}]} = 0.108\ \text{m/s}$$

Then, volumetric flow rate using the pipe cross-sectional area and Equation (2.17) is

$$\dot{V} = \frac{\pi \times 0.03^2\ [\text{m}^2]}{4} \times 0.108\ [\text{m/s}] - 7.63 \times 10^{-5}\ \text{m}^3/\text{s}$$

2. *Volume of tank*

$$= \frac{\pi(\text{diameter})^2(\text{height})}{4}$$

$$= \frac{\pi \times 1.5^2\ [\text{m}^2] \times 3\ [\text{m}]}{4}$$

$$= 5.3\ \text{m}^3$$

3. *The minimum time to fill the tank = (volume of tank)/(volumetric flow rate)*

$$= \frac{5.3\ [\text{m}^3]}{7.63 \times 10^{-5}\ [\text{m}^3/\text{s}]} = 6.95 \times 10^4\ \text{s} = 19.29\ \text{h}$$

Part (b)

4. *From Equation (2.19), minimum velocity under turbulent flow conditions is*

$$\bar{u} = \frac{4000\,\mu}{\rho D} = \frac{4000 \times 1600 \times 10^{-6}\ [\text{kg/ms}]}{1040\ [\text{kg/m}^3] \times 0.03\ [\text{m}]} = 0.205\ \text{m/s}$$

Then, volumetric flow rate using the pipe cross-sectional area and Equation (2.17) is

$$\dot{V} = \frac{\pi \times 0.03^3\ [\text{m}^2]}{4} \times 0.20\ [\text{m/s}] = 1.449 \times 10^{-4}\ \text{m}^3/\text{s}$$

5. *The maximum time to fill the tank = (volume of tank)/(volumetric flow rate)*

$$= \frac{5.3\ [m^3]}{1.449 \times 10^{-4}\ [m^3/s]} = 3.66 \times 10^4\ s = 10.16\ h$$

6. *The minimum time to fill the tank under laminar conditions is 19.29 hours, whereas the maximum time to fill the tank under turbulent conditions is 10.16 hours.*

Example 2.5

At what velocity does air and water flow convert from laminar to transitional in a 5 cm diameter pipe at 20°C?

Given

Pipe diameter = 5 cm = 0.05 m
Temperature = 20°C
From Table A.4.1 for water,
Density = 998.2 kg/m³
Viscosity = 993.414 × 10^{-6} Pa s
From Table A.4.4 for air,
Density = 1.164 kg/m³
Viscosity = 18.240 × 10^{-6} Pa s

Approach

We will use a Reynolds number of 2100 for change from laminar to transitional flow.

Solution

1. *From the Reynolds number and Equation (2.19), we obtain velocity as*

$$\bar{u} = \frac{N_{Re}\mu}{\rho D}$$

2. *For water*

$$\bar{u} = \frac{2100 \times 993.414 \times 10^{-6} \left[\frac{kg}{m\,s}\right]}{998.2 \left[\frac{kg}{m^3}\right] \times 0.05\ [m]}$$

$$\bar{u} = 0.042\ m/s$$

3. *For air*

$$\bar{u} = \frac{2100 \times 18.240 \times 10^{-6} \left[\frac{kg}{m\,s}\right]}{1.164 \left[\frac{kg}{m^3}\right] \times 0.05\ [m]}$$

$$\bar{u} = 0.658\ m/s$$

4. *To change from laminar to transitional flow, the calculated velocities of air and water for a 5 cm pipe are quite low; typically, much higher fluid velocities are used in commercial practice. Therefore, the fluid flow in industry is generally in the transition or turbulent region. Mostly, we encounter laminar flow in the case of more viscous liquids.*

2.3.3 Entrance Region and Fully Developed Flow

When a liquid enters a pipe, there is a certain initial length of the pipe, called the **entrance region**, where the flow characteristics of the liquid are quite different from those in the following length of pipe. As shown in Figure 2.14, immediately at the entrance to the pipe, the liquid has a uniform velocity profile, identified by the same length of arrows in the diagram. As it begins to move into the pipe, the liquid next to the inside wall of the pipe is held back by friction between the liquid and the wall surface. The velocity of the liquid is zero at the wall and increases toward the central axis of the pipe. Therefore, the boundary (or the wall surface) begins to influence the velocity profile. As shown in Figure 2.14, in the entrance region, the boundary layer develops from location X to Y. At location Y, the effect of the boundary layer on the velocity profile extends all the way to the central axis. The cross-sectional velocity profile at Y is parabolic in shape (as we will mathematically derive in the following section). From X to Y, the region is called the **entrance region**, and the liquid flow in the region beyond Y is commonly referred to as **fully developed flow**.

Using dimensional analysis, it has been shown that the dimensionless entrance length, L_e/D, is a function of the Reynolds number.

■ **Figure 2.14** Velocity profile in a fluid flowing in a pipe.

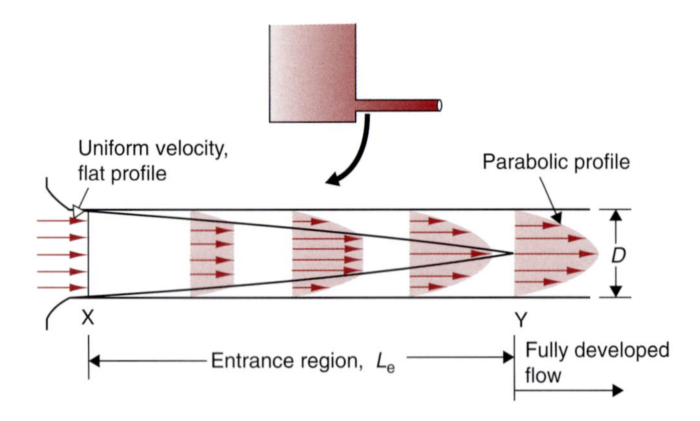

Therefore, the entrance length, L_e may be calculated from the following expressions:

For laminar flow,

$$\frac{L_e}{D} = 0.06\, N_{Re} \tag{2.21}$$

and for turbulent flow,

$$\frac{L_e}{D} = 4.4(N_{Re})^{1/6} \tag{2.22}$$

Example 2.6

A 2 cm diameter pipe is 10 m long and delivers wine at a rate of 40 L/min at 20°C. What fraction of the pipe represents the entrance region?

Given

Pipe diameter = 2 cm = 0.02 m
Length = 10 m
Flow rate = 40 L/min = 6.67×10^{-4} m³/s
Temperature = 20°C

Approach

Since the properties of wine are not given, we will approximate properties of wine as those of water obtained from Table A.4.2. First, we will determine the Reynolds number and then select an appropriate equation to calculate the entrance region from Equations (2.21) and (2.22).

Solution

1. *The average velocity is obtained from Equation (2.17)*

$$\bar{u} = \frac{0.000667 \left[\dfrac{m^3}{s}\right]}{\dfrac{\pi \times 0.02^2}{4} [m^2]}$$

$$= 2.12 \text{ m/s}$$

2. *The Reynolds number, using Equation (2.19), is*

$$N_{Re} = \frac{998.2 \left[\dfrac{kg}{m^3}\right] \times 2.12 \left[\dfrac{m}{s}\right] \times 0.02\ [m]}{993.414 \times 10^{-6}\ [Pa\ s]}$$

$$N_{Re} = 42{,}604$$

Therefore, the flow is turbulent, and we select Equation (2.22) to determine the entrance region.

3. *Using Equation (2.22)*

$$L_e = 0.02\ [m] \times 4.4 \times (42{,}604)^{1/6}$$
$$L_e = 0.52\ m$$

4. *The entrance region is 5% of the total length of the pipe.*

2.3.4 Velocity Profile in a Liquid Flowing Under Fully Developed Flow Conditions

Calculations to determine the velocity profile in a pipe depend upon whether the region of interest is at the entrance or further along the pipe where the flow is fully developed. In the entrance region, the calculations are complicated because the liquid velocity depends not only on the radial distance from the pipe centerline, r, but also on the axial distance from the entrance, x. On the other hand, the velocity profile in the fully developed region depends only on the radial distance from the central axis, r. In the entrance region, three forces—gravitational, pressure, and inertial—influence the flow. Largely due to inertial forces, the flow in the entrance region accelerates, and the velocity profile changes from location X to Y, as shown in Figure 2.14. A mathematical description of flow in the entrance region and under turbulent conditions is highly complex and beyond the scope of this book. Therefore, we will determine the velocity profile in a liquid flowing in a straight horizontal pipe of a constant diameter under fully developed laminar flow conditions.

Consider fluid flow taking place under steady and fully developed conditions in a constant-diameter pipe. Forces due to pressure and gravity cause the fluid flow in the fully developed region. In the case of a horizontal pipe, gravitational effects are negligible. Therefore, for the purpose of this analysis, we will consider only forces due to pressure. When a viscous liquid (a liquid with viscosity greater than zero) flows in a pipe, the viscous forces within the liquid oppose the pressure forces. Application of pressure is therefore necessary for flow to occur, as it overcomes the viscous forces opposing the flow. Furthermore, the flow takes place without accelerating; the velocity profile within the fully developed flow region does not change with location along the x-axis. For the flow to be steady, a balance must exist between the pressure and viscous forces in the liquid. We will conduct a force balance to analyze this case.

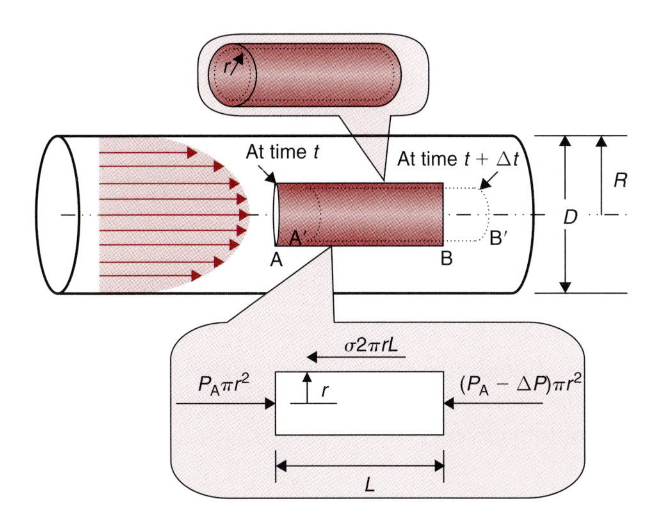

Figure 2.15 Force balance for a liquid flowing in a pipe.

Using Newton's second law of motion, we can describe forces acting on a small element of liquid as shown in Figure 2.15. The cylindrical element is of radius r and length L. The pipe diameter is D. Initially at time t, the location of the element is identified from A to B. After a small lapse of time Δt, the liquid element moves to a new location A′B′. The ends of the element, A′B′, indicating velocity profile, are shown distorted, indicating that under fully developed flow conditions, the velocity at the central axis is maximum and decreases with increasing r.

Since the pipe is horizontal, we neglect the gravitational forces. The pressure varies from one axial location to another, but remains constant along any vertical cross-section of the pipe. Let us assume that the pressure on the cross-sectional face at A is P_A and at B is P_B. If the decrease of pressure from A to B is ΔP, then $\Delta P = P_A - P_B$.

As seen in the force diagram for Figure 2.15, the pressure forces acting on the liquid element are as follows:

On the vertical cross-sectional area, πr^2,

$$\text{at location A, pressure forces} = P_A \pi r^2 \tag{2.23}$$

$$\text{at location B, pressure forces} = (P_A - \Delta P)\pi r^2 \tag{2.24}$$

and on the circumferential area, $2\pi rL$,

$$\text{forces opposing pressure forces due to viscous effects} = \sigma 2\pi rL \tag{2.25}$$

where σ is the shear stress.

According to Newton's second law of motion, the force in the x direction, $F_x = ma_x$. As noted earlier in this section, under fully developed flow conditions, there is no acceleration, or $a_x = 0$. Therefore, $F_x = 0$. Thus, for the liquid element, all forces acting on it must balance, or,

$$P_A \pi r^2 - (P_A - \Delta P)\pi r^2 - \sigma 2\pi r L = 0 \tag{2.26}$$

or simplifying,

$$\frac{\Delta P}{L} = \frac{2\sigma}{r} \tag{2.27}$$

For Newtonian liquids, the shear stress is related to viscosity as seen earlier in Equation (2.10). For pipe flow, we rewrite this equation in cylindrical coordinates as

$$\sigma = -\mu \frac{du}{dr} \tag{2.28}$$

Note that du/dr is negative in case of a pipe flow, the velocity decreasing with increasing radial distance, r, as seen in Figure 2.15. Therefore, we have introduced a negative sign in Equation (2.28) so that we obtain a positive value for shear stress, σ. Substituting Equation (2.27) in Equation (2.28),

$$\frac{du}{dr} = -\left(\frac{\Delta P}{2\mu L}\right) r \tag{2.29}$$

integrating,

$$\int du = -\frac{\Delta P}{2\mu L} \int r dr \tag{2.30}$$

or

$$u(r) = -\left(\frac{\Delta P}{4\mu L}\right) r^2 + C_1 \tag{2.31}$$

where C_1 is a constant.

For a viscous fluid flowing in a pipe, $u = 0$ at $r = R$; therefore

$$C_1 = \frac{\Delta P}{4\mu L} R^2 \tag{2.32}$$

Therefore, the velocity profile for a laminar, fully developed flow, in a horizontal pipe, is:

$$u(r) = \frac{\Delta P}{4\mu L} (R^2 - r^2) \tag{2.33}$$

or

$$u(r) = \frac{\Delta P R^2}{4\mu L}\left[1 - \left(\frac{r}{R}\right)^2\right] \tag{2.34}$$

Equation (2.34) is an equation of a parabola. Therefore, for fully developed flow conditions we obtain a parabolic velocity profile. Furthermore, from Equation (2.34), substituting $r = 0$, the maximum velocity, u_{max}, is obtained at the pipe centerline, or

$$u_{max} = \frac{\Delta P R^2}{4\mu L} \tag{2.35}$$

Next, let us determine the volumetric flow rate by integrating the velocity profile across the cross-section of the pipe. First, we will examine a small ring of thickness dr with an area dA, where $dA = 2\pi r dr$, as shown in Figure 2.15. The velocity, u, in this thin annular ring is assumed to be constant. Then the volumetric flow rate through the annular ring, $\dot{V}_{ring}$ is:

$$\dot{V}_{ring} = u(r)dA = u(r)2\pi r dr \tag{2.36}$$

The volumetric flow rate for the entire pipe cross-section is obtained by integration as follows:

$$\dot{V} = \int u(r)dA = \int_{r=0}^{r=R} u(r)2\pi r dr \tag{2.37}$$

or, substituting Equation (2.34) in (2.37),

$$\dot{V} = \frac{2\pi \Delta P R^2}{4\mu L}\int_0^R \left[1 - \left(\frac{r}{R}\right)^2\right] r dr \tag{2.38}$$

or

$$\dot{V} = \frac{\pi R^4 \Delta P}{8\mu L} \tag{2.39}$$

The mean velocity, $\bar{u}$, is defined as the volumetric flow rate divided by the cross-sectional area of the pipe, πR^2, or

$$\bar{u} = \frac{\dot{V}}{\pi R^2} \tag{2.40}$$

or, substituting Equation (2.39) in Equation (2.40), we obtain

$$\bar{u} = \frac{\Delta P R^2}{8\mu L} \tag{2.41}$$

Equation (2.39) is called Poiseuille's Law. The flow characteristics of fully developed laminar flow were described independently in experiments by two scientists, G. Hagen in 1839 and J. Poiseuille[2] in 1840.

If we divide Equation (2.41) by (2.35), we obtain

$$\frac{\bar{u}}{u_{max}} = 0.5 \quad \text{(Laminar Flow)} \tag{2.42}$$

From Equation (2.42), we observe that the average velocity is half the maximum velocity for fully developed laminar flow conditions. Furthermore, the radius (or diameter) of the pipe has a dramatic influence on the flow rate, as seen in Equation (2.39). Doubling the diameter increases the volumetric flow rate 16-fold.

For turbulent flow in the fully developed region, the mathematical analysis necessary to obtain a velocity profile is complex. Therefore, the following empirical expression is generally used.

$$\frac{\bar{u}(r)}{u_{max}} = \left(1 - \frac{r}{R}\right)^{1/j} \tag{2.43}$$

where j is a function of the Reynolds number. For most applications $j = 7$ is recommended. Under turbulent conditions, the velocity profile may be obtained from

$$u(r) = u_{max}\left(1 - \frac{r}{R}\right)^{1/7} \tag{2.44}$$

Equation (2.44) is also called the Blassius 1/7th power law.

A volumetric flow rate under turbulent conditions may be obtained in a similar manner as for laminar conditions. Substituting Equation (2.43) in Equation (2.37),

$$\dot{V} = \int_{r=0}^{r=R} u_{max}\left(1 - \frac{r}{R}\right)^{1/j} 2\pi r dr \tag{2.45}$$

Integrating Equation (2.45), we get

$$\dot{V} = 2\pi u_{max} \frac{R^2 j^2}{(j+1)(2j+1)} \tag{2.46}$$

[2] Jean-Louis-Marie Poiseuille (1799–1869), a French physiologist, studied the flow rate of fluids under laminar condition in circular tubes. The same mathematical expression also was determined by Gotthilf Hagen; therefore the relationship is called the Hagen–Poiseuille equation. Poiseuille also studied the circulation of blood and the flow of fluids in narrow tubes.

A relationship between average and maximum velocity may be obtained by substituting Equation (2.40) in Equation (2.46), and we get,

$$\frac{\bar{u}}{u_{\max}} = \frac{2j^2}{(j+1)\,(2j+1)} \tag{2.47}$$

Substituting the value 7 for j in Equation (2.47),

$$\frac{\bar{u}}{u_{\max}} = 0.82 \quad \text{(Turbulent Flow)} \tag{2.48}$$

Thus, in the case of turbulent flow, the average velocity is 82% of the maximum velocity. The maximum velocity occurs at the central axis of the pipe.

Example 2.7

A fluid is flowing under laminar conditions in a cylindrical pipe of 2 cm diameter. The pressure drop is 330 Pa, the viscosity of the fluid is 5 Pa s, and the pipe is 300 cm long. Calculate the mean velocity and velocity of fluid at different radial locations in the pipe.

Given

Diameter of pipe = 2 cm
Length of pipe = 300 cm
Pressure drop = 330 Pa
Viscosity = 5 Pa s

Approach

We will use Equation (2.33) to calculate velocity at different radial locations.

Solution

1. *From Equation (2.33)*

$$u = \frac{\Delta P}{4\mu L}(R^2 - r^2)$$

The velocity is calculated at r = 0, 0.25, 0.5, 0.75, and 1 cm.

r = 0 cm u = 0.055 cm/s
r = 0.25 cm u = 0.0516 cm/s
r = 0.5 cm u = 0.0413 cm/s
r = 0.75 cm u = 0.0241 cm/s
r = 1 cm u = 0 cm/s

2. *The mean velocity is calculated as 0.0275 cm/s; this value is half the maximum velocity.*

2.3.5 Forces Due to Friction

The forces that must be overcome in order to pump a liquid through a pipe derive from several sources. As we saw in Section 2.2.3, viscous forces are important in liquid flow; these forces occur due to the movement of one layer over another. The other important forces are due to friction between the liquid and the surface of the wall of a pipe. When a fluid flows through a pipe, some of its mechanical energy is dissipated due to friction. It is common to refer to this dissipated energy as a frictional energy loss. Though truly not a *loss*, some of the mechanical energy supplied to the liquid to cause flow is actually converted into heat, therefore all the mechanical energy is not available as useful energy in the liquid transport system.

The friction forces vary with conditions such as flow rates, as described with a Reynolds number, and surface roughness. The influence of the friction forces is expressed in the form of a **friction factor** f. The following mathematical development is for laminar flow conditions.

The friction factor is the ratio between the shear stress at the wall, σ_w, to the kinetic energy of the fluid per unit volume.

$$f = \frac{\sigma_w}{\rho \bar{u}^2/2} \tag{2.49}$$

Rewriting Equation (2.27) for shear stress at the wall, $r = D/2$

$$\sigma_w = \frac{D\Delta P}{4L} \tag{2.50}$$

Substituting Equation (2.50) in Equation (2.49) we obtain

$$f = \frac{\Delta PD}{2L\rho\bar{u}^2} \tag{2.51}$$

By rearranging terms in Equation (2.41), the pressure drop in fully developed laminar flow conditions is determined as

$$\Delta P = \frac{32\mu\bar{u}L}{D^2} \tag{2.52}$$

Substituting Equation (2.52) in Equation (2.51), we obtain

$$f = \frac{16}{N_{Re}} \tag{2.53}$$

where f is called the Fanning friction factor. Note that many civil and mechanical engineering textbooks refer to a different friction factor, called Darcy[3] friction factor, with the same symbol, f. The Darcy friction factor is four times the Fanning friction factor. In chemical engineering literature, the Fanning friction factor is used more commonly, and in this text we will use only the Fanning friction factor.

The previous calculations leading to Fanning friction factor are for laminar flow conditions only. In cases of transitional and turbulent flow conditions, the mathematical derivations become highly complex. For situations involving flow conditions other than laminar flow, we will use a graphical chart that presents friction factor as a function of the Reynolds number. This chart, called a **Moody chart**, is shown in Figure 2.16. The Moody chart presents the friction factor as a function of the Reynolds number for various magnitudes of relative roughness of the pipes. At a low Reynolds number ($N_{Re} \ll 2100$), the curve is described by Equation (2.53) and is not influenced by surface roughness, ε, of the pipe. In the transition from laminar to turbulent flow or critical region, either set of curves can be used. Most often, the friction factor is selected for turbulent flow, since it ensures that the pressure loss due to friction will not be underestimated. The Moody chart is accurate to ± 15 percent.

From the Moody chart, it is evident that the friction factor is never zero, even for smooth pipes. Because there is always some roughness at the microscopic level, a fluid will stick to the pipe surface regardless of how smooth it is. Thus, there is always a certain frictional loss when a fluid flows in a pipe.

An explicit equation to estimate the friction factor, f, was proposed by Haaland (1983). A slightly modified form of this equation for the Darcy friction factor, Equation (2.54) is recommended for calculating the Fanning friction factor for a turbulent region, preferably with the use of a spreadsheet.

$$\frac{1}{\sqrt{f}} \approx -3.6 \log \left[\frac{6.9}{N_{Re}} + \left(\frac{\varepsilon/D}{3.7} \right)^{1.11} \right] \tag{2.54}$$

[3] Henri-Philibert-Gaspard Darcy (1803–1858), a French hydraulic engineer, was the first person to develop a mathematical formulation of laminar flow of fluids in porous materials. His work laid the foundation for the subject of groundwater hydrology. In Dijon, his native city, he supervised the design and construction of the municipal water supply system. In his work he studied the flow of groundwater through granular material.

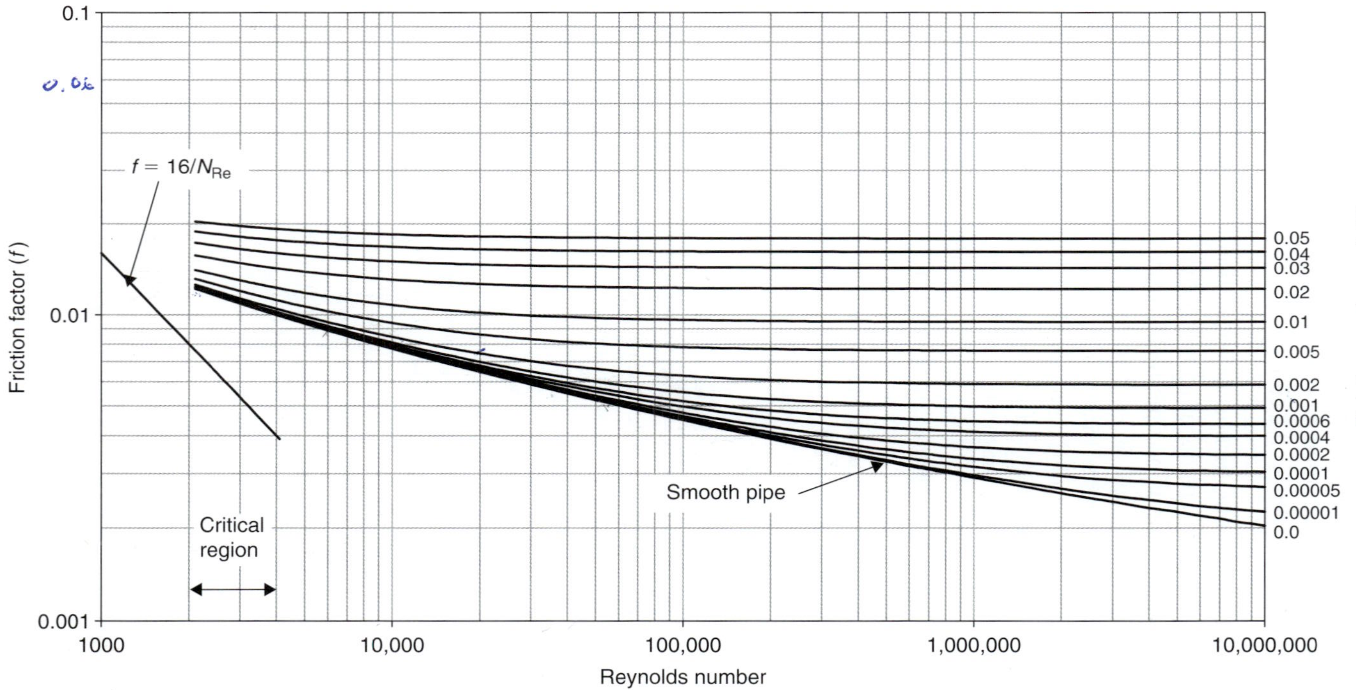

Figure 2.16 The Moody diagram for the Fanning friction factor. Equivalent roughness for new pipes (ε in meters): cast iron, 259×10^{-6}; drawn tubing, 1.5235×10^{-6}, galvanized iron, 152×10^{-6}; steel or wrought iron, 45.7×10^{-6}. (Based on L.F. Moody, 1944. Trans *ASME*, *66*, 671.)

Example 2.8

Water at 30°C is being pumped through a 30 m section of 2.5 cm diameter steel pipe at a mass flow rate of 2 kg/s. Compute the pressure loss due to friction in the pipe section.

Given

Density (ρ) = 995.7 kg/m^3, from Table A.4.1
Viscosity (μ) = 792.377 × 10^{-6} Pa s, from Table A.4.1
Length (L) of pipe = 30 m
Diameter (D) of pipe = 2.5 cm = 0.025 m
Mass flow rate ($\dot{m}$) = 2 kg/s

Approach

The pressure drop due to friction is computed using Equation (2.51) with the information given. Equation (2.51) requires knowledge of the friction factor f as obtained from Figure 2.16. Figure 2.16 can be used once the turbulence (N_{Re}) and relative roughness (ε/D) values have been determined.

Solution

1. Compute mean velocity $\bar{u}$ from Equation (2.15):

$$\bar{u} = \frac{(2\,\text{kg/s})}{(995.7\,\text{kg/m}^3)[\pi(0.025\,\text{m})^2/4]} = 4.092\,\text{m/s}$$

2. Compute the Reynolds number:

$$N_{Re} = \frac{(995.7\,\text{kg/m}^3)(0.025\,\text{m})(4.092\,\text{m/s})}{(792.377 \times 10^{-6}\,\text{Pa s})} = 128,550$$

3. Using the given information and Figure 2.16, relative roughness can be computed:

$$\varepsilon/D = \frac{45.7 \times 10^{-6}\,\text{m}}{0.025\,\text{m}} = 1.828 \times 10^{-3}$$

4. Using the computed Reynolds number and the computed relative roughness, friction factor f is obtained from Figure 2.16:

$$f = 0.006$$

5. Using Equation (2.51):

$$\frac{\Delta P}{\rho} = 2(0.006)\frac{(4.092\,\text{m/s})^2(30\,\text{m})}{(0.025\,\text{m})} = 241.12\,\text{m}^2/\text{s}^2$$

6. Note that (1 J = 1 kg m^2/s^2)

$$\frac{\Delta P}{\rho} = 241.12\,\text{m}^2/\text{s}^2 = 241.12\,\text{J/kg}$$

represents the energy consumed due to friction on a per-unit-mass basis.

7. *The pressure loss is calculated as (Note that 1 J = 1 kg m²/s²)*

$$\Delta P = (241.12\ \text{J/kg})(995.7\ \text{kg/m}^3) = 240.80 \times 10^3\ \text{kg/(m s}^2)$$

$$\Delta P = 240.08\ \text{kPa}$$

2.4 FORCE BALANCE ON A FLUID ELEMENT FLOWING IN A PIPE—DERIVATION OF BERNOULLI EQUATION

As noted in this chapter, fluids begin to move when a nonzero resultant force acts upon them. The resultant force brings about a change in the momentum of the fluid. We recall from physics that momentum is the product of mass and velocity. Under steady-flow conditions, the resultant force acting on a liquid must equal the net rate of change in momentum. We will use these concepts to derive one of the most widely used equations in fluid flow, called the Bernoulli equation.

Let us consider a particle of fluid moving along a stream line from location (1) to (2) as shown in Figure 2.17. The flow is assumed to be steady and the liquid has a constant density. The fluid is inviscid, meaning that its viscosity is zero. The x and z axes are shown; the y direction is in the perpendicular direction from the x–z horizontal plane. The s-direction is along the stream line and n-direction is normal to the s-direction. The velocity of the particle is u.

The forces acting on the particle, neglecting any forces due to friction, are the result of particle weight and pressure. Let us consider these forces separately.

a. A force component of the particle weight exerted in the s-direction is obtained as follows. The volume of the particle is $dn\ ds\ dy$. If the density of the liquid is ρ, then,

$$\text{weight of the particle} = \rho g\ dn\ ds\ dy \tag{2.55}$$

and,

$$\begin{array}{c}\text{the component of force}\\ \text{in the } s\text{-direction}\\ \text{due to weight}\end{array} = -\rho g \sin\theta\ dn\ ds\ dy$$

Since,

$$\sin\theta = \frac{\partial z}{\partial s}$$

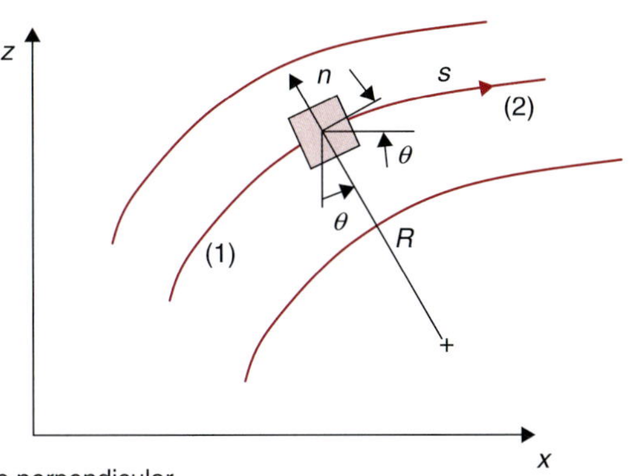

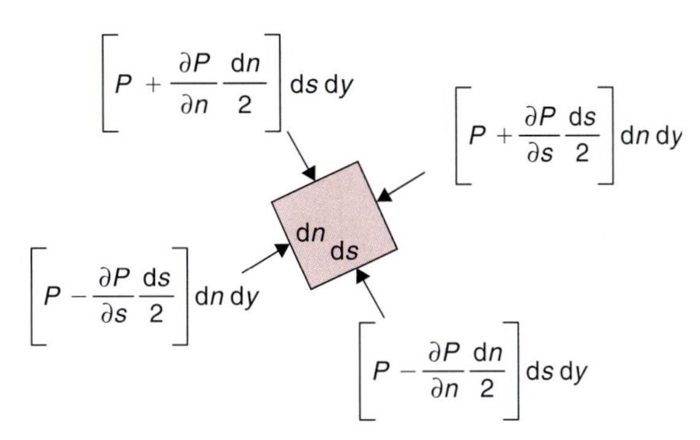

Figure 2.17 Force balance on a small volume of fluid.

force component in the s-direction

$$\text{due to weight} = -\rho g \frac{\partial z}{\partial s}\, dn\, ds\, dy \tag{2.56}$$

b. The second force acting on the particle due to pressure in the s-direction, as seen in Figure 2.17, is as follows.

$$\text{pressure force on particle} = \left(P - \frac{\partial P}{\partial s}\frac{ds}{2}\right) dn\, dy - \left(P + \frac{\partial P}{\partial s}\frac{ds}{2}\right) dn\, dy \tag{2.57}$$

Or, canceling and rearranging terms in the equation:

$$\text{pressure force on particle in } s\text{-direction} = -\frac{\partial P}{\partial s}\, dn\, dy\, ds \tag{2.58}$$

From Equations (2.56) and (2.58),

$$\text{the resultant force acting on the particle in the } s\text{-direction} = \left(-\frac{\partial P}{\partial s} - \rho g \frac{\partial z}{\partial s}\right) \mathrm{d}n\, \mathrm{d}s\, \mathrm{d}y$$

or,

$$\text{resultant force per unit volume} = -\frac{\partial P}{\partial s} - \rho g \frac{\partial z}{\partial s} \tag{2.59}$$

Because of this resultant force, the particle will accelerate as it moves along the stream line. Therefore, the velocity will change from u to $u + (\partial u/\partial s)\mathrm{d}s$ as the particle moves from s to $s + \mathrm{d}s$.

Recall from physics that momentum is mass multiplied with velocity. The rate of change of momentum of the particle due to the action of resultant force is then

$$\rho \left(\frac{u + \frac{\partial u}{\partial s}\mathrm{d}s - u}{\mathrm{d}t} \right)$$

or simply

$$\text{rate of change of momentum} = \rho \frac{\partial u}{\partial s} \frac{\mathrm{d}s}{\mathrm{d}t} \tag{2.60}$$

But

$$\frac{\mathrm{d}s}{\mathrm{d}t} = u$$

Then,

$$\text{rate of change of momentum} = \rho u \frac{\partial u}{\partial s} \tag{2.61}$$

The resultant force must equal the rate of change of momentum of the particle, or from Equation (2.59) and Equation (2.61):

$$\frac{\partial P}{\partial s} + \rho u \frac{\partial u}{\partial s} + \rho g \frac{\partial z}{\partial s} = 0 \tag{2.62}$$

This equation is also called the Euler equation of motion. If we multiply both sides by ds, we get

$$\frac{\partial P}{\partial s}\mathrm{d}s + \rho u \frac{\partial u}{\partial s}\mathrm{d}s + \rho g \frac{\partial z}{\partial s}\mathrm{d}s = 0 \tag{2.63}$$

The first term in Equation (2.63) expresses the change in pressure along the stream line; the second term is a change in velocity, and the third term is a change in elevation. Using laws of calculus, we can simply write this equation as

$$\frac{dP}{\rho} + udu + gdz = 0 \tag{2.64}$$

Integrating the preceding equation from location (1) to (2) along the stream line, we obtain

$$\int_{P_1}^{P_2} \frac{dP}{\rho} + \int_{u_1}^{u_2} udu + g\int_{z_1}^{z_2} dz = 0 \tag{2.65}$$

or, evaluating limits and multiplying by ρ, and rearranging,

$$P_1 + \frac{1}{2}\rho u_1^2 + \rho g z_1 = P_2 + \frac{1}{2}\rho u_2^2 + \rho g z_2 = \text{constant} \tag{2.66}$$

Equation (2.66) is called the **Bernoulli equation**, named after a Swiss mathematician, Daniel Bernoulli. It is one of the equations most widely used to solve problems in fluid dynamics. The application of this equation to many problems involving fluid flow provides great insight. However, if the assumptions used in deriving the equation are not followed, then erroneous results are likely to be obtained. Again, note the main assumptions used in deriving this equation:

- Locations 1 and 2 are on the same stream line.
- The fluid has a constant density, therefore the fluid is incompressible.
- The flow is inviscid; that is, the viscosity of the fluid is zero.
- The flow is steady.
- No shaft work is done on or by the fluid.
- No heat transfer takes place between the fluid and its surroundings.

As we will observe in some of the examples in this section, the Bernoulli equation may provide reasonable approximations even if the assumptions are not strictly followed. For example, fluids with low viscosities may approximate inviscid conditions.

Another frequently used form of the Bernoulli equation is written in terms of "head." If we divide Equation (2.66) by specific weight of the fluid, ρg, we obtain

$$\underbrace{\frac{P}{\rho g}}_{\text{pressure head}} + \underbrace{\frac{u^2}{2g}}_{\text{velocity head}} + \underbrace{z}_{\text{elevation head}} = \text{constant} = h_{\text{total}} \tag{2.67}$$

Each term on the left-hand side of Equation (2.67) is expressed in units of length, m. The three terms are pressure head, velocity head, and elevation head, respectively. The sum of these three heads is a constant, called total head, h_{total}. The total head of a fluid flowing in a pipe is measured with a Pitot-tube at the stagnation point, as we will discuss in Section 2.7.1. The use of the Bernoulli equation is illustrated in Examples 2.9 and 2.10.

Example 2.9

A 3 m diameter stainless steel tank contains wine. The tank is filled to 5 m depth. A discharge port, 10 cm in diameter, is opened to drain the wine. Calculate the discharge velocity of wine, assuming the flow is steady and frictionless, and the time required in emptying it.

Given

Height of tank = 5 m
Diameter of tank = 3 m

Approach

We will use the Bernoulli equation using the assumptions of steady and frictionless flow.

Solution

1. *We select location (1) as the wine free surface, and location (2) as the nozzle exit. The pressure at (1) is atmospheric. The velocity at location (1) is low enough for us to consider it as a quasi-steady state condition with zero velocity.*
2. *In the Bernoulli equation, Equation (2.66),* $P_1 = P_2 = P_{atm}$, $\rho_1 = \rho_2$ *and* $\bar{u}_1 = 0$; *therefore,*

$$gz_1 = \frac{1}{2}\bar{u}_2^2 + gz_2$$

or

$$\bar{u}_2 = \sqrt{2g(z_2 - z_1)}$$

This formula is named after Evangelista Torricelli, who discovered it in 1644.

3. *Substituting the known values in the Torricelli formula,*

$$\bar{u} = \sqrt{2 \times 9.81\left[\frac{m}{s^2}\right] \times 5[m]} = 9.9\,m/s$$

Then, volumetric flow rate from the discharge port, using Equation (2.17),

$$= \frac{\pi \times 0.10^2[m^2]}{4} \times 9.9\,[m/s] = 0.078\,m^3/s$$

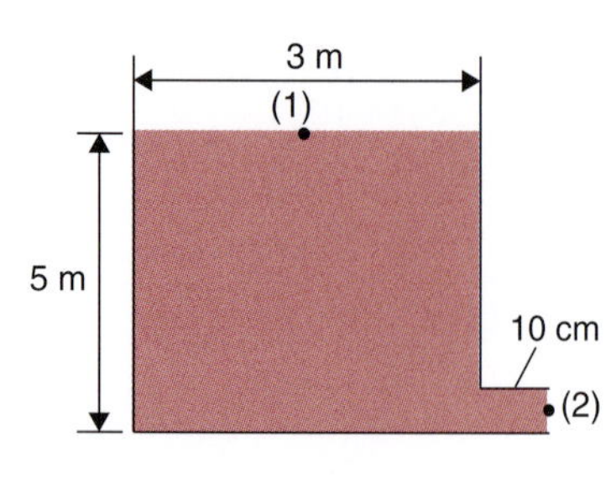

Figure E2.1 Discharge from a tank, for conditions given in Example 2.9.

4. *The volume of the tank is*

$$\frac{\pi \times 3^2\,[m^2]}{4} \times 5\ [m] = 35.3\,m^3$$

5. *Time to empty the tank equals*

$$\frac{35.3\ [m^3]}{0.078\ [m^3/s]} = 452.6\,s = 7.5\,min$$

Example 2.10

A 1.5 cm diameter tube is being used to siphon water out of a tank. The discharge end of the siphon tube is 3 m below the bottom of the tank. The water level in the tank is 4 m. Calculate the maximum height of the hill over which the tube can siphon the water without cavitation. The temperature of the water is 30°C.

Given

Siphon tube diameter = 1.5 cm = 0.015 m
Height of water in tank = 4 m
Discharge location below the bottom of tank = 3 m
Temperature of water = 30°C

Approach

Assuming an inviscid, steady, and incompressible flow, we will apply the Bernoulli equation at locations (1), (2), and (3). We will then calculate the pressure at location (2) and compare it with the vapor pressure of water at 30°C. Note that the atmospheric pressure is 101.3 kPa.

Solution

1. *To apply the Bernoulli equation at locations (1), (2), and (3), we note that*

$$P_1 = P_2 = P_{atm},\ \ \bar{u}_1 = 0,\ \ z_1 = 4\,m,\ \ z_3 = -3\,m.$$

2. *From the equation of continuity, Equation (2.16)*

$$A_2\bar{u}_2 = A_3\bar{u}_3$$

Therefore

$$\bar{u}_2 = \bar{u}_3$$

3. *Applying the Bernoulli equation, Equation (2.66), between locations (1) and (3) gives*

$$\rho g z_1 = \frac{1}{2}\rho\bar{u}_3^2 + \rho g z_3$$

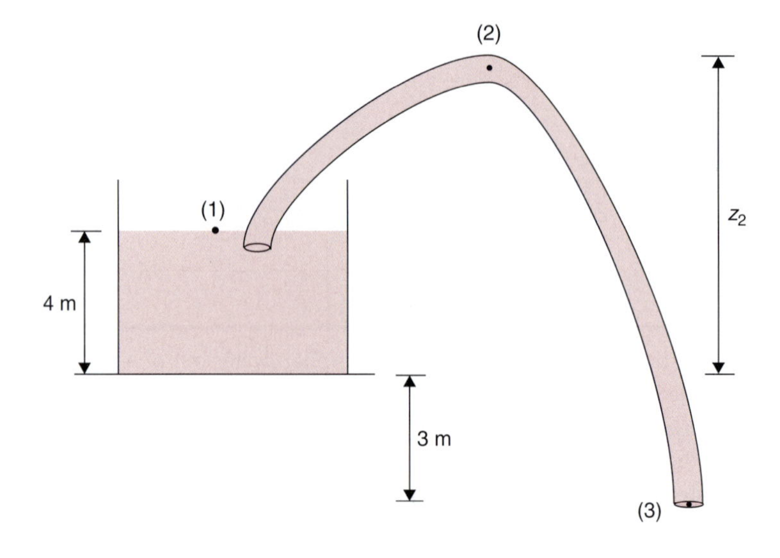

Figure E2.2 Siphoning water out of a tank, for conditions given in Example 2.10.

$$\bar{u}_3 = \sqrt{2g(z_1 - z_3)}$$

then

$$\bar{u}_3 = \sqrt{2 \times 9.81 \left[\frac{m}{s^2}\right] \times (4 - (-3))\ [m]}$$

$\bar{u}_3 = 11.72$ m/s, *same as at location (2), or,* $\bar{u}_2 = 11.72$ m/s.

4. *From Table A.4.2, at 30°C, the vapor pressure of water* = 4.246 kPa. *Again using the Bernoulli equation between locations (1) and (2), noting that* $\bar{u}_1 = 0$, *we get*

$$z_2 = z_1 + \frac{P_1}{\rho g} - \frac{P_2}{\rho g} - \frac{\bar{u}_2^2}{2g}$$

Substituting values and noting that 1 Pa = 1 kg/(ms^2),

$$z_2 = 4\ [m] + \frac{(101.325 - 4.246) \times 1000\ [Pa]}{995.7\ [kg/m^3] \times 9.81\ [m/s^2]} - \frac{1}{2} \times (11.72)^2 \left[\frac{m^2}{s^2}\right] \times \frac{1}{9.81}\left[\frac{s^2}{m}\right]$$

$$z_2 = 6.93\ m$$

5. *If* z_2 *is more than 6.93 m then cavitation will occur. If the discharge end of the siphon tube is lowered, say 5 m below the bottom of the tank, then the velocity will be higher and the value of* z_2 *will be lower.*

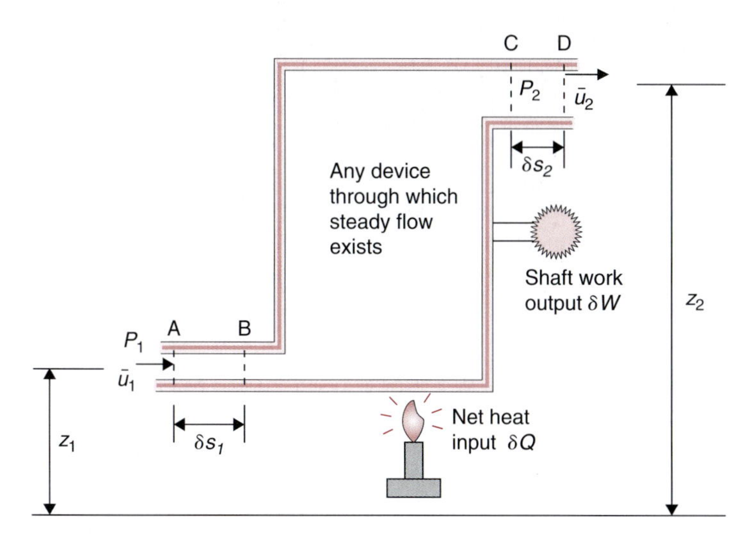

Figure 2.18 A device with a steady flow.

2.5 ENERGY EQUATION FOR STEADY FLOW OF FLUIDS

As noted in preceding sections, fluid flow occurs with the application of a force. Thus, a fluid transport system depends on a source of energy; for liquids we use pumps, whereas for gases we use blowers. In this section, we will develop mathematical expressions useful for determining the energy requirements for fluid flow. This mathematical development requires the application of the first law of thermodynamics and concepts presented in Chapter 1.

Consider a system involving fluid flow, such as the one shown in Figure 2.18. We assume that (a) the flow is continuous and under steady-state conditions, and the mass flow rate entering and exiting the system is constant; (b) fluid properties and conditions between inlet and outlet do not vary; (c) heat and shaft work between the fluid and the surroundings are transferred at a constant rate; and (d) energy transfer due to electricity, magnetism, and surface tension are negligible.

For the flow system shown in Figure 2.18, in a unit time, a constant amount of heat, δQ, is added to the system, and the system does a constant rate of work, δW, on the surroundings (e.g., it could rotate a shaft if it were a turbine or a steam engine; but if work is done by the surroundings on the fluid, as by a pump, then rate of work will carry a negative sign). At the inlet, the fluid velocity is $\bar{u}_1$, pressure

P_1, and elevation z_1. At the exit, the fluid velocity is $\bar{u}_2$, pressure P_2, and elevation z_2. At any moment, a certain packet of fluid is between locations A and C. After a short time duration, δt, this fluid packet moves to locations B and D. According to the Continuity equation, Equation (2.14), the mass of liquid, δm, in AB is the same as in CD. At the inlet, on a per unit mass basis, the fluid has a specific internal energy, E'_{i1}, kinetic energy, $\frac{1}{2}\bar{u}_1^2$, and potential energy, gz_1. Let the energy embodied in the system between B and C be $E_{\text{B-C}}$. Therefore, the energy embodied in the fluid between A and C is

$$E_{\text{A}-\text{C}} = E_{\text{A}-\text{B}} + E_{\text{B}-\text{C}} \tag{2.68}$$

or,

$$E_{\text{A}-\text{C}} = \delta m\left(E'_{i1} + \frac{1}{2}\bar{u}_1^2 + gz_1\right) + E_{\text{B}-\text{C}} \tag{2.69}$$

After a short time duration, δt, as the fluid packet moves from A–C to B–D, the energy embodied in the fluid between B and D will be

$$E_{\text{B}-\text{D}} = E_{\text{B}-\text{C}} + E_{\text{C}-\text{D}} \tag{2.70}$$

or,

$$E_{\text{B}-\text{D}} = E_{\text{B}-\text{C}} + \delta m(E'_{i2} + \frac{1}{2}\bar{u}_2^2 + gz_2) \tag{2.71}$$

Therefore, the increase in energy for the selected packet of fluid as it moves from A–C to B–D is

$$\delta E_{\text{increase}} = E_{\text{B}-\text{D}} - E_{\text{A}-\text{C}} \tag{2.72}$$

$$\begin{aligned}\delta E_{\text{increase}} = &\left[E_{\text{B}-\text{C}} + \delta m\left(E'_{i2} + \frac{1}{2}\bar{u}_2^2 + gz_2\right)\right] \\ &- \left[\delta m\left(E'_{i1} + \frac{1}{2}\bar{u}_1^2 + gz_1\right) + E_{\text{B}-\text{C}}\right]\end{aligned} \tag{2.73}$$

or simplifying,

$$\delta E_{\text{increase}} = \delta m\left[(E'_{i2} - E'_{i1}) + \frac{1}{2}(\bar{u}_2^2 - \bar{u}_1^2) + g(z_2 - z_1)\right] \tag{2.74}$$

During the time interval δt, when the liquid packet moves from A–C to B–D, the work done *by* the flowing fluid *on* the surroundings is δW. (Note that it will be $-\delta W$ if we were using a pump that does work *on* the fluid.) The heat transfer into the fluid system is δQ. Furthermore, there is work associated with pressure forces (as noted in Section 2.4). The work done *by* the fluid at the exit is $P_2A_2\delta x_2$, and at the inlet, the work done *on* the fluid is $-P_1A_1\delta x_1$, where areas A_1 and A_2 are cross-sectional areas at the inlet and exit, and P_1 and P_2 are pressures at the inlet and exit. Therefore, total work done *by* the flowing fluid is

$$\delta W_{\text{total}} = \delta W + P_2A_2\delta x_2 - P_1A_1\delta x_1 \tag{2.75}$$

From the energy balance, we note that the change in energy of the system is the heat added minus the total work done by the fluid on the surroundings, or,

$$\delta E_{\text{increase}} = \delta Q - \delta W_{\text{total}} \tag{2.76}$$

Substituting Equation (2.74) and Equation (2.75) in Equation (2.76), and rearranging, we obtain,

$$\begin{aligned}\delta Q = \delta m\left[(E'_{i2} - E'_{i1}) + \frac{1}{2}(\bar{u}_2^2 - \bar{u}_1^2) + g(z_2 - z_1)\right] \\ + \delta W + P_2A_2\delta x_2 - P_1A_1\delta x_1\end{aligned} \tag{2.77}$$

From the mass balance, we know

$$\delta m = \rho_1A_1\delta x_1 = \rho_2A_2\delta x_2 \tag{2.78}$$

Dividing Equation (2.77) by δm and substituting Equation (2.78),

$$\frac{\delta Q}{\delta m} = (E'_{i2} - E'_{i1}) + \frac{1}{2}(\bar{u}_2^2 - \bar{u}_1^2) + g(z_2 - z_1) + \frac{\delta W}{\delta m} + \frac{P_2}{\rho_2} - \frac{P_1}{\rho_1} \tag{2.79}$$

Rearranging terms,

$$Q_m = \left(\frac{P_2}{\rho_2} + \frac{1}{2}\bar{u}_2^2 + gz_2\right) - \left(\frac{P_1}{\rho_1} + \frac{1}{2}\bar{u}_1^2 + gz_1\right) + (E'_{i2} - E'_{i1}) + W_m \tag{2.80}$$

Equation (2.80) is the general energy equation for a system involving steady fluid flow, where Q_m is the heat added *to* the fluid system per unit mass, and W_m is the work done per unit mass *by* the fluid system on its surroundings (such as by a turbine).

Note that for an incompressible and inviscid fluid (viscosity = 0) if no transfer of heat or work takes place ($W_m = 0$, $Q_m = 0$) and the internal energy of the flowing fluid remains constant, then Equation (2.80) reduces to the Bernoulli equation, as presented in Section 2.4.

However, in case of a real fluid, we cannot ignore its viscosity. A certain amount of work is done to overcome viscous forces, commonly referred to as fluid friction. Due to frictional work, there is a transfer of energy into heat, with an increase in temperature. However, the temperature rise is usually very small and of little practical value, and the frictional work is often referred to as a loss of useful energy. Therefore, in Equation (2.80), we may express the terms $(E'_{i2} - E'_{i1})$ as E_f, the frictional loss of energy. Furthermore, in problems involving pumping of a liquid we may replace W_m with the work done *by* the pump, E_p. Note a change in sign will be required, since W_m was work done by a fluid on the surroundings. Assuming no transfer of heat with the surroundings, or $Q_m = 0$, Equation (2.80) is rewritten as:

$$\frac{P_2}{\rho_2} + \frac{1}{2}\bar{u}_2^2 + gz_2 + E_f = \frac{P_1}{\rho_1} + \frac{1}{2}\bar{u}_1^2 + gz_1 + E_p \tag{2.81}$$

Rearranging terms in Equation (2.81) to obtain an expression for the energy requirements of a pump, E_p, per unit mass, and noting that for an incompressible fluid, $\rho_2 = \rho_1 = \rho$,

$$E_p = \frac{P_2 - P_1}{\rho} + \frac{1}{2}(\bar{u}_2^2 - \bar{u}_1^2) + g(z_2 - z_1) + E_f \tag{2.82}$$

Equation (2.82) contains terms for pressure energy, kinetic energy, potential energy, and energy loss associated with frictional forces, respectively. We will now consider these items individually, to note any necessary modifications and practical implications.

2.5.1 Pressure Energy

The first term on the right-hand side of Equation (2.82) denotes energy dissipation related to the change in pressure between locations (1) and (2). If the transport system (Fig. 2.19a) connects two tanks, both of which are exposed to the atmosphere, then there is no change in pressure, or $P_1 - P_2 = 0$. However, in situations where one or both tanks are under pressure or vacuum (Fig. 2.19b), the pressure difference needs to be accounted for. Changes in pressure may add to energy requirements as

$$\frac{\Delta P}{\rho} = \frac{P_2 - P_1}{\rho} \tag{2.83}$$

Note that liquid density does not change in the type of systems being analyzed. In other words, the flow is incompressible.

2.5.2 Kinetic Energy

The second term in the right-hand side of Equation (2.82) accounts for the change in velocity of the flowing fluid from location (1) to (2),

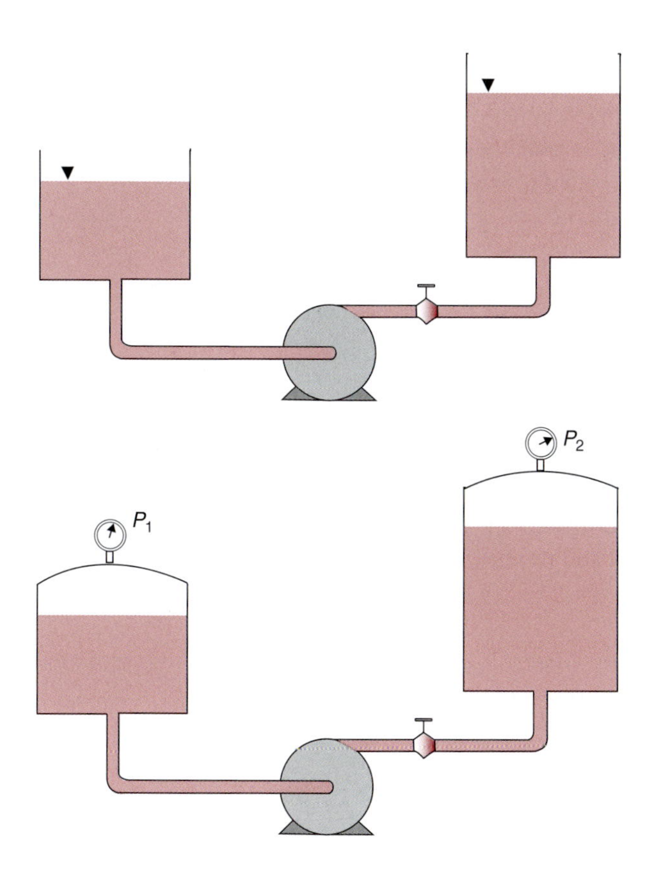

Figure 2.19 Pumping liquid between two tanks.

resulting in a change in kinetic energy. In deriving the energy equation (2.82), we assumed that the fluid velocity is uniform across the entire cross-section. However, due to viscous effects, the velocity is never uniform across the pipe cross-section, but varies as we observed in Section 2.3.4. Therefore, we must use a correction factor, α, and modify the kinetic energy term in Equation (2.82) as follows:

$$\text{Kinetic Energy} = \frac{\bar{u}_2^2 - \bar{u}_1^2}{2\alpha} \tag{2.84}$$

where for laminar flow, $\alpha = 0.5$, and for turbulent flow, $\alpha = 1.0$. Note that the units of the kinetic energy term in Equation (2.84) are expressed as J/kg, as follows:

$$\text{Kinetic Energy} = \frac{\bar{u}^2}{2} \equiv \frac{\text{m}^2}{\text{s}^2} \equiv \frac{\text{kg m}^2}{\text{kg s}^2} \equiv \left(\frac{\text{kg m}}{\text{s}^2}\right)\frac{\text{m}}{\text{kg}} \equiv \frac{\text{N m}}{\text{kg}} \equiv \frac{\text{J}}{\text{kg}}$$

2.5.3 Potential Energy

The energy required to overcome a change in elevation during liquid transport is potential energy. The general expression for change in potential energy per unit mass is

$$= g(z_2 - z_1) \tag{2.85}$$

where z_2 and z_1 are the elevations indicated in Figure 2.18, and the acceleration due to gravity (g) converts the elevation to energy units (J/kg).

$$\text{Potential Energy} = gz \equiv \frac{m^2}{s^2} \equiv \frac{kg\,m^2}{kg\,s^2} \equiv \left(\frac{kg\,m}{s^2}\right)\frac{m}{kg} \equiv \frac{N\,m}{kg} \equiv \frac{J}{kg}$$

2.5.4 Frictional Energy Loss

The frictional energy loss for a liquid flowing in a pipe is composed of major and minor losses. Or,

$$E_f = E_{f,\,major} + E_{f,\,minor} \tag{2.86}$$

The **major losses**, $E_{f,major}$ are due to the flow of viscous liquid in the straight portions of a pipe. Equation (2.51) may be rearranged to give an expression for pressure drop per unit density to account for energy loss due to friction per unit mass, $E_{f,major}$, as

$$E_{f,major} = \frac{\Delta P}{\rho} = 2f\,\frac{\bar{u}^2 L}{D} \tag{2.87}$$

where f is the friction factor obtained from the Moody diagram or Equation (2.54).

The second type of frictional losses, the **minor losses**, $E_{f,minor}$, are due to various components used in pipeline systems—such as valves, tees, and elbows—and to contraction of fluid when it enters from a tank into a pipe or expansion of a fluid when it empties out from a pipe into a tank. Although these types of losses are called *minor* losses, they can be quite significant. For example, if a valve installed in a pipeline is fully closed, then it offers an infinite resistance to flow, and the loss is certainly not minor. The minor losses have three components:

$$E_{f,minor} = E_{f,contraction} + E_{f,expansion} + E_{f,fittings} \tag{2.88}$$

We will consider each of these components separately.

2.5.4.1 *Energy Loss Due to Sudden Contraction, $E_{f,contraction}$*

When the diameter of a pipe suddenly decreases or, in a limiting case, when a liquid held in a tank enters a pipe, there is a contraction in flow (Fig. 2.20). A sudden contraction in the cross-section of the pipe causes energy dissipation. If $\bar{u}$ is the upstream velocity, the energy loss due to sudden contraction is evaluated as

$$\frac{\Delta P}{\rho} = C_{fc}\frac{\bar{u}^2}{2} \tag{2.89}$$

where,

$$C_{fc} = 0.4\left[1.25 - \left(\frac{A_2}{A_1}\right)\right] \quad \text{where } \frac{A_2}{A_1} < 0.715$$
$$C_{fc} = 0.75\left[1 - \left(\frac{A_2}{A_1}\right)\right] \quad \text{where } \frac{A_2}{A_1} > 0.715 \tag{2.90}$$

A limiting case of sudden contraction is when a pipe is connected to a large reservoir. As seen in Figure 2.20, for this case, the diameter A_1 is much larger than A_2, therefore, $A_2/A_1 = 0$ and $C_{fc} = 0.5$.

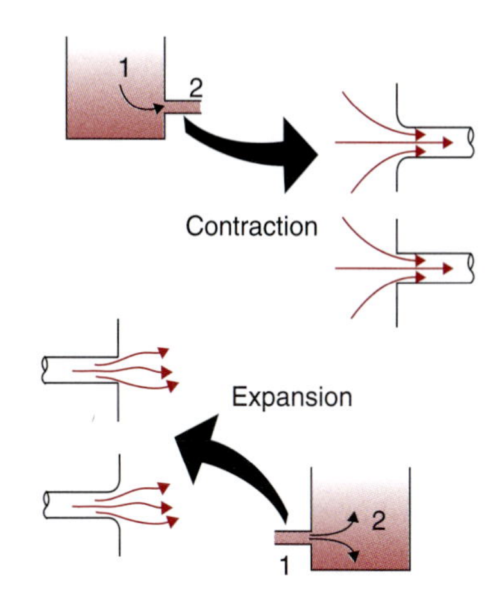

Figure 2.20 Liquid flow through a contraction and an expansion.

2.5.4.2 Energy Loss Due to Sudden Expansion

In a similar manner to sudden contraction, a sudden increase in the cross-section of a pipe will contribute to energy loss due to friction. The energy loss is

$$\frac{\Delta P}{\rho} = C_{fe}\frac{\bar{u}^2}{2} \tag{2.91}$$

and coefficient, C_{fe}, in this case is,

$$C_{fe} = \left(1 - \frac{A_1}{A_2}\right)^2 \tag{2.92}$$

where parameters having a subscript of one are located upstream from the expansion joint. For the limiting case, when a pipe exits into a reservoir, A_2 is much larger than A_1, and $A_1/A_2 = 0$, and $C_{fe} = 1.0$.

2.5.4.3 Energy Losses Due to Pipe Fittings

All pipe fittings such as elbows, tees, and valves will contribute to energy losses due to friction. The energy loss associated with pipe fittings is

$$\frac{\Delta P}{\rho} = C_{ff}\frac{\bar{u}^2}{2} \tag{2.93}$$

Table 2.2 Friction Losses for Standard Fittings

Type of Fitting	C_{ff}
Elbows	
Long radius 45°, flanged	0.2
Long radius 90°, threaded	0.7
Long radius 90°, flanged	0.2
Regular 45°, threaded	0.4
Regular 90°, flanged	0.3
Regular 90°, threaded	1.5
180° Return bends	
180° return bend, flanged	0.2
180° return bend, threaded	1.5
Tees	
Branch flow, flanged	1.0
Branch flow, threaded	2.0
Line flow, flanged	0.2
Line flow, threaded	0.9
Union threaded	0.8
Valves	
Angle, fully open	2
Ball valve, ⅓ closed	5.5
Ball valve, ⅔ closed	210
Ball valve, fully open	0.05
Diaphragm valve, open	2.3
Diaphragm valve, ¼ closed	2.6
Diaphragm valve, ½ closed	4.3
Gate, ¾ closed	17
Gate, ¼ closed	0.26
Gate, ½ closed	2.1
Gate, fully open	0.15
Globe, fully open	10
Swing check, backward flow	∞
Swing check, forward flow	2

Typical values of loss coefficient, C_{ff}, for various fittings are given in Table 2.2. Depending upon how many fittings are used in a fluid transport system, the C_{ff} is the summed up value for all the fittings in Equation (2.93). We will illustrate this procedure in Example 2.11.

Other processing equipment that may be installed in the liquid transport system, such as a heat exchanger, will usually have some assigned pressure drop due to friction. If not, a value for the pressure drop should be obtained by measurement. The measured pressure drop value is then divided by the liquid density to obtain the appropriate energy units.

2.5.5 Power Requirements of a Pump

We can compute the power requirements of a pump by knowing all the changes in energy associated with pumping liquid from one location to another. The energy requirements for pumping a liquid may be expressed by expanding Equation (2.82), as follows:

$$E_p = \frac{P_2 - P_1}{\rho} + \frac{1}{2}(\bar{u}_2^2 - \bar{u}_1^2) + g(z_2 - z_1) + E_{f,\text{major}} + E_{f,\text{minor}} \quad (2.94)$$

or

$$E_p = \frac{P_2 - P_1}{\rho} + \frac{1}{2}(\bar{u}_2^2 - \bar{u}_1^2) + g(z_2 - z_1) + \frac{2f\bar{u}^2L}{D} + C_{fe}\frac{\bar{u}^2}{2} + C_{fc}\frac{\bar{u}^2}{2} + C_{ff}\frac{\bar{u}^2}{2} \quad (2.95)$$

or, dividing each term by g, we may determine the pump requirements in terms of head, as

$$h_{\text{pump}} = \underbrace{\frac{P_2 - P_1}{\rho g}}_{\text{pressure head}} + \underbrace{\frac{1}{2g}(\bar{u}_2^2 - \bar{u}_1^2)}_{\text{velocity head}} + \underbrace{(z_2 - z_1)}_{\text{elevation head}} + \underbrace{\frac{2f\bar{u}^2L}{gD}}_{\text{major losses head}} + \underbrace{C_{fe}\frac{\bar{u}^2}{2g} + C_{fc}\frac{\bar{u}^2}{2g} + C_{ff}\frac{\bar{u}^2}{2g}}_{\text{minor losses head}} \quad (2.96)$$

We can compute the power requirements for the pump, Φ, by noting that power is the rate of doing work; if the mass flow rate $\dot{m}$ is known, then

$$\text{Power} = \Phi = \dot{m}(E_p) \quad (2.97)$$

where E_p is the work done per unit mass *by* the pump *on* the fluid, as given by Equation (2.95).

To calculate pump sizes, we need to incorporate accurate sizes of the pipes being used into computations. The information in Table 2.3 provides the type of values needed for this purpose. Note the variations in diameters of steel pipe as compared with sanitary pipe for the same nominal size.

Table 2.3 Pipe and Heat-Exchanger Tube Dimensions

	Steel pipe (Schedule 40)		Sanitary pipe		Heat-exchanger tube (18 gauge)	
Nominal size (in)	**ID in/(m)**	**OD in/(m)**	**ID in/(m)**	**OD in/(m)**	**ID in/(m)**	**OD in/(m)**
0.5	0.622 (0.01579)[a]	0.840 (0.02134)	–	–	0.402 (0.01021)	0.50 (0.0127)
0.75	0.824 (0.02093)	1.050 (0.02667)			0.652 (0.01656)	0.75 (0.01905)
1	1.049 (0.02644)	1.315 (0.03340)	0.902 (0.02291)	1.00 (0.0254)	0.902 (0.02291)	1.00 (0.0254)
1.5	1.610 (0.04089)	1.900 (0.04826)	1.402 (0.03561)	1.50 (0.0381)	1.402 (0.03561)	1.50 (0.0381)
2.0	2.067 (0.0525)	2.375 (0.06033)	1.870 (0.04749)	2.00 (0.0508)	–	–
2.5	2.469 (0.06271)	2.875 (0.07302)	2.370 (0.06019)	2.5 (0.0635)		
3.0	3.068 (0.07793)	3.500 (0.08890)	2.870 (0.07289)	3.0 (0.0762)	–	–
4.0	4.026 (0.10226)	4.500 (0.11430)	3.834 (0.09739)	4.0 (0.1016)		

Source: Toledo (1991)
[a]*Numbers in parentheses represent the dimension in meters.*

Example 2.11

A 20° Brix (20% sucrose by weight) apple juice is being pumped at 27°C from an open tank through a 1-in nominal diameter sanitary pipe to a second tank at a higher level, illustrated in Figure E2.3. The mass flow rate is 1 kg/s through 30 m of straight pipe with two 90° standard elbows and one angle valve. The supply tank maintains a liquid level of 3 m, and the apple juice leaves the system at an elevation of 12 m above the floor. Compute the power requirements of the pump.

Given

Product viscosity (μ) = 2.1×10^{-3} Pa s, assumed to be the same as for water, from Table A.2.4

Product density (ρ) = 997.1 kg/m^3, estimated from the density of water at 25°C

Pipe diameter (D) = 1 in nominal = 0.02291 m, from Table 2.3

Mass flow rate ($\dot{m}$) = 1 kg/s

Pipe length (L) = 30 m

90° standard elbow friction, from Table 2.2

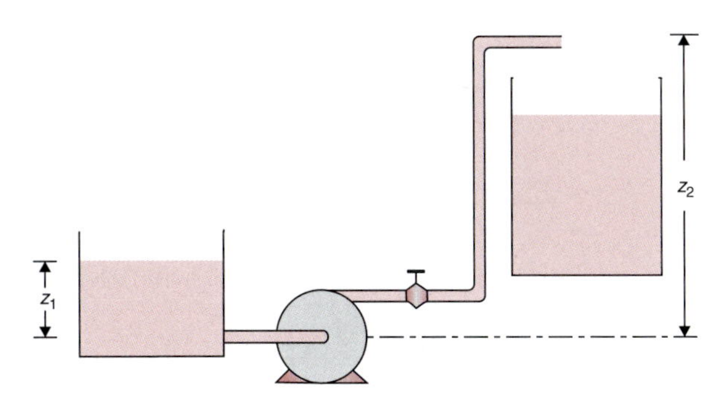

■ **Figure E2.3** Pumping water from one tank to another, for Example 2.11.

Angle valve friction, from Table 2.2
Liquid level $z_1 = 3\,m$, $z_2 = 12\,m$

Approach

Power requirements for the pump can be computed using the mechanical energy balance.

Solution

1. *First, compute mean velocity using mass flow rate equation.*

$$\bar{u} = \frac{\dot{m}}{\rho A} = \frac{(1\,kg/s)}{(997.1\,kg/m^3)[\pi(0.02291\,m)^2/4]} = 2.433\,m/s$$

2. *By computation of the Reynolds number,*

$$N_{Re} = \frac{(997.1\,kg/m^3)(0.02291\,m)(2.433\,m/s)}{(2.1 \times 10^{-3}\,Pa\,s)}$$
$$= 26{,}465$$

it is established that flow is turbulent.

3. *By using the energy equation (2.82) and identification of reference points, the following expression is obtained:*

$$g(3) + E_p = g(12) + \frac{(2.433)^2}{2} + E_f$$

where reference 1 is at the upper level of the supply tank, $\bar{u}_1 = 0$, and $P_1 = P_2$.

4. *By computing E_f, the power can be determined. Based on $N_{Re} = 2.6465 \times 10^4$ and smooth pipe, $f = 0.006$ from Figure 2.16.*
5. *The entrance from the tank to the pipeline can be accounted for by Equation (2.90), where*

$$C_{fc} = 0.4(1.25 - 0) \quad \text{since} \quad D_2^2/D_1^2 = 0$$
$$= 0.5$$

and

$$\frac{\Delta P}{\rho} = 0.5\frac{(2.433)^2}{2} = 1.48\,J/kg$$

6. *The contribution of the two elbows and the angle valve to friction is determined by using C_{ff} factors from Table 2.2. C_{ff} for 90° threaded regular elbows is 1.5 and for angle valve, fully open, is 2. Then by using Equation (2.93), we get*

$$\frac{\Delta P}{\rho} = \frac{(2 \times 1.5 + 2) \times 2.433^2}{2} = 14.79$$

For the 30 m length, the friction loss is obtained from Equation (2.87):

$$E_f = \frac{2 \times 0.006 \times 2.433^2 \times 30}{0.02291} = 93.01$$

7. *Then the total friction loss is*

$$E_f = 93.01 + 14.79 + 1.48 = 109.3 \text{ J/kg}$$

8. *Using the expression obtained in Equation (2.95),*

$$E_p = 9.81(12 - 3) + \frac{2.433^2}{2} + 109.3$$
$$E_p = 200.5 \text{ J/kg}$$

This represents the energy requirements of the pump.

9. *Since power is energy use per unit time,*

$$\text{Power} = (200.5 \text{ J/kg})(1 \text{ kg/s}) = 200.5 \text{ J/s}$$

10. *This answer must be considered theoretical, since delivery of power to pump may be only 60% efficient; then actual power is*

$$\text{Power} = 200.5 / 0.6 = 334.2 \text{ W}$$

Example 2.12

Develop a spreadsheet using the data given in Example 2.11. Rework the problem using the spreadsheet. Determine the influence on the power requirements of changing the pipe length to 60, 90, 120, and 150 m. Also, determine the influence on the power requirement of changing the pipe diameter to 1.5 in, 2 in, and 2.5 in nominal diameters.

Given

The conditions are the same as in Example 2.11.

Approach

We will develop a spreadsheet using Excel™. The mathematical expressions will be the same as those used in Example 2.11. For the friction factor we will use Equation (2.54).

	A	B	C	D	E	F	G	H
1	Given							
2	Viscosity (Pa s)	0.0021						
3	Density (kg/m^3)	997.1						
4	Diameter (m)	0.02291						
5	Mass flow rate (kg/s)	1						
6	Pipe length (m)	30						
7	C_{ff} Elbows from Table 2.2	1.5						
8	C_{ff} Angle Valve from Table 2.2	2						
9	Low liquid level (m)	3						
10	High liquid level (m)	12						
11								
12	Mean velocity	2.432883	=B5/(B3*PI()*B4^2/4)					
13	Reynolds Number	26464.62	=B3*B12*B4/B2					
14	Entrance Losses	1.47973	=0.5*B12^2/2					
15	Friction factor	0.006008	=(−1/(3.6*LOG(6.9/B13)))^2					
16	Friction Loss_pipe length	93.12623	=2*B15*B6*B12^2/B4					
17	Friction loss_fittings	14.7973	=(2*B7+B8)*B12^2/2					
18	Total Friction loss	109.4033	=B14+B16+B17					
19	Energy for Pump	200.6527	=9.81*(B10−B9)+B12^2/2+B18					
20	Power	334.4212	=B19*B5/0.6					
21								
22	Pipe length	Power		Diameter	Power			
23	30	334		0.02291	334			
24	60	490		0.03561	172			
25	90	645		0.04749	154			
26	120	800		0.06019	149			
27	150	955						
28								
29								
30								
31								
32								
33								
34								
35								
36								
37								
38								

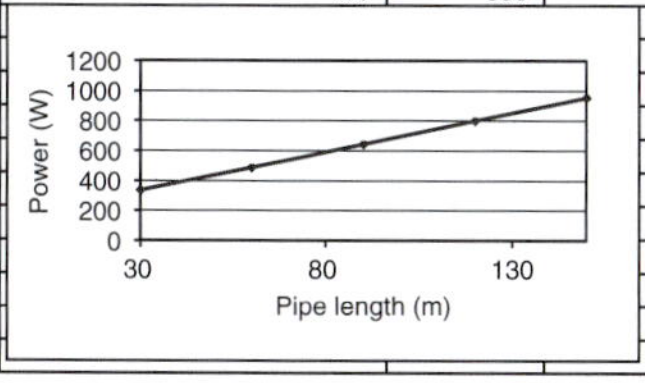

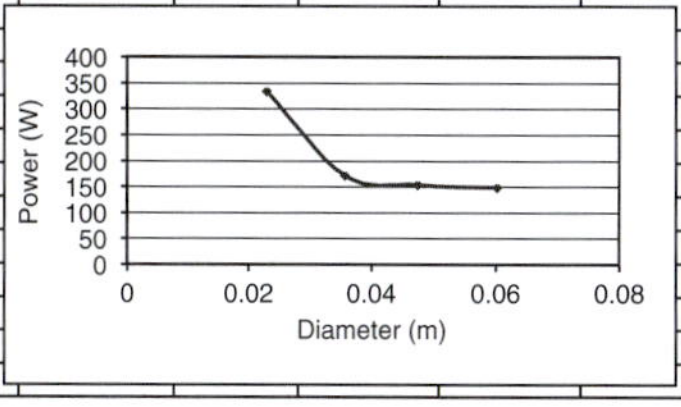

Figure E2.4 A spreadsheet solution for Example 2.12.

f0540

Solution

The spreadsheet is developed as shown in Figure E2.4. All mathematical equations are the same as those used in Example 2.11. The influence of changing length and diameter is seen in the plots. As is evident, for the conditions used in this example, there is a dramatic influence on power requirement when the pipe diameter is decreased from 2.5 in to 1.5 in.

2.6 PUMP SELECTION AND PERFORMANCE EVALUATION

2.6.1 Centrifugal Pumps

In Section 2.1.2, we described some salient features of different type of pumps. We will now consider centrifugal pumps in greater detail, as these are the most commonly used pumps for pumping water and a variety of low-viscosity Newtonian liquids.

As seen in Figure 2.21, a centrifugal pump has two components: an impeller firmly attached to a rotating shaft, and a volute-shaped housing, called casing, that encloses the impeller. The impeller contains a number of blades, called vanes, that usually are curved backward. The shaft of the pump is rotated using either an electric motor

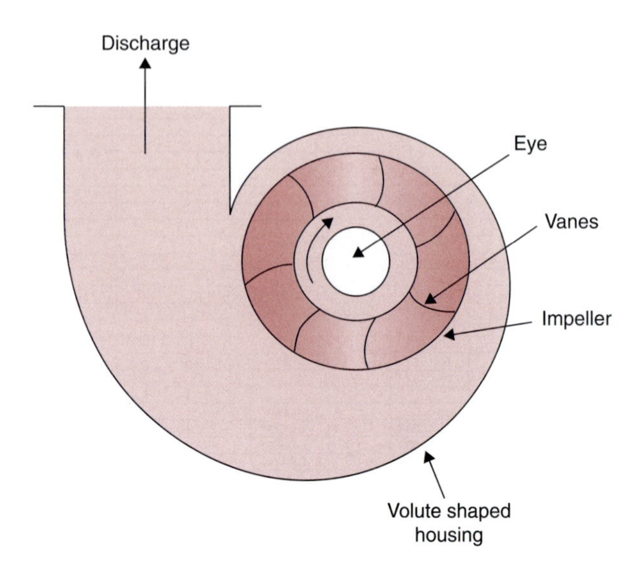

■ **Figure 2.21** A centrifugal pump.

or an engine. As the shaft rotates, the impeller, fixed to the shaft, also rotates. The liquid is sucked through an opening in the casing, called the eye. Due to the rotating impeller, and the direction of the vanes, the liquid flows from the eye to the periphery of the impeller. Therefore, work is done on the liquid by the rotating vanes. In moving from the eye to the periphery, the velocity of the liquid increases, raising its kinetic energy. However, as the liquid enters the peripheral zone, or the volute-shaped casing, the liquid velocity decreases. This decrease in velocity occurs because of the increasingly larger area in the volute. The decrease in liquid velocity causes its kinetic energy to decrease, which is converted into an increase in pressure. Thus, the discharging liquid has a higher pressure when compared with the liquid entering the pump at the suction eye. In summary, the main purpose of the pump is to raise the pressure of the liquid as it moves from suction to discharge.

Although considerable theory has been developed regarding the operation of centrifugal pumps, the mathematical complexity prevents theory alone from being an adequate basis for selection of pumps for a given flow system. Therefore, experimental data are usually obtained for actual pump performance. These data, obtained by manufacturers, are supplied with the pump as pump performance curves. The engineer's task is then to select an appropriate pump based on its performance characteristics. In calculations involving fluid flow and pump selection, a commonly used term is the head. We will first develop a general understanding of this term.

2.6.2 Head

In designing pumps, a common term used to express the energy of a fluid is the **head**. As noted previously, in Chapter 1, head is expressed in meters of liquid.

In Equation (2.96) if we sum all the energy terms into head for various items connected to the suction side of the pump, the summed up value of head is called **suction head**. Similarly, on the discharge side, if we convert all the energy terms to head and add them together, we obtain the **discharge head**.

Let us consider a pump being used to lift water from tank A to tank B, as shown in Figure 2.22a. First let us assume that there is no loss in energy due to friction within the pipes or any of the fittings. The height of water in tank A is 5 m from the pump centerline and it is 10 m in tank B. The pressure gauges on the discharge and suction side read 0.49 bar and 0.98 bar, respectively. This is in accordance with Equation (1.20). The total head is then the discharge head minus the suction head, or $10 - 5 = 5$ m.

The same tanks in Figure 2.22a are shown in Figure 2.22b, except the pressure gauge on the suction side reads a pressure of 0.39 bar. This decrease, of 0.1 bar (equivalent to 1.02 m of water) is due to the frictional losses in the pipe and other fittings used to convey the water from tank A to the pump. This is a more realistic situation because there will always be some frictional loss due to flow of a viscous liquid.

In Figure 2.22c, tank A is located below the pump centerline. In this case, first the pump must lift water from a lower level to the pump centerline. This is called the **suction lift**. The total head is then calculated as suction plus the discharge head or $7 + 10 = 17$ m.

The fourth case, Figure 2.22d involves water in a tank on the suction side where it is held under pressure (0.30 bar), as indicated by the gauge in the headspace of the tank. The pressure gauge in the suction line indicates 0.79 bar. The suction head is greater than the actual height of the water because of the applied pressure.

2.6.3 Pump Performance Characteristics

In designing liquid transport systems that involve pumps, two items are necessary: (1) quantitative information about a pump being considered, and (2) the energy requirements associated with liquid flow through various components of the transport system such as pipes, tanks, processing equipment, and fittings. The information on the pump should inform us about the energy that the pump will add to

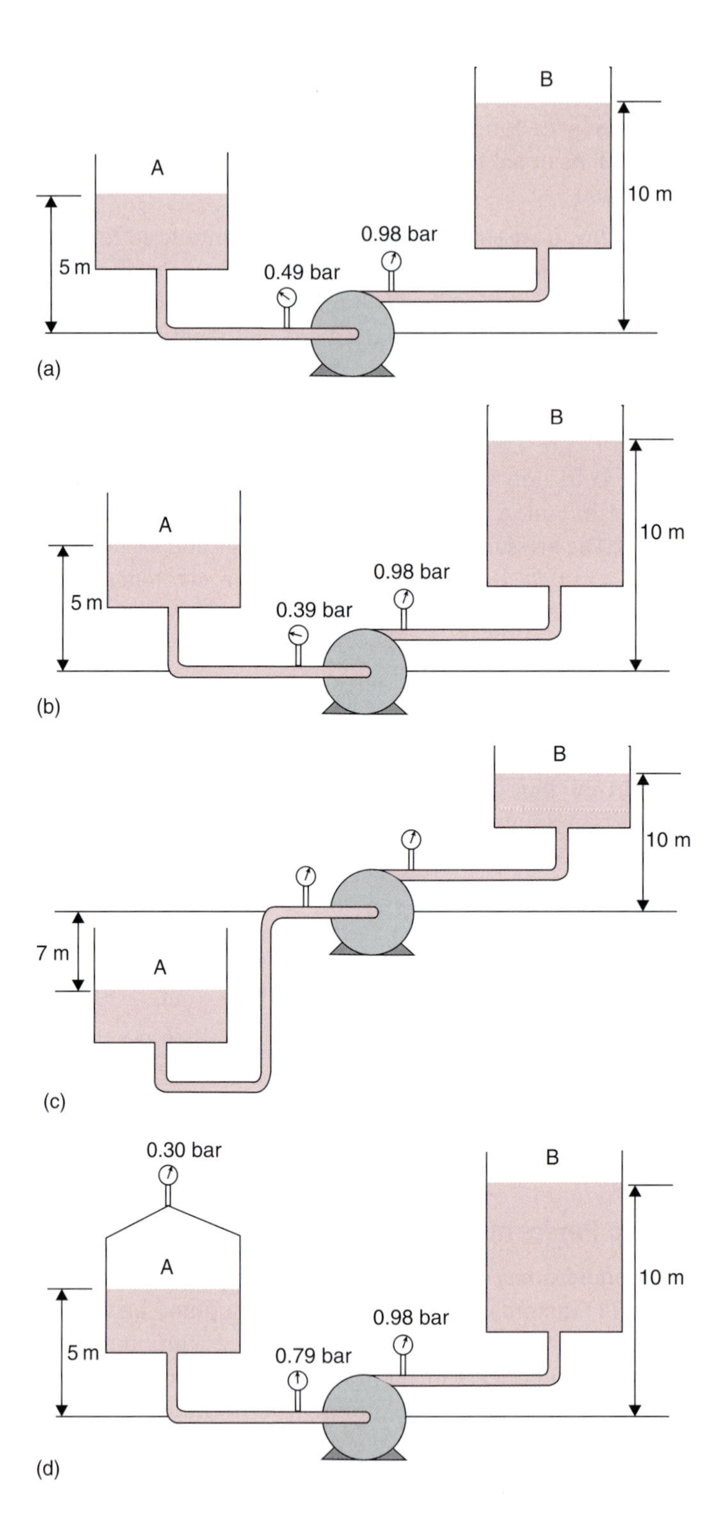

Figure 2.22 Suction and discharge pressures in a pumping system under different conditions.

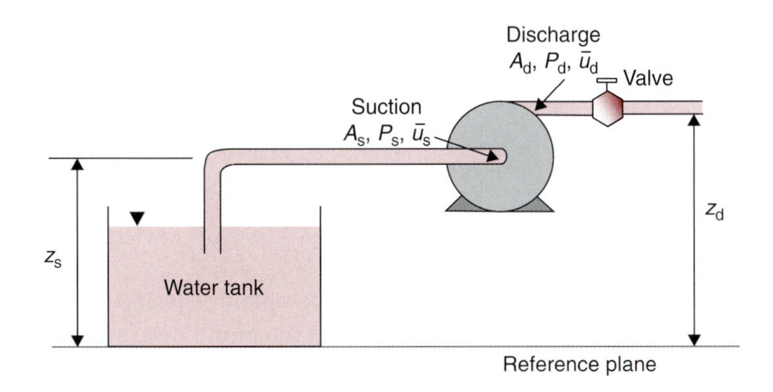

■ **Figure 2.23** A test unit to determine performance of a pump.

the liquid flowing at a certain flow rate. In other words, we need to know the performance of a pump under certain operating conditions. This information, in the form of a pump characteristic diagram, is determined by the manufacturer and is supplied with the pump so that an engineer can make a proper judgment on the suitability of the pump for a given transport system.

Standard procedures have been developed to test the performance of industrial pumps (Hydraulic Institute, 1975). Testing of a pump is conducted by running the pump at a constant speed when set up on a test stand, as shown in Figure 2.23. The height of the suction port from a reference plane, z_s, and the height of the discharge port, z_d, are recorded. Initially, the valve is kept fully open and measurements of pressure on the suction and discharge sides, volumetric flow rates, and torque supplied to turn the pump shaft are measured. Then the discharge valve is slightly closed and the measurements are repeated. This procedure is continued until the valve is nearly closed. The valve is never completely closed, or else the pump may be damaged.

The pump performance test involves measurement of volumetric flow rate, $\dot{V}$, areas of suction and discharge ports, A_s and A_d, heights of the suction and discharge ports, pressures at suction, P_s and discharge, P_d. The data are then used in the following calculations.

The velocity at the suction is calculated as

$$\bar{u}_s = \frac{\dot{V}}{A_s} \tag{2.98}$$

Similarly, the velocity at the pump discharge is obtained as

$$\bar{u}_d = \frac{\dot{V}}{A_d} \tag{2.99}$$

The suction head, h_s, and discharge head, h_d, based on discussion presented in Section 2.5, are obtained as

$$h_s = \frac{\bar{u}_s^2}{2\alpha g} + z_s + \frac{P_s}{\rho g} \quad (2.100)$$

$$h_d = \frac{\bar{u}_d^2}{2\alpha g} + z_d + \frac{P_d}{\rho g} \quad (2.101)$$

The values of suction and discharge heads obtained from Equations (2.100) and (2.101) are used in calculating the **pump head** as

$$h_{pump} = h_d - h_s \quad (2.102)$$

Note that in Equation (2.102) we do not consider friction losses in pipes, since our interest at this point is primarily in the performance of the pump, not the system.

The power output of the pump is called the fluid power, Φ_{fl}. It is the product of the mass flow rate of the fluid and the pump head

$$\Phi_{fl} = \dot{m} g h_{pump} \quad (2.103)$$

The fluid power may also be expressed in terms of volumetric flow rate, $\dot{V}$, as,

$$\Phi_{fl} - \rho g \dot{V} h_{pump} \quad (2.104)$$

The power required to drive the pump is called the break power, Φ_{bk}. It is obtained from the torque supplied to the pump shaft, Ω, and the angular velocity of the shaft, ω,

$$\Phi_{bk} = \omega \Omega \quad (2.105)$$

The efficiency of the pump is calculated from these two values of power. It is the ratio between the power gained by the fluid and the power supplied by the shaft driving the pump, or,

$$\eta = \frac{\Phi_{fl}}{\Phi_{bk}} \quad (2.106)$$

The calculated values of pump head, efficiency, and break power are used to develop a pump characteristic diagram as discussed in the following section.

Example 2.3

The following data were collected while testing a centrifugal pump for water at 30°C. Suction pressure = 5 bar, discharge pressure = 8 bar, volumetric flow rate = 15,000 L/h. Calculate the pump head at the given flow rate and power requirements.

Given

Suction pressure = 5 bar = 5×10^5 Pa = 5×10^5 N/m^2 = 5×10^5 kg/(m s^2)
Discharge pressure = 8 bar = 8×10^5 Pa = 8×10^5 N/m^2 = 8×10^5 kg/(m s^2)
Volumetric flow rate = 15,000 L/h = 0.0042 m^3/s

Approach

We will use Equation (2.102) to obtain pump head and Equation (2.104) to calculate the fluid power requirements.

Solution

1. *In Equation (2.102), for the pump shown in Figure 2.23, the suction and discharge velocities are about the same and the difference in elevation z_2–z_1 may be neglected, then*

$$h_{pump} = \frac{(P_d - P_s)}{\rho g}$$

$$h_{pump} = \frac{(8-5) \times 10^5 \ [kg/(m\,s^2)]}{995.7\ [kg/m^3] \times 9.81\ [m/s^2]}$$

$$h_{pump} = 30.7\ m$$

2. *From Equation (2.104)*

$$\Phi_{fl} = 995.7\ [kg/m^3] \times 9.81\ [m/s^2] \times 0.0042\ [m^3/s] \times 30.7\ [m]$$
$$\Phi_{fl} = 1259\ W = 1.26\ kW$$

3. *The fluid power requirement is 1.26 kW at a flow rate of 15,000 L/h. The pump head is 30.7 m.*

2.6.4 Pump Characteristic Diagram

The calculated values of the pump head, efficiency, and break power plotted against volumetric flow rate (also called capacity) constitute the characteristic diagram of the pump, as shown in Figure 2.24. Typically, the pump characteristic diagrams are obtained for water. Therefore, if a pump is to be used for another type of liquid, the curves must be adjusted for the different properties of the liquid. As seen in Figure 2.24, a centrifugal pump can deliver flow rate from zero to maximum, depending upon the head and conditions at the suction. These curves depend upon the impeller diameter and the casing size. The relationship between the head and volumetric flow rate may be rising, drooping, steep, or flat. As seen in the figure, a rising head curve is shown, since the head increases with decreasing flow rate. The shape of the curve depends upon the impeller type and its design characteristics. At zero capacity, when the discharge valve is completely shut, the efficiency is zero, and the power supplied to the pump is converted to heat.

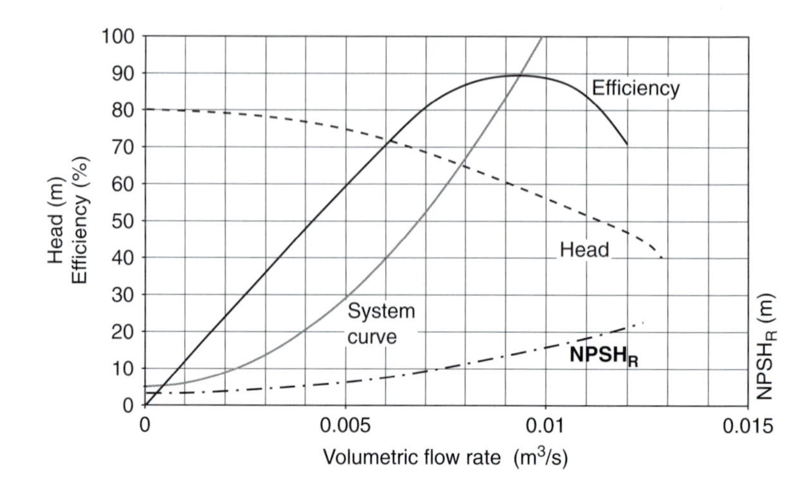

Figure 2.24 Performance characteristic curves for a pump.

We can draw several conclusions by examining the characteristic diagram of the pump. As the total head decreases, the volumetric flow rate increases. When the fluid level in the tank on the suction side decreases, the total head increases and the volumetric flow rate decreases. The efficiency of a pump is low both at low and high volumetric flow rates. The break power increases with the flow rate; however, it decreases as the maximum flow rate is reached.

The peak of efficiency curve represents the volumetric flow rate where the pump is most efficient. The flow rate at the peak efficiency is the design flow rate. The points on the head and power curve corresponding to the maximum efficiency are called best efficiency points, or BEP. With increasing volumetric flow rate, the power required to operate the pump increases. If a different impeller diameter is used, the head curve is shifted; increasing the diameter raises the curve. Thus, by using an impeller of a larger diameter pump, we can pump liquid to a higher head. Figure 2.24 also shows the net positive suction head (NPSH), which we will discuss in the following section.

2.6.5 Net Positive Suction Head

An important issue that requires our careful attention in designing pumps is to prevent conditions that may encourage vaporization of the liquid being transported. In a closed space, a certain pressure on the liquid surface is necessary to prevent vapors from escaping from the liquid. This pressure is the vapor pressure of the liquid. In a pumping system, it is important that the pressure of the liquid does not decrease below the vapor pressure of the liquid at that temperature. If it does so, a phenomenon called cavitation may occur at the eye of the impeller.

As the liquid enters the eye of the impeller, pressure at this location is the lowest in the entire liquid handling system. If the pressure at that location is lower than the liquid vapor pressure, then vaporization of liquid will begin.

Any formation of vapors will lower the efficiency of a pump. Furthermore, as the vapors travel further along the impeller toward the periphery, the pressure increases, and the vapors condense rapidly. Cavitation may be recognized as a crackling sound produced as vapor bubbles form and collapse on the impeller surface. With cavitation occurring at high frequency and extremely high local pressures, any brittle material such as the impeller surface may be damaged. To avoid cavitation, the pressure on the suction side must not be allowed to drop below the vapor pressure. The pump manufacturers specify the *required* net positive suction head ($NPSH_R$) as suction head minus the vapor pressure head, or:

$$NPSH_R = h_s - \frac{P_v}{\rho g} \tag{2.107}$$

where the total head on the suction side of a pump is

$$h_s = \frac{P_s}{\rho g} + \frac{u_s^2}{2g} \tag{2.108}$$

Then,

$$NPSH_R = \frac{P_s}{\rho g} + \frac{u_s^2}{2g} - \frac{P_v}{\rho g} \tag{2.109}$$

where P_v = vapor pressure of the liquid being pumped.

The $NPSH_R$ must be exceeded to ensure that cavitation will be prevented. The manufacturers test their pumps experimentally to determine the value of $NPSH_R$, and these values are provided graphically as shown in Figure 2.24. The pump user must ensure that the *available* net positive suction head ($NPSH_A$) for a given application is greater than the required $NPSH_R$ as specified by the manufacturer.

In using a pump for a given application, a calculation is first made to determine the $NPSH_A$, which depends on the given flow system. For example, in the case of the flow system shown in Figure 2.25 between locations (1) and (2), we may apply Equation (2.96) to obtain

$$\frac{P_{atm}}{\rho g} - z_1 = \frac{P_2}{\rho g} + \frac{\bar{u}_2^2}{2g} + h_{1-2} \tag{2.110}$$

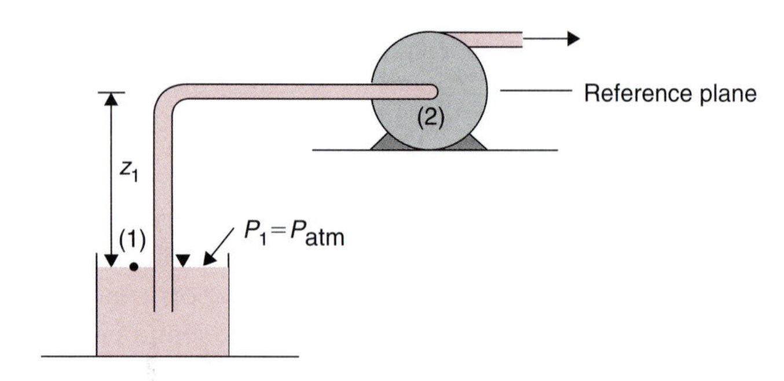

■ **Figure 2.25** Suction side of a pumping system.

where h_{1-2} represents the major and minor losses between locations (1) and (2). The head available at the suction side of the pump (the pump impeller inlet) is

$$\frac{P_2}{\rho g} + \frac{\bar{u}_2^2}{2g} = \frac{P_{atm}}{\rho g} - z_1 - h_{1-2} \tag{2.111}$$

Then, the $NPSH_A$ is the suction head minus the vapor pressure head, or

$$NPSH_A = \frac{P_{atm}}{\rho g} - z_1 - h_{1-2} - \frac{P_v}{\rho g} \tag{2.112}$$

To avoid cavitation, an engineer must ensure that $NPSH_A$ is equal to or greater than $NPSH_R$. Note that in Equation (2.112) the $NPSH_A$ decreases if the height of the pump above the liquid surface in the reservoir, z_1, is increased, or if installing more fittings on the suction side increases the friction head loss, h_{1-2}.

Example 2.14

A centrifugal pump is to be located 4 m above the water level in a tank. The pump will operate at a rate of 0.02 m^3/s. The manufacturer suggests a pump with a $NPSH_R$ at this flow rate as 3 m. All frictional losses may be neglected except a heat exchanger between the pipe inlet and the pump suction that has a loss coefficient $C_f = 15$. The pipe diameter is 10 cm and the water temperature is 30°C. Is this pump suitable for the given conditions?

Given

Pipe diameter = 10 cm = 0.1 m
Pump location above water level in tank = 4 m
Volumetric flow rate = 0.02 m^3/s
Loss coefficient due to heat exchanger $C_f = 15$
Temperature of water = 30°C
$NPSH_R = 3$ m

Approach

We will first determine the head losses and then using Equation (2.112) determine the $NPSH_A$. From steam tables (Table A.4.2), we will determine vapor pressure at 30°C.

Solution

1. *Velocity is obtained from the volumetric flow rate using Equation (2.17):*

$$\bar{u} = \frac{0.02 \left[\frac{m^3}{s}\right]}{\frac{\pi \times (0.1)^2 \ [m^2]}{4}}$$

$$\bar{u} = 2.55 \ m/s$$

2. *The frictional head loss due to the heat exchanger is obtained using an equation similar to Equation (2.93):*

$$h_L = C_{heat\ exchanger} \frac{u^2}{2g}$$

$$h_L = \frac{15 \times (2.55)^2 \left[\frac{m^2}{s^2}\right]}{2 \times 9.81 \left[\frac{m}{s^2}\right]}$$

$$h_L = 4.97 \ m$$

3. *From steam tables, at 30°C, vapor pressure is 4.246 kPa; then $NPSH_A$ is, from Equation (2.112),*

$$\frac{101.3 \times 1000 \ [Pa]}{9.81 \ [m/s^2] \times 995.7 \ [kg/m^3]} - 4 \ [m] - 4.97 \ [m] - \frac{4.246 \times 1000 \ [Pa]}{9.81 [m/s^2] \times 995.7 [kg/m^3]}$$

(Note that 1 Pa = 1 kg/(m s²))

$$NPSH_A = 10.37 - 4 - 4.97 - 0.43$$

$$NPSH_A = 0.97 \ m$$

4. *The $NPSH_A$ is less than the $NPSH_R$. This suggests that cavitation will occur. Therefore the recommended pump is unsuitable for the given conditions. Another pump with an $NPSH_R$ of less than 0.97 m should be chosen to prevent cavitation.*

2.6.6 Selecting a Pump for a Liquid Transport System

In Section 2.6.3 we noted the two requirements in designing a liquid transport system – information about the pump and the system.

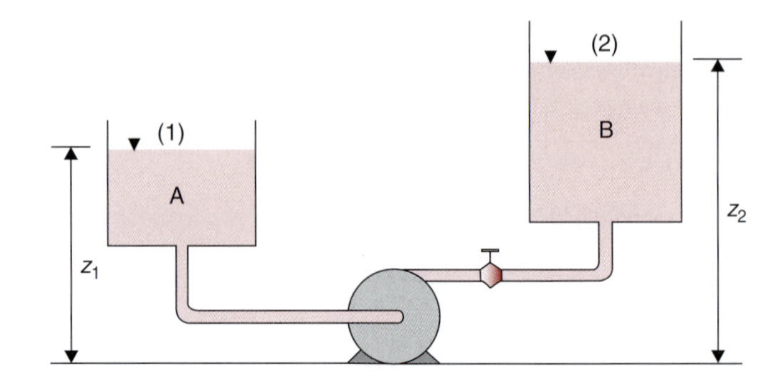

Figure 2.26 Pumping liquid from one tank to another.

So far we have examined the requirements of a pump. We will now consider a total liquid transport system containing pipes, valves, fittings, and other process equipment. Remember that the purpose of installing a pump in a liquid transport system is to increase the energy of the liquid so that it can be moved from one location to another. For example, in Figure 2.26, a pump is being used to pump liquid from tank A to tank B. The system contains a certain length of pipe, elbows, and a valve. The liquid level in tank A is z_1 from the ground. In tank B, the top of the liquid is z_2 from the ground. The velocity of the liquid surfaces at locations 1 and 2 is negligible, and in both tanks the surfaces are exposed to atmospheric pressure. Therefore, for this system,

$$h_{system} = z_2 - z_1 + h_{1-2} \tag{2.113}$$

From Equation (2.96), we observe that the friction losses, h_{1-2}, are proportional to the square of velocity. Since velocity is proportional to volumetric flow rate, the frictional losses are proportional to the square of the volumetric flow rate. Or,

$$h_{1-2} = C_{system}\dot{V}^2 \tag{2.114}$$

where C_{system} is a system constant. Thus, substituting Equation (2.114) in Equation (2.113),

$$h_{system} = z_2 - z_1 + C_{system}\dot{V}^2 \tag{2.115}$$

The system head as a function of the volumetric flow rate is shown in Figure 2.27a. The upward increasing curve is due to the quadratic function in Equation (2.115). The system head, h_{system}, depends upon the change in elevation (total static head) and any of the major and minor losses. In Figure 2.27b, two system head curves are shown, indicating situations where the static head may be changing. Similarly, if the frictional losses change, for example if a valve in a

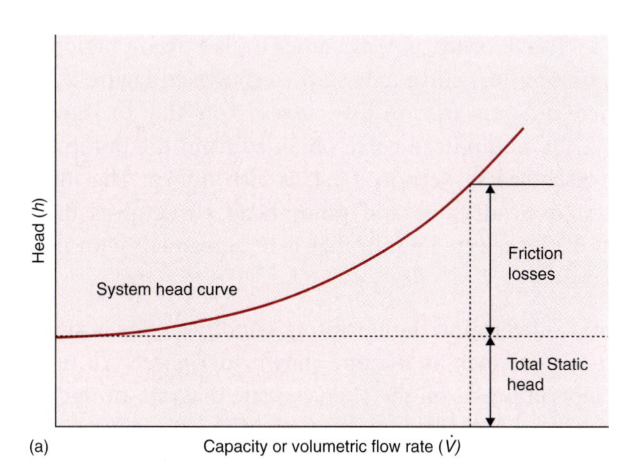

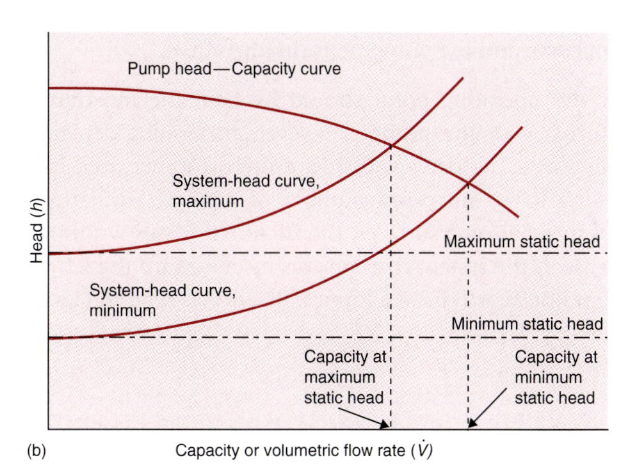

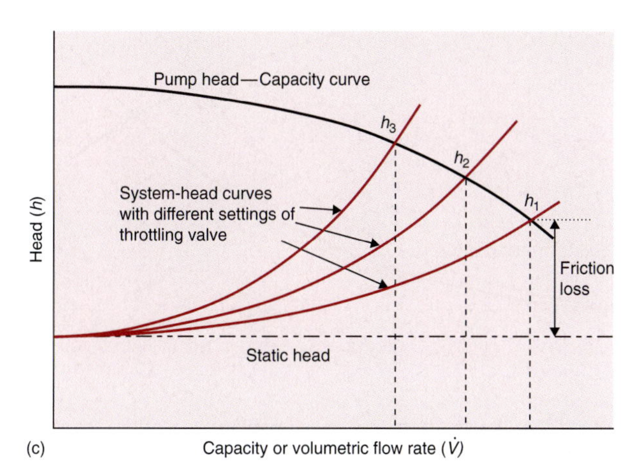

Figure 2.27 (a) System-head curve for a pump. (b) System-head curves for minimum and maximum static head. (c) System-head curves with different settings of throttling valve.

pipeline is closed, or the pipe becomes fouled over a period of time, then the friction loss curve may shift as shown in Figure 2.27c, indicating three different friction loss curves. Note that in Figures 2.27b and 2.27c, the pump-head curve obtained from the pump manufacturer, as discussed in Section 2.6.4, is also drawn. The intersection of the system-head curve and pump-head curve gives the operating point of the selected pump that is in agreement with the system requirements.

Therefore, to determine the operating conditions for a given liquid transport system, such as the one shown in Figure 2.26, the system curve is superimposed on the characteristic diagram of the pump, as shown in Figure 2.28. The intersection of the system curve and the pump performance curve, A, called the operating point, gives the operating values of flow rate and head. These two values satisfy both the system curve and the pump performance curve.

Typically, the operating point should be near the maximum value of the efficiency of the pump. However, this point depends upon the system curve. The curve will shift if there are increased losses, for example due to an increased number of fittings. Similarly, due to fouling of internal pipe surfaces, the frictional losses within the pipe may increase. If the system curve moves more toward the left, the new operating point, B, will be at a lower efficiency as seen in Figure 2.28.

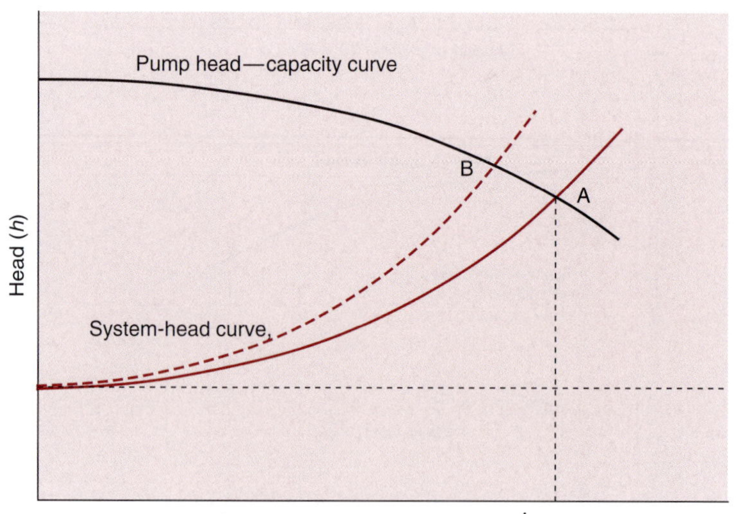

Figure 2.28 Pump head—capacity curve and system-head curve.

Example 2.15

A centrifugal pump is being considered for transporting water from tank A to tank B. The pipe diameter is 4 cm. The friction factor is 0.005. The minor losses include a contraction at the pipe inlet, expansion at the pipe outlet, four pipe bends, and a globe valve. The total length is 25 m and the elevation difference between the levels of water in tank A and B is 5 m. The performance characteristics of the pump supplied by the manufacturer are given in Figure E2.5.

Given

Pipe diameter = 4 cm = 0.04 m
Pipe length = 25 m
Friction factor = 0.005
C_f*(elbow) = 1.5 (from Table 2.2)*
C_f*(globe valve, fully opened) = 10 (from Table 2.2)*
$C_{fc} = 0.5$ *(from Equation (2.90) for* $D_1 >> D_2$*)*
$C_{fe} = 1.0$ *(from Equation (2.92) for* $D_2 >> D_1$*)*

Approach

We will apply the energy expression, Equation (2.96), between locations (1) and (2). Then we will express it in terms of pump head vs flow rate. We will plot it on the performance curve and determine the intersection point.

Solution

1. *Applying the energy equation (2.96), noting that* $P_1 = P_2 = 0$, $\bar{u}_1 = \bar{u}_2$, $z_2 - z_1 = 5\,m$, *we obtain*

$$h_{pump} = z_2 - z_1 + \text{major losses} + \text{minor losses}$$

$$h_{pump} = 5\,[m] + \left\{\frac{4 \times 0.005 \times 25\,[m]}{0.04\,[m]} + 0.5 + 1.0 + 4(1.5) + 10\right\} \times \frac{\bar{u}^2\,[m^2/s^2]}{2 \times 9.81\,[m/s^2]}$$

$$h_{pump} = 5 + 1.5291 \times \bar{u}^2$$

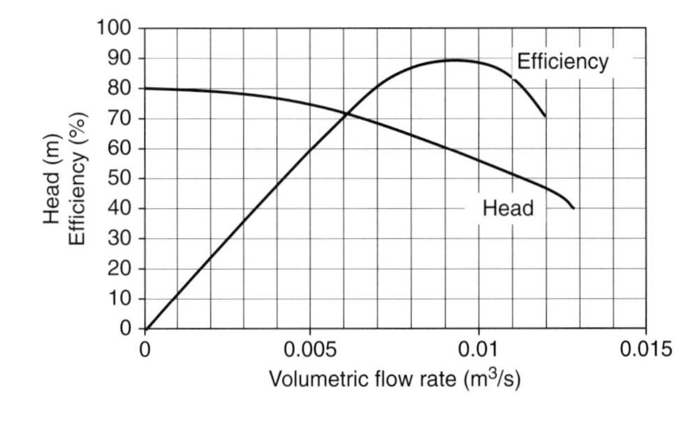

Figure E2.5 Performance characteristic of a centrifugal pump.

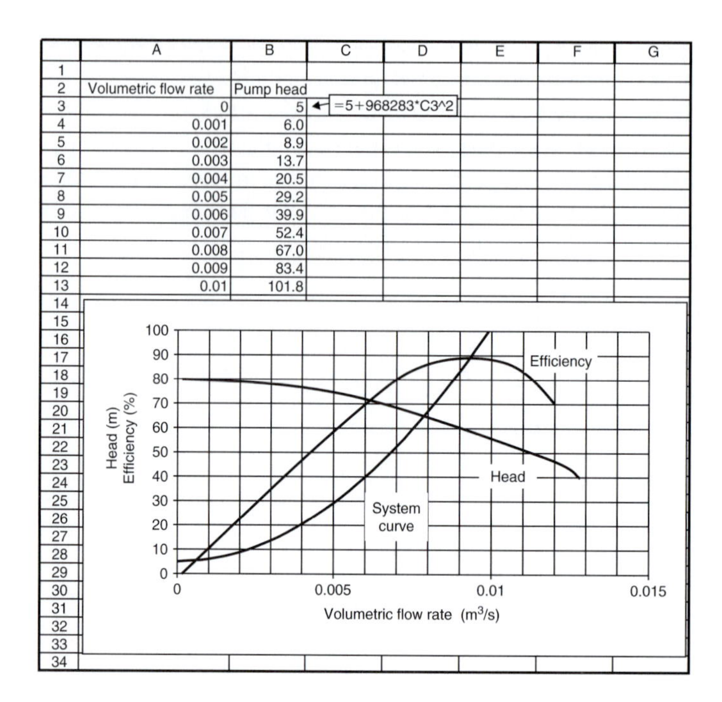

	A	B	C
1			
2	Volumetric flow rate	Pump head	
3	0	5	=5+968283*C3^2
4	0.001	6.0	
5	0.002	8.9	
6	0.003	13.7	
7	0.004	20.5	
8	0.005	29.2	
9	0.006	39.9	
10	0.007	52.4	
11	0.008	67.0	
12	0.009	83.4	
13	0.01	101.8	

■ Figure E2.6 System curve plotted with pump performance curves for Example 2.15.

2. Velocity may be expressed in terms of volumetric flow rate, using Equation (2.17) as

$$\bar{u} = \frac{4\dot{V}[m^3/s]}{\pi(0.04)^2[m^2]}$$

3. Substituting $\bar{u}$ in the expression for h_{pump} in step (1),

$$h_{pump} = 5 + 9{,}68{,}283 \times \dot{V}^2$$

4. Plotting the expression for h_{pump} obtained in step (3) in Figure E2.6 we determine the operating point where the system curve intersects the head curve. The volumetric flow rate at the operating point is 0.0078 m³/s with a head of 65 m, and an efficiency of 88%. This operating efficiency is close to the peak efficiency of 90%.
5. The pump head needed at the pump shaft:

$$= \frac{65\ [m]}{0.88} = 73.9\,m$$

6. The break power needed to drive the pump is obtained from Equations (2.104) and (2.106) as

$$= \frac{990\ [kg/m^3] \times 9.81\ [m/s^2] \times 0.0078\ [m^3/s] \times 65\ [m]}{0.88}$$

$$= 5.6\,kW$$

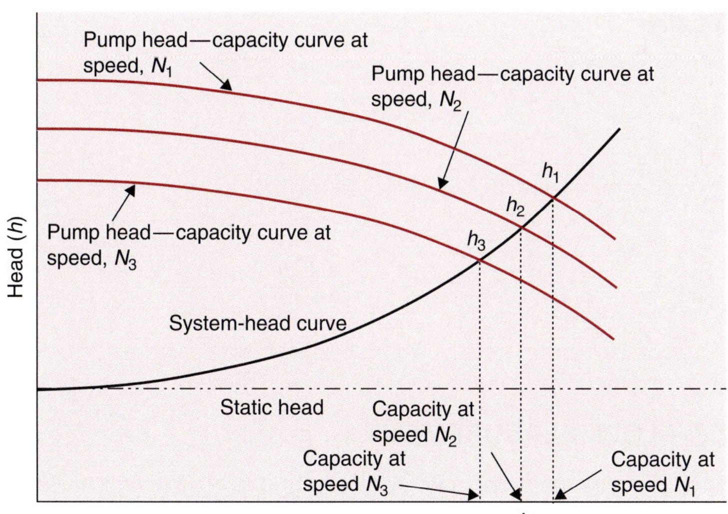

Figure 2.29 Pump head—capacity curve at different pump speeds.

2.6.7 Affinity laws

A set of formulas known as affinity laws govern the performance of centrifugal pumps at various impeller speeds. These formulas are as follows:

$$\dot{V}_2 = \dot{V}_1(N_2/N_1) \tag{2.116}$$

$$h_2 = h_1(N_2/N_1)^2 \tag{2.117}$$

$$\Phi_2 = \Phi_1(N_2/N_1)^3 \tag{2.118}$$

where N is impeller speed, $\dot{V}$ is volumetric flow rate, Φ is power, and h is head.

These equations can be used to calculate the effects of changing impeller speed on the performance of a given centrifugal pump. For example, Figure 2.29 shows the pump head curve for three different impeller speeds that may be obtained with the use of a variable speed motor operating the pump. Example 2.16 illustrates the use of these formulas.

Example 2.16

A centrifugal pump is operating with the following conditions:

volumetric flow rate = $5\,m^3/s$
total head = 10 m
power = 2 kW
impeller speed = 1750 rpm

Calculate the performance of this pump if it is operated at 3500 rpm.

Solution

The ratio of speeds is

$$\frac{N_2}{N_1} = \frac{3500}{1750} = 2$$

Therefore, using Equations (2.116), (2.117), and (2.118),

$$\dot{V}_2 = 5 \times 2 = 10 \text{ m}^3/\text{s}$$
$$h_2 = 10 \times 2^2 = 40 \text{ m}$$
$$P'_2 = 2 \times 2^3 = 16 \text{ kW}$$

2.7 FLOW MEASUREMENT

The measurement of flow rate in a liquid transport system is an essential component of the operation. As illustrated in previous sections, knowledge of flow rate and/or fluid velocity is important in design calculations. In addition, periodic measurements during actual operations are required to ensure that system components are performing in an expected manner.

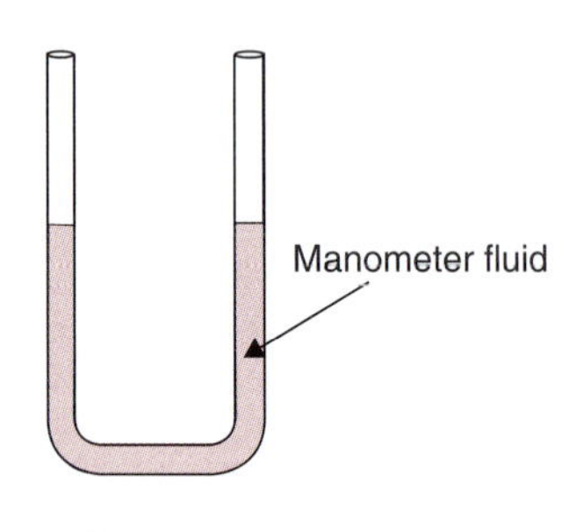

■ **Figure 2.30** A manometer.

There are several types of flow measurement devices that are inexpensive and lead to direct quantification of the mass flow rate or velocity. These methods include (a) Pitot tube, (b) orifice meter, and (c) venturi tube. With all three of these methods, a portion of the measurement involves pressure difference. The device most often used for this purpose is a U-tube manometer. Let us first consider how a U-tube manometer is used in measuring pressure, then we will examine its use in flow measuring devices.

A U-tube manometer is a small-diameter tube of constant diameter shaped as a "U" as shown in Figure 2.30. The tube is partially filled with a fluid called the manometer fluid, rising to a certain height in each of its arms. This fluid must be different from the fluid whose pressure is to be measured. For example, mercury is a commonly used manometer fluid.

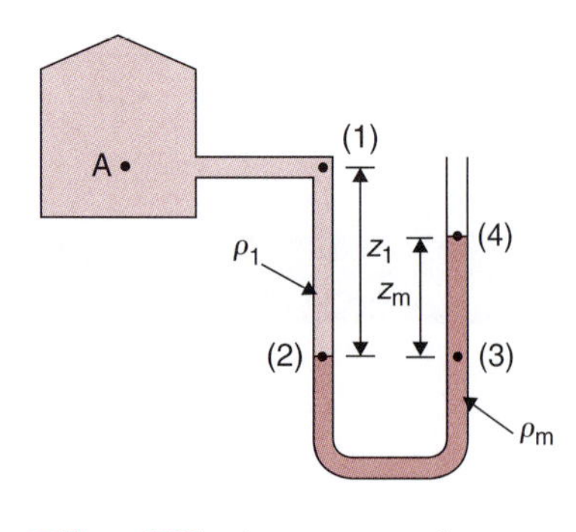

■ **Figure 2.31** A manometer used to measure pressure in a chamber.

Let us consider a case where we want to measure pressure at location A in a vessel containing some fluid, as shown in Figure 2.31. For this purpose, a hole is drilled in the side of the vessel at the same elevation as location A, and one arm of the U-tube manometer is connected to this hole. As shown in the figure, the pressure of the fluid in the vessel pushes the manometer fluid down in the left-hand arm while raising it by an equal distance in the right-hand arm. After the initial displacement, the manometer fluid comes to rest. Therefore, we can apply the expressions developed in Section 1.9 for static head.

An easy approach to analyze the pressures at various locations within the U-tube manometer is to account for pressures at selected locations starting from one side of the manometer and continuing to the other. Using this approach, we note that the pressure at location (1) is same as at A because they are at the same elevation. From location (1) to (2), there is an increase in pressure equivalent to $\rho_1 g z_1$. Pressures at locations (2) and (3) are the same because they are at the same elevation and the fluid between locations (2) and (3) is the same. From location (3) to (4) there is a decrease in pressure equal to $\rho_m g z_m$. The manometer fluid at location (4) is exposed to atmosphere. Therefore, we may write an expression as follows

$$P_A + \rho_1 g z_1 - \rho_m g z_m = P_{atm} \tag{2.119}$$

or

$$P_A = \rho_m g z_m - \rho_1 g z_1 + P_{atm} \tag{2.120}$$

If the density of manometer fluid, ρ_m, is much larger than the fluid in the vessel, ρ_1, then pressure at location A in the vessel is simply

$$P_A = \rho_m g z_m + P_{atm} \tag{2.121}$$

Therefore, knowing the difference between the heights of the manometer fluid in the two arms, z_m, and the density of the manometer fluid, we can determine the pressure at any desired location in the vessel. Note that the lengths of the manometer arms have no influence on the measured pressure. Furthermore, the term $\rho_m g z_m$ in Equation (2.121) is the gauge pressure according to Section 1.9.

Next, let us consider a case where a U-tube manometer is connected to two vessels containing fluids of different densities ρ_A and ρ_B, and under different pressures (Fig. 2.32). Assume that pressure in vessel A is greater than that in vessel B. Again we will approach this problem by tracking pressures from one arm of the manometer to another. Pressure at location (1) is the same as at A. From (1) to (2) there is an increase in pressure equivalent to $\rho_A g z_1$. Pressures at locations (2) and (3) are the same since they are at the same elevation and contain the same fluid. From (3) to (4), there is a decrease in pressure equal to $\rho_m g z_m$. From location (4) to (5) there is another decrease in pressure equal to $\rho_B g z_3$. The pressures at location B and (5) are the same. Thus we may write an expression for pressure as follows:

$$P_A + \rho_A g z_1 - \rho_m g z_m - \rho_B g z_3 = P_B \tag{2.122}$$

or

$$P_A - P_B = g(\rho_m z_m - \rho_A z_1) + \rho_B g z_3 \tag{2.123}$$

As is evident from these derivations, the manometer fluid must have a higher density than the fluid whose pressure is being measured. Furthermore, the two fluids must be immiscible. The common manometer fluids are mercury and water, depending upon the application. An extension of the preceding analysis involves determining pressure difference in a flow-measuring device using a U-tube manometer. We will consider this analysis for different types of flow-measuring devices.

In pressure measurements involving fluid flow, it is important to note that there are typically three kinds of pressures involved: static, dynamic, and stagnation pressure.

Static pressure is actual thermodynamic pressure of a moving fluid depicted by the first term in the Bernoulli equation (Equation (2.67)). As shown in Figure 2.33, the pressure of the moving fluid measured at location (1) is the static pressure. If the pressure sensor were moving with the fluid at the same velocity as the fluid, then the fluid would appear "static" to the sensor, hence the name. A common procedure to measure static pressure is to drill a hole in the duct, making sure that there are no imperfections in the hole such as burrs, so that the fluid moving in the duct is not disturbed. A pressure-measuring device such as a piezometer tube (a) is connected to the hole at location (2) to measure the static pressure as shown in Figure 2.33. Using the same approach as for a U-tube manometer to track pressures at various locations, as discussed in the previous section, the pressure at location (1) is

$$P_1 = P_3 + \rho g z_2 + \rho g z_1 \tag{2.124}$$

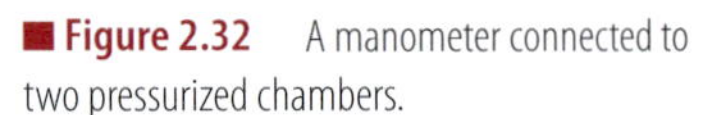

Figure 2.32 A manometer connected to two pressurized chambers.

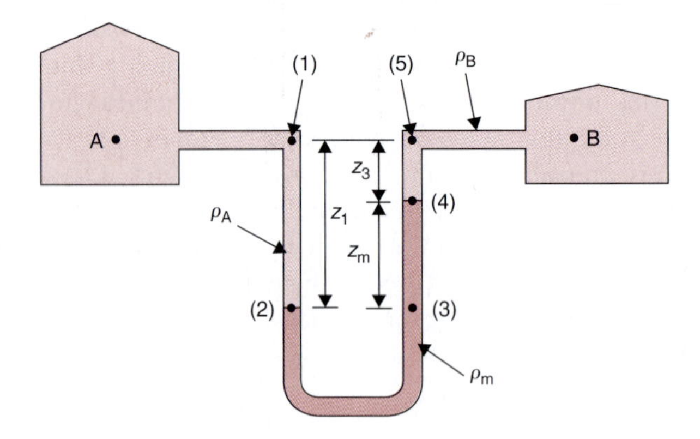

Since $P_3 = 0$, because it is the gauge atmospheric pressure, then

$$P_1 = \rho g(z_1 + z_2) = \rho g z_3 \tag{2.125}$$

If we insert a thin tube (b) into the duct, as shown in Figure 2.33, some of the fluid will be pushed through the tube to height z_4 in the tube. After a short transient period, the fluid inside the tube will come to rest and its velocity will be zero. This would imply that at the entrance of the tube, at location (4), the fluid velocity is zero and it is stagnant. Therefore, the pressure of the fluid at (4) will be the **stagnation pressure**. Applying the Bernoulli equation to locations (1) and (4), assuming they are at the same elevation, we have

$$\frac{P_1}{\rho} + \frac{u_1^2}{2g} = \frac{P_4}{\rho} + \frac{u_4^2}{2g}^{\,0} \tag{2.126}$$

Therefore, the stagnation pressure, P_4, is

$$P_4 = P_1 + \frac{\rho u_1^2}{2g} \tag{2.127}$$

In Equation (2.127), the term $\rho u_1^2/2g$ is called the **dynamic pressure** because it represents the pressure due to the fluid's kinetic energy. The stagnation pressure, P_4, is the sum of static and dynamic pressures, and it is the highest pressure obtainable along a given stream line assuming that elevation effects are negligible. Note the levels of fluid shown in the two tubes in Figure 2.33; the difference in the levels between tubes (a) and (b) is the kinetic energy term. We will

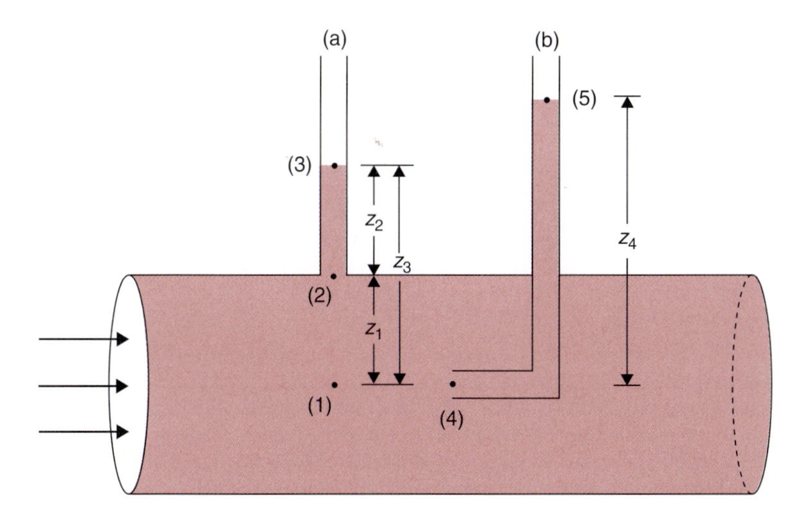

Figure 2.33 Measuring static and velocity head in fluid flow.

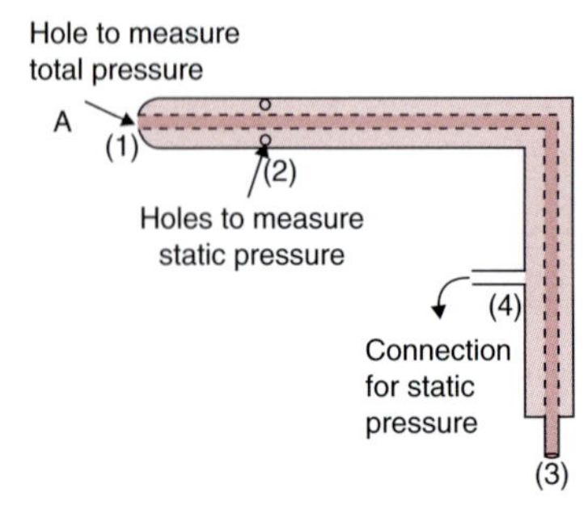

■ Figure 2.34 A pitot tube.

use these definitions of pressure in developing a design equation for a Pitot tube sensor that is commonly used to measure fluid velocity.

2.7.1 The Pitot Tube

A Pitot tube is a widely used sensor to measure velocity of a fluid. The design principle is based on the existence of stagnation and static pressures when an object is placed in the flowing fluid. A schematic of a Pitot[4] tube is presented in Figure 2.34. As indicated, the system is designed with two small concentric tubes, each leading to a separate outlet. The inlet hole of the inner tube is oriented directly into the fluid flow, whereas the inlet to the outer tube is through one or more holes located on the circumference of the outer tube. The outlets of the Pitot tube are connected to a U-tube manometer to measure the differential pressure. The inlet hole at location (1) measures the stagnation pressure. If the pressure and velocity in the fluid, at location A, upstream of (1), are P_A and u_A, and elevation difference between (1) and (3) is negligible, then

$$P_3 = P_A + \frac{\rho_f u_A^2}{2} \tag{2.128}$$

At location (2), the static pressure is measured. If the elevation difference between locations (2) and (4) is negligible then

$$P_4 = P_2 = P_A \tag{2.129}$$

Then, from Equations (2.128) and (2.129),

$$P_3 - P_4 = \frac{\rho_f u_A^2}{2} \tag{2.130}$$

or, rearranging,

$$u_A = \sqrt{\frac{2(P_3 - P_4)}{\rho_f}} \tag{2.131}$$

We obtained Equation (2.131) using the Bernoulli equation, which requires the fluid to be inviscid (viscosity = 0). This equation may be modified for real fluids by introducing a tube coefficient, C,

$$u_A = C\sqrt{\frac{2(P_3 - P_4)}{\rho_f}} \tag{2.132}$$

[4]Henri Pitot (1695–1771) was a French hydraulic engineer who began his career as a mathematician. In 1724, he was elected to the French Academy of Sciences. In Montpellier, he was in charge of constructing an aqueduct that included a 1 km long Romanesque stone arch section. His studies included flow of water in rivers and canals, and he invented a device to measure fluid velocities.

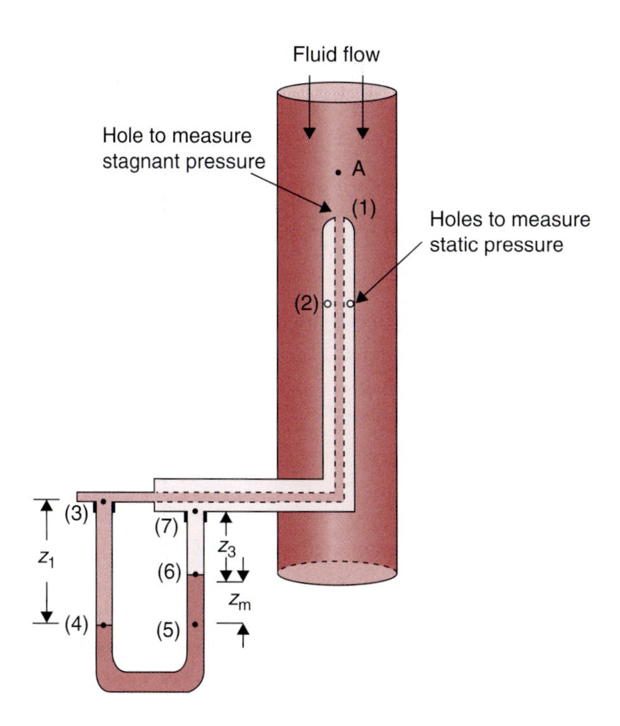

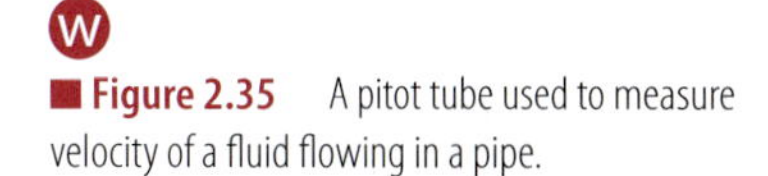

Figure 2.35 A pitot tube used to measure velocity of a fluid flowing in a pipe.

Equation (2.132) indicates that the fluid velocity in a stream at any desired location can be determined using a Pitot tube by measurement of pressure difference $P_3 - P_4$. The density ρ_f of the fluid and a tube coefficient C must be known. In most cases, $C \leq 1.0$. The velocity measured by the Pitot tube is the fluid velocity at the location A, upstream of the tip of the Pitot tube. To obtain an average velocity in a duct, several measurements are necessary.

If a U-tube manometer is used with the Pitot tube as shown in Figure 2.35, then we can follow the same approach as given earlier in this section to account for the pressures at various locations. Therefore, in Figure 2.35, from location (3) to (4), there will be an increase in pressure equal to $\rho_f g z_1$. Pressure at locations (4) and (5) will be the same as they are at the same elevation. There will be a decrease in pressure from (5) to (6) equal to $\rho_m g z_m$. From locations (6) to (7) there is additional decrease in pressure equal to $\rho_f g z_3$. Thus we may write the following:

$$P_3 + \rho_f g z_1 - \rho_m g z_m - \rho_f g z_3 = P_7 \tag{2.133}$$

or

$$P_3 + \rho_f g (z_1 - z_3) - \rho_m g z_m = P_7 \tag{2.134}$$

or rearranging terms,

$$P_3 - P_7 = gz_m(\rho_m - \rho_f) \tag{2.135}$$

By introducing Equation (2.135) into Equation (2.132), and noting that $P_3 - P_4$ in Figure 2.34 is analogous to $P_3 - P_7$ in Figure 2.35, we obtain

$$u_A = C\sqrt{\frac{2g(\rho_m - \rho_f)z_m}{\rho_f}} \tag{2.136}$$

We can measure the velocity directly from the change in height of a manometer fluid (z_m) when the two sides of the U-tube manometer are connected to the two outlets from the Pitot tube. The only other requirements for Equation (2.136) are knowledge of the fluid densities ρ_m and ρ_f, acceleration due to gravity (g), and the tube coefficient C.

Example 2.17

A Pitot tube is being used to measure maximum velocity for water flow in a pipe. The tube is positioned with the inlet to the inner tube along the center axis of the pipe. A U-tube manometer gives a reading of 20 mm Hg. Calculate the velocity of water, assuming a tube coefficient of 1.0. The density of mercury is 13,600 kg/m^3.

Given

Manometer reading = 20 mm Hg = 0.02 m Hg
Density (ρ_m) of mercury = 13,600 kg/m^3
Density (ρ) of water = 998 kg/m^3
Tube coefficient (C) = 1.0

Approach

By using Equation (2.136), the velocity of water can be computed.

Solution

1. *Using Equation (2.136) with C = 1,*

$$\bar{u}_2 = 1.0\left[\frac{2(9.81\ \text{m/s}^2)}{998\ \text{kg/m}^3}(13{,}600\ \text{kg/m}^3 - 998\ \text{kg/m}^3)(0.02\ \text{m})\right]^{1/2}$$

$$\bar{u}_2 = 2.226\ \text{m/s}$$

2.7.2 The Orifice Meter

By introducing a restriction of known dimensions into flow within a pipe or tube, we can use the relationship between pressure across the restriction and velocity through the restriction to measure fluid

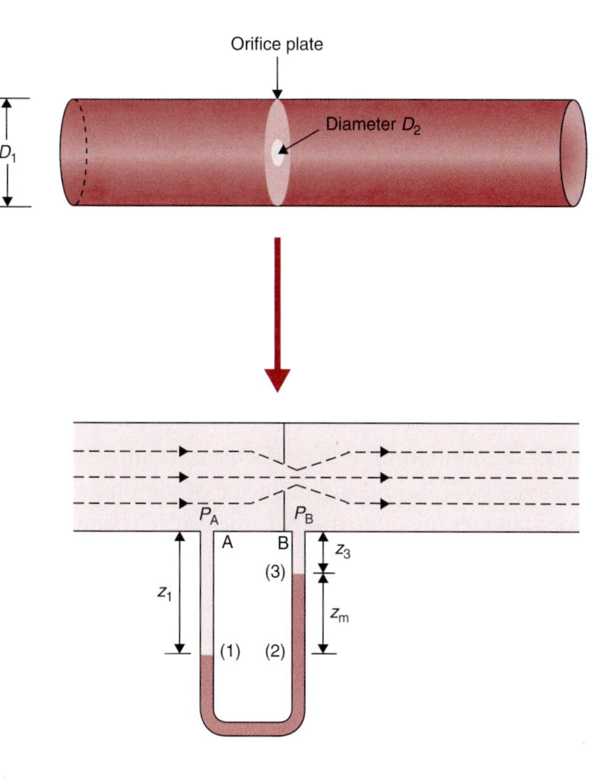

Figure 2.36 An orifice plate used to measure fluid flow.

flow rate. An orifice meter is a ring introduced into a pipe or tube that reduces the cross-sectional area of tube by a known amount. By attaching pressure taps or transducers at locations upstream and downstream from the orifice, the pressure changes can be measured.

We again use Equation (2.66) to analyze flow characteristics in the region near the orifice. Reference location A should be at sufficient distance upstream that the orifice does not influence flow characteristics. Reference location B is just slightly downstream from the orifice, where velocity is the same as within the orifice. Figure 2.36 illustrates the flow stream profile around the orifice meter and the reference locations. The pipe diameter is D_1 and the orifice diameter is D_2. Using Equation (2.66)

$$\frac{\bar{u}_A^2}{2} + \frac{P_A}{\rho_f} = \frac{\bar{u}_B^2}{2} + \frac{P_B}{\rho_f} \tag{2.137}$$

and

$$\bar{u}_A = \frac{A_2}{A_1}\bar{u}_B = \frac{D_2^2}{D_1^2}\bar{u}_B \tag{2.138}$$

By combining Equations (2.137) and (2.138),

$$\frac{\bar{u}_B^2}{2} + \frac{P_B}{\rho_f} = \left(\frac{D_2}{D_1}\right)^4 \frac{\bar{u}_B^2}{2} + \frac{P_A}{\rho_f} \tag{2.139}$$

or

$$\bar{u}_B = C\left\{\frac{2(P_A - P_B)}{\rho_f\left[1 - \left(\frac{D_2}{D_1}\right)^4\right]}\right\}^{1/2} \tag{2.140}$$

which allows computation of the velocity at location A from the pressure difference $P_A - P_B$ and the diameter of the pipe or tube D_1 and the orifice diameter D_2. Note that C is the orifice coefficient.

If we use a U-tube manometer to determine pressure drop, then we can use the same approach as shown earlier in Section 2.7.1 as follows. From Figure 2.36 we account for pressures. Moving from location A to (1), there is a pressure increase of $\rho_f g z_1$. Pressures at locations (1) and (2) are the same. From location (2) to (3), there is a pressure decrease equal to $\rho_m g z_m$. From location (3) to B there is a pressure decrease of $\rho_f g z_3$. Thus, we may write

$$P_A + \rho_f g z_1 - \rho_m g z_m - \rho_f g z_3 = P_B \tag{2.141}$$

Rearranging terms,

$$P_A - P_B = \rho_f g(z_3 - z_1) + \rho_m g z_m \tag{2.142}$$

or

$$P_A - P_B = z_m g(\rho_m - \rho_f) \tag{2.143}$$

By introducing Equation (2.140), the following relationship is obtained:

$$\bar{u}_B = C\left\{\frac{2g\left(\frac{\rho_m}{\rho_f} - 1\right)z_m}{\left[1 - \left(\frac{D_2}{D_1}\right)^4\right]}\right\}^{1/2} \tag{2.144}$$

which allows computation of average velocity in the fluid stream from the change in manometer fluid height and the density of the manometer fluid.

The magnitude of the orifice coefficient C is a function of exact location of the pressure taps, the Reynolds number, and the ratio of pipe diameter to orifice diameter. At $N_{Re} = 30{,}000$, the coefficient C will have a value of 0.61, and the magnitude will vary with N_{Re} at lower values. It is recommended that orifice meters be calibrated in known flow conditions to establish the exact values of the orifice coefficient.

Example 2.18

An orifice meter is being designed to measure steam flow in a food processing plant. The steam has a mass flow rate of approximately 0.1 kg/s in a 7.5 cm diameter (ID) pipe with a pressure of 198.53 kPa. Determine the density of the manometer fluid to be used so that pressure differences can be detected accurately and reasonably. A manometer of less than 1 m in height can be considered reasonable.

Given

Mass flow rate ($\dot{m}$) of steam = 0.1 kg/s
Pipe diameter (D_1) = 7.5 cm = 0.075 m
Steam density (ρ) = 1.12 kg/m^3 from Table A.4.2 at pressure of 198.53 kPa
Orifice coefficient (C) = 0.61 at N_{Re} = 30,000

Approach

To use Equation (2.144) to compute density of manometer fluid (ρ_m), the orifice diameter D_2 and manometer fluid height z_m must be assumed.

Solution

1. *By assuming an orifice diameter D_2 of 6 cm or 0.06 m,*

$$\bar{u} = \frac{\dot{m}}{\rho A} = \frac{(0.1\,\text{kg/s})}{(1.12\,\text{kg/m}^3)[\pi(0.06\,\text{m})^2/4]} = 31.578\,\text{m/s}$$

2. *Since the manometer fluid height (z_m) must be less than 1 m, a value of 0.1 m will be assumed. Using Equation (2.144),*

$$31.578\,\text{m/s} = 0.61\left[\frac{2(9.81\,\text{m/s}^2)\left(\dfrac{\rho_m}{1.12\,\text{kg/m}^3}-1\right)(0.1\,\text{m})}{1-(0.06/0.075)^4}\right]^{1/2}$$

$$\rho_m = 904.3\,\text{kg/m}^3$$

3. *This density could be obtained by using light oil with a density of 850 kg/m^3.*

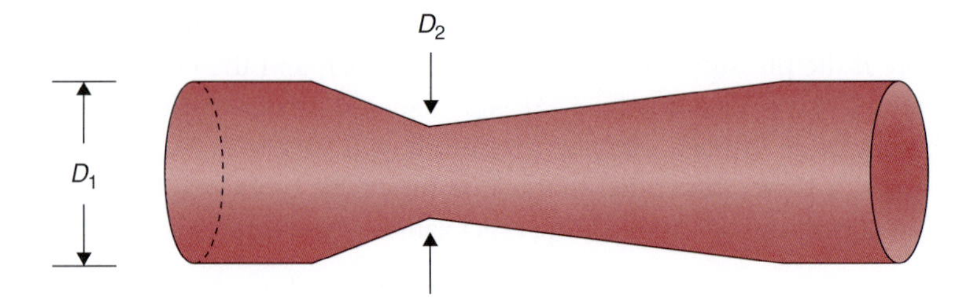

■ Figure 2.37 A venturi tube flow meter.

2.7.3 The Venturi Meter

To reduce energy loss due to friction created by the sudden contraction in flow in an orifice meter, a venturi tube of the type illustrated in Figure 2.37 can be used. An analysis similar to that presented for the orifice meter leads to the following equation:

$$\bar{u}_B = C\left\{\frac{2g\left(\dfrac{\rho_m}{\rho_f}-1\right)z_m}{\left[1-\left(\dfrac{D_2}{D_1}\right)^4\right]}\right\}^{1/2} \tag{2.145}$$

where the average velocity $\bar{u}_2$ is at reference location 2, where diameter D_2 is the smallest value for the venturi. The venturi meter requires careful construction to ensure proper angles of entrance to and exit from the venturi. The meter requires a significant length of pipe for installation compared with the orifice meter. In general, the orifice meter is less costly and simpler to design than the venturi meter.

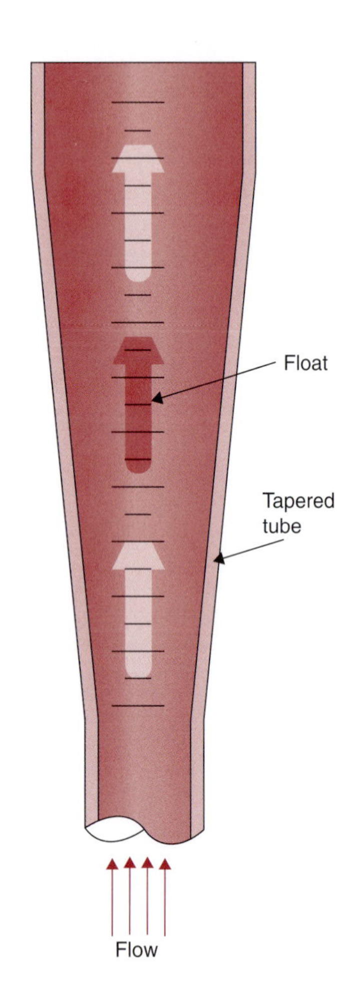

■ Figure 2.38 A variable flow meter.

2.7.4 Variable-Area Meters

The flow meters considered in the preceding sections, namely the orifice meter and venturi tube, involve a change in flow rate through a constant cross-sectional area that generates a variable pressure drop as a function of flow rate. In a variable-area meter, the fluid stream is throttled by a constriction arranged in a manner such that the cross-sectional area is varied. This allows a variation of flow while maintaining a nearly constant pressure drop. The cross-sectional area in these devices is related to the flow rate by proper calibration.

A popular type of variable flow meter is a **rotameter**, shown in Figure 2.38. In this device, the height of a plummet, also called bob or float, in a tapered tube indicates the flow. The float moves up or down in a vertically mounted tapered tube, with the largest diameter of the tube at the top. The fluid moves up from the bottom and lifts the float. Because of the higher density of the float, the passage remains blocked until the pressure builds up and the buoyant effect of the fluid lifts the float; the fluid then flows between the float and the tube wall. As the passage for flow increases, a dynamic equilibrium is established between the position of the float and the pressure difference across the float and the buoyant forces. A scale mounted on the outside of the tube provides a measurement of the vertical displacement of the float from a reference point. The fluid flow in the tube can thus be measured. As the float is raised higher in the tapered tube, a greater area becomes available for the fluid to pass through; this is why this meter is called a *variable-area flow meter*.

The tube material is typically glass, acrylic, or metal. For measuring low flow rates, a ball or plumb-bob float shape is used, whereas for high flow capacities and applications that require high accuracy with constant viscosity, a streamlined float shape is used. The float materials commonly used include black glass, red sapphire, stainless steel, and tungsten. The capacities of rotameters are usually given in terms of two standard fluids: water at 20°C and air at 20°C at 101.3 kPa. A proper flow meter can be selected based on the float selection curves and capacity tables supplied by the manufacturers. A single instrument can cover a wide range of flow, up to tenfold; using floats of different densities, ranges up to 200-fold are possible. Unlike orifice meters, the rotameter is not sensitive to velocity distribution in the approaching stream. The installation of rotameters requires no straight section of pipe either upstream or downstream.

Industrial rotameters offer excellent repeatability over a wide range of flows. Their standard accuracy is $\pm 2\%$ of full scale, with capacities of 6×10^{-8} to $1 \times 10^{-2}\,m^3/s$ of water and 5×10^{-7} to $0.3\,m^3/s$ of air at standard temperature and pressure. Rotameters are also available for special requirements such as low volume, and high pressure. These instruments are calibrated when obtained from the manufacturer with a given size and shape bob and for a given bob density for a fluid of specified specific gravity.

2.7.5 Other Measurement Methods

In addition to methods in which pressure drop caused by a restriction in flow is measured, several methods have been developed for unique applications in the food industry. These methods vary considerably in operation principles but meet the needs for sanitary design.

Volumetric displacement as a flow measurement principle involves the use of a measuring chamber of known volume and containing a rotating motor. As flow is directed through the chamber, the rotor turns and displaces known volume magnitudes. The flow rate is detected by monitoring the number of revolutions of the rotor and accounting for the volume in each revolution.

Several flow measurement methods use ultrasound as a flow-sensing mechanism. Generally, these methods use the response from a high-frequency wave directed at the flow as an indication of flow rate. As the flow changes, the frequency changes. One method of flow detection uses the Doppler shift; changes in flow rate cause shifts in the wave frequency as the wave passes through the flow.

Another method of flow measurement is the use of a vortex created by inserting an object of irregular shape into the flow stream. Since the vortices move downstream with a frequency that is a function of flow rate, this frequency can be used as an indicator of flow rate. Typically, the frequencies are measured by placing heated thermistors in the vortex stream, followed by detection of cooling rates.

The flow rate of a fluid in a tube can be measured by placing a turbine wheel into the flow stream. As flow rate changes, the rotation speed changes in some proportional manner. Measurement of rotation is accomplished by using small magnets attached to the rotating part of the turbine. The magnets generate a pulse to be detected by a coil circuit located on the outside tube wall.

Each of the flow measurement methods described have unique features, and their use should be determined by the circumstances of the application. All have been used in various situations in the food industry.

2.8 MEASUREMENT OF VISCOSITY

Viscosity of a liquid can be measured using a variety of approaches and methods. The capillary tube and the rotational viscometer are the more common types of instruments used.

2.8.1 Capillary Tube Viscometer

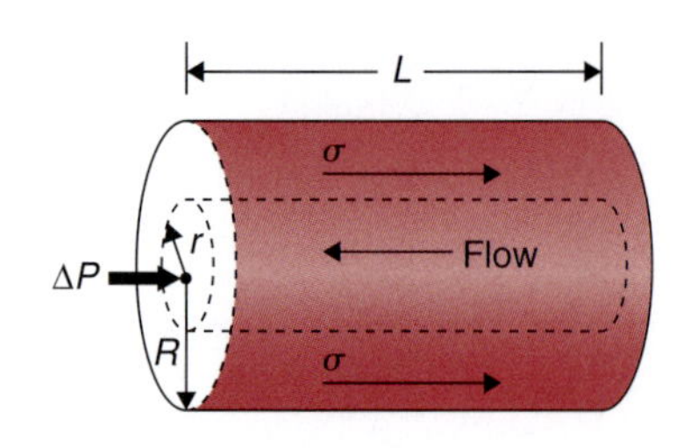

Figure 2.39 Force balance for a section of capillary tube.

Capillary tube measurement is based on the scheme shown in Figure 2.39. As shown, pressure (ΔP) is sufficient to overcome the shear forces within the liquid and produce flow of a given rate. The shear forces are operating on all internal liquid surfaces for the entire length L of tube and distance r from the tube center.

Equation (2.39) provides the basis for design and operation of any capillary tube viscometer. For a tube with length L and radius R, measurement of a volumetric flow rate $\dot{V}$ at a pressure ΔP will allow determination of viscosity μ:

$$\mu = \frac{\pi \Delta P R^4}{8L\dot{V}} \tag{2.146}$$

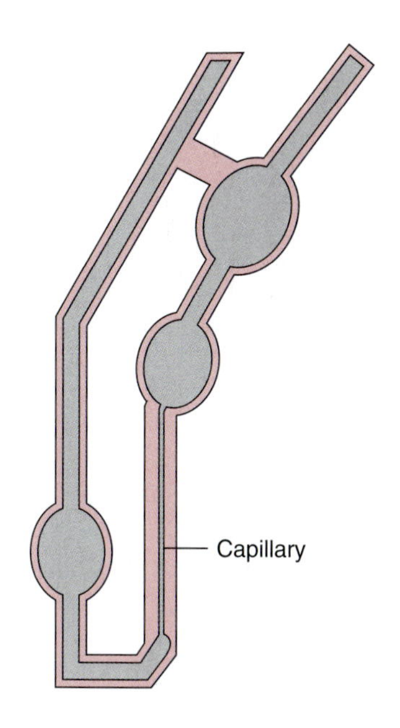

■ **Figure 2.40** A Cannon-Fenske viscometer.

Since Equation (2.146) is derived for a Newtonian fluid, any combination of flow rate and pressure drop will give the same viscosity value.

In a Cannon–Fenske type capillary viscometer, shown in Figure 2.40, we allow gravitational force to provide the pressure for liquid flow through the glass capillary tube. We can use a simple variation of the mathematical formulation developed for the capillary tube viscometer. By recognizing that

$$\Delta P = \frac{\rho V g}{A} \tag{2.147}$$

and the volumetric flow rate through the capillary tube is

$$\dot{V} = \frac{\text{volume of bulb}}{\text{discharge time}} = \frac{V}{t} \tag{2.148}$$

then, Equation (2.146) becomes

$$\mu = \frac{\pi \rho g R^4 t}{8V} \tag{2.149}$$

Equation (2.149) illustrates that viscosity of a liquid measured by a glass capillary tube will be a function of the liquid volume in the bulb, fluid density, the acceleration due to gravity ($g = 9.8\,\text{m/s}^2$), and tube length L. We can determine viscosity by measuring the length of time for the liquid to drain from the bulb.

Example 2.19

A capillary tube viscometer is being used to measure the viscosity of honey at 30°C. The tube radius is 2.5 cm and the length is 25 cm.The following data have been collected:

ΔP (Pa)	$\dot{V}$ (cm^3/s)
10.0	1.25
12.5	1.55
15.0	1.80
17.5	2.05
20.0	2.55

Determine the viscosity of honey from the data collected.

Given

Data required to compute viscosity values from Equation (2.146), for example,

$$\Delta P = 12.5\,Pa$$
$$R = 2.5\,cm = 0.025\,m$$
$$L = 25\,cm = 0.25\,m$$
$$\dot{V} = 1.55\,cm^3/s = 1.55 \times 10^{-6}\,m^3/s$$

Approach

The viscosity for each pressure difference (ΔP) and flow rate ($\dot{V}$) combination can be computed from Equation (2.146).

Solution

1. *Using Equation (2.146), a viscosity value can be computed for each $\Delta P - \dot{V}$ combination; for example,*

$$\mu = \frac{\pi(12.5\,Pa)(0.025\,m)^4}{8(0.25\,m)(1.55 \times 10^{-6}\,m^3/s)} = 4.948\,Pa\,s$$

2. *By repeating the same calculation at each $\Delta P - \dot{V}$ combination, the following information is obtained:*

ΔP (Pa)	$\dot{V}$ ($\times 10^{-6}\,m^3/s$)	μ (Pa s)
10	1.25	4.909
12.5	1.55	4.948
15	1.8	5.113
17.5	2.05	5.238
20	2.55	4.812

3. *Although there is some variability with pressure (ΔP), there is no indication of a consistent trend, and the best estimate of the viscosity would be the arithmetic mean*

$$\mu = 5.004\,Pa\,s$$

2.8.2 Rotational Viscometer

The second type of viscometer is the rotational viscometer illustrated in Figure 2.41. This illustration is more specific for a coaxial-cylinder viscometer with the liquid placed in the space between the inner and outer cylinders. The measurement involves recording of torque Ω required to turn the inner cylinder at a given number of revolutions per unit time. To calculate viscosity from the measurements, the relationships between torque Ω and shear stress σ, as well as revolutions per unit second N_r and shear rate zm $\dot{\gamma}$ must be established.

The relationship between torque Ω and shear stress σ can be shown as

$$\Omega = 2\pi r^2 L\sigma \tag{2.150}$$

where the length L of the cylinder and the radial location r between the inner and outer cylinder are accounted for.

The angular velocity at r is

$$u = r\omega \tag{2.151}$$

Using differential calculus,

$$\frac{du}{dr} = \omega + \frac{r d\omega}{dr} \tag{2.152}$$

We note that ω does not contribute to shear. And the shear rate $\dot{\gamma}$ for a rotational system becomes a function of angular velocity ω, as follows:

$$\dot{\gamma} = -\frac{du}{dr} = r\left(-\frac{d\omega}{dr}\right) \tag{2.153}$$

By substitution of these relationships into Equation (2.150),

$$\frac{\Omega}{2\pi L r^2} = -\mu\left(r\frac{d\omega}{dr}\right) \tag{2.154}$$

To obtain the desired relationship for viscosity, an integration between the outer and inner cylinders must be performed:

$$\int_{\mathrm{o}}^{\omega_i} d\omega = -\frac{\Omega}{2\pi\mu L}\int_{R_o}^{R_i} r^{-3} dr \tag{2.155}$$

where the outer cylinder (R_o) is stationary ($\omega = 0$) and the inner cylinder (R_i) has an angular velocity $\omega = \omega_i$. The integration leads to

$$\omega_i = \frac{\Omega}{4\pi\mu L}\left(\frac{1}{R_i^2} - \frac{1}{R_o^2}\right) \tag{2.156}$$

and since

$$\omega_i = 2\pi N_r \tag{2.157}$$

Note that ω is in units of radian/s and N_r is revolution/s. Then

$$\mu = \frac{\Omega}{8\pi^2 N_r L}\left(\frac{1}{R_i^2} - \frac{1}{R_o^2}\right) \tag{2.158}$$

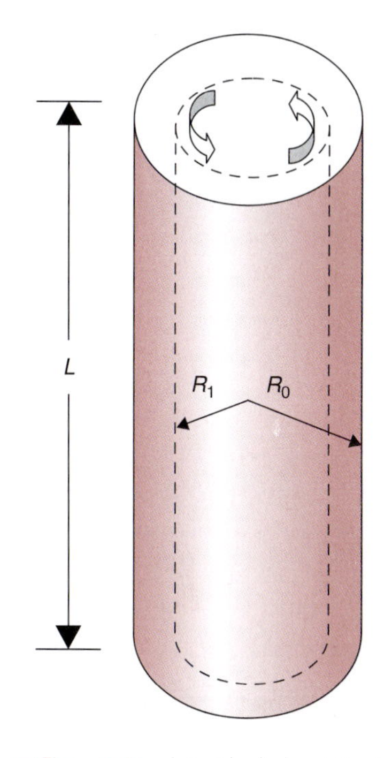

Figure 2.41 A coaxial-cylinder rotational viscometer.

Equation (2.158) illustrates that liquid viscosity can be determined using a coaxial-cylinder viscometer with an inner cylinder radius R_i, length L, and outer cylinder radius R_o by measurement of torque Ω at a given N_r (revolutions per second).

A variation of the coaxial-cylinder viscometer is the single-cylinder viscometer. In this device, a single cylinder of radius R_i is immersed in a container with the test sample. Then the outer cylinder radius R_o approaches infinity, and Equation (2.158) becomes

$$\mu = \frac{\Omega}{8\pi^2 N_r L R_i^2} \tag{2.159}$$

Several rotational viscometers operate using the single-cylinder principle, which assumes that the wall of the vessel containing the liquid during measurement has no influence on the shear stresses within the liquid. This may be a relatively good assumption for Newtonian liquids but should be evaluated carefully for each liquid to be measured.

Example 2.20

A single-cylinder rotational viscometer with a 1 cm radius and 6 cm length is being used to measure liquid viscosity. The following torque readings were obtained at several values of revolutions per minute (rpm):

N_r (rpm)	Ω ($\times 10^{-3}$ N cm)
3	1.2
6	2.3
9	3.7
12	5.0

Compute the viscosity of the liquid based on the information provided.

Given

Equation (2.159) requires the following input data (for example):

$$\Omega = 2.3 \times 10^{-3}\,\text{N cm} = 2.3 \times 10^{-5}\,\text{N m}$$
$$N_r = 6\,\text{rpm} = 0.1\,\text{rev/s}$$
$$L = 6\,\text{cm} = 0.06\,\text{m}$$
$$R_i = 1\,\text{cm} = 0.01\,\text{m}$$

Approach

Use Equation (2.159) to calculate viscosity from each rpm–torque reading combination.

Solution

1. *Using Equation (2.159) and the given data,*

$$\mu = \frac{(2.3 \times 10^{-5}\ \mathrm{N\,m})}{8\pi^2 (0.1\ \mathrm{rev/s})(0.06\ \mathrm{m})(0.01\ \mathrm{m})^2} = 0.485\ \mathrm{Pa\,s}$$

2. *Using the same approach, values of viscosity are obtained for each N_r–Ω combination.*

N_r (rev/s)	Ω ($\times 10^{-5}$ N m)	μ (Pa s)
0.05	1.2	0.507
0.1	2.3	0.485
0.15	3.7	0.521
0.2	5.0	0.528

3. *Since it is assumed that the liquid is Newtonian, the four values can be used to compute an arithmetic mean of*

$$\mu = 0.510\ \mathrm{Pa\,s}$$

2.8.3 Influence of Temperature on Viscosity

The magnitude of the viscosity coefficient for a liquid is influenced significantly by temperature. Since temperature is changed dramatically during many processing operations, it is important to obtain appropriate viscosity values for liquids over the range of temperatures encountered during processing. This temperature dependence of viscosity also requires that during measurements of viscosity, extra care must be taken to avoid temperature fluctuations. In the case of water, the temperature sensitivity of viscosity from Table A.4.1 is estimated to be 3%/°C at room temperature. This means that ±1% accuracy in its measurement requires the sample temperature to be maintained within ±0.3°C.

There is considerable evidence that the influence of temperature on viscosity for a liquid food may be described by an Arrhenius-type relationship:

$$\ln \mu = \ln B_A + \frac{E_a}{R_g T_A} \tag{2.160}$$

where B_A is the Arrhenius constant, E_a is an activation energy constant, and R_g is the gas constant. Equation (2.160) can be used to reduce the number of measurements required to describe the influence of temperature on viscosity of a liquid food. If we obtain values at three or more temperatures within the desired range and establish the magnitudes of the constants (B_A and E_a/R_g), we can predict with reasonable accuracy the viscosity coefficient at other temperatures within the range.

Example 2.21

For concentrated orange juice, Vitali and Rao (1984) obtained viscosity at a shear rate of $100\,s^{-1}$ at eight different temperatures as shown in the following table. Determine the activation energy and the preexponential factor. Calculate the viscosity at 5°C.

Given

The values of viscosity at a shear rate of $100\,s^{-1}$ at eight different temperatures are given in the table.

Temperature	Viscosity
−18.8	8.37
−14.5	5.32
−9.9	3.38
−5.4	2.22
0.8	1.56
9.5	0.77
19.4	0.46
29.2	0.28

Approach

We will use a spreadsheet first to convert the temperature to 1/T, where T is in absolute, K. Then we will plot ln (viscosity) vs 1/T. Using the trend line we will calculate the slope of the line, intercept, and regression coefficient. We will use the coefficients in the Arrhenius equation to obtain the viscosity at 5°C.

Solution

1. *Prepare a spreadsheet as shown in Figure E2.7.*
2. *Using the trendline feature of the Excel software, obtain the slope and intercept.*
3. *From the slope, $5401.5 \times R_g = 10{,}733.7$ cal/mol. Note that Gas Constant, $R_g = 1.98717$ cal/(mol K),*

$$\text{intercept},\ B_A = 4.3 \times 10^{-9},$$
$$\text{and the regression coefficient} = 0.99.$$

The high regression coefficient indicates a high degree of fit.

	A	B	C	D	E	F	G
1	Temperature	1/Tabs	Viscosity	ln (viscosity)			
2	−18.8	0.00393391	8.37	2.12465388			
3	−14.5	0.003868472	5.32	1.6714733			
4	−9.9	0.003800836	3.38	1.21787571			
5	−5.4	0.003736921	2.22	0.7975072			
6	0.8	0.003652301	1.56	0.44468582			
7	9.5	0.003539823	0.77	−0.26136476			
8	19.4	0.003419973	0.46	−0.77652879			
9	29.2	0.003309067	0.28	−1.27296568			
10							
11							
12							
13							
14							
15							
16							
17							
18							
19							
20							
21							
22							
23							
24							
25							
26							
27							
28							
29							
30	R	1.98717	=5401.5*B30				
31	Slope	10733.70					
32	Intercept	4.3028E-09	=EXP(−19.264)				
33							

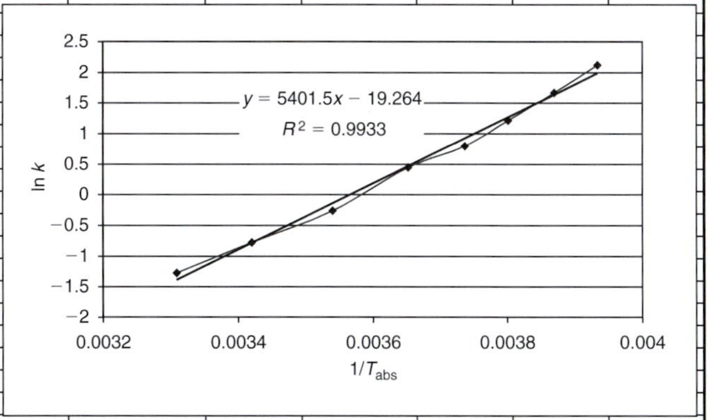

$y = 5401.5x - 19.264$

$R^2 = 0.9933$

ln k

$1/T_{abs}$

2.5, 2, 1.5, 1, 0.5, 0, −0.5, −1, −1.5, −2

0.0032, 0.0034, 0.0036, 0.0038, 0.004

■ **Figure E2.7** Spreadsheet solution for Example 2.21.

4. *Using the calculated coefficient of Arrhenius equation, we can write the equation for the viscosity value at 5°C as*

$$\mu = 4.3 \times 10^{-9}\, e^{\left(\frac{5401}{5+273}\right)}$$
$$= 1.178 \text{ Pa s}$$

5. *The viscosity at 5°C is 1.178 Pa s. This value is between the given values at temperatures 0.8° and 9.5°C.*

2.9 FLOW CHARACTERISTICS OF NON-NEWTONIAN FLUIDS

2.9.1 Properties of Non-Newtonian Fluids

It should be evident from the preceding discussion in this chapter that liquids offer interesting properties. They flow under gravity and do not retain their shape. They may exist as solids at one temperature, and liquid at another (e.g., ice cream and shortenings). Products such as applesauce, tomato purée, baby foods, soups, and salad dressings

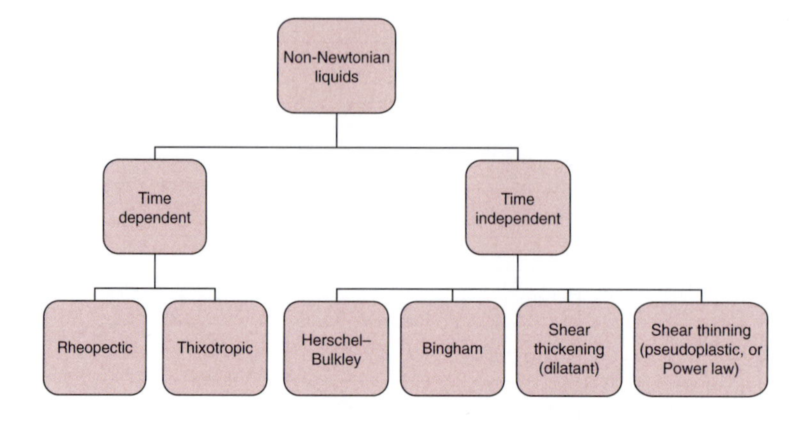

Figure 2.42 Classification of non-Newtonian liquids.

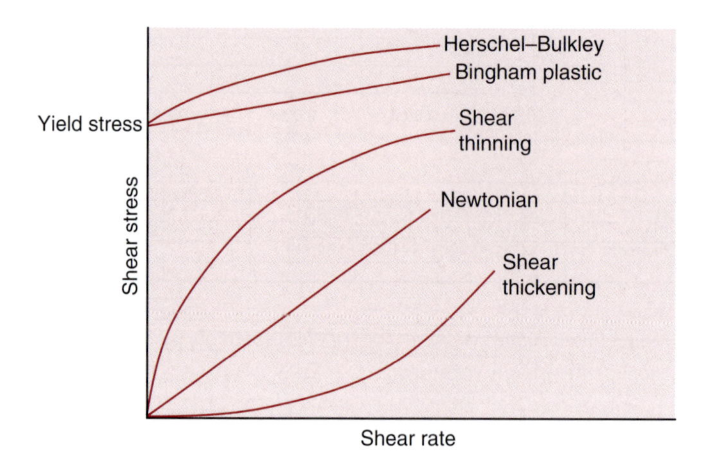

Figure 2.43 Relationship between shear stress and shear rate for Newtonian and non-Newtonian liquids.

are suspensions of solid matter in liquid. When droplets of one liquid are submerged in another, we obtain emulsions—milk, for example.

The properties of non-Newtonian liquids can be classified as time-independent and time-dependent (Fig. 2.42). The **time-independent non-Newtonian liquids** respond immediately with a flow as soon as a small amount of shear stress is applied. Unlike Newtonian liquids, the relationship between shear stress and shear rate is nonlinear, as shown in Figure 2.43. There are two important types of time-independent non-Newtonian liquids, namely, shear-thinning liquids and shear-thickening liquids. The differences between these two types of liquids can be understood easily by considering another commonly used term, **apparent viscosity**.

An apparent viscosity is calculated by using a gross assumption that the non-Newtonian liquid is obeying Newton's law of viscosity

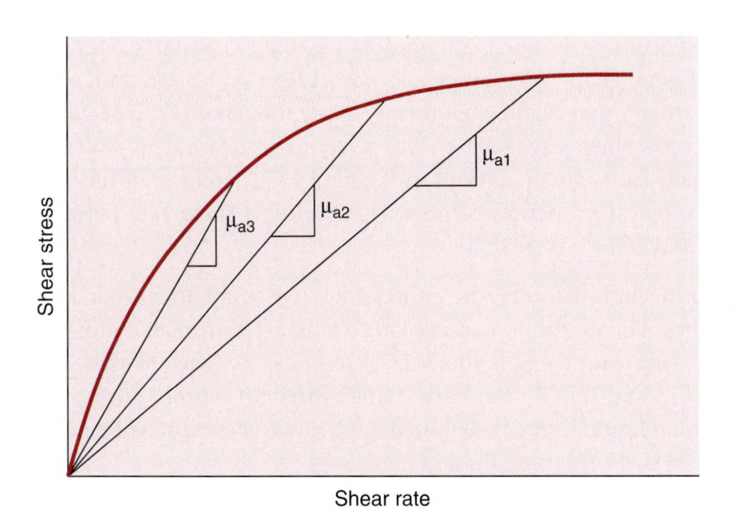

Figure 2.44 Determination of apparent viscosity from plot of shear stress vs shear rate.

(Equation (2.10)). Thus, at any selected shear rate, a straight line is drawn from the selected point on the curve to the origin (Fig. 2.44). The slope of this straight line gives a value for the apparent viscosity. Using this method, it should be evident that the value obtained for apparent viscosity is dependent on the selected shear rate. Therefore the apparent viscosity must always be expressed along with the value of shear rate used to calculate it; otherwise, it is meaningless. For a **shear-thinning liquid**, as the shear rate increases, the apparent viscosity decreases; thus, the name *shear thinning* is used to describe the behavior of these liquids.

Shear-thinning liquids also are called **pseudoplastic** or **power law** liquids. Some common examples of shear-thinning liquids are condensed milk, fruit purées, mayonnaise, mustard, and vegetable soups. When shear-thinning products are shaken in a jar, they become more "fluid." Similarly, if these products are mixed at high intensity in a mixer, their viscosity decreases, which may aid in their mixing. There are several reasons to explain shear-thinning behavior. A liquid that appears homogenous to the naked eye may actually contain microscopic particulates submerged in it. When these liquids are subjected to a shear, the randomly distributed particles may orient themselves in the direction of flow; similarly, coiled particulates may deform and elongate in the direction of flow. Any agglomerated particles may break up into smaller particles. These types of modifications due to shearing action improve the flow of such fluids, and an increased "fluidity" is observed. They are also usually reversible. Thus, when the shearing action is stopped, after a time lag, the particulates return to their original shape—the elongated particulates coil back, and separated particles may again agglomerate. Note that changes in viscosity at a very low shear rate ($<0.5\,s^{-1}$) or a very high shear rate ($>100\,s^{-1}$) are usually quite small, as seen in Figure 2.45. Therefore, in measuring rheological properties of power law fluids, a shear rate between $0.5\,s^{-1}$ and $100\,s^{-1}$ is used.

With some liquid foods, the processing steps may alter their flow properties. For example, raw egg at 21°C is a Newtonian fluid, but when frozen whole egg is thawed, its response changes to that of a shear-thinning liquid. Similarly, single-strength apple juice is a Newtonian liquid, but concentrated apple juice (depectinized and filtered) is a shear-thinning liquid.

If the increase in shear rate results in an increase in apparent viscosity, then the liquid is called a **shear-thickening liquid** (or sometimes referred to as **dilatant liquid**). Examples of shear-thickening liquids include 60% suspension of corn starch in water. With shear-thickening liquids, the apparent viscosity increases with increasing shear rate. These liquids become "stiffer" at higher shear rates. Mostly, these liquids are suspensions—solid particles in a liquid that acts as a plasticizer. At low shear rates, the liquid is sufficient to keep the solid particles well lubricated, and the suspension flows almost as a Newtonian liquid. But as shear rate increases, the solid particles begin to separate out, forming wedges while increasing the overall volume.

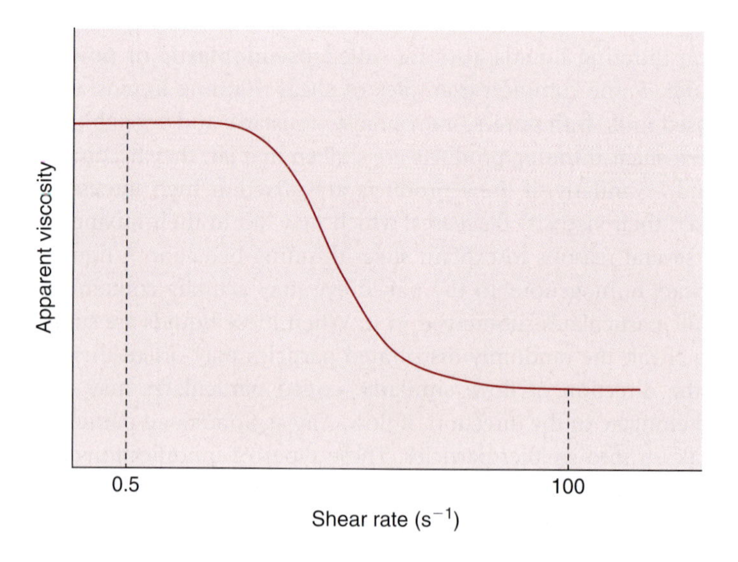

Figure 2.45 Apparent viscosity vs shear rate.

Hence, they are called dilatant liquids. The liquid is then unable to act as a plasticizer. As a result, the overall suspension becomes more resistant to flow.

Another important class of non-Newtonian liquids requires the application of **yield stress** prior to any response. For example, certain types of tomato catsup will not flow until a certain yield stress is applied. For these types of liquids, a plot of shear stress against shear rate does not pass through the origin, as shown in Figure 2.43. After the application of yield stress, the response of these liquids can be similar to a Newtonian liquid; in that case, they are called **Bingham plastic**. On the other hand, if the response of a liquid, after the yield stress is applied, is similar to a shear-thinning flow, then these liquids are called **Herschel–Bulkley** fluids. These liquids that require a yield stress to flow may be viewed as having an interparticle or intermolecular network that resists low-level shear force when at rest. Below the yield stress, the material acts like a solid and does not flatten out on a horizontal surface due to force of gravity. It is only when the applied stress exceeds the forces holding the network together that the material begins to flow.

Time-dependent non-Newtonian liquids obtain a constant value of apparent viscosity only after a certain finite time has elapsed after the application of shear stress. These types of liquids are also called **thixotropic** materials; examples include certain types of starch pastes. For more discussion of these types of liquids, refer to Doublier and Lefebvre (1989).

A common mathematical model may be used to express the non-Newtonian characteristics. This model is referred to as the **Herschel–Bulkley model** (Herschel and Bulkley, 1926):

$$\sigma = K\left(\frac{du}{dy}\right)^n + \sigma_0 \tag{2.161}$$

where the values for the different coefficients are given in Table 2.4.

Another model, used in interpreting the flow data of chocolate, is the **Casson model** (Casson, 1959), given as

$$\sigma^{0.5} = \sigma_0^{0.5} + K(\dot{\gamma})^{0.5} \tag{2.162}$$

The square root of shear stress is plotted against the square root of shear rate to obtain a straight line. The slope of the line gives the consistency coefficient, and the square of the intercept gives the yield stress.

Table 2.4 Values of Coefficients in Herschel-Bulkley Fluid Model

Fluid	K	n	σ_0	Typical examples
Herschel–Bulkley	>0	$0<n<\infty$	>0	minced fish paste, raisin paste
Newtonian	>0	1	0	water, fruit juice, honey, milk, vegetable oil
Shear-thinning (pseudoplastic)	>0	$0<n<1$	0	applesauce, banana purée, orange juice concentrate
Shear-thickening	>0	$1<n<\infty$	0	some types of honey, 40% raw corn starch solution
Bingham plastic	>0	1	>0	toothpaste, tomato paste

Source: Steffe (1996)

Example 2.22

Rheological data for Swedish commercial milk chocolate at 40°C is shown in the following. Using the Casson model, determine the consistency coefficient and the yield stress.

Shear rate	Shear stress
0.099	28.6
0.14	35.7
0.199	42.8
0.39	52.4
0.79	61.9
1.6	71.4
2.4	80.9
3.9	100
6.4	123.8
7.9	133.3
11.5	164.2
13.1	178.5
15.9	201.1
17.9	221.3
19.9	235.6

Given

Data on shear rate and shear stress given in the preceding table.

Approach

We will use a spreadsheet for this example.

Solution

1. *We will use Equation (2.162). Using a spreadsheet, first enter the shear stress and shear rate data into columns A and B. Then develop new columns, C and D, with square root of shear stress and shear rate. Then plot the data from*

	A	B	C	D	E
1	Shear rate	Shear stress	Shear rate^0.5	Shear stress^0.5	
2	0.099	28.6	0.315	5.348	
3	0.14	35.7	0.374	5.975	
4	0.199	42.8	0.446	6.542	
5	0.39	52.4	0.624	7.239	
6	0.79	61.9	0.889	7.868	
7	1.6	71.4	1.265	8.450	
8	2.4	80.9	1.549	8.994	
9	3.9	100	1.975	10.000	
10	6.4	123.8	2.530	11.127	
11	7.9	133.3	2.811	11.546	
12	11.5	164.2	3.391	12.814	
13	13.1	178.5	3.619	13.360	
14	15.9	201.1	3.987	14.181	
15	17.9	221.3	4.231	14.876	
16	19.9	235.6	4.461	15.349	
17					
18					
19					
20					
21					
22					
23					
24					
25					
26					
27					
28					
29					
30					
31					

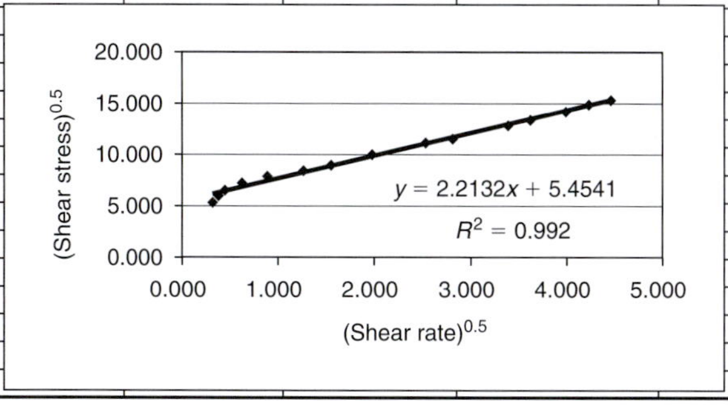

Figure E2.8 Spreadsheet solution for Example 2.22.

columns C and D as a scatter plot, as shown in Figure E2.8. If using Excel, then use "trend line" to determine the slope, intercept, and data fit.

2. *The slope is 2.213, and intercept is 5.4541.*
3. *The consistency coefficient is*

$$K = 2.213\ Pa^{0.5}\ s^{0.5}$$

and

$$\text{Yield stress } \sigma_0 = 5.4541^2 = 29.75\ Pa$$

4. *These coefficients were obtained when the entire range of shear rate 0–20 s^{-1} was used. For further analysis of these data see Steffe (1996).*

2.9.2 Velocity Profile of a Power Law Fluid

We recall from Equation (2.161) that the shear stress and shear rate are related by a power law equation, as follows:

$$\sigma = K\left(-\frac{du}{dr}\right)^n \qquad (2.163)$$

Note that du/dr is negative in a pipe flow where the velocity decreases from the center to the surface. Therefore, a negative sign is used to make the shear stress positive.

From Equations (2.27) and (2.163)

$$K\left(-\frac{du}{dr}\right)^n = \frac{\Delta Pr}{2L} \tag{2.164}$$

Rearranging and setting up integrals from the central axis to the wall of the pipe,

$$-\int_u^0 du = \left(\frac{\Delta P}{2LK}\right)^{1/n} \int_r^R r^{1/n}\, dr \tag{2.165}$$

Integrating between the limits,

$$u(r) = \left(\frac{\Delta P}{2LK}\right)^{1/n} \left(\frac{n}{n+1}\right)\left(R^{\frac{n+1}{n}} - r^{\frac{n+1}{n}}\right) \tag{2.166}$$

Equation (2.166) gives the velocity profile for a power law fluid.

2.9.3 Volumetric Flow Rate of a Power Law Fluid

The volumetric flow rate is obtained by integrating the velocity profile for a ring element as previously discussed for Newtonian fluid in Section 2.3.4:

$$\dot{V} = \int_{r=0}^{r=R} u(r) 2\pi r dr \tag{2.167}$$

Substituting Equation (2.166),

$$\dot{V} = \left(\frac{\Delta P}{2LK}\right)^{1/n} \left(\frac{n}{n+1}\right) 2\pi \int_{r=0}^{r=R} r\left(R^{\frac{n+1}{n}} - r^{\frac{n+1}{n}}\right) dr \tag{2.168}$$

Integrating between limits,

$$\dot{V} = \left(\frac{\Delta P}{2LK}\right)^{1/n} \left(\frac{n}{n+1}\right) 2\pi \left| \frac{r^2}{2} R^{\frac{n+1}{n}} - \frac{r^{\frac{2n+1}{n}+1}}{\frac{2n+1}{n}+1} \right|_0^R \tag{2.169}$$

Evaluating the limits,

$$\dot{V} = \left(\frac{\Delta P}{2LK}\right)^{1/n} 2\pi \left(\frac{n}{n+1}\right) \left[\frac{R^{\frac{3n+1}{n}}}{2} - \frac{R^{\frac{3n+1}{n}}}{\frac{3n+1}{n}}\right] \tag{2.170}$$

and simplifying,

$$\dot{V} = \pi \left(\frac{n}{3n+1}\right)\left(\frac{\Delta P}{2LK}\right)^{1/n} R^{\frac{3n+1}{n}} \tag{2.171}$$

Equation (2.171) gives the volumetric flow rate for a power law fluid.

2.9.4 Average Velocity in a Power Law Fluid

Recall from Equation (2.17) that

$$\bar{u} = \frac{\dot{V}}{\pi R^2} \tag{2.172}$$

Substituting Equation (2.171) in (2.172) we obtain the average velocity:

$$\bar{u} = \left(\frac{n}{3n+1}\right)\left(\frac{\Delta P}{2LK}\right)^{1/n} R^{\frac{n+1}{n}} \tag{2.173}$$

For Newtonian fluids the velocity ratio, u/u_{max}, was given by Equation (2.42) (for laminar flow), or Equation (2.48) (for turbulent flow). For non-Newtonian fluids, the velocity ratio can be obtained from Figure 2.46 based on the work of Palmer and Jones (1976).

2.9.5 Friction Factor and Generalized Reynolds Number for Power Law Fluids

In Section 2.3.5, we defined Fanning friction factor as

$$f = \frac{\Delta P D}{2\rho L \bar{u}^2} \tag{2.174}$$

We can rearrange the terms in Equation (2.173) as

$$\frac{\Delta P}{L} = \frac{4K\bar{u}^n}{D^{n+1}}\left(\frac{6n+2}{n}\right)^n \tag{2.175}$$

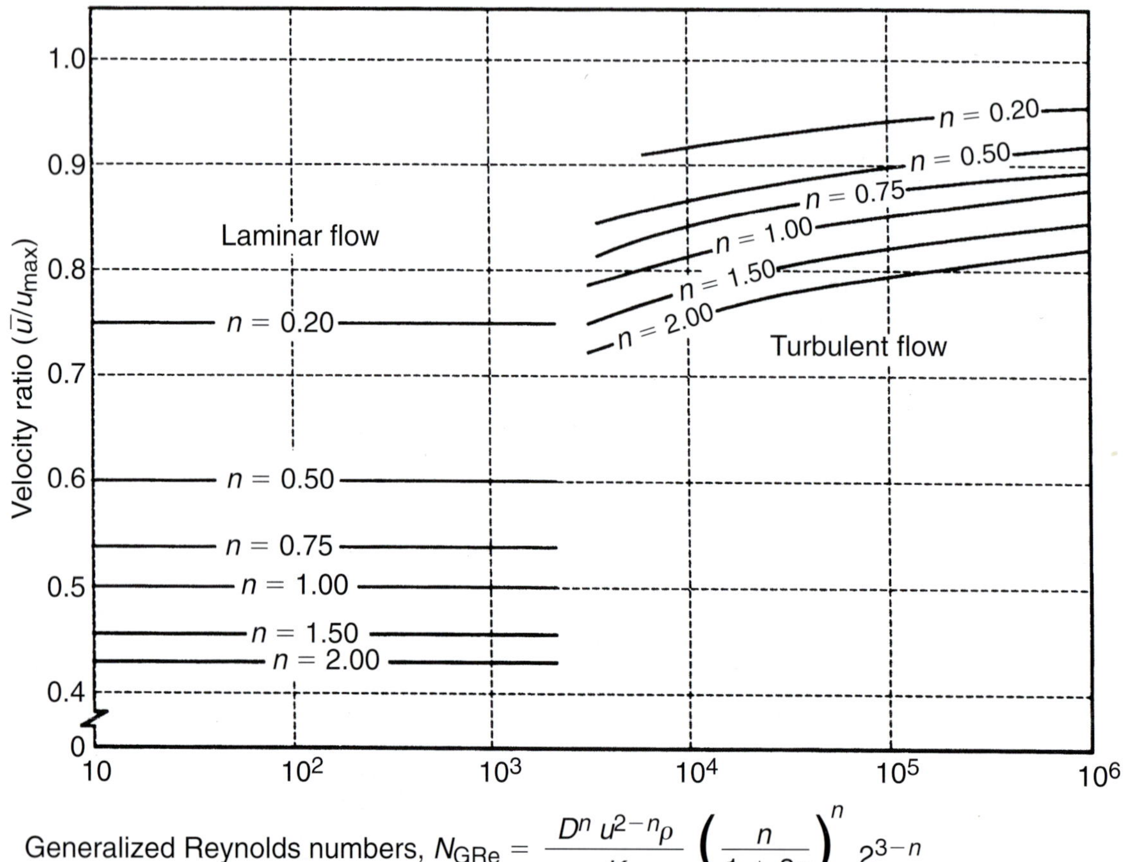

W

Figure 2.46 Plot of velocity ratio vs generalized Reynolds numbers. (From Palmer and Jones, 1976)

Substituting in Equation (2.174),

$$f = \frac{2K\bar{u}^n}{\rho \bar{u}^2 D^n}\left(\frac{6n+2}{n}\right)^n \tag{2.176}$$

Similar to Equation (2.53), the Fanning friction factor for laminar flow for a power law fluid is

$$f = \frac{16}{N_{GRe}} \tag{2.177}$$

where by comparison of Equation (2.176) and Equation (2.177), we obtain the generalized Reynolds number as

$$N_{GRe} = \frac{8D^n \bar{u}^{2-n} \rho}{K} \left(\frac{n}{6n+2} \right)^n \quad (2.178)$$

or, rearranging terms,

$$N_{GRe} = \frac{D^n \bar{u}^{2-n} \rho}{K 8^{n-1}} \left(\frac{4n}{3n+1} \right)^n \quad (2.179)$$

Note that if we substitute the values $n = 1$ and $K = \mu$, Equation (2.179) reduces to the Reynolds number for Newtonian fluids.

Example 2.23 illustrates the use of various expressions obtained in this section for the power law fluids.

Example 2.23

A non-Newtonian fluid is flowing in a 10 m long pipe. The inside diameter of the pipe is 3.5 cm. The pressure drop is measured at 100 kPa. The consistency coefficient is 5.2 and flow behavior index is 0.45. Calculate and plot the velocity profile, volumetric flow rate, average velocity, generalized Reynolds number, and friction factor.

Given

Pressure drop = 100 kPa
Inside diameter = 3.5 cm = 0.035 m
Consistency coefficient = 5.2
Flow behavior index = 0.45
Length = 10 m

Approach

We will program a spreadsheet with the given data and equations from Section 2.9 to obtain the plot and the results.

Solution

1. *The solution is shown in the spreadsheet. Note that as the flow behavior index increases, the velocity at the center decreases. Once a spreadsheet is programmed as shown in Figure E2.9, other values of pressure drop, consistency coefficient, and flow behavior index may be substituted in the appropriate cells to observe their influence on the calculated results. For example an increase in flow behavior index from 0.45 to 0.5 results in a significant change in the velocity profile, as shown in Figure E2.9.*

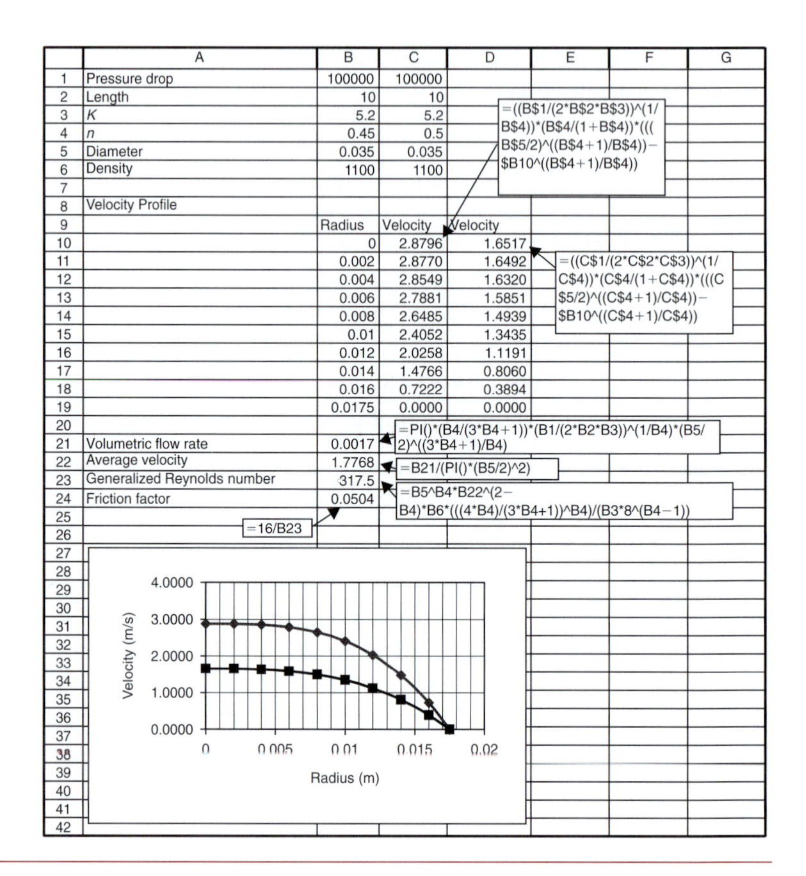

	A	B	C	D
1	Pressure drop	100000	100000	
2	Length	10	10	
3	*K*	5.2	5.2	
4	*n*	0.45	0.5	
5	Diameter	0.035	0.035	
6	Density	1100	1100	
7				
8	Velocity Profile			
9		Radius	Velocity	Velocity
10		0	2.8796	1.6517
11		0.002	2.8770	1.6492
12		0.004	2.8549	1.6320
13		0.006	2.7881	1.5851
14		0.008	2.6485	1.4939
15		0.01	2.4052	1.3435
16		0.012	2.0258	1.1191
17		0.014	1.4766	0.8060
18		0.016	0.7222	0.3894
19		0.0175	0.0000	0.0000
20				
21	Volumetric flow rate	0.0017		
22	Average velocity	1.7768		
23	Generalized Reynolds number	317.5		
24	Friction factor	0.0504		

■ **Figure E2.9** Spreadsheet solution for Example 2.23.

2.9.6 Computation of Pumping Requirement of Non-newtonian Liquids

The computation of pumping requirements of non-Newtonian liquids is similar to that shown earlier for Newtonian liquids in Section 2.5.5. Equation (2.95) is modified to incorporate the non-Newtonian properties as follows:

$$E_p = \frac{P_2 - P_1}{\rho} + \frac{1}{2\alpha'}(u_2^2 - u_1^2) + g(z_2 - z_1) + \frac{2f\bar{u}^2L}{D} + C'_{fe}\frac{\bar{u}^2}{2} + C'_{fc}\frac{\bar{u}^2}{2} + C'_{ff}\frac{\bar{u}^2}{2} \tag{2.180}$$

where, for laminar flow and shear-thinning liquids,

$$\alpha' = \frac{(2n+1)(5n+3)}{3(3n+1)^2} \tag{2.181}$$

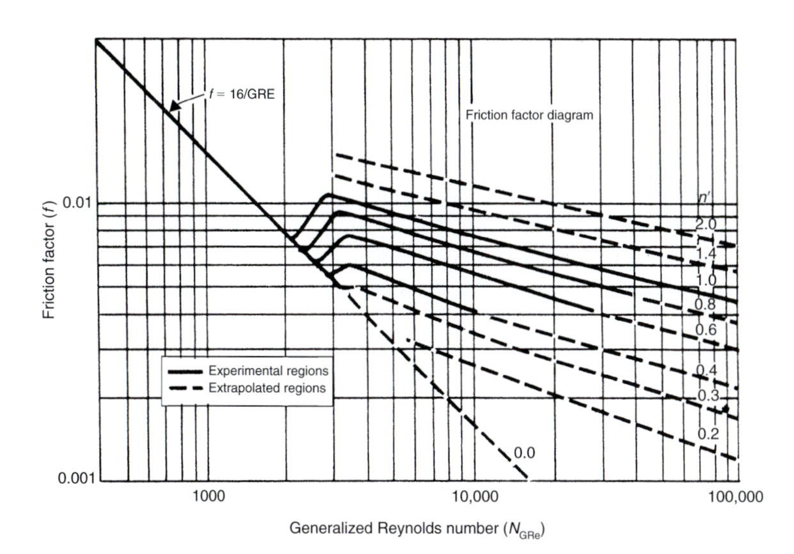

Figure 2.47 Friction factor-Generalized Reynolds number chart for non-Newtonian flow in cylindrical tubes. (From Dodge and Metzner, 1959)

and for turbulent flow

$$\alpha' = 1 \quad (2.182)$$

The friction factor, f, for laminar flow is given by Equation (2.177) and for turbulent flow, f is obtained from Figure 2.47.

For coefficients C'_{fe}, C'_{fc}, and C'_{ff}, experimental data for non-Newtonian fluids are limited. The following rule-of-thumb guidelines are proposed by Steffe (1996) to determine the three coefficients for non-Newtonian fluids.

a. For non-Newtonian fluids above a generalized Reynolds number of 500, use data for Newtonian fluids in turbulent flow (see Table 2.2 and Equations (2.90) and (2.92)).

b. For non-Newtonian fluids, for the range $20 < N_{GRe} < 500$, first determine coefficients C_{fe}, C_{fc}, C_{ff} as in step (a), then use the following expressions:

$$C'_{fe} = \frac{500 \times C_{fe}}{N_{GRe}};\ C'_{fc} = \frac{500 \times C_{fc}}{N_{GRe}};\ C'_{ff} = \frac{500 \times C_{ff}}{N_{GRe}} \quad (2.183)$$

The resulting values of the coefficients are then substituted in Equation (2.180).

Example 2.24 illustrates the determination of pumping requirements of a non-Newtonian liquid.

Example 2.24

A non-Newtonian fluid is being pumped from one tank to another in a 0.0348 m diameter pipe with a mass flow rate of 1.97 kg/s. The properties of the fluid are as follows: density 1250 kg/m^3, consistency coefficient 5.2 Pa s^n, flow behavior index 0.45. The total length of pipe between the tanks is 10 m. The difference of elevation from inlet to exit is 3 m. The fittings include three long-radius 90° flanged elbows, and one fully open angle valve. Furthermore, a filter present in the pipeline causes a 100 kPa pressure drop. Set up a spreadsheet to calculate the pumping requirements.

Given

Pipe diameter = 0.0348 m
Mass flow rate = 1.97 kg/s
Density = 1250 kg/m^3
Consistency coefficient = 5.2 Pa s^n
Flow behavior index = 0.45
Length of pipe = 10 m
Difference in elevation = 3 m
Pressure drop in filter = 100 kPa
C_{ff} of 90° flanged long radius elbows from Table 2.3 = 0.2
C_{ff} of fully open angle valve from Table 2.3 = 2

Approach

We will develop a spreadsheet using the given data and appropriate equations to calculate velocity from the mass flow rate, generalized Reynolds number, correction factor, α, coefficients C_{fe} and C_{ff} for various fittings, and friction factor. Finally we will substitute the calculated values in Equation (2.180) to determine the pumping requirements.

Solution

1. *Enter given data as shown in A1:B12 in Figure E2.10.*
2. *Enter the following equations to calculate required values.*

Equation number	To calculate	Cell
(2.15)	Velocity	B15
(2.179)	generalized Reynolds number	B16
(2.181)	correction factor	B17
(2.177)	friction factor	B21
(2.183)	coefficients for fittings, contraction	B24
(2.180)	energy for pumping	B25

3. *The spreadsheet may be used to calculate pumping requirements for different input values.*

	A	B	C	D	E	F	G
1	Given						
2	Pipe diameter (m)	0.03					
3	Mass flow rate (kg/s)	2					
4	Density (kg/m3)	1250					
5	Pressure drop (kPa)	100					
6	Angle valves	1					
7	Long radius elbows	3					
8	Length (m)	10					
9	Change in elevation (m)	3					
10	*K* (Pa s^n)	5.2					
11	*n*	0.45					
12	Pump efficiency	0.85					
13							
14							
15	Velocity	2.26	=B3/(B4*(PI()*B2^2/4))				
16	Generalized Reynolds number	489.92	=(B2^B11*B15^(2-B11)*B4/(B10*8^(B11−1)))*(4*B11/(3*B11=1))^B11				
17	Correction factor	1.20	=2*(2*B11+1)*(5*B11+3)/(3*(3*B11+1)^2)				
18	C_{fe} entrance	0.5					
19	C_{ff} long radius elbow	0.2					
20	C_{ff} angle valve	2					
21	Friction factor	0.0327	=16/B16				
22	Friction Loss (J/kg)	8.11	=(B18*500/B16+3*B19*500/B16+B20*500/B16)*B15^2/2				
23	Friction loss due to filter	80.00	=B5*1000/B4				
24	Total friction loss	88.11	=B22+B23				
25	Energy for pump (J/kg)	233.34	=9.81*B9+B15^2/B17+2*B21*B15^2*B8/B2+B24				
26	Power requirement (W)	549.04	=B26*B3/B12				
27							
28							

Figure E2.10 Spreadsheet solution for Example 2.24.

2.10 TRANSPORT OF SOLID FOODS

Many food products and ingredients are transported throughout a processing plant in the form of a solid, as opposed to liquids. The handling, storage, and movement of these materials require different types of equipment and process design. Many of the considerations are similar to liquids, in terms of requirements for sanitary design of product-contact surfaces within equipment and the attention to ensuring that designs reduce impact on product quality attributes. Ultimately, the transport of solid pieces, granular products, and/or food powders requires power inputs to an operation designed specifically for the product being handled. In many situations, gravity becomes a contributor to the transport, and introduces unique factors into the transport challenges.

The computation of material transport requirements depend directly on properties of the granular solid or powder. These properties include consideration for friction at the interface between the solid material and the surface of the transport conduit. After establishment of appropriate property magnitudes, the power requirements or similar process design configuration can be estimated.

2.10.1 Properties of Granular Materials and Powders

The physical properties of granular materials and powders have direct influence on the transport of these types of food within a food processing operation. In the following sections, the basic relationships used to predict or measure the properties of these types of food product will be presented and described.

2.10.1.1 *Bulk Density*

For granular materials or powders, there are different types of density to be described. As indicated in Section 1.5, bulk density can be defined as follows:

$$\rho_B = \frac{m}{V} \tag{2.184}$$

or the overall mass of the material divided by the volume occupied by the material. Depending on other characteristics of the food material, this property may require more specific description. For example, food powders will occupy variable volume depending on the size of individual particles and the volume of the space between the particles. Given these observations, the magnitude of the bulk density will vary with the extent of packing within the food particle structure. One approach to measurement includes *Loose Bulk Density*, obtained by careful placement of the granular material in a defined volumetric space without vibration, followed by measurement of the mass. *Packed Bulk Density* would be measured by vibration of a defined mass until the volume is constant, then compute bulk density according to Equation (2.184). In practice, bulk density will vary with the conditions within the operation, but the magnitude should fall between the extremes indicated by the measurement approach.

The bulk density can be predicted by the following relationship:

$$\rho_B = \epsilon_p \rho_p + \epsilon_a \rho_a \tag{2.185}$$

where ϵ_p is the volume fraction occupied by particles and ϵ_a is the volume fraction occupied by air. The parameter referred to as void (υ) or Interparticle Porosity (see Eq. 1.2) is defined as the ratio of volume of space not occupied by particles (ϵ_a) to the total volume. The following expression would apply:

$$\upsilon = 1 - \left(\frac{\rho_b}{\rho_p} \right) \tag{2.186}$$

and would reflect the variability in bulk density discussed in the previous paragraph.

2.10.1.2 Particle Density

As indicated in Section 1.5, the density of individual food particles is referred to as particle density, and is a function of gas phase (air) volume trapped within the particle structure. This property is measured by a picnometer; using a known-density solvent to replace the gas within the particle structure.

The particle density is predicted from the following relationship:

$$\rho_p = \rho_s e_{so} + \rho_a e_a \quad (2.187)$$

where the density of the gas phase or air (ρ_a) is obtained from standard tables (Table A.4), and the density of particle solids is based on product composition and the coefficients in Table A.2.9. The volume fractions of solids (e_{so}) and gas phase or air (e_a) would be evident in magnitude of the particle density.

Example 2.25

Estimate the particle density of a nonfat dry milk particle at 20°C, when 10% of the particle volume is air space and the moisture content is 3.5% (wet basis).

Solution

1. *Based on information presented in Table A.2, the composition of nonfat dry milk is 35.6% protein, 52% carbohydrate, 1% fat, 7.9% ash, and 3.5% water.*
2. *Using the relationships in Table A.2.9, the density of product components at 20°C are:*

   Protein = 1319.5 kg/m^3

   Carbohydrate = 1592.9 kg/m^3

   Fat = 917.2 kg/m^3

   Ash = 2418.2 kg/m^3

   Water = 995.7 kg/m^3

   Air = 1.164 kg/m^3 (from Table A.4.4)

3. *The density of the particle solids is a function of the mass of each component:*

$$\begin{aligned}\rho_{SO} &= (0.356)(1319.5) + (0.52)(1592.9) + (0.01)(917.2) \\ &\quad + (0.079)(2418.2) + (0.035)(995.7) \\ &= 1533.1\ \text{kg/m}^3\end{aligned}$$

4. *The particle density is estimated from Equation (2.187):*

$$\rho_p = (1533.1)(0.9) + (1.164)(0.1) = 1379.9\ \text{kg/m}^3$$

As indicated by the solution, the particle density is influenced primarily by the particle solids.

Porosity (Ψ) is the ratio of the air volume to total volume occupied by the product powder. As introduced in Section 1.5, the relationship of porosity to bulk density becomes:

$$\Psi = 1 - \left(\frac{\rho_b}{\rho_{so}}\right) \tag{2.188}$$

and the relationship to particle density is given by Equation (2.187).

All the properties of granular materials and powders are functions of moisture content and temperature. These relationships are accounted for in the dependence of individual components on moisture content and temperature, but other changes with the particle structure will occur. Details of these changes are discussed by Heldman (2001).

2.10.1.3 *Particle Size and Size Distribution*

An important property of a granular food or powder is particle size. This property has direct impact on the magnitude of the bulk density, as well as the porosity. Particle sizes are measured by using several different techniques, including a sieve, a microscope, or a light-scattering instrument (Coulter Counter). The results of such measurements reveal that a range of particle sizes exists within the typical food powder or granular material. These observations emphasize the importance of using at least two parameters to describe the particle size properties; a mean diameter and a standard deviation.

Mugele and Evans (1951) suggested a systematic approach to description of particle sizes for granular materials. The approach suggests the following model:

$$d_{qp}^{q-p} = \frac{\sum (d^q N)}{\sum (d^p N)} \tag{2.189}$$

where d is the particle diameter, N is the number of particles with a given diameter, and q and p are parameters defined in Table 2.5. For example, the arithmetic or linear mean diameter would become:

$$d_L = \frac{\sum dN}{N} \tag{2.190}$$

Table 2.5 Notations of Mean Particle Size for Use in Equation (2.189)

Symbol	Name of Mean Diameter	p	q	Order
x_L	Linear (arithmetic)	0	1	1
x_s	Surface	0	2	2
x_v	Volume	0	3	3
x_m	Mass	0	3	3
x_{sd}	Surface-diameter	1	2	3
x_{vd}	Volume-diameter	1	3	4
x_{vs}	Volume-surface	2	3	5
x_{ms}	Mass-surface	3	4	7

Source: Mugele and Evans (1951)

where the value is the straight forward mean of particle diameters within the distribution. Another commonly used expression is referred to as the Sauter mean diameter:

$$d_{vs} = \frac{\sum d^3N}{\sum d^2N} \qquad (2.191)$$

and expresses the mean diameter as a function of the ratio of particle volume to particle surface area.

Most often, the distribution of particle sizes in a granular food or powder is described by log-normal density function. The parameters include the arithmetic log-geometric mean:

$$\ln d_g = \frac{\sum (N \ln d)}{N} \qquad (2.192)$$

and the geometric standard deviation, as follows:

$$\ln s_{dg} = \left\{ \frac{\sum \left[N(\ln d - \ln d_g)^2 \right]}{N} \right\}^{1/2} \qquad (2.193)$$

These parameters are estimated from measurement of particle size, and the number of particles in sizes ranges over the range of sizes in the granular food or powder.

Example 2.26

The particle size distribution of a sample dry coffee product has been measured, and the results are as follows:

Portion (%)	Particle Size (micron)
2	40
8	30
50	20
40	15
10	10

Estimate the mean particle size based on the ratio of volume to surface area.

Solution

1. *Given Equation (2.191):*

$$d_{vs} = \frac{(40)^3(2) + (30)^3(8) + (20)^3(50) + (15)^3(40) + (10)^3(10)}{(40)^2(2) + (30)^2(8) + (20)^2(50) + (15)^2(40) + (10)^2(10)}$$

2. *The volume/surface area or Sauter mean particle size diameter is:*

$$d_{vs} = 22 \text{ micron}$$

2.10.1.4 *Particle Flow*

The movement or flow of a granular food or powder is influenced by several different properties of the powder, including several of the particle properties previously described. A property of the granular food that is directly related to flow is *angle of repose*. This is a relatively simple property to measure by allowing the powder to flow from a container over a horizontal surface. The height (H) of the mound of powder and the circumference of the mound (S) are measured, and the angle of repose is computed using the following equation:

$$\tan \beta = \frac{2\pi H}{S} \tag{2.194}$$

The magnitude of this property will vary with density, particle size, moisture content, and particle size distribution of the granular powder.

A more fundamental property of powder flow is the *angle of internal friction*. This property is defined by the following expression:

$$\tan \varphi = \frac{\sigma}{\sigma_n} \tag{2.195}$$

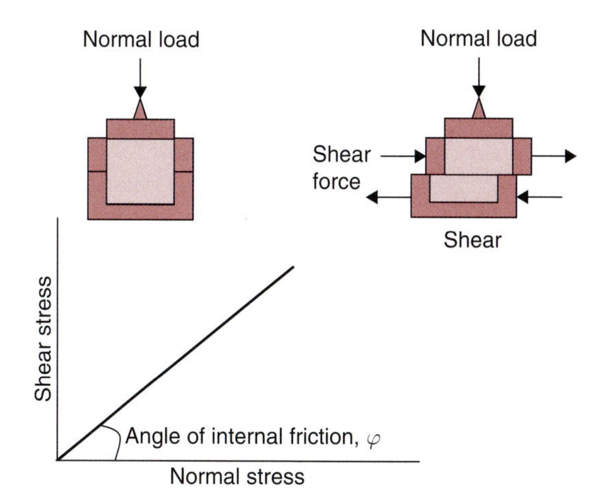

Figure 2.48 The relationship of shear stress to normal stress for a noncohesive powder.

indicating that the angle of internal friction (φ) is proportional to a ratio of the shear stress (σ) to the normal stress (σ_n). The magnitude of this property is determined by measurement of the forces required to move one layer of powder over another layer at various magnitudes of normal stress, as indicated in Figure 2.48.

A fundamental approach to measurement of granular material flow properties is based on a method proposed by Jenike (1970). Measurements of shear stress and normal stress are obtained using a specific shear cell, but are used to generate parameters, such as the unconfined yield stress (f_c) and the major unconsolidated stress (σ_1). These two parameters are then used to define the flow function (f'_c), as follows:

$$f'_c = \frac{\sigma_1}{f_c} \qquad (2.196)$$

This property (f'_c) characterizes the flowability of powders, and with specific applications in the design of bins and hoppers for granular foods. Details on measurement of the flow function (f'_c) are provided by Rao (2006). The relationships of the magnitude of the flow function (f'_c) to flowability of powders are presented in Table 2.6.

2.10.2 Flow of Granular Foods

A significant application of food powder flow properties is the description of product flow from a storage vessel or container. The properties of the granular material have direct influence on the design of the vessel, and specifically the configuration of the vessel at the exit

Table 2.6 Flowability of Powders According to Jenike's Flow Function

Magnitude of Flow Function, f_c'	Flowability of powders	Type of powder
$f_c' < 2$	Very cohesive, nonflowing	Cohesive powders
$2 < f_c' < 4$	Cohesive	
$4 < f_c' < 10$	Easy flowing	Noncohesive powders
$10 < f_c'$	Free flowing	

opening. The mass flow rate of a granular food from a storage vessel can be estimated from the following expression:

$$m_g = \frac{C_g \pi \rho_b \left[\frac{D^5 g \tan\beta}{2}\right]^{0.5}}{4} \tag{2.197}$$

where the mass flow rate (m_g) is a function of bulk density of the powder (ρ_b), diameter of the opening at the exit from the vessel (D), the angle of repose for the powder (β), and a discharge coefficient (C_g). The discharge coefficients will vary in magnitude with container configuration; usually between 0.5 and 0.7. Equation (2.197) is limited to situations where the ratio of particle diameter to opening diameter is less than 0.1.

A critical problem in the use of gravity to induce granular flow from a storage vessel is the tendency of arching or bridging to occur. This blockage of flow from the container is directly related to the powder properties, specifically the density and angle of internal friction. A closely related concern during gravity flow from a storage container is the occurrence of plug flow as opposed to mass flow, as illustrated in Figure 2.49. As illustrated, plug flow results in the collection of powder in regions of the container near the outlet, and the material is not removed without assistance. The vessel design feature is the cone angle (θ_c), or the angle between the vessel surface leading to the exit and the vertical sides of the container. This angle is dependent on the angle of internal friction for the powder.

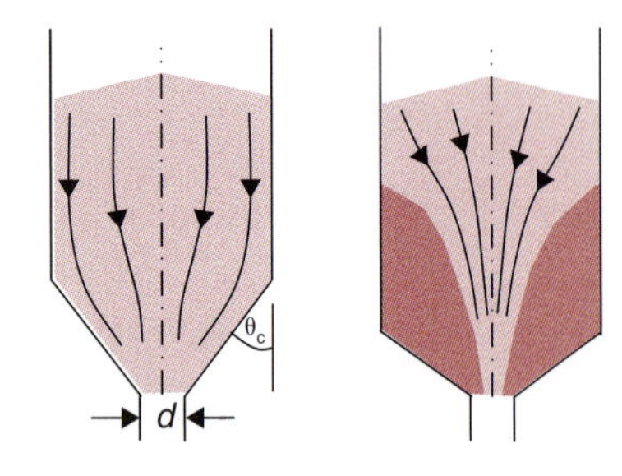

■ **Figure 2.49** An illustration of food powder flow from a storage vessel.

An expression proposed to assist in the design of the outlet from a storage vessel for a granular food is as follows:

$$D_b = \left(\frac{C_b}{\rho_b}\right)[1 + \sin\varphi] \tag{2.198}$$

where D_b is the minimum diameter of opening needed to prevent bridging of the powder above the opening, and is dependent on the natural

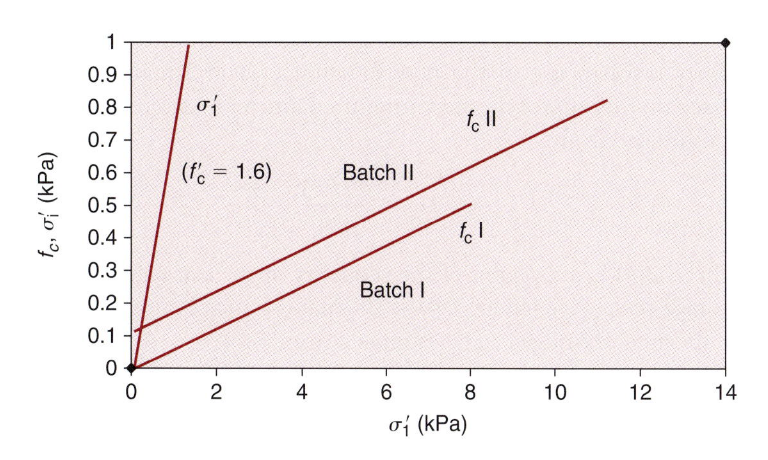

Figure 2.50 The relationship of shear stress to normal stress for instant cocoa powder. (From Schubert, H. (1987a). *J. Food Eng.* 6: 1–32; Schubert, H. (1987b). *J. Food Eng.* 6: 83–102. With permission.)

cohesiveness parameter (C_b) as determined during measurement of the angle of internal friction. More specifically, the natural cohesiveness parameter is the intercept on the shear stress axis of the shear stress versus normal stress plot, as illustrated in Figure 2.50. The angle of internal friction (φ) is the angle of the shear stress (τ) to the normal stress (σ).

Example 2.27

Estimate the minimum diameter of an opening from a storage vessel for an instant cocoa powder. The bulk density of the food powder is 450 kg/m^3.

Solution

1. *Based on the measurements of shear stress versus normal stress presented in Figure 2.50, the angle of internal friction (for Batch II) is determined by using Equation (2.195) to obtain:*

$$\tan\varphi = \frac{(1-0.11)}{14} = 0.064$$

$$\varphi = 3.66^\circ$$

2. *The same measurements indicate that the natural cohesiveness parameter (C_b) for the same batch (based on the intercept on the shear stress axis) is:*

$$C_b = 0.11\,kPa = 110\,Pa$$

3. *Using Equation (2.198):*

$$D_b = \left(\frac{110}{450}\right)[1+\sin(3.66)]$$

$$D_b = 0.26\,m = 26\,cm$$

4. *The result indicates that the diameter of the outlet for the storage vessel would need to exceed 26 cm to avoid bridging for this product.*

A more thorough analysis of the granular flow from a storage container involves use of the flow function (f_c') in Equation (2.196). Based on this approach, the minimum diameter to avoid arching (D_b) is estimated from:

$$D_b = \frac{\sigma' H(\theta)}{g\rho_b} \tag{2.199}$$

where $H(\theta)$ is a function of the geometry of the exit region from the storage vessel (see Jenike, 1970). The major principle stress (σ') acting in the arch or bridge can be estimated from:

$$\sigma' = \frac{\sigma_1}{f_c'} \tag{2.200}$$

where σ_1 is the unconsolidated stress involved in measurement of the flow function (f_c'). Once flowability of the powder has been established, Table 2.6 can be used to estimate the flow function (f_c') and the minimum diameter of the opening from the granular food storage vessel from Equation (2.199).

PROBLEMS

2.1 Calculate the Reynolds number for 25°C water flow in a 1-in nominal diameter sanitary pipe at 0.5 kg/s. What are the flow characteristics?

2.2 A pipe discharges wine into a 1.5 m diameter tank. Another pipe (15 cm diameter), located near the base of the tank, is used to discharge wine out of the tank. Calculate the volumetric flow rate into the tank if the wine level remains constant at 2.5 m.

2.3 Apple juice at 20°C is siphoned from a large tank using a constant-diameter hose. The end of the siphon is 1 m below the bottom of the tank. Calculate the height of the hill over which the hose may be siphoned without cavitation. Assume properties of apple juice are same as water. The atmospheric pressure is 101.3 kPa. Height of the juice in the tank is 2 m.

2.4 Sulfuric acid with a density of 1980 kg/m^3 and a viscosity of 26.7 cP is flowing in a 35 mm diameter pipe. If the acid flow rate is 1 m^3/min, what is the pressure loss due to friction for a 30 m length of smooth pipe?

2.5 Compute the mean and maximum velocities for a liquid with a flow rate of 20 L/min in a 1.5-in nominal diameter sanitary pipeline. The liquid has a density of 1030 kg/m^3 and viscosity of 50 cP. Is the flow laminar or turbulent?

2.6 Calculate the total equivalent length of 1-in wrought iron pipe that would produce a pressure drop of 70 kPa due to fluid friction, for a fluid flowing at a rate of 0.05 kg/s, a viscosity of 2 cP, and density of 1000 kg/m^3.

***2.7** A solution of ethanol is pumped to a vessel 25 m above a reference level through a 25 mm inside diameter steel pipe at a rate of 10 m^3/h. The length of pipe is 30 m and contains two elbows with friction equivalent to 20 diameters each. Compute the power requirements of the pump. Solution properties include density of 975 kg/m^3 and viscosity of 4×10^{-4} Pa s.

2.8 The flow of a liquid in a 2-in nominal diameter steel pipe produces a pressure drop due to friction of 78.86 Pa. The length of pipe is 40 m and the mean velocity is 3 m/s. If the density of the liquid is 1000 kg/m^3, then

a. Determine the Reynolds number.
b. Determine if the flow is laminar or turbulent.
c. Compute viscosity of the liquid.
d. Estimate the temperature, if the liquid is water.
e. Compute the mass flow rate.

***2.9** A pump is being used to transport a liquid food product ($\rho = 1000$ kg/m^3, $\mu = 1.5$ cP) from a holding tank to a filling machine at a mass flow rate of 2 kg/s. The liquid level in the holding tank is 10 m above the pump, and the filling machine is 5 m above the pump. There is 100 m of 2-in nominal diameter sanitary pipeline between the holding tank and the filling machine, with one open globe valve and four regular 90° flanged elbows in the system. The product is being pumped through a heat exchanger with 100 kPa of pressure drop due to friction before filling. Determine the theoretical power requirement for the pump.

2.10 A centrifugal pump, located above an open water tank, is used to draw water using a suction pipe (8 cm diameter). The pump is to deliver water at a rate of 0.02 m^3/s. The pump manufacturer has specified a $NPSH_R$ of 3 m. The water temperature is 20°C and atmospheric pressure is 101.3 kPa.

---

*Indicates an advanced level in solving.

Calculate the maximum height the pump can be placed above the water level in the tank without cavitation. A food process equipment located between the suction and the pump causes a loss of $C_f = 15$. All other losses may be neglected.

2.11 A centrifugal pump is operating at 1800 rpm against 30 m head with a flow rate of 1500 L/min. If the pump speed is doubled, calculate the new flow rate and developed head.

2.12 Edible oil (specific gravity = 0.83) flows through a venturi meter with a range of flow rates from 0.002 to 0.02 m^3/s. Calculate the range in pressure differences required to measure these flow rates. Pipe diameter is 15 cm and diameter of the venturi throat is 5 cm.

2.13 A pipe (inside diameter 9 cm) is being used to pump a liquid (density = 1100 kg/m^3) with a maximum mass flow rate of 0.015 kg/s. An orifice plate ($C = 0.61$) is to be installed to measure the flow rate. However, it is desired that the pressure drop due to the orifice plate must not exceed 15 kPa. Determine the diameter of the orifice plate that will ensure that the pressure drop will remain within the stated limit.

2.14 A capillary tube is being used to measure the viscosity of a Newtonian liquid. The tube has a 4 cm diameter and a length of 20 cm. Estimate the viscosity coefficient for the liquid if a pressure of 2.5 kPa is required to maintain a flow rate of 1 kg/s. The liquid density is 998 kg/m^3.

2.15 Calculate the viscosity of a fluid that would allow a pressure drop of 35 kPa over a 5 m length of 0.75-in stainless steel sanitary pipe if the fluid is flowing at 0.12 m^3/h and has a density of 1010 kg/m^3. Assume laminar flow.

2.16 A 2 cm diameter, 5 cm long capillary-tube viscometer is being used to measure viscosity of a 10 Pa s liquid food. Determine the pressure required for measurement when a flow rate of 1 kg/min is desired and $\rho = 1000$ kg/m^3.

***2.17** A capillary-tube viscometer is being selected to measure viscosity of a liquid food. The maximum viscosity to be measured will be 230 cP, and the maximum flow rate that can be measured accurately is 0.015 kg/min. If the tube length is 10 cm and a maximum pressure of 25 Pa can be measured,

*Indicates an advanced level in solving.

determine the tube diameter to be used. The density of the product is 1000 kg/m^3.

***2.18** A single-cylinder rotational viscometer is used to measure a liquid with viscosity of 100 cP using a spindle with 6 cm length and 1 cm radius. At maximum shear rate (rpm = 60), the measurements approach a full-scale reading of 100. Determine the spindle dimensions that will allow the viscometer to measure viscosities up to 10,000 cP at maximum shear rate.

2.19 A dry food powder with a bulk density of 650 kg/m^3 and an angle of repose of 55° is discharged by gravity through a circular opening at the bottom of a large storage vessel. Estimate the diameter of the opening needed to maintain a mass flow rate of 5 kg/min. The discharge coefficient is 0.6.

2.20 A dry food powder is being stored in a bin with a 10 cm diameter outlet for gravity flow during discharge. The angle of internal friction is 5° and the bulk density is 525 kg/m^3. Estimate the cohesiveness parameter of a powder that would result in bridging when the product is discharged from the storage bin.

***2.21** Zigrang and Sylvester (1982) give several explicit empirical equations for friction factor as a function of the Reynolds number. The equation attributed to Churchill is said to be applicable to all values of NRe and ε/D. Churchill's model is for the Darcy friction factor, fD, which is four times the value of the Fanning friction factor, *f*.

$$f_D = 8\left[\left(\frac{8}{N_{Re}}\right)^{12} + \frac{1}{(A+B)^{3/2}}\right]^{1/12}$$

where

$$A \equiv \left[2.457 \ln\left\{\left(\frac{7}{N_{Re}}\right)^{0.9} + 0.27\frac{\varepsilon}{D}\right\}\right]^{16} \qquad B \equiv \left(\frac{37{,}530}{N_{Re}}\right)^{16}$$

Using MATLAB® , evaluate Churchill's equation for $\varepsilon/D = 0.0004$ and $N_{Re} = 1 \times 10^4$, 1×10^5, 1×10^6, and 1×10^7. Compare results to readings from the Fanning friction factor chart (Fig. 2.16) and to Haaland's explicit friction factor for turbulent flow given by Equation (2.54).

---

* Indicates an advanced level in solving.

***2.22** The turbulent portion ($N_{Re} > 2100$) of the Moody diagram given in Figure 2.16 is constructed using the Colebrook equation:

$$\frac{1}{\sqrt{f_D}} = -2.0 \log_{10}\left(\frac{\varepsilon/D}{3.7} + \frac{2.51}{N_{Re}\sqrt{f_D}}\right)$$

where f_D is the Darcy friction factor, which is four times the value of the Fanning friction factor, f.

Use the MATLAB® function f_0 to solve for f_D (and f) in the Colebrook equation for $\varepsilon/D = 0.0004$ and the Reynolds numbers of 2100, 1×10^4, 1×10^5, 1×10^6, and 1×10^7.

Compare these results to values read from Figure 2.16.

***2.23** A thin-plate orifice has been installed to measure water flow in a pipe with an inner diameter (D_1) of 0.147m and an orifice diameter (D_2) of 0.0735m. The orifice pressure taps are located a distance D_1 upstream and $D_1/2$ downstream of the orifice. Write a MATLAB® script to calculate the flow rate, Q, if the measured pressure drop ($P_A - P_B$) is 9000 Pa when the water temperature is 20°C. Use the procedure outline by White (2008) as shown.

1. Guess a value for $C = 0.61$.
2. Calculate flow $Q(m^3/s)$ using the value of C.

$$Q = \frac{CA_2}{\sqrt{(1-\beta^4)}}\sqrt{\frac{2}{\rho}(p_1 - p_2)}$$

3. Calculate velocity in pipe, u_A, and the Reynolds number, N_{Re}.
4. Calculate a new value for C using the empirical equation for the orifice coefficient for D and $D/2$ taps:

$$C \approx 0.5899 + 0.05\beta^2 - 0.08\beta^6 + (0.0037\beta^{1.25} + 0.011\beta^8)\left(\frac{10^6}{N_{Re}}\right)^{1/2}$$

where

$$N_{Re} = \frac{u_A D_1 \rho_f}{\mu_f} \qquad \beta = \frac{D_2}{D_1}$$

5. Repeat steps (2) through (4) until you converge to constant Q.

---

*Indicates an advanced level in solving.

LIST OF SYMBOLS

A area (m^2)
α correction factor for Newtonian fluid
α' correction factor for non-Newtonian fluid
β angle of repose
B_A Arrhenius constant
C coefficient
C_{fe} coefficient of friction loss during expansion
C_{fc} coefficient of friction loss due to contraction
C_{ff} coefficient of friction loss due to fittings in a pipe
C_b cohesiveness parameter (Pa)
C_g discharge coefficient in Equation (2.197)
D pipe or tube diameter (m)
d particle diameter (micron)
d_c characteristic dimension (m)
e volume fraction within a particle
E internal energy (J/kg)
E_f frictional loss of energy
$\in$ volume fraction in particle bed
ε surface roughness (m)
E energy (J)
E' energy per unit mass (J/kg)
E_a activation energy (J/kg)
E_p energy supplied by the pump (J/kg)
F force (N)
f friction factor
f_c unconfined yield stress (Pa)
f_c' flow function
Φ power (watts)
g acceleration due to gravity (m/s^2)
$\dot{\gamma}$ rate of shear (1/s)
H height (m)
$H(\theta)$ exit geometry function
h head (m)
j exponent in Equation (2.43)
K consistency coefficient (Pa s)
L length (m)
L_e entrance length (m)
μ coefficient of viscosity (Pa s)
$\dot{m}$ mass flow rate (kg/s)
m mass (kg)

N	rotational speed, revolutions per second
N_p	number of particles
N_{Re}	Reynolds number
N_{GRe}	Generalized Reynolds number
n	flow behavior index
η	efficiency of pump
P	pressure (Pa)
Q	heat added or removed from a system (kJ)
θ	angle
R	radius (m)
ρ	density (kg/m^3)
r	radial coordinate
R_g	gas constant (cal/mol K)
s	distance coordinate along stream line
S	circumference
s_d	standard deviation
σ	shear stress (Pa)
σ_n	normal stress (Pa)
σ'	major principle stress (Pa)
σ_w	shear stress at wall (Pa)
T	temperature (°C)
t	time (s)
u	fluid velocity (m/s)
$\bar{u}$	mean fluid velocity (m/s)
ν	kinematic viscosity (m^2/s)
V	volume (m^3)
$\dot{V}$	volumetric flow rate (m^3/s)
φ	angle of internal friction
ω	angular velocity (rad/s)
Ψ	porosity
W	work (kJ/kg)
W_m	work done by the pump (kJ/kg)
υ	void
x	distance coordinate in x-direction (m)
y	distance coordinate in y-direction (m)
z	vertical coordinate
Ω	torque (N m)
ΔP	pressure drop (Pa)

Subscripts: a, air ; A, absolute; b, minimum; B, bulk; d, discharge; g, granular; i, inner; m, manometer; n, normal; o, outer; p, particle; s, suction; so, solids; w, wall.

BIBLIOGRAPHY

Brennan, J. G., Butters, J. R., Cowell, N. D., and Lilly, A. E. V. (1990). *Food Engineering Operations*, 3rd ed. Elsevier Science Publishing Co., New York.

Casson, N. (1959). A flow equation for pigmented-oil suspension of the printing ink type. In *Rheology of Dispersed Systems*, C. C. Mill, ed., 84–104. Pergamon Press, New York.

Charm, S. E. (1978). *The Fundamentals of Food Engineering*, 3rd ed. AVI Publ. Co., Westport, Connecticut.

Colebrook, C. F. (1939). Friction factors for pipe flow. *Inst. Civil Eng.* **11**: 133.

Dodge, D. W. and Metzner, A. B. (1959). Turbulent flow of non-Newtonian systems. *AIChE J.* **5**(7): 189–204.

Doublier, J. L. and Lefebvre, J. (1989). Flow properties of fluid food materials. In *Food Properties and Computer-Aided Engineering of Food Processing Systems*, R. P. Singh and A. G. Medina, eds., 245–269. Kluwer Academic Publishers, Dordrecht, The Netherlands.

Earle, R. L. (1983). *Unit Operations in Food Processing*, 2nd ed. Pergamon Press, Oxford.

Farrall, A. W. (1976). *Food Engineering Systems*, Vol. 1. AVI Publ. Co., Westport, Connecticut.

Farrall, A. W. (1979). *Food Engineering Systems*, Vol. 2. AVI Publ. Co., Westport, Connecticut.

Haaland, S. E. (1983). Simple and explicit formulas for the friction factor in turbulent pipe flow. *Fluids Eng.* March, 89–90.

Heldman, D. R. and Seiberling, D. A. (1976). In *Dairy Technology and Engineering,*, W. J. Harper and C. W. Hall, eds., 272–321. AVI Publ. Co., Westport, Connecticut.

Heldman, D. R. and Singh, R. P. (1981). *Food Process Engineering*, 2nd ed. AVI Publ. Co., Westport, Connecticut.

Heldman, Dennis R. (2001). Prediction of models for thermophysical properties of foods. Chapter 1. In *Food Processing Operation Modeling: Design and Analysis*, J. Irudayaraj, ed. Marcel-Dekker, Inc, New York.

Herschel, W. H. and Bulkley, R. (1926). Konsistenzmessungen von gummi-benzollusungen. *Kolloid-Zeitschr.* **39**: 291.

Hydraulic Institute (1975). *Hydraulic Institute Standards for Centrifugal, Rotary and Reciprocating Pumps*. Hydraulic Institute, Cleveland, Ohio.

Jenike, A. W. (1970). *Storage and Flow of Solids, Bulletin 123 of the Utah Engineering Experiment Station*, 4th printing (revised). University of Utah, Salt Lake City.

Loncin, M. and Merson, R. L. (1979). *Food Engineering; Principles and Selected Applications*. Academic Press, New York.

Mugele, K. A. and Evans, H. D. (1951). Droplet size distribution in sprays. *Ind. Eng. Chem.* **43**: 1317.

Munson, B. R., Young, D. F., and Okiishi, T. H. (1998). *Fundamentals of Fluid Mechanics*. John Wiley and Sons, New York.

Palmer, J. and Jones, V. (1976). Reduction of holding times for continuous thermal processing of power law fluids. *J. Food Sci.* **41**(5): 1233.

Peleg, M. (1977). Flowability of food powders and methods for evaluation. *J. Food Process Engr.* **1**: 303–328.

Rao, M. A. (2006). Transport and Storage of Food Products. Chapter 4. In *Handbook of Food Engineering*, D. R. Heldman and D. B. Lund, eds. CRC Press, Taylor & Francis Group, Boca Raton, Florida.

Reynolds, O. (1874). *Papers on Mechanical and Physical Subjects*. The University Press, Cambridge, England.

Rotstein, E., Singh, R. P., and Valentas, K. (1997). *Handbook of Food Engineering Practice*. CRC Press, Inc, Boca Raton, Florida.

Schubert, H. (1987a). Food particle technology: Properties of particles and particulate food systems. *J. Food Engr.* **6**: 1–32.

Schubert, H. (1987b). Food particle technology: Some specific cases. *J. Food Engr.* **6**: 83–102.

Slade, F. H. (1967). *Food Processing Plant*, Vol. 1. CRC Press, Cleveland, Ohio.

Slade, F. H. (1971). *Food Processing Plant*, Vol. 2. CRC Press, Cleveland, Ohio.

Smits, A. J. (2000). *A Physical Introduction to Fluid Mechanics*. John Wiley and Sons, Inc, New York.

Steffe, J. F. (1996). *Rheological Methods in Food Process Engineering*, 2nd ed. Freeman Press, East Lansing, Michigan.

Steffe, J. F. and Daubert, C. R. (2006). *Bioprocessing Pipelines, Rheology and Analysis*. Freeman Press, East Lansing, Michigan.

Toledo, R. T. (2007). *Fundamentals of Food Process Engineering*, 3rd ed. Springer-Science+Business Media, New York.

Van Wazer, J. R. (1963). *Viscosity and Flow Measurement*. Interscience Publishers, New York.

Vitali, A. A. and Rao, M. A. (1984). Flow properties of low-pulp concentrated orange juice: Effect of temperature and concentration. *J. Food Sci.* **49**(3): 882–888.

Watson, E. L. and Harper, J. C. (1988). *Elements of Food Engineering*, 2nd ed. Van Nostrand Reinhold, New York.

White, F. M. (2008). *Fluid Mechanics*, 6th edition. McGraw-Hill Book Co, New York.

Zigrang, D. J. and Sylvester, N. D. (1982). Explicit approximations of the solution of Colebrook's friction factor equation. *AIChE Journal* **28**(3): 514–515.

Chapter 3

Energy and Controls in Food Processes

The modern food processing plant cannot function without adequate supplies of basic utilities. The use of large quantities of water is not unexpected due to the handling of food in water and the need for water as a cleaning medium. Electricity is used as a utility to power many motors and related equipment throughout food processing. Heated air and water are used for a variety of purposes, with energy provided from several fuel sources, including natural gas, coal, or oil. Refrigeration is a much used utility throughout the food industry, with most applications involving conversion of electrical energy into cold air. Steam is a utility similar to refrigeration, in that its availability is dependent on generating facilities at a location near the point of use.

Within this chapter, three of the utilities used in food processing will be analyzed in some detail. These utilities include (1) generation and utilization of steam, (2) natural gas utilization, and (3) electric power utilization. Water utilization will not be analyzed, except as a part of steam generation, since it is not viewed as a source of energy in most applications. The subject of refrigeration will be described in a separate chapter to adequately reflect the importance of this subject.

All icons in this chapter refer to the author's web site, which is independently owned and operated. Academic Press is not responsible for the content or operation of the author's web site. Please direct your web site comments and questions to the author: Professor R. Paul Singh, Department of Biological and Agricultural Engineering, University of California, Davis, CA 95616, USA.
Email: rps@rpaulsingh.com.

3.1 GENERATION OF STEAM

Steam represents the vapor state of water and becomes a source of energy when the change-of-state is realized. This energy can be used for increasing the temperature of other substances, such as food products, and results in production of a water condensate as the energy is released. The vapor state of water or steam is produced by addition of energy from a more basic source, such as fuel oil or natural gas, to convert water from a liquid to a vapor state.

This section will first describe typical systems used in the food industry for conversion of water to steam. The thermodynamics of phase change will be discussed and will be used to explain steam tables. The values tabulated in steam tables will be used to illustrate energy requirements for steam generation, as well as availability of energy from steam to use in food processing. The efficient conversion of energy from the source used to generate steam to some food processing application will be emphasized.

3.1.1 Steam Generation Systems

The systems for generation of steam can be divided into two major classifications: fire-tube and water-tube. Both systems are used in the food industry, but water-tube systems are designed for the more modern applications. The steam generation system or boiler is a vessel designed to bring water into contact with a hot surface, as required to convert liquid to vapor. The hot surface is maintained by using hot gases, usually combustion gases from natural gas or other petroleum products. The boiler vessel is designed to contain the steam and to withstand the pressures resulting from the change of state for water.

Fire-tube steam generators (Fig. 3.1) utilize hot gases within tubes surrounded by water to convert the water from liquid to vapor state.

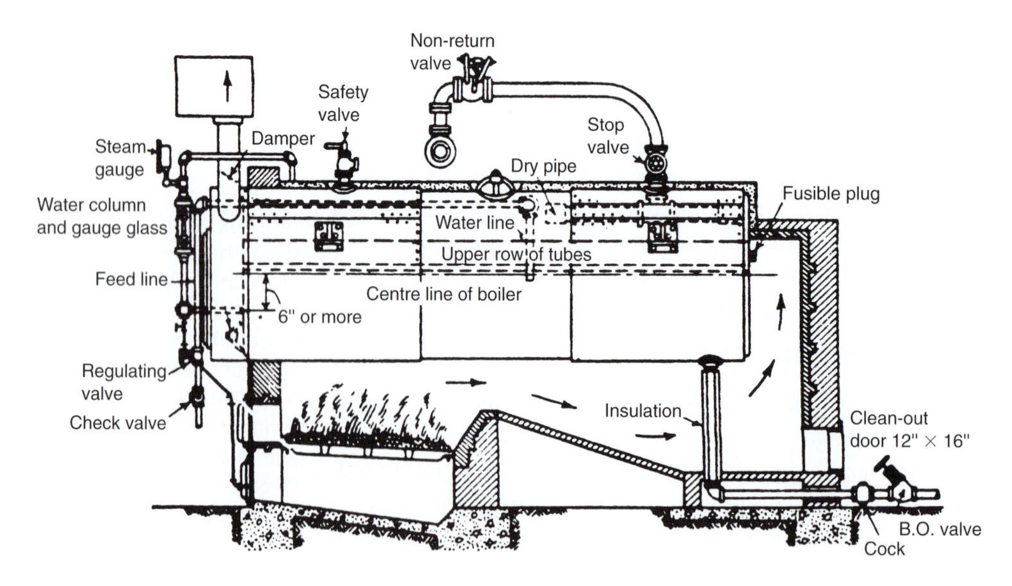

Figure 3.1 The horizontal return tubular (HRT) fire-tube boiler. (From Farrall, 1979)

The resulting heat transfer causes the desired change of state, with the vapors generated contained within the vessel holding the water. A water-tube steam generator (Fig. 3.2) utilizes heat transfer from hot gas surrounding the tubes to the water flowing through the tubes to produce steam. The heat transfer in the water-tube system tends to be somewhat more rapid because of the ability to maintain turbulent flow within the liquid flow tube.

Water-tube boilers generally operate with larger capacities and at higher pressures. These systems have greater flexibility and are considered safer to operate than the counterpart fire-tube systems. The safety feature is associated most closely with the change-of-phase occurring within small tubes in a water-tube system rather than in a large vessel in a fire-tube system. The latter system does have an advantage when the load on the system varies considerably with time. Nearly all modern installations in the food industry are of the water-tube design.

One of the more recent developments is the utilization of alternate fuels as a source of energy for steam generation. In particular, combustible waste materials from processing operations have become a viable alternative. In many situations, these materials are available in large quantities and may present a disposal problem.

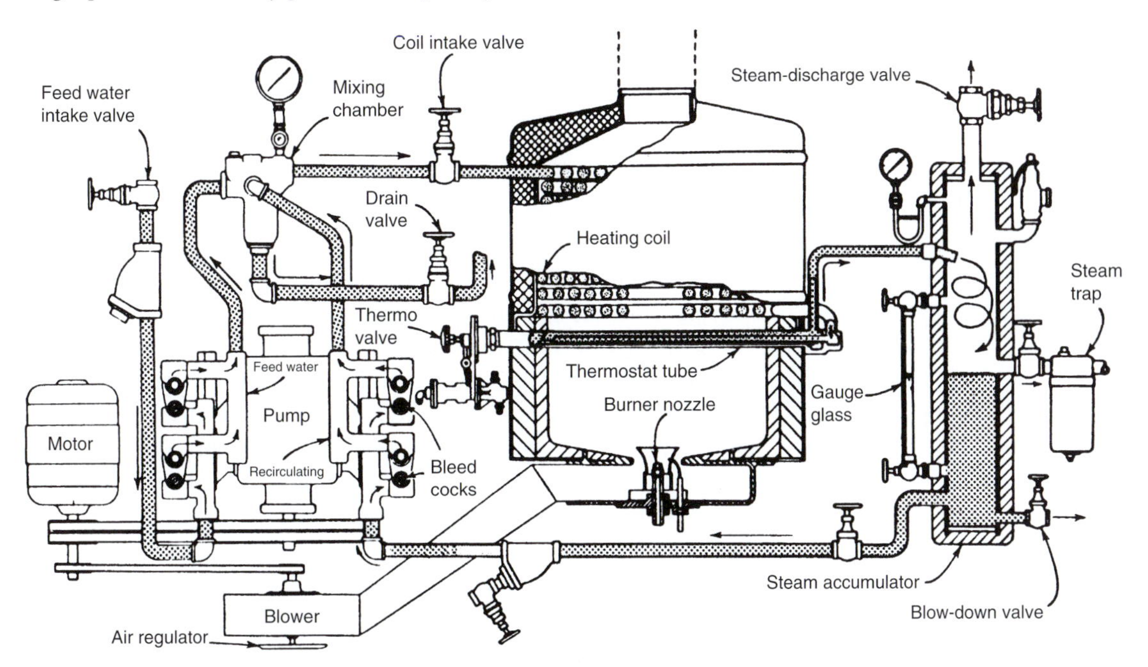

■ **Figure 3.2** Water-tube steam generator. (Courtesy of Cherry-Burrell Corporation)

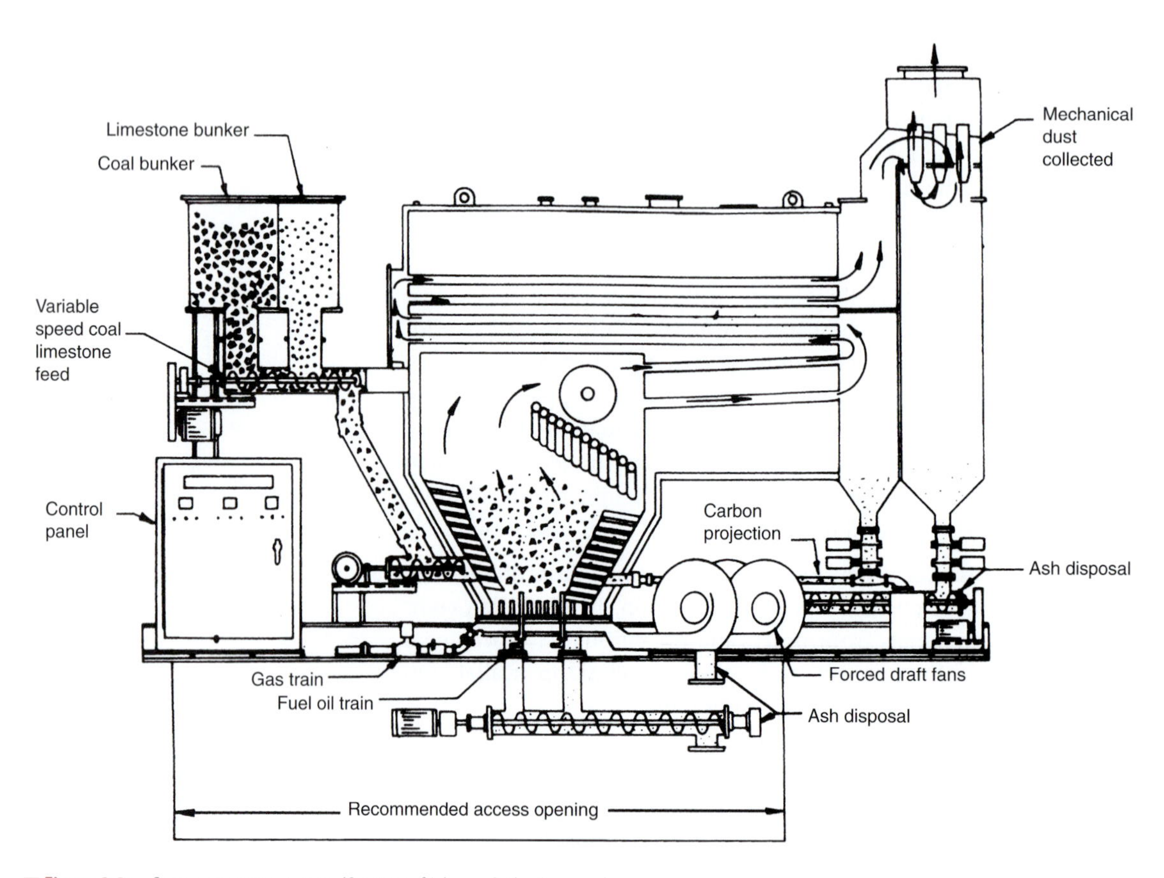

■ **Figure 3.3** Steam generation system. (Courtesy of Johnson Boiler Company)

Steam generation systems do require modifications in design to accommodate different combustion processes, as illustrated in Figure 3.3. The advantage of these systems is the opportunity to establish cogeneration, as sketched in Figure 3.4. This arrangement utilizes steam generated by burning waste materials to generate electric power, as well as to provide steam for processing operations. Depending on the availability of waste materials, significant percentages of electric power demand can be met in this way.

3.1.2 Thermodynamics of Phase Change

The conversion of water from a liquid to vapor state can be described in terms of thermodynamic relationships. If the phase change for water is presented as a pressure–enthalpy relationship, it appears as

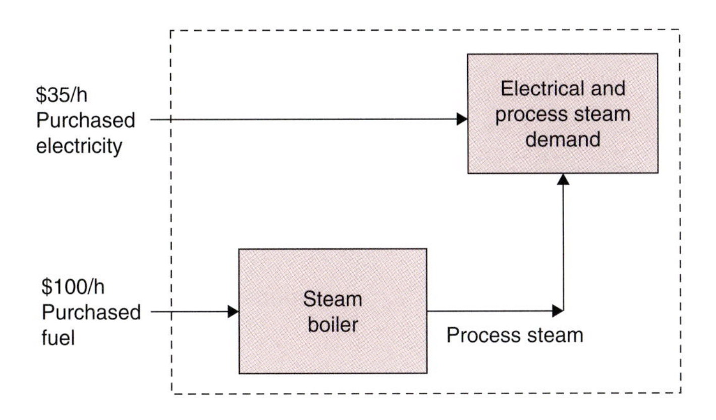

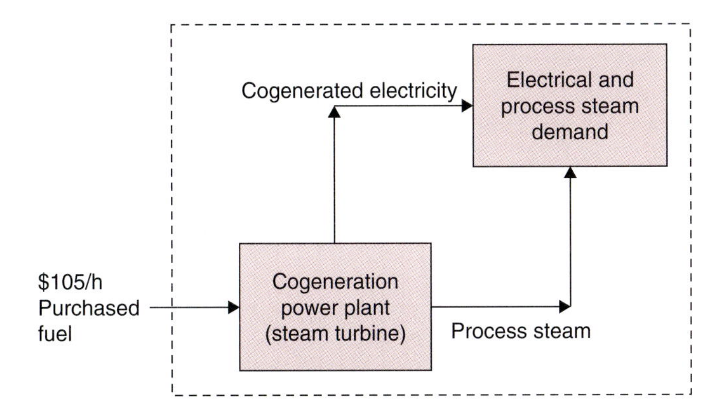

Figure 3.4 Steam generation systems with and without cogeneration. (From Teixeira, 1980)

shown in Figure 3.5. The bell-shaped curve represents the pressure, temperature, and enthalpy relationships of water at its different states.

The left-side curve is the saturated liquid curve, whereas the right-side curve is the saturated vapor curve. Inside the bell-shaped curve any location indicates a mixture of liquid and vapor. The region to the right side of the saturated vapor curve indicates superheated vapors. And the region to the left side of the saturated liquid curve indicates subcooled liquid. At atmospheric pressure, the addition of sensible heat increases the heat content of liquid water until it reaches the saturated liquid curve.

As an illustration, consider a process ABCD on Figure 3.5. Point A represents water at 90°C and 0.1 MPa pressure. The enthalpy content is about 375 kJ/kg of water. As heat is added to the water, the

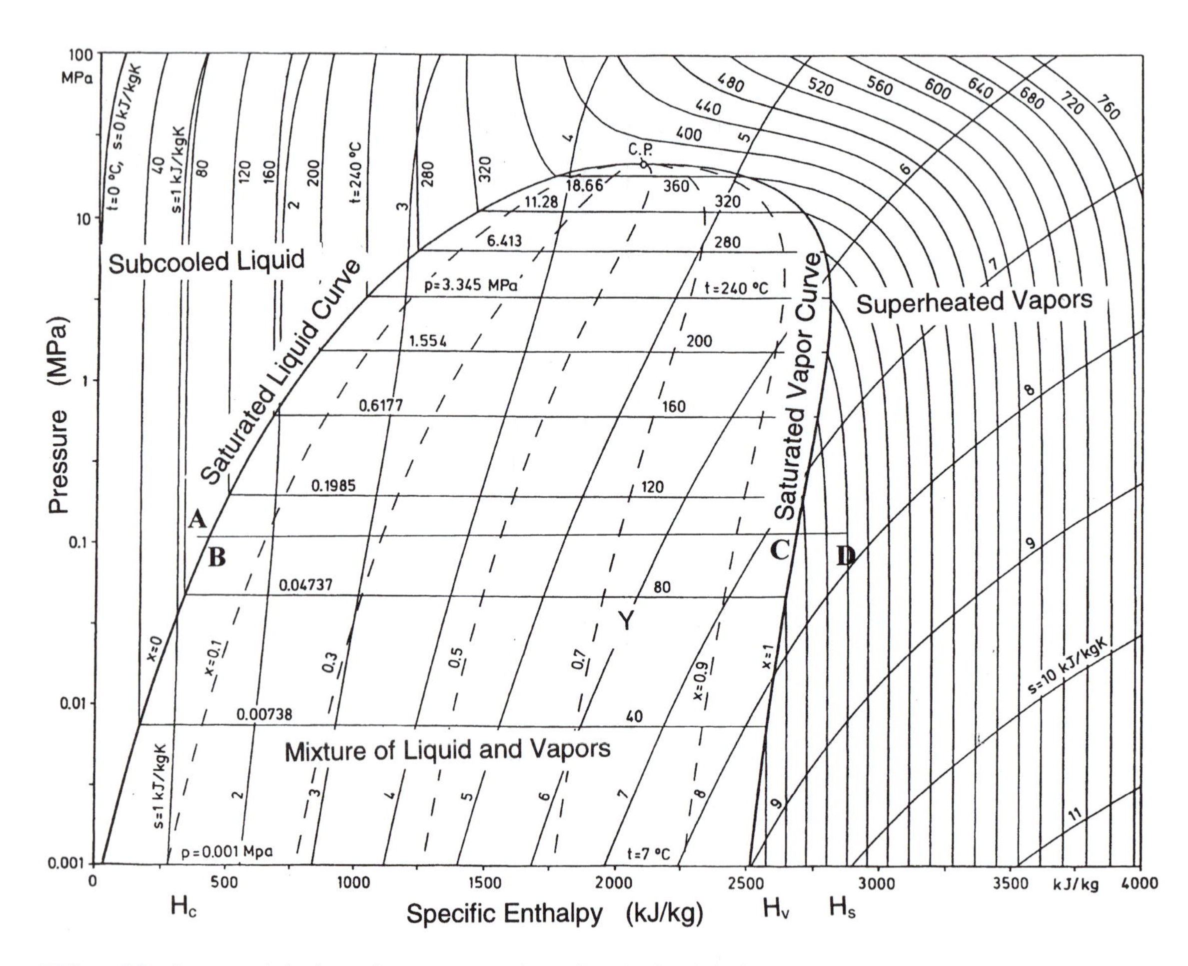

Figure 3.5 Pressure–enthalpy diagram for steam–water and vapor. (From Straub and Scheibner, 1984)

temperature increases to 100°C at point B on the saturated liquid curve. The enthalpy content of saturated water at point B is H_c (referring to enthalpy of condensate), which can be read off the chart as 420 kJ/kg. Further addition of thermal energy (in the form of latent heat) causes a phase change. As additional heat is added, more liquid water changes to vapor state. At point C, all the water has changed into vapors, thus producing saturated steam at 100°C. The enthalpy of saturated steam at point C is H_v (referring to enthalpy of saturated vapors) or 2675 kJ/kg. Further addition of thermal energy results in superheated steam at the same pressure but higher temperatures.

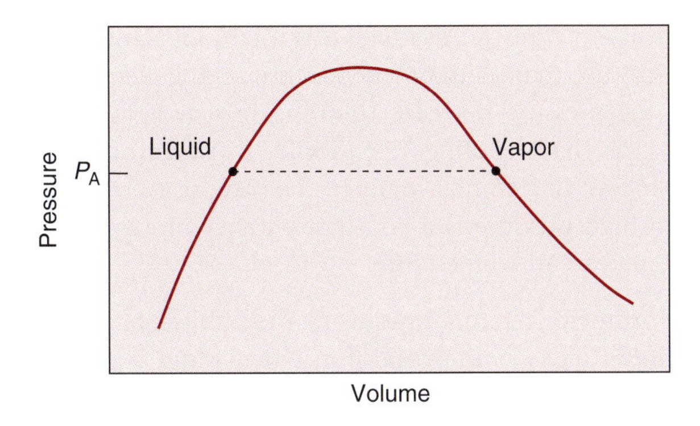

■ **Figure 3.6** Pressure–volume relationships for water liquid and vapor during phase change.

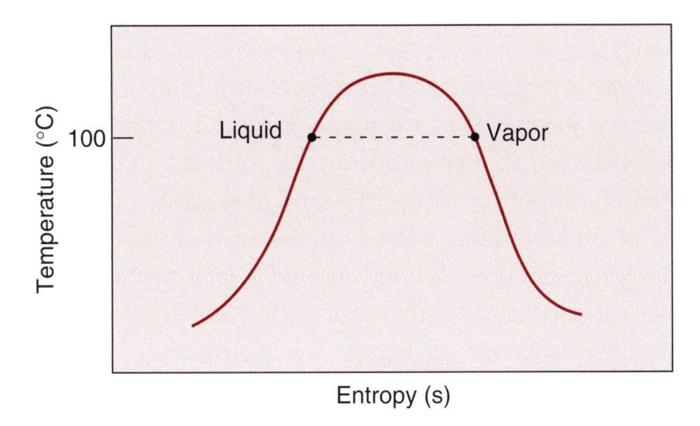

■ **Figure 3.7** Temperature–entropy relationships for water liquid and vapor during phase change.

Point D represents superheated steam at 200°C with an enthalpy content H_s (referring to superheated steam) of 2850 kJ/kg. Although Figure 3.5 provides a conceptual understanding of the steam generation processes, steam tables (to be described in the following section) give more accurate values.

By plotting the water phase-change process on pressure–volume coordinates, Figure 3.6 is obtained. This illustrates that a significant increase in volume occurs during the conversion of water from a liquid to vapor state. In practice, this conversion occurs within a constant-volume vessel, resulting in an increase in pressure as a result of the phase-change process. In a continuous steam generation process, the pressure and corresponding temperature of the steam to be used for processing operations are established by the magnitude of thermal energy added from the fuel source.

The third thermodynamic relationship would be on temperature–entropy coordinates, as illustrated in Figure 3.7. This relationship indicates that the phase change from liquid to vapor is accompanied by an increase in entropy. Although this thermodynamic property has less practical use than enthalpy, it has interesting characteristics. For example, the pressure decrease resulting in a temperature decrease (referred to as "flash cooling") is, ideally, an isoentropic or constant entropy process. In a similar manner, the compression of steam from a low to a high pressure is a constant entropy process with a corresponding increase in temperature.

There are numerous terms unique to the subject of steam generation. *Saturated liquid* is the condition when **water** is at equilibrium with its vapor. This condition exists at any pressure and corresponding temperature when the liquid is at the boiling point. *Saturated vapor* is **steam** at equilibrium with liquid water. Likewise, the condition exists at any pressure and temperature at the boiling point. *Superheated vapor* is steam at any pressure and temperature when the heat content is greater than saturated vapor. A continuous range of states exists between that of a saturated liquid and that of a saturated vapor, in which the proportions of liquid and vapor vary according to the degree of phase change transition. The extent to which the phase change has progressed is defined as *steam quality*. Normally, steam quality is expressed as a percentage indicating the heat content of the vapor–liquid mixture. In Figure 3.5, point Y indicates a mixture of liquid and vapor. The steam quality of the mixture represented by this point is 0.7 or 70%, meaning 70% of the mixture is vapor and the remaining 30% is in a liquid state. The enthalpy of steam with a steam quality less than 100% is expressed by the following equation:

$$H = H_c + x_s(H_v - H_c) \tag{3.1}$$

The preceding equation may be rearranged into the following alternative form:

$$H = (1 - x_s)H_c + x_s H_v \tag{3.2}$$

The specific volume of steam with a steam quality of x_s can be expressed by

$$V' = (1 - x_s)V_c' + x_s V_v' \tag{3.3}$$

3.1.3 Steam Tables

In the previous section, we saw the use of diagrams to obtain thermodynamic properties of steam. A more accurate procedure to obtain these values is by using tables (see Tables A.4.2 and A.4.3).

Table A.4.2 presents the properties of saturated steam. The properties include specific volume, enthalpy, and entropy, all presented as a function of temperature and pressure. Each property is described in terms of a magnitude for saturated liquid, an additional value for saturated vapor, and a value representing the difference between vapor and liquid. For example, the latent heat of vaporization, as given in Table A.4.2, is the difference between the enthalpy of saturated vapor and saturated liquid.

The properties of superheated steam are presented in Table A.4.3. The specific volume, enthalpy, and entropy are presented at several temperatures above saturation at each pressure. The property values represent the influence of temperature on the magnitude of specific volume, enthalpy, and entropy.

Another procedure to obtain thermodynamic properties of steam is with the use of mathematical equations. These mathematical equations are available in literature. When programmed into a computer, these equations allow determination of enthalpy values. A set of these empirical equations, suggested by Martin (1961) and Steltz and Silvestri (1958) are presented in Example 3.3. This example uses a spreadsheet for the determination of the thermodynamic properties.

Example 3.1

Determine the volume and enthalpy of steam at 120°C with 80% quality.

Given (from Table A.4.2)

Specific volume of liquid $(V'_c) = 0.0010603\ m^3/kg$
Specific volume of vapor $(V'_v) = 0.8919\ m^3/kg$
Enthalpy of liquid $(H_c) = 503.71\ kJ/kg$
Enthalpy of vapor $(H_v) = 2706.3\ kJ/kg$

Approach

The volume and enthalpy of the 80% quality steam can be determined from the saturation conditions using appropriate proper proportions based on steam quality expressed as a fraction.

Solution

1. *For enthalpy*

$$H = H_c + x_s(H_v - H_c) = (1 - x_s)H_c + x_s H_v$$

$$= 0.2(503.7) + 0.8(2706.3)$$

$$= 2265.78\ kJ/kg$$

2. *For specific volume*

$$V' = (1 - x_s)V'_c + x_s V'_v$$

$$= 0.2(0.0010603) + 0.8(0.8919)$$

$$= 0.7137\ m^3/kg$$

3. *Note that a small error results from ignoring the volume of saturated liquid.*

$$V' = x_s V'_v = 0.8(0.8919) = 0.7135\ m^3/kg$$

Example 3.2

Fluid milk is being heated from 60 to 115°C at a rate of 500 kg/h using steam as a heating medium. The heat exchanger being utilized has an efficiency of 85%, and the steam quality is 90%. The system is designed to allow condensate to be released at 115°C. The mass and volume flow rates of steam required for this process are to be determined.

Given

Product flow rate ($\dot{m}$) = 500 kg/h
Specific heat of milk (c_p) = 3.86 kJ/(kg °C) (Table A.2.1)
Initial product temperature (T_i) = 60°C
Final product temperature (T_o) = 115°C
Steam quality (x_s) = 90%
Steam temperature (T_s) = 120°C; selected to assure a minimum temperature gradient of 5°C between steam and product
For steam temperature of 120°C, pressure will be 198.55 kPa, and (from Table A.4.2):

$$H_c = 503.71\ kJ/kg \qquad V'_c = 0.0010603\ m^3/kg$$

$$H_v = 2706.3\ kJ/kg \qquad V'_v = 0.8919\ m^3/kg$$

Approach

The thermal energy requirements for the product will be used to establish the mass flow rate of steam required. The volumetric flow rate is computed from the mass flow rate and specific volume of steam.

Solution

1. *Thermal energy requirement*

$$q = \dot{m}c_p(T_o - T_i) = (500\ kg/h)(3.86\ kJ/kg\,°C)(115°C - 60°C)$$

$$= 106{,}150\ kJ/h$$

or for 85% heat exchanger efficiency

$$q = \frac{106{,}150}{0.85} = 124{,}882\ kJ/h$$

2. *For steam quality of 90%,*

$$H = (0.1)503.71 + (0.9)2706.3 = 2486.04 \text{ kJ/kg}$$

3. *The thermal energy content of condensate leaving the heat exchanger will be (the specific heat for water is obtained from Table A.4.1)*

$$H_c = (4.228 \text{ kJ/[kg °C]})(115°\text{C}) = 486 \text{ kJ/kg}$$

4. *Since the thermal energy provided by steam will be*

$$q_s = \dot{m}_s(H - H_c)$$

and this magnitude must match the steam requirements, then

$$\dot{m}_s = \frac{124{,}882 \text{ kJ/h}}{(2486.04 - 486) \text{ kJ/kg}} = 62.44 \text{ kg/h}$$

5. *For 90% quality,*

$$V' = (0.1)0.0010603 + (0.9)0.8919 = 0.8028 \text{ m}^3/\text{kg}$$

6. *Volumetric flow rate = (62.35 kg/h)(0.8028 m³/kg) = 50.05 m³/h*
7. *Steam generation system capacity*

$$\begin{aligned} 124{,}882 \text{ kJ/h} = 34{,}689 \text{ J/s} &= 34{,}689 \text{ W} \\ &= 34.7 \text{ kW} \\ &= 46.5 \text{ hp} \end{aligned}$$

Example 3.3

Develop a spreadsheet program for predicting enthalpy values of saturated and superheated steam.

Approach

We will use the numeric equations given by Martin (1961) and Steltz and Silvestri (1958) to program an Excel™ spreadsheet.

Solution

The spreadsheets with equations and a sample calculation for steam at a temperature of 120°C are as shown in Figures E3.1 and E3.2. The results are given in cells B45 through B48, as follows:

$$V_V = 0.89 \text{ m}^3/\text{kg}$$

$$H_c = 503.4 \text{ kJ/kg}$$

$$H_v = 2705.6 \text{ kJ/kg}$$

$$H_{evap} = 2202.2 \text{ kJ/kg}$$

	A	B
1	Temperature °C?	120
2		248
3		7.46908269
4		=−0.00750675994
5		−0.0000000046203229
6		−0.001215470111
7		0
8		=B2−705.398
9		=(EXP(8.0728362+B8*(B3+B4*B8+B5*B8^3+B7*B8^4)/(1+B6*B8)/(B2+459.688)))*6.89
10	Pressure kPa?	=B9
11		=B10*0.1450383
12	Temperature °C?	=B1
13		=B12*1.8+32
14		=(B13+459.688)/2.84378159
15		=0.0862139787*B14
16		=LN(B15)
17		=−B16/0.048615207
18		=0.73726439−0.0170952671*B17
19		=0.1286073*B11
20		=LN(B19)
21		=B20/9.07243502
22		=14.3582702+45.4653859*B21
23		=(B15)^2/0.79836127
24		=0.00372999654/B23
25		=186210.0562*B24
26		=EXP(B25+B20−B16+4.3342998)
27		=B26−B19
28		=B24*B27^2
29		=B28^2
30		=3464.3764/B15
31		=−1.279514846*B30
32		=B28*(B31+41.273)
33		=B29*(B15+0.5*B30)
34		=2*(B32+2*B33)
35		=B28*(B30*B28−B31)
36		=18.8131323+B22*B21
37		=B26+2*(B26*B25)
38		=B37*B34/B27+B34−B35−B37
39		−32.179105
40		1.0088084
41		−0.00011516996
42		0.00000048553836
43		−0.00000000073618778
44		9.6350315E−13
45	V_v	=(0.0302749643*(B34−B27+83.47150448*B15)/B19)*0.02832/0.45359
46	H_c	=(B39+B40*B13+B41*B2^2+B42*B2^3+B43*B2^4+B44*B2^5)*2.3258
47	H_v or H_s	=(835.417534−B17+B14+0.04355685*(B32+B23−B27+B38))*2.3258
48	H_{evap}	=B47−B46

■ **Figure E3.1** Spreadsheet for predicting enthalpy values of saturated and superheated steam in Example 3.3.

	A	B	C
1	Temperature °C?	120	
2		248	For calculations involving saturated vapors: Enter temperature in cell B1, =B9 in cell B10, and =B1 in cell B12
3		7.46908269	
4		−0.00750676	
5		−4.62032E−09	
6		−0.00121547	
7		0	
8		−457.398	
9		198.558129	For calculations involving superheated vapors: Enter pressure in cell B10 and temperature in cell B12
10	Pressure kPa?	198.558129	
11		28.79853348	
12	Temperature °C?	120	
13		248	
14		248.8545543	
15		21.45474124	
16		3.065945658	
17		−63.06556831	
18		1.815387125	
19		3.703701635	
20		1.309332761	
21		0.144319883	
22		20.91982938	
23		576.5634419	
24		6.46936E−06	
25		1.204659912	
26		43.91899067	
27		40.21528903	
28		0.010462699	
29		0.000109468	
30		161.4736976	
31		−206.6079934	
32		−1.729850209	
33		0.011186715	
34		−3.414953557	
35		2.179353383	
36		21.83227963	
37		149.7338856	
38		−168.0431145	
39		−32.179105	
40		1.0088084	
41		−0.00011516996	
42		4.8553836E−07	
43		−7.361878E−10	
44		9.6350315E−13	
45	V_v	0.89172	For superheated vapors, Hs is given by cell B47
46	H_c	503.41	
47	H_v or H_s	2705.61	
48	H_{evap}	2202.20	

Figure E3.2 Sample calculation spreadsheet for Example 3.3.

3.1.4 Steam Utilization

The capacity of the steam generation system in a food processing plant is established by requirements of the individual operations using steam. The requirements are expressed in two ways: (1) the temperature of steam needed as a heating medium, and (2) the quantity of steam required to supply the demands of the operation. Since the temperature requirement is a function of pressure, this establishes one of the operating conditions of the system. In addition, the steam properties are a function of pressure (and temperature), which in turn influences the quantity of steam utilized.

The steps involved in determining the capacity of a steam generation system include the following. The thermal energy requirements of all operations utilizing steam from a given system are determined. In most situations, those requirements will establish maximum temperature required and therefore the pressure at which the steam generation system must operate. After the operating pressure of the system is established, the properties of steam are known and the thermal energy available from each unit of steam can be determined. This information can then be used to compute quantities of steam required for the process. An important consideration in sizing the pipe connecting the process to the steam generation system is the volume of steam required. Using the quantity of steam required as expressed in mass units, and the specific volume of the steam being used, the volumetric flow rate for steam leading to the process is computed.

The use of steam by various processes in a food processing plant requires a transport system. The steam generation system is connected by a network of pipelines to the processes using steam. The transport system must account for two factors: (1) the resistance to flow of steam to the various locations, and (2) the loss of thermal energy or heat content during transport.

The transport of steam involves many of the considerations presented in Chapter 2. The flow of steam through a processing plant pipeline can be described by factors in the mechanical energy balance equation, Equation (2.81). In many situations, the steam generation system and the process using the steam will not be at the same elevation, and the third term on each side of the equation must be considered. Since the steam velocity within the steam generation system will be essentially zero, the kinetic energy term on the left side of the equation will be zero, at least in comparison to the same term on the right side of the equation. The pressure terms in Equation (2.81) are very important, since the left side represents the pressure at the steam generation system and the right side will be the pressure at the point of use. Since no work E_P is being done on the steam during transport, this term is zero; but the energy loss due to friction will be very important. In many situations, the energy loss due to friction can be translated directly into the loss of pressure between the steam generation system and the point of steam use.

Example 3.4

Steam is being transported from a steam generation system to a process at the rate of 1 kg/min through a 2-inch (nominal diameter) steel pipe. The distance is 20 m and there are five 90° standard elbows in the pipeline. If the steam is being generated at 143.27 kPa, compute the pressure of the steam at the point of use. The viscosity of steam is 10.335×10^{-6} Pa s.

Given

Steam flow rate $(\dot{m}_s) = 1$ *kg/min*
Steam pressure = *143.27 kPa*
Pipe diameter (D) = 2 inches (nominal) = 0.0525 m (Table 2.3)
Pipe length (L) = 20 m
Fittings include five 90° standard elbows
Steam viscosity $(\mu) = 10.335 \times 10^{-6}$ *Pa s*

Approach

By using the mechanical energy balance and computation of energy losses due to friction, the pressure losses for the 20 m length of pipe will be determined.

Solution

1. *To use Equation (2.51), the friction factor f must be determined from the Reynolds number and the relative roughness, for a steam density of 0.8263 kg/m³ obtained at 143.27 kPa.*

$$\bar{u} = \frac{(1\ \text{kg/min})(1/60\ \text{min/s})}{(0.8263\ \text{kg/m}^3)[\pi(0.0525\ \text{m})^2/4]} = 9.32\ \text{m/s}$$

and

$$N_{Re} = \frac{(0.8263\ \text{kg/m}^3)(0.0525\ \text{m})(9.32\ \text{m/s})}{(10.335 \times 10^{-6}\ \text{Pa s})} = 39{,}120$$

2. *For steel pipe (using Fig. 2.16),*

$$\frac{\varepsilon}{D} = \frac{45.7 \times 10^{-6}\ \text{m}}{0.0525\ \text{m}} = 0.00087$$

3. *The friction factor f is determined from Figure 2.16.*

$$f = 0.0061$$

4. *The energy loss due to friction is computed using Equation (2.51).*

$$\frac{\Delta P}{\rho} = 2(0.0061)\frac{(9.32\,\text{m/s})^2(20\,\text{m})}{(0.0525\,\text{m})} = 403.7\ \text{J/kg}$$

5. *Energy loss due to friction in five standard elbows: From Table 2.2, for standard elbow, $C_{ff} = 1.5$*

$$\frac{\Delta P}{\rho} = \frac{5 \times 1.5 \times (9.32)^2}{2}$$

$$\frac{\Delta P}{\rho} = 325.7\ \text{J/kg}$$

6. *Using the mechanical energy balance, Equation (2.81), without elevation and work terms and with velocity of zero at steam generation system,*

$$\frac{143,270\ \text{Pa}}{0.8263\ \text{kg/m}^3} = \frac{(9.32\,\text{m/s})^2}{2} + \frac{P_2}{\rho} + (403.7 + 325.7)$$

or

$$\frac{P_2}{\rho} = 173,387.4 - 43.4 - 729.4 = 172,614.6\ \text{J/kg}$$

7. *By assuming that steam density has not changed and noting that 1 Pa = 1 J/m³*

$$P_2 = (172,614.6\ \text{J/kg})(0.8263\ \text{kg/m}^3) = 142.63\ \text{kPa}$$

indicating that change in steam pressure due to friction losses during flow is relatively small.

Example 3.5

A liquid food with 12% total solids is being heated by steam injection using steam at a pressure of 232.1 kPa (Fig. E3.3). The product enters the heating system at 50°C at a rate of 100 kg/min and is being heated to 120°C. The product specific heat is a function of composition as follows:

$$c_p = c_{pw}(\text{mass fraction } H_2O) + c_{ps}(\text{mass fraction solid})$$

and the specific heat of product at 12% total solids is 3.936 kJ/(kg °C). Determine the quantity and minimum quality of steam to ensure that the product leaving the heating system has 10% total solids.

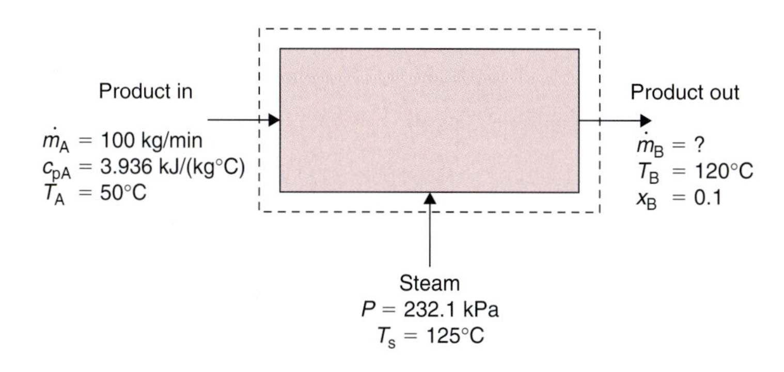

Figure E3.3 System diagram for Example 3.5.

Given

Product total solids in (X_A) = 0.12
Product mass flow rate ($\dot{m}_A$) = 100 kg/min
Product total solids out (X_B) = 0.1
Product temperature in (T_A) = 50°C
Product temperature out (T_B) = 120°C
Steam pressure = 232.1 kPa at (T_s) = 125°C
Product specific heat in (c_{PA}) = 3.936 kJ/(kg °C)

Approach

1. *Set up mass balance equations.*

$$\dot{m}_a + \dot{m}_S = \dot{m}_B$$

$$\dot{m}_A X_A = \dot{m}_B X_B$$

2. *Set up energy balance equation using reference temperature of 0°C.*

$$\dot{m}_A c_{PA}(T_A - 0) + \dot{m}_S H_S = \dot{m}_B c_{PB}(T_B - 0)$$

3. *By solving the mass balance equations for $\dot{m}_B$ and $\dot{m}_S$, the enthalpy of steam (H_S) required can be computed.*

Solution

1. *Mass and solid balance*

$$100 + \dot{m}_s = \dot{m}_B$$

$$100(0.12) + 0 = \dot{m}_B(0.1)$$

$$\dot{m}_B = \frac{12}{0.1} = 120\ \text{kg/min}$$

2. *Then*

$$\dot{m}_s = 120 - 100 = 20 \text{ kg/min}$$

3. *From energy balance*

$$(100)(3.936)(50 - 0) + (20)H_s = (120)c_{PB}(120 - 0)$$

where

$$c_{PB} = (4.232)(0.9) + c_{PS}(0.1)$$
$$3.936 = (4.178)(0.88) + c_{PS}(0.12)$$
$$c_{PS} = 2.161$$

then

$$c_{PB} = 4.025 \text{ kJ/(kg °C)}$$

4. *Solving for enthalpy* (H_s),

$$H_s = \frac{(120)(4.025)(120) - (100)(3.936)(50)}{20}$$

$$H_s = 1914.0 \text{ kJ/kg}$$

5. *From properties of saturated steam at 232.1 kPa,*

$$H_c = 524.99 \text{ kJ/kg}$$

$$H_v = 2713.5 \text{ kJ/kg}$$

then

$$\%\text{Quality} = \frac{1914 - 524.99}{2713.5 - 524.99}(100)$$
$$= 63.5\%$$

6. *Any steam quality above 63.5% will result in higher total solids in heated product.*

3.2 FUEL UTILIZATION

The energy requirements for food processing are met in a variety of ways. In general, the traditional energy sources are utilized to generate steam as well as to provide for other utilities used in the processing

Table 3.1 Energy Use by Fuel Type for 14 Leading Energy-Using Food and Kindred Products Industries for 1973

	Energy use by type of fuel (%)					
Industry	**Natural Gas**	**Purchased Electricity**	**Petroleum Products**	**Coal**	**Other**	**Total**
Meat packing	46	31	14	9	0	100
Prepared animal feeds	52	38	10	<1	0	100
Wet corn milling	43	14	7	36	0	100
Fluid milk	33	47	17	3	0	100
Beet sugar processing	65	1	5	25	4	100
Malt beverages	38	37	18	7	0	100
Bread and related products	34	28	38	0	0	100
Frozen fruits and vegetables	41	50	5	4	0	100
Soybean oil mills	47	28	9	16	0	100
Canned fruits and vegetables	66	16	15	3	0	100
Cane sugar refining	66	1	33	0	0	100
Sausage and other meat	46	38	15	1	0	100
Animal and marine fats and oils	65	17	17	1	0	100
Manufactured ice	12	85	3	0	0	100

Source: Unger (1975)

plant. As illustrated in Table 3.1, the energy types include natural gas, electricity, petroleum products, and coal. Although the information presented was collected in 1973 and percentages of natural gas utilization have declined somewhat, it seems evident that food processing has a definite dependence on petroleum products and natural gas.

To release the energy available from natural gas and petroleum products, they are exposed to a combustion process. This is a rapid chemical reaction involving fuel components and oxygen. The primary fuel components involved in the reaction include carbon, hydrogen, and sulfur, with the last being an undesirable component. The oxygen for the reaction is provided by air, which must be mixed with fuel in the most efficient manner.

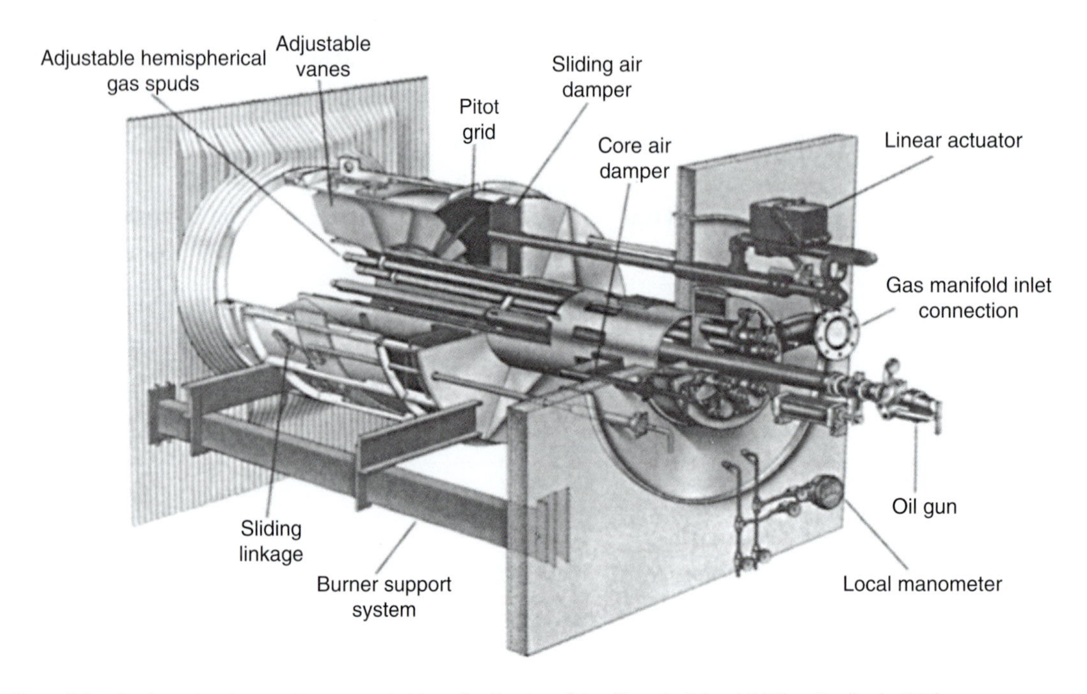

■ **Figure 3.8** Circular register burner with water-cooled throat for oil and gas firing. (From the Babcock & Wilcox Handbook, 1978)

3.2.1 Systems

The burner is the primary component of the system required for combustion of natural gas or petroleum products. Burners are used to produce the hot gases required in steam generation or the heated air for space heating in a building. Burners are designed to introduce fuel and air into the combustion chamber in a manner leading to maximally efficient generation of energy.

A typical burner is illustrated by Figure 3.8, a single circular register burner for both natural gas and oil. The orientation of doors in the air register provides turbulence needed to create mixing of fuel and air, as well as producing a desirable short flame. Burners are designed to minimize maintenance by keeping exposure of the burner to a minimum and allowing the replacement of vulnerable components while the unit continues to operate.

Safety is a definite concern in operating any system involving combustion. Ignition of a burner should occur at a location close to the burner, even at much higher air flows than required. Safety precautions should apply during starting and stopping of the system, as well as during load changes and variations in fuel.

3.2.2 Mass and Energy Balance Analysis

The combustion process can be described by equations involving the reaction between methane and oxygen as follows:

$$CH_4 + 2\,O_2 + 7.52\,N_2 = CO_2 + 2\,H_2O + 7.52\,N_2 \tag{3.4}$$

where 3.76 mol N_2 per mol O_2 are included in air for combustion and in the reaction processes. Actual fuel gas will contain 85.3% CH_4 (by volume), and the reaction will appear as follows:

$$\begin{aligned} 0.853\,CH_4 &+ 0.126\,C_2H_6 + 0.001\,CO_2 \\ &+ 0.017\,N_2 + 0.003\,O_2 + 2.147\,O_2 + 8.073\,N_2 \\ &= 1.106\,CO_2 + 2.084\,H_2O + 8.09\,N_2 \end{aligned} \tag{3.5}$$

where the theoretical balance has been established to indicate that 10.22 m^3 air would be needed for each cubic meter of fuel gas.

In actual combustion reactions, as much as 10% excess air will be provided and the reaction equation will appear as

$$\begin{aligned} 0.853\,CH_4 &+ 0.126\,C_2H_6 + 0.001\,CO_2 \\ &+ 0.017\,N_2 + 0.003\,O_2 + 2.362\,O_2 + 8.88\,N_2 \\ &= 1.106\,CO_2 + 0.218\,O_2 + 2.084\,H_2O + 8.897\,N_2 \end{aligned} \tag{3.6}$$

and indicate that the excess air produces excess oxygen and nitrogen in the flue gas from the combustion process. On a dry basis, the composition of the flue gas would be 87.1% nitrogen, 10.8% carbon dioxide, and 2.1% oxygen, where percentages are in volume basis.

The use of excess air is important to assure efficient combustion. Without sufficient oxygen, the reaction will be incomplete, resulting in the production of carbon monoxide (CO) along with associated safety hazards. In addition, the inefficient combustion has nearly 70% less heat released by the reaction. The amount of excess air must be controlled, however, since the air that is not involved in the reaction absorbs heat energy and decreases the amount of heat released by the combustion process.

The heat of combustion for a given reaction is dependent on the mixture of gases within the fuel. For the fuel previously described, the heat of combustion will be approximately 36,750 kJ/m^3. The losses with flue gas can be compared to this value, which would represent the maximum achievable from the process. The flue gas losses would be dependent on heat content of each component in the flue gas, and these values are a function of gas temperature, as indicated by Figure 3.9. Using this information, the energy losses associated with the previously described situation can be estimated (based on 1 m^3 fuel with 370°C flame gas).

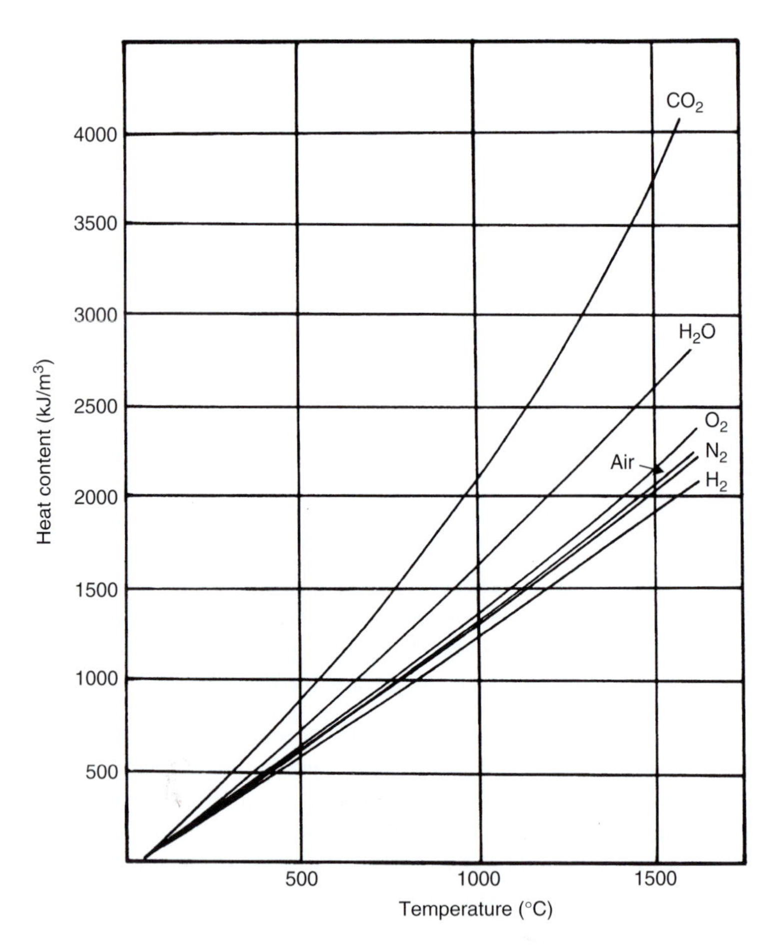

■ Figure 3.9 Heat content of gases found in flue products.

CO_2	$1.106\,m^3$	$\times$ 652 kJ/m^3	=	721.1 kJ
O_2	$0.2147\,m^3$	$\times$ 458 kJ/m^3	=	98.3 kJ
H_2O	$2.084\,m^3$	$\times$ 522 kJ/m^3	=	1087.9 kJ
N_2	$8.897\,m^3$	$\times$ 428 kJ/m^3	=	3807.9 kJ
				5715.2 kJ

This estimate indicates that flue gas energy losses are 5715.2 kJ/m^3 of fuel gas used. This represents 15.6% of the total energy available from the combustion process.

3.2.3 Burner Efficiencies

As indicated in Section 3.2.1, one of the primary purposes of the fuel burner is to assure optimum mixing of fuel and air. Without mixing, the combustion process will be incomplete and will produce the same results as insufficient oxygen. The burner is the key component of the combustion system in assuring that minimum amounts of excess air are required to maintain efficient combustion while minimizing the losses of energy in flue gas.

Example 3.6

Natural gas is being burned to produce thermal energy required to convert water to steam in a steam generator. The natural gas composition is 85.3% methane, 12.6% ethane, 0.1% carbon dioxide, 1.7% nitrogen, and 0.3% oxygen. An analysis of the flue gas indicates that the composition is 86.8% nitrogen, 10.5% carbon dioxide, and 2.7% oxygen. Determine the amount of excess air being utilized and the percentage of energy loss in the flue gas leaving at 315°C.

Given

Composition of natural gas

Composition of flue gas after combustion

All CO_2 in flue must originate with natural gas: 1.106 m^3 CO_2/m^3 fuel

Approach

The amount of excess air in the reaction can be evaluated by writing the reaction equation in a balanced manner and determining the extra oxygen in the reaction. The energy loss in the flue gas is based on thermal energy content in the flue gas, as determined from Figure 3.9.

Solution

1. *Based on the flue gas composition and the observation that the reaction must produce 1.106 m^3 CO_2/m^3 fuel,*

$$10.5\%\ CO_2 = 1.106\ m^3$$
$$2.7\%\ O_2 = 0.284\ m^3$$
$$86.8\%\ N_2 = 9.143\ m^3$$

2. *The reaction equation becomes*

$$\begin{aligned} &0.853\ CH_4 + 0.126\ C_2H_6 + 0.001\ CO_2 + 0.017\ N_2 \\ &\quad + 0.003\ O_2 + 2.428\ O_2 + 9.126\ N_2 \\ &= 1.106\ CO_2 + 0.284\ O_2 + 2.084\ H_2O + 9.143\ N_2 \end{aligned}$$

3. *Based on the analysis presented,*

$$\text{Excess air} = \frac{0.284}{2.428 - 0.284} \times 100 = 13.25\%$$

where the percentage of excess air is reflected in the amount of oxygen in the flue gas as compared with oxygen associated with the air involved in the reaction.

4. *Using the composition of flue gas and the heat content of various components from Figure 3.9, the following computations are obtained:*

CO_2	1.106 m^3	×	577.4 kJ/m^3	=	638.6 kJ
O_2	0.284 m^3	×	409.8 kJ/m^3	=	116.4 kJ
H_2O	2.084 m^3	×	465.7 kJ/m^3	=	970.5 kJ
N_2	9.143 m^3	×	372.5 kJ/m^3	=	3405.8 kJ
					5131.3 kJ

The analysis indicates that 5131.3 kJ are lost with flue gas per cubic meter of fuel used in the process.

5. *Using the heat combustion of 36,750 kJ/m^3 for the fuel, the loss of energy with flue gas represents 14% of the energy available from the fuel.*

3.3 ELECTRIC POWER UTILIZATION

Electric power has become so commonplace in the food industry that modern plants could not operate without this power source. In fact, most plants of significant size have acquired "back-up" electrical power generators to use in case disruptions occur in the primary supply. It is quite evident that electric power represents the most versatile and flexible power source available. In addition, the cost of electric power is very attractive when compared with other sources. In Figure 3.10,

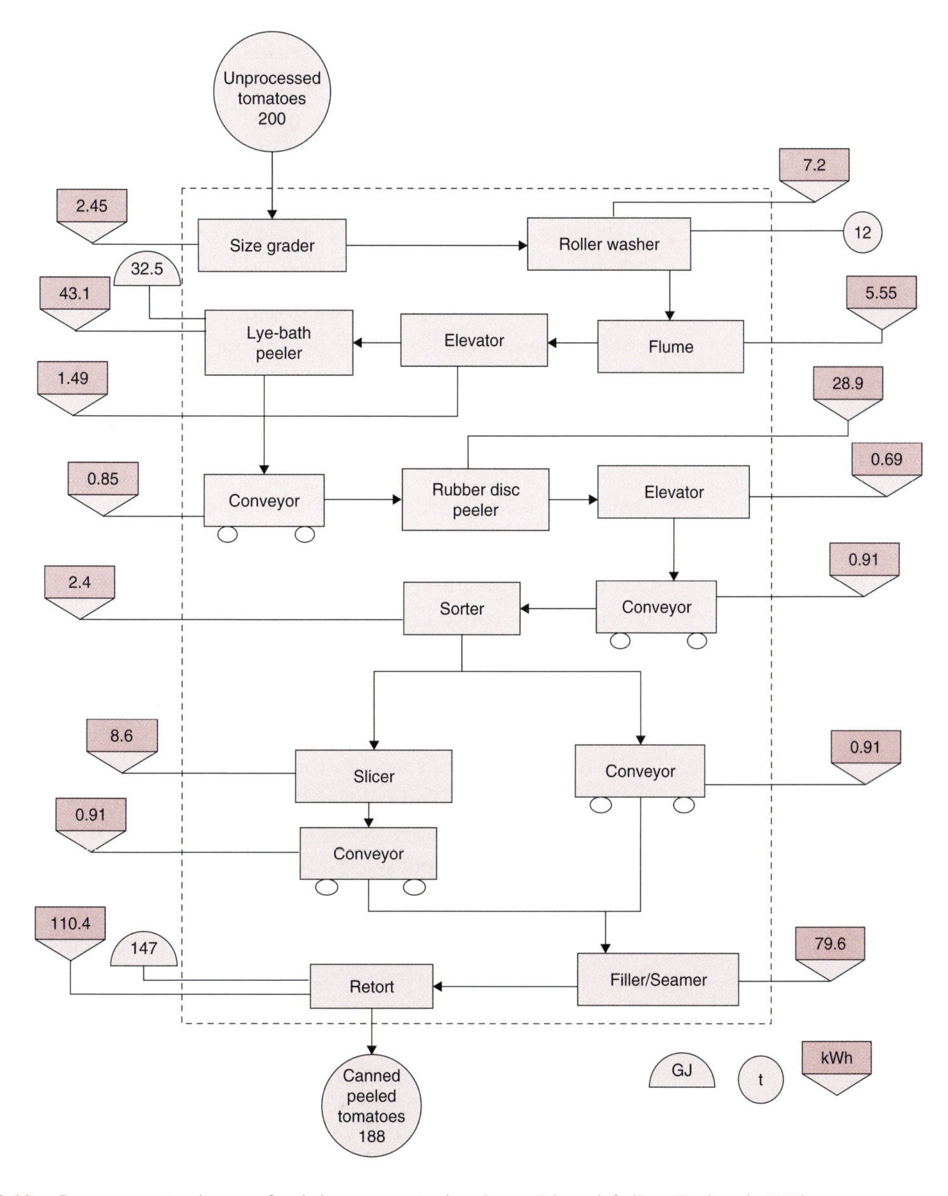

■ **Figure 3.10** Energy accounting diagram of peeled tomato canning based on an 8-hour shift. (From Singh et al., 1980)

a tomato processing line is shown along with energy requirements to operate each unit. As seen in this figure, most of the process equipment requires electrical energy for their operation.

3.3.1 Electrical Terms and Units

As in most physical systems, electricity has its own set of terms and units. These terms and units are entirely different from most physical systems, and it requires careful analysis to relate the terms to applications. This presentation is elementary and is intended to be a brief introduction to the subject. The following terms are essential.

Electricity can be defined as the flow of electrons from atom to atom in an electrical conductor. Most materials can be considered conductors, but will vary in the ability to conduct electricity.

Ampere is the unit used to describe the magnitude of electrical current flowing in a conductor. By definition, 1 ampere (A) is 6.06×10^{18} electrons flowing past a given point per second.

Voltage is defined as the force causing current flow in an electrical circuit. The unit of voltage is the volt (V).

Resistance is the term used to describe the degree to which a conductor resists current flow. The ohm (Ω) is the unit of electrical resistance.

Direct current is the type of electrical current flow in a simple electrical circuit. By convention, current is considered to flow from a positive to a negative terminal of a voltage generator.

Alternating current describes the type of voltage generated by an AC (alternating current) generator. Measurement of the actual voltage generated would indicate that the magnitude varies with time and a uniform frequency. The voltage ranges from positive to negative values of equal magnitudes. Most electrical service in the United States operates at 60 cycles per second (60 Hz).

Single-phase is the type of electrical current generated by a single set of windings in a generator designed to convert mechanical power to electrical voltage. The rotor in the generator is a magnet that produces magnetic lines as it rotates. These magnetic lines produce a voltage in the iron frame (stator) that holds the windings. The voltage produced becomes the source of alternating current.

Three-phase is the type of electrical current generated by a stator with three sets of windings. Since three AC voltages are generated simultaneously, the voltage can be relatively constant. This type of system has several advantages compared with single-phase electricity.

Watt is the unit used to express electrical power or the rate of work. In a direct current (DC) system, power is the product of voltage and current, whereas computation of power from an alternating current (AC) system requires use of a power factor.

Power factors are ratios of actual power to apparent power from an alternating current system. These factors should be as large as possible to ensure that excessive current is not carried through motors and conductors to achieve power ratings.

Conductors are materials used to transmit electrical energy from source to use. Ratings of conductors are on the basis of resistance to electrical flow.

3.3.2 Ohm's Law

The most basic relationship in electrical power use is Ohm's[1] law, expressed as

$$E_{\mathrm{v}} = IR_{\mathrm{E}} \tag{3.7}$$

where the voltage E_{v} is equal to the product of current I and resistance R_{E}. As might be expected, this relationship illustrates that for a given voltage, the current flow in a system will be inversely proportional to the resistance in the conductor.

As indicated earlier, the power generated is the product of voltage and current.

$$\text{Power} = E_{\mathrm{v}}I \tag{3.8}$$

or

$$\text{Power} = I^2R_{\mathrm{E}} \tag{3.9}$$

or

$$\text{Power} = \frac{E_{\mathrm{v}}^2}{R_{\mathrm{E}}} \tag{3.10}$$

These relationships can be applied directly to direct current (DC) systems and to alternating current (AC) with slight modifications.

[1]George Simon Ohm (1789–1854). A German physicist, in 1817 he was appointed as a professor of mathematics at the Jesuit College, Cologne. In 1827, he wrote the paper *Die galvanische Kette, mathamatisch bearbeitet* (The Galvanic Circuit Investigated Mathematically), but he remained unrecognized for his contribution. He resigned from his professorship to join the Polytechnic School in Nürnberg. Finally, in 1841, he was awarded the Copley medal by the Royal Society of London.

Example 3.7

A 12-volt battery is being used to operate a small DC motor with an internal resistance of 2 Ω. Compute the current flow in the system and the power required to operate the motor.

Given

Battery with voltage $E_v = 12\,V$
DC motor with resistance $R_E = 2\,\Omega$

Approach

The current flow in the motor can be computed using Equation (3.7), and power required can be determined from Equations (3.8), (3.9), or (3.10).

Solution

1. *Using Equation (3.7),*

$$I = \frac{E_v}{R_E} = \frac{12}{2} = 6\,A$$

indicating a current flow of 6 A in the system.

2. *The power required can be computed from Equation (3.10)*

$$\text{Power} = \frac{(12)^2}{2} = 72\,W$$

or 0.072 kW for the motor.

3.3.3 Electric Circuits

The manner in which conductors are used to connect the electric power source to the point of use is the electrical circuit. There are three basic types of circuits, with the series circuit being the simplest. As indicated by Figure 3.11, this type of circuit is recognized as having the resistances connected in series with the power source. In this type of situation, each resistance would probably represent the points at which the electrical power is used. Often, these points are referred to as electrical loads. Application of Ohm's law to this situation leads to

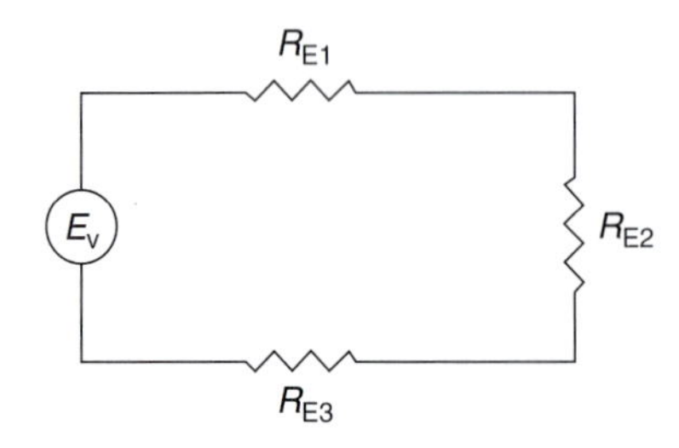

■ **Figure 3.11** Electrical circuit with resistance in series.

$$E_v = I(R_{E1} + R_{E2} + R_{E3}) \tag{3.11}$$

indicating that resistances in series are additive. In addition, the voltage is often expressed as the sum of the voltage drop across each resistance in the circuit.

A parallel electrical circuit has the resistance or loads connected in parallel with the power source, as illustrated in Figure 3.12. When Ohm's law is applied to the parallel circuit, the following relationship applies:

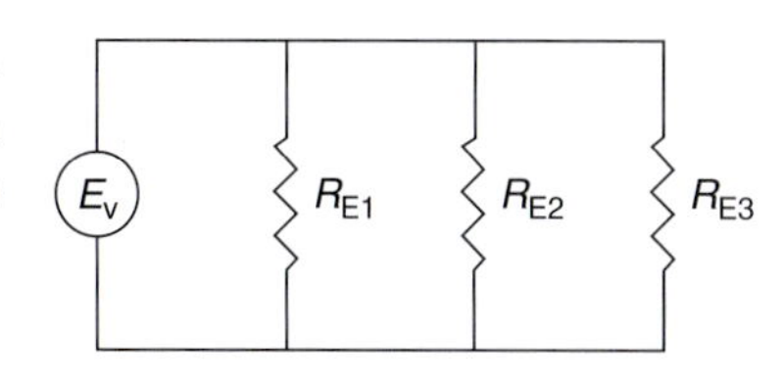

Figure 3.12 Electrical circuit with resistances in parallel.

$$E_v = I \Big/ \left(\frac{1}{R_{E1}} + \frac{1}{R_{E2}} + \frac{1}{R_{E3}} \right) \tag{3.12}$$

with the inverse of each resistance being additive. The most complex basic electrical circuit has a combination of series and parallel resistances, as illustrated in Figure 3.13. To analyze relationships between voltage and resistances, the combination circuit must be treated in two parts. First, the three resistances (R_{E1}, R_{E2}, R_{E3}) must be resolved as an equivalent R_e.

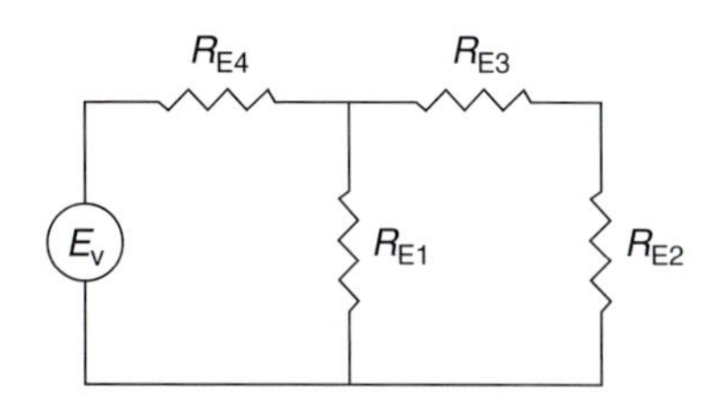

Figure 3.13 Electrical circuit with resistances in series and in parallel.

$$\frac{1}{R_e} = \frac{1}{R_{E1}} + \frac{1}{R_{E2} + R_{E3}} \tag{3.13}$$

Then the circuit can be analyzed by applying Ohm's law in the following manner:

$$E_v = I(R_{E4} + R_e) \tag{3.14}$$

since in the modified circuit, the resistance R_{E4} and R_e are in series.

Example 3.8

The four resistances in Figure 3.13 are $R_{E1} = 25\ \Omega$, $R_{E2} = 60\ \Omega$, $R_{E3} = 20\ \Omega$, $R_{E4} = 20\ \Omega$. Determine the voltage source E_v required to maintain a voltage drop of 45 V across the resistance R_{E2}.

Given

Four resistances as identified in Figure 3.13.
Voltage (E_{v2}) = 45 V

Approach

The voltage source E_v required can be evaluated by analysis of the circuit in terms of individual components and equivalent resistances.

Solution

1. *By using Ohm's law, the current flow through the resistance R_{E2} will be*

$$I_2 = \frac{45}{60} = 0.75\ A$$

2. *Since the current flow through R_{E3} must be the same as R_{E2}, then*

$$E_{v3} = (0.75)(20) = 15\,V$$

3. *Due to the circuit design, the voltage drop across R_{E1} must be the same as across R_{E2} plus R_{E3}; therefore,*

$$E_{v2} + E_{v3} = 45 + 15 = 60 = I_1(25)$$

$$I_1 = \frac{60}{25} = 2.4\,A$$

4. *The current flow through R_{E4} must be the total for the circuit, or*

$$I_4 = 0.75 + 2.4 = 3.15\,A$$

which is the current drawn from the voltage source E_v as well.

5. *The equivalent resistance for the circuit will be*

$$\frac{1}{R_e} = \frac{1}{25} + \frac{1}{60 + 20}$$
$$R_e = 19.05\,\Omega$$

3.3.4 Electric Motors

The basic component of an electric energy utilization system is the electric motor. This component converts electrical energy into mechanical energy to be used in operation of processing systems with moving parts.

The majority of the motors used in food processing operations operate with alternating current (AC), and their operation depends on three basic electrical principles. These principles include the electromagnet, formed by winding insulated wire around a soft iron core. Current flow through the wire produces a magnetic field in the iron core; orientation of the field is dependent on the direction of current flow.

The second electrical principle involved in the operation of a motor is electromagnetic induction. This phenomenon occurs when an electric current is induced in a circuit as it moves through a magnetic force field. The induced electric current produces a voltage within the circuit, with magnitude that is a function of the strength of the magnetic field, the speed at which the current moves through the field, and the number of conductor circuits in the magnetic field.

The third electrical principle is alternating current. As indicated earlier, this term refers to a current that changes direction of flow in a consistent manner. Normal electric service in the United States is 60 Hz, indicating that the change in current flow direction occurs 60 times per second.

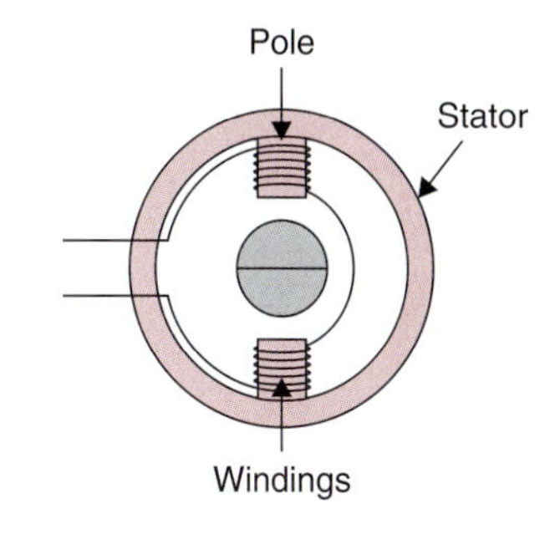

■ **Figure 3.14** Schematic diagram of a stator. (From Merkel, 1983)

An electric motor contains a stator: a housing that has two iron cores wound with insulated copper wire. The two cores or windings are located opposite one another, as illustrated in Figure 3.14, and the leads from the windings are connected to a 60 Hz alternating current source. With this arrangement, the stator becomes an electromagnet with reversing polarity as the current alternates.

A second component of an electric motor is the rotor: a rotating drum of iron with copper bars. The rotor is placed between the two poles or windings of the stator (Fig. 3.15). The current flow to the stator and the resulting electromagnetic field produces current flow within the copper bars of the rotor. The current flow within the rotor creates magnetic poles, which in turn react with the magnetic field of the stator to cause rotation of the rotor. Due to the 60 Hz alternating current to the stator, the rotation of the rotor should be 3600 revolutions per minute (rpm), but it typically operates at 3450 rpm.

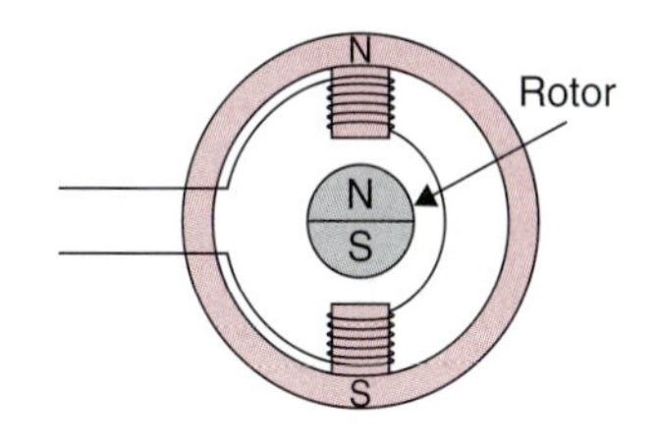

■ **Figure 3.15** Schematic diagram of a stator with rotor. (From Merkel, 1983)

Although there are numerous types of electric motors, they operate on these same basic principles. The most popular motor in the food processing plant is the single-phase, alternating current motor. There are different types of single-phase motors; the differences are related primarily to the starting of the motor.

The selection of the proper motor for a given application is of importance when ensuring that efficient conversion of electrical to mechanical energy occurs. The selection process takes into account the type of power supply available, as well as the use of the motor. The type and size of load must be considered, along with the environmental conditions of operation and the available space.

3.3.5 Electrical Controls

The efficient use of electrical energy and the equipment that is operated by this energy source is related to the opportunity for automatic control. Since the operation of processes and equipment in a food processing plant depends on responses to physical parameters, automatic control involves conversion of the physical parameter into an electrical response or signal. Fortunately, these conversions can be achieved rather easily using a variety of electrical transducers.

The control of electrical circuits is accomplished by using several different types of transducers. A magnetic relay utilizes a coil that produces electromagnetism to move a contact mechanically and complete the primary circuit. Thermostats and humidistats are controllers that use some physical change in temperature or humidity to provide the mechanical movement required to complete an electrical circuit. A timing device utilizes movement of the clock mechanism to mechanically bring two points into contact and complete an electrical circuit. Photoelectric controls use a photocell to produce a small current required to bring the points in the primary circuit into contact and allow for current flow. Time-delay relays, pressure switches, and limit switches are other types of controls used to accomplish the same type of electrical power utilization.

3.3.6 Electric Lighting

Another primary use of electric power in food processing plants is to provide illumination of work spaces. Often the work productivity of workers within the plant will be dependent on the availability of proper lighting. The design of a lighting system for a workspace will depend on several factors. The light must be distributed properly within the space, and the light source must be of sufficient size and efficiency. The light source must be supported properly and easily replaced or serviced. Finally, the cost of the entire system will be a factor to consider.

Light can be defined as visually evaluated radiant energy. Light is a small portion of the electromagnetic spectrum and varies in color depending on the wavelength. The intensity of a light at a point location is measured in the unit *lux*: the magnitude of illumination at a distance of one meter from a standard candle. A light source can be expressed in lumens: the amount of light on one square meter of surface when the intensity is one lux.

Two types of light sources are used in food processing plants: the incandescent lamp and the fluorescent lamp. The incandescent lamp uses a tungsten filament through which current flows. Due to the high electrical resistance of the filament, the flow of current through it causes it to glow white-hot. These types of lamps will provide efficiencies of approximately 20 lumens per watt.

A fluorescent lamp uses an inductance coil to create a current discharge within the tube. The heat from the discharge causes electrons to be removed from mercury vapor within the tube. The return of the electrons to the shell of mercury vapor causes emission of ultraviolet rays. These rays react with phosphor crystals at the tube surface to produce light. Fluorescent lamps are two to three times more efficient than comparable incandescent lamps. Although there are other factors to consider when comparing incandescent and fluorescent lamps, the efficiency and the longer life of fluorescent lamps are the most important.

One of the basic decisions related to lighting system design is determining the number of light sources required to maintain a desired level of illumination. An expression for illumination can be

$$\text{Illumination} = \frac{(\text{lumens/lamp}) \times \text{CU} \times \text{LLF}}{\text{area/lamp}} \tag{3.15}$$

where CU is the coefficient of utilization and LLF is the light loss factor.

The preceding equation indicates that the illumination maintained in a given space is a function of the magnitude of the light source and the number of lamps in the space. The coefficient of utilization CU accounts for various factors within the space, such as room size proportions, location of lamps, and workspace light. Light loss factors LLF account for room surface dust, lamp dust, and lamp lumen depreciation.

Example 3.9

A work area within a food processing plant is to be maintained at a light intensity of 800 lux. The room is 10 by 25 m, and 500-watt incandescent lamps (10,600 lumens/lamp) are to be utilized. A CU of 0.6 and LLF of 0.8 have been established. Determine the number of lamps required.

Given

Desired light intensity = 800 lux
Room size = 10 m by 25 m = 250 m^2
Lamps are 500 W, or 10,600 lumens/lamp
Coefficient of utilization CU = 0.6
Light loss factor LLF = 0.8

Approach

Equation (3.15) can be used to determine area/lamp, and the result is combined with the given room area to calculate the number of lamps required.

Solution

1. *Equation (3.15) can be used to compute the area per lamp allowed for the desired illumination.*

$$\text{Area/lamp} = \frac{10{,}600 \times 0.6 \times 0.8}{800} = 6.36\,\text{m}^2$$

2. *Based on the preceding,*

$$\text{Number of lamps} = \frac{10 \times 25}{6.36} = 39.3 \quad \text{or} \quad 40.$$

3.4 PROCESS CONTROLS IN FOOD PROCESSING

A typical food processing factory involves a number of unit operations that are carried out with different processing equipment: pumping, mixing, heating, cooling, freezing, drying, and packaging. Often the processing equipment operates in a continuous mode, which results in higher processing efficiencies than the batch mode. In designing a food processing plant, the processing equipment is arranged in a logical manner so that as raw food enters the plant, it is conveyed from one piece of equipment to the next while undergoing the desired processing.

Figure 3.16 is a flow diagram of typical operations used in the manufacture of canned tomatoes. The operations include unloading

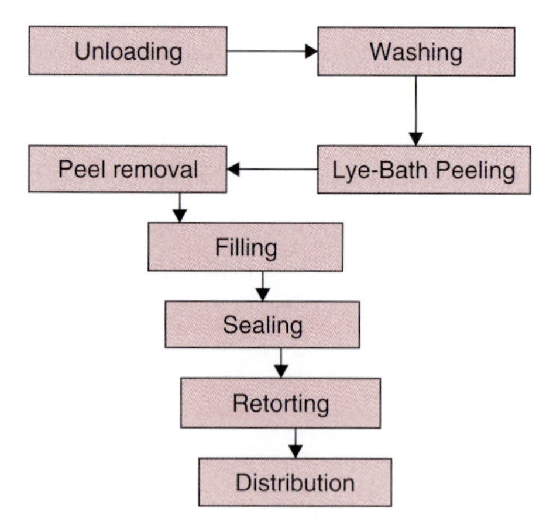

■ **Figure 3.16** Manufacturing steps in canning tomatoes.

tomatoes from the trucks, washing, cleaning, grading, peeling, filling into cans, and retorting. Most of the equipment used to carry out these processing operations is linked with different types of conveyors that allow the entire process to proceed in a continuous system. Some of the operations require human intervention, namely, inspecting incoming tomatoes to ensure that any undesirable foreign objects such as dirt clods and severely damaged fruit are removed. However, most of the processing equipment is operated without much human intervention with the use of automatic controllers and sensors.

The following criteria must be addressed in food manufacturing:

- Required production capacity
- Quality and hygiene of the end product
- Flexible manufacturing
- Optimum use of labor
- Economical operation that maximizes profit
- Compliance with the environmental regulations imposed by the local, national, or international laws
- Providing a safe working environment
- Meeting any special constraints imposed by the processing equipment.

These criteria require that food processing operations should be monitored continually and any deviations are promptly addressed.

In this chapter, we will examine some of the underlying principles of automatic control of processing equipment. The design and implementation of process controls requires advanced mathematics, which is beyond the scope of this book. However, you will be introduced to some of the common terminology and approaches used in designing these controls.

Let us consider a simple case where the task assigned to an operator is to pump wine into different tanks (Fig 3.17). For this task, the operator uses certain logic: choosing a precleaned empty tank, opening a valve to direct wine to that tank, knowing when it is full, and then directing wine to the next available tank, and so on. The operator keeps a check on the level of wine in the tank being filled to ensure that it does not overfill and cause loss of product. A similar logic may be programmed into an automatic control system to accomplish the desired task without significant human intervention.

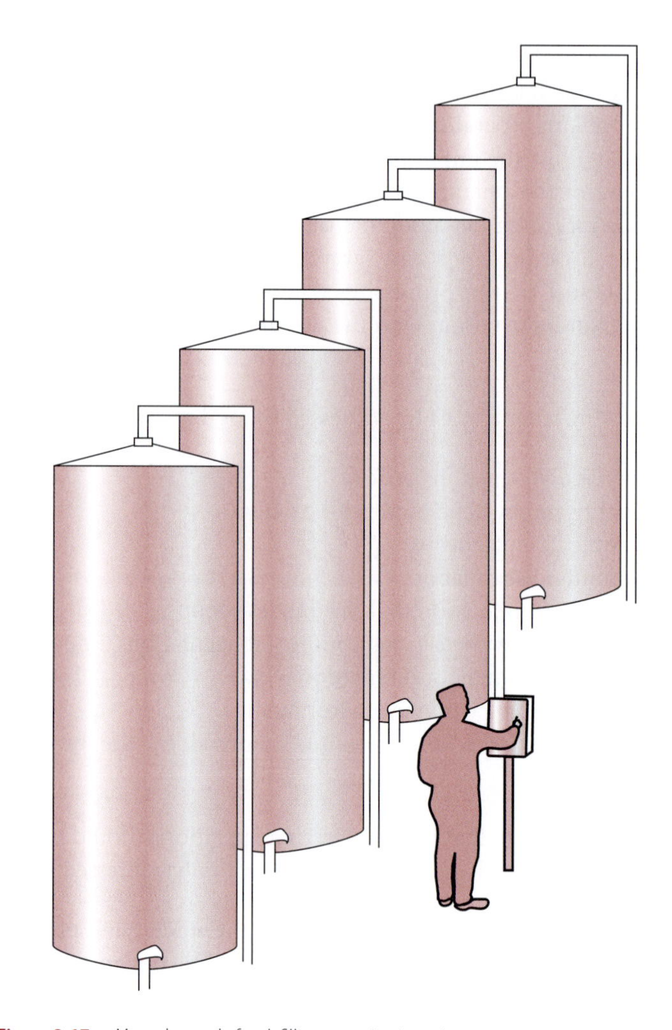

■ **Figure 3.17** Manual control of tank filling operation in a winery.

3.4.1 Processing Variables and Performance Indicators

In operating food processing equipment, an operator is often concerned with a variety of different process variables. For example, in a heating system, the temperature of a product may require careful monitoring. When milk is pasteurized using a heat exchanger, the temperature of the milk must reach 71°C and be held there for 16 seconds to destroy harmful pathogens. Therefore, the operator must ensure that the temperature reaches the desired value for the specified time or the milk will be either underprocessed (resulting in an unsafe product) or overprocessed with impairment to quality. Similarly, in operating different processing equipment, the flow rate, level, pressure, or weight may be an important variable that requires careful monitoring and control.

Controlled variables are simply those variables that can be controlled in a system. For example, the steam composition, steam flow rate, temperature of a water stream, and level of water in a tank are all variables that can be controlled. When heating milk, temperature is a controlled variable. Other examples of a controlled variable include pressure, density, moisture content, and measurable quality attributes such as color.

Uncontrolled variables are those variables that cannot be controlled when a process is carried out. For example, during operation of an extruder, the impact of operation on the extruder screw surface is an uncontrolled variable.

Manipulated variables are dependent variables that can be changed to bring about a desired outcome. For example, by changing the flow rate of steam to a tank of water, the water temperature will change. This variable may be manipulated by either a human operator or a control mechanism. When heating water in a chamber, the feed rate of water is a manipulated variable. A measured variable is used to alter the manipulated variable. Examples of measured variables are temperature, pH, or pressure, whereas the manipulated variable is the flow rate of a certain material or energy (such as electricity and steam).

Disturbances in the variables are those changes that are not caused by an operator or a control mechanism but result from some change outside the boundaries of the system. Disturbances cause undesirable changes in the output of a system. For example, the temperature of water in a tank is a controlled variable that may be influenced by the inlet flow rate, temperature of the inlet flow, and exit flow rate of water.

Robustness describes how tolerant the system is to changes in process parameters. When the robustness of a control system decreases, a small change in a process parameter makes the system unstable.

Performance communicates the effectiveness of the control system. There is a tradeoff between robustness and performance.

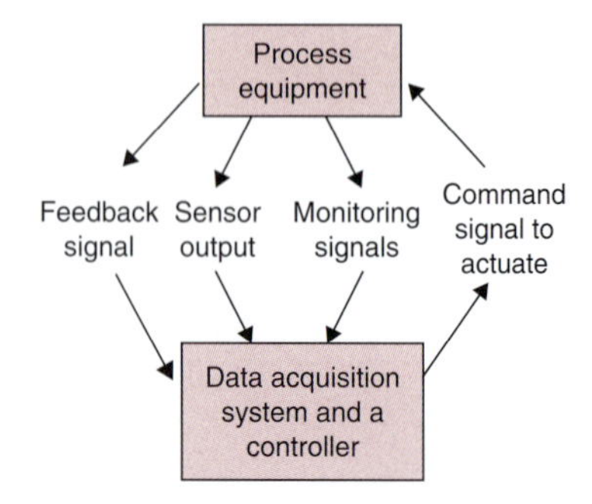

Figure 3.18 Communication between process equipment and data acquisition system/controller.

3.4.2 Input and Output Signals to Control Processes

A variety of signals are transmitted between the control system and processing equipment. These signals include:

- **Output signals**, which send a command to actuate components of a processing equipment such as valves and motors
- **Input signals** to the controller that
  a. Provide feedback from the processing components when a certain valve or motor has been actuated
  b. Measure selected process variables including temperature, flow, and pressure
  c. Monitor the processing equipment and detect completion of a certain process

Signals received by a controller are analyzed by following certain logic that is programmed in the control system, similar to the logic a human operator may follow in controlling a process. The overall aims of controlling a processing system are to minimize the effect of any external disturbances, operate the process under stable conditions, and achieve optimal performance. We will examine different strategies used in design of a control system in the following section.

3.4.3 Design of a Control System

3.4.3.1 *Control Strategy*

A control system may be designed to provide digital or analog control, or monitor tasks. For example, processing equipment may be set up under digital control so that it can be turned on or off from a control panel located at a remote location. Similarly, valves may be opened or closed, or the operation of multiple pieces of equipment may be carried out in some desired sequence.

An analog control is obtained via analog signals sent from the control unit. When combined with a feedback signal, analog controls are useful to operate valves that may be partially closed or opened. For example, the flow of steam or hot or cold water to processing equipment may be controlled.

Monitoring allows for checking critical aspects of the process for any major faults. Upon receiving a signal indicating a fault, the equipment can be shut down or the process stopped until the fault is corrected. Data acquisition is another feature of the automation system where the collected data can be used by the plant management to improve process efficiencies, or conduct scheduled maintenance, quality assurance, and cost analysis.

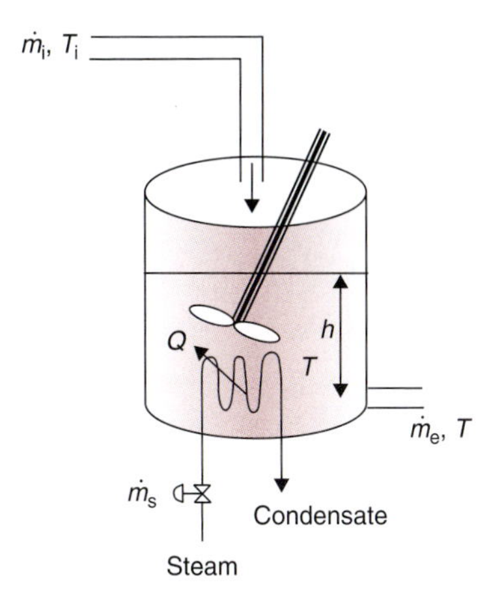

Figure 3.19 A tank of water heated with an indirect steam heat exchanger.

3.4.3.2 *Feed Backward Control System*

Consider a simple case of heating water in a tank. The tank is equipped with a steam coil and an agitator. The purpose of the agitator is to provide good mixing so that the temperature of water is uniform inside the tank—that is, it does not vary from one location to another inside the tank. Steam is conveyed to the coil by first passing it through a control valve. Steam condenses inside the coil and the heat of condensation, Q, is discharged into the water surrounding the coil while the condensate exits the coil. Water at a rate of $\dot{m}_i$ (kg/s) at a temperature of T_i is pumped into the tank and exits the tank at a rate of $\dot{m}_e$ (kg/s) with an exit temperature T (same as the temperature of water in the tank). The height of water in the tank is h. When operating this water heater, it is important to ensure that the volume of the water in the tank is maintained at some predetermined level; it should not overflow or run empty. Likewise, the temperature of the water exiting the tank must be maintained at some desired value.

Under steady state conditions, this water heating system should operate well if there are no changes in the inlet flow rate of water ($\dot{m}_i$) or its temperature (T_i). What if there is a change in either $\dot{m}_i$ or T? This will cause a disturbance in the process and will require intervention. If the process is supervised by an operator who is checking the temperature and notices a change (disturbance), the operator will try to change the steam flow rate by closing or opening the steam valve. This simple description of the process implies that we cannot leave this system to operate on its own. It requires some type of either manual supervision or an automatic control.

The objective of a control system is to determine and continuously update the valve position as the load condition changes. In a feedback control loop, the value of a controlled variable is measured and compared against a desired value usually called the **set** value. The difference between the **desired** and **set** value is called **controller error**. The output from the controller, which is a function of the controller error, is used to adjust the manipulated variable.

Next, let us consider a temperature control that may substitute for manual supervision.

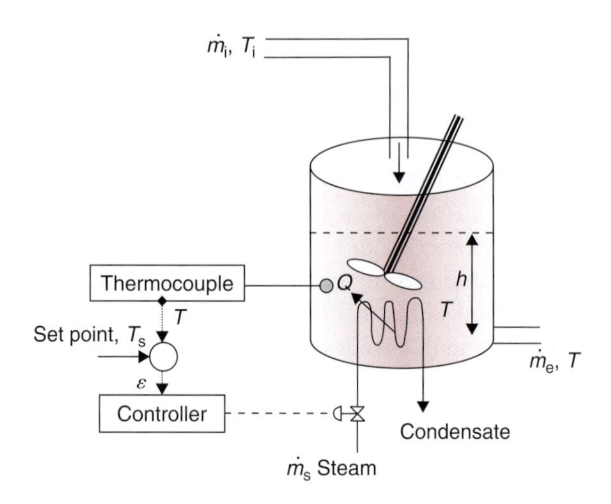

■ **Figure 3.20** A feedback control system used in a tank of water heated with an indirect steam heat exchanger.

As seen in Figure 3.20, a temperature sensor (thermocouple) is installed in the tank. The steam valve and the thermocouple are connected by a controller. The objective of this controller is to keep the temperature of the water constant at T_s (a set point temperature) whenever there is a change in the flow rate $\dot{m}_i$ or the inlet temperature T_i. In this arrangement, when the thermocouple senses a change in temperature, say ε, where

$$\varepsilon = T_s - T \tag{3.16}$$

the deviation ε is conveyed to the controller. If the deviation ε is greater than zero, meaning that the temperature of the water has fallen and should be increased, the controller sends a signal to the steam valve, opening it to allow steam to be conveyed to the coil. When the temperature of water reaches the desired value T_s, the controller shuts off the steam valve.

The control shown in Figure 3.20 is called a feedback control because the signal conveyed by the controller to the steam valve occurs after measuring the water temperature and comparing it with the set point temperature.

3.4.3.3 *Feedforward Control System*

Another type of control is a feedforward control, as shown in Figure 3.21. In this case the thermocouple is installed in the inlet feed pipe. When the temperature of the feed water (T_i) decreases below a set point (T_s), it implies that deviation ε will be greater than $T_s - T_i$,

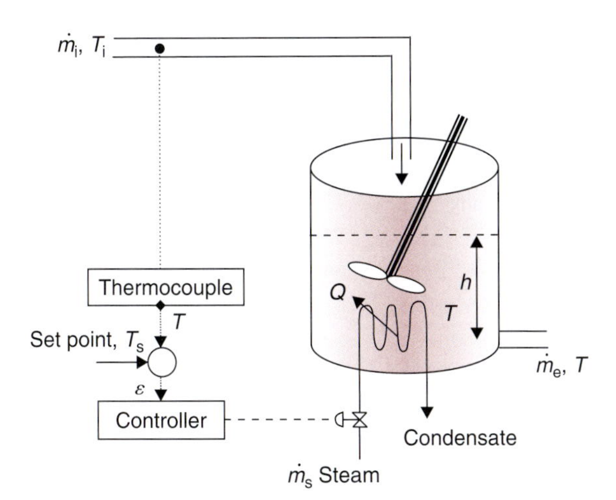

Figure 3.21 A feedback control system used in a tank of water heated with an indirect steam heat exchanger.

and it will cause the water temperature in the tank to decrease. In this case, the controller sends a signal to the steam valve, causing the valve to open and resulting in more steam flowing to the coil.

The difference between the feedforward and feedback controls should be clear by observing that in a feedforward control the controller anticipates that there will be a change of water temperature in the tank because of the change in the inlet water temperature, and a corrective action is taken prior to any observed change of water temperature in the tank.

Similar controls may be installed to monitor the height of water (h) in the tank and maintain it at a set point by controlling the feed rate of water into the tank.

3.4.3.4 *Stability and Modes of Control Functions*

Another important reason for using automatic controls is to maintain the stability of a system or process. In a stable system, a disturbance (i.e., change) in any variable should decrease with time, as shown in Figure 3.22a. For example, if the temperature goes above a certain set point and a system is able to bring it back to the initial value, an external means of correction is not required. In an unstable system (Fig. 3.22b), a disturbance in a variable would continue to increase to the point where the system is unable to bring the variable back to its original value on its own, requiring some external corrective action. Thus a controller is required to prevent instability in a system.

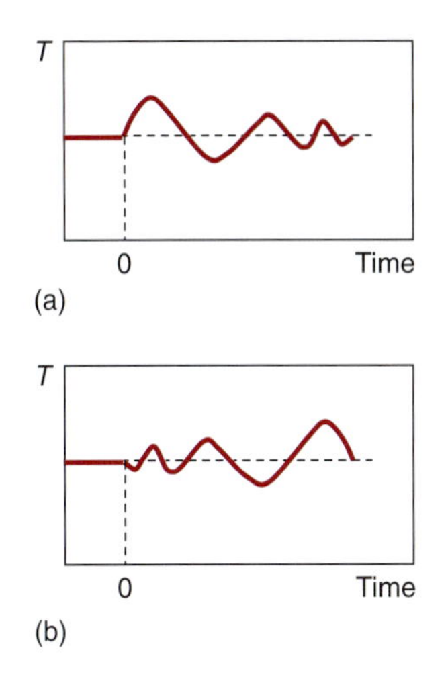

Figure 3.22 Response of a system: (a) stable, (b) unstable.

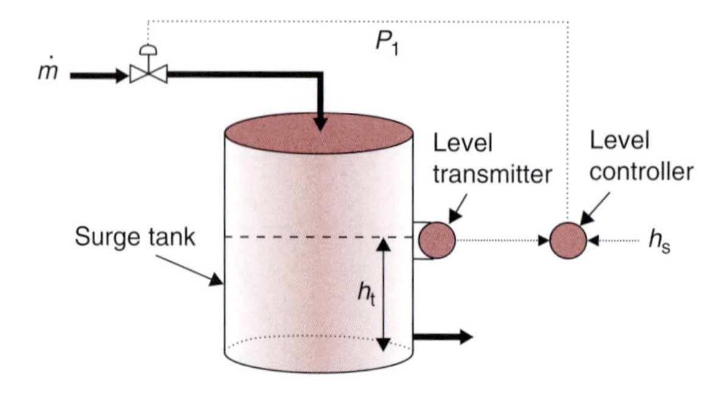

■ **Figure 3.23** Control of water level in a surge tank.

To understand the operating principles of different types of controllers, let us consider controlling the height of liquid in an intermediate storage tank. Intermediate storage tanks are often used to act as a buffer between two different processes. As shown in Figure 3.23, water is pumped through a control valve into the tank, and it exits at the bottom. To ensure that the tank does not overflow, a level sensor is installed to measure the height of the water in the tank, h_t. The sensor is connected to a level controller. The desired height in the tank, called the set point, is h_s. The difference between the desired height and measured height is the error. If the error is zero, then no control action is required. However, if an error is present, then the level controller sends a signal to a valve installed at the inlet pipe. A common signal to operate the valve is pneumatic pressure, P_1, required to operate the valve to open it. With this set up, we will now consider different types of controllers that can be used.

3.4.3.5 *On-Off Control*

An on-off control is similar to that used on a household thermostat for an air-conditioning system. The control is either maximum or zero flow. In our application, shown in Figure 3.23, the on-off controller sends a signal to the control valve when the error indicates that the level in the tank has decreased. The signal results in the application of pneumatic pressure P_1 on the control valve, and the valve opens. This is called a *fail-closed* valve since it remains closed until pressure is applied to open it. Such a valve will prevent water from flowing into the tank if the control system breaks down.

Many of the control valves used in the industry operate with signals between 3 and 15 psig. This means that a *fail-close* valve will be fully closed at 3 psig and fully open when a pressure of 15 psig is applied.

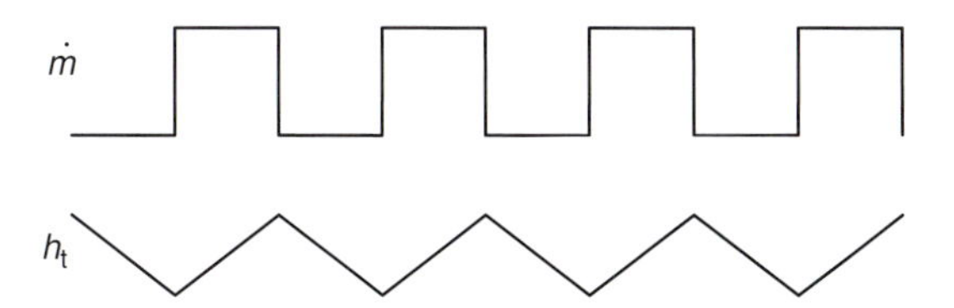

Figure 3.24 A cyclic response to on-off control signal.

An on-off controller is generally avoided for a continuous operation because it yields a cycling response (shown in Fig. 3.24), which is not desirable. There is also more wear on the control valves in this system, due to excessive operation.

3.4.3.6 *Proportional Controller*

In a proportional controller, the actuating output of a controller (c) is proportional to the input error (ε), the difference between the set point and the measured values. Thus,

$$c(t) = G_p \varepsilon(t) + c_s \qquad (3.17)$$

where G_p is the proportional gain of the controller and c_s is the controller's bias signal. The bias signal is the actuating signal when error is zero. Thus, in our example the bias signal is the steady state pressure applied on the valve when the error is zero.

The proportional gain (G_p), may be viewed as an amplification of the signal. In our example, a positive error, or when the set point is greater than the measured height, results in an increase in the flow rate into the tank. The flow rate will increase only if the pressure applied on the valve is increased. For a given amount of error, as the proportional gain is increased, it will result in more of the control action.

In some controllers, a proportional band is used instead of proportional gain. The proportional band represents the range of error that causes the output of the controller to change over its entire range. The proportional band is given by the following equation:

$$\text{Proportional Band} = \frac{100}{G_p} \qquad (3.18)$$

In a proportional controller, an offset between the set point and the actual output is created when the controller output and the process output arrive at a new equilibrium value even before the error diminishes to zero. To prevent the offset, a proportional integral controller is used.

3.4.3.7 *Proportional Integral (PI) Controller*

In a proportional integral (PI) controller, both error and the integral of the error are included in determining the actuating signal. Thus, for a PI controller, the actuating signal is

$$c(t) = G_p\varepsilon(t) + G_i\int_0^t \varepsilon(t)dt + c_s \tag{3.19}$$

The first term in the right-hand side of the equation is proportional to the error and the second term is proportional to the integral of the error. In a PI controller, an error signal will cause the controller output to change in a continuous manner, and the integral of the error will reduce the error to zero. In the integral action, the output of a controller is related to the time integral of the controller error, making the controller output dependent upon the size and duration of the error. Thus, with the integral function the past history of the controller response is accounted for in determining the actuating signal. For this reason, PI controllers are preferred in many applications.

3.4.3.8 *Proportional–Integral–Derivative Controller*

In a proportional–integral–derivative (PID) controller, the current rate of change or the derivative of the error is incorporated. The output of a PID controller is as follows:

$$c(t) = G_p\varepsilon(t) + G_i\int_0^t \varepsilon(t)dt + G_d\frac{d\varepsilon(t)}{dt} + c_s \tag{3.20}$$

A PID controller can be viewed as an anticipatory controller. The current rate of change of error is used to predict whether the future error is going to increase or decrease, and this information is then used to determine the actuating output. If there is a constant nonzero error, then the derivative control action is zero. PID controllers are not suitable when there is noise in the input signal, since a small nonzero error can result in an unnecessary large control action.

3.4.3.9 *Transmission Lines*

Transmission lines are used to carry the signal from the measurement sensor to the controller and from the controller to the control element. Transmission lines are electric or pneumatic using compressed air or liquid.

3.4.3.10 *Final Control Element*

The final control element carries out the implementing action. Upon receiving a signal, the control element will adjust. The most common control element used in the food industry is a pneumatic valve, as shown in Figure 3.25. This valve is operated with air and controls the flow by positioning a plug in an orifice. The plug is connected to a stem that is attached to a diaphragm. With a change in the air pressure due to the control signal sent by the controller, the stem moves and the plug restricts the flow through the orifice. This is the principle behind an air-to-close valve. If, due to some failure, the air supply is lost, the valve will fail open because the spring would push the stem and the plug upward. In addition, there are air-to-open with fail-closed valves. These valves will be fully open or closed as the air pressure on the side of the diaphragm changes from 3 to 15 psig (20 to 100 kPa).

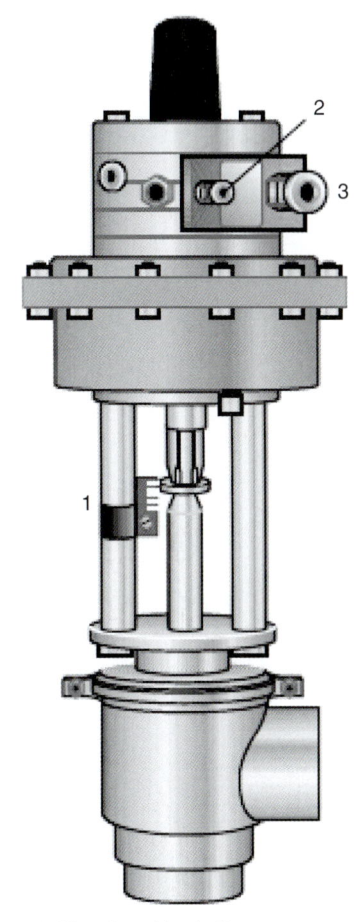

1. Visual position indicator
2. Connection for electrical signal
3. Connection for compressed air

Figure 3.25 A pneumatic valve. (Courtesy Alpha Laval)

An accurate measurement of controlled output is necessary for any feedback control system. The most common controlled variables in process controls are as follows: temperature, pressure, flow rate, composition, and liquid level. A large number of commercial sensors, based on numerous operating principles, are available to measure these variables. Some of the variables are measured directly, such as pressure. Temperature is measured indirectly; for example, in a thermocouple, temperature change is converted into voltage. The output signal of a sensor is often converted into another signal for transmission. For signal transmission the common standard systems are shown in Table 3.2.

The output of sensors may be changed to the same type of signal; for example, both temperature and pressure signals may be changed to a current of 4 to 20 mA carried on different pairs of wires. When a signal arrives at the processing unit, it may be necessary to amplify it to operate a solenoid valve or to change the position of a control valve.

During transmission, electrical signals may be corrupted by extraneous causes, such as by the operation of nearby large electrical equipment. External noise can be eliminated using appropriate filters. Low signal strength (e.g., mV signal) is more prone to extraneous noise

Table 3.2 Standard Systems Used for Signal Transmission

Pneumatic	Pneumatic (3–15 psig or 20–100 kPa)
Electric Current (direct current)	4–20 mA
Voltage (d.c.)	0–10 V
	0–5 V

than higher signal strength (V signal). Voltage signals are less robust than electric current signals, and digital signals are less prone to external electrical noise.

3.5 SENSORS

As we have learned in the preceding discussion, sensors play an important role in measuring process variables. Next, we will review the operating principles of some common sensors used in food processing.

3.5.1 Temperature

Temperature sensors used in the industry may be classified broadly as indicating and recording type. The common indicating sensors used in measuring temperature are a bimetallic strip thermometer and filled thermal systems. For recording and process control purposes, the most common types of temperature sensors are thermocouples, thermistors, and resistance temperature detectors. These three sensors provide an electrical signal based on the measurement.

A thermocouple is a simple device containing an electrical circuit made with two wires of dissimilar materials. The wires are joined at the ends to create two junctions, as shown in Figure 3.26a. If the two junctions are held at different temperatures, a difference in electric potential is created, resulting in a flow of current in the circuit (called the Seebeck effect). The potential difference (typically on the order of mV) measured in this circuit is a function of the temperature difference between the two junctions. In the circuit shown in Figure 3.26a, only temperature difference can be determined. To determine the unknown temperature we need to know the temperature of one of the junctions. This can be accomplished by keeping one of the junctions in ice, as shown in Figure 3.26b. Alternatively, an electronic ice junction

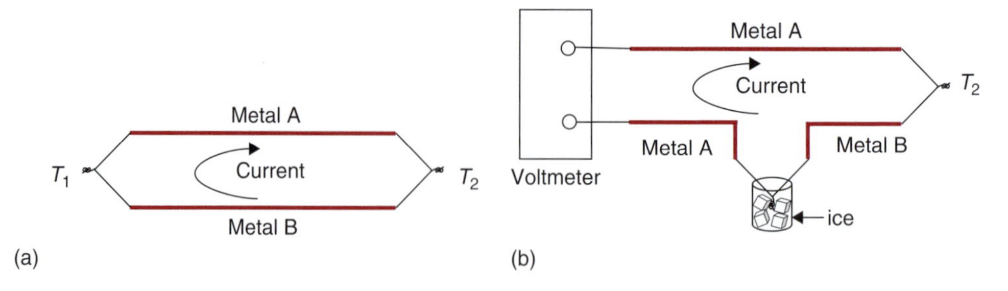

Figure 3.26 Thermocouple circuit: (a) Current is generated when temperature $T_1 \neq T_2$; (b) When one junction is placed in ice then the temperature of second junction can be measured.

is built into the circuit to determine the absolute temperature. Table 3.3 lists some common combinations of metals and alloys used in thermocouples.

A resistance temperature detector (RTD) provides high accuracy and is useful when a difference in temperatures is to be measured. Either a thin film or wire resistor is used. Standard resistance of 100, 500, or 1000 ohms is most common. Typically, RTDs are made of either platinum or nickel. Thus, a Pt100 refers to an RTD made of platinum with 100 ohm resistance. Since an RTD is a very thin film or wire, it is embedded in another material to make it suitable for mechanical handling. Typically ceramic is used for this purpose, as shown in Figure 3.27. Performance features of thermocouples and RTDs are presented in Table 3.4.

Table 3.3 Some Common Combinations of Metals and Alloys Used in Thermocouples

Thermocouple type	Composition of positive wire	Composition of negative wire	Suitability (Environment, limits of temperature)
J	Iron	Constantan	Oxidizing, reducing, inert, up to 700°C
K	90% Ni 10% Cr	95% Ni 5% Al	Oxidizing, inert, up to 1260°C
T	Copper	Constantan	Oxidizing, vacuum, reducing or inert, up to 370°C, suitable for −175°C.

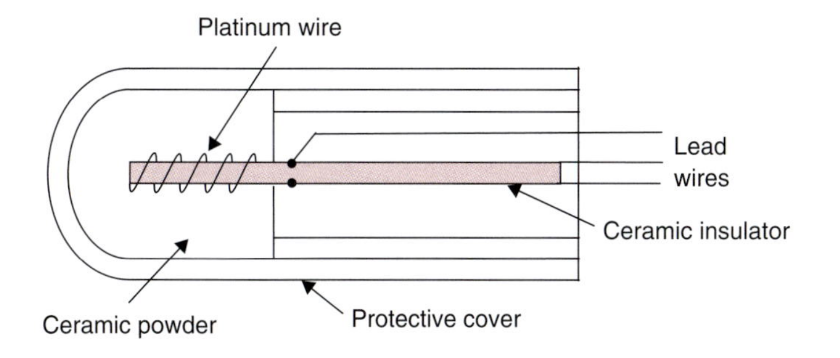

Figure 3.27 A resistance temperature device.

Table 3.4 Performance of Thermocouples and Resistance Temperature Detectors

Performance	Thermocouple	Resistance temperature detector (RTD)
Operating temperature range	Wide	Wide, −200 to +850°C
Time response	Fast (faster than RTD)	Slow
Functional relation with temperature	Nonlinear	Linear, well-defined relationship
Durability	Simple and rugged	Sensitive to vibration, shock, and mechanical handling
Sensitivity to electromagnetic interference	Sensitive to surrounding electromagnetic interference	
Reference	Reference junction is required	No
Cost	Inexpensive	Expensive

In process equipment such as tanks and pipes, temperature is measured by inserting the sensor in a thermowell, as shown in Figure 3.28. Thermowells isolate the temperature sensor from the surrounding environment. They are designed to prevent performance degradation of the temperature sensor. The length of the thermowell should be at least 15 times the diameter of the thermowell tip. The dynamic response of a temperature sensor when located in a thermowell can be quite long and therefore problematic in undermining the control performance.

3.5.2 Liquid Level in a Tank

The level of liquid in a tank may be measured using different methods. One method is to use a float that is lighter than the fluid in the tank, similar to the valve used in a bathroom toilet. In another system, the apparent weight of a cylinder is measured as it is buoyed up or down in a tank of liquid. Yet another method involves measurement of the differential pressure between two locations, one in

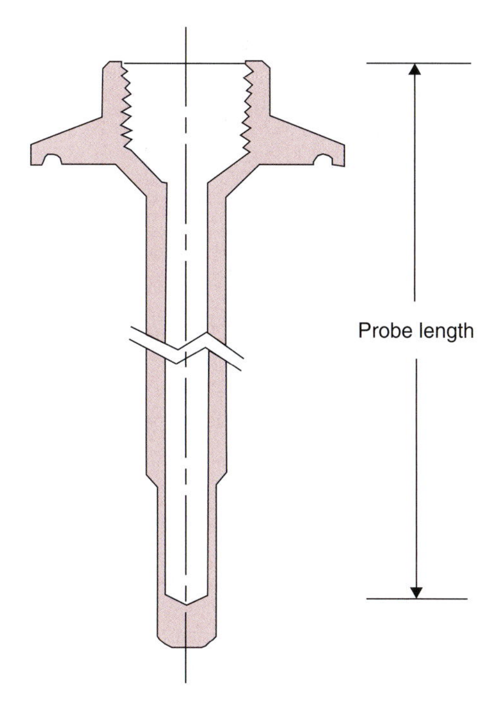

■ **Figure 3.28** A thermowell used to isolate a temperature sensor from the process environment.

the vapor and the other in the liquid. There are also ultrasonic level detectors that are used to measure liquid level in a tank.

3.5.3 Pressure Sensors

Pressure sensors are used to measure pressure in the equipment and also to measure pressure differences to determine the height of a liquid in a tank or flow rates of liquids and gases. In a variable capacitance differential pressure transducer, the pressure difference causes displacement of the diaphragm, which is sensed by capacitor plates located on both sides of the diaphragm. The differential capacitance between the sensing diaphragm and the surrounding capacitor plates is converted into DC voltage. For indicating purposes, manometers and a Bourdon gauge are commonly used. However, for recording pressures, a pressure sensor capable of generating an electrical signal is preferred. One type of pressure sensor with electrical signal output is shown in Figure 3.29. Inside this instrument, a flexible diaphragm is the pressure transmitting

■ **Figure 3.29** A differential pressure transmitter.

element. The pressure acting on the diaphragm causes a rod to push a flexible beam. Two strain gages are mounted on the beam. The deflection of the beam is sensed by the strain gages, and the reading is converted to a pressure reading.

3.5.4 Flow Sensors

Liquid and gas flow are commonly measured by passing the fluid through a constriction that causes a pressure drop. The pressure drop is measured and the flow rate is determined using a Bernoulli equation as previously described in Chapter 2. Typical constrictions used in flow measurement are an orifice plate and venturi tube. Another type of flow sensor is a turbine flow meter, where the number of turbine revolutions is determined to calculate the flow rate. For air flow, a vane-type anemometer is used to count the number of revolutions. For measuring air flow, hot-wire anemometers are commonly used.

In flow meters installed in-line, the material of the meter comes into direct contact with the food medium. Therefore, hygienic design and construction (that allow ease of cleaning) and minimization of any dead spaces are vital in food applications. If the flow measuring device can be cleaned *in situ*, then it must withstand the conditions employed during clean-in-place procedures. If the device cannot be cleaned *in situ*, then it must be easy to dismantle for cleaning purposes.

Typical flow meters used in the food industry include turbine, positive displacement, electromagnetic, and sensors based on the Coriolis principle.

In positive displacement meters, fixed volume chambers rotate around an axis. The liquid fills in the inlet chamber and causes the rotation. The rotation of these meters is linked to some type of mechanical counter or electromagnetic signal. The advantage of these meters is that they do not require an auxiliary source of energy as the moving fluid provides the movement of the sensor, and they can be installed without requiring any special length of upstream straight section. They operate well with a range of viscosities and are good at low flow rates. Typical fluids using positive displacement meters are edible oils and sugar syrups. Turbine meters are more suitable for low viscosity fluids such as milk, beer, and water. Their rotary motion is obtained by conversion of the free stream energy. A turbine is installed so that it forms a helix, and every revolution of helix is equivalent to the length of the screw. Turbine meters should be installed away from upstream fittings.

3.5.5 Glossary of Terms Important in Data Acquisition

Many of the following terms are encountered in experimental measurements and data analysis. Brief definitions are provided so that these terms are correctly used in reporting results from experimental measurements.

Accuracy: The difference between an indicated and an actual value.

Drift: A change in the reading of an instrument of a fixed variable over time.

Error Signal: The difference between the set point and the amplitude of measured value.

Hysteresis: The difference in readings obtained when an instrument approaches a signal from opposite directions.

Offset: A reading of instrument with zero input.

Precision: The limit within which a signal can be read.

Range: The lowest and highest value an instrument is designed to operate or that it can measure.

Repeatability: A measure of closeness of agreement between a number of readings taken consecutively of a variable.

Resolution: The smallest change in a variable to which the instrument will respond.

Sensitivity: A measure of change of output of an instrument for a change in the measured variable.

3.6 DYNAMIC RESPONSE CHARACTERISTICS OF SENSORS

As we learned in the previous sections, a variety of sensors are used in control and measurement of variables in food processing. For example, a thermocouple, thermometer, or a thermistor is used for measuring temperature. Similarly, other properties such as pressure, velocity, and density are frequently measured with appropriate sensors. In selecting a sensor, a key criterion is its dynamic response. We will briefly review the dynamic response characteristics of sensors.

The dynamic response characteristic of any sensor is measured by determining its time constant, which provides us with a measure of how fast or slow a given sensor responds to a change in input. For example, if we are measuring the temperature of a liquid food as it is being heated, we need to know how much time lag exists between the actual temperature of the liquid and the temperature indicated by a sensor placed in the liquid food. If the liquid temperature is changing rather slowly, then this time lag may not be a major problem. But, if the liquid is heating rapidly, we should select a sensor that can respond without much lag time. Quantitatively, we describe this lag period by determining the *time constant* of the sensor.

The aforementioned terms "rapid" and "slow" in instrument response are subjective and are not very useful in selecting sensors. What we need is an objective method to determine whether a selected sensor will be suitable for a given job. For this purpose, we determine the time constant of a temperature sensor. Furthermore, the same methodology may be used for other types of sensors, such as for measuring pressure, velocity, and mass.

The time constant is expressed in units of time, such as seconds. The time constant is a function of a given sensor.

To determine the time constant, note that the response of a given sensor to a sudden change in input is exponential. As shown in Figure 3.30, the temperature response of a thermometer follows an exponential plot when the surrounding temperature is suddenly changed from 0°C to 25°C. The time constant, τ, is calculated using the following equation that describes such an exponential plot:

$$T = T_u - (T_u - T_0)e^{-\frac{t}{\tau}} \tag{3.21}$$

where T is the sensor temperature (°C), T_u is the ambient temperature (°C), T_0 is the initial temperature (°C), t is time (s), and τ is time constant (s).

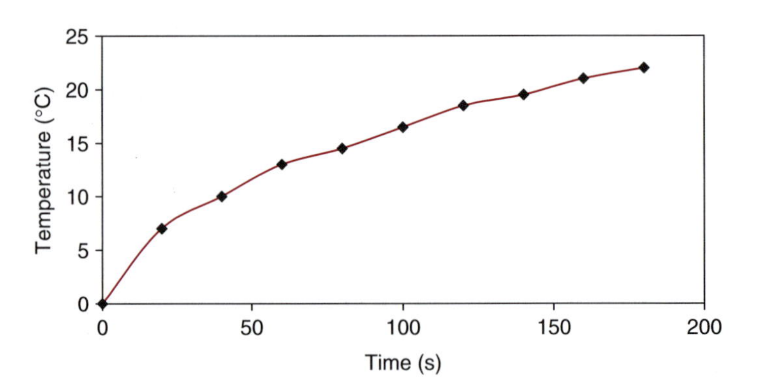

Figure 3.30 Exponential response of a temperature sensor.

Rearranging Equation (3.21)

$$\frac{T_u - T}{T_u - T_0} = e^{-\frac{t}{\mathrm{t}}} \tag{3.22}$$

Taking natural logarithm of both sides, we get

$$\ln \frac{T_u - T}{T_u - T_0} = -\frac{t}{\mathrm{t}} \tag{3.23}$$

Equation (3.23) is that of a straight line, $y = mx + c$ where y-ordinate is

$$\ln \frac{T_u - T}{T_u - T_0}$$

and x-axis is t, and the time constant, τ, is obtained from the inverse of the negative slope.

Subtracting both left- and right-hand sides of Equation (3.22) from 1, we obtain

$$1 - \frac{T_u - T}{T_u - T_0} = 1 - e^{-\frac{t}{\mathrm{t}}} \tag{3.24}$$

Rearranging,

$$\frac{T - T_0}{T_u - T_0} = 1 - e^{-\frac{t}{\tau}} \tag{3.25}$$

Therefore, at time equal to one time constant, that is, $t = \tau$,

$$\frac{T - T_0}{T_u - T_0} = 1 - e^{-\frac{\tau}{\tau}} = 1 - e^{-1} = 1 - 0.3679 = 0.6321$$

or

$$T = T_0 + 0.6321(T_u - T_0)$$

This implies that at a time equal to one time constant, the temperature of the sensor would have increased by 63.21% of the step temperature change, $T_u - T_0$. Example 3.10 illustrates how the time constant is obtained from experimental data.

Example 3.10

An experiment was conducted to measure the time constant of a bimetallic sensor. The sensor was initially equilibrated in ice water. The experiment began by removing the sensor from ice water. Any residual water from the sensor was quickly wiped and it was held in a room with an ambient temperature of 23°C. The following data on temperature vs time were obtained at 20 s intervals. Using these data, estimate the time constant.

Time (s)	Temperature (°C)
0	0
20	7
40	10
60	13
80	14.5
100	16.5
120	18.5
140	19.5
160	21
180	22

Given

Initial temperature, $T_0 = 0°C$

Ambient temperature, $T_u = 23°C$

Approach

We will use a spreadsheet to solve this problem. Equation (3.23) will be programmed and the slope of a straight line will give the time constant value.

Solution

1. *Program a spreadsheet as shown in Figure E3.4.*
2. *Create a plot between* $\ln\frac{(T_u - T)}{(T_u - T_0)}$ *and time.*
3. *Use trend line feature to calculate slope. As shown in the figure, the slope is*

$$\text{Slope} = -0.0149$$

$$\text{Time constant} = -\left(\frac{1}{\text{slope}}\right)$$

$$= 67.1\,\text{s}$$

4. *The time constant of the given bimetallic sensor is 67.11 s.*

	A	B	C	D	E	F
1	Initial Temperature, To	23				
2						
3	Time(s)	Temperature (C)	(Tu-T)	(Tu-T)/(Tu-To)	ln((Tu-T)/Tu-To))	
4						
5	0	0	23	1.0000	0.00000	
6	20	7	16	0.6957	-0.36291	
7	40	10	13	0.5652	-0.57054	
8	60	13	10	0.4348	-0.83291	
9	80	14.5	8.5	0.3696	-0.99543	
10	100	16.5	6.5	0.2826	-1.26369	
11	120	18.5	4.5	0.1957	-1.63142	
12	140	19.5	3.5	0.1522	-1.88273	
13	160	21	2	0.0870	-2.44235	
14	180	22	1	0.0435	-3.13549	

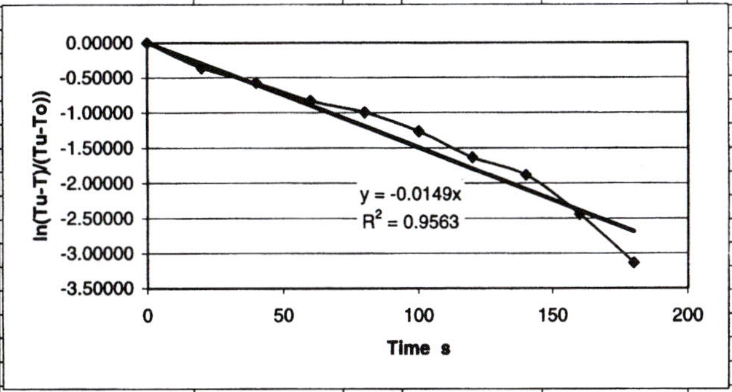

ln(Tu-T)/(Tu-To))
0.00000
-0.50000
-1.00000
-1.50000
-2.00000
-2.50000
-3.00000
-3.50000
0
50
100
150
200
Time s
y = -0.0149x
R^2 = 0.9563

■ **Figure E3.4** Spreadsheet for calculating time constant for data given in Example 3.10.

PROBLEMS

3.1 Compute the energy requirements to convert 50°C water to superheated steam at 170°C when the pressure is 210.82 kPa.

3.2 Determine the quality of steam at 169.06 kPa when 270 kJ/kg of energy are lost from saturated steam. What is the steam temperature?

3.3 Calculate the amount of energy (kJ/kg) required to convert saturated water at 150 kPa to superheated steam at 170°C and at the same pressure.

3.4 Determine the quality of steam at 143.27 kPa after a loss of 270 kJ/kg from the saturated steam conditions. What is the steam temperature?

3.5 A fruit juice is being heated in an indirect heat exchanger using steam as a heating medium. The product flows through the heat exchanger at a rate of 1500 kg/h and the inlet temperature is 20°C. Determine the quantity of steam required to heat the product to 100°C when only latent heat of vaporization (2200 kJ/kg) is used for heating. The specific heat of the product is 4 kJ/(kg °C).

3.6 A pudding mix is being formulated to achieve a total solids content of 20% in the final product. The initial product has a temperature of 60°C and is preheated to 90°C by direct steam injection using saturated steam at 105°C. If there is no additional gain or loss of moisture from the product, what is the total solids content of the initial product?

***3.7** Steam with 80% quality is being used to heat a 40% total solids tomato purée as it flows through a steam injection heater at a rate of 400 kg/h. The steam is generated at 169.06 kPa and is flowing to the heater at a rate of 50 kg/h. Assume that the heat exchanger efficiency is 85%. If the specific heat of the product is 3.2 kJ/(kg K), determine the temperature of the product leaving the heater when the initial temperature is 50°C. Determine the total solids content of the product after heating. Assume the specific heat of the purée is not influenced by the heating process.

3.8 Natural gas combustion is being used for steam generation and 5% excess air is incorporated into the combustion. Estimate the composition of the flue gas and compute the percent thermal energy loss if the flue gas temperature is 20 °C.

3.9 An electrical circuit includes a voltage source and two resistances (50 and 75 Ω) in parallel. Determine the voltage source required to provide 1.6 A of current flow through the 75 Ω resistance and compute current flow through the 50 Ω resistance.

***3.10** The manufacturing of pie filling involves blending of concentrated product with liquid sugar and heating by steam injection. The product being manufactured will contain 25% product solids and 15% sugar solids and will be heated to 115 °C. The process has input streams of concentrated product with 40% product solids at 40 °C and 10 kg/s and liquid sugar with 60% sugar solids at 50 °C. Heating is accomplished using steam at 198.53 kPa. The concentrated product entering the process and the final product have specific heats of 3.6 kJ/(kg °C), whereas the liquid sugar has a specific heat of 3.8 kJ/(kg °C). Determine (a) the rate of product manufacturing;

---

*Indicates an advanced level of difficulty in solving.

(b) the flow rate of liquid sugar into the process; (c) the steam requirements for the process; and (d) the quality of steam required for the process.

***3.11** Balan et al. (1991) present an empirical equation for the enthalpy of saturated steam vapor:

$$H_g = -484.836273 + 3.741550922T + 1.3426566x10^{-3}T^2 + 97.21546936(647.3 - T)^{1/2} - 1.435427715(1.0085)^T$$

The enthalpy is in (kJ/kg) and the temperature, T, is in degrees K.

Write a MATLAB® script to compare the results predicted by this equation to those for saturated steam from a set of steam tables (see Table A.4.2).

3.12 Irvine and Liley (1984) present equations for the estimation of vapor enthalpy, H_v, of saturated steam given the temperature in degrees K, T_{sat}.

$$Y = A + BT_C^{1/3} + CT_C^{5/6} + DT_C^{7/8} + \sum_{N=1}^{7} E(N)T_C^N$$

$$Y = \frac{H_v}{H_{v\,CR}} \qquad H_{v\,CR} = 2099.3 \text{ kJ/kg}$$

$$T_C = \frac{T_{CR} - T_{sat}}{T_{CR}} \qquad T_{CR} = 647.3 \text{ K}$$

	273.16 ≤ T ≤ 647.3 K
A	1.0
B	0.457874342
C	5.08441288
D	−1.48513244
E(1)	−4.81351884
E(2)	2.69411792
E(3)	−7.39064542
E(4)	10.4961689
E(5)	−5.46840036
E(6)	0.0
E(7)	0.0
$H_{v\,CR}$	2099.3

*Indicates an advanced level of difficulty in solving.

Write a MATLAB® script to evaluate the enthalpy and compare them to values given in saturated steam tables such as in Table A.4.2.

***3.13** Irvine and Liley (1984) present equations for the estimation of liquid enthalpy, H_c, of saturated steam given the temperature in degrees K, T_{sat}.

$$Y = A + BT_C^{1/3} + CT_C^{5/6} + DT_C^{7/8} + \sum_{N=1}^{7} E(N)T_C^N$$

$$Y = \frac{H_C}{H_{c\,CR}} \qquad H_{c\,CR} = 2099.3 \text{ kJ/kg}$$

$$T_C = \frac{T_{CR} - T_{sat}}{T_{CR}} \qquad T_{CR} = 647.3 \text{ K}$$

The parameter values (A, B, C, D, E(N)) vary with temperature.

	273.16 ≤ *T*, 300 K	**300 ≤ *T*, 600 K**	**600 ≤ *T* ≤ 647.3 K**
A	0.0	0.8839230108	1.0
B	0.0	0.0	−0.441057805
C	0.0	0.0	−5.52255517
D	0.0	0.0	6.43994847
E(1)	624.698837	−2.67172935	−1.64578795
E(2)	−2343.85369	6.22640035	−1.30574143
E(3)	−9508.12101	−13.1789573	0.0
E(4)	71628.7928	−1.91322436	0.0
E(5)	−163535.221	68.7937653	0.0
E(6)	166531.093	−124.819906	0.0
E(7)	−64785.4585	72.1435404	0.0
$H_{f\,CR}$	2099.3	2099.3	2099.3

Write a MATLAB® script to evaluate the enthalpy and compare results to values given in saturated steam tables such as in Table A.4.2.

LIST OF SYMBOLS

c	actuating output of a controller
c_p	specific heat (kJ/[kg K])
c_s	bias signal
CU	coefficient of utilization

*Indicates an advanced level of difficulty in solving.

D	diameter (m)
ΔP	pressure drop (Pa)
ε	difference between set point and measured value
ε	surface roughness factor (m) in Example 3.4
E_v	voltage (V)
f	friction factor
G_d	differential gain
G_i	integral gain
G_p	proportional gain
h	height, (m)
H	enthalpy (kJ/kg)
H_{evap}	latent heat of evaporation (kJ/kg)
I	current (A)
L	length (m)
L_e	equivalent length (m)
LLF	light loss factor
$\dot{m}$	mass flow rate
μ	viscosity (Pa s)
N_{Re}	Reynolds number, dimensionless
P	pressure (Pa)
q	heat transfer rate (kJ/s)
Q	heat energy, (kJ)
ρ	density
R_E	electrical resistance (Ω)
R_e	equivalent electrical resistance (Ω)
s	entropy (kJ/[kg K])
T	temperature (°C or K)
$\bar{u}$	average fluid velocity (m/s)
V'	specific volume (m^3/kg)
x_s	steam quality

Subscripts: c, liquid/condensate; e, exit; v, vapor; i, initial or inlet; o, outer; s, steam.

BIBLIOGRAPHY

Babcock & Wilcox Handbook (1978). *Steam—Its Generation and Use.* Babcock & Wilcox Co., New York, New York.

Balan, G. P., Hariarabaskaran, A. N., and Srinivasan, D. (1991). Empirical formulas calculate steam properties quickly. *Chemical Engineering* Jan, 139–140.

Farrall, A. W. (1979). *Food Engineering Systems,* Vol. 2. AVI Publ. Co., Westport, Connecticut.

Gustafson, R. J. (1980). *Fundamentals of Electricity for Agriculture*. AVI Publ. Co., Westport, Connecticut.

Irvine, T. F. and Liley, P. E. (1984). *Steam and Gas Tables with Computer Equations*. Academic Press, Inc., Orlando, Florida.

Martin, T. W. (1961). Improved computer oriented methods for calculation of steam properties. *J. Heat Transfer* **83**: 515–516.

Merkel, J. A. (1983). *Basic Engineering Principles*, 2nd ed. AVI Publ. Co., Westport, Connecticut.

Morgan, M. T. and Haley, T. A. (2007). Design of food process control systems. In *Handbook of Farm, Dairy, and Food Machinery*. M. Kurz, ed. 485-552, William Andrew Inc., Norwich, New York.

Singh, R. P., Carroad, P. A., Chinnan, M. S., Rose W. W., and Jacob, N. L. (1980). Energy accounting in canning tomato products. *J. Food Sci.* **45**: 735–739.

Steltz, W. G. and Silvestri, G. J. (1958). The formulation of steam properties for digital computer application. *Trans. ASME* **80**: 967–973.

Straub, U. G. and Scheibner, G. (1984). *Steam Tables in SI Units*, 2nd ed. Springer-Verlag, Berlin.

Teixeira, A. A. (1980). Cogeneration of electricity in food processing plants. *Agric. Eng.* **61**(1): 26–29.

Unger, S. G. (1975). Energy utilization in the leading energy-consuming food processing industries. *Food Technol.* **29**(12): 33–43.

Watson, E.L. and Harper, J.C. (1988). *Elements of Food Engineering*, 2nd ed. Van Nostrand Reinhold, New York, New York.

Webster, J. G. (1999). *The Measurement, Instrumentation, and Sensors Handbook*. CRC Press, Boca Raton, Florida.

Chapter 4

Heat Transfer in Food Processing

The most common processes found in a food processing plant involve heating and cooling of foods. In the modern industrialized food industry, we commonly find unit operations such as refrigeration, freezing, thermal sterilization, drying, and evaporation. These unit operations involve the transfer of heat between a product and some heating or cooling medium. Heating and cooling of food products is necessary to prevent microbial and enzymatic degradation. In addition, desired sensorial properties—color, flavor, texture—are imparted to foods when they are heated or cooled.

The study of heat transfer is important because it provides a basis for understanding how various food processes operate. In this chapter, we will study the fundamentals of heat transfer and learn how they are related to the design and operation of food processing equipment.

We will begin by studying heat-exchange equipment. We will observe that there is a wide variety of heat-exchange equipment available for food applications. This description will identify the need to study properties of foods that affect the design and operation of heat exchangers. Thereafter, we will examine various approaches to obtaining thermal properties of foods. We will consider basic modes of heat transfer such as conduction, convection, and radiation. Simple mathematical equations will be developed to allow prediction of heat transfer in solid as well as liquid foods. These mathematical equations will provide us with sufficient tools to design and evaluate the performance of simple heat exchangers. Next, we will consider more complicated situations arising from heat transfer under unsteady-state conditions, when temperature changes with time. A good understanding of the various concepts presented in this chapter is important, since they will be the basis for topics in the following chapters.

All icons in this chapter refer to the author's web site, which is independently owned and operated. Academic Press is not responsible for the content or operation of the author's web site. Please direct your web site comments and questions to the author: Professor R. Paul Singh, Department of Biological and Agricultural Engineering, University of California, Davis, CA 95616, USA. Email: rps@rpaulsingh.com.

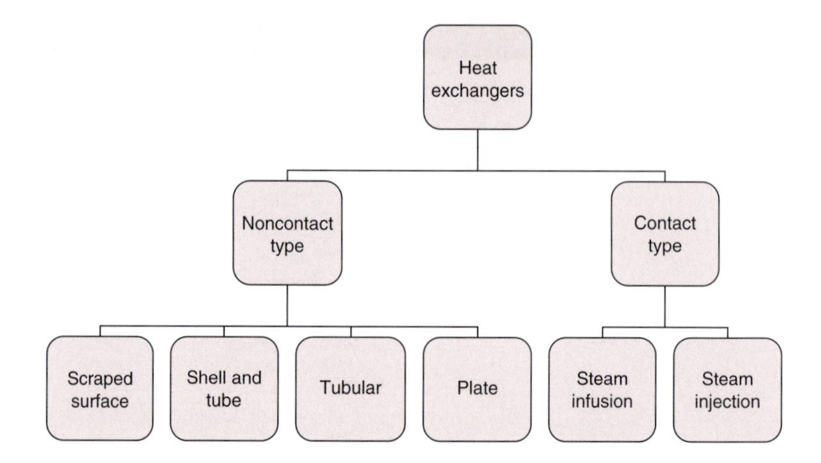

■ **Figure 4.1** Classification of commonly used heat exchangers.

4.1 SYSTEMS FOR HEATING AND COOLING FOOD PRODUCTS

In a food processing plant, heating and cooling of foods is conducted in equipment called heat exchangers. As shown in Figure 4.1, heat exchangers can be broadly classified into noncontact and contact types. As the name implies, in noncontact-type heat exchangers, the product and heating or cooling medium are kept physically separated, usually by a thin wall. On the other hand, in contact-type heat exchangers, there is direct physical contact between the product and the heating or cooling streams.

For example, in a steam-injection system, steam is directly injected into the product to be heated. In a plate heat exchanger, a thin metal plate separates the product stream from the heating or cooling stream while allowing heat transfer to take place without mixing. We will discuss some of the commonly used heat exchangers in the food industry in the following subsections.

4.1.1 Plate Heat Exchanger

The plate heat exchanger invented more than 70 years ago has found wide application in the dairy and food beverage industry. A schematic of a plate heat exchanger is shown in Figure 4.2. This heat exchanger consists of a series of parallel, closely spaced stainless-steel plates pressed in a frame. Gaskets, made of natural or synthetic rubber, seal the plate edges and ports to prevent intermixing of liquids. These gaskets help to direct the heating or cooling and the product streams

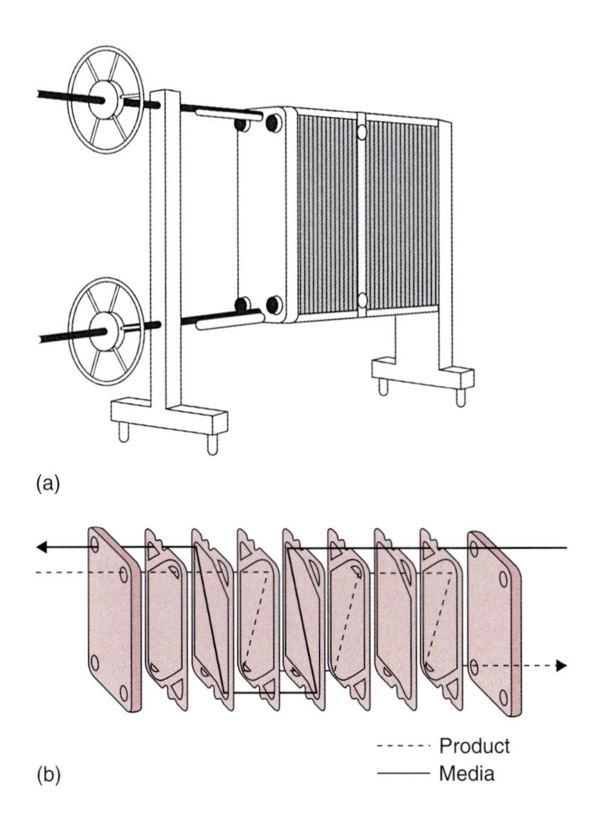

■ **Figure 4.2** (a) Plate heat exchanger. (b) Schematic view of fluid flow between plates. (Courtesy of Cherry-Burrell Corporation)

into the respective alternate gaps. The direction of the product stream versus the heating/cooling stream can be either parallel flow (same direction) or counterflow (opposite direction) to each other. We will discuss the influence of flow direction on the performance of the heat exchanger later in Section 4.4.7.

The plates used in the plate heat exchanger are constructed from stainless steel: Special patterns are pressed on the plates to cause increased turbulence in the product stream, thus achieving better heat transfer. An example of such a pattern is a shallow herringbone-ribbed design, as shown in Figure 4.3.

Plate heat exchangers are suitable for low-viscosity (<5 Pa s) liquid foods. If suspended solids are present, the equivalent diameter of the particulates should be less than 0.3 cm. Larger particulates can bridge across the plate contact points and "burn on" in the heating section.

In industrial-size plate heat exchangers, product flow rates from 5000 to 20,000 kg/h often are obtained. When using plate heat exchangers,

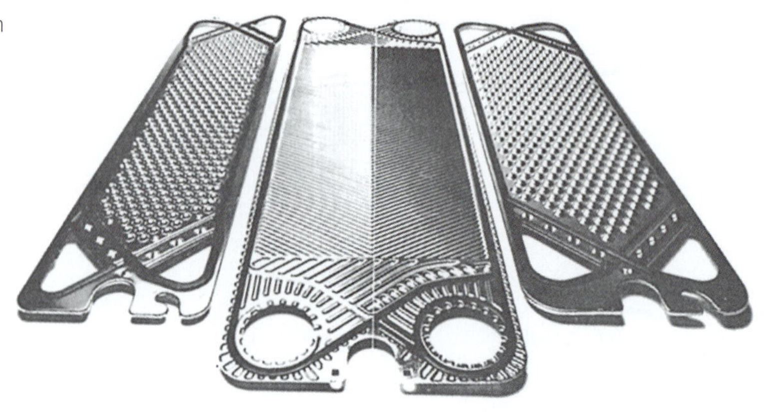

■ **Figure 4.3** Patterns pressed on plates used on a plate heat exchanger. (Courtesy of Cherry-Burrell Corporation)

care should be taken to minimize the deposition of solid food material such as milk proteins on the surface of the plates. This deposition, also called *fouling*, will decrease the heat transfer rate from the heating medium to the product; in addition, the pressure drop will increase over a period of time. Eventually, the process is stopped and the plates are cleaned. For dairy products, which require ultra-high-temperature applications, the process time is often limited to 3 to 4 h. Plate heat exchangers offer the following advantages:

- The maintenance of these heat exchangers is simple, and they can be easily and quickly dismantled for product surface inspection.
- The plate heat exchangers have a sanitary design for food applications.
- Their capacity can easily be increased by adding more plates to the frame.
- With plate heat exchangers, we can heat or cool product to within 1°C of the adjacent media temperature, with less capital investment than other noncontact-type heat exchangers.
- Plate heat exchangers offer opportunities for energy conservation by regeneration.

As shown in a simple schematic in Figure 4.4, a liquid food is heated to pasteurization or other desired temperature in the heating section; the heated fluid then surrenders part of its heat to the incoming raw fluid in the regeneration section. The cold stream is heated to a temperature where it requires little additional energy to bring it up to the desired temperature. For regeneration, additional plates are required; however, the additional capital cost may be recovered quickly by lowered operating costs.

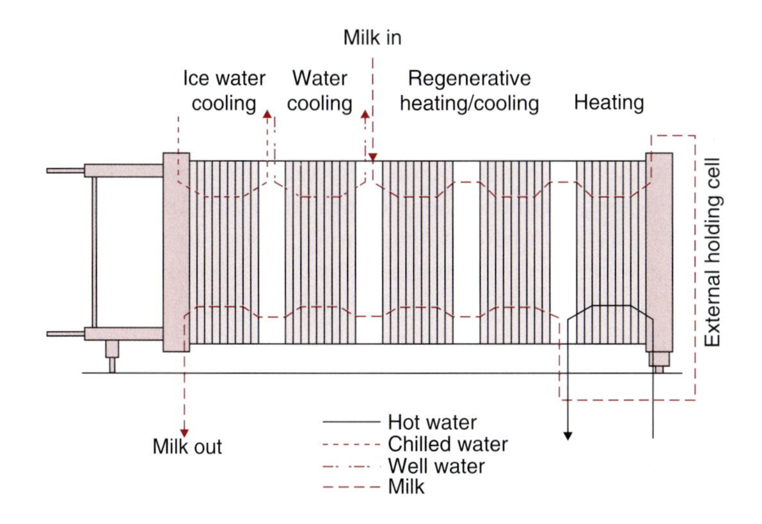

Figure 4.4 A five-stage plate pasteurizer for processing milk. (Reprinted with permission of Alfa-Laval AB, Tumba, Sweden, and Alfa-Laval, Inc., Fort Lee, New Jersey)

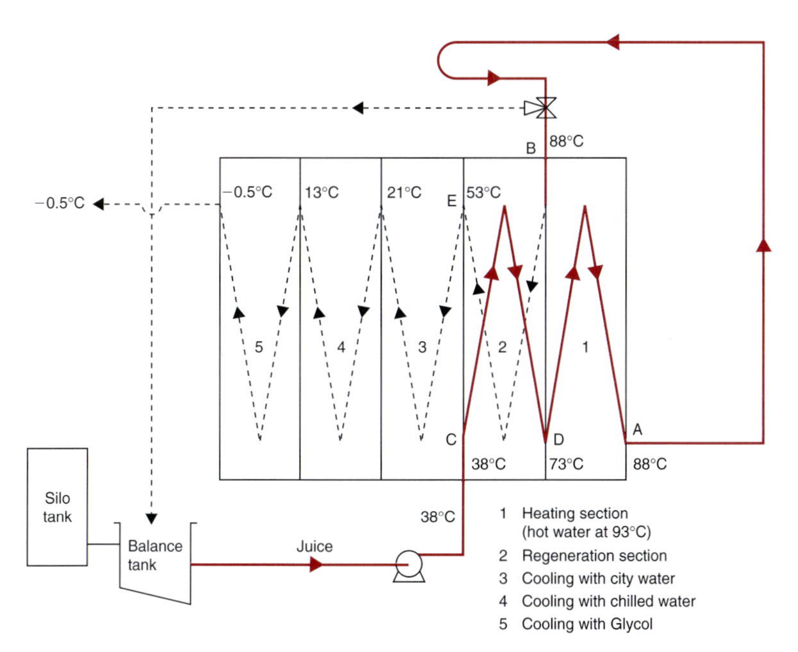

Figure 4.5 A two-way regeneration system used in processing grape juice. (Courtesy of APV Equipment, Inc.)

An actual two-way regeneration process is shown in Figure 4.5 for pasteurizing grape juice. After the "starter" juice has been heated to 88°C (at location A), it is passed through a holding loop and into the regenerative section (entering at location B). In this section, the juice releases its heat to incoming raw juice entering (at location C) into the exchanger at 38°C. The temperature of raw juice increases to 73°C (at location D), and the "starter" juice temperature decreases to 53°C (at location E). In this example, the regeneration is [(73 − 38)/(88 − 38)] × 100 or 70%, since the incoming raw juice was heated to 70% of its eventual pasteurization temperature without the use of an external heating medium. The juice heated to 73°C passes through the heating section, where its temperature is raised to 88°C by using 93°C hot water as the heating medium. The heated juice is then pumped to the regeneration section, where it preheats the incoming raw juice, and the cycle continues. The cooling of hot pasteurized juice is accomplished by using city water, chilled water, or recirculated glycol. It should be noted that, in this example, less heat needs to be removed from the pasteurized juice, thus decreasing the cooling load by the regeneration process.

4.1.2 Tubular Heat Exchanger

The simplest noncontact-type heat exchanger is a double-pipe heat exchanger, consisting of a pipe located concentrically inside another pipe. The two fluid streams flow in the annular space and in the inner pipe, respectively.

The streams may flow in the same direction (parallel flow) or in the opposite direction (counterflow). Figure 4.6 is a schematic diagram of a counterflow double-pipe heat exchanger.

A slight variation of a double-pipe heat exchanger is a triple-tube heat exchanger, shown in Figure 4.7. In this type of heat exchanger, product flows in the inner annular space, whereas the heating/cooling medium flows in the inner tube and outer annular space. The innermost tube may contain specially designed obstructions to create

W

Figure 4.6 Schematic illustration of a tubular heat exchanger.

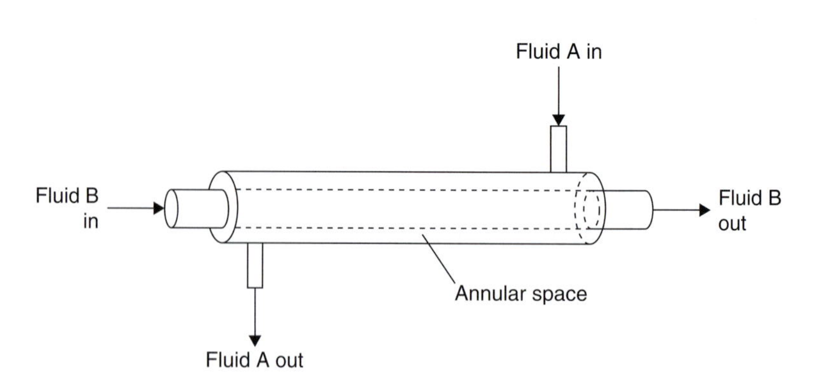

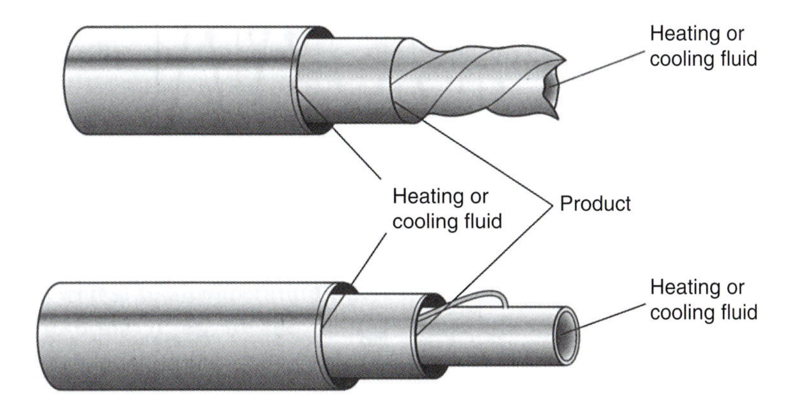

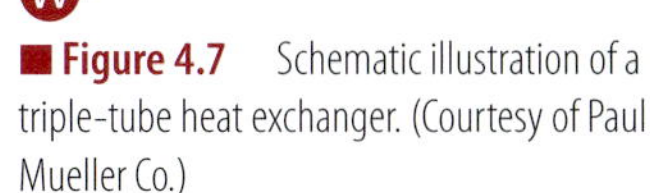

■ **Figure 4.7** Schematic illustration of a triple-tube heat exchanger. (Courtesy of Paul Mueller Co.)

turbulence and better heat transfer. Some specific industrial applications of triple-tube heat exchangers include heating single-strength orange juice from 4 to 93°C and then cooling to 4°C; cooling cottage cheese wash water from 46 to 18°C with chilled water; and cooling ice cream mix from 12 to 0.5°C with ammonia.

Another common type of heat exchanger used in the food industry is a shell-and-tube heat exchanger for such applications as heating liquid foods in evaporation systems. As shown in Figure 4.8, one of the fluid streams flows inside the tube while the other fluid stream is pumped over the tubes through the shell. By maintaining the fluid stream in the shell side to flow over the tubes, rather than parallel to the tubes, we can achieve higher rates of heat transfer. Baffles located in the shell side allow the cross-flow pattern. One or more tube passes can be accomplished, depending on the design. The shell-and-tube heat exchangers shown in Figure 4.8 are one shell pass with two tube passes, and two shell passes with four tube passes.

4.1.3 Scraped-Surface Heat Exchanger

In conventional types of tubular heat exchangers, heat transfer to a fluid stream is affected by hydraulic drag and heat resistance due to film buildup or fouling on the tube wall. This heat resistance can be minimized if the inside surface of the tube wall is scraped continuously by some mechanical means. The scraping action allows rapid heat transfer to a relatively small product volume. A scraped-surface heat exchanger, used in food processing, is shown schematically in Figure 4.9.

■ Figure 4.8 A shell-and-tube heat exchanger.

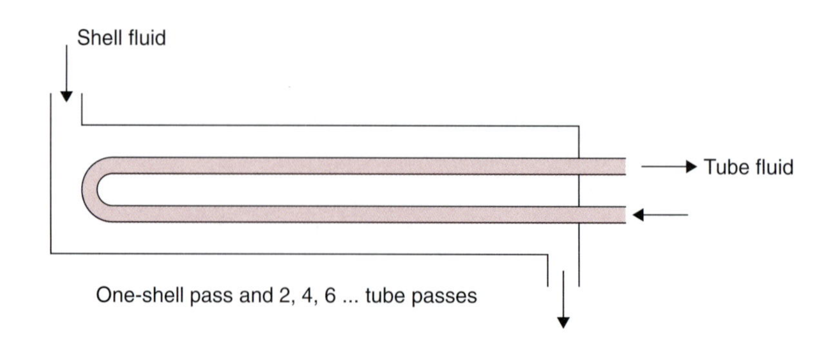

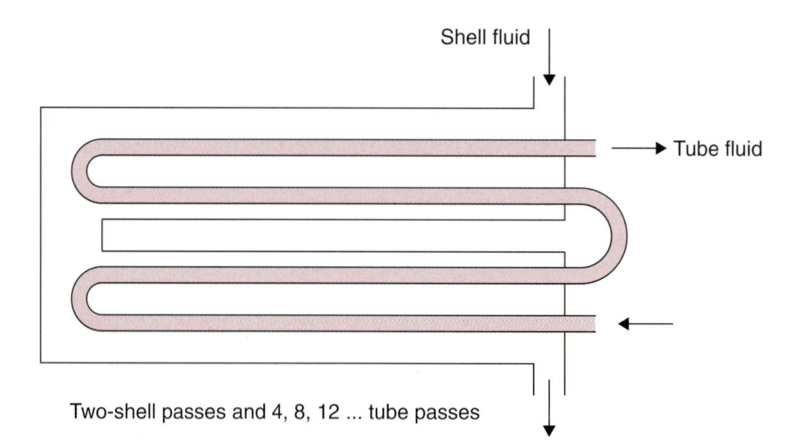

■ Figure 4.9 A scraped-surface heat exchanger with a cutaway section illustrating various components. (Courtesy of Cherry-Burrell Corporation)

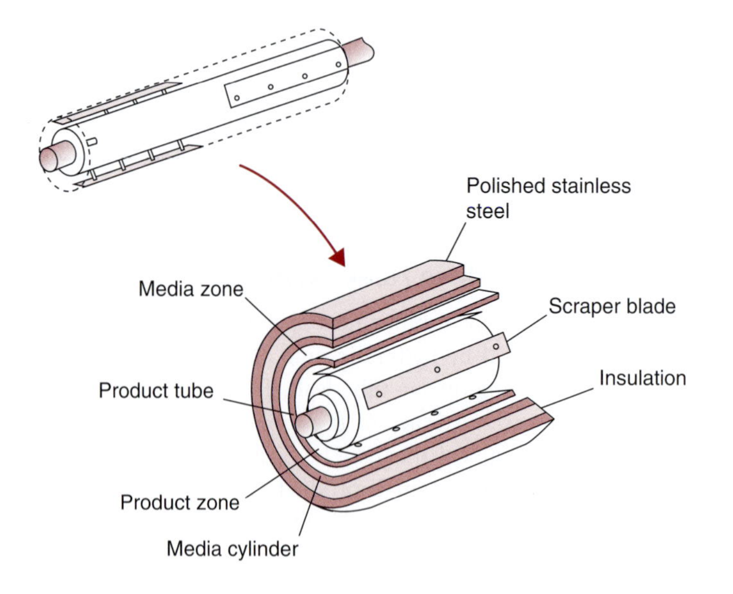

The food contact areas of a scraped-surface cylinder are fabricated from stainless steel (type 316), pure nickel, hard chromium-plated nickel, or other corrosion-resistant material. The inside rotor contains blades that are covered with plastic laminate or molded plastic (Fig. 4.9). The rotor speed varies between 150 and 500 rpm. Although higher rotation speed allows better heat transfer, it may affect the quality of the processed product by possible maceration. Thus, we must carefully select the rotor speed and the annular space between the rotor and the cylinder for the product being processed.

As seen in Figure 4.9, the cylinder containing the product and the rotor is enclosed in an outside jacket. The heating/cooling medium is supplied to this outside jacket. Commonly used media include steam, hot water, brine, or a refrigerant. Typical temperatures used for processing products in scraped-surface heat exchangers range from −35 to 190°C.

The constant blending action accomplished in the scraped-surface heat exchanger is often desirable to enhance the uniformity of product flavor, color, aroma, and textural characteristics. In the food processing industry, the applications of scraped-surface heat exchangers include heating, pasteurizing, sterilizing, whipping, gelling, emulsifying, plasticizing, and crystallizing. Liquids with a wide range of viscosities that can be pumped are processed in these heat exchangers; examples include fruit juices, soups, citrus concentrate, peanut butter, baked beans, tomato paste, and pie fillings.

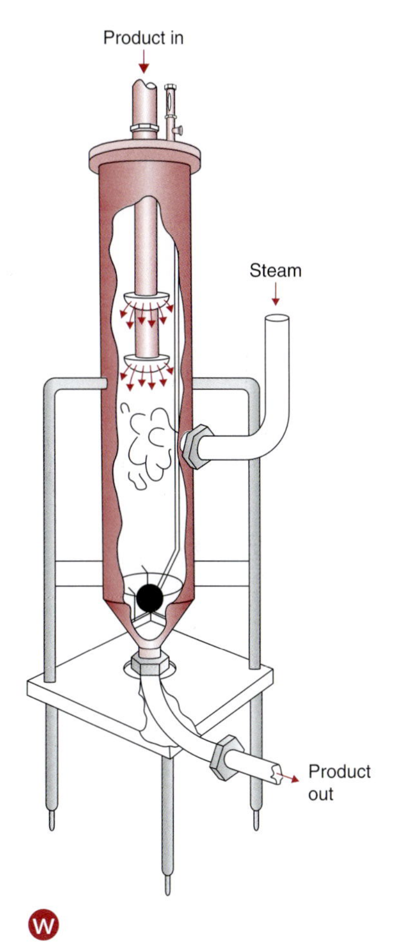

W

■ **Figure 4.10** A steam-infusion heat exchanger. (Courtesy of CREPACO, Inc.)

4.1.4 Steam-Infusion Heat Exchanger

A steam-infusion heat exchanger provides a direct contact between steam and the product. As shown in Figure 4.10, product in liquid state is pumped to the top of the heat exchanger and then allowed to flow in thin sheets in the heating chamber. The viscosity of the liquid determines the size of the spreaders. Products containing particulates, such as diced vegetables, meat chunks, and rice, can be handled by specially designed spreaders. High rates of heat transfer are achieved when steam contacts tiny droplets of the food. The temperature of the product rises very rapidly due to steam condensation. The heated products with condensed steam are released from the chamber at the bottom. A specific amount of liquid is retained in the bottom of the chamber to achieve desired cooking.

The temperature difference of the product between the inlet and the outlet to the heating chamber may be as low as 5.5°C, such as for deodorizing milk (76.7 to 82.2°C), or as high as 96.7°C, such as for sterilizing puddings for aseptic packaging (48.9 to 145.6°C).

The water added to the product due to steam condensation is sometimes desirable, particularly if the overall process requires addition of water. Otherwise, the added water of condensation can be "flashed off" by pumping the heated liquid into a vacuum cooling system. The amount of water added due to condensation can be computed by measuring the temperature of the product fed to the heat exchanger and the temperature of the product discharged from the vacuum cooler.

This type of heat exchanger has applications in cooking and/or sterilizing a wide variety of products, such as concentrated soups, chocolate, processed cheese, ice cream mixes, puddings, fruit pie fillings, and milk.

4.1.5 Epilogue

In the preceding subsections, we discussed several types of commonly used heat exchangers. It should be evident that a basic understanding of the mechanisms of heat transfer, both in the food and the materials used in construction of the food processing equipment, is necessary before we can design or evaluate any heat exchange equipment. A wide variety of food products is processed using heat exchangers. These products present unique and often complex problems related to heat transfer. In the following sections, we will develop quantitative descriptions emphasizing the following:

1. *Thermal properties*. Properties such as specific heat, thermal conductivity, and thermal diffusivity of food and equipment materials (such as metals) play an important role in determining the rate of heat transfer.
2. *Mode of heat transfer*. A mathematical description of the actual mode of heat transfer, such as conduction, convection, and/or radiation is necessary to determine quantities, such as total amount of heat transferred from heating or cooling medium to the product.
3. *Steady-state and unsteady-state heat transfer*. Calculation procedures are needed to examine both the unsteady-state and steady-state phases of heat transfer.

We will develop an analytical approach for cases involving simple heat transfer. For more complex treatment of heat transfer, such as for non-Newtonian liquids, the textbook by Heldman and Singh (1981) is recommended.

4.2 THERMAL PROPERTIES OF FOODS

4.2.1 Specific Heat

Specific heat is the quantity of heat that is gained or lost by a unit mass of product to accomplish a unit change in temperature, without a change in state:

$$c_{\mathrm{p}} = \frac{Q}{m(\Delta T)} \tag{4.1}$$

where Q is heat gained or lost (kJ), m is mass (kg), ΔT is temperature change in the material (°C), and c_{p} is specific heat (kJ/[kg °C]).

Specific heat is an essential part of the thermal analysis of food processing or of the equipment used in heating or cooling of foods. With food materials, this property is a function of the various components that constitute a food, its moisture content, temperature, and pressure. The specific heat of a food increases as the product moisture content increases. For a gas, the specific heat at constant pressure, c_{p}, is greater than its specific heat at constant volume, c_{v}. In most food processing applications, we use specific heat at constant pressure c_{p}, since pressure is generally kept constant except in high-pressure processing.

For processes where a change of state takes place, such as freezing or thawing, an apparent specific heat is used. Apparent specific heat incorporates the heat involved in the change of state in addition to the sensible heat.

In designing food processes and processing equipment, we need numerical values for the specific heat of the food and materials to be used. There are two ways to obtain such values. Published data are available that provide values of specific heat for some food and nonfood materials, such as given in Tables A.2.1, A.3.1, and A.3.2 (in the appendix). Comprehensive databases are also available to obtain published values (Singh, 1994). Another way to obtain a specific heat value is to use a predictive equation. The predictive equations are empirical expressions, obtained by fitting experimental data into mathematical models. Typically these mathematical models are based on one or more constituents of the food. Since water is a major component of many foods, a number of models are expressed as a function of water content.

One of the earliest models to calculate specific heat was proposed by Siebel (1892) as,

$$c_p = 0.837 + 3.349\,X_w \tag{4.2}$$

where X_w is the water content expressed as a fraction. This model does not show the effect of temperature or other components of a food product. The influence of product components was expressed in an empirical equation proposed by Charm (1978) as

$$c_p = 2.093\,X_f + 1.256\,X_s + 4.187\,X_w \tag{4.3}$$

where X is the mass fraction; and subscripts f is fat, s is nonfat solids, and w is water. Note that in Equation (4.3), the coefficients of each term on the right-hand side are specific heat values of the respective food constituents. For example, 4.187 is the specific heat of water at 70°C, and 2.093 is the specific heat of liquid fat.

Heldman and Singh (1981) proposed the following expression based on the components of a food product:

$$c_p = 1.424X_h + 1.549X_p + 1.675X_f + 0.837X_a + 4.187X_w \tag{4.4}$$

where X is the mass fraction; the subscripts on the right-hand side are h, carbohydrate; p, protein; f, fat; a, ash; and w, moisture.

Note that these equations do not include a dependence on temperature. However, for processes where temperature changes, we must use predictive models of specific heat that include temperature dependence. Choi and Okos (1986) presented a comprehensive model to predict specific heat based on composition and temperature. Their model is as follows:

$$c_p = \sum_{i=1}^{n} c_{pi} X_i \tag{4.5}$$

where X_i is the fraction of the ith component, n is the total number of components in a food, and c_{pi} is the specific heat of the ith component. Table A.2.9 gives the specific heat of pure food components as a function of temperature. The coefficients in this table may be programmed in a spreadsheet for predicting specific heat at any desired temperature, as illustrated in Example 4.1.

The units for specific heat are

$$c_{\mathrm{p}} = \frac{\mathrm{kJ}}{\mathrm{kg\ K}}$$

Note that these units are equivalent to kJ/(kg °C), since 1° temperature *change* is the same in Celsius or Kelvin scale.

Food composition values may be obtained from *Agriculture Handbook No. 8* (Watt and Merrill, 1975). Values for selected foods are given in Table A.2.8.

Example 4.1

Predict the specific heat for a model food with the following composition: carbohydrate 40%, protein 20%, fat 10%, ash 5%, moisture 25%.

Given

$$X_h = 0.4 \quad X_p = 0.2 \quad X_f = 0.1 \quad X_a = 0.05 \quad X_m = 0.25$$

Approach

Since the product composition is given, Equation (4.4) will be used to predict specific heat. Furthermore, we will program a spreadsheet with Equation (4.5) to determine a value for specific heat.

	A	B	C	D	E	F	G	H	I
1	Temperature (°C)	20							
2	Water	0.25							
3	Protein	0.2							
4	Fat	0.1							
5	Carbohydrate	0.4							
6	Fiber	0							
7	Ash	0.05							
8									
9		Coefficients							
10	Water	4.1766	←	=4.1762-0.000090864*B1+0.0000054731*B1^2					
11	Protein	2.0319	←	=2.0082+0.0012089*B1-0.0000013129*B1^2					
12	Fat	2.0117	←	=1.9842+0.0014733*B1-0.0000048008*B1^2					
13	Carbohydrate	1.5857	←	=1.5488+0.0019625*B1-0.0000059399*B1^2					
14	Fiber	1.8807	←	=1.8459+0.0018306*B1-0.0000046509*B1^2					
15	Ash	1.1289	←	=1.0926+0.0018896*B1-0.0000036817*B1^2					
16									
17		Eq(4.5)							
18	Water	1.044	←	=B2*B10					
19	Protein	0.406	←	=B3*B11					
20	Fat	0.201	←	=B4*B12					
21	Carbohydrate	0.634	←	=B5*B13					
22	Fiber	0.000	←	=B6*B14					
23	Ash	0.056	←	=B7*B15					
24	Result	2.342	←	=SUM(B18:B23)					

Figure E4.1 Spreadsheet for data given in Example 4.1.

Solution

1. *Using Equation (4.4)*

$$
\begin{aligned}
c_p &= (1.424 \times 0.4) + (1.549 \times 0.2) + (1.675 \times 0.1) \\
&\quad + (0.837 \times 0.05) + (4.187 \times 0.25) \\
&= 2.14\,\text{kJ/(kg °C)}
\end{aligned}
$$

2. *We can program a spreadsheet using Equation (4.5) with coefficients given in Table A.2.9 as shown in Figure E4.1.*
3. *Specific heat predicted using Equation (4.4) is 2.14 kJ/(kg °C) whereas using Equation (4.5) is slightly different as 2.34 kJ/(kg °C). Equation (4.5) is preferred since it incorporates information about the temperature.*

4.2.2 Thermal Conductivity

The thermal conductivity of a food is an important property used in calculations involving rate of heat transfer. In quantitative terms, this property gives the amount of heat that will be conducted per unit time through a unit thickness of the material if a unit temperature gradient exists across that thickness.

In SI units, thermal conductivity is

$$
k \equiv \frac{\text{J}}{\text{s m °C}} \equiv \frac{\text{W}}{\text{m °C}} \tag{4.6}
$$

Note that W/(m °C) is same as W/(m K).

There is wide variability in the magnitude of thermal conductivity values for commonly encountered materials. For example:

- Metals: 50–400 W/(m °C)
- Alloys: 10–120 W/(m °C)
- Water: 0.597 W/(m °C) (at 20°C)
- Air: 0.0251 W/(m °C) (at 20°C)
- Insulating materials: 0.035–0.173 W/(m °C)

Most high-moisture foods have thermal conductivity values closer to that of water. On the other hand, the thermal conductivity of dried, porous foods is influenced by the presence of air with its low value. Tables A.2.2, A.3.1, and A.3.2 show thermal conductivity values obtained numerically for a number of food and nonfood materials. In addition to the tabulated values, empirical predictive equations are useful in process calculations where temperature may be changing.

For fruits and vegetables with a water content greater than 60%, the following equation has been proposed (Sweat, 1974):

$$k = 0.148 + 0.493\,X_{\mathrm{w}} \qquad (4.7)$$

where k is thermal conductivity (W/[m°C]), and X_w is water content expressed as a fraction. For meats and fish, temperature 0–60°C, water content 60–80%, wet basis, Sweat (1975) proposed the following equation:

$$k = 0.08 + 0.52X_{\mathrm{w}} \qquad (4.8)$$

Another empirical equation developed by Sweat (1986) is to fit a set of 430 data points for solid and liquid foods, as follows:

$$k = 0.25X_{\mathrm{h}} + 0.155X_{\mathrm{p}} + 0.16X_{\mathrm{f}} + 0.135X_{\mathrm{a}} + 0.58X_{\mathrm{w}} \qquad (4.9)$$

where X is the mass fraction, and subscript h is carbohydrate, p is protein, f is fat, a is ash, and w is water.

The coefficients in Equation (4.9) are thermal conductivity values of the pure component. Note that the thermal conductivity of pure water at 25°C is 0.606 W/(m °C). The coefficient of 0.58 in Equation (4.9) indicates that there is either a bias in the data set used for regression, or the effective thermal conductivity of water in a food is different from that of pure water.

Equations (4.7) to (4.9) are simple expressions to calculate the thermal conductivity of foods, however they do not include the influence of temperature. Choi and Okos (1986) gave the following expression that includes the influence of product composition and temperature:

$$k = \sum_{i=1}^{n} k_i Y_i \qquad (4.10)$$

where a food material has n components, k_i is the thermal conductivity of the ith component, Y_i is the volume fraction of the ith component, obtained as follows:

$$Y_i = \frac{X_i/\rho_i}{\sum_{i=1}^{n}(X_i/\rho_i)} \qquad (4.11)$$

where X_i is the weight fraction and ρ_i is the density (kg/m^3) of the ith component.

The coefficients for k_i for pure components are listed in Table A.2.9. They may be programmed into a spreadsheet, as illustrated later in Example 4.2.

For the additive models, Equations (4.10) and (4.11), the food composition values may be obtained from Table A.2.8. These equations predict thermal conductivity of foods within 15% of experimental values.

In the case of anisotropic foods, the properties of the material are direction dependent. For example, the presence of fibers in beef results in different values of thermal conductivity when measured parallel to the fibers (0.476 W/[m °C]) versus perpendicular to them (0.431 W/[m °C]). Mathematical models to predict the thermal conductivity of anisotropic foods are discussed in Heldman and Singh (1981).

4.2.3 Thermal Diffusivity

Thermal diffusivity, a ratio involving thermal conductivity, density, and specific heat, is given as,

$$\alpha = \frac{k}{\rho c_p} \tag{4.12}$$

The units of thermal diffusivity are

$$\alpha \equiv \frac{m^2}{s}$$

Thermal diffusivity may be calculated by substituting values of thermal conductivity, density, and specific heat in Equation (4.12). Table A.2.3 gives some experimentally determined values of thermal diffusivity. Choi and Okos (1986) provided the following predictive equation, obtained by substituting the values of k, ρ, and c_p in Equation (4.12):

$$\alpha = \sum_{i=1}^{n} \alpha_i X_i \tag{4.13}$$

where n is the number of components, α_i is the thermal diffusivity of the ith component, and X_i is the mass fraction of each component. The values of α_i are obtained from Table A.2.9.

Example 4.2

Estimate the thermal conductivity of hamburger beef that contains 68.3% water.

Given

$X_m = 0.683$

Approach

We will use Equation (4.8), which is recommended for meats. We will also program a spreadsheet using Equations (4.10) and (4.11) at 20°C to calculate thermal conductivity.

Solution

1. *Using Equation (4.8)*

$$\begin{aligned} k &= 0.08 + (0.52 \times 0.683) \\ &= 0.435\ W/(m\,°C) \end{aligned}$$

2. *Next we will program a spreadsheet as shown in Figure E4.2 using the composition of hamburger beef from Table A.2.8 and coefficients of Equations (4.10) and (4.11) given in Table A.2.9. We will use a temperature of 20°C.*

	A	B	C	D	E	F	G	H	I
1									
2	Given								
3	Temperature (C)	20							
4	Water	0.683							
5	Protein	0.207							
6	Fat	0.1							
7	Carbohydrate	0							
8	Fiber	0							
9	Ash	0.01							
10									
11		density coeff	Xi/ri	Yi					
12	Water	995.739918	0.000686	0.717526					
13	Protein	1319.532	0.000157	0.164102					
14	Fat	917.2386	0.000109	0.114046					
15	Carbohydrate	1592.8908	0.000000	0					
16	Fiber	1304.1822	0.000000	0					
17	Ash	2418.1874	0.000004	0.004326					
18		sum	0.000956						
19									
20		k Coeff							
21	Water	0.6037	0.4331						
22	Protein	0.2016	0.0331						
23	Fat	0.1254	0.0143						
24	Carbohydrate	0.2274	0.0000						
25	Fiber	0.2070	0.0000						
26	Ash	0.3565	0.0015						
27									
28	Result		0.4821						

Figure E4.2 Spreadsheet for data given in Example 4.2.

3. *The thermal conductivity predicted by Equation (4.8) is 0.435 W/(m °C), whereas using Equation (4.10) it is 0.4821 W/(m °C). Although Equation (4.8) is easier to use, it does not include the influence of temperature.*

4.3 MODES OF HEAT TRANSFER

In Chapter 1, we reviewed various forms of energy, such as thermal, potential, mechanical, kinetic, and electrical. Our focus in this chapter will be on thermal energy, commonly referred to as heat energy or heat content. As noted in Section 1.19, heat energy is simply the sensible and latent forms of internal energy. Recall that the heat content of an object such as a tomato is determined by its mass, specific heat, and temperature. The equation for calculating heat content is

$$Q = mc_p \Delta T \quad (4.14)$$

where m is mass (kg), c_p is specific heat at constant pressure (kJ/[kg K]), and ΔT is the temperature difference between the object and a reference temperature (°C). Heat content is always expressed relative to some other temperature (called a datum or reference temperature).

Although determining heat content is an important calculation, the knowledge of how heat may *transfer* from one object to another or within an object is of even greater practical value. For example, to thermally sterilize tomato juice, we raise its heat content by transferring heat from some heating medium such as steam into the juice. In order to design the sterilization equipment, we need to know how much heat is necessary to raise the temperature of tomato juice from the initial to the final sterilization temperature using Equation (4.14). Furthermore, we need to know the rate at which heat will transfer from steam into the juice first passing through the walls of the sterilizer. Therefore, our concerns in heating calculations are twofold: the quantity of heat transferred, Q, expressed in the units of joule (J); and the rate of heat transfer, q, expressed as joule/s (J/s) or watt (W).

We will first review some highlights of the three common modes of heat transfer—conduction, convection, and radiation—and then examine selected topics of rates of heat transfer important in the design and analysis of food processes.

4.3.1 Conductive Heat Transfer

Conduction is the mode of heat transfer in which the transfer of energy takes place at a molecular level. There are two commonly accepted theories that describe conductive heat transfer. According to one theory, as molecules of a solid material attain additional thermal energy, they become more energetic and vibrate with increased amplitude of vibration while confined in their lattice. These vibrations are transmitted from one molecule to another without actual translatory motion of the molecules. Heat is thus conducted from regions of higher temperature to those at lower temperature. The second theory states that conduction occurs at a molecular level due to the drift of free electrons. These free electrons are prevalent in metals, and they carry thermal and electrical energy. For this reason, good conductors of electricity such as silver and copper are also good conductors of thermal energy.

Note that in conductive mode, there is no physical movement of the object undergoing heat transfer. Conduction is the common mode of heat transfer in heating/cooling of opaque solid materials.

From everyday experience, we know that on a hot day, heat transfer from the outside to the inside through the wall of a room (Fig. 4.11) depends on the surface area of the wall (a wall with larger surface area will conduct more heat), the thermal properties of construction materials (steel will conduct more heat than brick), wall thickness (more heat transfer through a thin wall than thick), and temperature difference (more heat transfer will occur when the outside temperature is much hotter than the inside room temperature). In other words, the rate of heat transfer through the wall may be expressed as

$$q \propto \frac{\text{(wall surface area)(temperature difference)}}{\text{(wall thickness)}} \tag{4.15}$$

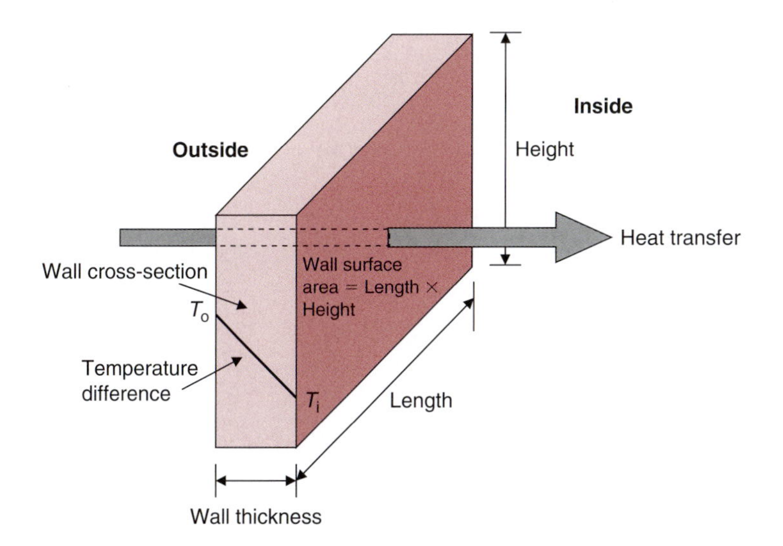

Figure 4.11 Conductive heat flow in a wall.

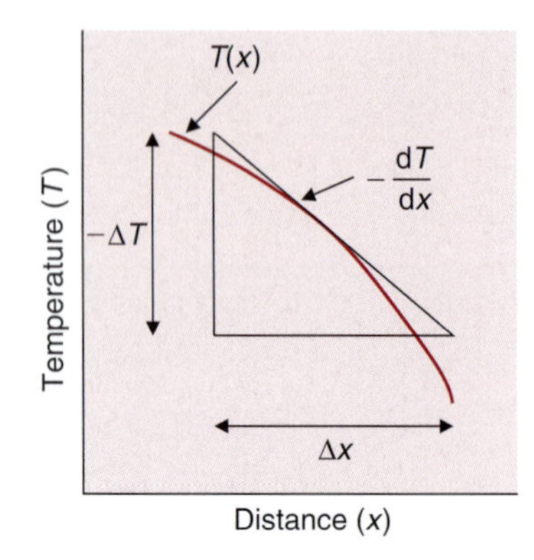

Figure 4.12 Sign convention for conductive heat flow.

or

$$q_x \propto \frac{A\,dT}{dx} \tag{4.16}$$

or, by inserting a constant of proportionality,

$$q_x = -kA\frac{dT}{dx} \tag{4.17}$$

where q_x is the rate of heat flow in the direction of heat transfer by conduction (W); k is thermal conductivity (W/[m °C]); A is area (normal to the direction of heat transfer) through which heat flows (m^2); T is temperature (°C); and x is length (m), a variable.

Equation (4.17) is also called the Fourier's law for heat conduction, after Joseph Fourier, a French mathematical physicist. According to the second law of thermodynamics, heat will always conduct from higher temperature to lower temperature. As shown in Figure 4.12, the gradient dT/dx is negative, because temperature decreases with increasing values of x. Therefore, in Equation (4.17), a negative sign is used to obtain a positive value for heat flow in the direction of decreasing temperature.

Example 4.3

One face of a stainless-steel plate 1 cm thick is maintained at 110°C, and the other face is at 90°C (Fig. E4.3). Assuming steady-state conditions, calculate the rate of heat transfer per unit area through the plate. The thermal conductivity of stainless steel is 17 W/(m °C).

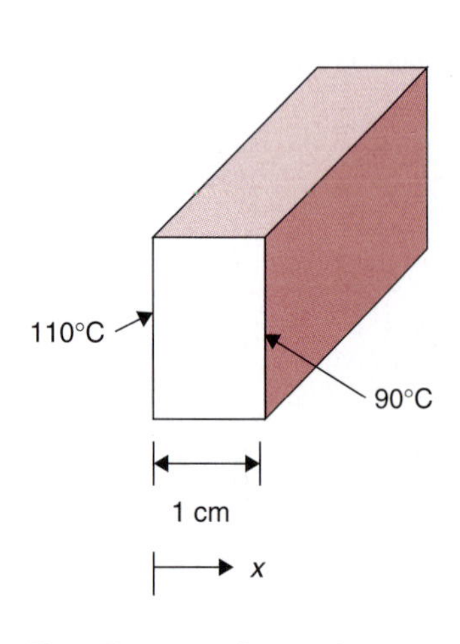

Figure E4.3 Heat flow in a plate.

Given

Thickness of plate = 1 cm = 0.01 m
Temperature of one face = 110°C
Temperature of other face = 90°C
Thermal conductivity of stainless steel = 17 W/(m °C)

Approach

For steady-state heat transfer in rectangular coordinates we will use Equation (4.17) to compute rate of heat transfer.

Solution

1. *From Equation (4.17)*

$$q = -\frac{17[W/(m\,°C)] \times 1[m^2] \times (110-90)[°C]}{(0-0.01)[m]}$$
$$= 34{,}000\,W$$

2. *Rate of heat transfer per unit area is calculated to be 34,000 W. A positive sign is obtained for the heat transfer, indicating that heat always flows "downhill" from 110°C to 90°C.*

4.3.2 Convective Heat Transfer

When a fluid (liquid or gas) comes into contact with a solid body such as the surface of a wall, heat exchange will occur between the solid and the fluid whenever there is a temperature difference between the two. During heating and cooling of gases and liquids the fluid streams exchange heat with solid surfaces by convection.

The magnitude of the fluid motion plays an important role in convective heat transfer. For example, if air is flowing at a high velocity past a hot baked potato, the latter will cool down much faster than if the air velocity was much lower. The complex behavior of fluid flow next to a solid surface, as seen in velocity profiles for laminar and turbulent flow conditions in Chapter 2, make the determination of convective heat transfer a complicated topic.

Depending on whether the flow of the fluid is artificially induced or natural, there are two types of convective heat transfer: **forced** convection and **free** (also called **natural**) convection. Forced convection involves the use of some mechanical means, such as a pump or a fan, to induce movement of the fluid. In contrast, free convection occurs due to density differences caused by temperature gradients within the system. Both of these mechanisms may result in either laminar or turbulent flow of the fluid, although turbulence occurs more often in forced convection heat transfer.

Consider heat transfer from a heated flat plate, PQRS, exposed to a flowing fluid, as shown in Figure 4.13. The surface temperature of the plate is T_s, and the temperature of the fluid far away from the plate surface is T_∞. Because of the viscous properties of the fluid, a velocity profile is set up within the flowing fluid, with the fluid velocity decreasing to zero at the solid surface. Overall, we see that the rate of heat transfer from the solid surface to the flowing fluid is proportional to the surface area of solid, A, in contact with the fluid, and the difference between the temperatures T_s and T_∞. Or,

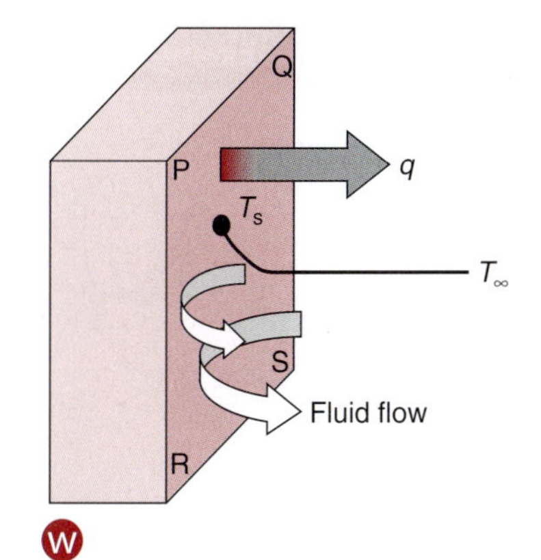

W

Figure 4.13 Convective heat flow from the surface of a flat plate.

$$q \propto A(T_s - T_\infty) \tag{4.18}$$

or,

$$q = hA(T_s - T_\infty) \tag{4.19}$$

Table 4.1 Some Approximate Values of Convective Heat-Transfer Coefficient

Fluid	Convective heat-transfer coefficient (W/[m^2 K])
Air	
Free convection	5–25
Forced convection	10–200
Water	
Free convection	20–100
Forced convection	50–10,000
Boiling water	3000–100,000
Condensing water vapor	5000–100,000

The area is A (m^2), and h is the convective heat-transfer coefficient (sometimes called surface heat-transfer coefficient), expressed as W/(m^2 °C). This equation is also called Newton's law of cooling.

Note that the convective heat transfer coefficient, h, is not a property of the solid material. This coefficient, however, depends on a number of properties of fluid (density, specific heat, viscosity, thermal conductivity), the velocity of fluid, geometry, and roughness of the surface of the solid object in contact with the fluid. Table 4.1 gives some approximate values of h. A high value of h reflects a high rate of heat transfer. Forced convection offers a higher value of h than free convection. For example, you feel cooler sitting in a room with a fan blowing air than in a room with stagnant air.

Example 4.4

The rate of heat transfer per unit area from a metal plate is 1000 W/m^2. The surface temperature of the plate is 120°C, and ambient temperature is 20°C (Fig. E4.4). Estimate the convective heat transfer coefficient.

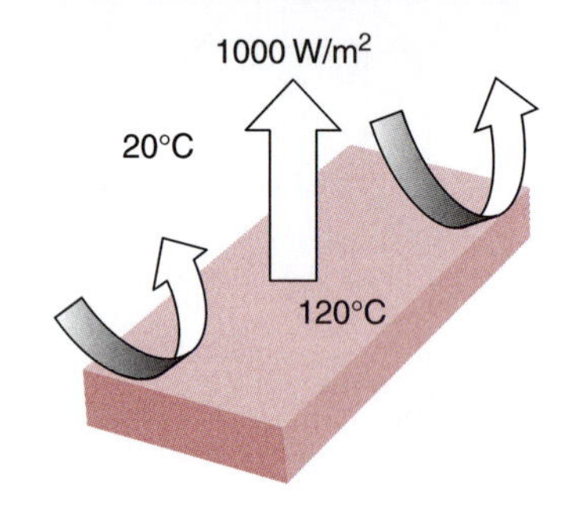

■ **Figure E4.4** Convective heat transfer from a plate.

Given

Plate surface temperature = 120°C
Ambient temperature = 20°C
Rate of heat transfer per unit area = 1000 W/m^2

Approach

Since the rate of heat transfer per unit area is known, we will estimate the convective heat transfer coefficient directly from Newton's law of cooling, Equation (4.19).

Solution

1. *From Equation (4.19),*

$$h = \frac{1000[W/m^2]}{(120 - 20)[°C]}$$

$$= 10\,W/(m^2\,°C)$$

2. *The convective heat transfer coefficient is found to be 10 W/(m² °C).*

4.3.3 Radiation Heat Transfer

Radiation heat transfer occurs between two surfaces by the emission and later absorption of electromagnetic waves (or photons). In contrast to conduction and convection, radiation requires no physical medium for its propagation—it can even occur in a perfect vacuum, moving at the speed of light, as we experience everyday solar radiation. Liquids are strong absorbers of radiation. Gases are transparent to radiation, except that some gases absorb radiation of a particular wavelength (for example, ozone absorbs ultraviolet radiation). Solids are opaque to thermal radiation. Therefore, in problems involving thermal radiation with solid materials, such as with solid foods, our analysis is concerned primarily with the surface of the material. This is in contrast to microwave and radio frequency radiation, where the wave penetration into a solid object is significant.

All objects at a temperature above 0 Absolute emit thermal radiation. Thermal radiation emitted from an object's surface is proportional to the absolute temperature raised to the fourth power and the surface characteristics. More specifically, the rate of heat emission (or radiation) from an object of a surface area A is expressed by the following equation:

$$q = \sigma \varepsilon A T_A^4 \tag{4.20}$$

where σ is the Stefan–Boltzmann[1] constant, equal to 5.669×10^{-8} W/(m^2 K^4); T_A is temperature, Absolute; A is the area (m^2); and ε is

---

[1]Josef Stefan (1835–1893). An Austrian physicist, Stefan began his academic career at the University of Vienna as a lecturer. In 1866, he was appointed director of the Physical Institute. Using empirical approaches, he derived the law describing radiant energy from blackbodies. Five years later, another Austrian, Ludwig Boltzmann, provided the thermodynamic basis of what is now known as the Stefan–Boltzmann law.

emissivity, which describes the extent to which a surface is similar to a blackbody. For a blackbody, the value of emissivity is 1. Table A.3.3 gives values of emissivity for selected surfaces.

Example 4.5

Calculate the rate of heat energy emitted by $100\,m^2$ of a polished iron surface (emissivity = 0.06) as shown in Figure E4.5. The temperature of the surface is 37°C.

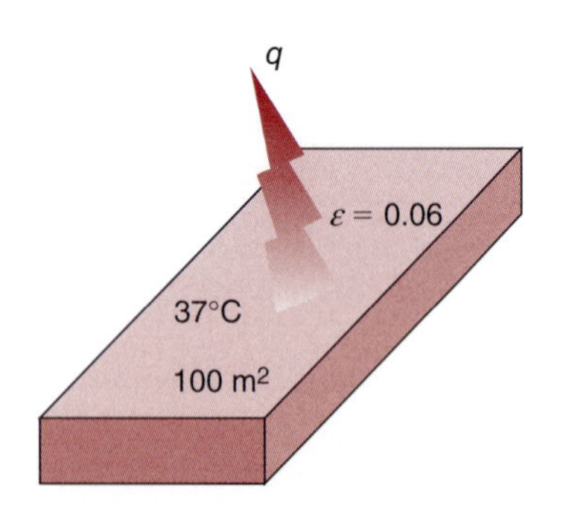

■ **Figure E4.5** Heat transfer from a plate.

Given

Emissivity $\varepsilon = 0.06$
Area $A = 100\,m^2$
Temperature $= 37°C = 310\,K$

Approach

We will use the Stefan–Boltzmann law, Equation (4.20), to calculate the rate of heat transfer due to radiation.

Solution

1. *From Equation (4.20)*

$$q = (5.669 \times 10^{-8}\,W/[m^2\,K^4])(0.06)(100\,m^2)(310\,K)^4$$
$$= 3141\,W$$

2. *The total energy emitted by the polished iron surface is 3141 W.*

4.4 STEADY-STATE HEAT TRANSFER

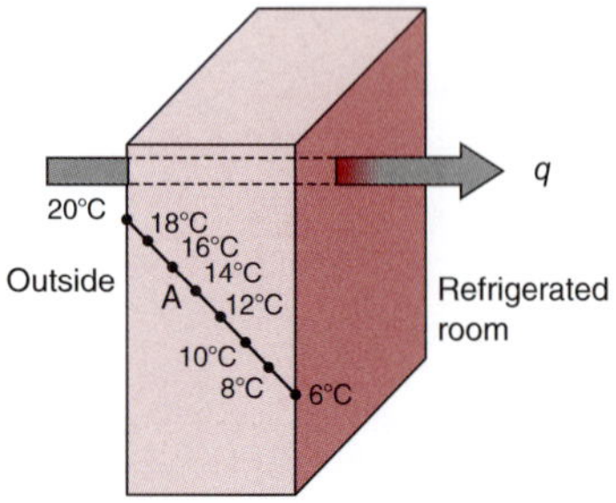

In problems involving heat transfer, we often deal with steady state and unsteady state (or transient) conditions. **Steady-state** conditions imply that time has no influence on the temperature distribution within an object, although temperature may be different at different locations within the object. Under **unsteady-state** conditions, the temperature changes with location and time. For example, consider the wall of a refrigerated warehouse as shown in Figure 4.14. The inside wall temperature is maintained at 6°C using refrigeration, while the outside wall temperature changes throughout the day and night. Assume that for a few hours of the day, the outside wall temperature is constant at 20°C, and during that time duration the rate of heat transfer into the warehouse through the wall will be under steady-state conditions. The temperature at any location inside the wall cross-section (e.g., 14°C at location A) will remain constant,

■ **Figure 4.14** Steady state conductive heat transfer in a wall.

although this temperature is different from other locations along the path of heat transfer within the wall, as shown in Figure 4.14. If, however, the temperature of the outside wall surface changes (say, increases above 20°C), then the heat transfer through the wall will be due to unsteady-state conditions, because now the temperature within the wall will change with time and location. Although true steady-state conditions are uncommon, their mathematical analysis is much simpler. Therefore, if appropriate, we assume steady-state conditions for the analysis of a given problem to obtain useful information for designing equipment and processes. In certain food processes such as in heating cans for food sterilization, we cannot use steady-state conditions, because the duration of interest is when the temperature is changing rapidly with time, and microbes are being killed. For analyzing those types of problems, an analysis involving unsteady-state heat transfer is used, as discussed later in Section 4.5.

Another special case of heat transfer involves change in temperature inside an object with time but not with location, such as might occur during heating or cooling of a small aluminum sphere, which has a high thermal conductivity. This is called a **lumped system**. We will discuss this case in more detail in Section 4.5.2.

In the following section, we will examine several applications of steady-state conduction heat transfer.

4.4.1 Conductive Heat Transfer in a Rectangular Slab

Consider a slab of constant cross-sectional area, as shown in Figure 4.15. The temperature, T_1, on side X is known. We will develop an equation to determine temperature, T_2, on the opposite side Y and at any location inside the slab under steady-state conditions.

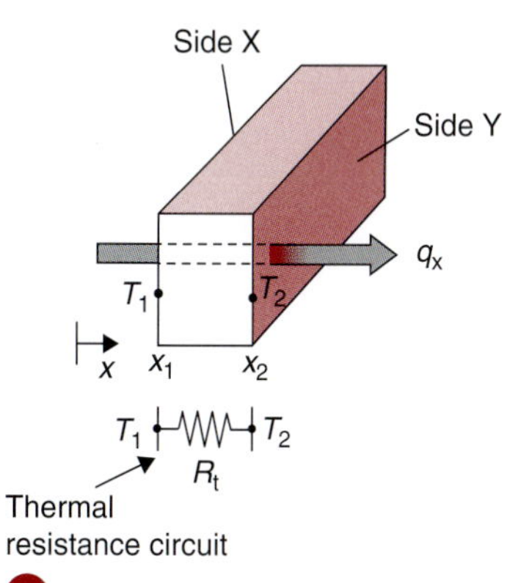

Figure 4.15 Heat transfer in a wall, also shown with a thermal resistance circuit.

This problem is solved by first writing Fourier's law,

$$q_x = -kA\frac{dT}{dx} \tag{4.21}$$

The boundary conditions are

$$\begin{aligned} x &= x_1 \quad T = T_1 \\ x &= x_2 \quad T = T_2 \end{aligned} \tag{4.22}$$

Separating variables in Equation (4.21), we get

$$\frac{q_x}{A}dx = -kdT \tag{4.23}$$

Setting up integration and substituting limits, we have

$$\int_{x_1}^{x_2} \frac{q_x}{A} \mathrm{d}x = -\int_{T_1}^{T_2} k \mathrm{d}T \tag{4.24}$$

Since q_x and A are independent of x, and k is assumed to be independent of T, Equation (4.24) can be rearranged to give

$$\frac{q_x}{A} \int_{x_1}^{x_2} \mathrm{d}x = -k \int_{T_1}^{T_2} \mathrm{d}T \tag{4.25}$$

Finally, integrating this equation, we get

$$\frac{q_x}{A}(x_2 - x_1) = -k(T_2 - T_1) \tag{4.26}$$

or

$$q_x = -kA \frac{(T_2 - T_1)}{(x_2 - x_1)} \tag{4.27}$$

Temperature on face Y is T_2; thus, rearranging Equation (4.27),

$$T_2 = T_1 - \frac{q_x}{kA}(x_2 - x_1) \tag{4.28}$$

To determine temperature, T, at any location, x, within the slab, we may replace T_2 and x_2 with unknown T and distance variable x, respectively, in Equation (4.28) and obtain,

$$T = T_1 - \frac{q_x}{kA}(x - x_1) \tag{4.29}$$

4.4.1.1 *Thermal Resistance Concept*

We noted in Chapter 3 that, according to Ohm's Law, electrical current, I, is directly proportional to the voltage difference, E_V, and indirectly proportional to the electrical resistance R_E. Or,

$$I = \frac{E_V}{R_E} \tag{4.30}$$

If we rearrange the terms in Equation (4.27), we obtain

$$q_x = \frac{(T_1 - T_2)}{\left[\dfrac{(x_2 - x_1)}{kA}\right]} \tag{4.31}$$

or,

$$q_x = \frac{T_1 - T_2}{R_t} \tag{4.32}$$

Comparing Equations (4.30) and (4.32), we note an analogy between rate of heat transfer, q_x, and electrical current, I, temperature difference, $(T_1 - T_2)$ and electrical voltage, E_V, and thermal resistance, R_t, and electrical resistance, R_E. From Equations (4.31) and (4.32), thermal resistance may be expressed as

$$R_t = \frac{(x_2 - x_1)}{kA} \tag{4.33}$$

A thermal resistance circuit for a rectangular slab is also shown in Figure 4.15. In solving problems involving conductive heat transfer in a rectangular slab using this concept, we first obtain thermal resistance using Equation (4.33) and then substitute it in Equation (4.32). The rates of heat transfer across the two surfaces of a rectangular slab are thus obtained. This procedure is illustrated in Example 4.6. The advantage of using the thermal resistance concept will become clear when we study conduction in multilayer walls. Moreover, the mathematical computations will be much simpler compared with alternative procedures used in solving these problems.

Example 4.6

a. Redo Example 4.3 using the thermal resistance concept.

b. Determine the temperature at 0.5 cm from the 110°C temperature face.

Given

See Example 4.3

Location at which temperature is desired = 0.5 cm = 0.005 m

Approach

We will use Equation (4.33) to calculate thermal resistance, and then Equation (4.32) to determine the rate of heat transfer. To determine temperature within the slab,

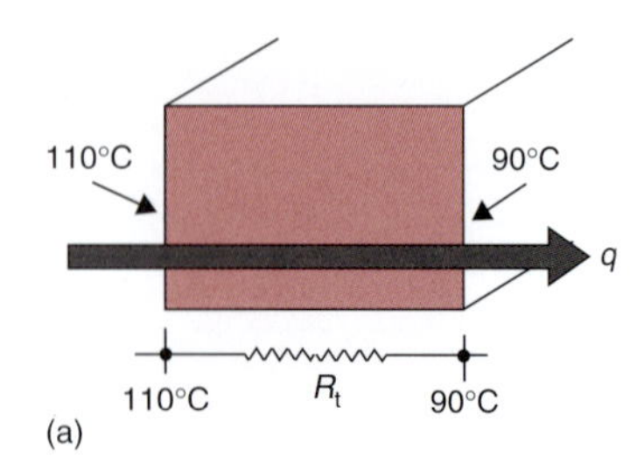

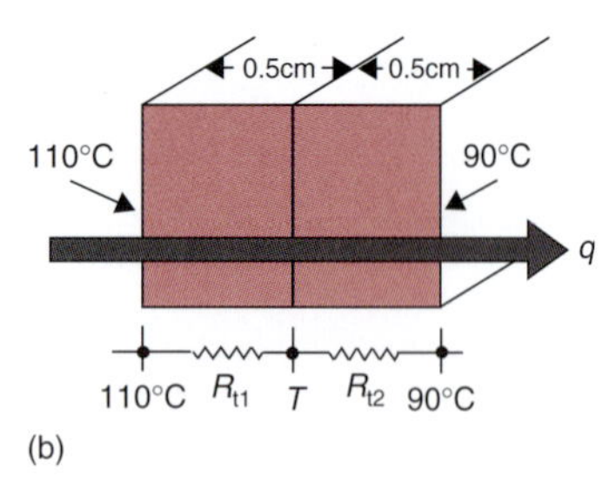

Figure E4.6 Thermal resistance circuits for heat transfer through a wall.

we will calculate the thermal resistance for the thickness of the slab bounded by 110°C and the unknown temperature (Fig. E4.6). Since the steady-state heat transfer remains the same throughout the slab, we will use the previously calculated value of q to determine the unknown temperature using Equation (4.32).

Solution

Part (a)

1. Using Equation (4.33), the thermal resistance R_t is

$$R_t = \frac{0.01[m]}{17[W/(m\,°C)] \times 1[m^2]}$$

$$R_t = 5.88 \times 10^{-4}\ °C/W$$

2. Using Equation (4.32), we obtain rate of heat transfer as

$$q = \frac{110[°C] - 90[°C]}{5.88 \times 10^{-4}[°C/W]}$$

or

$$q = 34{,}013 W$$

Part (b)

3. Using Equation (4.33) calculate resistance R_{t1}

$$R_{t1} = \frac{0.005[m]}{17[W/(m°C)] \times 1[m^2]}$$

$$R_{t1} = 2.94 \times 10^{-4}\ °C/W$$

4. Rearranging terms in Equation (4.32) to determine the unknown temperature T

$$T = T_1 - (q \times R_{t1})$$

$$T = 110[°C] - 34{,}013[W] \times 2.94 \times 10^{-4}[°C/W]$$

$$T = 100°C$$

5. The temperature at the midplane is 100°C. This temperature was expected, since the thermal conductivity is constant, and the temperature profile in the steel slab is linear.

4.4.2 Conductive Heat Transfer through a Tubular Pipe

Consider a long, hollow cylinder of inner radius r_i, outer radius r_o, and length L, as shown in Figure 4.16. Let the inside wall temperature be T_i and the outside wall temperature be T_o. We want to calculate the

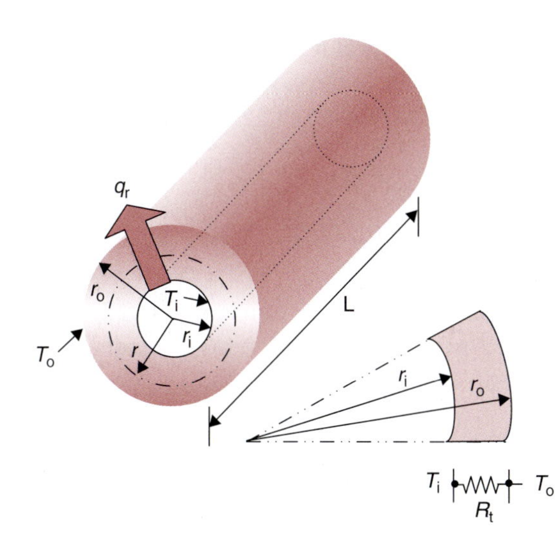

Figure 4.16 Heat transfer in a radial direction in a pipe, also shown with a thermal resistance circuit.

rate of heat transfer along the radial direction in this pipe. Assume thermal conductivity of the metal remains constant with temperature.

Fourier's law in cylindrical coordinates may be written as

$$q_r = -kA\frac{dT}{dr} \tag{4.34}$$

where q_r is the rate of heat transfer in the radial direction.

Substituting for circumferential area of the pipe,

$$q_r = -k(2\pi rL)\frac{dT}{dr} \tag{4.35}$$

The boundary conditions are

$$\begin{aligned} T &= T_i \quad r = r_i \\ T &= T_o \quad r = r_o \end{aligned} \tag{4.36}$$

Rearranging Equation (4.35), and setting up the integrals,

$$\frac{q_r}{2\pi L}\int_{r_i}^{r_o}\frac{dr}{r} = -k\int_{T_i}^{T_o} dT \tag{4.37}$$

Equation (4.37) gives

$$\frac{q_r}{2\pi L}\left|\ln r\right|_{r_i}^{r_o} = -k\left|T\right|_{T_i}^{T_o} \tag{4.38}$$

$$q_r = \frac{2\pi Lk(T_i - T_o)}{\ln(r_o/r_i)} \tag{4.39}$$

Again, we can use the electrical resistance analogy to write an expression for thermal resistance in the case of a cylindrical-shaped object. Rearranging the terms in Equation (4.39), we obtain

$$q_r = \frac{(T_i - T_o)}{\left[\dfrac{\ln(r_o/r_i)}{2\pi Lk}\right]} \tag{4.40}$$

Comparing Equation (4.40) with Equation (4.32), we obtain the thermal resistance in the radial direction for a cylinder as

$$R_t = \frac{\ln(r_o/r_i)}{2\pi Lk} \tag{4.41}$$

Figure 4.16 shows a thermal circuit to obtain R_t. An illustration of the use of this concept is given in Example 4.7.

Example 4.7

A 2 cm thick steel pipe (thermal conductivity = 43 W/[m °C]) with 6 cm inside diameter is being used to convey steam from a boiler to process equipment for a distance of 40 m. The inside pipe surface temperature is 115°C, and the outside pipe surface temperature is 90°C (Fig. E4.7). Calculate the total heat loss to the surroundings under steady-state conditions.

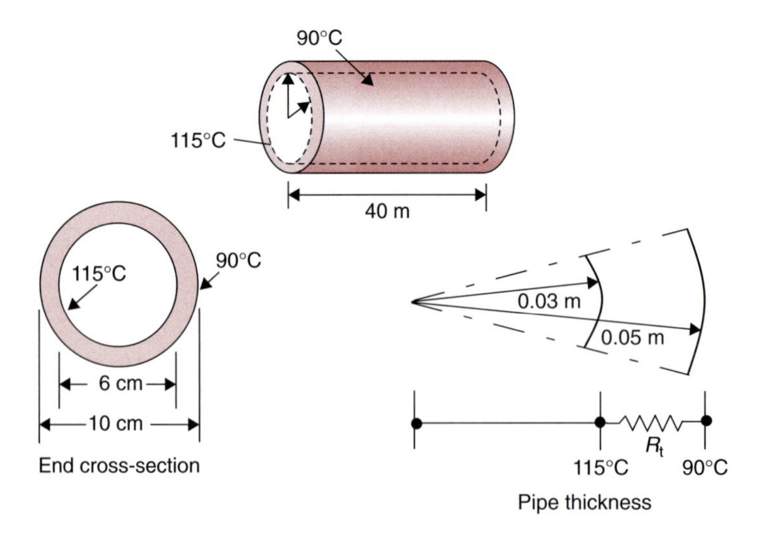

■ **Figure E4.7** Thermal resistance circuit for heat transfer through a pipe.

Given

Thickness of pipe = 2 cm = 0.02 m
Inside diameter = 6 cm = 0.06 m
Thermal conductivity $k = 43\ W/(m\ °C)$
Length $L = 40\ m$
Inside temperature $T_i = 115°C$
Outside temperature $T_o = 90°C$

Approach

We will determine the thermal resistance in the cross-section of the pipe and then use it to calculate the rate of heat transfer, using Equation (4.40).

Solution

1. *Using Equation (4.41)*

$$R_t = \frac{\ln(0.05/0.03)}{2\pi \times 40[m] \times 43[W/(m°C)]}$$
$$= 4.727 \times 10^{-5}\ °C/W$$

2. *From Equation (4.40)*

$$q = \frac{115[°C] - 90[°C]}{4.727 \times 10^{-5}[°C/W]}$$
$$= 528,903\ W$$

3. *The total heat loss from the 40 m long pipe is 528,903 W.*

4.4.3 Heat Conduction in Multilayered Systems

4.4.3.1 *Composite Rectangular Wall (in Series)*

We will now consider heat transfer through a composite wall made of several materials of different thermal conductivities and thicknesses. An example is a wall of a cold storage, constructed of different layers of materials of different insulating properties. All materials are arranged in series in the direction of heat transfer, as shown in Figure 4.17.

From Fourier's law,

$$q = -kA\frac{dT}{dx}$$

This may be rewritten as

$$\Delta T = -\frac{q\Delta x}{kA} \tag{4.42}$$

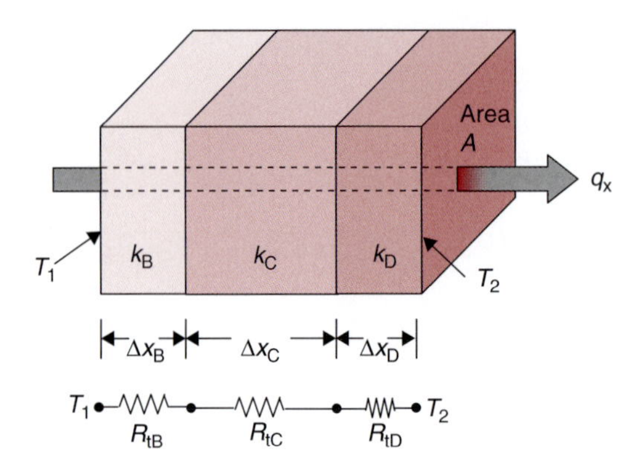

■ **Figure 4.17** Conductive heat transfer in a composite rectangular wall, also shown with a thermal resistance circuit.

Thus, for materials B, C, and D, we have

$$\Delta T_B = -\frac{q\Delta x_B}{k_B A} \qquad \Delta T_C = -\frac{q\Delta x_C}{k_C A} \qquad \Delta T_D = -\frac{q\Delta x_D}{k_D A} \tag{4.43}$$

From Figure 4.17,

$$\Delta T = T_1 - T_2 = \Delta T_B + \Delta T_C + \Delta T_D \tag{4.44}$$

From Equations (4.42), (4.43), and (4.44),

$$T_1 - T_2 = -\left(\frac{q\Delta x_B}{k_B A} + \frac{q\Delta x_C}{k_C A} + \frac{q\Delta x_D}{k_D A}\right) \tag{4.45}$$

or, rearranging the terms,

$$T_1 - T_2 = -\frac{q}{A}\left(\frac{\Delta x_B}{k_B} + \frac{\Delta x_C}{k_C} + \frac{\Delta x_D}{k_D}\right) \tag{4.46}$$

We can rewrite Equation (4.46) for thermal resistance as

$$q = \frac{T_2 - T_1}{\left(\dfrac{\Delta x_B}{k_B A} + \dfrac{\Delta x_C}{k_C A} + \dfrac{\Delta x_D}{k_D A}\right)} \tag{4.47}$$

or, using thermal resistance values for each layer, we can write Equation (4.47) as,

$$q = \frac{T_2 - T_1}{R_{tB} + R_{tC} + R_{tD}} \tag{4.48}$$

where

$$R_{tB} = \frac{\Delta x_B}{k_B A} \qquad R_{tC} = \frac{\Delta x_C}{k_C A} \qquad R_{tD} = \frac{\Delta x_D}{k_D A}$$

The thermal circuit for a multilayer rectangular system is shown in Figure 4.17. Example 4.8 illustrates the calculation of heat transfer through a multilayer wall.

Example 4.8

A cold storage wall (3 m × 6 m) is constructed of 15 cm thick concrete (thermal conductivity = 1.37 W/[m °C]). Insulation must be provided to maintain a heat transfer rate through the wall at or below 500 W (Fig. E4.8). If the thermal conductivity of the insulation is 0.04 W/(m °C), compute the required thickness of the insulation. The outside surface temperature of the wall is 38°C, and the inside wall temperature is 5°C.

Given

Wall dimensions = 3 m × 6 m

Thickness of concrete wall = 15 cm = 0.15 m

$k_{concrete}$ *= 1.37 W/(m °C)*

Maximum heat gain permitted, $q = 500$ *W*

$k_{insulation}$ *= 0.04 W/(m °C)*

Outer wall temperature = 38°C

Inside wall (concrete/insulation) temperature = 5°C

Approach

In this problem we know the two surface temperatures and the rate of heat transfer through the composite wall, therefore, using this information we will first calculate the thermal resistance in the concrete layer. Then we will calculate the thermal resistance in the insulation layer, which will yield the thickness value.

Solution

1. *Using Equation (4.48)*

$$q = \frac{(38-5)[°C]}{R_{t1} + R_{t2}}$$

2. *Thermal resistance in the concrete layer,* R_{t2} *is*

$$R_{t2} = \frac{0.15[m]}{1.37[W/(m°C)] \times 18[m^2]}$$

$$R_{t2} = 0.0061°C/W$$

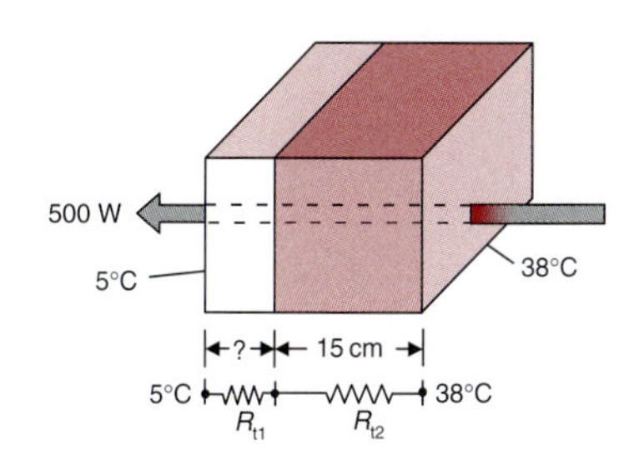

Figure E4.8 Heat transfer through a two-layered wall.

3. *From Step 1,*

$$\frac{(38-5)[°C]}{R_{t1}+0.0061[°C/W]}=500$$

or,

$$R_{t1}=\frac{(38-5)[°C]}{500[W]}-0.0061[°C/W]$$

$$R_{t1}=0.06\,°C/W$$

4. *From Equation (4.48)*

$$\Delta x_B = R_{tB}k_BA$$

$$\text{Thickness of insulation} = 0.06[°C/W]\times 0.04[W/(m\,°C)]\times 18[m^2]$$

$$= 0.043\,m = 4.3\,cm$$

5. *An insulation with a thickness of 4.3 cm will ensure that heat loss from the wall will remain below 500 W. This thickness of insulation allows a 91% reduction in heat loss.*

4.4.3.2 *Composite Cylindrical Tube (in Series)*

Figure 4.18 shows a composite cylindrical tube made of two layers of materials, A and B. An example is a steel pipe covered with a layer of insulating material. The rate of heat transfer in this composite tube can be calculated as follows.

In Section 4.4.2 we found that rate of heat transfer through a single-wall cylinder is

$$q_r = \frac{(T_i - T_o)}{\left[\dfrac{\ln(r_o/r_i)}{2\pi Lk}\right]}$$

The rate of heat transfer through a composite cylinder using thermal resistances of the two layers is

$$q_r = \frac{(T_1 - T_3)}{R_{tA} + R_{tB}} \tag{4.49}$$

or, substituting the individual thermal resistance values,

$$q_r = \frac{(T_1 - T_3)}{\dfrac{\ln(r_2/r_1)}{2\pi Lk_A} + \dfrac{\ln(r_3/r_2)}{2\pi Lk_B}} \tag{4.50}$$

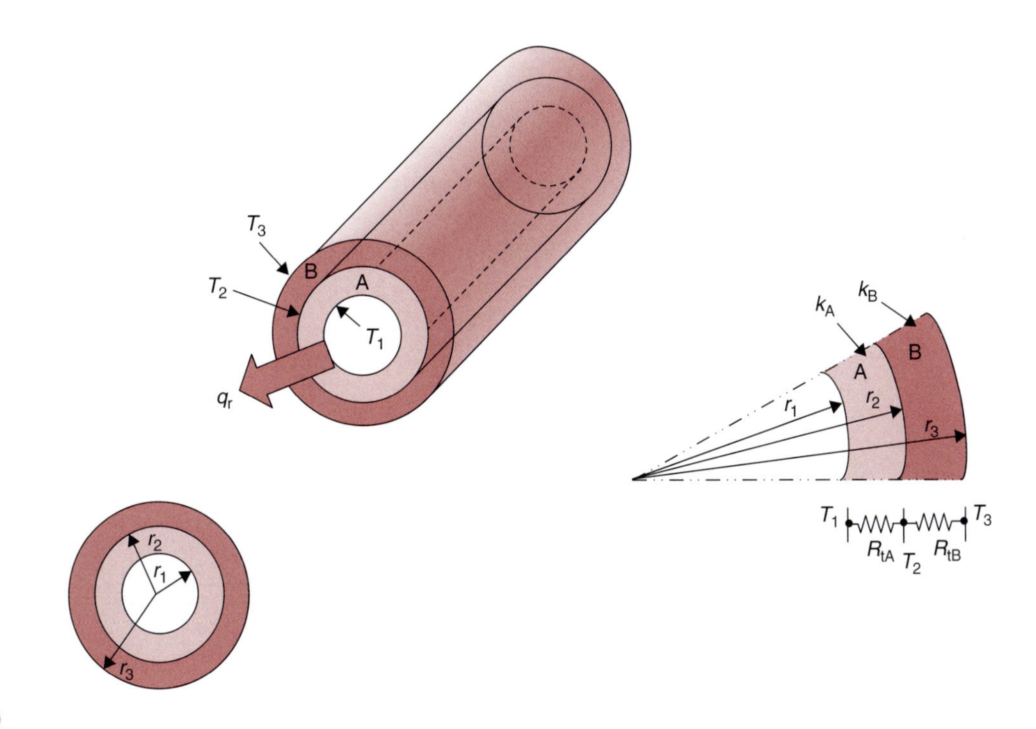

W

Figure 4.18 Conductive heat transfer in concentric cylindrical pipes, also shown with a thermal resistance circuit.

The preceding equation is useful in calculating the rate of heat transfer through a multilayered cylinder. Note that if there were three layers present between the two surfaces with temperatures T_1 and T_3, then we just add another thermal resistance term in the denominator.

Suppose we need to know the temperature at the interface between two layers, T_2, as shown in Figure 4.18. First, we calculate the steady-state rate of heat transfer using Equation (4.50), noting that under steady-state conditions, q_r has the same value through each layer of the composite wall. Then, we can use the following equation, which represents the thermal resistance between the known temperature, T_1, and the unknown temperature, T_2.

$$T_2 = T_1 - q\left(\frac{\ln(r_2/r_1)}{2\pi L k_A}\right) \qquad (4.51)$$

This procedure to solve problems for unknown interfacial temperatures is illustrated in Example 4.9.

Example 4.9

A stainless-steel pipe (thermal conductivity = 17 W/[m °C]) is being used to convey heated oil (Fig. E4.9). The inside surface temperature is 130°C. The pipe is 2 cm thick with an inside diameter of 8 cm.The pipe is insulated with 0.04 m thick insulation (thermal conductivity = 0.035 W/[m °C]). The outer insulation temperature is 25°C. Calculate the temperature of the interface between steel and insulation, assume steady-state conditions.

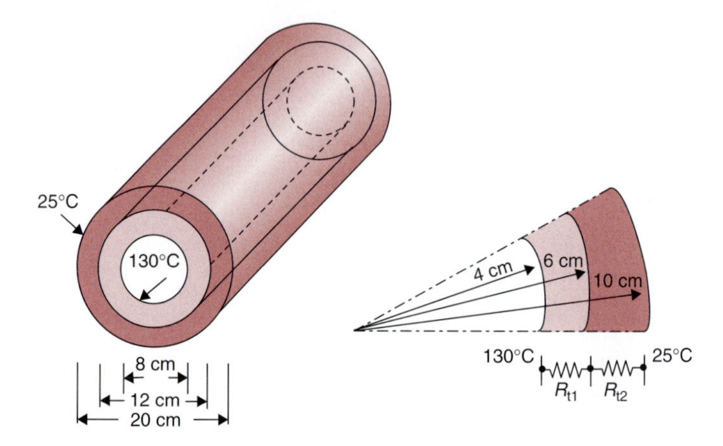

■ **Figure E4.9** Heat transfer through a multilayered pipe.

Given

Thickness of pipe = 2 cm = 0.02 m
Inside diameter = 8 cm = 0.08 m
$k_{steel} = 17\ W/(m\ °C)$
Thickness of insulation = 0.04 m
$k_{insulation} = 0.035\ W/(m\ °C)$
Inside pipe surface temperature = 130°C
Outside insulation surface temperature = 25°C
Pipe length = 1 m (assumed)

Approach

We will first calculate the two thermal resistances, in the pipe and the insulation. Then we will obtain the rate of heat transfer through the composite layer. Finally, we will use the thermal resistance of the pipe alone to determine the temperature at the interface between the pipe and insulation.

Solution

1. *Thermal resistance in the pipe layer is, from Equation (4.41),*

$$R_{t1} = \frac{\ln(0.06/0.04)}{2\pi \times 1[m] \times 17[W/(m°C)]}$$

$$= 0.0038\ °C/W$$

2. *Similarly, the thermal resistance in the insulation layer is,*

$$R_{t2} = \frac{ln(0.1/0.06)}{2\pi \times 1[m] \times 0.035[W/(m\ °C)]}$$
$$= 2.3229\ °C/W$$

3. *Using Equation (4.49), the rate of heat transfer is*

$$q = \frac{(130 - 25)[°C]}{0.0038[°C/W] + 2.3229[°C/W]}$$
$$= 45.13\ W$$

4. *Using Equation (4.40)*

$$45.13[W] = \frac{(130 - T)[°C]}{0.0038[°C/W]}$$
$$T = 130\ [°C] - 0.171\ [°C]$$
$$T = 129.83\ °C$$

5. *The interfacial temperature is 129.8°C. This temperature is very close to the inside pipe temperature of 130°C, due to the high thermal conductivity of the steel pipe. The interfacial temperature between a hot surface and insulation must be known to ensure that the insulation will be able to withstand that temperature.*

Example 4.10

A stainless-steel pipe (thermal conductivity = 15 W/[m K]) is being used to transport heated oil at 125°C (Fig. E4.10). The inside temperature of the pipe is 120°C. The pipe has an inside diameter of 5 cm and is 1 cm thick. Insulation is necessary to keep the heat loss from the oil below 25 W/m length of the pipe. Due to space limitations, only 5 cm thick insulation can be provided. The outside surface temperature of the insulation must be above 20°C (the dew point temperature of surrounding air) to avoid condensation of water on the surface of insulation. Calculate the thermal conductivity of insulation that will result in minimum heat loss while avoiding water condensation on its surface.

Given

Thermal conductivity of steel = 15 W/(m K)
Inside pipe surface temperature = 120°C
Inside diameter = 0.05 m
Pipe thickness = 0.01 m
Heat loss permitted in 1 m length of pipe = 25 W
Insulation thickness = 0.05 m
Outside surface temperature >20°C = 21°C (assumed)

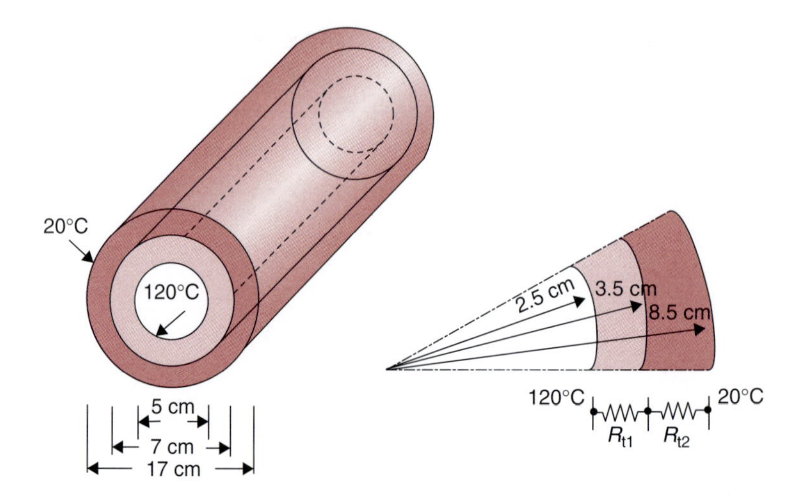

■ **Figure E4.10** Heat transfer through a multilayered pipe.

Approach

We will first calculate the thermal resistance in the steel layer, and set up an equation for the thermal resistance in the insulation layer. Then we will substitute the thermal resistance values into Equation (4.50). The only unknown, thermal conductivity, k, will be then calculated.

Solution

1. *Thermal resistance in the steel layer is*

$$R_{t1} = \frac{\ln(3.5/2.5)}{2\pi \times 1[m] \times 15[W/(m\,°C)]} = 0.0036\,°C/W$$

2. *Thermal resistance in the insulation layer is*

$$R_{t2} = \frac{\ln(8.5/3.5)}{2\pi \times 1[m] \times k[W/(m\,°C)]} = \frac{0.1412[1/m]}{k[W/(m\,°C)]}$$

3. *Substituting the two thermal resistance values in Equation (4.50)*

$$25[W] = \frac{(120-21)[°C]}{0.0036[°C/W] + \dfrac{0.1412[1/m]}{k[W/(m\,°C)]}}$$

or,

$$k = 0.0357\,W/(m\,°C)$$

4. *An insulation with a thermal conductivity of 0.0357 W/(m °C) will ensure that no condensation will occur on its outer surface.*

4.4.4 Estimation of Convective Heat-Transfer Coefficient

In Section 4.3.1 on the conduction mode of heat transfer, we observed that any material undergoing conduction heating or cooling remains stationary. Conduction is the main mode of heat transfer within solids. Now we will consider heat transfer between a solid and a surrounding fluid, a mode of heat transfer called convection. In this case, the material experiencing heating or cooling (a fluid) also moves. The movement of fluid may be due to the natural buoyancy effects or caused by artificial means, such as a pump in the case of a liquid or a blower for air.

Determination of the rate of heat transfer due to convection is complicated because of the presence of fluid motion. In Chapter 2, we noted that a velocity profile develops when a fluid flows over a solid surface because of the viscous properties of the fluid material. The fluid next to the wall does not move but "sticks" to it, with an increasing velocity away from the wall. A boundary layer develops within the flowing fluid, with a pronounced influence of viscous properties of the fluid. This layer moves all the way to the center of a pipe, as was shown in Figure 2.14. The parabolic velocity profile under laminar flow conditions indicates that the drag caused by the sticky layer in contact with the solid surface influences velocity at the pipe center.

Similar to the velocity profile, a temperature profile develops in a fluid as it flows through a pipe, as shown in Figure 4.19. Suppose the temperature of the pipe surface is kept constant at T_s, and the fluid enters with a uniform temperature, T_i. A temperature profile develops because the fluid in contact with the pipe surface quickly reaches the wall temperature, thus setting up a temperature gradient as shown in the figure. A thermal boundary layer develops. At the end of the thermal entrance region, the boundary layer extends all the way to the pipe centerline.

Therefore, when heating or cooling a fluid as it flows through a pipe, two boundary layers develop—a hydrodynamic boundary layer and a thermal

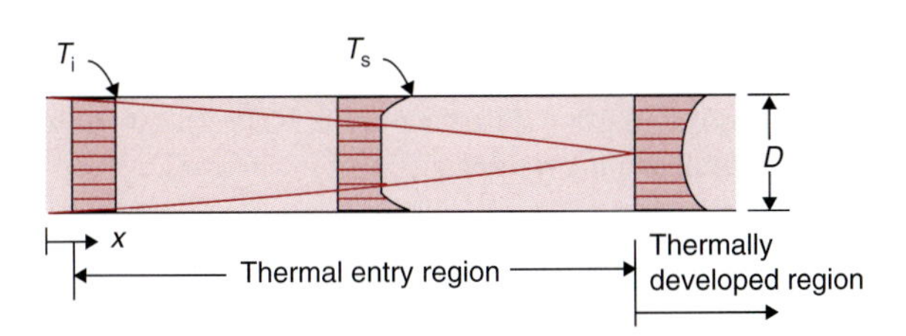

Figure 4.19 Thermal entry region in fluid flowing in a pipe.

boundary layer. These boundary layers have a major influence on the rate of heat transfer between the pipe surface and the fluid. The mathematics involved in an analytical treatment of this subject is complicated and beyond the scope of this book. However, there is an equally useful procedure called the *empirical approach*, which is widely used to determine the rate of convective heat transfer. A drawback of the empirical approach is that it requires a large number of experiments to obtain the required data. We overcome this problem and keep the data analysis manageable by using dimensionless numbers. To formulate this approach, first we will identify and review the required dimensionless numbers: Reynolds number, N_{Re}, Nusselt number, N_{Nu}, and Prandtl number, N_{Pr}.

The Reynolds number was described in Section 2.3.2. It provides an indication of the inertial and viscous forces present in a fluid. The Reynolds number is calculated using Equation (2.20).

The second required dimensionless number for our data analysis is Nusselt number—the dimensionless form of convective heat transfer coefficient, h. Consider a fluid layer of thickness l, as shown in Figure 4.20. The temperature difference between the top and bottom of the layer is ΔT. If the fluid is stationary, then the rate of heat transfer will be due to conduction, and the rate of heat transfer will be

$$q_{\text{conduction}} = -kA\frac{\Delta T}{l} \tag{4.52}$$

However, if the fluid layer is moving, then the heat transfer will be due to convection, and the rate of heat transfer using Newton's law of cooling will be

$$q_{\text{convection}} = hA\Delta T \tag{4.53}$$

Dividing Equation (4.53) by (4.52), we get

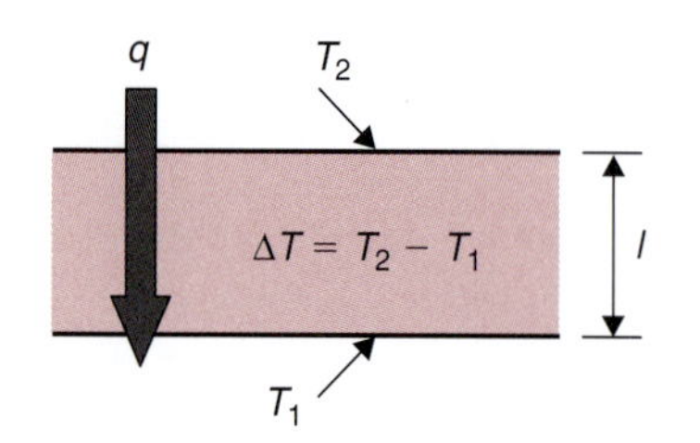

■ **Figure 4.20** Heat transfer through a fluid layer.

$$\frac{q_{\text{convection}}}{q_{\text{conduction}}} = \frac{hA\Delta T}{kA\Delta T/l} = \frac{hl}{k} \equiv N_{Nu} \tag{4.54}$$

Replacing thickness l with a more general term for dimension, the characteristic dimension d_c, we get

$$N_{Nu} \equiv \frac{hd_c}{k} \tag{4.55}$$

Nusselt number may be viewed as an enhancement in the rate of heat transfer caused by convection over the conduction mode. Therefore, if $N_{Nu} = 1$, then there is no improvement in the rate of heat transfer due to convection. However, if $N_{Nu} = 5$, the rate of convective heat transfer due to fluid motion is five times the rate of heat transfer if the fluid in contact with the solid surface is stagnant. The fact that by blowing air over a hot surface we can cool it faster is due to increased Nusselt number and consequently to an increased rate of heat transfer.

The third required dimensionless number for the empirical approach to determine convective heat transfer is Prandtl number, N_{Pr}, which describes the thickness of the hydrodynamic boundary layer compared with the thermal boundary layer. It is the ratio between the molecular diffusivity of momentum to the molecular diffusivity of heat. Or,

$$N_{Pr} = \frac{\text{molecular diffusivity of momentum}}{\text{molecular diffusivity of heat}} \tag{4.56}$$

or,

$$N_{Pr} = \frac{\text{kinematic viscosity}}{\text{thermal diffusivity}} = \frac{\nu}{\alpha} \tag{4.57}$$

Substituting Equations (2.11) and (4.12) in Equation (4.57),

$$N_{Pr} = \frac{\mu c_p}{k} \tag{4.58}$$

If $N_{Pr} = 1$, then the thickness of the hydrodynamic and thermal boundary layers will be exactly the same. On the other hand, if $N_{Pr} << 1$, the molecular diffusivity of heat will be much larger than that of momentum. Therefore, the heat will dissipate much faster, as in the case of a liquid metal flowing in a pipe. For gases, N_{Pr} is about 0.7, and for water it is around 10.

With a basic understanding of these three dimensionless numbers, we will now plan the following experiment to determine convective rate of heat transfer. Assume that a fluid is flowing in a heated pipe. We are interested in determining convective rate of heat transfer from the inside surface of the heated pipe into the fluid flowing inside the pipe, as shown in Figure 4.21. We carry out this experiment by pumping a fluid such as water, entering at a velocity of u_i at a temperature of T_i and flowing parallel to the inside surface of the pipe. The pipe

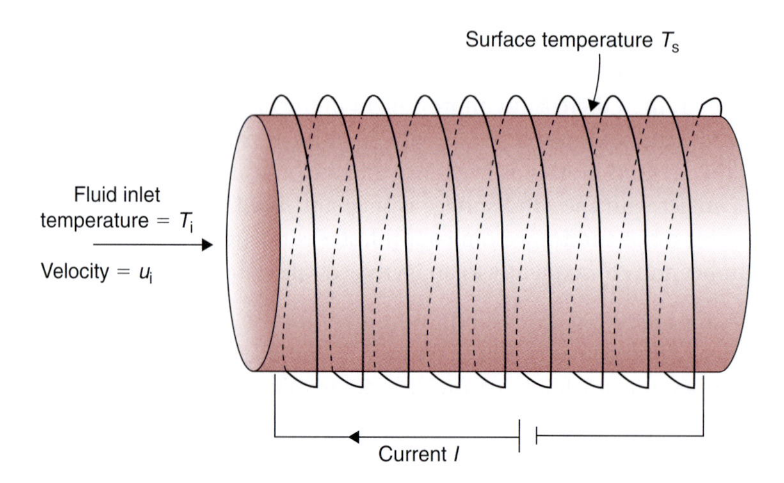

■ **Figure 4.21** Heating of fluid with electrically heated pipe surface.

is heated using an electrical heater so that the inside pipe surface is maintained at temperature T_s, which is higher than the inlet fluid temperature, T_i. We measure the electric current, I, and electrical resistance, R_E, and calculate the product of the two to determine the rate of heat transfer, q. The pipe is well insulated so that all the electrically generated heat transfers into the fluid. Thus, we can experimentally determine values of q, A, T_i, u_i, and T_s. Using Equation (4.53), we can calculate the convective heat transfer coefficient, h.

If we repeat this experiment with a different diameter pipe or different temperature of pipe surface, a new value of h will be obtained. It should become clear that we can perform a series of experiments to obtain h values that are a function of the operating variables, q, A, u_i, T_i, and T_s. The disadvantage of this experiment is that a large amount of experimental data are generated, and organizing these data for meaningful applications is a daunting task. However, the data analysis can be greatly simplified if we combine various properties and operating variables into the three dimensionless numbers, N_{Re}, N_{Nu}, and N_{Pr}, which will accommodate all the properties and variables that are important to our experiment.

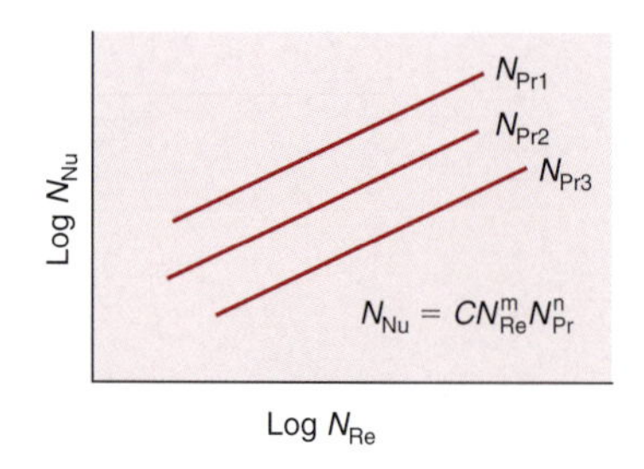

■ **Figure 4.22** A plot of Nusselt and Reynolds numbers on a log-log scale.

Thus, for each experimental set, we calculate the respective dimensionless numbers and, using log-log scale, plot the Nusselt number as a function of the Reynolds number for different values of the Prandtl number. Figure 4.22 shows a typical plot. It has been experimentally determined that for a given fluid with a fixed Prandtl number, straight line plots are obtained on the log-log scale, as shown in Figure 4.22.

This type of graphical relationship may be conveniently expressed with an equation as

$$N_{Nu} = CN_{Re}^{m}N_{Pr}^{n} \tag{4.59}$$

where C, m, and n are coefficients.

By substituting the experimentally obtained coefficients in Equation (4.59), we obtain *empirical correlations* specific for a given condition. Several researchers have determined these empirical correlations for a variety of operating conditions, such as fluid flow inside a pipe, over a pipe, or over a sphere. Different correlations are obtained, depending on whether the flow is laminar or turbulent.

A suggested methodology to solve problems requiring the calculation of convective heat transfer coefficients using empirical correlations is as follows:

1. *Identify flow geometry*. The first step in a calculation involving convection heat transfer is to clearly identify the geometrical shape of the solid surface in contact with the fluid and its dimensions. For example, is it a pipe, sphere, rectangular duct, or a rectangular plate? Is the fluid flowing inside a pipe or over the outside surface?

2. *Identify the fluid and determine its properties*. The second step is to identify the type of fluid. Is it water, air, or a liquid food? Determine the average fluid temperature far away from the solid surface, T_∞. In some cases the average inlet and outlet temperatures may be different, for example, in a heat exchanger; in that case calculate the average fluid temperature as follows:

$$T_\infty = \frac{T_i + T_e}{2} \tag{4.60}$$

   where T_i is the average inlet fluid temperature and T_e is the average exit fluid temperature. Use the average fluid temperature, T_∞, to obtain physical and thermal properties of the fluid, such as viscosity, density, and thermal conductivity, from appropriate tables (such as Table A.4.1 for water, Table A.4.4 for air), paying careful attention to the units of each property.

3. *Calculate the Reynolds number*. Using the velocity of the fluid, fluid properties and the characteristic dimension of the object

in contact with the fluid, calculate the Reynolds number. The Reynolds number is necessary to determine whether the flow is laminar, transitional, or turbulent. This information is required to select an appropriate empirical correlation.

4. *Select an appropriate empirical correlation*. Using the information from steps (1) and (3), select an empirical correlation of the form given in Equation (4.59) for the conditions and geometry of the object that resembles the one being investigated (as presented later in this section). For example, if the given problem involves turbulent water flow in a pipe, select the correlation given in Equation (4.67). Using the selected correlation, calculate Nusselt number and finally the convective heat transfer coefficient.

The convective heat-transfer coefficient h is predicted from empirical correlations. The coefficient is influenced by such parameters as type and velocity of the fluid, physical properties of the fluid, temperature difference, and geometrical shape of the physical system under consideration.

The empirical correlations useful in predicting h are presented in the following sections for both forced and free convection. We will discuss selected physical systems that are most commonly encountered in convective heat transfer in food processing. For other situations refer to handbooks such as Rotstein et al. (1997) or Heldman and Lund (1992). All correlations apply to Newtonian fluids only. For expressions for non-Newtonian fluids, the textbook by Heldman and Singh (1981) is recommended.

4.4.4.1 *Forced Convection*

In forced convection, a fluid is forced to move over a solid surface by external mechanical means, such as an electric fan, pump, or a stirrer (Fig. 4.23). The general correlation between the dimensionless numbers is

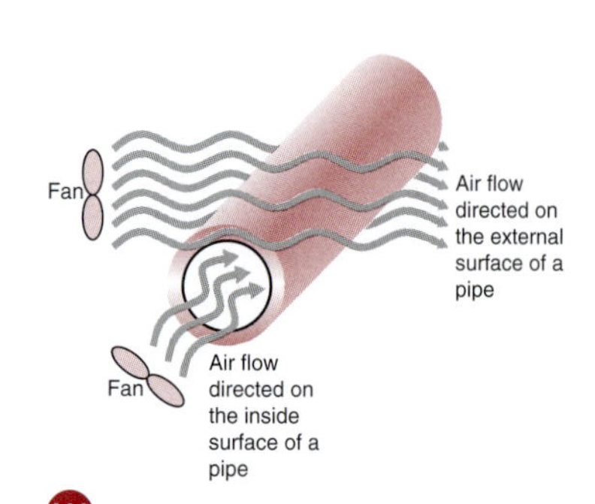

Figure 4.23 Forced convective heat transfer from a pipe with flow inside and outside the pipe.

$$N_{Nu} = \Phi(N_{Re}, N_{Pr}) \tag{4.61}$$

where N_{Nu} is Nusselt number $= hd_c/k$; h is convective heat-transfer coefficient (W/[m^2 °C]); d_c is the characteristic dimension (m); k is thermal conductivity of fluid (W/[m °C]); N_{Re} is Reynolds number $= \rho \bar{u} d_c/\mu$; ρ is density of fluid (kg/m^3); $\bar{u}$ is velocity of fluid (m/s); μ is viscosity (Pa s); N_{Pr} is Prandtl number $= \mu c_p/k$; c_p is specific heat (kJ/[kg °C]); and Φ stands for "function of".

Laminar flow in pipes

1. Fully developed conditions with constant surface temperature of the pipe:

$$N_{Nu} = 3.66 \tag{4.62}$$

where thermal conductivity of the fluid is obtained at average fluid temperature, T_∞, and d_c is the inside diameter of the pipe.

2. Fully developed conditions with uniform surface heat flux:

$$N_{Nu} = 4.36 \tag{4.63}$$

where thermal conductivity of the fluid is obtained at average fluid temperature, T_∞, and d_c is the inside diameter of the pipe.

3. For both entry region and fully developed flow conditions:

$$N_{Nu} = 1.86\left(N_{Re} \times N_{Pr} \times \frac{d_c}{L}\right)^{0.33}\left(\frac{\mu_b}{\mu_w}\right)^{0.14} \tag{4.64}$$

where L is the length of pipe (m); characteristic dimension, d_c, is the inside diameter of the pipe; all physical properties are evaluated at the average fluid temperature, T_∞, except μ_w, which is evaluated at the surface temperature of the wall.

Transition flow in pipes

For Reynolds numbers between 2100 and 10,000,

$$N_{Nu} = \frac{(f/8)(N_{Re} - 1000)N_{Pr}}{1 + 12.7(f/8)^{1/2}(N_{Pr}^{2/3} - 1)} \tag{4.65}$$

where all fluid properties are evaluated at the average fluid temperature, T_∞, d_c is the inside diameter of the pipe, and the friction factor, f, is obtained for smooth pipes using the following expression:

$$f = \frac{1}{(0.790 \ln N_{Re} - 1.64)^2} \tag{4.66}$$

Turbulent flow in pipes

The following equation may be used for Reynolds numbers greater than 10,000:

$$N_{Nu} = 0.023 N_{Re}^{0.8} \times N_{Pr}^{0.33} \times \left(\frac{\mu_b}{\mu_w}\right)^{0.14} \tag{4.67}$$

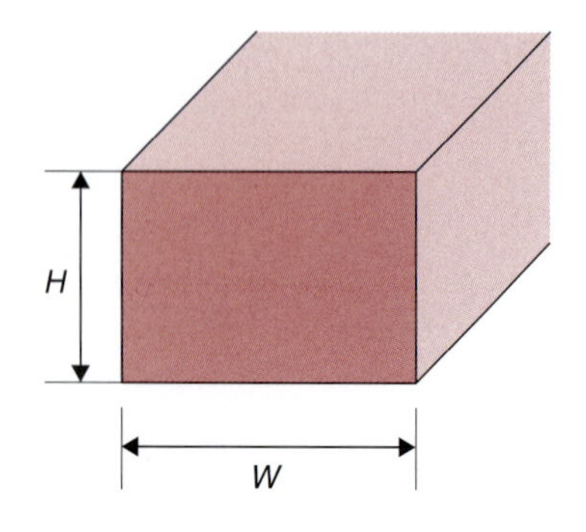

Figure 4.24 Cross-section of a rectangular duct.

Fluid properties are evaluated at the average film temperature, T_∞, except μ_w, which is evaluated at the wall temperature; d_c is the inside diameter of the pipe. Equation (4.67) is valid both for constant surface temperature and uniform heat flux conditions.

Convection in noncircular ducts

For noncircular ducts, an equivalent diameter, D_e, is used for the characteristic dimension:

$$D_e = \frac{4 \times \text{free area}}{\text{wetted perimeter}} \tag{4.68}$$

Figure 4.24 shows a rectangular duct with sides of length W and H. The equivalent diameter in this case will be equal to $2WH/(W + H)$.

Flow past immersed objects

In several applications, the fluid may flow past immersed objects. For these cases, the heat transfer depends on the geometrical shape of the object, relative position of the object, proximity of other objects, flow rate, and fluid properties.

For a flow past a single sphere, when the single sphere may be heated or cooled, the following equation will apply:

$$N_{Nu} = 2 + 0.60 N_{Re}^{0.5} \times N_{Pr}^{1/3} \quad \text{for} \begin{cases} 1 < N_{Re} < 70{,}000 \\ 0.6 < N_{Pr} < 400 \end{cases} \tag{4.69}$$

where the characteristic dimension, d_c, is the outside diameter of the sphere. The fluid properties are evaluated at the film temperature T_f where

$$T_f = \frac{T_s + T_\infty}{2}$$

For heat transfer in flow past other immersed objects, such as cylinders and plates, correlations are available in Perry and Chilton (1973).

Example 4.11

Water flowing at a rate of 0.02 kg/s is heated from 20 to 60°C in a horizontal pipe (inside diameter = 2.5 cm). The inside pipe surface temperature is 90°C (Fig. E4.11). Estimate the convective heat-transfer coefficient if the pipe is 1 m long.

Given

Water flow rate = 0.02 kg/s
Inlet temperature = 20°C
Exit temperature = 60°C
Inside diameter = 2.5 cm = 0.025 m
Inside pipe surface temperature = 90°C
Length of pipe = 1 m

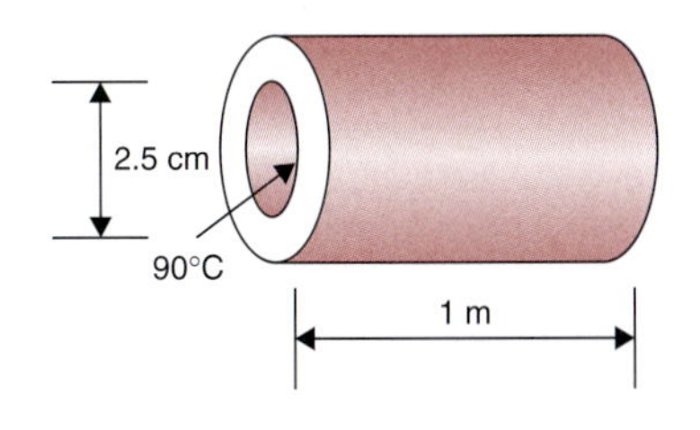

Figure E4.11 Convective heat transfer inside a pipe.

Approach

Since water is flowing due to some external means, the problem indicates forced convective heat transfer. We will first determine if the flow is laminar by calculating the Reynolds number. If the Reynolds number is less than 2100, we will use Equation (4.64) to calculate the Nusselt number. From the Nusselt number we will calculate the h value.

Solution

1. *Physical properties of water are needed to calculate the Reynolds number. All physical properties except μ_w must be evaluated at average bulk fluid temperature, (20 + 60)/2 = 40°C. From Table A.4.1 at 40°C,*
   Density $\rho = 992.2$ kg/m^3
   Specific heat $c_p = 4.175$ kJ/(kg °C)
   Thermal conductivity $k = 0.633$ W/(m °C)
   Viscosity (absolute) $\mu = 658.026 \times 10^{-6}$ Pa s
   Prandtl number $N_{Pr} = 4.3$

   Thus,

$$N_{Re} = \frac{\rho \bar{u} D}{\mu} = \frac{4\dot{m}}{\pi \mu D}$$

$$= \frac{(4)(0.02\ \text{kg/s})}{(\pi)(658.026 \times 10^{-6}\ \text{Pa s})(0.025\ \text{m})}$$

$$= 1547.9$$

   Note that 1 Pa = 1 kg/(m s^2). Since the Reynolds number is less than 2100, the flow is laminar.

2. *We select Equation (4.64), and using $\mu_w = 308.909 \times 10^{-6}$ Pa s at 90°C,*

$$N_{Nu} = 1.86(1547.9 \times 4.3 \times 0.025)^{0.33} \left(\frac{658.016 \times 10^{-6}}{308.909 \times 10^{-6}} \right)^{0.14}$$

$$= 11.2$$

3. *The convective heat-transfer coefficient can be obtained from the Nusselt number.*

$$h = \frac{N_{Nu}k}{D} = \frac{(11.2)(0.633\, W/[m\, °C])}{(0.025\, m)}$$
$$= 284\, W/(m^2\, °C)$$

4. *The convective heat-transfer coefficient is estimated to be 284 W/(m² °C).*

Example 4.12

If the rate of water flow in Example 4.11 is raised to 0.2 kg/s from 0.02 kg/s while all other conditions are kept the same, calculate the new convective heat-transfer coefficient.

Given

See Example 4.11

New mass flow rate of water = 0.2 kg/s

Approach

We will calculate the Reynolds number to find whether the flow is turbulent. If the flow is turbulent, we will use Equation (4.67) to compute the Nusselt number. The surface heat-transfer coefficient will be computed from the Nusselt number.

Solution

1. *First, we compute the Reynolds number using some of the properties obtained in Example 4.11.*

$$N_{Re} = \frac{(4)(0.2\, kg/s)}{(\pi)(658.026 \times 10^{-6}\, Pa\, s)(0.025\, m)} = 15{,}479$$

Thus, flow is turbulent.

2. *For turbulent flow, we select Equation (4.67).*

$$N_{Nu} = (0.023)(15479)^{0.8}(4.3)^{0.33}\left(\frac{658.026 \times 10^{-6}}{308.909 \times 10^{-6}}\right)^{0.14}$$
$$= 93$$

3. *The convective heat transfer can be computed as*

$$h = \frac{N_{Nu}k}{D} = \frac{(93)(0.633\, W/[m\, °C])}{(0.025\, m)}$$
$$= 2355\, W/(m^2\, °C)$$

4. *The convective heat-transfer coefficient for turbulent flow is estimated to be 2355 W/(m² °C). This value is more than eight times higher than the value of h for laminar flow calculated in Example 4.11.*

Example 4.13

What is the expected percent increase in convective heat-transfer coefficient if the velocity of a fluid is doubled while all other parameters are kept the same for turbulent flow in a pipe?

Approach

We will use Equation (4.67) to solve this problem.

Solution

1. *For turbulent flow in a pipe*

$$N_{Nu} = 0.023 N_{Re}^{0.8} \times N_{Pr}^{0.33} \times \left(\frac{\mu_b}{\mu_w}\right)^{0.14}$$

We can rewrite this equation as

$$N_{Nu_1} = f(\bar{u}_1)^{0.8}$$

$$N_{Nu_2} = f(\bar{u}_2)^{0.8}$$

$$\frac{N_{Nu_2}}{N_{Nu_1}} = \left(\frac{\bar{u}_2}{\bar{u}_1}\right)^{0.8}$$

2. *Since* $\bar{u}_2 = 2\bar{u}_1$,

$$\frac{N_{Nu_2}}{N_{Nu_1}} = (2)^{0.8} = 1.74$$

$$N_{Nu_2} = 1.74 N_{Nu_1}$$

3. *This expression implies that* $h_2 = 1.74h_1$. *Thus,*

$$\%\text{Increase} = \frac{1.74h_1 - h_1}{h_1} \times 100 = 74\%$$

4. *As expected, velocity has a considerable effect on the convective heat-transfer coefficient.*

Example 4.14

Calculate convective heat-transfer coefficient when air at 90°C is passed through a deep bed of green peas. Assume surface temperature of a pea to be 30°C. The diameter of each pea is 0.5 cm. The velocity of air through the bed is 0.3 m/s.

Given

Diameter of pea = 0.005 m
Temperature of air = 90°C
Temperature of a pea = 30°C
Velocity of air = 0.3 m/s

Approach

Since the air flows around a spherically immersed object (green pea), we will estimate N_{Nu} from Equation (4.69). The Nusselt number will give us the value for h.

Solution

1. *The properties of air are evaluated at T_f, where*

$$T_f = \frac{T_s + T_\infty}{2} = \frac{30 + 90}{2} = 60°C$$

From Table A.4.4,

$$\rho = 1.025 \text{ kg/m}^3$$
$$c_p = 1.017 \text{ kJ/(kg °C)}$$
$$k = 0.0279 \text{ W/(m °C)}$$
$$\mu = 19.907 \times 10^{-6} \text{ Pa s}$$
$$N_{Pr} = 0.71$$

2. *The Reynolds number is computed as*

$$N_{Re} = \frac{(1.025 \text{ kg/m}^3)(0.3 \text{ m/s})(0.005 \text{ m})}{(19.907 \times 10^{-6} \text{ Pa s})}$$
$$= 77.2$$

3. *From Equation (4.69),*

$$N_{Nu} = 2 + 0.6(77.2)^{0.5}(0.71)^{0.33}$$
$$= 6.71$$

4. *Thus*

$$h = \frac{6.71(0.0279\,W/[m\,°C])}{(0.005\,m)} = 37\,W/(m^2\,°C)$$

5. *The convective heat-transfer coefficient is 37 W/(m² °C).*

4.4.4.2 Free Convection

Free convection occurs because of density differences in fluids as they come into contact with a heated surface (Fig. 4.25). The low density of fluid at a higher temperature causes buoyancy forces, and as a result, heated fluid moves upward and colder fluid takes its place.

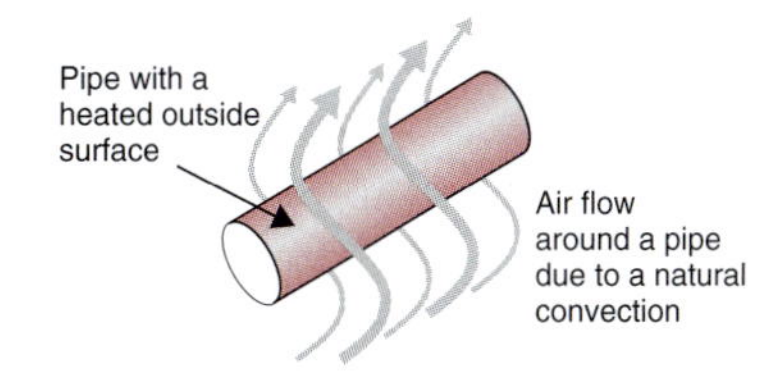

■ **Figure 4.25** Heat transfer from the outside of a heated pipe due to natural convection.

Empirical expressions useful in predicting convective heat-transfer coefficients are of the following form:

$$N_{Nu} = \frac{hd_c}{k} = a(N_{Ra})^m \tag{4.70}$$

where a and m are constants; N_{Ra}, is the Rayleigh number. Rayleigh number is a product of two dimensionless numbers, Grashof number and Prandtl number.

$$N_{Ra} = N_{Gr} \times N_{Pr} \tag{4.71}$$

The Grashof number, N_{Gr}, is defined as follows:

$$N_{Gr} = \frac{d_c^3 \rho^2 g \beta \Delta T}{\mu^2} \tag{4.72}$$

where d_c is characteristic dimension (m); ρ is density (kg/m³); g is acceleration due to gravity (9.80665 m/s²); β is coefficient of volumetric expansion (K^{-1}); ΔT is temperature difference between wall and the surrounding bulk (°C); and μ is viscosity (Pa s).

A Grashof number is a ratio between the buoyancy forces and viscous forces. Similar to the Reynolds number, the Grashof number is useful for determining whether a flow over an object is laminar or turbulent. For example, a Grashof number greater than 10^9 for fluid flow over vertical plates signifies a turbulent flow.

In the case of heat transfer due to free convection, physical properties are evaluated at the film temperature, $T_f = (T_s + T_\infty)/2$.

Table 4.2 gives various constants that may be used in Equation (4.70) for natural convection from vertical plates and cylinders, and from horizontal cylinders and plates.

Table 4.2 Coefficients for Equation (4.70) for Free Convection

Geometry	Characteristic Length	Range of N_{Ra}	a	m	Equation
Vertical plate 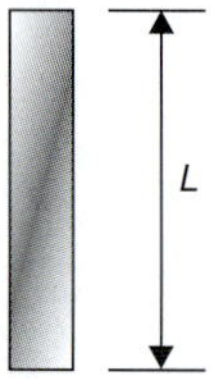	L	10^4–10^9 10^9–10^{13}	0.59 0.1	0.25 0.333	$N_{Nu} = a(N_{Ra})^m$
Inclined plate 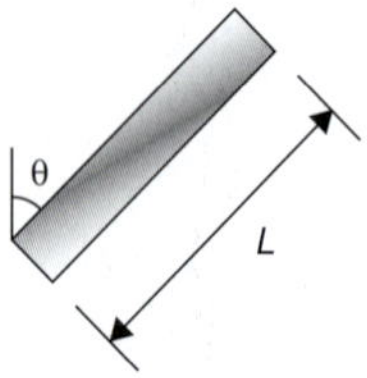	L				Use same equations as vertical plate, replace g by $g \cos\theta$ for $N_{Ra} < 10^9$
Horizontal plate Surface area = A Perimeter = p (a) Upper surface of a hot plate (or lower surface of a cold plate) 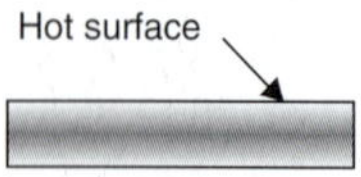	A/p	10^4–10^7 10^7–10^{11}	0.54 0.15	0.25 0.333	$N_{Nu} = a(N_{Ra})^m$
Horizontal plate Surface area = A Perimeter = p (b) Lower surface of a hot plate (or upper surface of a cold plate) 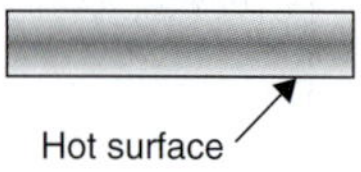	A/p	10^5–10^{11}	0.27	0.25	$N_{Nu} = a(N_{Ra})^m$

(Continued)

Table 4.2 Continued

Geometry	Characteristic Length	Range of N_{Ra}	a	m	Equation
Vertical cylinder	L				A vertical cylinder can be treated as a vertical plate when $D \geq \frac{35L}{N_{Gr}^{0.25}}$
Horizontal cylinder	D	10^{-5}–10^{12}			$N_{Nu} = \left\{ 0.6 + \frac{0.387 N_{Ra}^{1/6}}{\left[1 + \left(\frac{0.559}{N_{Pr}}\right)^{9/16}\right]^{8/27}} \right\}^2$
Sphere	$\frac{1}{2}\pi D$	$N_{Ra} \leq 10^{11}$ ($N_{Pr} \geq 0.7$)			$N_{Nu} = 2 + \frac{0.589 N_{Ra}^{1/4}}{\left[1 + \left(\frac{0.469}{N_{Pr}}\right)^{9/16}\right]^{4/9}}$

Example 4.15

Estimate the convective heat-transfer coefficient for convective heat loss from a horizontal 10 cm diameter steam pipe. The surface temperature of the uninsulated pipe is 130°C, and the air temperature is 30°C (Fig. E4.12).

Given

Diameter of pipe = 10 cm = 0.1 m

Pipe surface temperature T_w = 130°C

Ambient temperature T_∞ = 30°C

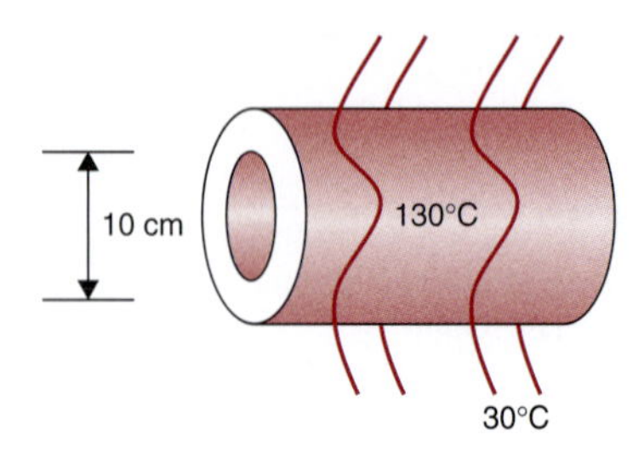

■ **Figure E4.12** Convective heat transfer from a horizontal pipe.

Approach

Since no mechanical means of moving air are indicated, heat loss from the horizontal pipe is by free convection. After finding the property values of air at film temperature, we will calculate the Grashof number. The product of the Grashof number and Prandtl number will allow determination of a and m parameters from Table 4.2; these parameters will be used in Equation (4.72). We will then compute the surface heat-transfer coefficient from the Nusselt number.

Solution

1. *Since no mechanical means of moving the air are indicated, heat loss is by free convection.*
2. *The film temperature is obtained as*

$$T_f = \frac{T_s + T_\infty}{2} = \frac{130 + 30}{2} = 80\,°C$$

3. *The properties of air at 80°C are obtained from Table A.4.4.*

$$\rho = 0.968\ kg/m^3$$

$$\beta = 2.83 \times 10^{-3}\ K^{-1}$$

$$c_p = 1.019\ kJ/(kg\ °C)$$

$$k = 0.0293\ W/(m\ °C)$$

$$\mu = 20.79 \times 10^{-6}\ N\ s/m^2$$

$$N_{Pr} = 0.71$$

$$g = 9.81\ m/s^2$$

4. *We calculate Rayleigh number, N_{Ra}, the product of N_{Gr} and N_{Pr}. The characteristic dimension is the outside diameter of the pipe.*

$$N_{Gr} = \frac{d_c^3 \rho^2 g \beta \Delta T}{\mu^2}$$

$$= \frac{(0.1\,m)^3 (0.968\ kg/m^3)^2 (9.81\ m/s^2)(2.83 \times 10^{-3}\ K^{-1})(130\,°C - 30\,°C)}{(20.79 \times 10^{-6}\ N\ s/m^2)^2}$$

$$= 6.019 \times 10^6$$

(Note that $1\ N = kg\ m/s^2$.) Thus,

$$N_{Gr} \times N_{Pr} = (6.019 \times 10^6)(0.71) = 4.27 \times 10^6$$

5. *From Table 4.2, for horizontal cylinder*

$$N_{Nu} = \left(0.6 + \frac{0.387(4.27 \times 10^6)^{1/6}}{\left(1 + \left(\frac{0.559}{0.71}\right)^{9/16}\right)^{8/27}} \right)^2$$

6. $N_{Nu} = 22$
7. *Thus*

$$h = \frac{(22)(0.0293\,W/[m\,°C])}{(0.1\,m)} = 6.5\,W/(m^2\,°C)$$

4.4.4.3 *Thermal Resistance in Convective Heat Transfer*

A thermal resistance term for convective heat transfer may be defined in a similar manner as in conductive heat transfer (Section 4.4). From Equation (4.19), we know that

$$q = hA(T_s - T_\infty) \tag{4.73}$$

or, rearranging terms in Equation (4.73),

$$q = \frac{T_s - T_\infty}{\left(\frac{1}{hA}\right)} \tag{4.74}$$

where the thermal resistance due to convection $(R_t)_{convection}$ is

$$(R_t)_{convection} = \frac{1}{hA} \tag{4.75}$$

In problems involving conduction and convection heat transfer in series, along the path of heat transfer, the thermal resistance due to convection is added to the thermal resistance due to conduction to obtain the total thermal resistance. We will discuss this further in the context of overall heat transfer involving both conduction and convection heat transfer.

4.4.5 Estimation of Overall Heat-Transfer Coefficient

In many heating/cooling applications, conductive and convective heat transfer may occur simultaneously. An example shown in Figure 4.26 involves heat transfer in a pipe that carries a fluid at a temperature greater than the temperature of the environment surrounding the outside of the pipe. In this case, heat must first transfer from the inside fluid by forced convection to the inside surface of the pipe, then by conduction through the pipe wall material, and finally by free convection from the outer pipe surface to the surrounding environment. Thus, heat transfer is through three layers in a series.

Using the approach of thermal resistance values, we can write:

$$q = \frac{T_i - T_\infty}{R_t} \tag{4.76}$$

where R_t is a combination of the thermal resistances in the inside convective layer, the conductive layer in the pipe material, and the outside convective layer, or

$$R_t = (R_t)_{\text{inside convection}} + (R_t)_{\text{conduction}} + (R_t)_{\text{outside convection}} \tag{4.77}$$

where

$$(R_t)_{\text{inside convection}} = \frac{1}{h_i A_i} \tag{4.78}$$

where h_i is the inside convective heat transfer coefficient, and A_i is the inside surface area of the pipe.

Resistance to heat transfer in the pipe wall is

$$(R_t)_{\text{conduction}} = \frac{\ln\left(\dfrac{r_o}{r_i}\right)}{2\pi k L} \tag{4.79}$$

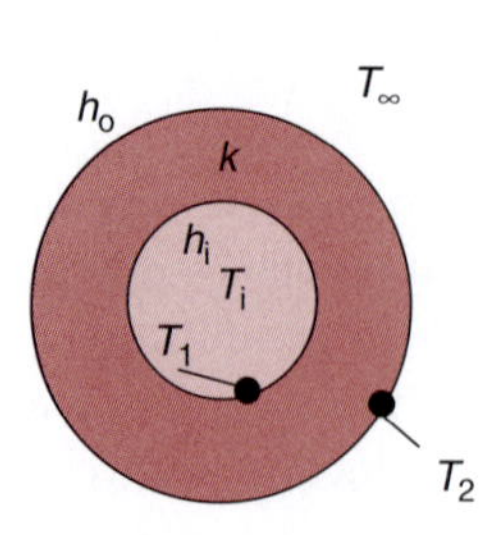

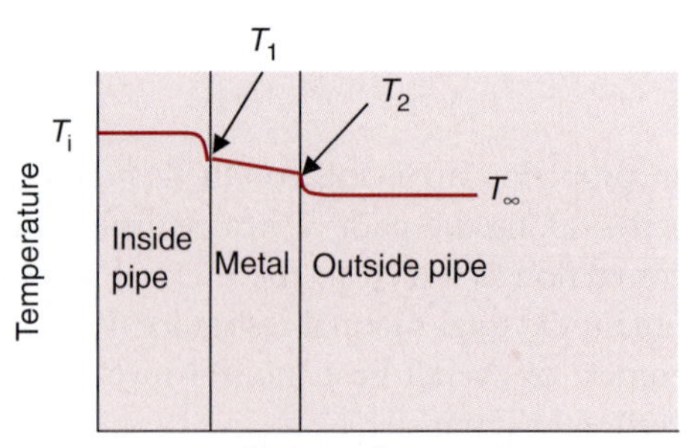

■ **Figure 4.26** Combined conductive and convective heat transfer.

where k is the thermal conductivity of the pipe material (W/[m K]), r_i is the inside radius (m), and r_o is the outside radius (m). Resistance to heat transfer due to convection at the outside pipe surface is

$$(R_t)_{\text{outside convection}} = \frac{1}{h_o A_o} \tag{4.80}$$

where h_o is the convective heat transfer coefficient at the outside surface of the pipe (W/[m^2 K]), and A_o is the outside surface area of the pipe. Substituting Equations (4.78), (4.79), and (4.80) in Equation (4.76), we obtain

$$q = \frac{T_i - T_\infty}{\dfrac{1}{h_i A_i} + \dfrac{\ln(r_o/r_i)}{2\pi Lk} + \dfrac{1}{h_o A_o}} \tag{4.81}$$

We can also write an expression for the overall heat transfer for this example as

$$q = U_i A_i (T_i - T_\infty) \tag{4.82}$$

where A_i is the inside area of the pipe, and U_i is the overall heat-transfer coefficient based on the inside area of the pipe. From Equation (4.82),

$$q = \frac{T_i - T_\infty}{\left(\dfrac{1}{U_i A_i}\right)} \tag{4.83}$$

From Equations (4.83) and (4.81) we obtain

$$\frac{1}{U_i A_i} = \frac{1}{h_i A_i} + \frac{\ln \dfrac{r_o}{r_i}}{2\pi Lk} + \frac{1}{h_o A_o} \tag{4.84}$$

Equation (4.84) is used to calculate the overall heat-transfer coefficient. The selection of area over which to calculate the overall heat transfer is quite arbitrary. For example, if U_o is selected as the overall heat-transfer coefficient based on outside area of the pipe, then Equation (4.84) is written as

$$\frac{1}{U_o A_o} \quad \frac{1}{h_i A_i} \quad \frac{\ln \dfrac{r_o}{r_i}}{2\pi Lk} + \frac{1}{h_o A_o} \tag{4.85}$$

and Equation (4.82) is modified to

$$q = U_o A_o (T_i - T_\infty) \tag{4.86}$$

Both Equations (4.82) and (4.86) yield the same value of the rate of heat transfer, q. This is shown in Example 4.16.

Example 4.16

A 2.5 cm inside diameter pipe is being used to convey a liquid food at 80°C (Fig. E4.13). The inside convective heat transfer coefficient is 10 W/(m^2 °C). The pipe (0.5 cm thick) is made of steel (thermal conductivity = 43 W/[m °C]). The outside ambient temperature is 20°C. The outside convective heat-transfer coefficient is 100 W/(m^2 °C). Calculate the overall heat transfer coefficient and the heat loss from 1 m length of the pipe.

Given

Inside diameter of pipe = 0.025 m
Bulk temperature of liquid food = 80°C
Inside convective heat-transfer coefficient = 10 W/(m^2 °C)
Outside convective heat-transfer coefficient = 100 W/(m^2 °C)
$k_{steel} = 43$ *W/(m °C)*
Outside ambient temperature = 20°C

Approach

The overall heat-transfer coefficient can be computed by using a basis of either the inside area of the pipe or the outside area of the pipe. We will use Equation (4.84) to find U_i and then use a modification of Equation (4.84) to find U_o. We will prove that the computed rate of heat flow will remain the same regardless of whether U_i or U_o is selected.

Solution

1. *Calculate the overall heat-transfer coefficient based on inside area using Equation (4.84):*

$$\frac{1}{U_i A_i} = \frac{1}{h_i A_i} + \frac{\ln\left(\frac{r_o}{r_i}\right)}{2\pi KL} + \frac{1}{h_o A_o}$$

2. *By canceling area terms and noting that $A_i = 2\pi r_i L$,*

$$\frac{1}{U_i} = \frac{1}{h_i} + \frac{r_i \ln\left(\frac{r_o}{r_i}\right)}{k} + \frac{r_i}{h_o r_o}$$

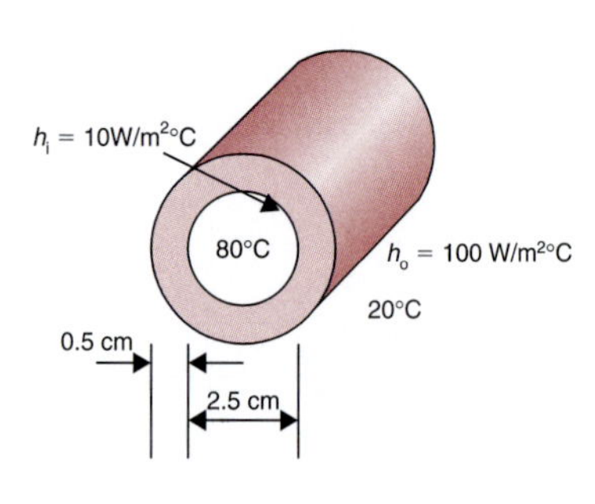

■ **Figure E4.13** Overall heat transfer in a pipe.

3. Substituting,

$$\frac{1}{U_i} = \frac{1}{10[W/(m^2\,°C)]} + \frac{0.0125[m] \times \ln\left(\frac{0.0175}{0.0125}\right)\left[\frac{m}{m}\right]}{43[W/(m\,°C)]}$$

$$+ \frac{0.0125[m]}{100[W/(m^2\,°C)] \times 0.0175[m]}$$

$$= 0.1 + 0.0001 + 0.00714 = 0.10724\,m^2\,°C/W$$

Thus, $U_i = 9.32\,W/(m^2\,°C)$.

4. Heat loss

$$q = U_i A_i (80 - 20)$$

$$= 9.32[W/(m^2\,°C)] \times 2\pi \times 1[m] \times 0.0125[m] \times 60[°C]$$

$$= 43.9\,W$$

5. Overall heat transfer coefficient based on outside area may be computed as

$$\frac{1}{U_o A_o} = \frac{1}{h_i A_i} + \frac{\ln\left(\frac{r_o}{r_i}\right)}{2\pi k L} + \frac{1}{h_o A_o}$$

6. By canceling area terms and noting that $A_o = 2\pi r_o L$,

$$\frac{1}{U_o} = \frac{r_o}{h_i r_i} + \frac{r_o \ln\left(\frac{r_o}{r_i}\right)}{k} + \frac{1}{h_o}$$

Substituting,

$$\frac{1}{U_o} = \frac{0.0175[m]}{10[W/(m^2\,°C)] \times 0.0125[m]}$$

$$+ \frac{0.0175[m] \times \ln\left(\frac{0.0175}{0.0125}\right)\left[\frac{m}{m}\right]}{43[W/(m\,°C)]}$$

$$+ \frac{1}{100[W/(m^2\,°C)]}$$

$$= 0.14 + 0.00014 + 0.01$$

$$= 0.1501\,m^2\,°C/W$$

$$U_o = 6.66\,W/(m^2\,°C)$$

7. *Heat loss*

$$q = U_o A_o (80 - 20)$$
$$= 6.66[W/(m^2\ °C)] \times 2\pi \times 0.0175[m] \times 1[m] \times 60[°C]$$
$$= 43.9 W$$

8. *As expected, the rate of heat loss remains the same regardless of which area was selected for computing overall heat-transfer coefficient.*
9. *It should be noted from steps (3) and (6) that the resistance offered by the metal wall is considerably smaller than the resistance offered in the convective layers.*

4.4.6 Fouling of Heat Transfer Surfaces

In heating equipment, when a liquid food comes into contact with a heated surface, some of its components may deposit on the hot surface, causing an increase in the resistance to heat transfer. This phenomenon of product buildup on the heat transfer surface is called *fouling*. A similar phenomenon is observed when a liquid is brought into contact with subcooled surfaces. Fouling of heat transfer surfaces not only increases thermal resistance but may also restrict fluid flow. Furthermore, valuable components of the food are lost to the fouled layer. Fouling is remedied by cleaning heating surfaces with strong chemicals that are also environmental pollutants.

Fouling is of major concern in the chemical process industries. Its role is even more pronounced in the food industry where many heat-sensitive components of foods can easily deposit on a heat-transfer surface. As a result, factory operations involving heating or cooling require frequent cleaning, often on a daily basis. Some of the common types of fouling and their underlying mechanisms are shown in Table 4.3.

The fouling layer often has a composition different from the liquid stream that causes fouling. With milk, which has a protein content of around 3%, the fouling deposits resulting at temperatures less than 110°C contain 50 to 60% protein and 30 to 35% minerals. About half of the protein in the fouled layer is β-lactoglobulin. When the temperature of the milk increases above 70 to 74°C, protein denaturation increases. The protein (β-lactoglobulin) first unfolds and the reactive sulphydryl groups are exposed. This is followed by polymerization (or aggregation) of the molecule with itself or other proteins including α-lactoglobulin.

Table 4.3 Common Mechanisms in Fouling of Heat Exchange Surfaces

Type of Fouling	Fouling Mechanism
Precipitation	Precipitation of dissolved substances. Salts such as $CaSO_4$, $CaCO_3$ cause scaling.
Chemical Reaction	Surface material acts as a reactant; chemical reactions of proteins, sugars, and fats.
Particulate	Accumulation of fine particles suspended in the processed fluids on the heat transfer surface.
Biological	Attachment of organisms both macro and micro on heat transfer surface.
Freezing	Solidification of liquid components on subcooled surfaces.
Corrosion	Heat transfer surface reacts with ambient and corrodes.

Fouling results from a complex series of reactions, and in heating processes these reactions are accelerated with temperature. To compensate for the reduced rate of heat transfer due to the increased thermal resistance, a larger heat-transfer surface area is required, which increases the cost of heat exchange equipment. For an operating heat exchanger with a fouled surface, the reduced rate of heat transfer is compensated for by using higher temperature gradients across the heat transfer medium. Consequently, the energy requirements to operate heat exchange equipment increase significantly. It has been estimated that the annual worldwide cost of fouling to process industries is several billion dollars.

Let us examine the role of fouling on heat transfer by considering the rates of heat transfer in a clean pipe and in one that has been fouled on both the inside and outside surfaces (Fig. 4.27). Assuming that the deposited layers are thin, the convective heat transfer coefficient on the inside surface of the fouled pipe, h_{fi}, will be same as that on the inside surface of the clean pipe h_{ci}. The same will hold true for the convective heat transfer coefficients on the outside surfaces; that is, $h_{fo} = h_{co}$. Similarly, the inside surface area of the fouled pipe, $A_{fi} = A_{ci} = A_i$, and, for the outside surface area, $A_{fo} = A_{co} = A_o$.

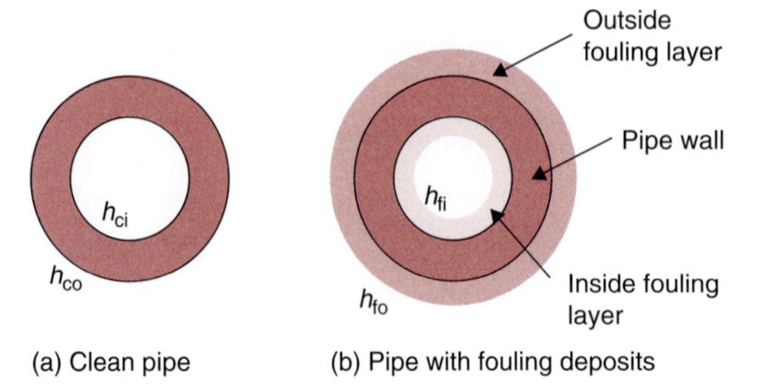

Figure 4.27 A pipe with fouling deposits on the inside and outside surfaces

Using Equation (4.85), we write an equation for the overall heat transfer coefficient based on the outside area for the clean pipe,

$$\frac{1}{U_{co}A_o} = \frac{1}{h_i A_i} + \frac{\ln \frac{r_o}{r_i}}{2\pi Lk} + \frac{1}{h_o A_o} \tag{4.87}$$

or,

$$\frac{1}{U_{co}} = \frac{A_o}{h_i A_i} + \frac{A_o \ln \frac{r_o}{r_i}}{2\pi Lk} + \frac{1}{h_o} \tag{4.88}$$

Next, we will consider a pipe that is fouled on both the inside and outside surface. As shown in Figure 4.27b, the fouling resistance due to the deposited layer on the inside of the pipe is R_{fi} (m^2°C/W) and for the outside pipe it is R_{fo} (m^2°C/W). Then for the fouled pipes,

$$\frac{1}{U_{fo}A_o} = \frac{1}{h_i A_i} + \frac{R_{fi}}{A_i} + \frac{\ln \frac{r_o}{r_i}}{2\pi Lk} + \frac{R_{fo}}{A_o} + \frac{1}{h_o A_o} \tag{4.89}$$

or,

$$\frac{1}{U_{fo}} = \frac{A_o}{h_i A_i} + \frac{R_{fi} A_o}{A_i} + \frac{A_o \ln \frac{r_o}{r_i}}{2\pi Lk} + R_{fo} + \frac{1}{h_o} \tag{4.90}$$

Since R_{fi} and R_{fo} are difficult to determine separately, the two terms for fouling resistance in Equations (4.88) are combined to get the total resistance due to fouling, R_{ft},

$$R_{ft} = \frac{A_o}{A_i} R_{fi} + R_{fo} \tag{4.91}$$

Then

$$\frac{1}{U_{fo}} = \frac{A_o}{h_i A_i} + \frac{A_o \ln \frac{r_o}{r_i}}{2\pi L k} + R_{ft} + \frac{1}{h_o} \tag{4.92}$$

Combining Equations (4.88) and (4.92), we get

$$\frac{1}{U_{fo}} = \frac{1}{U_{co}} + R_{ft} \tag{4.93}$$

In industrial practice, a term called Cleaning factor (C_F) is defined as a ratio between the two overall heat transfer coefficients, or,

$$C_F = \frac{U_{fo}}{U_{co}} \tag{4.94}$$

Note that the value of C_F is less than 1. Substituting Equation (4.93) in Equation (4.94), and rearranging terms we can write the fouling resistance in terms of the cleaning factor.

$$R_{ft} = \frac{1}{U_{co}}\left(\frac{1}{C_F} - 1\right) \tag{4.95}$$

In designing heat exchangers, we may want to determine the extra area required due to expected fouling of the heat transfer area. For this purpose, we consider the rates of heat transfer in the clean and fouled surface to be the same. Thus,

$$q = U_{co}A_{co}\Delta T_m = U_{fo}A_{fo}\Delta T_m \tag{4.96}$$

Then eliminating ΔT_m in the preceding equation,

$$U_{co}A_{co} = U_{fo}A_{fo} \tag{4.97}$$

or,

$$\frac{U_{co}}{U_{fo}} = \frac{A_{fo}}{A_{co}} \tag{4.98}$$

Combining Equations (4.94), (4.95), and (4.98), and rearranging terms, we get

$$R_{ft} = \frac{1}{U_{co}}\left(\frac{A_{fo}}{A_{co}} - 1\right) \tag{4.99}$$

In Equation (4.99) the term $\left(\frac{A_{fo}}{A_{co}} - 1\right)$ multiplied by 100 is the percent extra area required for the fouled surface compared with the clean surface. In the following example, we will use the previous derived relations to develop a graph of percent increase in area required to compensate for fouling for different values of overall heat transfer coefficients.

Example 4.17

Using a spreadsheet, develop a chart that shows the relationship between the total fouling resistance as a function of overall heat transfer coefficients of 1000 to 5000 W/m^{2}K for cleanliness factors of 0.8, 0.85, 0.9, 0.95. Also, develop a chart that demonstrates the increase in surface area required for fouling resistances of 0.0001, 0.001, 0.01, and 0.05 m^{2}K/W for an overall heat transfer coefficient varying from 1 to 10,000 W/m^{2}K.

Given

Part (a)

Overall heat transfer coefficient = 1000, 2000, 3000, 4000 and 5000 W/m^{2}K

Cleanliness factor = 0.8, 0.85, 0.9, 0.95

	A	B	C	D	E
1			Cleanliness Factor		
2	Overall Heat Transfer Coefficient	0.80	0.85	0.90	0.95
3	1000	2.50	1.76	1.11	0.53
4	2000	1.25	0.88	0.56	0.26
5	3000	0.83	0.59	0.37	0.18
6	4000	0.63	0.44	0.28	0.13
7	5000	0.50	0.35	0.22	0.11

1) Enter values as shown in cells A3:A7 and B2:E2
2) Enter=(1/$A3)*(1/B$2-1)*10000 in cell B3 and copy in cells B4:B7 and C3:E7

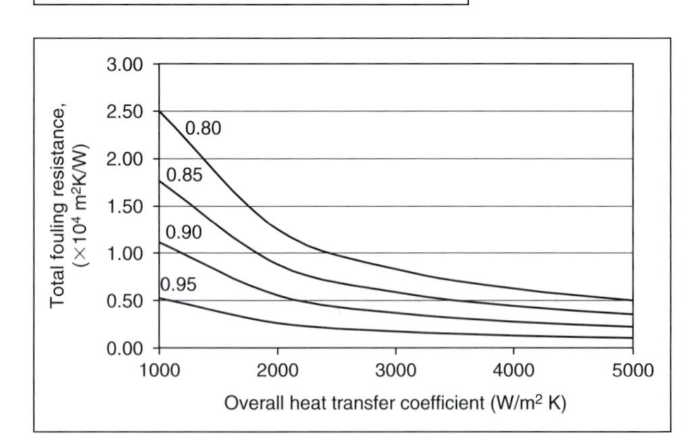

■ **Figure E4.14** Total fouling resistance as a function of overall heat transfer coefficient for different cleanliness factors.

Part (b)

Fouling resistance = 0.0001, 0.001, 0.01 and 0.05 m^2K/W
Overall heat transfer coefficient = 1, 10, 100, 1000, 10,000 W/m^2K

Approach

We will develop two spreadsheets using Excel™. First, we will calculate the fouling resistance for part (a) and the required increase in surface area for part (b). Then we will create a plot from the calculated data.

Solution

The first spreadsheet is developed using Equation (4.95) for the fouling resistance R_f as a function of the overall heat transfer coefficient for a clean pipe. The second spreadsheet is developed using Equation (4.99), and we will plot $\left(\frac{A_{fo}}{A_{co}} - 1\right) \times 100$ against overall heat transfer coefficient on a log-log scale. As we observe, from Figure E4.14, as the cleanliness factor decreases, the fouling resistance increases for the same value of clean-pipe overall heat transfer coefficient. This effect is more pronounced at low values of overall heat transfer coefficient. Similarly, from Figure E4.15,

	A	B	C	D	E
1			Fouling resistance (m^2K/W)		
2	Overall Heat Transfer Coefficient	0.0001	0.001	0.01	0.05
3	1	0.01	0.1	1	5
4	10	0.1	1	10	50
5	100	1	10	100	500
6	1000	10	100	1000	5000
7	10000	100	1000	10000	50000

1) Enter values as shown in cells A3:A7 and B2:E2
2) Enter = \$A3*B\$2*100 in cell B3 and
paste in cells B4:B7 and C3:E7

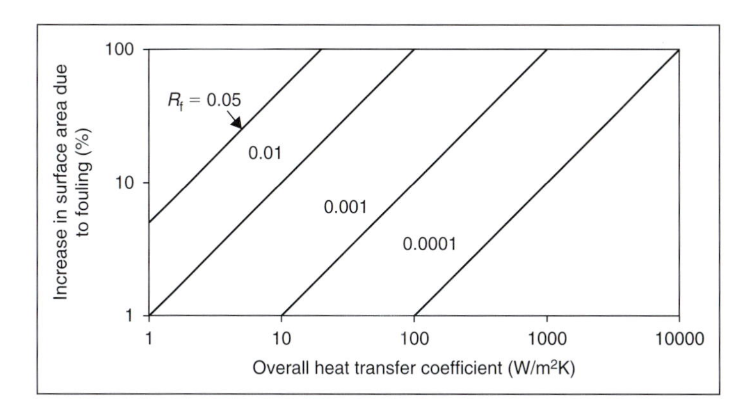

Figure E4.15 Increase in surface area due to fouling.

for a small increase in the resistance factor, for the same overall heat transfer coefficient, a much larger surface area is required. These graphs show that the resistance due to fouling has a substantial effect on heat transfer, and larger surface areas are required with increasing fouling.

4.4.7 Design of a Tubular Heat Exchanger

In Section 4.1, we examined a variety of heat exchange equipment used in the food process industry. Recall that a number of different geometrical configurations are used in the design of heat exchange equipment, such as tubular, plate, and scraped surface heat exchangers. The primary objective in using a heat exchanger is to transfer thermal energy from one fluid to another. In this section we will develop calculations necessary to design a tubular heat exchanger.

One of the key objectives in calculations involving a heat exchanger is to determine the required heat transfer area for a given application. We will use the following assumptions:

1. Heat transfer is under steady-state conditions.
2. The overall heat-transfer coefficient is constant throughout the length of pipe.
3. There is no axial conduction of heat in the metal pipe.
4. The heat exchanger is well insulated. The heat exchange is between the two liquid streams flowing in the heat exchanger. There is negligible heat loss to the surroundings.

Recall from Chapter 1 that change in heat energy in a fluid stream, if its temperature changes from T_1 to T_2, is expressed as:

$$q = \dot{m}c_p(T_1 - T_2) \tag{4.100}$$

where $\dot{m}$ is mass flow rate of a fluid (kg/s), c_p is specific heat of a fluid (kJ/[kg °C]), and the temperature change of a fluid is from some inlet temperature T_1 to an exit temperature T_2.

Consider a tubular heat exchanger, as shown in Figure 4.28. A *hot* fluid, H, enters the heat exchanger at location (1) and it flows through the inner pipe, exiting at location (2). Its temperature decreases from $T_{H,inlet}$ to $T_{H,exit}$. The second fluid, C, is a *cold* fluid that enters the annular space between the outer and inner pipes of the tubular heat exchanger at location (1) and exits at location (2). Its temperature

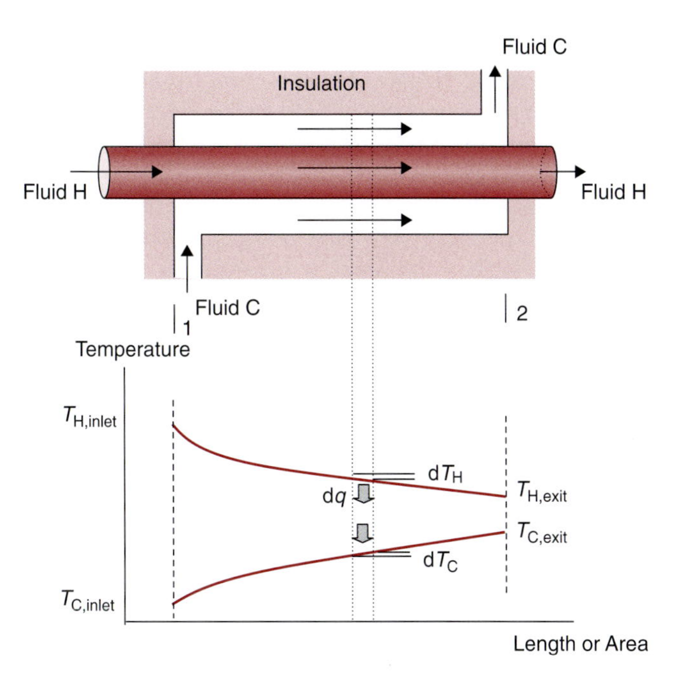

Figure 4.28 A concurrent flow heat exchanger and temperature plots.

increases from $T_{C,inlet}$ to $T_{C,exit}$. The outer pipe of the heat exchanger is covered with an insulation to prevent any heat exchange with the surroundings. Because the heat transfer occurs only between fluids H and C, the decrease in the heat energy of fluid H must equal the increase in the energy of fluid C. Therefore, conducting an energy balance, the rate of heat transfer between the fluids is:

$$q = \dot{m}_{H} c_{pH} (T_{H,inlet} - T_{H,exit}) = \dot{m}_{C} c_{pC} (T_{C,exit} - T_{C,inlet}) \qquad (4.101)$$

where c_{pH} is the specific heat of the hot fluid (kJ/[kg °C]), c_{pC} is the specific heat of the cold fluid (kJ/[kg °C]), $\dot{m}_H$ is the mass flow rate of the hot fluid (kg/s), and $\dot{m}_C$ is the mass flow rate of the cold fluid (kg/s).

Equation (4.101) is useful if we are interested in determining the inlet and exit temperatures of the two fluid streams. Furthermore, we may use this equation to determine the mass flow rate of either fluid stream, provided all other conditions are known. But, this equation does not provide us with any information about the size of the heat exchanger required for accomplishing a desired rate of heat transfer, and we cannot use it to determine how much thermal resistance to heat transfer exists between the two fluid streams. For those questions, we need to determine heat transfer perpendicular to the flow of the fluid streams, as discussed in the following.

Consider a thin slice of the heat exchanger, as shown in Figure 4.28. We want to determine the rate of heat transfer from fluid H to C, perpendicular to the direction of the fluid streams. For this thin slice of the heat exchanger, the rate of heat transfer, dq, from fluid H to fluid C may be expressed as:

$$dq = U\Delta T\, dA \tag{4.102}$$

where ΔT is the temperature difference between fluid H and fluid C. Note that this temperature difference, ΔT, varies from location (1) to (2) of the heat exchanger. At the inlet of the fluid streams, location (1), the temperature difference, ΔT, is $T_{H,inlet} - T_{C,inlet}$ and on the exit side, location (2), it is $T_{H,exit} - T_{C,exit}$ (Fig. 4.28). To solve Equation (4.102) we can substitute only one value of ΔT, or its average value that represents the temperature gradient perpendicular to the direction of the flow. Although it may be tempting to take an arithmetic average of the two ΔT values from locations (1) and (2), the arithmetic average value will be incorrect because, as seen in Figure 4.28, the temperature plots are nonlinear. Therefore, we will develop the following mathematical analysis to determine a value of ΔT that will correctly identify the "average" temperature difference between the fluids H and C as they flow through the heat exchanger.

The temperature difference, ΔT, between the two fluids H and C is

$$\Delta T = T_H - T_C \tag{4.103}$$

where T_H is the temperature of the hot stream and T_C is that of the cold stream. For a small differential ring element as shown in Figure 4.28, using energy balance for the hot stream H we get

$$dq = -\dot{m}_H c_{pH} dT_H \tag{4.104}$$

and, for cold stream C in the differential element,

$$dq = \dot{m}_C c_{pC} dT_C \tag{4.105}$$

In Equation (4.104), dT_H is a negative quantity; therefore, we added a negative sign to obtain a positive value for dq. Solving for dT_H and dT_C, we obtain

$$dT_H = -\frac{dq}{\dot{m}_H c_{pH}} \tag{4.106}$$

and

$$dT_C = \frac{dq}{\dot{m}_C c_{pC}} \tag{4.107}$$

Then, subtracting Equation (4.107) from Equation (4.106),

$$dT_H - dT_C = d(T_H - T_C) = -dq\left(\frac{1}{\dot{m}_H c_{pH}} + \frac{1}{\dot{m}_C c_{pC}}\right) \tag{4.108}$$

Using Equations (4.102) and (4.103), and substituting in Equation (4.108),

$$\frac{d(T_H - T_C)}{(T_H - T_C)} = -U\left(\frac{1}{\dot{m}_H c_{pH}} + \frac{1}{\dot{m}_C c_{pC}}\right) dA \tag{4.109}$$

Integrating Equation (4.109) from locations (1) to (2) shown in Figure 4.28,

$$\ln\frac{(T_{H,exit} - T_{C,exit})}{(T_{H,inlet} - T_{C,inlet})} = -UA\left(\frac{1}{\dot{m}_H c_{pH}} + \frac{1}{\dot{m}_C c_{pC}}\right) \tag{4.110}$$

Noting that

$$\begin{aligned} T_{H,inlet} - T_{C,inlet} &= \Delta T_1 \\ T_{H,exit} - T_{C,exit} &= \Delta T_2 \end{aligned} \tag{4.111}$$

we get

$$\ln\frac{\Delta T_2}{\Delta T_1} = -UA\left(\frac{1}{\dot{m}_H c_{pH}} + \frac{1}{\dot{m}_C c_{pC}}\right) \tag{4.112}$$

Substituting Equation (4.101) in Equation (4.112),

$$\ln\left(\frac{\Delta T_2}{\Delta T_1}\right) = -UA\left(\frac{T_{H,inlet} - T_{H,exit}}{q} + \frac{T_{C,exit} - T_{C,inlet}}{q}\right) \tag{4.113}$$

Rearranging terms in Equation (4.113),

$$\ln\left(\frac{\Delta T_2}{\Delta T_1}\right) = -\frac{UA}{q}[(T_{H,inlet} - T_{C,inlet}) - (T_{H,exit} - T_{C,exit})] \tag{4.114}$$

Substituting Equation (4.111) in Equation (4.114), we obtain

$$\ln\left(\frac{\Delta T_2}{\Delta T_1}\right) = -\frac{UA}{q}(\Delta T_1 - \Delta T_2) \tag{4.115}$$

Rearranging terms,

$$q = UA\frac{\Delta T_2 - \Delta T_1}{\ln\dfrac{\Delta T_2}{\Delta T_1}} \tag{4.116}$$

$$q = UA(\Delta T_{lm}) \tag{4.117}$$

where

$$\Delta T_{lm} = \frac{\Delta T_2 - \Delta T_1}{\ln\dfrac{\Delta T_2}{\Delta T_1}} \tag{4.118}$$

ΔT_{lm} is called the log mean temperature difference (LMTD). Equation (4.117) is used to design a heat exchanger and determine its area and the overall resistance to heat transfer, as illustrated in Examples 4.18 and 4.19.

Example 4.18

A liquid food (specific heat = 4.0 kJ/[kg °C]) flows in the inner pipe of a double-pipe heat exchanger. The liquid food enters the heat exchanger at 20°C and exits at 60°C (Fig. E4.16). The flow rate of the liquid food is 0.5 kg/s. In the annular section, hot water at 90°C enters the heat exchanger and flows countercurrently at a flow rate of 1 kg/s. The average specific heat of water is 4.18 kJ/(kg °C). Assume steady-state conditions.

1. Calculate the exit temperature of water.
2. Calculate log-mean temperature difference.
3. If the average overall heat transfer coefficient is 2000 W/(m^2 °C) and the diameter of the inner pipe is 5 cm, calculate the length of the heat exchanger.
4. Repeat these calculations for parallel-flow configuration.

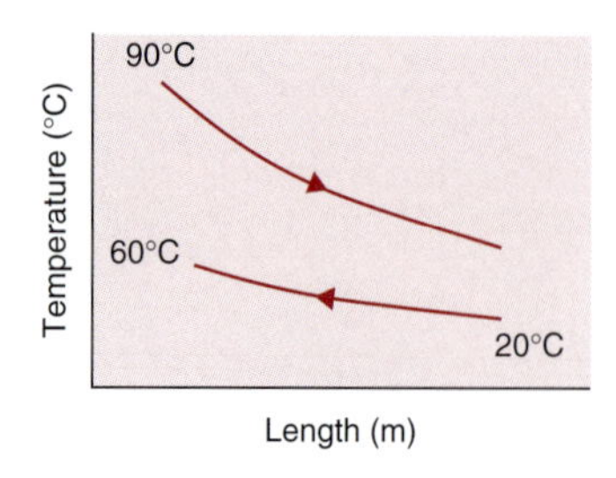

Figure E4.16 A countercurrent flow heat exchanger with unknown exit temperatures.

Given

Liquid food:

Inlet temperature = 20°C

Exit temperature = 60°C

Specific heat = 4.0 kJ/(kg °C)
Flow rate = 0.5 kg/s

Water:

Inlet temperature = 90°C
Specific heat = 4.18 kJ/(kg °C)
Flow rate = 1.0 kg/s

Heat exchanger:

Diameter of inner pipe = 5 cm
Flow = countercurrent

Approach

We will first calculate the exit temperature of hot water by using a simple heat balance equation. Then we will compute log-mean temperature difference. The length of the heat exchanger will be determined from Equation (4.117). The solution will be repeated for parallel-flow configuration to obtain a new value for log-mean temperature difference and length of the heat exchanger.

Solution

1. *Using a simple heat balance,*

$$q = \dot{m}_C c_{pC} \Delta T_C = \dot{m}_h c_{ph} \Delta T_h$$
$$= (0.5\ kg/s)(4\ kJ/[kg\,°C])(60°C - 20°C)$$
$$= (1\ kg/s)(4.18\ kJ/[kg\,°C])(90°C - T_e\,°C)$$
$$T_e = 70.9°C$$

2. *The exit temperature of water is 70.9°C.*
3. *From Equation (4.118)*

$$(\Delta T)_{lm} = \frac{\Delta(T)_1 - \Delta(T)_2}{\ln\left[\dfrac{\Delta(T)_1}{\Delta(T)_2}\right]} = \frac{(70.9 - 20) - (90 - 60)}{\ln\left(\dfrac{50.9}{30}\right)}$$
$$= 39.5°C$$

4. *The log-mean temperature difference is 39.5°C.*
5. *From Equation (4.117),*

$$q = UA(\Delta T)_{lm} = U\pi D_i L(\Delta T)_{lm}$$

where q, from step (1), is

$$q = (0.5\ kg/s)(4\ kJ/[kg\,°C])(60°C - 20°C) = 80\ kJ/s$$

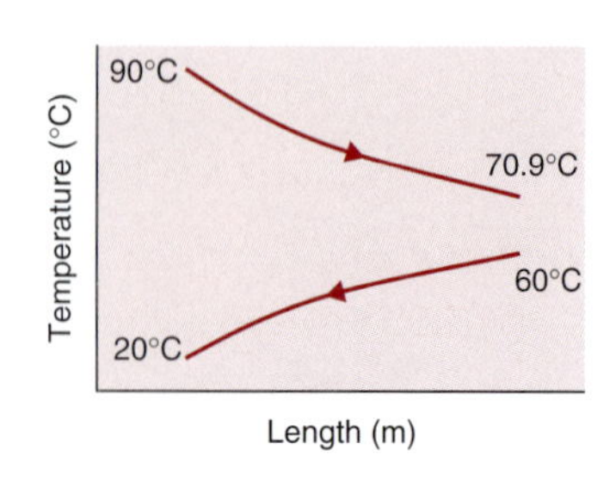

■ **Figure E4.17** Temperature plots for a counterflow heat exchanger.

Thus,

$$L = \frac{(80\ kJ/s)(1000\ J/kJ)}{(\pi)(0.05\ m)(39.5°C)(2000\ W/[m^2\ °C])} = 6.45\ m$$

6. *The length of the heat exchanger, when operated counter-currently, is 6.5 m.*
7. *For parallel-flow operation, the system diagram will be as shown in Figure E4.16.*
8. *Assuming that for parallel flow the exit temperature will be the same as for counterflow, $T_e = 70.9°C$.*
9. *Log-mean temperature difference is calculated from Equation (4.118).*

$$(\Delta T)_{lm} = \frac{(90-20)-(70.9-60)}{\ln\left(\dfrac{90-20}{70.9-60}\right)} = 31.8°C$$

10. *The log-mean temperature difference for parallel flow is 31.8°C, about 8°C less than that for the countercurrent flow arrangement.*
11. *The length can be computed as in step (5).*

$$L = \frac{(80\ kJ/s)(1000\ J/kJ)}{(\pi)(0.05\ m)(31.8°C)(2000\ W/[m^2\ °C])} = 8\ m$$

12. *The length of the heat exchanger, when operated as parallel flow, is 8 m. This length of heat exchanger is longer by 1.55 m to obtain the same exit temperature of hot-water stream as for the counterflow arrangement.*

Example 4.19

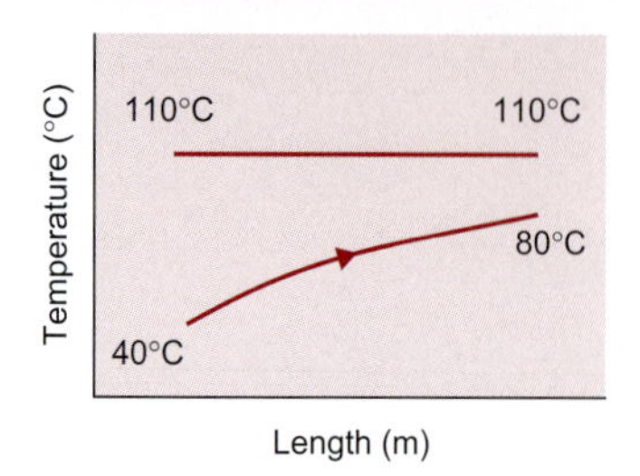

■ **Figure E4.18** Temperature plots for a parallel-flow heat exchanger.

Steam with 90% quality, at a pressure of 143.27 kPa, is condensing in the outer annular space of a 5 m-long double-pipe heat exchanger (Fig. E4.18). Liquid food is flowing at a rate of 0.5 kg/s in the inner pipe. The inner pipe has an inside diameter of 5 cm. The specific heat of the liquid food is 3.9 kJ/(kg °C). The inlet temperature of the liquid food is 40°C and the exit temperature is 80°C.

a. Calculate the average overall heat-transfer coefficient.

b. If the resistance to conductive heat transfer caused by the inner steel pipe is negligible, and the convective heat-transfer coefficient on the steam side is very large (approaches infinity), estimate the convective heat-transfer coefficient for the liquid food in the inside pipe.

Given

Steam pressure = 143.27 kPa
Length = 5 m
Flow rate of liquid = 0.5 kg/s
Inside diameter = 0.05 m
Specific heat = 3.9 kJ/(kg °C)
Product inlet temperature = 40°C
Product exit temperature = 80°C

Approach

We will obtain steam temperature from Table A.4.2. We also note that steam quality has no effect on the steam condensation temperature. We will calculate the heat required to raise the liquid food temperature from 40 to 80°C. Next, we will calculate log-mean temperature difference. Then we will obtain the overall heat-transfer coefficient by equating the heat gain of the liquid food and heat transfer across the pipe wall, from steam to the liquid food.

Solution

Part (a)

1. *From steam table (Table A.4.2), steam temperature = 110°C.*
2. $$q = \dot{m}c_p\Delta T$$
$$= (0.5\ kg/s)(3.9\ kJ/[kg\,°C])(1000\ J/kJ)(80°C - 40°C)$$
$$= 78{,}000\ J/s$$
3. $$q = UA(\Delta T)_{lm} = \dot{m}c_p\Delta T$$
$$(\Delta T)_{lm} = \frac{(110-40)-(110-80)}{\ln\left(\dfrac{110-40}{110-80}\right)} = 47.2°C$$
and
$$A = \pi(0.05)(5) = 0.785\ m^2$$
4. $$U = \frac{\dot{m}c_p\Delta T}{A(\Delta T)_{lm}} = \frac{(78{,}800\ J/s)}{(0.785\ m^2)(47.2°C)} = 2105\ W/(m^2\,°C)$$
5. *Overall heat-transfer coefficient = 2105 W/(m^2 °C)*

Part (b)

The overall heat-transfer equation may be written as follows:

$$\frac{1}{U_i A_i} = \frac{1}{h_i A_i} + \frac{\ln\left(\frac{r_o}{r_i}\right)}{2\pi kL} + \frac{1}{h_o A_o}$$

The second term on the right-hand side of the above equation is zero, since the resistance offered by steel to conductive heat transfer is considered negligible. Likewise, the third term is zero, because the convective heat-transfer coefficient on the steam side is very large.

Therefore,

$$U_i = h_i$$

or

$$h_i = 2105 \text{ W/(m}^2 \text{ °C)}$$

4.4.8 The Effectiveness-NTU Method for Designing Heat Exchangers

In the preceding section, we used the log-mean-temperature-difference (LMTD) approach to design a heat exchanger. The LMTD approach works well when we are designing a new heat exchanger, where the temperatures of the fluid streams at the inlet and exit are known and we are interested in determining the size of the heat exchanger (in terms of the heat transfer area, length, and diameter of the pipe). However, in other situations when the size of the heat exchanger, and the inlet temperatures of the product and heating/cooling streams are known but the exit temperatures of the two streams are unknown, the LMTD approach can be used but the solution requires iterative procedures and becomes tedious. For this purpose, another calculation technique called the effectiveness-NTU method is easier to use. This method involves three dimensionless quantities, namely the heat capacity rate ratio, heat exchanger effectiveness, and number of transfer units (NTU).

4.4.8.1 *Heat Capacity Rate Ratio, C**

A heat capacity rate of a liquid stream is obtained as a product of the mass flow rate and specific heat capacity. Thus, for the hot and cold streams the heat capacity rates are, respectively:

$$C_H = \dot{m}_H c_{pH} \tag{4.119}$$

$$C_C = \dot{m}_C c_{pC} \tag{4.120}$$

These two quantities are evaluated using the given data in a problem; the smaller of the two quantities is called C_{min} and the larger C_{max}.

The heat capacity rate ratio, C^*, is defined as

$$C^* = \frac{C_{min}}{C_{max}} \tag{4.121}$$

4.4.8.2 *Heat Exchanger Effectiveness, ε_E*

The heat exchanger effectiveness is a ratio of the actual rate of heat transfer accomplished and the maximum attainable rate of heat transfer for a given heat exchanger. The heat exchanger effectiveness, ε_E, is defined as

$$\varepsilon_E = \frac{q_{actual}}{q_{max}} \tag{4.122}$$

The actual rate of heat transfer can be determined for both hot and cold streams as

$$q_{actual} = C_H(T_{H,inlet} - T_{H,exit}) = C_C(T_{C,exit} - T_{C,inlet}) \tag{4.123}$$

And, the maximum rate of attainable heat transfer is obtained by observing that in any heat exchanger the maximum possible temperature difference is between the temperatures of hot and cold streams at the inlet. This temperature difference is multiplied with the minimum heat capacity rate, C_{min}, to obtain q_{max}:

$$q_{max} = C_{min}(T_{H,inlet} - T_{C,inlet}) \tag{4.124}$$

Thus, from Equation (4.122),

$$q_{actual} = \varepsilon_E q_{max} = \varepsilon_E C_{min}(T_{H,inlet} - T_{C,inlet}) \tag{4.125}$$

Table 4.4 Effectiveness-NTU Relations for Heat Exchangers

Type of Heat Exchanger	Effectiveness Relation
Double pipe Concurrent flow	$\varepsilon_E = \dfrac{1-\exp[-NTU(1+C^*)]}{1+C^*}$
Double pipe Countercurrent flow	$\varepsilon_E = \dfrac{1-\exp[-NTU(1-C^*)]}{1-C^*\exp[-NTU(1-C^*)]}$
Shell and tube: One-shell pass 2, 4, 6… tube passes	$\varepsilon_E = \dfrac{2}{1+C^*+\sqrt{1+C^{*2}}\dfrac{1+\exp\left[-NTU\sqrt{1+C^{*2}}\right]}{1-\exp\left[-NTU\sqrt{1+C^{*2}}\right]}}$
Plate heat exchanger	$\varepsilon_E = \dfrac{\exp\left[(1-C^*)\times NTU\right]-1}{\exp\left[(1-C^*)\times NTU\right]-C^*}$
All heat exchangers, $C^* = 0$	$\varepsilon_E = 1-\exp(-NTU)$

4.4.8.3 *Number of Transfer Units, NTU*

The number of transfer units provides a measure of the heat transfer surface area for a given overall heat transfer coefficient and minimum heat capacity rate. It is expressed as

$$NTU = \frac{UA}{C_{min}} \tag{4.126}$$

where A is the heat transfer area (m^2), U is the overall heat transfer coefficient ($W/m^2\,°C$) based on the selected area (see Section 4.4.5), and C_{min} is the minimum heat capacity rate ($W/°C$).

Relationships between NTU and effectiveness can be obtained for different types of heat exchangers with prescribed flow conditions (such as counterflow or concurrent flow). These relationships include the heat capacity rate ratios. Some of these relationships for commonly used heat exchangers are shown in Tables 4.4 and 4.5. In Table 4.4, the exchanger effectiveness is given in terms of NTU, and in Table 4.5, the NTU values are given as a function of the exchanger effectiveness. The steps involved in using the effectiveness-NTU method are illustrated in the following example.

Table 4.5 NTU-Effectiveness Relations for Heat Exchangers

Type of Heat Exchanger	NTU Relation
Double pipe Concurrent flow	$NTU = -\frac{\ln[1-\varepsilon_E(1+C^*)]}{1+C^*}$
Double pipe Countercurrent flow	$NTU = \frac{1}{1-C^*}\ln\left[\frac{1-C^*\varepsilon_E}{1-\varepsilon_E}\right] \quad (C^* < 1)$ $NTU = \frac{\varepsilon_E}{1-\varepsilon_E} \quad (C^* = 1)$
Shell and tube: One-shell pass 2, 4, 6... tube passes	$NTU = \frac{1}{\sqrt{1+C^{*2}}}\ln\frac{2-\varepsilon_E\left[1+C^* - \sqrt{1+C^{*2}}\right]}{2-\varepsilon_E\left[1+C^* + \sqrt{1+C^{*2}}\right]}$
Plate heat exchanger	$NTU = \frac{ln\left[\frac{(1-C^*)}{(1-\varepsilon_E)}\right]}{(1-C^*)}$
All heat exchangers, $C^* = 0$	$NTU = -ln(1-\varepsilon_E)$

Example 4.20

We will use some of the data from Example 4.18 to show the use of the effectiveness-NTU method in solving problems when both exit temperatures are unknown. A liquid food (specific heat = 4.0 kJ/[kg °C]) flows in the inner pipe of a double-pipe heat exchanger. The liquid food enters the heat exchanger at 20°C. The flow rate of the liquid food is 0.5 kg/s. In the annular section, hot water at 90°C enters the heat exchanger and flows in countercurrent direction at a flow rate of 1 kg/s. The average specific heat of water is 4.18 kJ/(kg °C). The average overall heat transfer coefficient based on the inside area is 2000 W/(m^2 °C), and the diameter of the inner pipe is 5 cm and length is 6.45 m. Assume steady state conditions. Calculate the exit temperature of liquid food and water.

Given

Liquid food:

Inlet temperature = 20°C

Specific heat = 4.0 kJ/(kg °C)

Flow rate = 0.5 kg/s

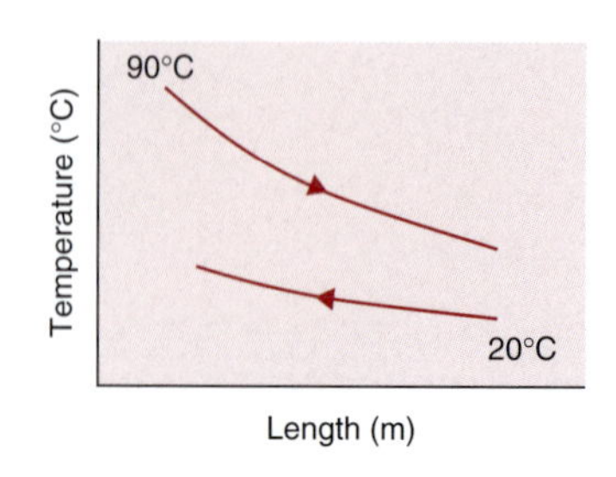

Figure E4.19 Temperature plots for a double-pipe heat exchanger.

Water:

Inlet temperature = 90°C

Specific heat = 4.18 kJ/(kg °C)

Flow rate = 1.0 kg/s

Heat exchanger:

Diameter of inner pipe = 5 cm

Length of inner pipe = 6.45 m

Overall heat transfer coefficient = 2000 W/m^2 °C

Flow = countercurrent

Approach

Since exit temperatures of both streams are unknown, we will use the effectiveness-NTU method. We will first calculate the maximum and minimum heat capacity rates. The two heat capacity rates will be used to calculate the heat capacity rate ratio, C^*. Next, we will obtain NTU from the given overall heat transfer coefficient, heat exchanger area, and the calculated value of C_{min}. Using the calculated value of NTU, we will use an appropriate table to determine the exchanger effectiveness. The definition of the exchanger effectiveness will be used to determine q_{actual}, and the unknown temperatures, $T_{H, exit}$ and $T_{L,exit}$.

Solution

1. *The heat capacity rates for the hot water and liquid food are, respectively,*

$$C_H = \dot{m}_H c_{pH}$$
$$= (1.0\ kg/s)(4.18\ kJ/[kg\ °C])$$
$$= 4.18\ kW/°C$$
$$C_L = \dot{m}_L c_{pL}$$
$$= (0.5\ kg/s)(4\ kJ/[kg\ °C])$$
$$= 2\ kW/°C$$

Therefore, using the smaller value of C between the above calculated values C_H and C_L,

$$C_{min} = 2\ kW/°C$$

Then, C^* is obtained as

$$C^* = \frac{2}{4.18} = 0.4785$$

2. *The NTU value is obtained from Equation (4.126) as*

$$NTU = \frac{UA}{C_{min}} = \frac{(2000\ W/m^2\ °C)(\pi)(0.05m)(6.45m)}{(2\ kW/°C)(1000\ W/kW)}$$

$$NTU = 1.0132$$

3. *From Table 4.4 we select an expression for ε_E for tubular heat exchanger, for counterflow, and substitute known terms as*

$$\varepsilon_E = \frac{1 - e^{(-1.0132(1-0.4785))}}{1 - 0.4785e^{(-1.0132(1-0.4785))}}$$

$$\varepsilon_E = 0.5717$$

According to Equation (4.124)

$$q_{max} = (2kW/°C)(90 - 20)(°C) = 140\ kW$$

$$\text{Since } \varepsilon_E = \frac{q_{actual}}{q_{max}}$$

$$q_{actual} = 0.5717 \times 140(kW) = 80.038\ kW$$

For hot water stream:

$$q_{actual} = 4.18(kW/°C) \times (90 - T_{H,exit})(°C) = 80.038\ kW$$

$$T_{H,exit} = 90 - 19.15 = 70.85°C$$

Similarly, for liquid food stream:

$$q_{actual} = 2(kW/°C) \times (T_{L,exit} - 20)(°C) = 80.038\ kW$$

$$T_{L,exit} = 40.019 + 20 = 60°C$$

The calculated exit temperatures of hot water and product streams are 70.85°C and 60°C, respectively. These values are comparable to those given in Example 4.18.

4.4.9 Design of a Plate Heat Exchanger

As discussed in Section 4.1.1, plate heat exchangers are commonly used in the food industry. The design of a plate heat exchanger requires certain relationships that are often unique to the plates used in the heat exchanger. Some of the required information for design purposes is closely guarded by the equipment manufacturers and is not readily available. We will consider a general design approach with some of the key relationships required in designing plate heat exchangers. First, it is important to understand the type of flow pattern present inside a plate heat exchanger.

Figure 4.2 shows an arrangement of plates with two liquid streams, product and heating/cooling medium, entering and exiting through their respective ports. The assembly of a plate heat exchanger involves placing the required number of plates in a frame and tightening the end bolts so that a fixed gap space is created between the plates. The gaps between the plates create channels through which the fluid streams flow. The fluid streams enter and exit through the port holes (Fig. 4.2b).

Each plate contains a gasket that helps orient the direction of the flow stream through the channel. The assembly of plates with gaskets is done in such a manner that it allows the product stream to move in one channel while the heating/cooling stream moves in the neighboring channel. Thus, the flow of product and heating/cooling streams alternates through the channels and they never come into direct contact (Fig. 4.2).

The heat exchange between the two liquid streams in a plate heat exchanger is across the plates, normal to the direction of the flow. Therefore, the plates are made as thin as possible to minimize resistance to heat transfer while maintaining the physical integrity of the plates. The plates are corrugated with a pattern that promotes turbulence in the liquid stream. Several types of patterns are used for the corrugations stamped on the plates, the most common one being a herringbone type design called the Chevron design (Fig. 4.3). The plates used in food processing applications are made of stainless steel (ANSI 316), although the thermal conductivity of stainless steel is not as high as other metals that are used in nonfood applications. In the past, the material used for gaskets did not permit use of high temperatures. However, new heat-resistant materials for gaskets have considerably extended the use of plate heat exchangers for high-temperature applications.

To determine the rate of heat transfer across the plates, it is necessary to know the convective heat transfer coefficient on both sides of the plates. Since the mechanism is forced convection, we use a dimensionless correlation involving Nusselt number, Reynolds number, and Prandtl number. The dimensionless correlation depends upon the design of the corrugations used. An approximate expression suitable for plate heat exchanger is as follows:

$$N_{Nu} = 0.4N_{Re}^{0.64}N_{Pr}^{0.4} \qquad (4.127)$$

To evaluate the Reynolds number, it is necessary to determine the velocity of the liquid stream in the channel. This determination is complex because of the corrugations. We will consider a simplified method to estimate the fluid velocity.

In a plate heat exchanger, the two end plates do not take part in heat transfer. Thus, the number of "thermal plates" involved in heat transfer is obtained by subtracting two from the total number of plates in a heat exchanger.

The flow rate of a liquid stream through each channel is obtained as

$$\dot{m}_{Hc} = \frac{\dot{m}_H}{\left(\frac{N+1}{2}\right)} \qquad (4.128)$$

and

$$\dot{m}_{Pc} = \frac{2\dot{m}_P}{N+1} \qquad (4.129)$$

where $\dot{m}_H$ and $\dot{m}_P$ are the total mass flow rates of the heating/cooling stream and the product stream (kg/s), respectively; $\dot{m}_{Hc}$ and $\dot{m}_{Pc}$ are the channel flow rates for heating/cooling stream and product stream (kg/s), respectively; and N is the total number of thermal plates.

The cross-sectional area of a channel between two adjacent plates is obtained as,

$$A_c = bw \qquad (4.130)$$

where b is the gap between two adjacent plates and w is the width of the plate.

The velocities of the two streams are

$$\bar{u}_{Pc} = \frac{\dot{m}_{Pc}}{\rho_P A_c} \qquad (4.131)$$

$$\bar{u}_{Hc} = \frac{\dot{m}_{Hc}}{\rho_H A_c} \qquad (4.132)$$

The equivalent diameter (or the hydraulic diameter), D_e, is calculated from the following expression:

$$D_e = \frac{4 \times \text{channel free-flow area for fluid stream}}{\text{wetted perimeter for the fluid}} \tag{4.133}$$

For simplicity, we use the projected area (disregarding the corrugations) to determine the wetted perimeter as

$$\text{Wetted perimeter for the fluid} = 2(b + w) \tag{4.134}$$

$$\text{Channel free-flow area} = bw \tag{4.135}$$

Then,

$$D_e = \frac{4bw}{2(b + w)} \tag{4.136}$$

Since in a plate heat exchanger, $b << w$, we may neglect b in the denominator to obtain,

$$D_e = 2 \times b \tag{4.137}$$

The Reynolds number for each stream is obtained as

Product stream:

$$N_{Re,P} = \frac{\rho_P \bar{u}_{Pc} D_e}{\mu_P} \tag{4.138}$$

Heating/cooling stream:

$$N_{Re,H} = \frac{\rho_H \bar{u}_{Hc} D_e}{\mu_H} \tag{4.139}$$

Knowing the Reynolds number, the Nusselt number is calculated for each stream using Equation (4.127). The heat transfer coefficient is obtained as

$$h_P = \frac{N_{Nu} \times k_P}{D_e} \tag{4.140}$$

$$h_H = \frac{N_{Nu} \times k_H}{D_e} \tag{4.141}$$

where k_P and k_H are the thermal conductivities of the product and heating/cooling streams, (W/m °C), respectively.

The overall heat transfer coefficient is determined from the two values of the convective heat transfer coefficients, assuming that the conductive resistance of the thin metal plate is negligible,

$$\frac{1}{U} = \frac{1}{h_P} + \frac{1}{h_H} \tag{4.142}$$

Once the overall heat transfer coefficient is obtained using Equation (4.142), the remaining computations for the design of the plate heat exchanger are done using the effectiveness-NTU method similar to the procedure described in Section 4.4.7.2. For a plate heat exchanger, the relevant expressions between NTU and effectiveness are given in Tables 4.4 and 4.5. The calculation procedure for designing a plate heat exchanger is illustrated in the following example.

Example 4.21

A counterflow plate heat exchanger is being used to heat apple juice with hot water. The heat exchanger contains 51 plates. Each plate is 1.2 m high and 0.8 m wide. The gap between the plates is 4 mm. The heating characteristics for this heat exchanger have been previously determined to follow the following relationship, $N_{Nu} = 0.4 N_{Re}^{0.64} N_{Pr}^{0.4}$. Hot water enters the heat exchanger at 95°C at a rate of 15 kg/s and apple juice enters at 15°C at a rate of 10 kg/s. It is assumed that the physical properties of water and apple juice are the same. The properties of water at 55°C are density = 985.7 kg/m^3, specific heat = 4179 J/kgK, thermal conductivity = 0.652 W/m°C, dynamic viscosity = 509.946 × 10^{-6} Ns/m^2, Prandtl Number = 3.27. Calculate the exit temperatures of water and apple juice.

Given

Number of plates = 51
Plate height = 1.2 m
Plate width = 0.8 m
Gap between plates = 4 mm = 0.004 m
Inlet temperature of hot water = 95°C
Mass flow rate of hot water = 15 kg/s
Inlet temperature of apple juice = 15°C
Mass flow rate of apple juice = 10 kg/s
Properties of water at (95 + 15)/2 = 55°C from Table A.4.1:
Density = 985.7 kg/m^3
Specific heat = 4.179 kJ/kg K

Thermal conductivity = 0.652 W/m°C
Viscosity = 509.946×10^{-6} Ns/m^2
Prandtl number = 3.27

Approach

We will assume that the heat exchanger is operating under steady state conditions. First, we will determine the channel velocity of each fluid stream using the given flow rates and gap dimensions. Next, we will determine the Reynolds number, and using the dimensionless correlation, we will calculate Nusselt number and convective heat transfer coefficient. The convective heat transfer coefficients from both sides of a plate will be used to determine the overall heat transfer coefficient. Once the overall heat transfer coefficient is known, then we will use the effectiveness-NTU method to determine the exit temperatures of each stream.

Solution

Heat transfer coefficient on the hot water side:

1. *Equivalent diameter for the channel created between the plates is*

$$D_e = 2 \times b = 2 \times 0.004 = 0.008\ m$$

2. *Flow rate of hot water in each channel:*
   Total number of plates = 51. Since the two end plates do not take part in heat exchange, the number of thermal plates = 49. Then

$$\dot{m}_{Hc} = \frac{2 \times 15}{(49+1)} = 0.60\ kg/s$$

3. *The cross-sectional area of each channel is*

$$A_c = 0.004 \times 0.8 = 0.0032\ m^2$$

4. *Velocity of hot water in each channel is*

$$\bar{u}_c = \frac{0.60\ (kg/s)}{985.7 (kg/m^3) \times 0.0032\ (m^2)} = 0.19 m/s$$

5. *Reynolds number:*

$$N_{Re} = \frac{985.7 \times 0.19 \times 0.008}{509.946 \times 10^{-6}} = 2941$$

6. *Nusselt number:*

$$N_{Nu} = 0.4 \times 2941^{0.64} \times 3.27^{0.4} = 106.6$$

Then, heat transfer coefficient on the hot water side is

$$h_H = \frac{106.6 \times 0.652}{0.008} = 8{,}688\,W/m^2\,°C$$

Heat transfer coefficient on the apple juice side:

7. *The flow rate of apple juice in each channel is*

$$\dot{m}_A = \frac{2 \times 10}{(49+1)} = 0.4\,kg/s$$

8. *The velocity of apple juice in each channel is*

$$\bar{u}_c = \frac{0.4(kg/s)}{985.7(kg/m^3) \times 0.0032(m^2)} = 0.127\,m/s$$

9. *The Reynolds number is*

$$N_{Re} = \frac{985.7 \times 0.127 \times 0.008}{509.946 \times 10^{-6}} = 1961$$

10. *The Nusselt number is*

$$N_{Nu} = 0.4 \times 1961^{0.64} \times 3.27^{0.4} = 82.2$$

11. *Convective heat transfer coefficient on the apple juice side is*

$$h_H = \frac{82.2 \times 0.652}{0.008} = 6702\ W/m^2\,°C$$

12. *The overall heat transfer coefficient, U, is*

$$\frac{1}{U} = \frac{1}{8688} + \frac{1}{6702}$$

$$U = 3784\,W/m^2\,°C$$

13. *The total area of the heat exchanger is*

$$A_h = 49 \times 1.2 \times 0.8 = 47.04\ m^2$$

14. *NTU for the heat exchanger is*

$$NTU = \frac{3784 \times 47.04}{10 \times 4179} = 4.259$$

15. *The capacity rate ratio is*

$$C^* = \frac{10}{15} = 0.67$$

16. *The exchanger effectiveness is obtained from the relationship given in Table 4.4 as*

$$\varepsilon_E = \frac{\exp[(1-0.67)\times 4.259]-1}{\exp[(1-0.67)\times 4.259]-0.67}$$

$$\varepsilon_E = 0.9$$

17. *The exit temperature of apple juice is determined by noting that $C_{juice} = C_{min}$, or*

$$\frac{q_{actual}}{q_{max}} = 0.9 = \frac{C_{juice}(T_{ae}-15)}{C_{min}(95-15)} = \frac{(T_{ae}-15)}{(95-15)}$$

Then, $T_{ae} = 87°C$

18. *The exit temperature of hot water is determined as*

$$0.9 = \frac{C_{water}}{C_{min}}\frac{(95-T_{we})}{(95-15)} = 1.5\frac{(95-T_{we})}{(95-15)}$$

or, $T_{we} = 47°C$

In the given counter flow plate heat exchanger operating under steady state conditions, the hot water stream will exit at 47°C and the apple juice will be heated to 87°C.

4.4.10 Importance of Surface Characteristics in Radiative Heat Transfer

All materials in the universe emit radiation of an electromagnetic nature based on their surface temperature. At a temperature of 0°K, the emission of radiation ceases. The characteristics of the radiation are also dependent on temperature. As temperature increases, the wavelength decreases. For example, radiation emitted by the sun is shortwave compared with the longwave radiation emitted by the surface of a hot coffee mug.

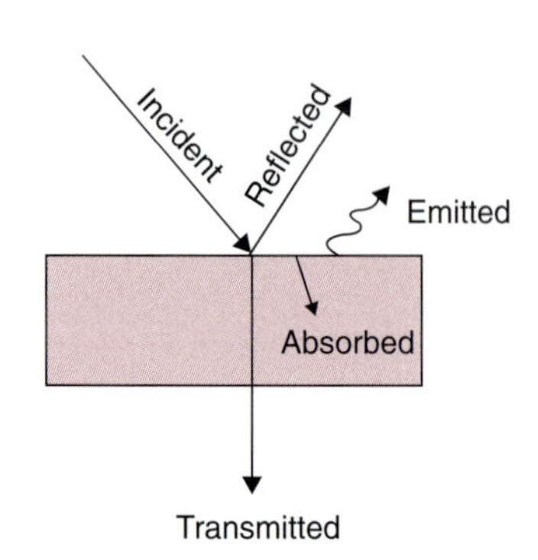

■ **Figure 4.29** Radiant energy incident on a semi-opaque object.

When radiation of a given wavelength, as shown in Figure 4.29, is incident on an object, some of the incident radiation is reflected, some transmitted, and some absorbed. The following expression holds true:

$$\phi + \chi + \psi = 1 \tag{4.143}$$

where ϕ is absorptivity, χ is reflectivity, and ψ is transmissivity. The absorbed radiation will result in an increase of temperature.

To compare the absorption of radiation for different materials, an ideal reference called a blackbody is used. For a blackbody, the absorptivity value is 1.0. Note that nothing in the universe is a true blackbody; even lampblack has $\phi = 0.99$ and $\chi = 0.01$. Regardless, blackbody is a useful concept for comparing radiative properties of different materials.

The absolute magnitudes of ϕ, χ, and ψ depend on the nature of the incident radiation. Thus, a brick wall of a house is opaque to visible light but transparent to radio waves.

Energy emitted (also called radiated) and energy reflected must be clearly distinguished. These are quite different terms. A material, depending on its surface absorptivity value, will reflect some of the incident radiation. In addition, based on its own temperature, it will emit radiation, as shown in Figure 4.29. The amount of radiation emitted can be computed from Equation (4.20).

Kirchoff's law states that the emissivity of a body is equal to its absorptivity for the same wavelength. Thus, mathematically,

$$\varepsilon = \phi \tag{4.144}$$

This identity is discussed in Example 4.22.

Example 4.22

Compare the selection of white paint versus black paint for painting the rooftop of a warehouse. The objective is to allow minimum heat gain from the sun during summer.

Given

White paint from Table A.3.3:

$\varepsilon_{shortwave} = 0.18$

$\varepsilon_{longwave} = 0.95$

Black paint from Table A.3.3:

$\varepsilon_{shortwave} = 0.97$

$\varepsilon_{longwave} = 0.96$

Approach

From the emissivity values, we will examine the use of white or black paint for both short- and longwave radiation.

Solution

1. *White paint: $\varepsilon_{shortwave} = 0.18$, therefore from Kirchoff's law $\phi_{shortwave} = 0.18$, so $\chi_{shortwave} = 1 - 0.18$, assuming $\psi = 0$, thus $\chi_{shortwave} = 0.82$. Thus, of the total shortwave radiation incident on the rooftop, 18% is absorbed and 82% reflected.*
2. *White paint, $\varepsilon_{longwave} = 0.95$. The white-painted surface, in terms of longwave radiation, emits 95% of radiation emitted by a blackbody.*
3. *Black paint: $\varepsilon_{shortwave} = 0.97$, therefore using Kirchoff's Law $\phi_{shortwave} = 0.97$. Thus, of the total shortwave radiation incident on the rooftop, 97% of radiation incident on it is absorbed and 3% is reflected.*
4. *Black paint: $\varepsilon_{longwave} = 0.96$. The black-painted surface, in terms of longwave radiation, emits 96% of radiation emitted by a blackbody.*
5. *White paint should be selected, as it absorbs only 18% shortwave (solar) radiation compared to 97% by black paint. Both black and white paint are similar in emitting longwave radiation to the surroundings.*

4.4.11 Radiative Heat Transfer between Two Objects

The transfer of heat by radiation between two surfaces is dependent on the emissivity of the radiating surface and the absorptivity of that same surface. The expression normally used to describe this type of heat transfer is as follows:

$$q_{1-2} = A\sigma(\varepsilon_1 T_{A1}^4 - \phi_{1-2} T_{A2}^4) \quad (4.145)$$

where ε_1 is the emissivity of the radiating surface at temperature T_{A1}, and ϕ_{1-2} is the absorptivity of the surface for radiation emitted at temperature T_{A2}.

Although the basic expression describing radiative heat transfer is given by Equations (4.20) and (4.145), one of the important factors requiring attention is the shape of the object. The shape factor accounts for the fraction of the radiation emitted by the high-temperature surface that is not absorbed by the low-temperature surface. For example, Equation (4.145) assumes that all radiation emitted at temperature T_{A1} is absorbed by the surface at temperature T_{A2}. If both the surfaces are blackbodies, then the expression describing heat transfer and incorporating the shape factor would be as follows:

$$q_{1-2} = \sigma F_{1-2} A_1 (T_{A1}^4 - T_{A2}^4) \quad (4.146)$$

where F_{1-2} is the shape factor, and it physically represents the fraction of the total radiation leaving the surface A_1 that is intercepted by the surface A_2. The values of shape factors have been tabulated and presented in the form of curves of the type shown in Figures 4.30 and 4.31. In the first case, the shape factors deal only with adjacent rectangles, which are in perpendicular planes. Figure 4.31 can be utilized for various shapes, including disks, squares, and rectangles.

Equation (4.146) does not account for nonblackbodies, and Equation (4.20) does not account for the shape factor; therefore, an expression that combines the two must be used. Such an expression would be

$$q_{1-2} = \sigma A_1 \xi_{1-2}(T_{A1}^4 - T_{A2}^4) \tag{4.147}$$

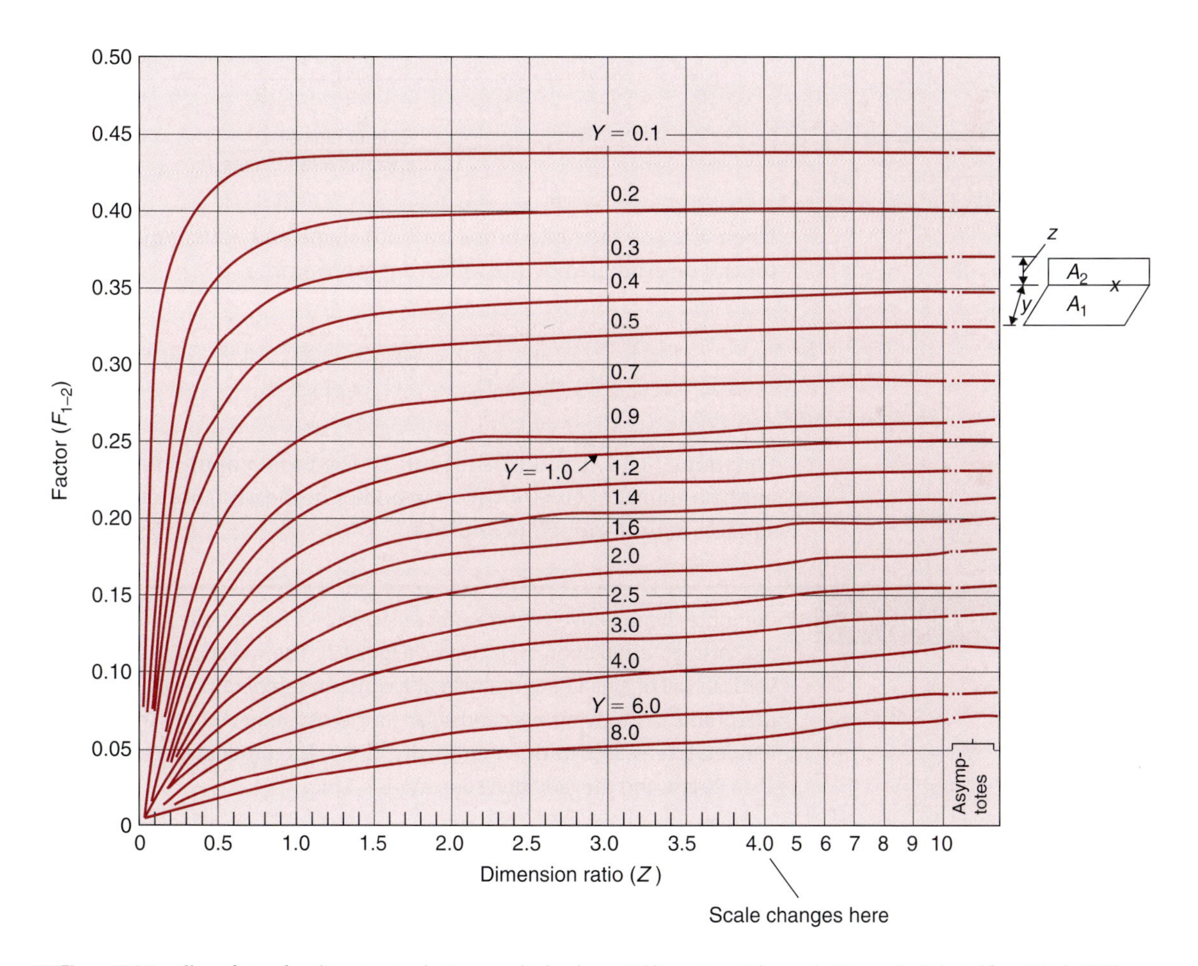

Figure 4.30 Shape factors for adjacent rectangles in perpendicular planes. Y (dimension ratio) $= y/x$; $Z = z/x$. (Adapted from Hottel, 1930)

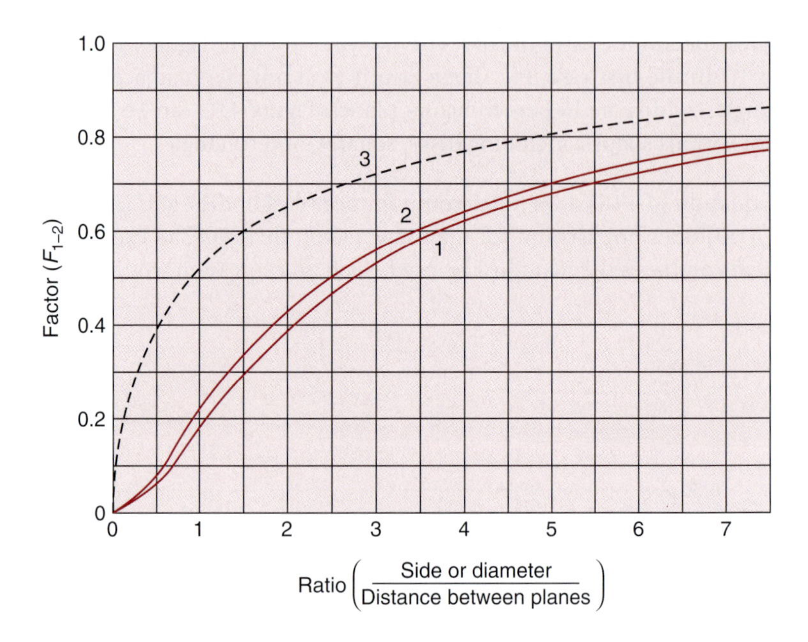

Figure 4.31 Shape factors for equal and parallel squares, rectangles, and disks: (1) direct radiation between disks; (2) direct radiation between squares; (3) total radiation between squares or disks connected by nonconducting but reradiating walls. (Adapted from Hottel, 1930)

where the ξ_{1-2} factor accounts for both shape and emissivity. This factor can be evaluated by the following expression:

$$\xi_{1-2} = \frac{1}{\dfrac{1}{F_{1-2}} + \left(\dfrac{1}{\varepsilon_1} - 1\right) + \dfrac{A_1}{A_2}\left(\dfrac{1}{\varepsilon_2} - 1\right)} \tag{4.148}$$

Equations (4.147) and (4.148) can be used to compute the net radiant heat transfer between two gray bodies in the presence of radiating surfaces at uniform temperatures.

Example 4.23

Compute the radiative heat transfer received by a rectangular product moving through a radiation-type heater (Fig. E4.2). The radiation source is one vertical wall of a heater and is held at a constant temperature of 200°C while the product is moving perpendicular to the radiation source. The product temperature is 80°C with an emissivity of 0.8. The product dimensions are 15 × 20 cm, and the radiation source is 1 × 5 m.

Given

Temperature of heater = 200°C
Product temperature = 80°C
Emissivity of product = 0.8
Product dimensions = 0.15 m × 0.2 m
Heater dimensions = 1 m × 5 m

Approach

To compute the ξ_{1-2} factor from Equation (4.148), we will use Figure 4.30, for objects perpendicular to each other, for F_{1-2} value. Then we will use Equation (4.147) to calculate radiative heat received by the rectangular product.

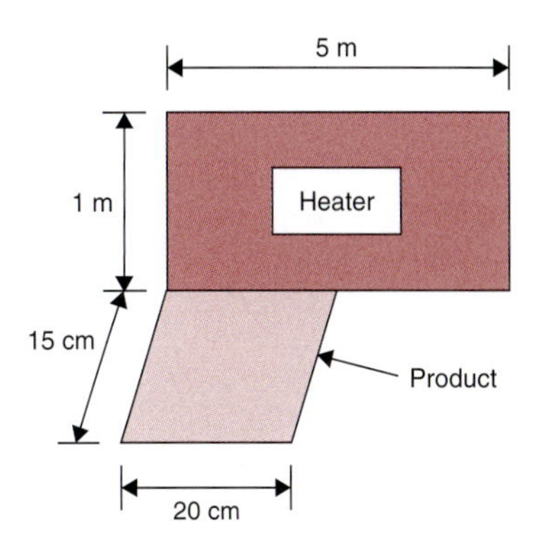

Figure E4.20 Radiation heater.

Solution

1. *To use Equation (4.147), the ξ factor must be computed from Equation (4.148).*

$$\xi_{1-2} = \frac{1}{\frac{1}{F_{1-2}} + \left(\frac{1}{1} - 1\right) + \frac{5}{0.03}\left(\frac{1}{0.8} - 1\right)}$$

where F_{1-2} must be obtained from Figure 4.30 for $z/x = 5.0$ and $y/x = 0.75$.

$$F_{1-2} = 0.28$$

Then

$$\xi_{1-2} = \frac{1}{(3.57 + 0 + 41.67)} = \frac{1}{45.24} = 0.0221$$

2. *From Equation (4.147),*

$$q_{1-2} = (5.669 \times 10^{-8}\ \mathrm{W/[m^2\ K^4]})(0.0221)(5\ \mathrm{m^2}) \times [(473\ \mathrm{K})^4 - (353\ \mathrm{K})^4] = 216\ \mathrm{W}$$

4.5 UNSTEADY-STATE HEAT TRANSFER

Unsteady-state (or transient) heat transfer is that phase of the heating and cooling process when the temperature changes as a function of both location and time. By contrast, in steady-state heat transfer, temperature varies only with location. In the initial unsteady-state period, many important reactions in the food may take place. With thermal processes, the unsteady-state phase may even dominate the entire process; for example, in several pasteurization and food-sterilization processes, the unsteady-state period is an important component of the process. Analysis of temperature variations with time during the unsteady-state period is essential in designing such a process.

Since temperature is a function of two independent variables, time and location, the following partial differential equation is the governing equation for a one-dimensional case:

$$\frac{\partial T}{\partial t} = \frac{k}{\rho c_p r^n} \frac{\partial}{\partial r}\left(r^n \frac{\partial T}{\partial r}\right) \tag{4.149}$$

where T is temperature (°C), t is time (s), and r is distance from center location (m). We can make this equation specific for a particular geometrical shape using $n = 0$ for a slab, $n = 1$ for a cylinder, and $n = 2$ for a sphere. The combination of properties $k/\rho c_p$ is defined as thermal diffusivity, α. If the rate of heat transfer at the surface of the object is due to convection, then

$$k\frac{\partial T}{\partial r}\Big|_{r=R} = h(T_a - T_s) \tag{4.150}$$

where h is the convective heat transfer coefficient (W/[m^2 °C]), T_a is the temperature of heating or cooling medium far away from the surface (°C), and T_s is the temperature at the surface (°C).

The procedure we use to solve the governing equation, Equation (4.149), involves the use of advanced mathematics, which is beyond the scope of this book. Myers (1971) gives the complete derivation for various types of boundary-value problems encountered in unsteady-state heat transfer. Due to mathematical complexity, the analytical solution of Equation (4.149) is possible only for objects of simplified geometrical shapes, such as a sphere, an infinite cylinder, or an infinite slab.

An infinite cylinder is a "long" cylinder of radius r, a sphere is of radius r, an infinite plate is a "large" plane wall of thickness $2z$. These three objects are geometrically and thermally symmetric about their centerline (for a cylinder), center plane (for a slab), or center point (for a sphere), as shown in Figure 4.32.

Consider heating an infinitely long cylindrical object. Initially, the object is assumed to have a uniform temperature, T_i. At time $t = 0$, we place this object in a heating medium maintained at a constant temperature T_a. A constant heat transfer coefficient, h, describes the convective heat transfer at the surface of the object. Temperature profiles at different time intervals inside the object are shown in Figure 4.33. At time $t = 0$, the temperature is uniform at T_i. At time, $t = t_1$, the temperature along the wall increases, establishing a temperature gradient within the object that promotes heat conduction. At time $t = t_2$, the temperature at the center is still at T_i. However, with the passage of time, at time $t = t_3$, the centerline temperature begins to increase, and eventually at $t = t_4$ the temperature of the cylinder becomes uniform at T_a. At this time, the cylinder is in thermal equilibrium with the surrounding medium, and heat transfer ceases. Note that there is no heat transfer from the axial ends of the cylinder. Because the cylinder is infinitely long, the implication of "infinite" length in this

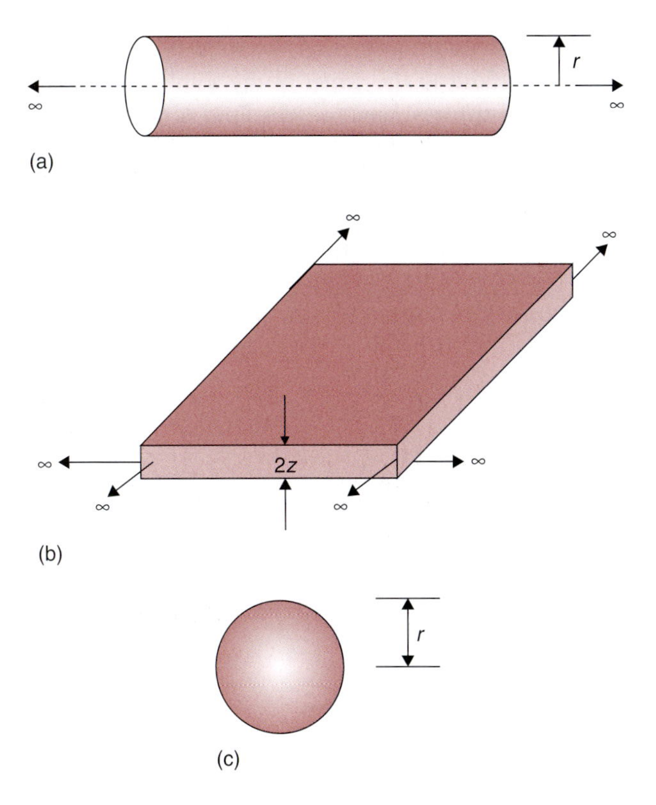

■ Figure 4.32 (a) An infinite cylinder, (b) an infinite plate, (c) a sphere.

case is that the heat transfer is only in the radial direction and not axial. Similarly, for an infinite slab of thickness, $2z$, the heat transfer is only from the two faces around the thickness $2z$ of the slab. The slab extends to infinity along the other four faces, and no heat transfer takes place through those sides. We will explore these concepts in greater detail in the following sections.

4.5.1 Importance of External versus Internal Resistance to Heat Transfer

In transient heat transfer analysis, one of the first steps is to consider the relative importance of heat transfer at the surface and interior of an object undergoing heating or cooling. Consider an object that is suddenly immersed in a fluid (Fig. 4.34). If the fluid is at a temperature different from the initial temperature of the solid, the temperature inside the solid will increase or decrease until it reaches a value in equilibrium with the temperature of the fluid.

During the unsteady-state heating period, the temperature inside the solid object (initially at a uniform temperature) will vary with

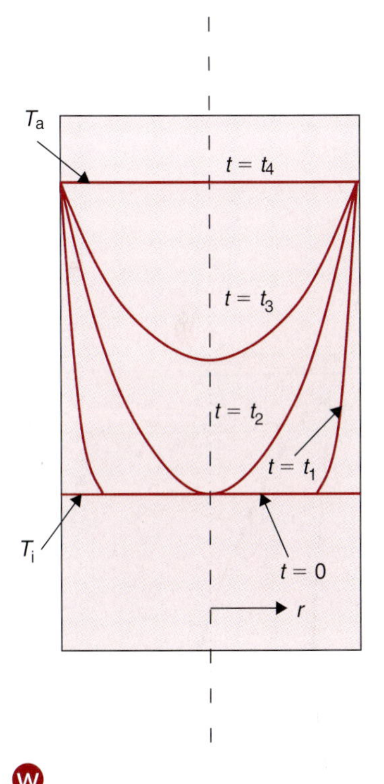

W

■ Figure 4.33 Temperature profiles as a function of time in an infinitely long cylinder.

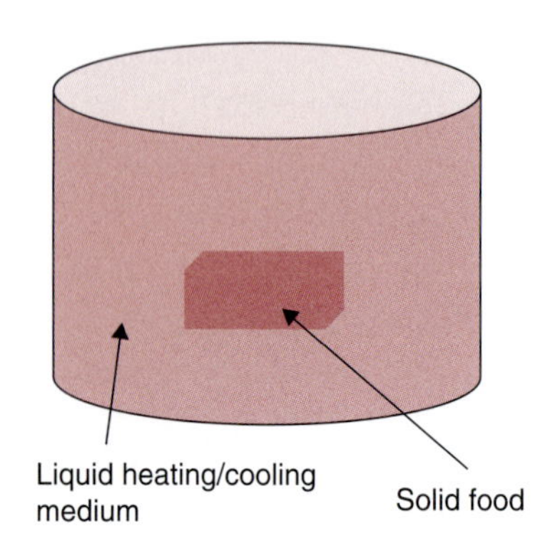

■ **Figure 4.34** A solid object suddenly exposed to a heating/cooling medium.

location and time. Upon immersing the solid in the fluid, the heat transfer from the fluid to the center of the solid will encounter two resistances: convective resistance in the fluid layer surrounding the solid, and conductive resistance inside the solid. The ratio of the internal resistance to heat transfer in the solid to the external resistance to heat transfer in the fluid is defined as the Biot number, N_{Bi}.

$$\frac{\text{internal conductive resistance within the body}}{\text{external convective resistance at the surface of the body}} = N_{\mathrm{Bi}} \tag{4.151}$$

or

$$N_{\mathrm{Bi}} = \frac{d_{\mathrm{c}}/k}{1/h} \tag{4.152}$$

or

$$N_{\mathrm{Bi}} = \frac{hd_{\mathrm{c}}}{k} \tag{4.153}$$

where d_{c} is a characteristic dimension.

According to Equation (4.151), if the convective resistance at the surface of a body is much smaller than the internal conductive resistance, then the Biot number will be high. For Biot numbers greater than 40, there is negligible surface resistance to heat transfer. On the other hand, if internal conductive resistance to heat transfer is small, then the Biot number will be low. For Biot numbers less than 0.1, there is negligible internal resistance to heat transfer. Between a Biot number of 0.1 and 40, there is a finite internal and external resistance to heat transfer. Steam condensing on the surface of a broccoli stem will result in negligible surface resistance to heat transfer ($N_{\mathrm{Bi}} > 0$). On the other hand, a metal can containing hot tomato paste, being cooled in a stream of cold air, will present finite internal and surface resistance to heat transfer, and a small copper sphere placed in stagnant heated air will have a Biot number of less than 0.1. In the following subsections, we will consider these three cases separately.

4.5.2 Negligible Internal Resistance to Heat Transfer ($N_{\mathrm{Bi}} < 0.1$)—A Lumped System Analysis

For a Biot number smaller than 0.1, there is a negligible internal resistance to heat transfer. This condition will occur in heating and cooling of most solid metal objects but not with solid foods, because the thermal conductivity of a solid food is relatively low.

Negligible internal resistance to heat transfer also means that the temperature is nearly uniform throughout the interior of the object. For this reason, this case is also referred to as a "lumped" system. This condition is obtained in objects with high thermal conductivity when they are placed in a medium that is a poor conductor of heat, such as motionless air. In these cases heat is transferred instantaneously into the object, thus avoiding temperature gradients with location. Another way to obtain such a condition is a well-stirred liquid food in a container. For this special case, there will be no temperature gradient with location, as the product is well mixed.

A mathematical expression to describe heat transfer for a negligible internal resistance case may be developed as follows.

Consider an object at low uniform temperature T_i, immersed in a hot fluid at temperature Ta, as shown in Figure 4.34. During the unsteady-state period, a heat balance across the system boundary gives

$$q = \rho c_p V \frac{dT}{dt} = hA(T_a - T) \tag{4.154}$$

where T_a is the temperature of the surrounding medium, and A is the surface area of the object.

By separating variables,

$$\frac{dT}{(T_a - T)} = \frac{hA\,dt}{\rho c_p V} \tag{4.155}$$

Integrating, and setting up limits,

$$\int_{T_i}^{T} \frac{dT}{T_a - T} = \frac{hA}{\rho c_p V} \int_0^t dt \tag{4.156}$$

$$-\ln(T_a - T)\,\Big|_{T_i}^{T} = \frac{hA}{\rho c_p V}(t - 0) \tag{4.157}$$

$$-\ln\left(\frac{T_a - T}{T_a - T_i}\right) = \frac{hAt}{\rho c_p V} \tag{4.158}$$

Rearranging terms,

$$\frac{T_a - T}{T_a - T_i} = e^{-(hA/\rho c_p V)t} \tag{4.159}$$

Rewriting Equation (4.159) as

$$\frac{T_a - T}{T_a - T_i} = e^{-bt} \tag{4.160}$$

where

$$b = \frac{hA}{\rho c_p V}$$

In Equation (4.160), the numerator $T_a - T$ on the left-hand side is the unaccomplished temperature difference between the heat transfer medium and the object. The denominator is the maximum temperature difference at the start of the heating/cooling process. Thus, the temperature ratio shown on the left-hand side of this equation is the unaccomplished temperature fraction. At the start of the heating/cooling process, the unaccomplished temperature fraction is one, and it decreases with time. The right-hand side of Equation (4.160) shows an exponentially decreasing (or decaying) function. This implies that with increasing passage of time the unaccomplished temperature fraction decreases but it never reaches a value of zero, it approaches zero asymptotically. Furthermore, when an object is being heated, with a higher value of b, the object temperature increases more rapidly (with a greater decay in temperature difference). The b value is influenced directly by the convective conditions on the surface described by h, its thermal properties, and the size. Small objects with low specific heat take a shorter time to heat or cool.

Example 4.24

Calculate the temperature of tomato juice (density = 980 kg/m^3) in a steam-jacketed hemispherical kettle after 5 min. of heating. (See Fig. E4.21.) The radius of the kettle is 0.5 m. The convective heat-transfer coefficient in the steam jacket is 5000 W/(m^2 °C). The inside surface temperature of the kettle is 90°C. The initial temperature of tomato juice is 20°C. Assume specific heat of tomato juice is 3.95 kJ/(kg °C).

Given

Kettle:

Surface temperature $T_a = 90°C$
Radius of kettle = 0.5 m

Tomato juice:

Initial temperature $T_i = 20°C$
Specific heat $c_p = 3.95$ kJ/(kg °C)
Density $\rho = 980\,kg/m^3$
Time of heating t = 5 min

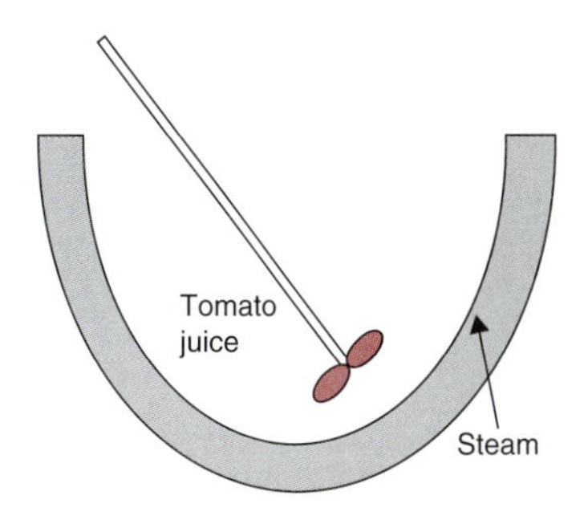

Figure E4.21 Heating tomato juice in a hemispherical steam jacketed kettle.

Approach

Since the product is well mixed, there are no temperature gradients inside the vessel, and we have negligible internal resistance to heat transfer. We will use Equation (4.159) to find the temperature after 5 minutes.

Solution

1. *We will use Equation (4.159). First, the inside surface area and volume of the hemispherical kettle are computed:*

$$A = 2\pi r^2 = 2\pi(0.5)^2 = 1.57\,m^2$$
$$V = \frac{2}{3}\pi r^3 = \frac{2}{3}\pi(0.5)^3 = 0.26\,m^3$$

2. *Using Equation (4.159):*

$$\frac{90 - T}{90 - 20} = \exp\frac{-(5000\,W/[m^2\,°C])(1.57\,m^2)(300\,s)}{(980\,kg/m^3)(3.95\,kJ/[kg\,°C])(1000\,J/kJ)(0.26\,m^3)}$$

$$\frac{90 - T}{90 - 20} = 0.096$$

$$T = 83.3°C$$

3. *The product temperature will rise to 83.3°C in 5 minutes of heating.*

Example 4.25

An experiment was conducted to determine surface convective heat-transfer coefficient for peas being frozen in an air-blast freezer. For this purpose, a metal analog of peas was used. The analog was a solid copper ball with a diameter of 1 cm. A small hole was drilled to the center of the copper ball, and a thermocouple junction was located at the center using a high-conductivity epoxy. The density of copper is $8954\,kg/m^3$, and its specific heat is 3830 J/(kg K). The copper ball (at a uniform initial temperature of 10°C) was hung in the path of air flow (at −40°C) and the center temperature indicated by the thermocouple was recorded. The following table lists the temperature at one-minute intervals for 14 minutes. Determine the surface heat-transfer coefficient from these data.

Time (s)	**Temperature (°C)**
0	10.00
60	9.00
120	8.00
180	7.00
240	6.00
300	5.00
360	4.00
420	3.50
480	2.50
540	1.00
600	1.00
660	0.00
720	−2.00
780	−2.00
840	−3.00

Given

Diameter of copper ball $D = 1$ cm
Density of copper $\rho = 8954$ kg/m^3
Specific heat of copper $c_p = 3830$ J/(kg K)
Initial temperature of copper $T_i = 10$°C
Temperature of cold air = −40°C

Approach

We will use a modified form of Equation (4.159) to plot the temperature–time data. If these data are plotted on a semilog paper, we can obtain the h value from the slope. An alternative approach is to use statistical software to develop a correlation and determine the slope.

Solution

1. *Equation (4.159) may be rewritten as follows:*

$$ln(T - T_a) = ln(T_i - T_a) - \frac{hAt}{\rho c_p V}$$

2. *The tabulated data for temperature and time is converted to $ln(T - T_a)$.*

3. *Using a statistical software (e.g., StatView™), a correlation is obtained between t and $\ln(T - T_a)$. The results are*

$$slope = -3.5595 \times 10^{-4}\ 1/s$$

4. *Thus,*

$$\frac{hA}{\rho c_p V} = 3.5595 \times 10^{-4}$$

5. *Surface area of sphere, $A = 4\pi r^2$*
   Volume of sphere, $V = 4\pi r^3/3$
6. *Substituting given and calculated values in the expression in step (4), we get*

$$h = 20\ W/(m^2\ {}^\circ C)$$

7. *A pea held in the same place as the copper ball in the blast of air will experience a convective heat-transfer of 20 W/(m² °C).*

4.5.3 Finite Internal and Surface Resistance to Heat Transfer ($0.1 < N_{Bi} < 40$)

As noted previously, the solution of Equation (4.149) is complicated, and it is available only for well-defined shapes such as sphere, infinite cylinder, and infinite slab. In each case, the solution is an infinite series containing trigonometric and/or transcendental functions. These solutions are as follows.

Sphere:

$$T = T_a + (T_a - T_i)\frac{2}{\pi}\left(\frac{d_c}{r}\right) \times \sum_{n=1}^{\infty} \frac{(-1)^{n+1}}{n} e^{(-n^2\pi^2\alpha t/d_c^2)} \sin(n\pi r/d_c) \quad (4.161)$$

with root equation,

$$N_{Bi} = 1 - \beta_1 \cot \beta_1 \quad (4.162)$$

Infinite cylinder:

$$T = T_a + 2(T_i - T_a)\sum_{n=0}^{\infty} \frac{e^{-\lambda_n^2 \alpha t / d_c^2} J_0(\lambda_n r/d_c)}{\lambda_n J_1(\lambda_n)} \quad (4.163)$$

with root equation,

$$N_{\mathrm{Bi}} = \frac{\beta_1 J_1(\beta_1)}{J_0(\beta_1)} \tag{4.164}$$

Infinite slab:

$$T = T_a + (T_i - T_a) \sum_{n=0}^{\infty} \frac{[2(-1)^n]\, e^{-\lambda_n^2 \alpha t/d_c^2}}{\lambda_n} \cos(\lambda_n x/d_c) \tag{4.165}$$

with root equation,

$$N_{\mathrm{Bi}} = \beta_1 \tan \beta_1 \tag{4.166}$$

These analytical solutions with infinite series may be programmed into a spreadsheet for use on a computer. We will illustrate this procedure later in this section with an example. These solutions have also been reduced to simple temperature–time charts that are relatively easy to use. In constructing a temperature–time chart for a typical transient heat transfer problem, the variety of factors can be numerous: for example, r, t, k, ρ, c_p, h, T_i, and T_a. However, these factors may be combined into three dimensionless numbers, making it convenient to develop charts for universal use regardless of units used for measuring these factors. Temperature–time charts developed for the three geometric shapes (sphere, infinite cylinder, and infinite slab) are presented in Figures 4.35, 4.36, and 4.37. These charts are called Heisler charts, based on the work of Heisler (1944). The three dimensionless numbers shown on these charts are the unaccomplished temperature fraction, $(T_a - T)/(T_a - T_i)$, Biot number, N_{Bi}, and a dimensionless time expressed as the Fourier[2] number. The Fourier number is defined as follows:

$$\text{Fourier number} = N_{\mathrm{Fo}} = \frac{k}{\rho c_p} \frac{t}{d_c^2} = \frac{\alpha t}{d_c^2} \tag{4.167}$$

---

[2]Joseph Baron Fourier (1768–1830) was a French mathematician and highly regarded Egyptologist. In 1798, he accompanied Napoleon to Egypt and conducted extensive research on Egyptian antiques. From 1798 to 1801 he served as a Secretary of Institut d'Egypte in Cairo. His work *Theorie Analytique de la Chaleur* (The Analytical Theory of Heat) was started in 1807 in Grenoble and was completed in 1822 in Paris. He developed a mathematical basis for the conductive heat transfer in solids.

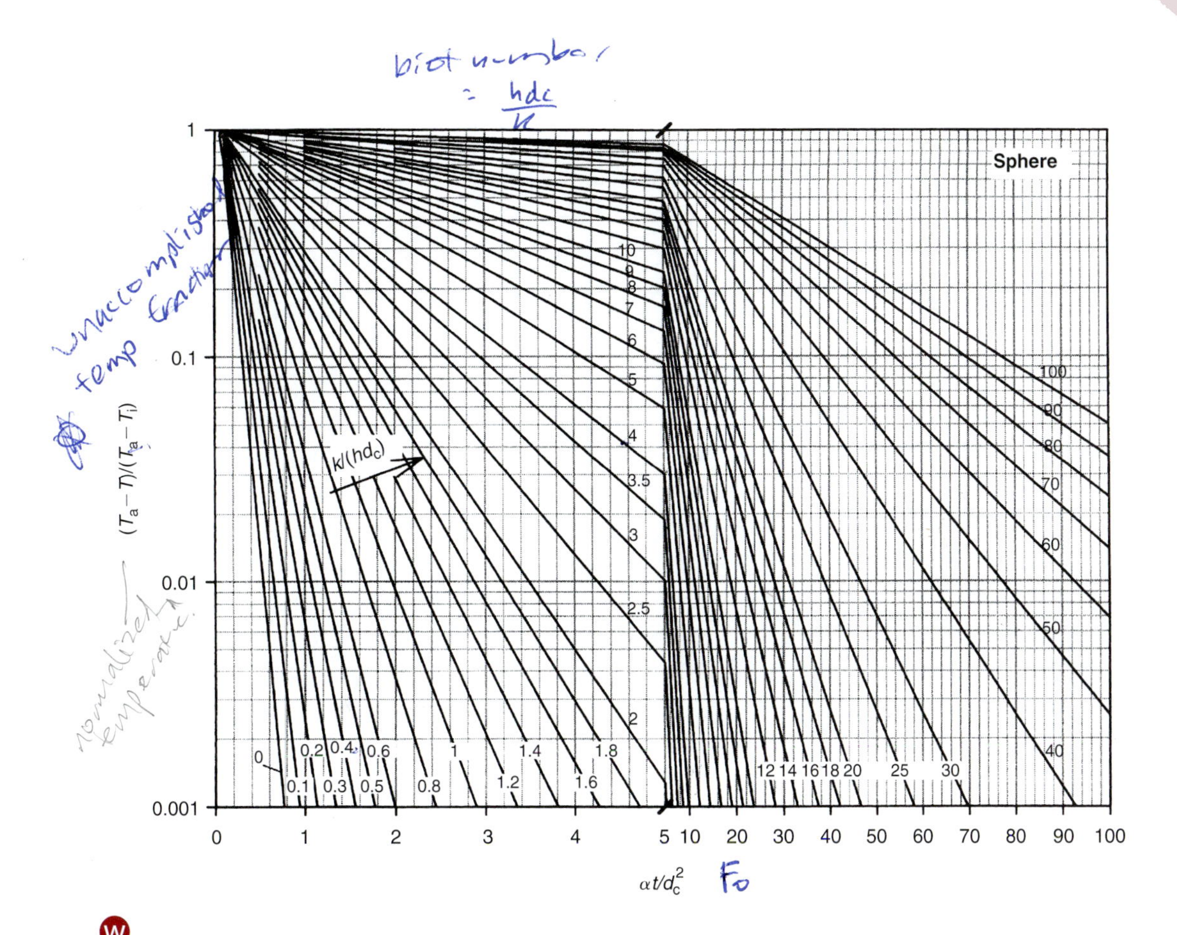

■ **Figure 4.35** Temperature at the geometric center of a sphere of radius d_c.

where d_c is a characteristic dimension. The value for d_c indicates the shortest distance from the surface to the center of the object. The characteristic dimension for both a sphere and an infinite cylinder is the radius; for an infinite slab, it is half the thickness of the slab.

We can examine the physical significance of the Fourier number, N_{Fo}, by rearranging Equation (4.167) as follows:

$$N_{Fo} = \frac{\alpha t}{d_c^2} = \frac{k(1/d_c)d_c^2}{\rho c_p d_c^3/t}$$

$$= \frac{\text{rate of heat conduction across } d_c \text{ in a body of volume } d_c^3 \text{ (W/°C)}}{\text{rate of heat storage in a body of volume } d_c^3 \text{ (W/°C)}}$$

For a given volume element, the Fourier number is a measure of the rate of heat conduction per unit rate of heat storage. Thus, a larger value of Fourier number indicates deep penetration of heat into the solid in a given period of time.

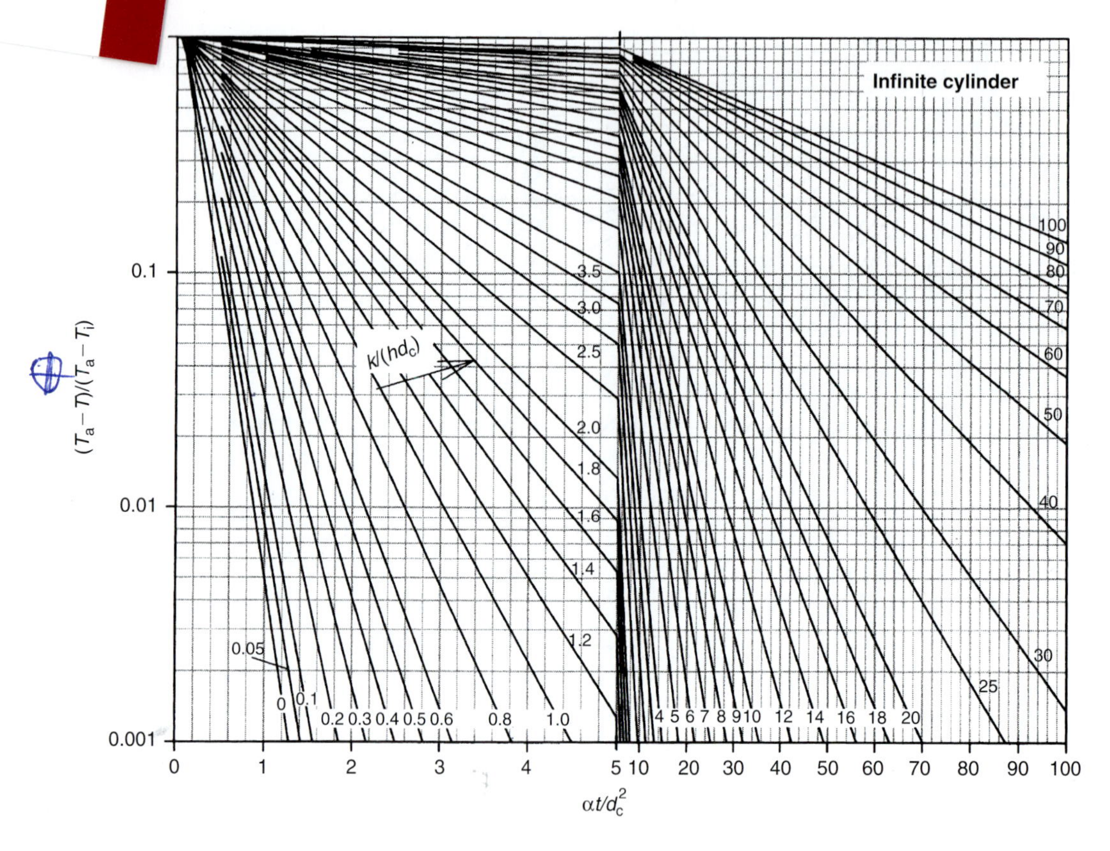

Figure 4.36 Temperature at the geometric center of an infinitely long cylinder of radius d_c.

Note that the Heisler charts shown in Figures 4.35, 4.36, and 4.37 are plotted on a log-linear scale.

4.5.4 Negligible Surface Resistance to Heat Transfer ($N_{Bi} > 40$)

For situations where the Biot number is greater than 40, indicating negligible surface resistance to heat transfer, we can use the charts in Figures 4.35, 4.36, and 4.37. In these figures, the lines for $k/hd_c = 0$ represent negligible surface resistance to heat transfer.

4.5.5 Finite Objects

Myers (1971) has shown mathematically that

$$\left(\frac{T_a - T}{T_a - T_i}\right)_{\text{finite cylinder}} = \left(\frac{T_a - T}{T_a - T_i}\right)_{\text{infinite cylinder}} \times \left(\frac{T_a - T}{T_a - T_i}\right)_{\text{infinite slab}} \tag{4.168}$$

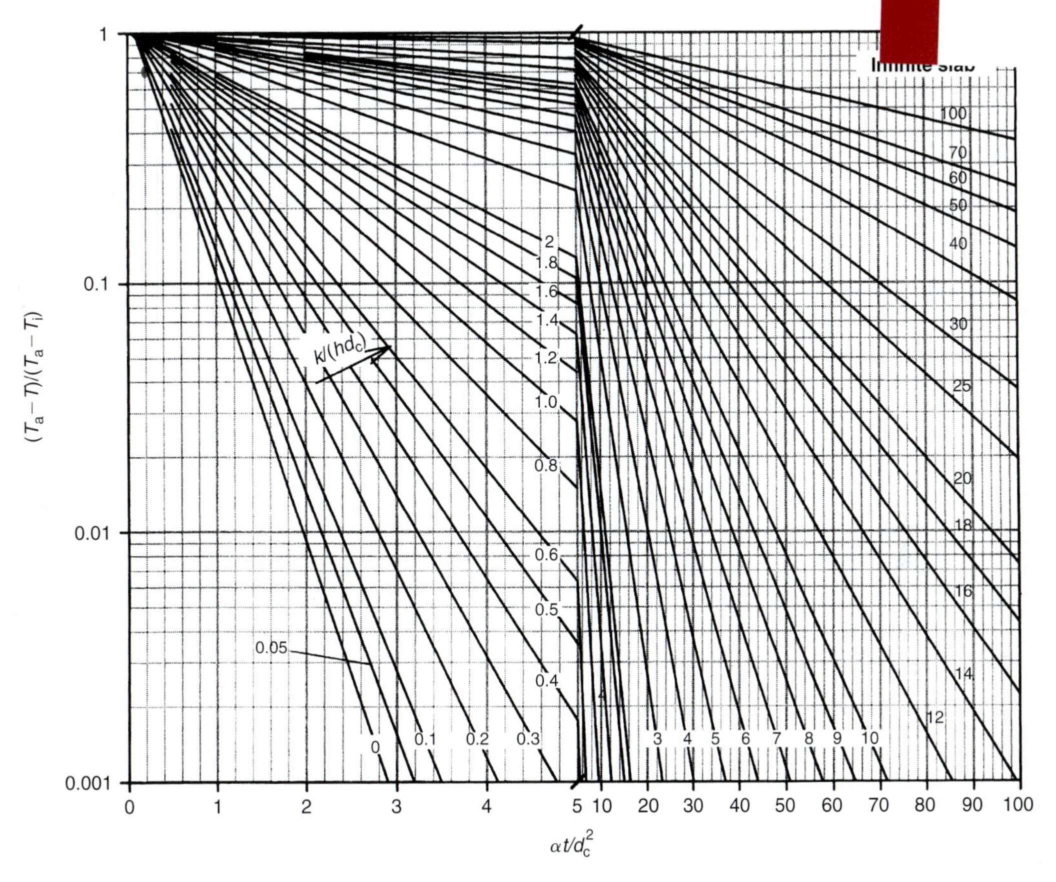

■ **Figure 4.37** Temperature at the midplane of an infinite slab of thickness $2d_c$.

and

$$\left(\frac{T_a - T}{T_a - T_i}\right)_{\text{finite brick shape}} = \left(\frac{T_a - T}{T_a - T_i}\right)_{\text{infinite slab, width}} \times \left(\frac{T_a - T}{T_a - T_i}\right)_{\text{infinite slab, depth}} \times \left(\frac{T_a - T}{T_a - T_i}\right)_{\text{infinite slab, height}} \tag{4.169}$$

These expressions allow us to determine temperature ratios for objects of finite geometry, such as a cylindrical can commonly used in heat sterilization of food. Although the mathematics to prove Equations (4.168) and (4.169) is beyond the scope of this text, Figure 4.38 may be studied as a visual aid.

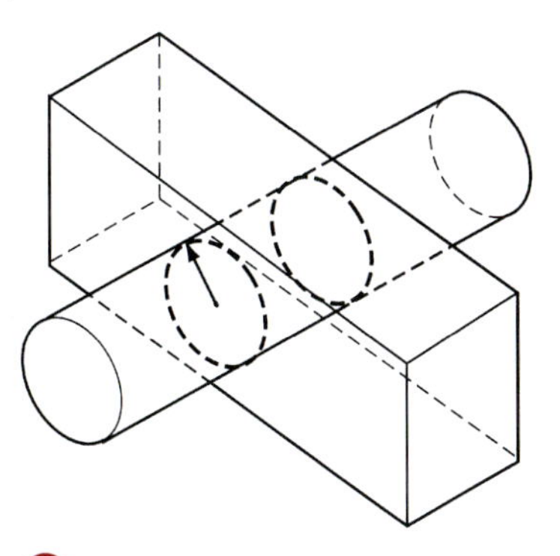

W

■ **Figure 4.38** A finite cylinder considered as part of an infinite cylinder and an infinite slab.

A finite cylinder (Fig. 4.38) can be visualized as a part of an infinite cylinder and a part of an infinite slab. Heat transfer in a radial direction is similar to heat transfer for an infinite cylinder. Invoking the infinite cylinder shape, we mean that heat transfer to the geometric center is only through the radial direction—through the circumferential area of the cylinder—whereas the ends of the cylinder are too far to have any measurable influence on heat transfer. Heat transfer from the two end surfaces is similar to heat transfer for an infinite slab. Considering the finite cylinder to be an infinite slab will account for all the heat transfer from the two ends of the cylindrical can while ignoring heat transfer in the radial direction. This approach allows us to include heat transfer both from the radial direction and from the cylinder ends. Similarly, a brick-shape object can be considered to be constructed from three infinite slabs maintaining width, depth, and height, respectively, as the finite thickness.

4.5.6 Procedures to Use Temperature–Time Charts

The following steps may be used to determine heat transfer in finite objects with the use of temperature–time charts.

Heat transfer to an object of finite cylindrical shape, such as a cylindrical can, requires the use of temperature–time charts for both infinite cylinder and infinite slab. Thus, if the temperature at the geometric center of the finite cylinder is required at a given time, the following steps may be used.

For an infinite cylinder:

1. Calculate the Fourier number, using the radius of the cylinder as the characteristic dimension.
2. Calculate the Biot number, using the radius of the cylinder as the characteristic dimension. Calculate the inverse of Biot number for use in the Heisler Charts.
3. Use Figure 4.36 to find the temperature ratio.

For an infinite slab:

1. Calculate the Fourier number, using the half-height as the characteristic dimension.
2. Calculate the Biot number, using the half-height as the characteristic dimension. Calculate the inverse of Biot number for use in the Heisler Charts.
3. Use Figure 4.37 to determine the temperature ratio.

The temperature ratio for a finite cylinder is then calculated from Equation (4.168). We can compute the temperature at the geometric center if the surrounding medium temperature T_a and initial temperature T_i are known.

Steps similar to the preceding may be used to compute temperature at the geometric center for a finite slab-shaped object (such as a parallelepiped or a cube). In the case of a spherical object such as an orange, Figure 4.35 for a sphere is used.

A major drawback of the Heisler charts is that they are difficult to use for situations when the Fourier number is small. For example, with problems involving transient heat transfer in foods, the Fourier numbers are often less than 1 because of the low thermal diffusivity of foods. In these cases, charts with expanded scales (see Appendix A.8), based on the work of Schneider (1963), are helpful. The procedure to use these expanded-scale charts is exactly the same as with the Heisler charts. Note that the expanded-scale charts use Biot numbers directly, without inverting them, and they are plotted on a linear-log scale, whereas the Heisler charts are plotted on a log-linear scale.

Examples 4.26 to 4.28 illustrate the procedures presented in this section.

Example 4.26

Estimate the time when the temperature at the geometric center of a 6 cm diameter apple held in a 2°C water stream reaches 3°C. The initial uniform temperature of the apple is 15°C. The convective heat transfer coefficient in water surrounding the apple is 50 W/(m^2 °C). The properties of the apple are thermal conductivity $k = 0.355$ W/(m °C); specific heat $c_p = 3.6$ kJ/(kg °C); and density $\rho = 820$ kg/m^3.

Given

Diameter of apple = 0.06 m
Convective heat-transfer coefficient h = 50 W/(m^2 °C)
Temperature of water stream $T_a = 2$°C
Initial temperature of apple $T_i = 15$°C
Final temperature of geometric center T = 3°C
Thermal conductivity k = 0.355 W/(m °C)
Specific heat $c_p = 3.6$ kJ/(kg °C)
Density $\rho = 820$ kg/m^3

Approach

Considering an apple to be a sphere in shape, we will use Figure 4.35 to find the Fourier number. The time of cooling will be computed from Equation (4.167).

Solution

1. *From given temperatures, we first calculate the temperature ratio.*

$$\left(\frac{T_a - T}{T_a - T_i}\right) = \frac{2-3}{2-15} = 0.077$$

2. *The Biot number is computed as*

$$N_{Bi} = \frac{hd_c}{k} = \frac{(50\,W/[m^2\ °C])(0.03\,m)}{(0.355\,W/[m\ °C])} = 4.23$$

Thus,

$$\frac{1}{N_{Bi}} = 0.237$$

3. *From Figure 4.35, for a temperature ratio of 0.077 and ($1/N_{Bi}$) of 0.237, the Fourier number can be read as*

$$N_{Fo} = 0.5$$

4. *The time is calculated from the Fourier number.*

$$\frac{k}{\rho c_p}\frac{t}{d_c^2} = 0.5$$

$$t = \frac{(0.5)(820\,kg/m^3)(3.6\,kJ/[kg\ °C])(0.03\,m)^2(1000\,J/kJ)}{(0.355\,W/[m\ °C])}$$

$$= 3742\,s$$

$$= 1.04\,h$$

Example 4.27

Estimate the temperature at the geometric center of a food product contained in a 303 × 406 can exposed to boiling water at 100°C for 30 min. The product is assumed to heat and cool by conduction. The initial uniform temperature of the product is 35°C. The properties of the food are thermal conductivity $k = 0.34\,W/(m\ °C)$; specific heat $c_p = 3.5\,kJ/(kg\ °C)$; and density

$\rho = 900\,kg/m^3$. The convective heat transfer coefficient for the boiling water is estimated to be $2000\,W/(m^2\,°C)$.

Given

Can dimensions:

$$Diameter = 3\tfrac{3}{16}\ inches = 0.081\,m$$

$$Height = 4\tfrac{6}{16}\ inches = 0.11\,m$$

Convective heat transfer coefficient $h = 2000\,W/(m^2\,°C)$
Temperature of heating media $T_a = 100°C$
Initial temperature of food $T_i = 35°C$
Time of heating $= 30\,min = 1800\,s$
Properties:

$$k = 0.34\ W/(m\ °C)$$

$$c_p = 3.5\ kJ/(kg\,°C)$$

$$\rho = 900\ kg/m^3$$

Approach

Since a finite cylindrical can may be considered a combination of infinite cylinder and infinite slab, we will use time–temperature figures for both these shapes to find respective temperature ratios. The temperature ratio for a finite cylinder will then be calculated from Equation (4.168).

Solution

1. *First, we estimate temperature ratio for an infinite cylinder.*
2. *Biot number* $= hd_c/k$ *where* d_c *is radius* $= 0.081/2 = 0.0405\,m$

$$N_{Bi} = \frac{(2000\,W/[m^2\ °C])(0.0405\,m)}{(0.34\,W/[m\ °C])} = 238$$

Thus,

$$\frac{1}{N_{Bi}} = 0.004$$

3. *The Fourier number for an infinite cylinder is*

$$N_{Fo} = \frac{k}{\rho c_p}\left(\frac{t}{d_c^2}\right)$$
$$= \frac{(0.34\,W/[m\,°C])(1800\,s)}{(900\,kg/m^3)(3.5\,kJ/[kg\,°C])(1000\,J/kJ)(0.0405\,m)^2}$$
$$= 0.118$$

4. *The temperature ratio can be estimated from Figure 4.36 for* $1/N_{Bi} = 0.004$ *and* $N_{Fo} = 0.118$ *as*

$$\left(\frac{T_a - T}{T_a - T_i}\right)_{\text{infinite cylinder}} = 0.8$$

5. *Next, we estimate the temperature ratio for an infinite slab.*
6. *Biot number* $= hd_c/k$ *where* d_c *is half-height* $= 0.11/2 = 0.055\,m$

$$N_{Bi} = \frac{(2000\,W/[m^2\,°C])(0.055\,m)}{(0.34\,W/[m\,°C])} = 323.5$$

Thus,

$$1/N_{Bi} = 0.003$$

7. *The Fourier number for an infinite slab is*

$$N_{Fo} = \frac{kt}{\rho c_p d_c^2}$$
$$= \frac{(0.34\,W/[m\,°C])(1800\,s)}{(900\,kg/m^3)(3.5\,kJ/[kg\,°C])(1000\,J/kJ)(0.055\,m)^2}$$
$$= 0.064$$

8. *The temperature ratio can be estimated from Figure 4.37 for* $1/N_{Bi} = 0.003$ *and* $N_{Fo} = 0.064$ *as*

$$\left(\frac{T_a - T}{T_a - T_i}\right)_{\text{infinite slab}} = 0.99$$

9. *The temperature ratio for a finite cylinder is computed using Equation (4.168)*

$$\left(\frac{T_a - T}{T_a - T_i}\right)_{\text{finite cylinder}} = (0.8)(0.99) = 0.792$$

Therefore,

$$
\begin{aligned}
T &= T_a - 0.792(T_a - T_i) \\
&= 100 - 0.792(100 - 35) \\
&= 48.4°C
\end{aligned}
$$

10. *The temperature at the geometric center of the can after 30 min. will be 48.4°C. Note that most of the heat transfers radially; only a small amount of heat transfers axially, since*

$$\left(\frac{T_a - T}{T_a - T_i}\right)_{\text{infinite slab}} = 0.99$$

or a value close to 1. If

$$\left(\frac{T_a - T}{T_a - T_i}\right) = 1$$

then $T = T_i$; this means that the temperature at the end of the heating period is still T_i, the initial temperature, indicating no transfer of heat. Conversely, if

$$\left(\frac{T_a - T}{T_a - T_i}\right) = 0$$

then $T = T_a$, indicating that the temperature of the end-of-heating period equals that of the surrounding temperature.

Example 4.28

Using Equations (4.161), (4.163), and (4.165), develop spreadsheet programs, and compare the calculated results with the values obtained from charts given in Figures 4.35, 4.36, and 4.37.

Approach

We will use spreadsheet EXCEL and program Equations (4.161), (4.163), and (4.165). Since these equations involve series solutions, we will consider the first 30 terms of each series, which should be sufficiently accurate for our purposes.

Solution

The spreadsheets written for a sphere, infinite cylinder, and infinite slab are shown in Figures E4.22, E4.23, and E4.24, respectively.

	A	B	C	D	E	F	G	H
1								
2		SPHERE - Negligible surface resistance to heat transfer						
3								
4								
5		Domain: 0 < r < d_c where d_c is characteristic dimension, or radius of sphere)						
6								
7						Terms of series	=((−1^(E9+1))/E9*EXP(−E9*E9*PI()*PI()*C8)*SIN(E9*PI()*C9))	
8		Fourier number	0.3		n	term_n		
9		r/d_c	0.00001		1	1.6265E−06		
10					2	−2.2572E−10		
11		Temperature ratio	0.104		3	8.3965E−17		
12					4	−8.3722E−26		
13		=SUM(F9:F38)*2/PI()*(1/C9)			5	2.2376E−37		
14					6	−1.6031E−51		
15					7	3.0784E−68		
16					8	−1.5846E−87		
17					9	2.186E−109		
18					10	−8.086E−134		
19					11	8.015E−161		
20					12	−2.13E−190		
21					13	1.517E−222		
22					14	−2.896E−257		
23					15	1.482E−294		
24					16	0		
25					17	0		
26					18	0		
27					19	0		
28					20	0		
29					21	0		
30					22	0		
31					23	0		
32					24	0		
33					25	0		
34					26	0		
35					27	0		
36					28	0		
37					29	0		
38					30	0		

Steps:

1) Enter numbers 1 though 30 in cells E9 to E38
2) Enter formula in cell F9, then copy it into cells F10 to F38.
3) Enter formula in cell C11
4) Enter any Fourier Number in cell C8 and radial location/(characteristic dimension) in cell C9
   For a sphere characteristic dimension is radius
5) If temperature ratio is desired at the center of a sphere, do not use r = 0, instead use a very small number in cell C9, e.g. 0.00001
6) The result is shown in cell C11

Figure E4.22 Spreadsheet solution (sphere) for Example 4.26.

	A	B	C	D	E	F	G	H	I	J	K	L
1												
2		INFINITE CYLINDER - Negligible Surface Resistance to Heat Transfer										
3												
4												
5		Domain: 0 < r < d_c (where d_c is the characteristic dimension, or radius of infinite cylinder)										
6												
7							Terms of Series	J0 for -3 < x < 3			J1 for 3 < x < inf	
8		Fourier Number	0.2		n	lambda_n	ArgJ0	J0			J1(lambda_n)	term_n
9		r/d_c	0		0	2.4048255577	0	1	0.820682791	0.195379932	0.519147809	0.251944131
10					1	5.5200781103	0	1	0.802645916	3.23090329	−0.340264805	−0.00120098.5
11		Temperature Ratio	0.5015		2	8.6537279129	0	1	0.799856138	6.340621261	0.271452299	1.33199E−07
12					3	11.7915344391	0	1	0.79895277	9.467043934	−0.232459829	−3.0563E−13
13					4	14.9309177086	0	1	0.798552578	12.5997901	0.206546432	1.4036E−20
14					5	18.0710639679	0	1	0.798341245	15.73559327	−0.187728803	−1.27219E−29
15					6	21.2116366299	0	1	0.798216311	18.87310399	0.173265895	2.25935E−40
16					7	24.3524715308	0	1	0.798136399	22.01166454	−0.161701553	−7.82877E−53
17					8	27.4934791320	0	1	0.798082224	25.15091632	0.152181217	5.27858E−67
18					9	30.6346064684	0	1	0.798043816	28.29064728	−0.144165981	−6.91297E−83
19												
20												
21												
22												
23												
24												
25												
26												
27												
28												
29												

Steps:

1) Enter numbers 0 to 9 in cells E9 to E18
2) Enter coefficients as indicated in cells F9 to F18
3) Enter formulas for cells G9, H9, I9, J9, K9 and L9, and copy into cells G10 to G18, H10 to H18, I10 to I18, J10 to J18, K10 to K18 and L10 to L18, respectively.
4) Enter formula for cell C11
5) Enter any Fourier Number in cell C8, and r/(characteristic dimension) in cell C9
6) Results are shown in cell C11

Formulas:
Cell C11 = 2*SUM(L9:L18)
Cell G9 = C9*F9
Cell H9 = 1−2.2499997*(G9/3)^2+1.2656208*(G9/3)^4−0.3163866*(G9/3)^6 + 0.0444479 *(G9/3)^8 − 0.0039444*(G9/3)^10 + 0.00021*(G9/3)^12
Cell I9 = 0.79788456 + 0.00000156*(3/F9) + 0.01659667*(3/F9)^2 + 0.00017105*(3/F9)^3− 0.00249511*(3/F9)^4 + 0.00113653*(3/F9)^5 − 0.00020033*(3/F9)^6
Cell J9 = F9−2.35619449 + 0.12499612*(3/F9) + 0.0000565*(3/F9)^2 − 0.00637879*(3/F9)^3 + 0.00074348*(3/F9)^4 + 0.00079824*(3/F9)^5 − 0.00029166*(3/F9)^6
Cell K9 = F9^(−1/2)*I9*COS(J9)
Cell L9 = EXP(−F9*F9*C8)*H9/(F9*K9)

Figure E4.23 Spreadsheet solution (infinite cylinder) for Example 4.26.

	A	B	C	D	E	F	G	H	I	J
1										
2		INFINITE SLAB - Negligible Surface Resistance to Heat Transfer								
3										
4										
5		Domain: −d_c<x<d_c where d_c is the characteristic dimension, or half-thickness of a slab								
6						=(2*E9+1)/2*PI()				
7							Terms of Series			
8		Fo	1		n	lambda_n	term_n	=(2*(−1)^E9)*EXP(−F9*F9*C8)/F9*COS(F9*C9)		
9		x/d_c	0		0	1.570796327	0.107977045			
10					1	4.71238898	−9.62899−11			
11		Temperature Ratio	0.108		2	7.853981634	4.13498E−28			
12					3	10.99557429	−5.65531E−54			
13		=+SUM(G9:G39)			4	14.13716694	2.25317E-88			
14		Steps:			5	17.27875959	−2.5264E−131			
15					6	20.42035225	7.8374E−183			
16		1) Enter numbers 0 to 30			7	23.5619449	−6.6622E−243			
17		in cells E9 to E39			8	26.70353756	0			
18		2) Enter formula in cell F9			9	29.84513021	0			
19		3) Copy formula from cell F9			10	32.98672286	0			
20		to cells F10 to F39			11	36.12831552	0			
21		4) Enter formula in cell G9			12	39.26990817	0			
22		5) Copy formula from cell G9 to cells G10 to G39			13	42.41150082	0			
23		6) Enter formula in cell C11			14	45.55309348	0			
24		7) Result is shown in cell C11			15	48.69468613	0			
25		8) Enter any Fourier number in			16	51.83627878	0			
26		cell C8 and x/d_c in cell C9			17	54.97787144	0			
27		and the results will be shown			18	58.11946409	0			
28		in cell C11			19	61.26105675	0			
29					20	64.4026494	0			
30					21	67.54424205	0			
31					22	70.68583471	0			
32					23	73.82742736	0			
33					24	76.96902001	0			
34					25	80.11061267	0			
35					26	83.25220532	0			
36					27	86.39379797	0			
37					28	89.53539063	0			
38					29	92.67698328	0			
39					30	95.81857593	0			

Figure E4.24 Spreadsheet solution (infinite slab) for Example 4.26.

These spreadsheets also include results for temperature ratios for arbitrarily selected Fourier numbers. The calculated results for temperature ratios compare favorably with those estimated from Figures 4.35, 4.36, and 4.37.

4.5.7 Use of f_h and j Factors in Predicting Temperature in Transient Heat Transfer

In many problems common to food processing, an unknown temperature is determined after the unaccomplished temperature fraction has decreased to less than 0.7. In such cases, the series solutions of the governing partial differential equation (Eq. (4.149)) is simplified. Since only the first term of the series is significant, all remaining terms become small and negligible. This was recognized by Ball (1923), who developed a mathematical approach to predicting temperatures in foods for calculations of thermal processes. We will consider Ball's method of thermal process calculations in more detail in Chapter 5. However, the approach to predict temperature for longer time durations is presented here.

In Equation (4.160) we noted that the unaccomplished temperature fraction decays exponentially. Thus, for a general case we may write

$$\frac{T_a - T}{T_a - T_i} = a_1\, e^{-b_1 t} + a_2\, e^{-b_2 t} + a_3\, e^{-b_3 t} \cdots \tag{4.170}$$

For long time durations, only the first term in the series is significant. Therefore,

$$\frac{T_a - T}{T_a - T_i} = a_1\, e^{-b_1 t} \tag{4.171}$$

or, rearranging,

$$\ln\left[\frac{(T_a - T)}{a_1(T_a - T_i)}\right] = -b_1 t \tag{4.172}$$

Ball used two factors to describe heat transfer equation, a time factor called f_h and a temperature lag factor called j_c. To be consistent with Ball's method, we will replace symbol a_1 with j_c, and b_1 with $2.303/f_h$. The factor 2.303 is due to the conversion of log scale from base 10 to base e. Substituting these symbols in Equation (4.172),

$$\ln\frac{(T_a - T)}{j_c(T_a - T_i)} = -\frac{2.303}{f_h}t \tag{4.173}$$

Rearranging,

$$\ln(T_a - T) = -\frac{2.303t}{f_h} + \ln[j_c(T_a - T_i)] \tag{4.174}$$

Converting to log10,

$$\log(T_a - T) = -\frac{t}{f_h} + \log[j_c(T_a - T_i)] \tag{4.175}$$

Ball used Equation (4.175) in developing his mathematical approach. He plotted the unaccomplished temperature ($T_a - T$) against time t on a log-linear graph rotated by 180°, as shown in Figure 4.39. From the plot, he obtained f_h as the time required for the straight line portion to traverse one log cycle. In other words, f_h is the time taken for the unaccomplished temperature to decrease by 90%. The j_c factor was obtained by extending the straight line to time 0 to obtain an intercept with the ordinate as $T_a - T_A$. Then j_c was defined as $(T_a - T_A)/(T_a - T_i)$. A similar procedure, used for a cooling curve, will be illustrated with an example at the end of this section.

The exact solutions of the first term of infinite series shown in Equations (4.161), (4.163), and (4.165) can either be programmed into a spreadsheet or plotted as originally done by Pflug et al. (1965) as f_h vs N_{Bi}, j_c vs N_{Bi}, and j_m vs N_{Bi}, shown in Figures 4.40, 4.41, and 4.42. Factor j_c is for the temperature lag at the center of an object, and j_m is the mean temperature lag for the object. The f_h value is the same for either center temperature or mean temperature. It is useful to know the mean temperature of an object when determining the heat load

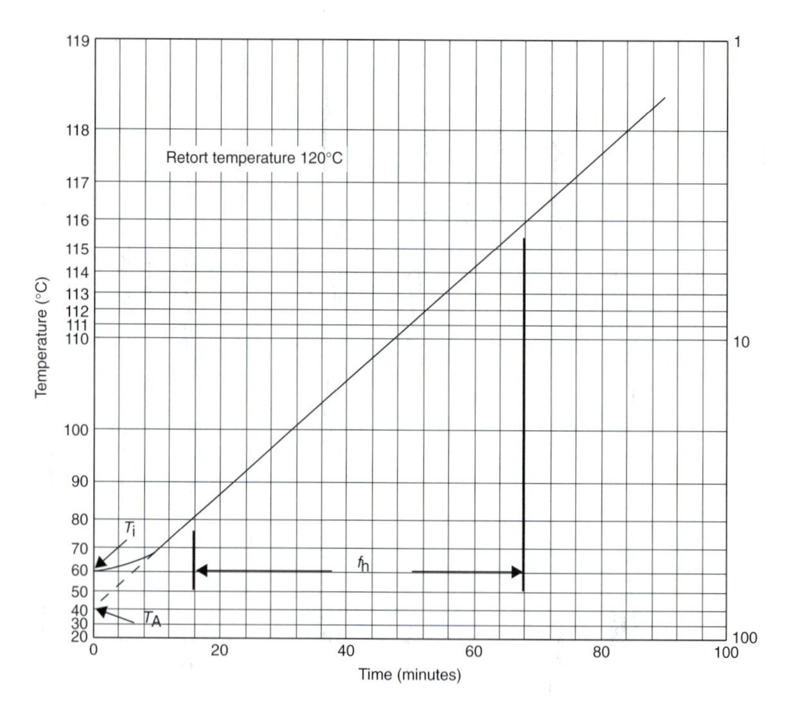

Figure 4.39 Heating curve plotted on a semi-log paper rotated 180°.

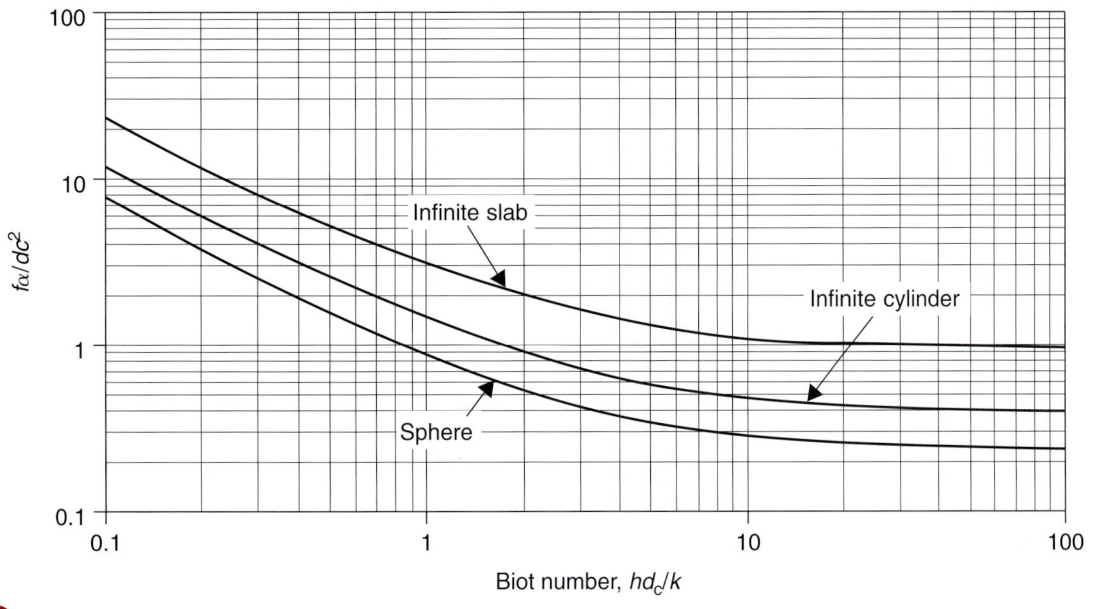

■ **Figure 4.40** Heating rate parameter, f_h, as a function of Biot number.

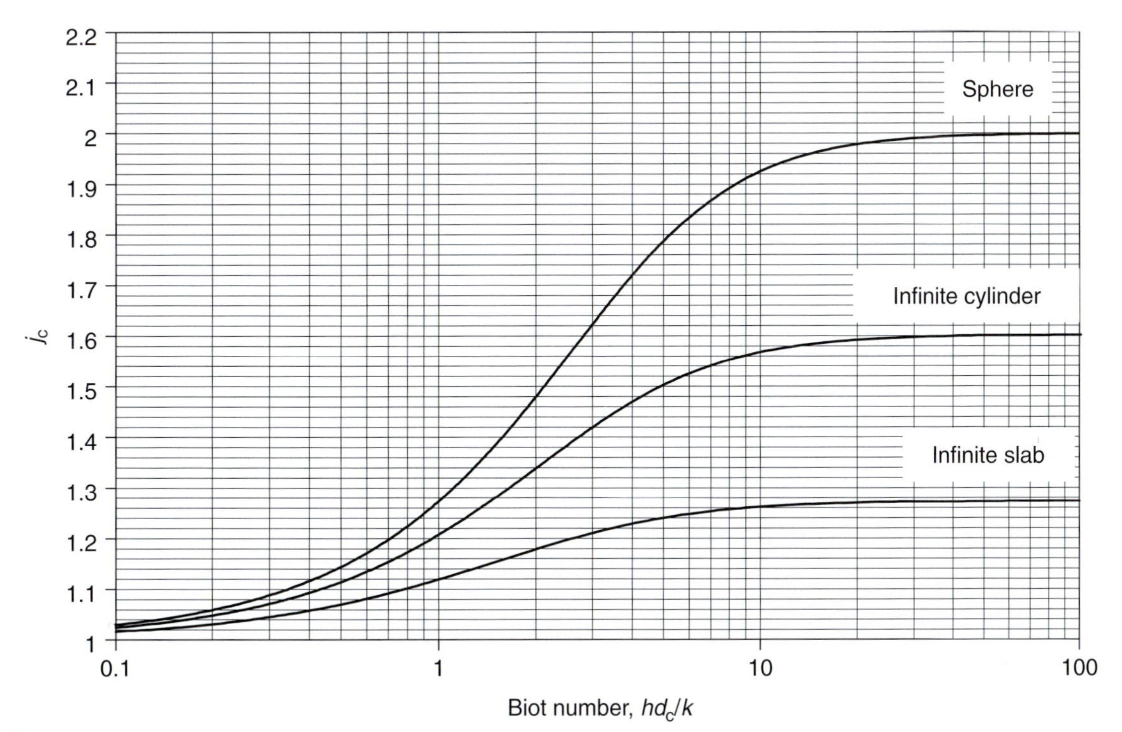

■ **Figure 4.41** Lag factor, j_c, at the geometric center of a sphere, infinite cylinder, and infinite slab as a function of Biot number.

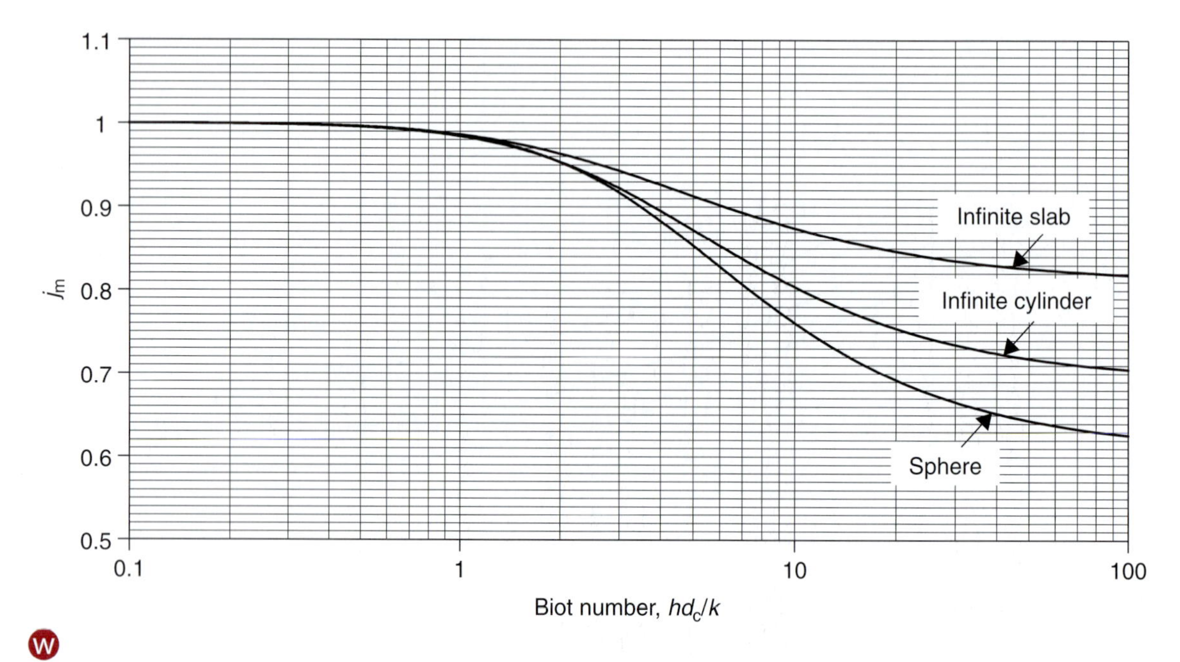

W

Figure 4.42 Average lag factor, j_m of a sphere, infinite cylinder, and infinite slab as a function of Biot number.

required for heating/cooling applications. Using Figures 4.40, 4.41, or 4.42, the f_h, j_c, or j_m factors are obtained for simplified geometrical shapes and used in Equation (4.174) to calculate the temperature at any time. If the shape of the object is a finite cylinder, then we can obtain the f_h and j_c factors using the following relations:

$$\frac{1}{f_{\text{finite cylinder}}} = \frac{1}{f_{\text{infinite cylinder}}} + \frac{1}{f_{\text{infinite slab}}} \tag{4.176}$$

and

$$j_{\text{c, finite cylinder}} = j_{\text{c, infinite cylinder}} \times j_{\text{c, infinite slab}} \tag{4.177}$$

For a brick-shaped object,

$$\frac{1}{f_{\text{brick}}} = \frac{1}{f_{\text{infinite slab}_1}} + \frac{1}{f_{\text{infinite slab}_2}} + \frac{1}{f_{\text{infinite slab}_3}} \tag{4.178}$$

and

$$j_{\text{c, brick}} = j_{\text{c, infinite slab}_1} \times j_{\text{c, infinite slab}_2} \times j_{\text{c, infinite slab}_3} \tag{4.179}$$

The usefulness of the product rule is limited for shapes with dimension ratios much greater than 1 (Pham, 2001). A gross overestimate of cooling time may occur. In those cases, f_h and j_c for finite shapes may be calculated using empirical methods given by Lin et al. (1996).

Example 4.29

A solid food in a cylindrical container is cooled in a 4°C immersion water cooler. Estimate the *f* and *j* factors if the following data were obtained for temperature at the geometric center of the container.

Time (minutes)	Temperature (°C)
0	58
5	48
10	40
15	26
20	25
25	19
30	15
35	12
40	10
45	9
50	7.5
55	7
60	6.5

Given

Cooling medium temperature = 4°C

Approach

We will use a two-cycle log paper and create a scale for the y-axis as shown in Figure E4.25. The straight line portion will be extended to intersect the y-axis to determine the pseudo initial temperature. The f_c factor will be obtained by determining the time traversed for one log cycle change in temperature.

Solution

1. *As shown in Figure E4.25 the y axis is labeled starting from the bottom with 1 + 4°C = 5°C scale; label the rest of the numbers on the y-axis according to the log scale.*
2. *Locate all the points for temperature and time on the graph. Extend the straight line to intersect the y-axis to determine the pseudo initial temperature. From the plot, this temperature is 69°C.*

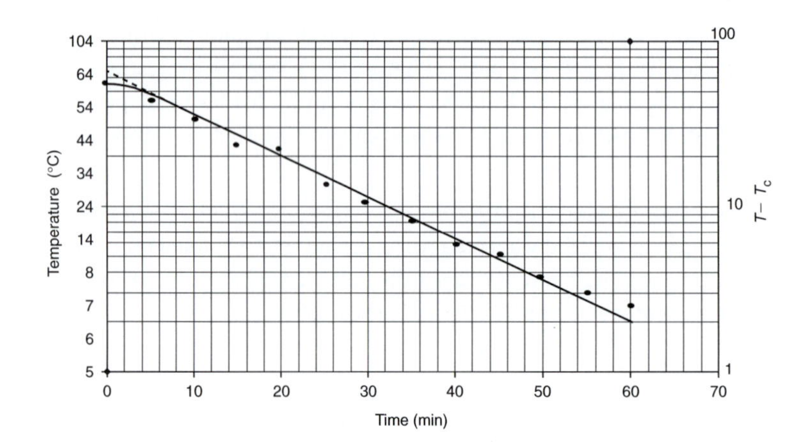

Figure E4.25 Cooling curve for data given in Example 4.27.

3. *Determine the time for the straight line to traverse one log cycle to obtain f_c. From the plot we get $f_c = 40$ minutes.*
4. *The j_c value is then obtained as*

$$j_c = \frac{69 - 4}{58 - 4} = 1.2$$

5. *For the given data, the f_c is 40 minutes and j_c is 1.2.*

Example 4.30

A hot dog is initially at 5°C and it is being heated in hot water at 95°C. The convective heat transfer coefficient is 300° W/(m² °C). The dimensions of the hot dog are 2 cm diameter and 15 cm long. Assuming that the heat transfer is largely in the radial direction, estimate the product temperature at the geometric center after 10 min. The properties of the hot dog are as follows: density = 1100 kg/m³, specific heat = 3.4 kJ/(kg °C), and thermal conductivity = 0.48 W/(m °C).

Given

Initial temperature = 5°C
Heating medium temperature = 95°C
Heat transfer coefficient = 300 W/(m² °C)
Hot dog length = 15 cm
Hot dog diameter = 2 cm
Heating time = 10 min

Density = 1100 kg/m³
Specific heat = 3.4 kJ/(kg °C)
Thermal conductivity = 0.48 W/(m °C)

Approach

We will consider Pflug's charts for solving this problem. First we will calculate the Biot number and use Figures 4.40 and 4.41 to obtain f_h and j_c factors. The required temperature will be obtained from Equation (4.176).

Solution

1. *Biot number for an infinite cylinder is obtained as*

$$N_{Bi} = \frac{300[W/(m^2\,°C)] \times 0.01[m]}{0.48[W/(m\,°C)]}$$

$$N_{Bi} = 6.25$$

2. *From Figure 4.40, we obtain for an infinite cylinder,*

$$\frac{f_h \alpha}{d_c^2} = 0.52$$

$$f_h = \frac{0.52 \times (0.01)^2 [m^2] \times 1100[kg/m^3] \times 3400[J/(kg\,°C)]}{0.48[W/(m\,°C)]}$$

$$f_h = 405.17\,s$$

3. *From Figure 4.41, we obtain for an infinite cylinder,*

$$j_c = 1.53$$

4. *Using Equation (4.176),*

$$\log(95 - T) = -\frac{10 \times 60[s]}{405.17[s]} + \log(1.53[95 - 5])$$

$$T = 90.45°C$$

5. *The temperature at the geometric center after 10 minutes of heating is 90.45°C. The validity of the method may be checked by calculating the unaccomplished temperature fraction at 10 minutes. For this example, the method is valid because the fraction is 0.05 ≪0.7, the assumption used in this method.*

4.6 ELECTRICAL CONDUCTIVITY OF FOODS

It is well known that when electrolytes are placed in an electric field, the ions present within the electrolyte move toward the electrodes with opposite charge. The movement of ions in the electrolyte generates heat. Similarly, when a food containing ions is placed between two electrodes and alternating current or any other wave form current is passed through the food, the food heats by internal heat generation.

Electrical conductance, κ_E, not to be confused with electrical conductivity, is the reciprocal of electrical resistance, or

$$\kappa_E = \frac{1}{R_E} \tag{4.180}$$

where R_E is the electrical resistance of the food material (ohms).

From Ohm's law, we know that

$$R_E = \frac{E_V}{I} \tag{4.181}$$

where E_V is the applied voltage (V), and I is the electric current (A).

Then, electrical conductance is

$$\kappa_E = \frac{I}{E_V} \tag{4.182}$$

Electrical conductivity, σ_E, is a measure of a material's ability to conduct electric current. It is equal to electrical conductance measured between the opposite faces of a 1-meter cube of the material.

$$\sigma_E = \frac{\kappa_E L}{A} = \frac{IL}{E_V A} \tag{4.183}$$

where A is the area (m^2) and L is the length (m). The SI units of electrical conductivity are siemens/m or S/m.

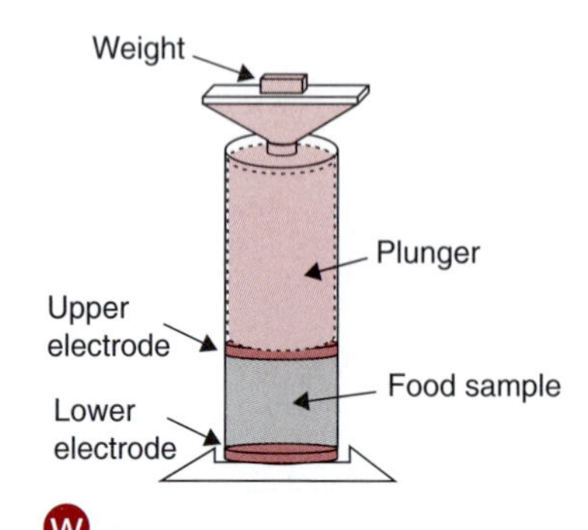

W

Figure 4.43 A setup for measurement of electrical conductivity of foods. (Adapted from Mitchell and Alvis, 1989)

Electrical conductivity of a food material is measured using an electrical conductivity cell, as shown in Figure 4.43. In this cell, a food sample is placed between two electrodes and the electrodes are connected to a power supply. Care is taken to ensure that the electrodes make a firm contact with the food sample.

Electrical conductivity of foods increases with temperature in a linear manner. The following equation may be used to calculate electrical conductivity of a food:

$$\sigma_E = \sigma_o(1 + m''T) \tag{4.184}$$

where σ_o is the electrical conductivity at 0°C (S/m), m'' is coefficient, (1/°C); T is temperature, (°C).

If a reference temperature other than 0°C is chosen then an alternate expression for estimating electrical conductivity is as follows:

$$\sigma_E = \sigma_{ref}[1 + K(T - T_{ref})] \tag{4.185}$$

Values of σ_o, σ_{ref} and coefficients m'' and K for different foods are given in Table 4.6.

The electrical conductivity of a food is a function of its composition—the quantity and type of various components present in the food. Foods containing electrolytes such as salts, acids, certain gums, and thickeners contain charged groups that have a notable effect on the value of electrical conductivity. Based on experimental studies, researchers have reported mathematical relationships to predict electrical conductivity

Table 4.6 Coefficients for Equations (4.184) and (4.185) to Estimate Electrical Conductivity

Product	σ_{25} (S/m)	K (°C^{-1})	σ_o (S/m)	m'' (°C^{-1})
Potato	0.32	0.035	0.04	0.28
Carrot	0.13	0.107	−0.218	−0.064
Yam	0.11	0.094	−0.149	−0.07
Chicken	0.37	0.019	0.194	0.036
Beef	0.44	0.016	0.264	0.027
Sodium Phosphate 0.025 M	0.189	0.027	0.614	0.083
Sodium Phosphate 0.05 M	0.361	0.022	0.162	0.048
Sodium Phosphate 0.1 M	0.676	0.021	0.321	0.0442

Source: Palaniappan and Sastry (1991)

based on selected food constituents. In fruit juices, the inert suspended solids in the form of pulp or other cellular material act as insulators and tend to decrease the electrical conductivity of the liquid media. Sastry and Palaniappan (1991) obtained the following relationship to describe the effect of solid concentration on electrical conductivity of orange and tomato juices:

$$\sigma_{T,\text{tomato}} = 0.863[1 + 0.174(T - 25)] - 0.101 \times M_s \qquad (4.186)$$

$$\sigma_{T,\text{orange}} = 0.567[1 + 0.242(T - 25)] - 0.036 \times M_s \qquad (4.187)$$

where M_s is the solid concentration (percent), and T is temperature (°C).

Example 4.31

Estimate the electrical conductivity of 0.1 M Sodium Phosphate solution at 30°C.

Given

Temperature = 30°C
0.1 M Sodium Phosphate solution

Approach

We will use Equation (4.184) to determine the electrical conductivity.

Solution

1. *Using Equation (4.184) and appropriate values for electrical conductivity at the reference temperature of 0°C and coefficient m″, we obtain*

$$\sigma_E = 0.321(1 + 0.0442 \times 30)$$

$$\sigma_E = 0.746 \text{ S/m}$$

2. *Note that if we use Equation (4.185) with appropriate values of electrical conductivity at reference temperature of 25°C and coefficient K, we obtain*

$$\sigma_E = 0.676(1 + 0.021(30 - 25))$$

$$\sigma_E = 0.747 \text{ S/m}$$

As expected, we get the same result using Equation (4.184) or (4.185).

4.7 OHMIC HEATING

In ohmic heating, main alternating current is passed directly through a conductive food, which causes heat generation within the food. Due to internal heat generation, the heating is rapid and more uniform than traditional systems used for heating foods where heat must travel from the outside surface to the inside of the food. The rapid and uniform heating of a food is advantageous in retaining many quality characteristics such as color, flavor, and texture. The efficiency of ohmic heating is dependent upon how well the electric current can pass through the food, as determined by its electrical conductivity. Therefore, the knowledge of electrical conductivity of foods is important in designing processes and equipment involving ohmic heating.

As an example of ohmic heating, we will consider heating a liquid food with Newtonian characteristics when pumped through an ohmic heater. We assume that the flow conditions through the tubular-shaped heater are similar to plug flow, and a constant voltage gradient exists along the heater. In this setup, heat is generated within the liquid due to ohmic heating, and heat loss from the fluid is in radial direction to the outside, if the heater pipe is uninsulated.

For this setup, conducting a heat balance we get,

$$\dot{m}c_p \frac{\mathrm{d}T}{\mathrm{d}t} = (|\Delta V|^2 \sigma_o(1 + m''T))\left(\frac{\pi d_c^2 L}{4}\right) - U\pi d_c L\left(T - T_\infty\right) \quad (4.188)$$

where $|\Delta V|$ is voltage gradient along the heater pipe length, (V/m); σ_o is the electrical conductivity at 0°C; m is the slope obtained from Equation (4.184); d_c is the characteristic dimension or diameter of the heater pipe (m); L is the length of heater pipe (m); U is the overall heat transfer coefficient based on the inside area of the heater pipe, (W/m^2°C); T_∞ is the temperature of the air surrounding the heater (°C).

The initial condition is

$$t = 0, \qquad T = T_o$$

Expanding the terms in Equation (4.188) and rearranging, we get

$$\frac{\mathrm{d}T}{\mathrm{d}t} = \frac{a\pi d_c LT}{\dot{m}c_p} + \frac{b\pi DL}{\dot{m}c_p} \quad (4.189)$$

where

$$a = \frac{|\Delta V|^2 \, d_c \sigma_o m''}{4} - U \tag{4.190}$$

$$b = \frac{d_c \, |\Delta V|^2 \, \sigma_o}{4} + UT_\infty \tag{4.191}$$

Integrating Equation (4.189), we obtain

$$\frac{aT + b}{aT_o + b} = e^{\left(\frac{a\pi d_c L}{\dot{m} c_p}\right)} \tag{4.192}$$

In the following example, we will use Equation (4.192) to determine the temperature of a liquid exiting an ohmic heater.

Example 4.32

A liquid food is being pumped through an ohmic heater at 0.5 kg/s. The inside diameter of the heater pipe is 0.05 m and it is 3 m long. The specific heat of the liquid food is 4000 J/kg °C. The applied voltage is 15,000V. The overall heat transfer coefficient based on the inside pipe area is 100 W/m^2 °C. The surrounding temperature of the air is 20 °C. The liquid food enters the ohmic heater at 50°C. Assume that the properties of the liquid food are similar to 0.05 M Sodium Phosphate solution. Calculate the temperature at which the liquid food exits.

Given

Flow rate = 0.5 kg/s
Inside diameter of ohmic heater = 0.05 m
Length of ohmic heater = 3 m
Specific heat capacity = 4000 J/kg °C
Voltage = 15000 V
Overall heat transfer coefficient = 100 W/m^2 °C
Surrounding air temperature = 20°C
Inlet temperature = 50°C

Approach

Using the given information, we will determine the voltage gradient, and obtain electrical properties using Table 4.6 for 0.05 M Sodium Phosphate. Using Equation (4.192) we will calculate the liquid temperature at the exit of the heater.

Solution

1. *Voltage gradient is obtained from the given information as*

$$|\Delta V| = \frac{15000}{3} = 5000\,V/m$$

2. *From Table 4.6, for 0.05 M Sodium Phosphate,*

$$\sigma_o = 0.162\,S/m$$

$$m'' = 0.048(°C^{-1})$$

3. *Using Equations (4.190) and (4.191), we obtain a and b values as follows:*

$$a = 2330\,W/m^2\,°C$$

$$b = 52625\,W/m^2\,°C$$

4. *Substituting a and b values in Equation (4.192)*

$$\frac{2330T + 52625}{2330 \times 50 + 52625} = e^{\left(\frac{\pi \times 2330 \times 0.05 \times 3}{0.5 \times 4000}\right)}$$

Solving for the unknown temperature, T, we get

$$T = 103°C$$

The temperature of the liquid food will increase from 50 to 103°C when heated in the ohmic heater.

4.8 MICROWAVE HEATING

Electromagnetic radiation is classified by wavelength or frequency. The electromagnetic spectrum between frequencies of 300 MHz and 300 GHz is represented by microwaves. Since microwaves are used in radar, navigational equipment, and communication equipment, their use is regulated by governmental agencies. In the United States, the Federal Communications Commission (FCC) has set aside two frequencies for industrial, scientific, and medical (ISM) apparatus in the microwave range, namely 915 ± 13 MHz, and 2450 ± 50 MHz. Similar frequencies are regulated worldwide through the International Telecommunication Union (ITU).

Microwaves have certain similarities to visible light. Microwaves can be focused into beams. They can transmit through hollow tubes. Depending on the dielectric properties of a material, they may be reflected or absorbed by the material. Microwaves may also transmit through materials without any absorption. Packaging materials such as glass, ceramics, and most thermoplastic materials allow microwaves to pass through with little or no absorption. When traveling from one material to another, microwaves may change direction, similar to the bending of light rays when they pass from air to water.

In contrast to conventional heating systems, microwaves penetrate a food, and heating extends within the entire food material. The rate of heating is therefore more rapid. Note that microwaves generate heat due to their interactions with the food materials. The microwave radiation itself is nonionizing radiation, distinctly different from ionizing radiation such as X-rays and gamma rays. When foods are exposed to microwave radiation, no known nonthermal effects are produced in food material (IFT, 1989; Mertens and Knorr, 1992).

The wavelength, frequency, and velocity of electromagnetic waves are related by the following expression

$$\lambda = u/f' \tag{4.193}$$

where, λ is wavelength in meters; f' is frequency in hertz; u is the speed of light (3×10^8 m/s).

Using Equation (4.193), the wavelengths of the permitted ISM frequencies in the microwave range can be calculated as

$$\lambda_{915} = \frac{3 \times 10^8 (\text{m/s})}{915 \times 10^6 (1/\text{s})} = 0.328\ \text{m}$$

and

$$\lambda_{2450} = \frac{3 \times 10^8 (\text{m/s})}{2450 \times 10^6 (1/\text{s})} = 0.122\ \text{m}$$

4.8.1 Mechanisms of Microwave Heating

The absorption of microwaves by a dielectric material results in the microwaves giving up their energy to the material, with a consequential rise in temperature. The two important mechanisms that explain heat generation in a material placed in a microwave field are ionic polarization and dipole rotation.

4.8.1.1 *Ionic Polarization*

When an electrical field is applied to food solutions containing ions, the ions move at an accelerated pace due to their inherent charge. The resulting collisions between the ions cause the conversion of kinetic energy of the moving ions into thermal energy. A solution with a high concentration of ions would have more frequent ionic collisions and therefore exhibit an increase in temperature.

4.8.1.2 *Dipole Rotation*

Food materials contain polar molecules such as water. These molecules generally have a random orientation. However, when an electrical field is applied, the molecules orient themselves according to the polarity of the field. In a microwave field, the polarity alternates rapidly (e.g., at the microwave frequency of 2450 MHz, the polarity changes at 2.45×10^9 cycles per second). The polar molecules rotate to maintain alignment with the rapidly changing polarity (Fig. 4.44). Such rotation of molecules leads to friction with the surrounding medium, and heat is generated. With increasing temperatures, the molecules try to align more rapidly with the applied field. Several factors influence the microwave heating of a material, including the size, shape, state (e.g., water or ice), and properties of the material, and the processing equipment.

4.8.2 Dielectric Properties

In microwave processing, we are concerned with the electrical properties of the material being heated. The important electrical properties are the relative dielectric constant ε' and the relative dielectric loss ε''.

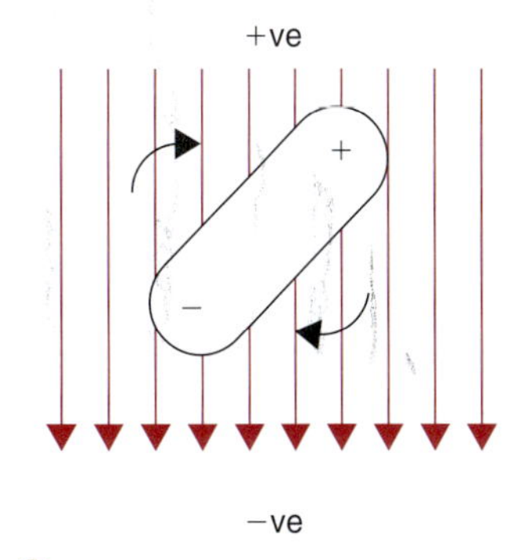

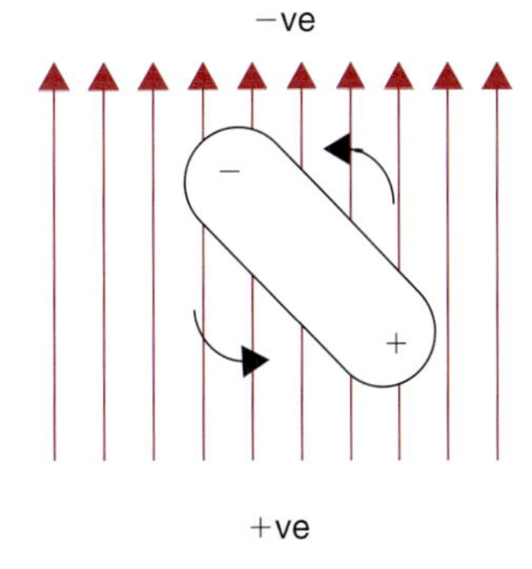

W

Figure 4.44 Movement of a dipole in an electrical field. (From Decareau and Peterson, 1986)

The term *loss* implies the conversion (or "loss") of electrical energy into heat, and the term *relative* means relative to free space.

The relative dielectric constant ε' expresses the ability of the material to store electrical energy, and the relative dielectric loss ε'' denotes the ability of the material to dissipate the electrical energy. These properties provide an indication of the electrical insulating ability of the material. Foods are in fact very poor insulators; therefore, they generally absorb a large fraction of the energy when placed in a microwave field, resulting in instantaneous heating (Mudgett, 1986). The dielectric loss factor for the material, ε'', which expresses the degree to which an externally applied electrical field will be converted to heat, is given by

$$\varepsilon'' = \varepsilon' \tan \delta \tag{4.194}$$

The loss tangent, $\tan \delta$, provides an indication of how well the material can be penetrated by an electrical field and how it dissipates electrical energy as heat.

4.8.3 Conversion of Microwave Energy into Heat

Microwave energy in itself is not thermal energy; rather, heating is a consequence of the interactions between microwave energy and a dielectric material. The conversion of the microwave energy to heat can be approximated with the following equation (Copson, 1975; Decareau and Peterson, 1986):

$$P_D = 55.61 \times 10^{-14} E^2 f' \varepsilon' \tan \delta \tag{4.195}$$

where P_D is the power dissipation (W/cm^3); E is electrical field strength (V/cm); f' is frequency (Hz); ε' is the relative dielectric constant, and $\tan \delta$ is the loss tangent.

In Equation (4.195), the dielectric constant ε' and the loss tangent $\tan \delta$ are the properties of the material, and the electrical field strength E and frequency f' represent the energy source. Thus, there is a direct relationship between the material being heated and the microwave system providing the energy for heating. It is evident in Equation (4.195) that increasing the electrical field strength has a dramatic effect on the power density, since the relationship involves a square term.

The governing heat transfer equation presented earlier in this chapter, Equation (4.149), can be modified for use in predicting heat transfer in a material placed in a microwave field. A heat generation term q''' equivalent to the power dissipation obtained from Equation (4.195) is introduced in Equation (4.149). Thus, for transient heat transfer in an infinite slab, we can obtain the following expression for a one-dimensional case:

$$\frac{\partial^2 T}{\partial x^2} + \frac{q'''}{k} = \frac{\rho c_p \partial T}{k \partial t} \tag{4.196}$$

Numerical techniques are used to solve the preceding equation (Mudgett, 1986).

4.8.4 Penetration Depth of Microwaves

The energy transfer between microwaves and the material exposed to the microwave field is influenced by the electrical properties of the material. The distribution of energy within a material is determined by the attenuation factor α'.

The attenuation factor α' is calculated from the values for the loss tangent, relative dielectric constant, and the frequency of the microwave field:

$$\alpha' = \frac{2\pi}{\lambda}\left[\frac{\varepsilon'}{2}\left(\sqrt{1+\tan^2\delta} - 1\right)\right]^{1/2} \tag{4.197}$$

The penetration of an electrical field can be calculated from the attenuation factor. As shown by von Hippel (1954), the depth Z below the surface of the material at which the electrical field strength is 1/e that of the electrical field in the free space, is the inverse of the attenuation factor. Thus,

$$Z = \frac{\lambda}{2\pi}\left[\frac{2}{\varepsilon'\left(\sqrt{1+\tan^2\delta} - 1\right)}\right]^{1/2} \tag{4.198}$$

Noting that frequency and wavelength are inversely related, it is evident from Equation (4.198) that microwave energy at 915 MHz penetrates more deeply than at 2450 MHz.

In addition to the foregoing description of penetration of a microwave field in a material, the depth of penetration for microwave power is usually described in two different ways. First, the penetration depth is the distance from the surface of a dielectric material where the incident power is decreased to 1/e of the incident power. Lambert's expression for power absorption gives

$$P = P_0\, e^{-2\alpha' d} \tag{4.199}$$

where P_0 is the incident power, P is the power at the penetration depth, d is the penetration depth, and α' is the attenuation factor.

If the power is reduced to 1/e of the incident power at depth d, we have $P/P_0 = 1/e$. Therefore, from Equation (4.199), $2\alpha' d = 1$ and $d = 1/2\alpha'$.

The second definition of the penetration depth is stated in terms of half-power depth (i.e., one-half of the incident power). Therefore, at half-power depth, $P/P_0 = 1/2$. From Equation (4.199), $e^{-2\alpha' d} = 1/2$, and solving for d we get $d = 0.347/\alpha'$.

Example 4.33

In a paper on microwave properties, Mudgett (1986) provides data on dielectric constants and loss tangents for raw potatoes. For a microwave frequency of 2450 MHz and at 20°C, the dielectric loss is 64 and the loss tangent is 0.23. Determine the attenuation factor, the field penetration depth, and the depth below the surface of a potato at which the microwave power is reduced to one-half of the incident power.

Approach

We will use Equations (4.197) and (4.198) to determine the attenuation factor and the penetration depth for the microwave field, respectively. The distance from the surface of the material at which the power is reduced to one-half of the incident power will be calculated using modifications of Equation (4.199).

Solution

1. *From Equation (4.197):*

$$\alpha' = \frac{2\pi \times 2450 \times 10^6 (1/s)}{3 \times 10^8 (m/s) \times 100 (cm/m)} \left[\frac{64}{2} \left(\sqrt{1 + (0.23)^2} - 1 \right) \right]^{1/2}$$

$$\alpha' = 0.469\ (cm^{-1})$$

2. *The penetration depth for the microwave field is the inverse of α' as seen in Equation (4.198). Therefore,*

$$\text{Field penetration depth} = Z = 1/\alpha' = 1/0.469 = 2.13\ cm$$

3. *To obtain the half-power depth of penetration, we use the modification of Lambert's expression, Equation (4.199), and solve for d:*

$$d = \frac{0.347}{\alpha'} = \frac{0.347}{0.469} = 0.74\,\text{cm}$$

4. *The half-power depth for potatoes at 2450 MHz and 20°C is calculated to be 0.74 cm.*

4.8.5 Microwave Oven

A typical microwave oven consists of the following major components (Fig. 4.45).

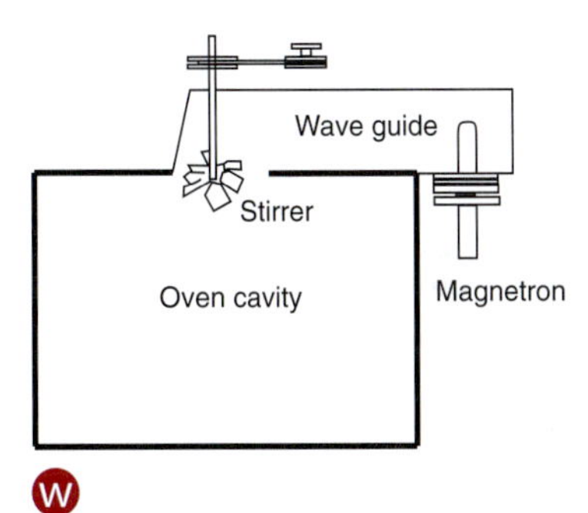

W

Figure 4.45 Major components of a microwave oven.

- *Power supply*. The purpose of the power supply is to draw electrical power from the line and convert it to the high voltage required by the magnetron. The magnetron usually requires several thousand volts of direct current.
- *Magnetron or power tube*. The magnetron is an oscillator capable of converting the power supplied into microwave energy. The magnetron emits high-frequency radiant energy. The polarity of the emitted radiation changes between negative and positive at high frequencies (e.g., 2.45×10^9 cycles per second for a magnetron operating at a frequency of 2450 MHz, the most common frequency used for domestic ovens).
- *Wave guide or transmission section*. The wave guide propagates, radiates, or transfers the generated energy from the magnetron to the oven cavity. In a domestic oven, the wave guide is a few centimeters long, whereas in industrial units it can be a few meters long. The energy loss in the wave guide is usually quite small.
- *Stirrer*. The stirrer is usually a fan-shaped distributor that rotates and scatters the transmitted energy throughout the oven. The stirrer disturbs the standing wave patterns, thus allowing better energy distribution in the oven cavity. This is particularly important when heating nonhomogenous materials like foods.
- *Oven cavity or oven*. The oven cavity encloses the food to be heated within the metallic walls. The distributed energy from the stirrer is reflected by the walls and intercepted by the food from many directions with more or less uniform energy density. The energy impinging on the food is absorbed and converted into heat. The size of the oven cavity is influenced by the wavelength. The length of the cavity wall should be greater than

one-half the wavelength and any multiple of a half-wave in the direction of the wave propagation. The wavelength at 2450 MHz frequency was calculated earlier to be 12.2 cm; therefore, the oven cavity wall must be greater than 6.1 cm. The oven cavity door includes safety controls and seals to retain the microwave energy within the oven during the heating process.

4.8.6 Microwave Heating of Foods

Heating of foods in a microwave field offers several advantages over more conventional methods of heating. The following are some of the important features of microwave heating that merit consideration.

4.8.6.1 *Speed of Heating*

The speed of heating of a dielectric material is directly proportional to the power output of the microwave system. In industrial units, the typical power output may range from 5 to 100 kW. Although high speed of heating is attainable in the microwave field, many food applications require good control of the rate at which the foods are heated. Very high-speed heating may not allow desirable physical and biochemical reactions to occur. The speed of heating in a microwave is governed by controlling the power output. The power required for heating is also proportional to the mass of the product.

4.8.6.2 *Frozen Foods*

The heating behavior of frozen foods is markedly influenced by the different dielectric properties of ice and water (Table 4.7). Due to its low dielectric loss factor, ice is more transparent to microwaves than water. Thus, ice does not heat as well as water. Therefore, when using microwaves to temper frozen foods, care is taken to keep the temperature of the frozen food just below the freezing point. If the ice melts,

Table 4.7 Dielectric Properties of Water and Ice at 2450 MHz

	Relative dielectric constant, ε'	Relative dielectric loss constant, ε''	Loss tangent, $\tan\delta$
Ice	3.2	0.0029	0.0009
Water (at 25°C)	78	12.48	0.16

Source: Schiffman (1986)

runaway heating may occur because the water will heat much faster due to the high dielectric loss factor of water.

4.8.6.3 *Shape and Density of the Material*

The shape of the food material is important in obtaining uniformity of heating. Nonuniform shapes result in local heating; similarly, sharp edges and corners cause nonuniform heating.

4.8.6.4 *Food Composition*

The composition of the food material affects how it heats in the microwave field. The moisture content of food directly affects the amount of microwave absorption. A higher amount of water in a food increases the dielectric loss factor ε''. In the case of foods of low moisture content, the influence of specific heat on the heating process is more pronounced than that of the dielectric loss factor. Therefore, due to their low specific heat, foods with low moisture content also heat at acceptable rates in microwaves. If the food material is highly porous with a significant amount of air, then due to low thermal conductivity of air, the material will act as a good insulator and show good heating rates in microwaves.

Another compositional factor that has a marked influence on heating rates in microwaves is the presence of salt. As stated previously, an increased concentration of ions promotes heating in microwaves. Thus, increasing the salt level in foods increases the rate of heating. Although oil has a much lower dielectric loss factor than water, oil has a specific heat less than half that of water. Since a product with high oil content will require much less heat to increase in temperature, the influence of specific heat becomes the overriding factor, and oil exhibits a much higher rate of heating than water (Ohlsson, 1983). More details on these and other issues important during the microwave heating of foods are elaborated by Schiffman (1986) and by Decareau and Peterson (1986).

The industrial applications of microwave processing of foods are mostly for tempering of frozen foods (increasing the temperature of frozen foods to -4 to $-20°C$), such as meat, fish, butter, and berries; for drying of pasta, instant tea, herbs, mushrooms, fish protein, bread crumbs, onions, rice cakes, seaweed, snack foods, and egg yolk; for cooking of bacon, meat patties, and potatoes; for vacuum drying of citrus juices, grains, and seeds; for freeze-drying of meat, vegetables, and fruits; for pasteurization and sterilization of prepared foods; for baking of bread and doughnuts; and for roasting of nuts, coffee beans, and cocoa beans (Decareau, 1992; Giese, 1992).

PROBLEMS

4.1 Calculate the rate of heat transfer per unit area through a 200 mm thick concrete wall when the temperatures are 20°C and 5°C on the two surfaces, respectively. The thermal conductivity of concrete is 0.935 W/(m°C).

4.2 A side wall of a storage room is 3 m high, 10 m wide, and 25 cm thick. The thermal conductivity is $k = 0.85$ W/(m °C). If, during the day, the inner surface temperature of the wall is 22°C and the outside surface temperature is 4°C:

a. Using the Thermal Resistance Concept, calculate the resistance to heat transfer for the wall.

b. Calculate the rate of heat transfer through the wall, assuming steady-state conditions.

4.3 An experiment was conducted to measure thermal conductivity of a formulated food. The measurement was made by using a large plane plate of the food material, which was 5 mm thick. It was found, under steady-state conditions, that when a temperature difference of 35°C was maintained between the two surfaces of the plate, a heat-transfer rate per unit area of 4700 W/m^2 was measured near the center of either surface. Calculate the thermal conductivity of the product, and list two assumptions used in obtaining the result.

4.4 Estimate the thermal conductivity of applesauce at 35°C. (Water content = 78.8% wet basis)

4.5 A 20 cm diameter cooking pan is placed on a stove. The pan is made of steel ($k = 15$ W/[m°C]), and it contains water boiling at 98°C. The bottom plate of the pan is 0.4 cm thick. The inside surface temperature of the bottom plate, in contact with water, is 105°C.

a. If the rate of heat transfer through the bottom plate is 450 W, determine the outside surface temperature of the bottom plate exposed to the heating stove.

b. Determine the convective heat-transfer coefficient for boiling water.

4.6 A 10 m long pipe has an inside radius of 70 mm, an outside radius of 80 mm, and is made of stainless steel ($k = 15$W/[m °C]). Its inside surface is held at 150°C, and its outside surface is at 30°C. There is no heat generation, and steady-state conditions hold. Compute the rate at which heat is being transferred across the pipe wall.

4.7 In a multilayered rectangular wall, the thermal resistance of the first layer is 0.005 °C/W, the resistance of the second layer is 0.2°C/W, and for the third layer it is 0.1 °C/W. The overall temperature gradient in the multilayered wall from one side to another is 70°C.

a. Determine the heat flux through the wall.
b. If the thermal resistance of the second layer is doubled to 0.4°C/W, what will be its influence in % on the heat flux, assuming the temperature gradient remains the same?

4.8 A plain piece of insulation board is used to reduce the heat loss from a hot furnace wall into the room. One surface of the board is at 100°C and the other surface is at 20°C. It is desired to keep the heat loss down to 120 W/m^2 of the insulation board. If the thermal conductivity of the board is 0.05 W/(m °C), calculate the required thickness of the board.

4.9 Consider an ice chest with the following dimensions: length = 50 cm, width = 40 cm, and height = 30 cm, made of a 3 cm thick insulating material (k = 0.033 W/[m °C]). The chest is filled with 30 kg of ice at 0°C. The inner wall surface temperature of the ice chest is assumed to be constant at 0°C. The latent heat of fusion of ice is 333.2 kJ/kg. The outside wall surface temperature of the chest is assumed to remain constant at 25°C. How long would it take to completely melt the ice? Assume negligible heat transfer through the bottom surface.

***4.10** Steam with 80% quality is being used to heat a 40% total solids tomato purée as it flows through a steam injection heater at a rate of 400 kg/h. The steam is generated at 169.06 kPa and is flowing to the heater at a rate of 50 kg/h. If the specific heat of the product is 3.2 kJ/(kg K), determine the temperature of the product leaving the heater when the initial temperature is 50°C. Determine the total solids content of the product after heating. Assume the specific heat of the heated purée is 3.5 kJ/(kg°C).

---

*Indicates an advanced level of difficulty in solving.

4.11 A stainless steel pipe ($k = 15\,W/[m\ ^\circ C]$) with a 2.5 cm inner diameter and 5 cm outer diameter is being used to convey high-pressure steam. The pipe is covered with a 5 cm layer of insulation ($k = 0.18\,W/[m\ ^\circ C]$). The inside steel pipe surface temperature is 300°C, and the outside insulation surface temperature is 90°C.

a. Determine the rate of heat transfer per meter length of the pipe.

b. The insulation selected for the purpose has a melting temperature of 220°C. Should you be concerned about the integrity of the insulation for the listed conditions?

4.12 Air at 25°C blows over a heated steel plate with its surface maintained at 200°C. The plate is 50×40 cm and 2.5 cm thick. The convective heat-transfer coefficient at the top surface is $20\,W/(m^2\,K)$. The thermal conductivity of steel is $45\,W/(m\,K)$. Calculate the heat loss per hour from the top surface of the plate.

4.13 A liquid food is being heated in a tubular heat exchanger. The inside pipe wall temperature is 110°C. The internal diameter of the pipe is 30 mm.The product flows at 0.5 kg/s. If the initial temperature of the product is 7°C, compute the convective heat-transfer coefficient. The thermal properties of the product are the following: specific heat = 3.7 kJ/(kg°C), thermal conductivity = $0.6\,W/(m\ ^\circ C)$, product viscosity = 500×10^{-6} Pa s, density = $1000\,kg/m^2$, product viscosity at 110°C = 410×10^{-6} Pa s.

4.14 Compute the convective heat-transfer coefficient for natural convection from a vertical, 100 mm outside diameter, 0.5 m long, stainless-steel pipe. The surface temperature of the uninsulated pipe is 145°C, and the air temperature is 40°C.

***4.15** A 30 m long pipe with an external diameter of 75 mm is being used to convey steam at a rate of 1000 kg/h. The steam pressure is 198.53 kPa. The steam enters the pipe with a dryness fraction of 0.98 and must leave the other end of the pipe with a minimum dryness fraction of 0.95. Insulation with a thermal conductivity of $0.2\,W/(m\,K)$ is available. Determine the minimum thickness of insulation necessary. The outside surface temperature of insulation is assumed to be

* Indicates an advanced level of difficulty in solving.

25 °C. Neglect the conductive resistance of the pipe material and assume no pressure drop across the pipe.

4.16 Estimate the convective heat-transfer coefficient for natural convection from a horizontal steam pipe. The outside surface temperature of the insulated pipe is 80 °C. The surrounding air temperature is 25 °C. The outside diameter of the insulated pipe is 10 cm.

4.17 A vertical cylindrical container is being cooled in ambient air at 25 °C with no air circulation. If the initial temperature of the container surface is 100 °C, compute the surface heat-transfer coefficient due to natural convection during the initial cooling period. The diameter of the container is 1 m, and it is 2 m high.

4.18 Water at a flow rate of 1 kg/s is flowing in a pipe of internal diameter 5 cm. If the inside pipe surface temperature is 90 °C and mean bulk water temperature is 50 °C, compute the convective heat-transfer coefficient.

4.19 A blower is used to move air through a pipe at a rate of 0.01 kg/s. The inside pipe surface temperature is 40 °C. The bulk temperature of the air reduces from 80 °C to 60 °C as it passes through a 5 m section of the pipe. The inside diameter of the pipe is 2 cm. Estimate the convective heat-transfer coefficient using the appropriate dimensionless correlation.

4.20 Estimate the convective heat-transfer coefficient on the outside of oranges (external diameter = 5 cm) when submerged in a stream of chilled water pumped around the orange. The velocity of water around an orange is 0.1 m/s. The surface temperature of the orange is 20 °C and the bulk water temperature is 0 °C.

***4.21** A flat wall is exposed to an environmental temperature of 38 °C. The wall is covered with a layer of insulation 2.5 cm thick whose thermal conductivity is 1.8 W/(m K), and the temperature of the wall on the inside of the insulation is 320 °C. The wall loses heat to the environment by convection. Compute the value of the convection heat-transfer coefficient that must be maintained on the outer surface of the insulation to ensure that the outer surface temperature does not exceed 40 °C.

---

*Indicates an advanced level of difficulty in solving.

4.22 Steam at 150 °C flows inside a pipe that has an inside radius of 50 mm and an outside radius of 55 mm. The convective heat-transfer coefficient between the steam and the inside pipe wall is 2500 W/(m^2 °C). The outside surface of the pipe is exposed to ambient air at 20 °C with a convective heat-transfer coefficient of 10 W/(m^2 °C). Assuming steady state and no heat generation, calculate the rate of heat transfer per meter from the steam to the air across the pipe. Assume thermal conductivity of stainless steel is 15 W/(m °C.)

4.23 The outside wall of a refrigerated storage room is 10 m long and 3 m high and is constructed with 100 mm concrete block ($k = 0.935$ W/[m °C]) and 10 cm of fiber insulation board ($k = 0.048$ W/[m °C]). The inside of the room is at −10 °C and the convective heat-transfer coefficient is 40 W/(m^2 K); the outside temperature is 30 °C with a convective heat-transfer coefficient of 10 W/(m^2 K) on the outside wall surface. Calculate the overall heat-transfer coefficient.

4.24 In a food processing plant, a steel pipe (thermal conductivity = 17 W/m °C, internal diameter = 5 cm; thickness = 3 mm) is being used to transport a liquid food. The inside surface temperature of the pipe is at 95 °C. A 4-cm thick insulation (thermal conductivity = 0.03 W/m °C) is wrapped around the pipe. The outside surface temperature of the insulation is 30 °C. Calculate the rate of heat transfer per unit length of the pipe.

4.25 A walk-in freezer with 4 m width, 6 m length, and 3 m height is being built. The walls and ceiling contain 1.7 mm thick stainless steel ($k = 15$ W/[m °C]), 10 cm thick foam insulation ($k = 0.036$ W/[m °C]), and some thickness of corkboard ($k = 0.043$ W/[m °C]) to be established, and 1.27 cm-thickness wood siding ($k = 0.104$ W/[m °C]). The inside of the freezer is maintained at −40 °C. Ambient air outside the freezer is at 32 °C. The convective heat-transfer coefficient is 5 W/(m^2 K) on the wood side of the wall and 2 W/(m^2 K) on the steel side. If the outside air has a dew point of 29 °C, calculate the thickness of corkboard insulation that would prevent condensation of moisture on the outside wall of the freezer. Calculate the rate of heat transfer through the walls and ceiling of this freezer.

4.26 A liquid food is being conveyed through an uninsulated pipe at 90 °C. The product flow rate is 0.25 kg/s and has a density of

1000 kg/m^3, specific heat of 4 kJ/(kg K), viscosity of 8×10^{-6} Pa s, and thermal conductivity of 0.55 W/(m K). Assume the viscosity correction is negligible. The internal pipe diameter is 20 mm with 3 mm thickness made of stainless steel ($k = 15$ W/[m °C]). The outside temperature is 15 °C. If the outside convective heat-transfer coefficient is 18 W/(m^2 K), calculate the steady-state heat loss from the product per unit length of pipe.

***4.27** A liquid food is being pumped in a 1 cm thick steel pipe. The inside diameter of the pipe is 5 cm. The bulk temperature of liquid food is 90 °C. The inside pipe surface temperature is 80 °C. The surface heat-transfer coefficient inside the pipe is 15 W/(m^2 K). The pipe has a 2 cm thick insulation. The outside bulk air temperature is 20 °C. The surface heat-transfer coefficient on the outside of insulation is 3 W/(m^2 K).

a. Calculate the insulation surface temperature exposed to the outside.
b. If the pipe length is doubled, how would it influence the insulation surface temperature? Discuss.

4.28 For a metal pipe used to pump tomato paste, the overall heat-transfer coefficient based on internal area is 2 W/(m^2 K). The inside diameter of the pipe is 5 cm. The pipe is 2 cm thick. The thermal conductivity of the metal is 20 W/(m K). Calculate the outer convective heat-transfer coefficient. The inside convective heat-transfer coefficient is 5 W/(m^2 K).

4.29 A cold-storage room is maintained at −18 °C. The internal dimensions of the room are 5 m × 5 m × 3 m high. Each wall, ceiling, and floor consists of an inner layer of 2.5 cm thick wood with 7 cm thick insulation and an 11 cm brick layer on the outside. The thermal conductivities of respective materials are wood 0.104 W/(m K), glass fiber 0.04 W/(m K), and brick 0.69 W/(m K). The convective heat-transfer coefficient for wood to still air is 2.5 W/(m^2 K), and from moving air to brick is 4 W/(m^2 K). The outside ambient temperature is 25 °C. Determine:

a. The overall heat-transfer coefficient.
b. The temperature of the exposed surfaces.
c. The temperatures of the interfaces.

---

*Indicates an advanced level of difficulty in solving.

4.30 Steam at 169.60 kPa is condensed inside a pipe (internal diameter = 7 cm, thickness = 3 mm). The inside and outside convective heat-transfer coefficients are 1000 and 10 W/(m^2 K), respectively. The thermal conductivity of the pipe is 45 W/(m K). Assume that all thermal resistances are based on the outside diameter of the pipe, and determine the following:

a. Percentage resistance offered by the pipe, by the steam, and by the outside.
b. The outer surface temperature of the pipe if the temperature of the air surrounding the pipe is 25°C.

4.31 A steel pipe (outside diameter 100 mm) is covered with two layers of insulation. The inside layer, 40 mm thick, has a thermal conductivity of 0.07 W/(m K). The outside layer, 20 mm thick, has a thermal conductivity of 0.15 W/(m K). The pipe is used to convey steam at a pressure of 700 kPa. The outside temperature of insulation is 24°C. If the pipe is 10 m long, determine the following, assuming the resistance to conductive heat transfer in steel pipe and convective resistance on the steam side are negligible:

a. The heat loss per hour.
b. The interface temperature of insulation.

***4.32** A 1 cm thick steel pipe, 1 m long, with an internal diameter of 5 cm is covered with 4 cm thick insulation. The inside wall temperature of the steel pipe is 100°C. The ambient temperature around the insulated pipe is 20°C. The convective heat-transfer coefficient on the outer insulated surface is 50 W/(m^2 K). Calculate the temperature at the steel insulation interface. The thermal conductivity of steel is 54 W/(m K), and the thermal conductivity of insulation is 0.04 W/(m K).

4.33 Calculate the overall heat-transfer coefficient of a steel pipe based on the inside area. The inside diameter of the pipe is 10 cm, and the pipe is 2 cm thick. The inside convective heat-transfer coefficient is 350 W/(m^2 °C), the outside convective heat-transfer coefficient is 25 W/(m^2 °C), the thermal conductivity of the steel pipe is 15 W/(m °C). If the pipe is used to convey steam at a bulk temperature of 110°C and the outside ambient temperature is 20°C, determine the rate of heat transfer from the pipe.

---

*Indicates an advanced level of difficulty in solving.

***4.34** Saturated refrigerant (Freon, R-12) at $-40°C$ flows through a copper tube of 20 mm inside diameter and wall thickness of 2 mm. The copper tube is covered with 40 mm thick insulation ($k = 0.02\,W/[m\ K]$). Determine the heat gain per meter of the pipe. The internal and external convective heat-transfer coefficients are 500 and $5\,W/(m^2\,K)$, respectively. The ambient air temperature is 25°C. Compare the amount of refrigerant vaporized per hour per meter length of pipe for insulated versus uninsulated pipe. The latent heat of the refrigerant at $-40°C$ is 1390 kJ/kg.

4.35 To cool hot edible oil, an engineer has suggested that the oil be pumped through a pipe submerged in a nearby lake. The pipe (external diameter = 15 cm) will be located in a horizontal direction. The average outside surface temperature of the pipe will be 130°C. The surrounding water temperature may be assumed to be constant at 10°C. The pipe is 100 m long. Assume there is no movement of water.

- **a.** Estimate the convective heat-transfer coefficient from the outside pipe surface into water.
- **b.** Determine the rate of heat transfer from the pipe into water.

4.36 In a concurrent-flow tubular heat exchanger, a liquid food, flowing in the inner pipe, is heated from 20 to 40°C. In the outer pipe the heating medium (water) cools from 90 to 50°C. The overall heat-transfer coefficient based on the inside diameter is $2000\,W/(m^2\ °C)$. The inside diameter is 5 cm and length of the heat exchanger is 10 m. The average specific heat of water is 4.181 kJ/(kg °C). Calculate the mass flow rate of water in the outer pipe.

***4.37** A countercurrent heat exchanger is being used to heat a liquid food from 15 to 70°C. The heat exchanger has a 23 mm internal diameter and 10 m length with an overall heat-transfer coefficient, based on the inside area, of $2000\,W/(m^2\,K)$. Water, the heating medium, enters the heat exchanger at 95°C, and leaves at 85°C. Determine the flow rates for product and water that will provide the conditions described. Use specific heats of 3.7 kJ/(kg K) for product and 4.18 kJ/(kg K) for water.

---

*Indicates an advanced level of difficulty in solving.

4.38 A 10 m long countercurrent-flow heat exchanger is being used to heat a liquid food from 20 to 80 °C. The heating medium is oil, which enters the heat exchanger at 150 °C and exits at 60 °C. The specific heat of the liquid food is 3.9 kJ/(kg K). The overall heat-transfer coefficient based on the inside area is 1000 W/(m^2 K). The inner diameter of the inside pipe is 7 cm.

a. Estimate the flow rate of the liquid food.

b. Determine the flow rate of the liquid food if the heat exchanger is operated in a concurrent-flow mode for the same conditions of temperatures at the inlet and exit from the heat exchanger.

4.39 Calculate the radiative heat gain in watts by a loaf of bread at a surface temperature of 100 °C. The surrounding oven surface temperature is 1000 °C. The total surface area of the bread is 0.15 m^2 and the emissivity of the bread surface is 0.80. Assume the oven is a blackbody radiator.

4.40 It is desired to predict the temperature after 30 min at the geometric center of a cylindrical can containing a model food. The dimensions of the can are 5 cm diameter and 3 cm height. The thermal conductivity of the food is 0.5 W/(m °C), specific heat = 3.9 kJ/(kg °C), and density = 950 kg/m^3. There is a negligible surface resistance to heat transfer. The surrounding medium temperature is 100 °C and the uniform initial temperature of food is 20 °C.

4.41 An 8 m^3 batch of oil with specific heat of 2 kJ/(kg K) and density of 850 kg/m^3 is being heated in a steam-jacketed, agitated vessel with 1.5 m^2 of heating surface. The convective heat-transfer coefficient on the oil side is 500 W/(m^2 K), and 10,000 W/(m^2 K) on the steam side. If the steam temperature is 130 °C and the initial temperature is 20 °C, estimate the oil temperature after 10 minutes.

4.42 Determine if a tubular heat exchanger can operate under the following conditions: Fluid A enters the heat exchanger at 120 °C and exits at 40 °C; fluid B enters the heat exchanger at 30 °C and exits at 70 °C. Calculate the log mean temperature difference.

4.43 Milk (c_p = 3.9 kJ/[kg K]) is cooled in a countercurrent flow heat exchanger at a rate of 1.5 kg/s from 70 °C to 30 °C. Cooling is done by using chilled water available at 5 °C with a flow rate of 2 kg/s. The inside diameter of the inner pipe is 2 cm. The overall heat transfer coefficient is 500 W/(m^2 °C). Determine the length of the heat exchanger.

4.44 In a double-pipe heat exchanger, made of stainless steel (k = 15 W/[m °C]), the inside pipe has an inner diameter of 2 cm and an outside diameter of 2.5 cm. The outer shell has an inner diameter of 4 cm. The convective heat transfer coefficient on the inside surface of the inner pipe is 550 W/(m^2 °C), whereas on the outside surface of the inner pipe it is 900 W/(m^2 °C). Over continuous use of the heat exchanger, fouling (depositing of solids from the liquids on the pipe surfaces) causes additional resistance to heat transfer. It is determined that the resistance to heat transfer due to fouling on the inside surface of the inner pipe is 0.00038 m^2 °C/W, and on the outside surface of the inner pipe is 0.0002 m^2 °C/W. Calculate:

a. The total thermal resistance of the heat exchanger per unit length.

b. The overall heat transfer coefficients U_i and U_o based on the inside and outside area of the inner pipe, respectively.

4.45 Water at 5 °C is being used to cool apples from an initial temperature of 20° to 8 °C. The water flow over the surface of the apple creates a convective heat-transfer coefficient of 10 W/(m^2 K). Assume the apple can be described as a sphere with an 8 cm diameter and the geometric center is to be reduced to 8 °C. The apple properties include thermal conductivity of 0.4 W/(m K), specific heat of 3.8 kJ/(kg K), and density of 960 kg/m^3. Determine the time that the apples must be exposed to the water.

4.46 A liquid food with density of 1025 kg/m^3 and specific heat of 3.77 kJ/(kg K) is being heated in a can with 8.5 cm diameter and 10.5 cm height. The heating will occur in a retort with temperature at 115 °C and convective heat-transfer coefficients of 50 W/(m^2 K) on the inside of the can and 5000 W/(m^2 K) on the outside surface. Determine the product temperature after 10 min if the initial temperature is 70 °C. Assume perfect mixing in the can.

4.47 Create a spreadsheet for Example 4.21. Keeping all conditions the same as given in the example, determine the exit temperatures of hot water and apple juice if the number of plates used in the heat exchanger are 21 or 31.

***4.48** A conduction-cooling food product with density of 1000 kg/m^3, specific heat of 4 kJ/(kg K), and thermal conductivity of 0.4 W/(m K) has been heated to 80 °C. The cooling of the product in a 10 cm high and 8 cm diameter can is accomplished using cold water with a convective heat-transfer coefficient of 10 W/(m^2 K) on the can surface. Determine the water temperature required to reduce the product temperature at geometric center to 50 °C in 7 h. Neglect conductive heat resistance through the can wall.

4.49 Cooked mashed potato is cooled on trays in a chilling unit with refrigerated air at 2 °C blown over the product surface at high velocity. The depth of product is 30 mm and the initial temperature is 95 °C. The product has a thermal conductivity of 0.37 W/(m K), specific heat of 3.7 kJ/(kg K), and density of 1000 kg/m^3. Assuming negligible resistance to heat transfer at the surface, calculate product temperature at the center after 30 min.

4.50 Program Example 4.9 on a spreadsheet. Determine the interfacial temperatures if the following thickness of insulation are used:

a. 2 cm
b. 4 cm
c. 6 cm
d. 8 cm
e. 10 cm

4.51 A liquid food at a flow rate of 0.3 kg/s enters a countercurrent flow double-pipe heat exchanger at 22 °C. In the annular section, hot water at 80 °C enters at a flow rate of 1.2 kg/s. The average specific heat of water is 4.18 kJ/(kg °C). The average overall heat transfer coefficient based on the inside area is 500 W/(m^2 °C). The diameter of the inner pipe is 7 cm, and length is 10 m. Assume steady state conditions. The specific heat of liquid food is assumed to be 4.1 kJ/[kg °C]. Calculate the exit temperature of liquid food and water.

4.52 A pure copper sphere of radius 1 cm is dropped into an agitated oil bath that has a uniform temperature of 130 °C. The initial temperature of the copper sphere is 20 °C. Using

---

*Indicates an advanced level of difficulty in solving.

a spreadsheet predict the internal temperature of the sphere at 5 min intervals until it reaches 130 °C for three different convective heat-transfer coefficients: 5, 10, and 100 W/(m^2 °C), respectively. Plot your results as temperature versus time.

4.53 Heated water with a bulk temperature of 90 °C is being pumped at a rate of 0.1 kg/s in a metal pipe placed horizontally in ambient air. The pipe has an internal diameter of 2.5 cm and it is 1 cm thick. The inside pipe surface temperature is 85 °C. The outside surface of the pipe is at 80 °C and exposed to the air. The bulk temperature of the air is 20 °C.

a. Determine the convective heat-transfer coefficient for water inside the pipe.

b. Determine the convective heat-transfer coefficient for air outside the pipe.

c. It is desired to double the convective heat-transfer coefficient inside the pipe. What operating conditions should be changed? By how much?

4.54 A solid food is being cooled in a cylindrical can of dimensions 12 cm diameter and 3 cm thickness. The cooling medium is cold water at 2 °C. The initial temperature of the solid food is 95 °C. The convective heat-transfer coefficient is 200 W/(m^2 °C).

a. Determine the temperature at the geometric center after 3 h. The thermal properties of the solid food are $k = 0.36$ W/(m °C), density of 950 kg/m^3, and specific heat of 3.9 kJ/(kg°C).

b. Is it reasonable to assume the cylindrical can to be an infinite cylinder (or an infinite slab)? Why?

4.55 A three-layered composite pipe with an inside diameter of 1 cm has an internal surface temperature of 120 °C. The first layer, from the inside to the outside, is 2 cm thick with a thermal conductivity of 15 W/(m°C), the second layer is 3 cm thick with a thermal conductivity of 0.04 W/(m °C), and the third layer is 1 cm thick and has a thermal conductivity of 164 W/(m °C). The outside surface temperature of the composite pipe is 60 °C.

a. Determine the rate of heat transfer through the pipe under steady-state conditions.

b. Can you suggest an approach that will allow you to quickly make an estimate for this problem?

4.56 It is known that raw eggs will become hard when heated to 72°C. To manufacture diced eggs, trays of liquid egg are exposed to steam for cooking.

a. How long will it take to cook the eggs given the following conditions? The tray dimension is 30 cm long, 30 cm wide, and 2 cm deep. The liquid egg inside the tray is 1 cm deep. The thermal conductivity of liquid egg is 0.45 W/(m °C); density is 800 kg/m^3, specific heat is 3.8 kJ/(kg °C); surface convective heat transfer coefficient is 5000 W/(m^2 °C); and the initial temperature of liquid egg is 2°C. Steam is available at 169.06 kPa. Ignore resistance to heat transfer caused by the metal tray.

b. What rate of steam flow per tray of liquid egg must be maintained to accomplish this? The latent heat of vaporization at 169.06 kPa is 2216.5 kJ/kg.

4.57 Determine the time required for the center temperature of a cube to reach 80°C. The cube has a volume of 125 cm^3. The thermal conductivity of the material is 0.4 W/(m °C); density is 950 kg/m^3; and specific heat is 3.4 kJ/(kg K). The initial temperature is 20°C. The surrounding temperature is 90°C. The cube is immersed in a fluid that results in a negligible surface resistance to heat transfer.

4.58 A tubular heat exchanger is being used for heating a liquid food from 30° to 70°C. The temperature of the heating medium decreases from 90° to 60°C.

a. Is the flow configuration in the heat exchanger countercurrent or concurrent flow?

b. Determine the log mean temperature difference.

c. If the heat transfer area is 20 m^2 and the overall heat transfer coefficient is 100 W/(m^2 °C), determine the rate of heat transfer from the heating medium to the liquid food.

d. What is the flow rate of the liquid food if the specific heat of the liquid is 3.9 kJ/(kg °C)? Assume no heat loss to the surroundings.

4.59 What is the flow rate of water in a heat exchanger if it enters the heat exchanger at 20°C and exits at 85°C? The heating medium is oil, where oil enters at 120°C and leaves at 75°C. The overall heat-transfer coefficient is 5 W/(m^2°C). The area of the heat exchanger is 30 m^2.

4.60 A thermocouple is a small temperature sensor used in measuring temperature in foods. The thermocouple junction, that senses the temperature, may be approximated as a sphere. Consider a situation where a thermocouple is being used to measure temperature of heated air in an oven. The convective heat-transfer coefficient is $400\,W/(m^2\,K)$. The properties of the thermocouple junction are $k = 25\,W/(m\,°C)$, $c_p = 450\,J/(kg\,K)$, and $\rho = 8000\,kg/m^3$. The diameter of the junction, considered as a sphere, is 0.0007 m. If the junction is initially at 25 °C, and it is placed in the oven where the estimated heated air temperature is 200 °C, how long will it take for the junction to reach 199 °C?

4.61 In an ohmic heater, the inside diameter of the pipe is 0.07 m and it is 2 m long. The applied voltage is 15,000 V. A liquid food is being pumped through the heater at 0.2 kg/s with an inlet temperature of 30 °C. The overall heat transfer coefficient based on the inside pipe area is $200\,W/m^2\,°C$. The specific heat of the liquid food is 4000 J/kg °C. The surrounding temperature of the air is 25 °C. Assume that the properties of the liquid food are similar to 0.1 M Sodium Phosphate solution. Determine the exit temperature of the liquid food.

4.62 A liquid food at 75 °C is being conveyed in a steel pipe ($k = 45\,W/[m\ °C]$). The pipe has an internal diameter of 2.5 cm and it is 1 cm thick. The overall heat-transfer coefficient based on internal diameter is $40\,W/(m^2\,K)$. The internal convective heat-transfer coefficient is $50\,W/(m^2\,K)$. Calculate the external convective heat-transfer coefficient.

4.63 Set up Example 4.16 on a spreadsheet. Determine the heat loss from 1 m length of the pipe if the inside diameter of the pipe is

a. 2.5 cm
b. 3.5 cm
c. 4.5 cm
d. 5.5 cm

***4.64** A plot of tomato juice density versus temperature was scanned and digitized from Choi and Okos (1983). Juice density

---

*Indicates an advanced level of difficulty in solving.

versus solids content for several temperatures are given in the following table.

Solids	$T = 30°C$	$T = 40°C$	$T = 50°C$	$T = 60°C$	$T = 70°C$	$T = 80°C$
0%	997	998	984	985	979	972
4.8%	1018	1018	1012	1006	1003	997
8.3%	1032	1032	1026	1026	1020	1017
13.9%	1070	1067	1064	1061	1058	1048
21.5%	1107	1108	1102	1102	1093	1086
40.0%	1190	1191	1188	1185	1179	1176
60.0%	1290	1294	1288	1289	1286	1276
80.0%	1387	1391	1385	1382	1379	1376

Create a MATLAB® script that plots the values of density as a function of percent solids at a temperature of 40 °C. Use the Basic Fitting option under the Tools menu of the Figure window to fit the data with an appropriate polynomial. Turn in a copy of your script, plot, and the equation that you determine.

***4.65** A plot of tomato juice density versus temperature was scanned and digitized by Choi and Okos (1983). Values of juice density versus solids content for several temperatures are shown in a table in Problem 4.64. Write a MATLAB® script to evaluate and plot the model for tomato juice density developed by Choi and Okos.

$$\rho = \rho_s X_s + \rho_w X_w$$

$$\rho_w = 9.9989 \times 10^2 - 6.0334 \times 10^{-2} T - 3.6710 \times 10^{-3} T^2$$

$$\rho_s = 1.4693 \times 10^3 + 5.4667 \times 10^{-1} T - 6.9646 \times 10^{-3} T^2$$

ρ = density (kg/m^3)
ρ_w = density of water (kg/m^3)
ρ_s = density of solids (kg/m^3)
T = temperature of juice (°C)
X_s = percent solids in juice (%)
X_w = percent water in juice (%)

Plot this model over the preceding experimental data for 40 °C.

*Indicates an advanced level of difficulty in solving.

***4.66** Telis-Romero et al. (1998) presented data for the specific heat of orange juice as a function of temperature and percent water. A portion of their data was digitized and is given in the following table.

	$T = 8°C$	$T = 18°C$	$T = 27°C$	$T = 47°C$	$T = 62°C$
X_w (w/w)	c_p (kJ/kg °C)	c_p (kJ/kg °C)	c_p (kJ/kg °C)	c_p (kJ/kg °C)	c_p (kJ/kg °C)
0.34	2.32	2.35	2.38	2.43	2.45
0.40	2.49	2.51	2.53	2.59	2.61
0.44	2.59	2.62	2.64	2.68	2.72
0.50	2.74	2.78	2.80	2.85	2.88
0.55	2.88	2.91	2.93	2.98	3.01
0.59	2.99	3.01	3.03	3.08	3.12
0.63	3.10	3.12	3.14	3.19	3.23
0.69	3.26	3.28	3.30	3.36	3.39
0.73	3.37	3.39	3.41	3.46	3.49

Notice that specific heat is a function of temperature as well as solids content for all juices.

Two general-purpose empirical equations are often used to estimate the specific heat for plant material and their juices. The first equation is known as Siebel's correlation equation, and is used to estimate the specific heat of "fat-free fruits and vegetables, purees, and concentrates of plant origin":

$$c_p = 3.349(X_w) + 0.8374$$

The second equation is given from the ASHRAE Handbook—Fundamentals (2005):

$$c_p = 4.187(0.6X_w + 0.4)$$

Using MATLAB®, create a plot of c_p vs. moisture content using data given in the table for all temperatures and compare with calculated values using the two empirical equations.

***4.67** The solution to temperature in an infinite slab with finite internal and surface resistance to heat transfer is given by Incropera et al. (2007) as

$$\frac{T - T_a}{T_i - T_a} = \sum_{n=1}^{\infty} C_n \exp(-\beta_n^2 N_{Fo}) \cos(\beta_n x/d_c)$$

*Indicates an advanced level of difficulty in solving.

$$C_n = \frac{4\sin(\beta_n)}{2\beta_n + \sin(2\beta_n)}$$

$$\beta_n \tan\beta_n = N_{Bi}$$

They give the first four roots of the transcendental equation for β for a range of Biot numbers. The first four roots for $N_{Bi} = 0.5$ are $\beta_1 = 0.6533$, $\beta_2 = 3.2923$, $\beta_3 = 6.3616$, and $\beta_4 = 9.4775$.

Write a MATLAB® script to evaluate the temperature for a slab with

$$d_c = 0.055 \text{ (m)}$$
$$k = 0.34 \text{ (W/m °C)}$$
$$c_p = 3500 \text{ (J/kg °C)}$$
$$\rho = 900 \text{ (kg/m}^3\text{)}$$
$$N_{Bi} = 0.5$$
$$T_a = 100 \text{ (°C)}$$
$$T_i = 35 \text{ (°C)}$$

Plot the temperatures $T(x,t)$ over the range of x from 0 to 0.055 m for times of 20, 40, and 60 minutes.

***4.68** Write a MATLAB® script to use the MATLAB® function *pdepe* to evaluate the temperature for an infinite slab with

$$d_c = 0.055 \text{ (m)}$$
$$k = 0.34 \text{ (W/m °C)}$$
$$c_p = 3500 \text{ (J/kg °C)}$$
$$\rho = 900 \text{ (kg/m}^3\text{)}$$
$$N_{Bi} = 0.5$$
$$T_a = 100 \text{ (°C)}$$
$$T_i = 35 \text{ (°C)}$$

Plot the temperatures $T(x,t)$ over the range of x from 0 to 0.055 m for times of 20, 40, and 60 minutes.

---

*Indicates an advanced level of difficulty in solving.

LIST OF SYMBOLS

A	area (m^2)
a	coefficient in Equation (4.70)
α	thermal diffusivity (m^2/s)
α'	attenuation factor (m^{-1})
b	gap between two adjacent plates (m)
β	coefficient of volumetric expansion (K^{-1})
C_F	cleaning factor, dimensionless
C_H	heat capacity rate (kJ/[s °C])
C_{min}	minimum heat capacity rate (kJ/[s °C])
C^*	heat capacity rate ratio, dimensionless
c_p	specific heat at constant pressure (kJ/[kg °C])
c_v	specific heat at constant volume (kJ/[kg °C])
χ	reflectivity, dimensionless
D	diameter (m)
d_c	characteristic dimension (m)
D_e	equivalent diameter (m)
d	penetration depth (m)
E	electrical field strength (V/cm)
E_V	voltage (V)
ε	emissivity, dimensionless
ε_E	heat exchanger effectiveness, dimensionless
ε'	relative dielectric constant, dimensionless
ε''	relative dielectric loss constant, dimensionless
F	shape factor, dimensionless
f_h	heating rate factor (s)
f	friction factor, dimensionless
f'	frequency (Hz)
g	acceleration due to gravity (m/s^2)
h	convective heat transfer coefficient ($W/[m^2 K]$)
I	electric current (A)
J_0	Bessel function of zero order
J_1	Bessel function of first order
j_c	temperature lag factor at center
j_m	temperature lag factor, mean
K	coefficient in Equation (4.185)
k	thermal conductivity (W/[m K])
κ_E	electrical conductance (siemens)
L	length (m)
l	thickness of fluid layer (m)
λ	wavelength (m)

λ_n	eigenvalue roots
M	mass concentration, percent
m	mass (kg); coefficient in Equations (4.59) and (4.70)
m''	coefficient in Equation (4.184)
$\dot{m}$	mass flow rate (kg/s)
μ	viscosity (Pa s)
N	number of plates
N_{Bi}	Biot number, dimensionless
N_{Fo}	Fourier number, dimensionless
N_{Gr}	Grashoff number, dimensionless
N_{Nu}	Nusselt number, dimensionless
N_{Pr}	Prandtl number, dimensionless
N_{Re}	Reynolds number, dimensionless
N_{Ra}	Raleigh number, dimensionless
n	coefficient in Equation (4.59)
v	kinematic viscosity (m^2/s)
P	power at the penetration depth (W)
P_D	power dissipation (W/cm^3)
P_o	incident power (W/cm^3)
ϕ	absorptivity, dimensionless
Φ	function
ψ	transmissivity, dimensionless
Q	heat gained or lost (kJ)
q	rate of heat transfer (W)
q'''	rate of heat generation (W/m^3)
R_E	electrical resistance (Ω)
R_f	fouling resistance, $[(m^2\,°C]/W)$
R_t	thermal resistance (°C/W)
r	radius or variable distance in radial direction (m)
r_c	critical radius
ρ	density (kg/m^3)
σ	Stefan–Boltzmann constant ($5.669 \times 10^{-8}\,W/[m^2\,K^4]$)
σ_E	electrical conductivity (S/m)
σ_0	electrical conductivity at 0°C (S/m)
T	temperature (°C)
t	time (s)
T_e	exit temperature (°C)
T_A	absolute temperature (K) or pseudo initial temperature for Ball's method (°C)
T_a	temperature of surrounding medium (°C)
T_f	film temperature (°C)
T_i	initial or inlet temperature (°C)

T_p	plate surface temperature (°C)
T_s	surface temperature (°C)
T_∞	fluid temperature far away from solid surface (°C)
$\tan\delta$	loss tangent, dimensionless
U	overall heat transfer coefficient (W/[m^2 °C])
$\bar{u}$	velocity (m/s)
V	volume (m^3)
w	width of a plate (m)
X	mass fraction, dimensionless
x	variable distance in x direction (m)
Y	volume fraction, dimensionless
Z	depth (m)
z	space coordinate
ξ	factor to account for shape and emissivity, dimensionless

Subscripts: a, ash; b, bulk; c, channel; ci, inside clean surface; co, outside clean surface; f, fat; fi, inside fouled surface; fo, outside fouled surface; h, carbohydrate; i, inside surface; lm, log mean; m, moisture; o, outside surface; p, protein; r, radial direction; s solid; x, x-direction; w, at wall (or water); H, hot stream; C, cold stream; P, product stream.

BIBLIOGRAPHY

American Society of Heating, Refrigerating and Air Conditioning Engineers, Inc. (2005). *ASHRAE Handbook—2005 Fundamentals*. ASHRAE, Atlanta, Georgia.

Brennan, J. G., Butters, J. R., Cowell, N. D., and Lilly, A. E. V. (1990). *Food Engineering Operations*, 3rd ed. Elsevier Science Publishing Co., New York.

Cengel, Y. A. (2007). *Heat Transfer. A Practical Approach*, 3rd ed. McGraw Hill, Boston.

Charm, S. E. (1978). *The Fundamentals of Food Engineering*, 3rd ed. AVI Publ. Co., Westport, Connecticut.

Choi, Y. and Okos, M. R. (1986). Effects of temperature and composition on the thermal properties of food. In *Food Engineering and Process Applications*, Vol. 1. "Transport Phenomena." M. Le Maguer and P. Jelen, eds., 93–101. Elsevier Applied Science Publishers, London.

Choi, Y. and Okos, M. R. (1983). The thermal properties of tomato juice concentrates. *Transactions of the ASAE* **26**(1): 305–315.

Copson, D. A. (1975). *Microwave Heating*. AVI Publ. Co., Westport, Connecticut.

Coulsen, J. M. and Richardson, J. F. (1999). *Chemical Engineering*, Vol. 1. Pergamon Press, Oxford.

Das, S. K. (2005). *Process Heat Transfer*. Alpha Science International Ltd., Harrow, UK.

Decareau, R. V. (1992). *Microwave Foods: New Product Development*. Food and Nutrition Press, Trumbull, Connecticut.

Decareau, R. V. and Peterson, R. A. (1986). *Microwave Processing and Engineering*. VCH Publ., Deerfield Beach, Florida.

Dickerson, R. W., Jr. (1969). Thermal properties of foods. In *The Freezing Preservation of Foods. 4th ed.*, D. K. Tressler, W. B. Van Arsdel, and M. J. Copley, eds., Vol. 2, 26–51. AVI Publ. Co., Westport, Connecticut.

Giese, J. (1992). Advances in microwave food processing. *Food Tech.* **46**(9): 118–123.

Green, D. W. and Perry, R. H. (2007). *Perry's Chemical Engineer's Handbook*, 8th ed. McGraw-Hill, New York.

Heldman, D. R. and Lund, D. B. (2007). *Handbook of Food Engineering*. CRC Press, Taylor & Francis Group, Boca Raton, Florida.

Heldman, D. R. and Singh, R. P. (1981). *Food Process Engineering*, 2nd ed. AVI Publ. Co., Westport, Connecticut.

Holman, J. P. (2002). *Heat Transfer*, 9th ed. McGraw-Hill, New York.

Hottel, H. C. (1930). Radiant heat transmission. *Mech. Eng.* **52**(7): 700.

Incropera, F. P., DeWitt, D. P., Bergman, T. L., and Lavine, A. S. (2007). *Fundamentals of Heat and Mass Transfer*, 6th ed. John Wiley and Sons, Inc., New York.

IFT (1989). Microwave food processing. A Scientific Status Summary by the IFT Expert Panel on Food Safety and Nutrition. *Food Tech.* **43**(1): 117–126.

Kakac, S. and Liu, H. (1998). *Heat Exchangers. Selection, Rating and Thermal Design*. CRC Press, Boca Raton, Florida.

Kreith, F. (1973). *Principles of Heat Transfer*. IEP-A Dun-Donnelley Publisher, New York.

Lin, Z., Cleland, A. C., Sellarach, G. F., and Cleland, D. J. (1996). A simple method for prediction of chilling times: Extension to three-dimensional irregular shapes. *Int. J. Refrigeration,* **19**: 107–114, Erratum, *Int. J. Refrigeration* **23**: 168.

Mertens, B. and Knorr, D. (1992). Developments of nonthermal processes for food preservation. *Food Tech.* **46**(5): 124–133.

Mudgett, R. E. (1986). Microwave properties and heating characteristics of foods. *Food Tech.* **40**(6): 84–93.

Myers, G. E. (1971). *Analytical Methods in Conduction Heat Transfer*. McGraw-Hill, New York.

Ohlsson, T. (1983). Fundamentals of microwave cooking. *Microwave World* **4**(2): 4.

Palaniappan, S. and Sastry, S. K. (1991). Electrical conductivity of selected solid foods during ohmic heating. *J. Food Proc. Engr.* **14**: 221–236.

Pflug, I. J., Blaisdell, J. L., and Kopelman, J. (1965). Developing temperature-time curves for objects that can be approximated by a sphere, infinite plate or infinite cylinder. *ASHRAE Trans.* **71**(1): 238–248.

Pham, Q. T. (2001). Prediction of cooling/freezing time and heat loads. In *Advances in Food Refrigeration*, D. W. Sun, ed. Leatherhead International, Leatherhead, Surrey, UK.

Rotstein, E., Singh, R. P., and Valentas, K. J. (1997). *Handbook of Food Engineering Practice*. CRC Press, Boca Raton, Florida.

Schiffman, R. F. (1986). Food product development for microwave processing. *Food Tech.* **40**(6): 94–98.

Schneider, P. J. (1963). *Temperature Response Charts*. Wiley, New York.

Siebel, J. E. (1892). Specific heat of various products. *Ice Refrigeration* **2**: 256.

Singh, R. P. (1994). *Food Properties Database*. RAR Press, Davis, California.

Sweat, V. E. (1974). Experimental values of thermal conductivity of selected fruits and vegetables. *J. Food Sci.* **39**: 1080.

Sweat, V. E. (1975). Modeling thermal conductivity of meats. *Trans. Am. Soc. Agric. Eng.* **18**(1): 564–565, 567, 568.

Sweat, V. E. (1986). Thermal properties of foods. In *Engineering Properties of Foods*, M. A. Rao and S. S. H. Rizvi, eds., 49–87. Marcel Dekker Inc., New York.

Telis-Romero, J., Telis, V. R. N., Gabas, A. L., and Yamashita, F. (1998). Thermophysical properties of Brazilian orange juice as affected by temperature and water content. *J. Food Engr.* **38**: 27–40.

von Hippel, A. R. (1954). *Dielectrics and Waves*. MIT Press, Cambridge, Massachusetts.

Whitaker, S. (1977). *Fundamental Principles of Heat Transfer*. Krieger Publishing Co., Melbourne, Florida.

Chapter 5

Preservation Processes

In the food industry, we refer to the processing steps required to eliminate the potential for foodborne illness as *preservation processes*. Pasteurization is one of the traditional preservation processes, and uses thermal energy to increase the product temperature and inactivate specific pathogenic microorganisms. Pasteurization results in a shelf-stable product with refrigeration. Commercial sterilization is a more intense thermal process to reduce the population of all microorganisms in the product and leads to shelf-stable products in cans and similar containers. Recently, technologies such as high pressure and pulsed electric fields have been investigated to reduce microbial populations in foods without the need for thermal energy. The subject of thermal processing has been described in detail in standard references such as Ball and Olson (1957), Stumbo (1973), NFPA (1980), Lopez (1969), and Teixiera (1992). The current status of nonthermal processes is presented by Barbosa-Canovas et al. (1998).

5.1 PROCESSING SYSTEMS

The systems used for preservation of foods will vary significantly with the type of process being used. Traditional thermal processing systems are designed to provide the desired increase in product temperature, followed by a period of holding time and cooling from the elevated temperature. Systems for alternative preservation processes involve bringing a treatment agent into contact with the food product for the period of time needed to reduce the action of deterioration reactions within the product. The design of all processing systems is unique for the specific food products being processed.

All icons in this chapter refer to the author's web site, which is independently owned and operated. Academic Press is not responsible for the content or operation of the author's web site. Please direct your web site comments and questions to the author: Professor R. Paul Singh, Department of Biological and Agricultural Engineering, University of California, Davis, CA 95616, USA. Email: rps@rpaulsingh.com.

5.1.1 Pasteurization and Blanching Systems

Many foods receive a mild thermal process designed to eliminate pathogenic microorganisms and other components causing deterioration of the product and to extend the shelf-life and the safety of the food product. The most recognized of these processes is pasteurization—a thermal process used to eliminate specific pathogenic microorganisms from a food. Blanching is a similar thermal process used to inactivate enzymes in foods and prevent the deterioration reactions in the product. Both processes accomplish the desired result without using the high temperatures normally associated with commercial sterilization.

Most pasteurization systems are designed for liquid foods, and with specific attention to achieving a specific time–temperature process. A typical system is presented in Figure 5.1. The continuous high-temperature-short-time (HTST) pasteurization system has several basic components, including the following:

- *Heat exchangers for product heating/cooling.* Most often, plate heat exchangers are used to heat the product to the desired temperature. The heating medium may be hot water or steam, and a regeneration section is used to increase efficiency of the process. In this section, hot product becomes the heating medium. Cold water is the cooling medium in a separate section of the heat exchanger.

- *Holding tube.* The holding tube is an important component of the pasteurization system. Although lethality accumulates in the heating, holding, and cooling sections, the Food and Drug Administration (FDA) will consider only the lethality accumulating in the holding section (Dignan et al., 1989). It follows that the design of the holding tube is crucial to achieve a uniform and sufficient thermal process.

- *Pumps and flow control.* A metering pump, located upstream from the holding tube, is used to maintain the required product flow rate. Usually a positive displacement pump is used for this application. Centrifugal pumps are more sensitive to pressure drop and should be used only for clean-in-place (CIP) applications.

- *Flow diversion valve.* An important control point in any pasteurization system is the flow diversion valve (FDV). This remotely activated valve is located downstream from the holding tube. A temperature sensor located at the exit to the holding tube

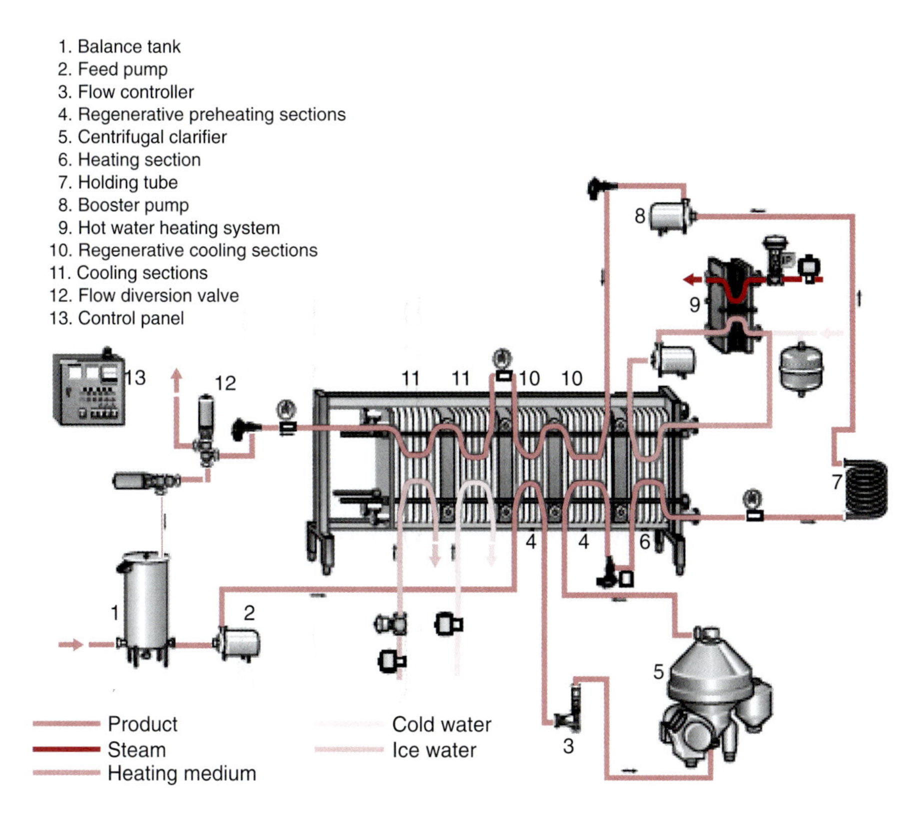

Figure 5.1 A milk pasteurization system. (Courtesy of Alpha Laval)

activates the FDV; when the temperature is above the established pasteurization temperature, the valve is maintained in a forward flow position. If the product temperature drops below the desired pasteurization temperature, the FDV diverts the product flow to the unheated product inlet to the system. The valve and sensor prevent product that has not received the established time–temperature treatment from reaching the product packaging system.

A blanching system achieves a process similar to pasteurization, but with application to a solid food and to inactivate an enzyme system. A schematic of a typical blanching system is presented in Figure 5.2. Since these systems are designed for solid foods, conveying systems are used to carry the product through the system. The environment within the system, where the product pieces are heated, is steam or

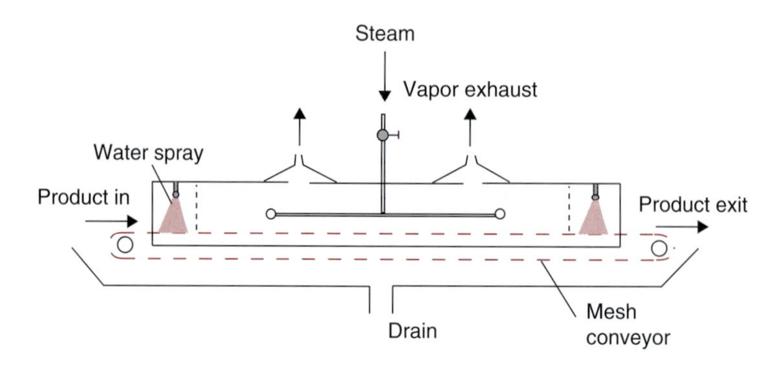

■ **Figure 5.2** A steam blanching system. (Adapted from Rumsey, 1982)

heated water. The design and speed of the conveyor ensures that the thermal treatment of the product provides the desired inactivation of the enzyme system for a given product. In general, the slowest heating location of the product pieces will increase in temperature during the initial stage of the process, while the product is exposed to the steam or hot water. During the second stage of the process, the product is exposed to a cold environment, usually cold air or water. The temperature-time profile at the slowest heating/cooling location of the product pieces are critical in ensuring the process, and are established by the speed of the conveyor through both stages of the system.

5.1.2 Commercial Sterilization Systems

The use of a thermal process to achieve a shelf-stable food product is referred to as commercial sterilization. The systems used to accomplish a thermal process are designed to increase the product temperature to rather high magnitudes for specified periods of time. Typically, the temperatures used for these processes exceed the boiling point of water, and the equipment must include this capability. The systems used for commercial sterilization are in three categories: batch, continuous, and aseptic. When considering these categories, the batch and continuous systems accomplish the thermal treatment after the product is placed in the container or package. For systems in the aseptic category, the process is accomplished before the product is placed in the container or package, and the container or package requires a separate process.

5.1.2.1 *Batch Systems*

The typical batch system for commercial sterilization is referred to as a still retort. As illustrated by Figure 5.3, a still retort is a vessel

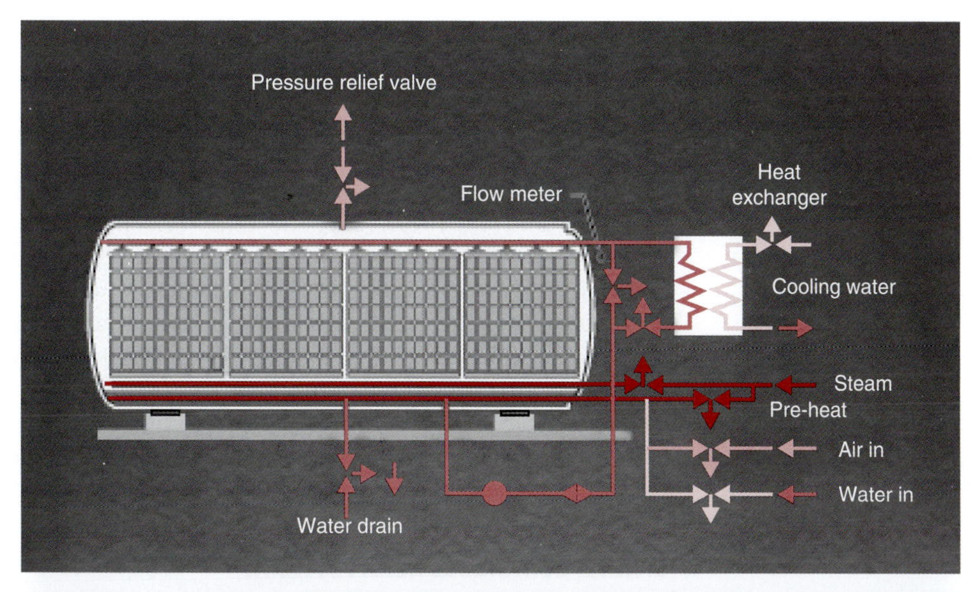

W

■ **Figure 5.3** A typical batch retort system for commercial sterilization of foods. (Courtesy of FMC)

designed to expose the product to temperatures above the boiling point of water. The vessel must maintain pressures up to 475 kPa, or the pressures needed to maintain steam temperature as high as 135° to 150°C. The control system is designed to allow the environment (pressure and temperature) within the vessel to be increased to some desired level for a specified time, held at the desired condition for a specified time, and then returned to ambient pressure and temperature conditions. For these systems, the product is introduced into the vessel after being placed into a container or package and sealed.

For a typical process, the product containers (with product) would be placed into the vessel. After the vessel is sealed, high-pressure steam would be released into the vessel until the desired pressure and temperature are reached. After the final pressure and temperature are reached, the product (in containers) is held for a predetermined period of time. The period of time, at the specified temperature, is established in advance, and delivers the desired thermal treatment for commercial sterilization of the product. Following the holding period, the pressure is released and the temperature surrounding the product containers decreases.

5.1.2.2 *Continuous Retort Systems*

Various systems have been developed to allow commercial sterilization to be accomplished in a retort and in a continuous manner. One approach is referred to as the *crateless* retort, and involves a design of the processing vessel to allow for automated filling and discharge of product containers. The approach is actually semicontinuous, in that the vessel is first filled with product containers, followed by sealing of the vessel, prior to administration of the thermal process in the same manner as the batch system. Following the process, the product containers are removed in an automated manner.

A truly continuous process is the hydrostatic sterilization system illustrated in Figure 5.4. The hydrostatic system uses a tower and two columns of water to maintain a high pressure steam environment for the product containers to move through. The height of the water columns is sufficient to maintain the desired steam pressure and temperature. The product conveyor carries the containers through the system in a continuous manner. The product containers enter the system through a column of hot water and heating of product is initiated. The heating of the product is completed as the product is conveyed through the steam environment. The final stage of the process is accomplished as the product containers are conveyed through a column of cold water. The residence time for product within the system is a function of the conveyor speed. Ultimately, the desired process for the product is a function of the steam temperature and the time required for product to be conveyed through the system.

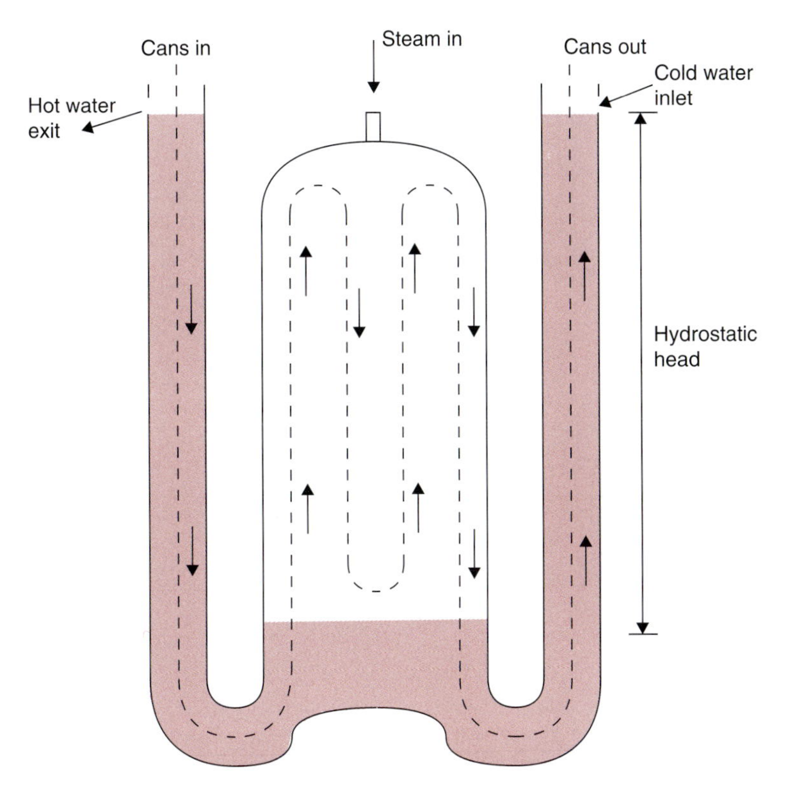

W

■ **Figure 5.4** Schematic diagram of a hydrostatic sterilization system for canned foods.

5.1.2.3 *Pouch Processing Systems*

Most commercial sterilization systems are designed for food product placed in metal cans. An alternative system for product placed in a flexible pouch requires an approach to suspend the pouch in the steam environment within the retort vessel. The design of this batch system includes a unique crate within the vessel to suspend the product pouches in the steam environment and allow the heating of the product in the most efficient manner. By exposing both sides of the thin pouch to the heating medium, the slowest heating location of the product reaches the desired temperature rapidly, and can be cooled rapidly to complete the thermal process.

5.1.2.4 *Aseptic Processing Systems*

Another approach to continuous sterilization of a food product is referred to as aseptic processing as illustrated in Figure 5.5. The unique aspect of these systems is that the product is thermally processed prior to being placed in a container. The systems require

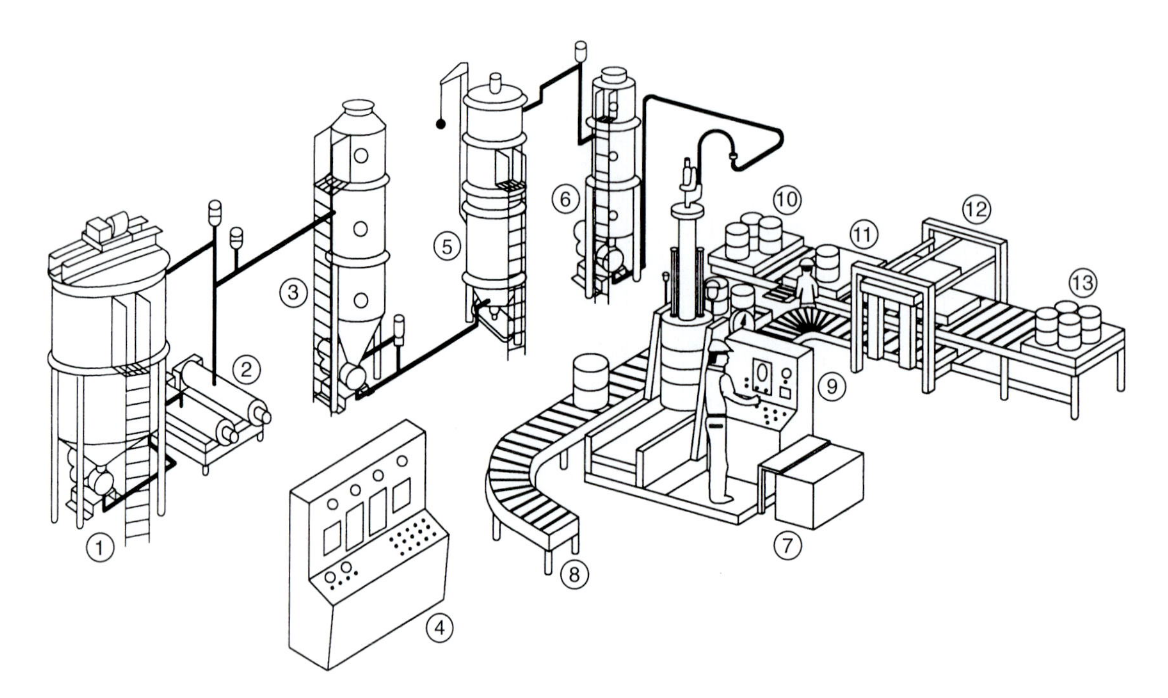

■ **Figure 5.5** An aseptic processing system. (1) feed tank with pump; (2) scraped surface heaters; (3) steam-pressurized hot-hold vessel with aseptic pump; (4) master process control panel; (5) aseptic scraped surface cooler; (6) nitrogen-pressurized cold surge; (7) aseptic low-pressure drum filler; (8) empty drum feed conveyor; (9) uppender; (10) palletized empty drum feed; (11) manual drum depalletizer; (12) semiautomatic drum palletizer; (13) palletized full drum discharge. (Adapted from Alpha Laval)

independent sterilization of the container, and placement of the product into the container while in an aseptic environment. These systems are limited to products that can be pumped through a heat exchanger for both heating and cooling. By using high pressure steam as a heating medium in a heat exchanger, the product can be heated to temperatures in excess of 100°C. Following the heating step, the product is pumped through a holding tube for residence time needed to achieve the desired thermal process. Product cooling is accomplished in a heat exchanger using cold water as a cooling medium. These systems cannot be used for solid foods, but have been adapted for high viscosity foods and liquid products containing solid particles.

5.1.3 Ultra-High Pressure Systems

The use of high pressures to achieve food preservation has evolved as a potential commercial process. Historically, there have been demonstrations of inactivation of microbial populations by using ultra-high pressures in the range of 300 to 800 MPa. More recently, systems have

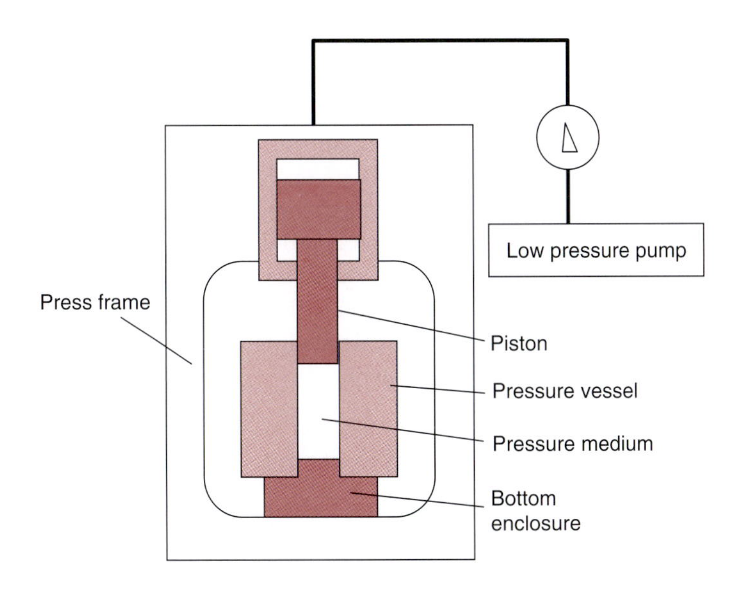

Figure 5.6 Schematic of an ultra-high pressure processing system. (Adapted from Mertens and Deplace, 1993)

been developed and used to expose food products and achieve significant reductions in microbial populations in the product.

The typical system for ultra-high pressure (UHP) processing of a food product is illustrated in Figure 5.6. The primary component of the system is a vessel designed to maintain the high pressures required for the process. A transmitting medium within the vessel is in contact with the product and delivers the impact of the agent to the product and microbial population within the product. Current systems operate in a batch or semicontinuous mode.

A typical UHP process for a solid food or product in a container would be accomplished by placing the product in the system vessel, followed by filling the space around the product with the transmitting medium. Pressure is increased by high pressure pumps for the transmitting liquid or by activating a piston to reduce the volume of the medium surrounding the product. After the desired pressure is reached in the vessel, the pressure is maintained for the period of time needed to accomplish the required reduction in microbial population in the product. At the end of the holding period, the pressure is released and the process is completed.

A system for pumpable product would use a high-pressure transfer value to introduce product into the vessel. For these types of semicontinuous systems, the pressure is increased by introducing water behind a free piston applied directly to the product. Following the process, the product is pumped out of the vessel, and the process cycle is repeated. The processed product must be filled into packages or containers in an aseptic environment.

During the UHP process, the product temperature will be increased further to adiabatic heating. A typical product would experience an increase of 3°C per 100 MPa, although the magnitude of increase would vary with product composition. This temperature increase may or may not influence the process depending on the temperature of the product entering the UHP system.

5.1.4 Pulsed Electric Field Systems

The exposure of a food product to a pulsed electric field results in a reduction in the microbial population within the product. The application of this process to a food product requires a component to generate the electric field and a treatment chamber to expose the product to the field in a controlled manner.

In general, the treatment chamber includes a method for conveying the product through an environment where it is exposed to the pulsed electric field for a controlled period of time. Most systems are designed for liquid foods that can be pumped through a tube or pipe, and the components used to generate the electric field are designed to surround the tube or pipe. The portion of the system designed to create the electric field has at least two electrodes: one high voltage and the other at ground level. The treatment involves exposing the product to pulses of voltage between the two electrodes.

Several different configurations for the electrodes and the product flow have been developed, as illustrated in Figure 5.7. These configurations include parallel plate, coaxial, and colinear. Although the parallel plate arrangement provides the most uniform electric field intensity, the intensity is reduced in boundary regions. At product flow rates for commercial operations, the frequency of pulses results in product temperature rise.

A primary process variable is the electric field intensity. Depending on the microbial population, the intensities may be as low as 2 kV/cm or as high as 35 kV/cm. For all microorganisms, the rate of reduction in the population will increase as the field intensity is increased. The pulse geometry is an additional factor influencing the process.

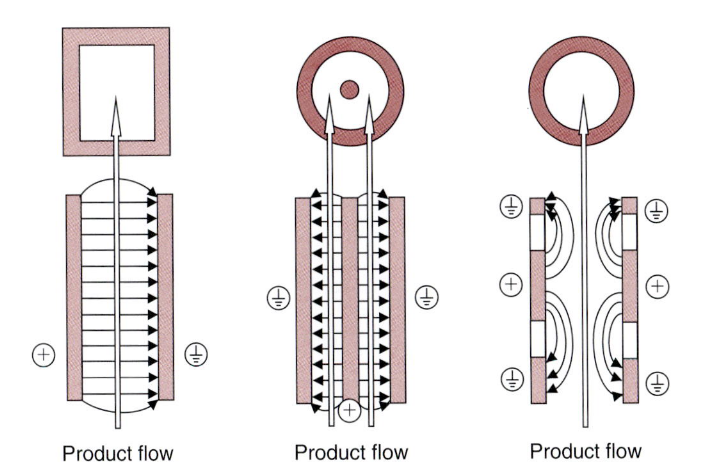

Figure 5.7 A schematic of three configurations of continuous flow pulsed-electric-field systems. (Adapted from Toepfl, Heinz, and Knorr, 2005)

Temperature is an additional variable, and has a highly synergistic influence on the effectiveness of the process. Product composition may influence the process, with products creating higher electrical conductivity, contributing to the effectiveness of the process. The existence of gas bubbles or particles in the product liquid require careful consideration as well.

5.1.5 Alternative Preservation Systems

A variety of other preservation technologies have been investigated. These technologies include ultraviolet light, pulsed light, and ultrasound. The applications of these technologies have not evolved to commercial-scale operations. Very limited information on the application of high voltage arc discharge, oscillating magnetic fields, and X-rays to foods is available. Microwave and radio frequency and ohmic-induction heating systems have been developed for foods, but depend on the increase of product temperature to achieve the process.

5.2 MICROBIAL SURVIVOR CURVES

During preservation processes for foods, an external agent is used to reduce the population of microorganisms present in the food. The population of vegetative cells such as *E. coli*, *Salmonella*, or *Listeria monocytogenes* will decrease in a pattern as shown in Figure 5.8.

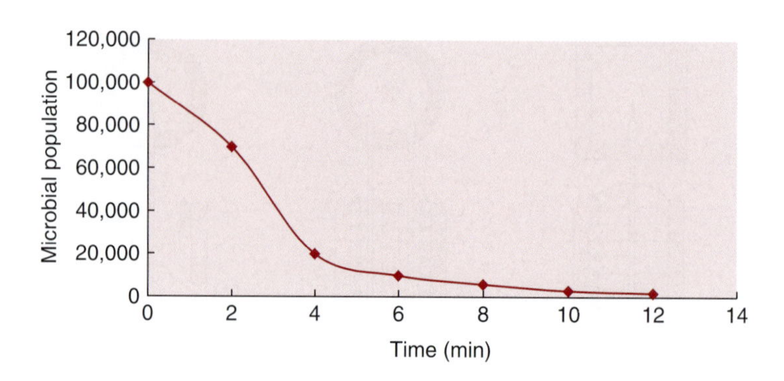

Figure 5.8 A survivor curve for a microbial population.

The population of microbial spores will decrease in a similar manner, but after an initial lag period. These curves are referred to as microbial survivor curves. Although the shape of these curves is often described by a first-order model, there is increasing evidence that alternate models are more appropriate when the application is the design of a preservation process.

A general model for description of the microbial curve would be:

$$\frac{dN}{dt} = -kN^n \quad (5.1)$$

where k is the rate constant and n is the order of the model. This general model describes the reduction in the microbial population (N) as a function of time. A special case of Equation (5.1) is:

$$\frac{dN}{dt} = -kN \quad (5.2)$$

where n is 1.0: a first-order kinetic model. This basic model has been used to describe survivor curves obtained when microbial populations are exposed to elevated temperatures. When survivor curve data are presented on semilog coordinates, a straight line is obtained, as shown in Figure 5.9. The slope of the straight line is the first-order rate constant (k), and is inversely related to the decimal reduction time, D.

The decimal reduction time D is defined as the time necessary for a 90% reduction in the microbial population. Alternatively, the D value is the time required for a one log-cycle reduction in the population of

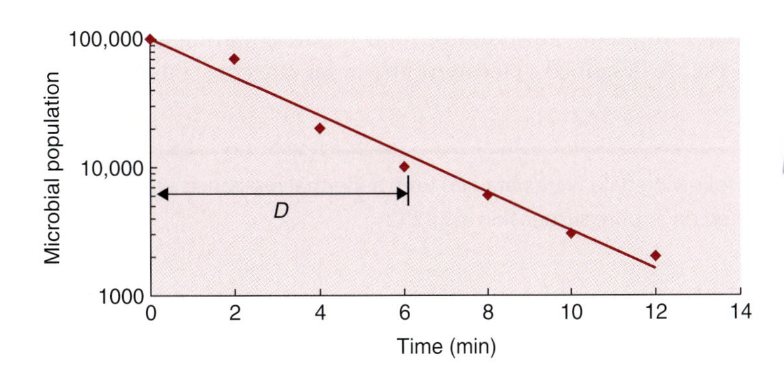

microorganisms. Based on the definition of decimal reduction time, the following equation would describe the survivor curve:

$$\log N_0 - \log N = \frac{t}{D} \tag{5.3}$$

or

$$D = \frac{t}{\log N_0 - \log N} \tag{5.4}$$

and

$$\frac{N}{N_0} = 10^{-t/D} \tag{5.5}$$

A solution to Equation (5.2), when the initial population is N_0 and the final population is N at time t, would be:

$$\frac{N}{N_0} = e^{-kt} \tag{5.6}$$

By comparison of Equations (5.5) and, (5.6), it is evident that:

$$k = \frac{2.303}{D} \tag{5.7}$$

The kinetics of a chemical reaction are more often described by Equation (5.1), and the rates of change in chemical components are expressed by first-order rate constants (k). In many situations, the changes in quality attributes of food products during a preservation process are described in terms of first-order rate constants (k).

Example 5.1

The following data were obtained from a thermal resistance experiment conducted on a spore suspension at 112°C:

Time (min)	Number of survivors
0	10^6
4	1.1×10^5
8	1.2×10^4
12	1.2×10^3

Determine the D value of the microorganism.

Approach

The microbial population will be plotted on a semilog plot to obtain the slope.

Solution

1. *On a semilog graph paper, plot the number of survivors as a function of time (Fig. E5.1). From the straight line obtained, determine the time for a one-log cycle reduction in the population of spores, which gives the D value as 4.1 minutes.*
2. *Alternatively, this problem can be solved using a spreadsheet by first taking the natural logarithm of the number of survivors and entering the data in a linear regression model. The D value of 4.1 minutes is obtained.*

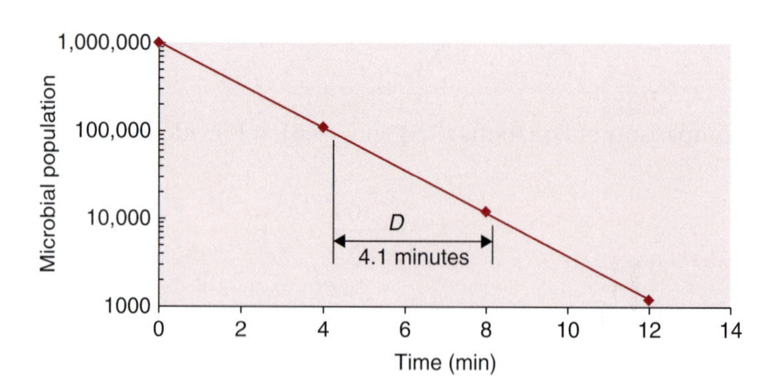

■ **Figure E5.1** Graphical determination of the decimal reduction time (D) in Example 5.1.

There are situations when the survivor curve for microbial populations may not follow a first-order relationship. For these situations, the following relationship is recommended:

$$\log N_o - \log N = \left[\frac{t}{D'}\right]^n \tag{5.8}$$

where D' is a parameter similar to the decimal reduction time, and n is a constant to account for deviation from the log-linear survivor curve. When survivor curves exhibit a concave upward relationship, the n-value will be less than 1.0. Concave downward survivor curves are described by n-values with magnitudes greater than 1.0.

Example 5.2

The following survivor curve data were obtained during an ultra-high pressure process of a food product at 300 MPa:

Time (s)	Microbial Numbers
0	1000
1	100
2	31
3	20
4	30
5	10
6	6
7	6
8	5
16	2
24	1

Determine the parameters needed to describe the survivor curve.

Approach

Equation (5.8) can be revised in the following manner:

$$\log[\log(N_o/N)] = n[\log t - \log D']$$

A plot of the log of the survivor curve ratio versus log of time results in evaluation of the two parameters.

Solution

1. *The plot of the survivor curve data is presented in Figure E5.2.*
2. *From the plot, slope = 0.3381, intercept = 0.0425.*
3. *Then* $D' = 10^{\left(-\frac{0.0425}{0.3381}\right)} = 0.74\,s$ *and* $n = 0.34$

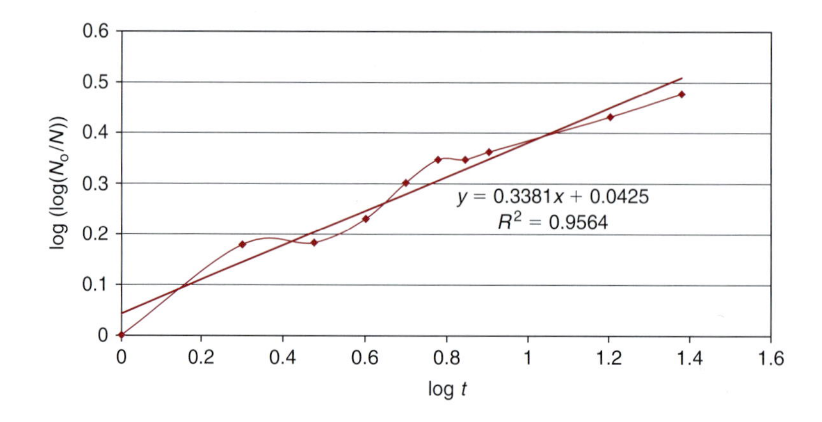

Figure E5.2 Graphical determination of D' in Example 5.2.

5.3 INFLUENCE OF EXTERNAL AGENTS

The survivor curves for microbial populations are influenced by external agents. As the magnitude of preservation agents such as temperature, pressure, and pulsed electric fields increase, the rate of the microbial population reduction increases. Exposure of a microbial population to an array of higher temperatures results in an increasing slope for the first-order curves.

In chemical kinetics, we use the Arrhenius equation to describe the influence of temperature on the rate constant. Thus,

$$k = Be^{-(E_a/R_g T_A)} \tag{5.9}$$

or

$$\ln k = \ln B - \frac{E_a}{R_g T_A} \tag{5.10}$$

where the influence of temperature on the rate constant (k) is expressed by the magnitude of the Activation Energy Constant (E_a). These constants are determined from experimental data by plotting ln k versus $1/T_A$, and the slope of the linear curve is equal to E_a/R_g, as illustrated in Figure 5.10.

Traditional thermal processing has used the thermal resistance constant z to describe the influence of temperature on decimal reduction time, D, for microbial populations. The thermal resistance constant

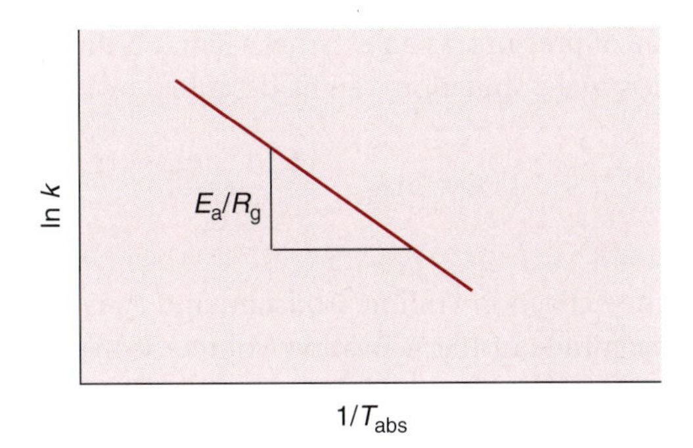

■ **Figure 5.10** An Arrhenius plot of the reaction rate constant (k) versus the inverse of absolute temperature.

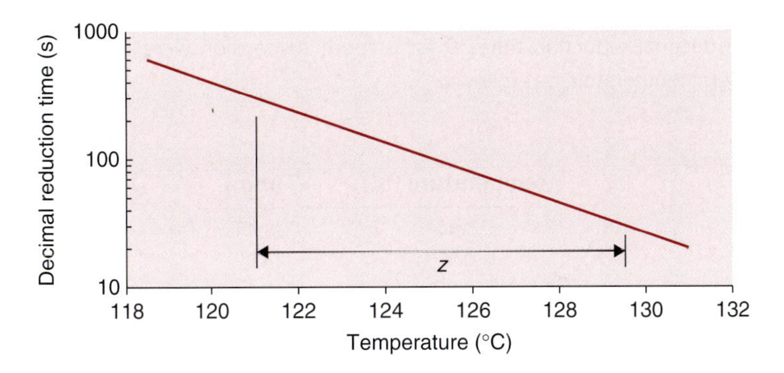

■ **Figure 5.11** A plot of the logarithm of D versus temperature, used to determine the thermal resistance constant (z).

(z) is defined as the increase in temperature necessary to cause a 90% reduction in the decimal reduction time D. The D values for different temperatures are plotted on semilog coordinates, and the temperature increase for a one log-cycle change in D values is the z value, as shown in Figure 5.11.

Based on the definition, z can be expressed by the following equation:

$$z = \frac{T_2 - T_1}{\log D_{T_1} - \log D_{T_2}} \tag{5.11}$$

By comparing Equations (5.10) and (5.11) and noting that $k = 2.303/D$,

$$E_a = \frac{2.303 R_g}{z} [T_{A_1} \times T_{A_2}] \tag{5.12}$$

The influence of pressure, as an external agent, on the rate of inactivation of a microbial population, can be described by the following:

$$\ln k = \ln k_R - \left[\frac{V(P - P_R)}{R_g T_A}\right] \quad (5.13)$$

where V is the Activation Volume Constant and P_R is a reference pressure. The magnitude of the Activation Volume Constant is evaluated from experimental data by plotting $\ln k$ versus $(P - P_R)$ at a constant temperature (T_A). The slope of the linear curve is $V/R_g T_A$.

Example 5.3

The decimal reduction times D for a spore suspension were measured at several temperatures, as follows:

Temperature (°C)	*D* (min)
104	27.5
107	14.5
110	7.5
113	4.0
116	2.2

Determine the thermal resistance constant z for the spores.

Approach

The D values will be plotted versus temperature on semilog coordinates and the z value is estimated from the slope of the established straight line.

Solution

1. *Using semilog graph paper, the D values are plotted versus temperature (Fig. E5.3).*
2. *A straight line through the plotted points is established.*
3. *Based on the straight-line curve, a temperature increase of 11°C is required for a one-log cycle reduction in the D value.*
4. *Based on this analysis, z = 11°C.*
5. *Alternately, z can be estimated using a linear regression program to analyze the relationship between the logarithm of D values versus temperature. The z value is 11.1°C.*

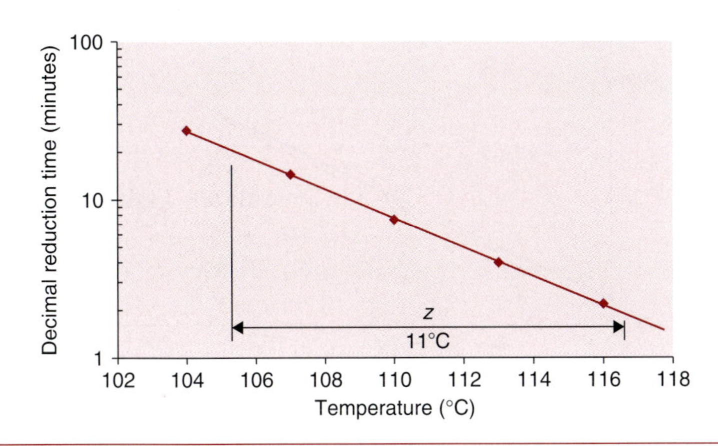

■ **Figure E5.3** Graphical determination of the thermal resistance constant (z) in Example 5.3.

Example 5.4

Data from Mussa et al. (1999) indicate that inactivation rates for L. monocytogenes Scott A as a function of pressure at 30°C were as follows:

Pressure (MPa)	Rate (k) ($\times 10^{-2}$ min^{-1})
200	4.34
250	7.49
300	14.3
350	27.0
400	65.2

Using the data presented, estimate the Activation Volume Constant (V).

Approach

The Activation Volume Constant (V) can be estimated using Equation (5.13). As indicated, a plot of Ln k versus ($P - P_R$) will provide linear relationship with a slope of V/R_gT_A.

Solution

1. *The data presented have been plotted as Ln k vs. (P-P_R), where P_R is 200 MPa, in Figure E.5.4.*
2. *The results indicate that the slope of the linear relationship is:*

$$V/R_gT_A = -0.0134 \times 10^{-6}\ Pa^{-1}$$

$$\text{And for } R_g = 8.314\ m^3Pa/(mol\ K) \quad \text{and} \quad T_A = 303\ K,$$

$$V = -3.38 \times 10^{-5}\ m^3/mol$$

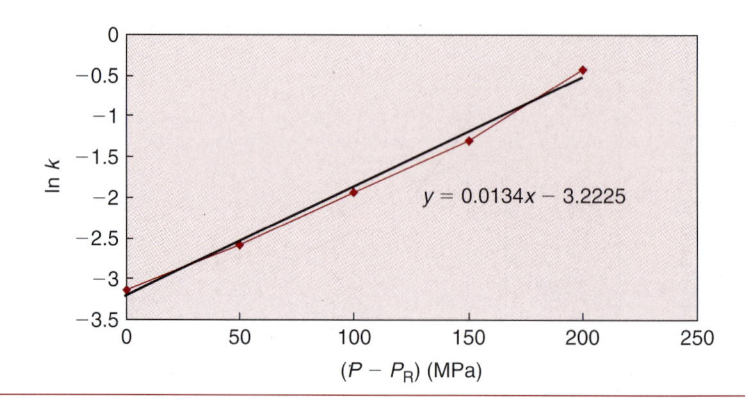

■ **Figure E5.4** Graphical determination of the activation volume constant (V) in Example 5.4.

5.4 THERMAL DEATH TIME *F*

The thermal death time F is the total time required to accomplish a stated reduction in a population of vegetative cells or spores. This time can be expressed as a multiple of D values, as long as the survivor curve follows a first-order model. For example, a 99.99% reduction in microbial population would be equivalent to four log-cycle reductions or $F = 4D$. A typical thermal death time in thermal processing of shelf-stable foods is $F = 12D$, with the D value for *Clostridium botulinum*. When there is uncertainty about the shape of the survivor curve, end-point data are needed to confirm the adequacy of the thermal death time.

In the food science literature for thermal processes, we traditionally express F with a subscript denoting the process temperature, and a superscript indicating the z value for the microorganism being considered. Thus, F_T^z is the thermal death time for a temperature T and a thermal resistance constant z. For reference purposes, a commonly used thermal death time is F_{250}^{18} in the Fahrenheit temperature scale, or F_{121}^{10} in the Celsius temperature scale. This reference thermal death time, simply written as F_0, represents the time for a given reduction in population of a microbial spore with a z value of 10°C (or 18°F) at 121°C (or 250°F).

Thermal death times for nonthermal processes would have the same relationship to the survivor curve, or rate of reduction in the microbial population. The magnitude of F would represent the total time of exposure to a preservation agent, as required to achieve the desired reduction in the defined microbial population.

5.5 SPOILAGE PROBABILITY

When considering shelf-stable food products, we can design the preservation process to eliminate spoilage in addition to ensuring microbial safety. The spoilage probability is used to estimate the number of spoiled containers within a total batch of processed product.

From Equation (5.3), with N representing the desired final microbial population for a thermal death time of F,

$$\log N_0 - \log N = \frac{F}{D} \tag{5.14}$$

Then, if r is the number of containers exposed to the preservation process, and N_0 is the initial population of spoilage microorganisms in each container, the total microbial load at the beginning of the process is rN_0, and

$$\log(rN_0) - \log(rN) = \frac{F}{D} \tag{5.15}$$

If the goal of the preservation process is to achieve a probability of one survivor from the microbial population for all containers processed, then

$$\log(rN_0) = \frac{F}{D} \tag{5.16}$$

or

$$rN_0 = 10^{F/D} \tag{5.17}$$

and

$$\frac{1}{r} = \frac{N_0}{10^{F/D}} \tag{5.18}$$

The ratio on the left side of Equation (5.18) represents the total number of containers processed (r) and resulting in one container with spoilage. The expression can be used to estimate the thermal death time required to accomplish a stated spoilage probability, based on knowledge of the initial population and the decimal reduction time, D, for the microbial population. It should be noted that the spoilage probability expression does assume that the survivor curve for the spoilage microorganism follows a first-order model.

Example 5.5

Estimate the spoilage probability of a 50-minute process at 113°C when $D_{113} = 4$ minutes and the initial microbial population is 10^4 per container.

Approach

Use Equation (5.18) to compute spoilage probability.

Solution

1. *From Equation (5.18),*

$$\frac{1}{r} = \frac{10^4}{10^{50/4}} = \frac{10^4}{10^{12.5}} = 10^{-8.5}$$

Therefore,

$$r = 10^{8.5} = 10^8 \times 10^{0.5} = 10^8 \times \sqrt{10}$$
$$= 3.16 \times 10^8$$

2. *Since $r = 3.16 \times 10^8$, then spoilage of one container in 3.16×10^8 can be expected, or approximately three containers from 10^9 containers processed.*

5.6 GENERAL METHOD FOR PROCESS CALCULATION

The General Method for Process Calculation is based on a classic paper by Bigelow et al. (1920). It established the basis of modern thermal process calculations. A major requirement of the general method is that the thermal death time, F, for the microbial population considered must be known at all temperatures to which the product is exposed during the preservation process. It should be noted that the thermal death time decreases as temperature increases.

The original method involved development of a sterility curve for the process. This curve is drawn with the ordinate scale equal to the sterilizing rate (F/t): the thermal death time at a given temperature divided by the actual time at that temperature. When this rate is plotted versus time, the area under the curve is the lethal effect of the process in time units. Assuming equal time increments at each temperature, the sterilizing rate becomes smaller at higher temperatures.

The General Method can be based on the relationship of the thermal death time to the survivor curve, as presented in Equation (5.14), and the relationship to the thermal resistance constant in Equation (5.11).

By using the relationship of thermal death time to the decimal reduction time, Equation (5.11) can be revised as:

$$\log\left(\frac{F_R}{F}\right) = \frac{(T - T_R)}{z} \tag{5.19}$$

or

$$\frac{F_R}{F} = 10^{(T-T_R)/z} \tag{5.20}$$

Equation (5.20) can be used to compute the thermal death time, F, at any temperature, T, when the thermal death time, F_R, is known at a reference temperature, T_R. According to Ball (1923), Equation (5.20) can be defined as lethal rate, or the proportion of the thermal death time at temperature T to the thermal death time at a reference temperature T_R. A plot of lethal rate (LR) versus time of process is a lethal rate curve as illustrated in Figure 5.12. The area under the lethal rate curve is defined as the **lethality**, or the integrated impact of time and temperature on the microbial population, expressed as time at the reference temperature. Often, the lethality is referred to as F, or the impact of a thermal process with a defined temperature profile and

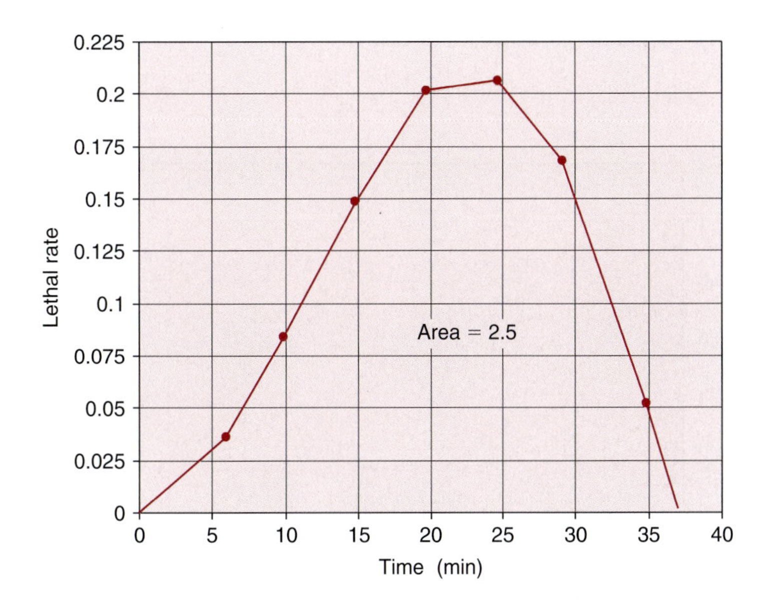

Figure 5.12 A plot of lethal rate versus time for a thermal process.

duration. To avoid confusion with thermal death time, lethality will be defined as:

$$L = \int 10^{(T-T_R)/z} dt \tag{5.21}$$

When estimating a process time, the process must create a lethality, L, sufficient to match a target thermal death time, as determined from Equation (5.18). This is a key step in the use of the General Method of Process Calculation.

5.6.1 Applications to Pasteurization

During pasteurization, the food is heated to a defined temperature, and held at that temperature for a defined time period. The lethality associated with the pasteurization process is based on the holding period only; impact of elevated temperatures on lethality during heating and cooling are not significant or are not considered. The pasteurization of milk (shown previously in Fig. 2.1) is based on reduction of a microbial pathogen with $D_{63} = 2.5$ min and $z = 4.1$°C; and a thermal death time of 12D, or 30 min. This process ensures that the probability of survival of the pathogen is negligible.

The traditional batch pasteurization process is accomplished using a holding time of 30 min at 63°C. By using a reference temperature of 63°C, the lethal rate is 1.0 for the entire holding period of 30 min. As illustrated in Figure 5.13, the lethality for the process is the area under the lethal rate curve, or 30 min.

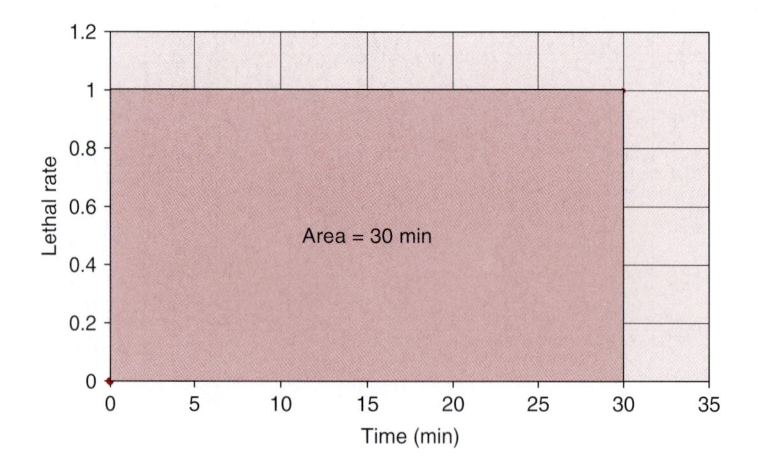

Figure 5.13 A lethal rate curve for a pasteurization process.

The continuous HTST process for pasteurization is unique in that a major portion of the lethality may be accumulated during a holding period at the temperature close to the heating medium. The extent to which the heating and cooling portions of the process will contribute to the lethality is dependent on the rates of heating and cooling and, in turn, on the method of heating and cooling.

The HTST pasteurization process is a continuous process accomplished by heating the product to 71.5 °C and passing through a holding tube at a rate that ensures the required holding time. When the lethal rate for the HTST process is based on the reference temperature of 63 °C, the magnitude is 120. The holding time needed to achieve the same lethality as the batch process is 15 s.

Example 5.6

A thermal process is accomplished by instantaneous heating to 138°C followed by a four-second hold and instantaneous cooling. Estimate the lethality at 121°C when the thermal resistance (z) for the microorganism is 8.5°C.

Approach

Use Equation (5.20) modified to express thermal death times and compute lethality. Figure E5.5 shows a plot of temperature with respect to time.

Solution

1. *Using a modified form of Equation (5.20),*

$$\frac{F_{138}}{F_{121}} = 10^{(121-138)/8.5}$$

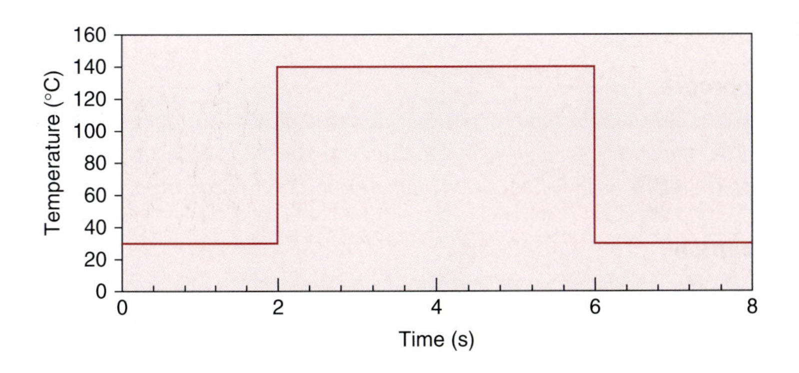

■ **Figure E5.5** Temperature history during the thermal process in Example 5.6.

or

$$F_{121} = (4\ s)10^{(138-121)/8.5} = 4 \times 10^2 = 400\ s$$

2. *The lethality at 121°C (F_{121}) is 100 times greater than at 138°C (F_{138}), in order to represent equivalent thermal processes.*

Under actual situations, the heating and cooling portions of the process will not be instantaneous. These circumstances lead to accumulation of lethality during the heating and cooling steps.

Example 5.7

An ultra-high pressure process is being used to reduce the population of spoilage microorganisms in a food product. The process occurs as follows:

Time (min)	Pressure (MPa)
0	0
1	100
2	200
3	300
4	400
5	400
6	400
7	400
8	200
9	100
10	0

Estimate the reduction in microbial population that occurs as a result of this process, when the initial population is 2×10^3. The response of the spoilage microorganism to the pressure treatment is described by $D_{200} = 60$ min and $z_p = 130$ MPa.

Approach

The change in microbial population occurring as a result of the high pressure process will be estimated by computing the lethal rate associated with each step of the process. The total lethality will be used to compute reduction in microbial population.

Solution

1. *The lethal rates for each step of the high pressure process are indicated in the following table.*

Time	Lethal Rate
0	0
1	0.1701
2	1.0000
3	5.8780
4	34.5511
5	34.5511
6	34.5511
7	34.5511
8	1.0000
9	0.1701
10	0
Total	146.4226

The computations indicate that the total lethality for the process is 146.4226 min. Using this magnitude and the survivor curve equation, the following result is obtained:

$$N = (2 \times 10^3)10^{(-146.42/60)}$$
$$N = 7$$

These calculations indicate that the number of surviving spoilage microorganisms from the process would be 7.

5.6.2 Commercial Sterilization

Commercial sterilization of foods can be accomplished by using retorts or continuous-flow systems. The purpose of these processes is the reduction of microbial populations by sufficient magnitudes to create a shelf-stable food, without refrigeration. Since the microbial populations in these products are spores, the processes are based on the appropriate kinetic parameters to account for the higher resistance of spores to thermal treatment. Although the resistance will vary with pH of the product, the most severe process is required for low-acid foods (pH > 4.5), where the kinetic parameters for *Cl. botulinum* are $D_{121} = 0.2$ min and $z = 10°$C. Since the thermal resistance of several spoilage microorganisms exceeds the magnitude for *Cl. botulinum*, the processes for shelf-stable foods may be based on kinetic parameters for these spoilage microorganisms.

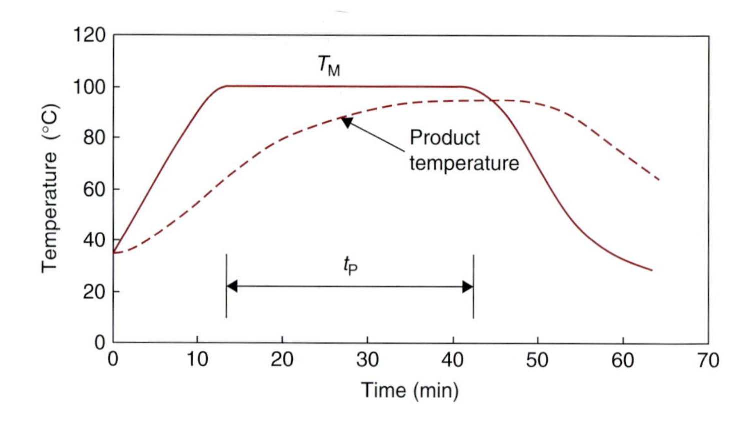

■ **Figure 5.14** Typical temperature history curves for the heating medium and the product during a thermal process.

When the thermal process is developed for a food product in a container to be exposed to high pressure steam in a retort, three factors influence the application of the process:

1. The physical characteristics of the product: solid foods will heat and cool primarily by conduction, whereas liquid foods will heat and cool primarily by convection.
2. The process must be based on the temperature history at the slowest-heating location within the product and container.
3. The time required for the temperature at the slowest-heating location to reach the heating-medium temperature may represent a significant portion of the process time; in conduction-heating products, the temperature at the end of the process may be less than the heating-medium temperature.

In order to express the process in terms of the application, the actual time for the product to be exposed to the retort environment must be defined. As illustrated in Figure 5.14, a finite amount of time is required for the temperature within the retort to reach the final and stable condition (T_M). The temperature history at the slowest-heating location would follow the curve indicated; the temperature does not reach T_M, and the decrease in temperature at the slowest-heating location occurs significantly after the container is exposed to the cooling medium. The impact of the thermal process is evaluated by using the measured temperature history to create the lethal rate curve as shown previously in Figure 5.12.

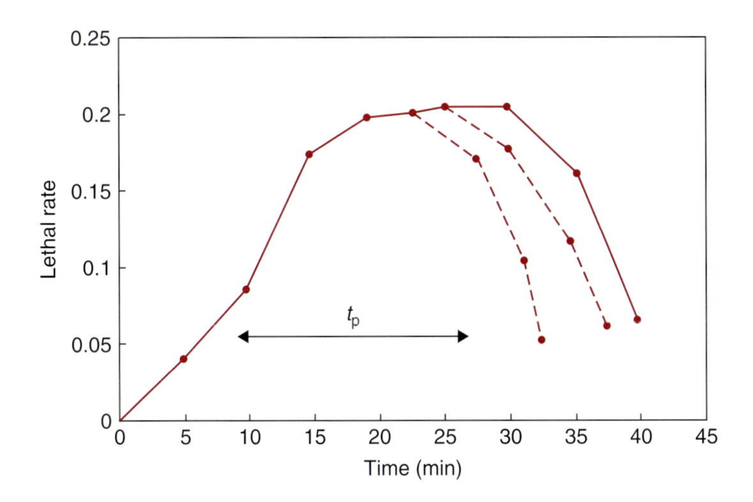

Figure 5.15 Determination of operator process time from a lethal rate curve.

In many situations, the purpose of the process calculation is to estimate the actual process time required to achieve a target thermal death time, F. In applications, the **operator time,** t_p, is the difference between the time at the end of product heating (beginning of cooling) and the time when the retort reaches a stable heating medium temperature. The process establishment is the selection of the t_p required for the lethality (area under the lethality curve) to equal the target thermal death time. As illustrated in Figure 5.15, the lethality for one of the three curves would equal the target thermal death time, and the operator time associated with that curve would be selected.

Example 5.8

The temperature at the slowest-heating location for a liquid food in a can is as follows:

Time (min)	Temperature (°C)
0	75
1	105
2	125
3	140
4	135
5	120
6	100

The thermal process is being applied to a microbial population with D_{121} of 1.1 min and $z = 11°C$. Estimate the lethality (F_{121}) for the thermal process.

Given

$$D_{121} = 1.1\,min$$
$$z = 11°C$$

Temperature–time data are given in the problem statement.

Approach

By computing the lethal rate at each time interval, the lethality can be determined by summation of lethal rates.

Solution

1. *The lethal rate is computed by using Equation (5.20):*

$$LR = 10^{(T-121)/11}$$

2. *When the lethal rate is computed at each time interval, the results are as follows:*

Time (min)	Lethal rate
0	–
1	0.035
2	2.310
3	53.367
4	18.738
5	0.811
6	0.012
	75.273

3. *As indicated, the total of the lethal rate values at 1 min intervals is 75.273 min.*
4. *The lethality is*

$$F_{121} = 75.27\,min$$

5.6.3 Aseptic Processing and Packaging

The heating of a liquid food during a continuous UHT process may require finite time for the product temperature to increase from the initial temperature to the holding-tube temperature. The time required for the product temperature to decrease after holding may be significant as well. Due to regulations, the lethality associated with the area under the lethal rate curve before and after holding may not be considered when evaluating continuous aseptic processes. The magnitude of these portions of the lethality curve will depend on heating time and cooling time, as well as the temperature during holding. Only when the heating is instantaneous, as would occur with steam injection or infusion, and the cooling is instantaneous, as would occur with "flash" cooling, will the lethality of the process be equal to the lethality accumulated during the holding period.

Example 5.9

In pilot plant testing of a liquid food subjected to an HTST process, one microorganism survived the heat treatment. Laboratory tests established that $D_{121} = 1.1$ minute and $z = 11°C$ for the microorganism. The maximum initial count of the microorganism in the food is assumed to be 10^5/g, and the largest container to be used is 1000 g. A process is desired that will assure that spoilage occurs in less than one container per 10,000. Temperatures at selected points in the process are as follows:

Process	Time (s)	Temperature (°C)
Heating	0.5	104
Heating	1.3	111
Heating	3.4	127
Heating	5.3	135
Heating	6.5	138
Holding	8.3	140
Holding	12.3	140
Cooling	12.9	127
Cooling	14.1	114
Cooling	16.2	106

Calculate the minimum holding time needed to achieve the desired result.

Approach

Compute the additional holding time required to achieve the desired spoilage rate.

Solution

1. *Since the desired process will reduce the microbial population to one survivor in 10,000 containers or one microorganism in 10^7 g, the process must be adequate to reduce the population from 10^5/g to one in 10^7 g, or 10^{-7}/g. This is equivalent to a reduction of 12 log cycles, and the process must be equivalent to $12D_{121}$ or $F_{121} = 12(1.1) = 13.2$ minutes.*

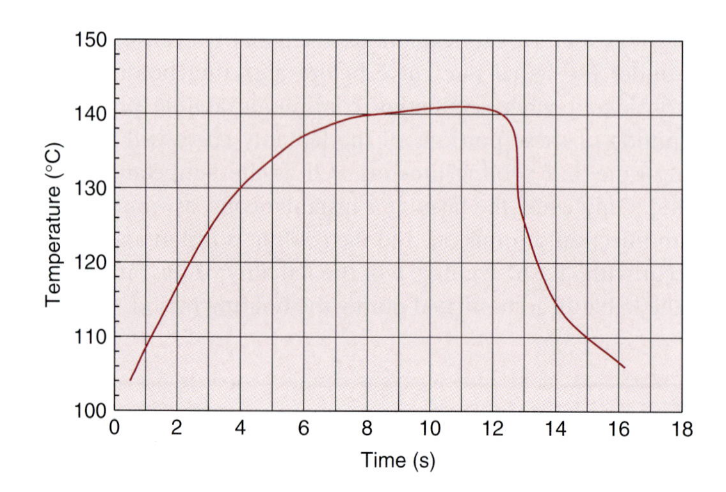

Figure E5.6 Temperature history for the thermal process in Example 5.9.

2. *To determine the adequacy of the process described, the lethality in terms of equivalent time at 121°C must be determined.*
3. *The time–temperature relationship as measured during the process has been plotted in Figure E5.6. In addition, the midpoint temperatures for one-second intervals throughout the process have been identified as illustrated in Table E5.1. Note that the midpoint temperatures have been chosen so that temperatures fall at the midpoint of each of the four seconds during holding. Since the midpoint of a one second interval is 0.5 s, the midpoints for the holding period would be 8.3 + 0.5 = 8.8, and 9.8, 10.8, and 11.8 s. All midpoint temperatures and corresponding times are listed in Table E5.1.*
4. *For this situation, the lethality using the temperature at holding as a reference would be computed as:*

$$\text{Lethal rate} = 10^{(T-140)/z}$$

Since one-second intervals have been chosen, a histogram to illustrate graphical integration used to obtain total lethality can be constructed (Fig. E5.7). The area under the histogram chart is 6.505 s, the same as the total for the lethal rate column in Table E5.1. The accumulated lethality represents a process of 140°C for 6.505 s.

5. *The computed process is equivalent to*

$$F_{121}^{11} = F_{140}^{11} \times 10^{(140-121)/11} = 6.505(53.4) = 347\ \text{s} = 5.79\ \text{min}$$

6. *Since the desired process is 13.2 minutes, an additional 7.41 minutes of process at 121°C is required.*
7. *The additional process can be accomplished by extending the holding time beyond the four seconds. By converting the additional process to an equivalent time at 140°C,*

$$F_{140}^{11} = F_{121}^{11} \times 10^{(121-140)/11} = 7.41(0.0187) = 0.139\ \text{min} = 8.31\ \text{s}$$

Table E5.1 Lethality Calculations for Example 5.9

Time (s)	Midpoint temperature (°C)	Lethal rate
0.8	107	0.001
1.8	114.8	0.005
2.8	122.4	0.025
3.8	128.7	0.094
4.8	132.9	0.226
5.8	136.25	0.456
6.8	138.3	0.701
7.8	139.4	0.882
8.8	140	1.000
9.8	140	1.000
10.8	140	1.000
11.8	140	1.000
12.8	129.2	0.104
13.8	117.25	0.008
14.8	111	0.002
15.8	108	0.001
	Total	6.505

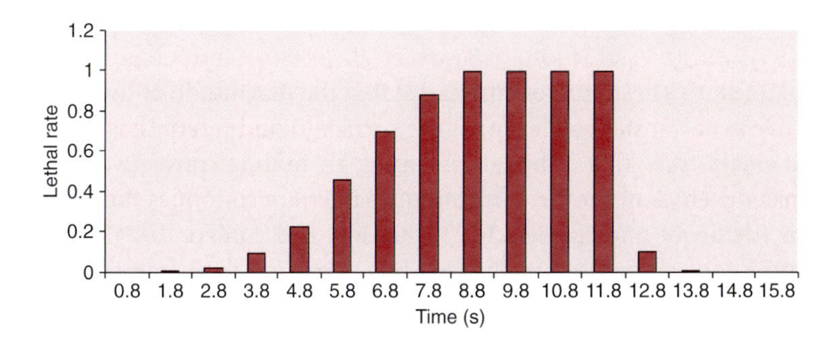

Figure E5.7 The lethal rate curve for the thermal process in Example 5.9.

8. *The total holding required to provide the desired process would become*

$$4 + 8.31 = 12.31\,s$$

Low-acid liquid foods that contain particles present an extra challenge to the heating and cooling system. In an indirect type of system, heating usually results in the carrier liquid being overprocessed to ensure adequate processing of the particles. The APV Jupiter system offers an alternative of heating particles and liquid in separate subsystems, but the system has not achieved commercial acceptance.

In a continuous aseptic processing system, heating or cooling of particulates occurs by conduction. The transient heat conduction equations presented in Chapter 4 can be used to describe heat transfer within a particle. As indicated by the solutions, knowledge of the magnitude of convective heat transfer coefficient (h_{fp}) at the fluid/particle interface has a significant effect on the heating rates for the particle.

For fluid flow over the surface of a sphere, a correlation developed by Ranz and Marshall (1952) can be used.

$$N_{Nu} = 2.0 + 0.6 N_{Re}^{0.5} N_{Pr}^{0.33} \tag{5.22}$$

If we assume that a particle and fluid are traveling at identical velocities, the Reynolds number is reduced to zero, and

$$N_{Nu} = 2.0 \tag{5.23}$$

Equation (5.23) gives the most conservative estimate of the convective heat transfer coefficient between the fluid and the particulate. Recent studies that measure the heat transfer coefficients include Heppell (1985), Zuritz et al. (1987), Sastry et al. (1989), and Chandarana et al. (1989). These studies emphasize that the magnitude of the convective heat transfer coefficient at the particle/liquid interface is finite and small. Note that although assuming an infinite convective heat transfer coefficient for the particle/liquid interface simplifies the solution (deRuyter and Burnet, 1973; Manson and Cullen, 1974), this assumption can predict significantly higher particle temperatures and underestimate lethality. The expressions found by Chandarana et al. (1990) were

$$\begin{aligned} &N_{Nu} = 2 + 2.82 \times 10^{-3} N_{Re}^{1.16} N_{Pr}^{0.89} \\ &1.23 < N_{Re} < 27.38 \\ &9.47 < N_{Pr} < 376.18 \end{aligned} \tag{5.24}$$

for starch solutions, and

$$N_{Nu} = 2 + 1.33 \times 10^{-3} N_{Re}^{1.08}$$
$$287.29 < N_{Re} < 880.76 \qquad (5.25)$$

for water. The expressions apply to convective heat transfer coefficients between 55 and 89 W/(m^2 °C) for starch solutions and between 65 and 107 W/(m^2 °C) for water. The holding tube is an important component of the aseptic processing system. Although lethality accumulates in the heating, holding, and cooling sections, the FDA will consider only the lethality accumulating in the holding section (Dignan et al., 1989). It follows that the design of the holding tube is crucial to achieve a uniform and sufficient thermal process.

Numerous systems can be used to package aseptically processed foods. Differences in systems are related primarily to the size and shape of the package as well as the type of material. The key design component of all packaging systems is that of the space where the product is introduced into the package. This space must be sterilized in a manner that prevents the product from postprocessing contamination during package filling. In addition, many packaging systems will contain a component to sterilize the product-contact surfaces of the container.

Example 5.10

An aseptic processing system is being designed for a vegetable soup in which the largest particulate is a 15 mm spherical potato. The carrier liquid is a starch solution that achieves the temperature of 140°C at the exit of the scraped-surface heating portion of the system and entrance to the holding tube. The particulates enter the holding tube at a uniform temperature of 80°C. The product flow rate is 1.5 m^3/h with laminar flow in the holding tube, where the inside diameter is 4.75 cm. The relative velocity between particulate and solution is 0.005 m/s. The thermal process must achieve a 12-log reduction in microbial population at the center of the particulates. The microorganism has a decimal reduction time (*D*) at 121°C of 1.665 s and a thermal resistance factor (*z*) of 10°C. The starch solution has a specific heat of 4 kJ/(kg °C), thermal conductivity of 0.6 W/(m °C), density of 1000 kg/m^3, and viscosity of 1.5×10^{-2} Pa s. Determine the length of holding tube required to complete the desired thermal process.

Approach

The temperature at the center of the largest particle is predicted based on conduction of heat within the particle and the convective heat transfer coefficient at the particle surface. Based on the temperature history at the particle center, the lethality is predicted and compared to the target thermal death time. The residence time for the particle is established for the particle velocity at the center of the holding tube and target thermal death time, then used to determine the holding tube length.

Solution

1. *The prediction of temperature history at the particle center is estimated by using Figure 4.35. The N_{Fo} and the inverse of N_{Bi} are needed.*
2. *To determine the inverse of Biot number:*

$$k = 0.554 \text{ W/(m K) (from Table A.2.2)}$$

$$d_c = 0.0075 \text{ m (characteristic dimension)}$$

For the convective heat transfer coefficient (h), we will use Equation (5.24):

$$N_{Re} = \frac{1000[kg/m^3] \times 0.015[m] \times 0.005[m/s]}{1.5 \times 10^{-2}[Pa\ s]} = 5$$

based on the particle diameter and the relative velocity between particle and carrier solution, and

$$N_{Pr} = \frac{4000[J/(kg\ °C)] \times 1.5 \times 10^{-2}[Pa\ s]}{0.6[W/(m\ °C)]} = 100$$

based on properties of the starch carrier solution.

Then

$$N_{Nu} = 2 + 2.82 \times 10^{-3}(5)^{1.16}(100)^{0.89} = 3.099$$

Therefore

$$h = \frac{3.099 \times 0.6[W/(m\ K)]}{0.0075[m]}$$

$$h = 248 \text{ W/(m}^2\text{ °C)}$$

Then

$$\frac{k}{hd_c} = \frac{0.554[W/(m\ °C)]}{248[W/(m^2\ °C)] \times 0.0075[m]} = 0.3$$

3. The values of N_{Fo} will depend on

$$N_{Fo} = \frac{\alpha t}{d_c^2}$$

$$\alpha = \frac{k}{\rho c_p} = \frac{0.554[W/(m\ °C)]}{950[kg/m^3] \times 3634[J/(kg\ °C)]} = 1.6 \times 10^{-7}\ m^2/s$$

and the density is based on values for water at 110°C from Table A.4.1, the thermal conductivity of potato is obtained from Table A.2.2, and the specific heat of potato from Table A.2.1.

4. Using Figure 4.35, the temperature history at the center of the spherical particle is illustrated in the following table.

Time (s)	N_{Fo}	Temperature ratio	Temperature (°C)
0	0	1.0	80
60	0.171	0.9	86
80	0.228	0.69	98.6
100	0.284	0.58	105.2
120	0.341	0.47	111.8
140	0.398	0.40	116.0
160	0.455	0.35	119.0

5. Before evaluation of lethality based on temperature history at the particle center, the largest thermal death time is computed:

$$F_{121} = 12 D_{121} = 12 \times 1.665 = 19.98\ s$$

6. Using the temperature history at the particle center, the lethal rate can be computed at 20 s intervals as illustrated in the following table.

$$LR = 10^{(T-121)/10}$$

Time (s)	Temperature (°C)	Lethal rate
0	80	–
60	86	0.000316
80	98.6	0.005754
100	105.2	0.026303
120	111.8	0.120226
140	116	0.316228
160	119	0.630957

7. *By recognizing that each increment of lethal rate represents a 20 s interval, the first 150 s (the end of the interval, with midpoint of 140 s and 116°C) contributes lethality of 9.37654 s. This value is obtained by summation of lethal rate and multiplied by 20 s.*
8. *By substituting the lethality from the target thermal death times,*

$$19.98 - 9.37654 = 10.60346\ \text{s}$$

is the lethality needed to be added during the time interval between 150 s and 170 s.

9. *Based on the lethal rate during the interval with midpoint at 160 s, the additional time becomes:*

$$\frac{10.60346}{0.630957} = 16.805\ \text{s}$$

10. *The total residence time needed for the process is*

$$150 + 16.805 = 166.805\ \text{s} \quad \text{or} \quad 2.78\ \text{min}$$

11. *Using the volumetric flow rate of 1.5 m^3/h and holding-tube diameter of 0.0475 m,*

$$\bar{u} = \frac{1.5}{\pi \times 0.02375^2} = 846\ \text{m/h} = 14.1\ \text{m/min}$$

12. *Based on laminar flow,*

$$u_{max} = \frac{\bar{u}}{0.5} = \frac{14.1}{0.5} = 28.2\ \text{m/min}$$

13. *Since the fastest moving particle will be at the maximum velocity, the particle velocity will be 28.2 m/min*
14. *Using the residence time of 2.78 min,*

$$\text{holding tube length} = 28.2 \times 2.78 = 78.4\ \text{m}$$

5.7 MATHEMATICAL METHODS

An obvious limitation to the General Method has been the indirect approach to establishing an operator time for commercial sterilization of a food container in a retort. This limitation was recognized and resulted in the development of the Formula Method by Ball (1923). The method used the equation of a heating curve as presented in Chapter 4, as follows:

$$\log(T_M - T) = -\frac{t}{f_h} + \log[j_h(T_M - T_i)] \tag{5.26}$$

Note that Equation (5.26) is valid after the unaccomplished temperature ratio becomes less than 0.7 as discussed in Chapter 4. However, this limitation is usually of less concern in thermal process calculations since lethality accumulation occurs only at longer time durations. Assume that the final desired temperature at a particular location in the food container (usually the slowest-heating point) is T_B. By defining the temperature difference at the end of product heating with a parameter g, as:

$$g = T_M - T_B \tag{5.27}$$

then

$$\log(g) = -\frac{t_B}{f_h} + \log[j_h(T_M - T_i)] \tag{5.28}$$

or

$$t_B = f_h \log\left[\frac{j_h(T_M - T_i)}{g}\right] \tag{5.29}$$

Equation (5.29) has been solved for the actual time (t_B) needed for a specific location in the product container to reach a desired temperature, T_B, as indicated by the parameter (g). Use of Equation (5.29) assumes that values of the heating rate constant (f_h) and the lag constant (j_h) are known for the product at the slowest-heating location in the container being processed.

Ball's Formula Method incorporates an analysis of an array of lethal rate curves for conditions existing during thermal sterilization conditions. The thermal death time, U, at the heating medium temperature, T_M, was defined as:

$$U = F_R 10^{(T_R - T_M)/z} \tag{5.30}$$

The results of the analysis of lethal rate curves was incorporated into charts of f_h/U versus log g charts, as illustrated in Figure 5.16. The chart is specific for microbial populations with $z = 10°C$ and a temperature difference between the heating medium and cooling medium ($T_M - T_{CM}$) of 100°C. The lethality data incorporated into the chart (Fig. 5.16) includes the lethality accumulated during cooling, as reflected by four different magnitudes of the lag constant (j_c) for product cooling at the slowest-heating location. The analysis also assumes that the cooling rate constant (f_c) has a magnitude equal to the heating rate constant (f_h). Charts for alternative values of z and

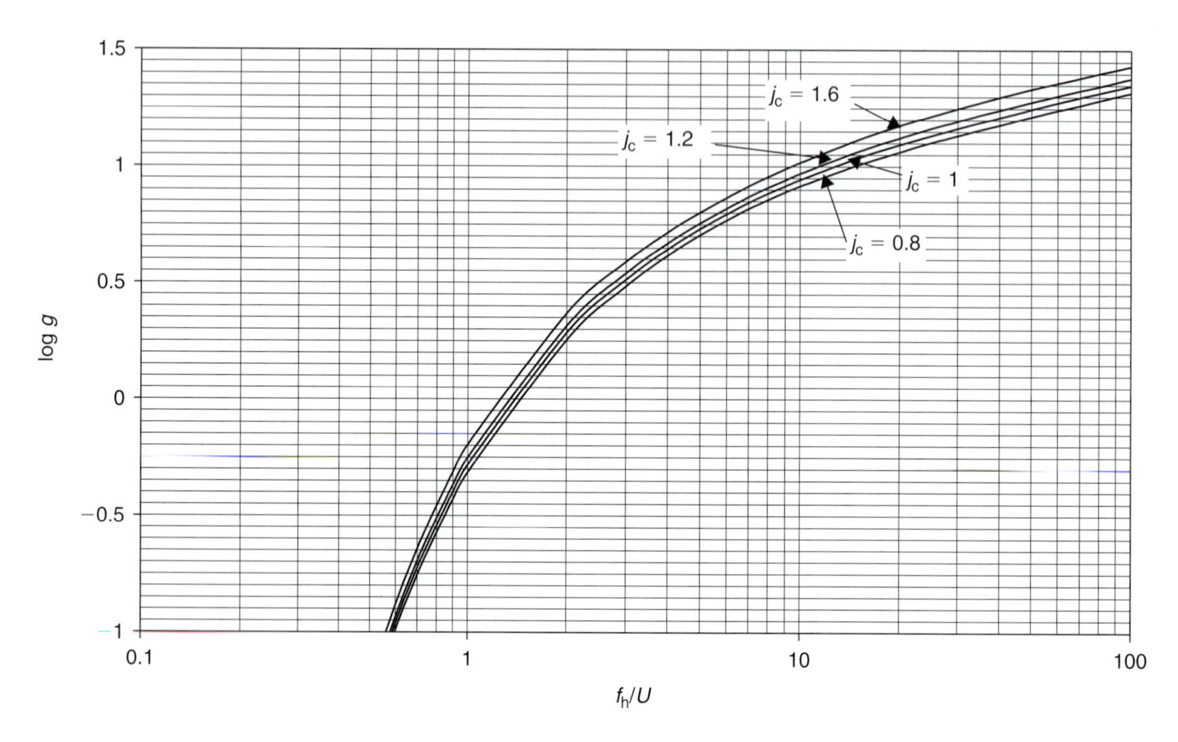

■ **Figure 5.16** A chart for evaluation of parameters in the mathematical method for process calculation.

$(T_M - T_{CM})$ have been developed, and should be used when conditions are appropriate. In addition, methods and charts/tables have been developed to account for situations when the cooling rate constant does not equal the heating rate constant.

Applications of mathematical methods require special attention to the conditions occurring during commercial sterilization of a food in a retort process. As indicated in Figure 5.14, a finite amount of time is required for a retort to reach and stabilize at the heating-medium temperature. This period of changing temperature in the heating medium has been defined as "come-up-time" (t_{cut}), and will influence the heating of the product, and require adjustments to the prediction of product temperature. Since the magnitude of the heating rate constant has been determined with a constant heating-medium temperature, the influence of t_{cut} on the predicted product temperature is to extend the time required for the slowest-heating location in the product to approach the heating-medium temperature. Based on experience, the influence of t_{cut} on a process evaluation can be expressed by the following relationship:

$$t_B = t_P + 0.42t_{cut} \tag{5.31}$$

Equation (5.31) provides a relationship between the actual process time (t_B) from Equation (5.29), and the operator time (t_P).

The mathematical method presented provides a direct approach to predicting the operator time for a retort sterilization process. After establishing the target thermal death time, based on microbial safety or acceptable spoilage rate, the thermal death time at the heating-medium temperature (U) is computed using Equation (5.30). The next step is to use the ratio (f_h/U) and the chart in Figure 5.16 to obtain log g. Equation (5.29) is then used to determine actual process time (t_B), given the value of g from Figure 5.16. Finally, the operator time (t_p) is computed using Equation (5.31).

Example 5.11

A conduction-heating food product is being thermally processed in a can; the slowest-heating location has $f_h = 40$ min, $j_h = 1.602$ and $j_c = 1.602$. The retort temperature is 125°C, the initial product temperature is 24°C, and the retort come-up-time is 5 minutes. The spoilage microorganism has a $D_{121} = 1.1$ min and $z = 10$°C. The acceptable spoilage rate has been established as one can per one million processed. The initial microbial population is 10^3 per can. Determine the operator time.

Given

$f_h = 40$ min

$j_h = 1.602$

$j_c = 1.602$

Retort temperature = 125°C

Initial product temperature = 24°C

Retort come-up-time = 5 min

$D_{121} = 1.1$ min

$z = 10$°C

Initial microbial load = 10^3 per can

Approach

After establishing the target thermal death time for the acceptable spoilage rate, the operator time will be determined using the mathematical method.

Solution

1. *Using Equation (5.18),*

$$\frac{1}{10^6} = \frac{10^3}{10^{F/1.1}}$$

Then $F_{121} = 9 \times 1.1 = 9.9$ *min.*

2. *In order to use Figure 5.16, the value of* f_h/U *must be determined at 125°C:*

$$U = F_{121} 10^{(T_R - T)/z}$$

$$U = 9.9 \times 10^{(121-125)/10} = 3.94 \text{ min}$$

$$\text{Then } f_h/U = 40/3.94 = 10.15$$

3. *From Figure 5.16,*

$$\log g = 1.02 \text{ at } f_h/U = 10.15$$

and

$$g = 10.47$$

4. *Using Equation (5.31),*

$$t_B = 40 \log\left[\frac{1.602(125 - 24)}{10.47}\right]$$

$$t_B = 47.56 \text{ min}$$

5. *Using Equation (5.31),*

$$t_P = 47.56 - 0.42(5) = 45.46 \text{ min}$$

5.7.1 Pouch Processing

The development of a process for pouch processing can be accomplished using either the General Method or mathematical methods. The unique aspects of pouch processing occur during commercial sterilization and the enhanced quality achieved when processing the product and pouch in the environment of a typical retort. The improvements in quality are achieved as a result of the ability to accomplish the desired process at the geometric center of the pouch without detrimental impacts to contents of the entire container.

Example 5.12

The process presented in Example 5.11 can be accomplished in a metal can or a pouch. Compare the impact of the process on retention of lycopene in concentrated tomato, with the following kinetic parameters: $k_{125} = 0.0189$/min, and $E_A = 88.7$ kJ/mole. The amount of product in the container is 1.125 kg, and the dimensions of the can are 8 cm diameter and 25 cm height. The pouch dimensions are 2 cm thickness and 25 cm length and width.

Approach

In order to compare the retention of lycopene from the same process in the two containers, the influence of the process on the entire product mass within the containers must be established. This can be estimated based on the average temperature achieved within the product during the process and the process time.

Solution

1. *As indicated in Example 5.11, the time required to accomplish the process in the can is 47.56 min.*
2. *To determine the average product temperature at the end of the process, the average temperature lag factor (j_m) (Fig. 4.42) is used in Equation (5.26):*

$$\log(125 - T_m) = -47.56/40 + \log[0.707(125 - 24)]$$
$$T_m = 120.34°C$$

Note that an infinite cylinder geometry has been used to describe the can geometry

$$[f_h = 40\ \text{min};\quad j_c = 1.602;\quad j_m = 0.707]$$

3. *Based on this average product temperature*

$$g = 125 - 120.34 = 4.66°C$$
$$\log g = 0.668$$

4. *Using Figure 5.16*

$$f_h/U = 4.7$$

then

$$U_{125} = 40/4.7 = 8.51\ \text{min}$$

5. *Based on the process time and temperature, the retention of lycopene for the process can be estimated*

$$\text{Retention} = 100e^{[-0.0189 \times 8.51]} = 0.851$$

6. *The calculations indicate 85.1% retention of the lycopene in the tomato concentrate in the can at the beginning of the process.*

7. *The relationship needed to provide an exact comparison of an infinite cylinder and an infinite plate geometry is as follows*

$$\alpha = 0.398\frac{d_c^2}{f_{hc}} = 0.933\frac{d_c^2}{f_{hp}}$$

where the thermal diffusivity of the product has been equated to the appropriate relationship for an infinite cylinder, and for an infinite plate.

8. *Based on the previous relationship*

$$f_{hp} = \frac{0.933 \times 1^2 \times 40}{0.398 \times 4^2} = 5.95 \text{ min}$$

to obtain the heating rate constant for an infinite plate equivalent to the same constant ($f_{hc} = 40$ min) for an infinite cylinder.

9. *The process for product in the pouch would be determined from Equation (5.30), starting with*

$$U_{125} = 3.94 \text{ min}$$

based on the process needed to achieve the target reduction in microbial population.

$$\text{Then } \frac{f_h}{U} = \frac{5.85}{3.94} = 1.485$$

and log g = 0.1 from Figure 5.16

$$g = 1.2589$$

then, using Equation (5.29) and $j_h = 1.273$ from Figure (4.41)

$$t_B = 5.85 \log\left[\frac{1.273(125-24)}{1.2589}\right]$$

$$t_B = 11.75 \text{ min}$$

This represents the actual processing time in order to achieve the desired reduction in microbial population.

10. *In order to estimate the retention of lycopene in tomato concentrate during processing in a pouch, using Equation (5.28),*

$$\log(125 - T_m) = -\frac{11.75}{5.85} + \log[0.808(125-24)]$$

where $j_m = 0.808$ for the infinite plate geometric

$$T_m = 124.2°C$$

11. *Using the average temperature,*

$$g = 125 - 124.2 = 0.8$$

and

$$\log g = -0.0968$$

12. *Using Figure 5.16,*

$$\frac{f_h}{U} = 1.25$$

and

$$U_{125} = \frac{5.85}{1.25} = 4.68 \text{ min}$$

13. *Based on the preceding, the retention of lycopene becomes*

$$\text{Retention} = 100e^{[-0.0189 \times 4.68]} = 0.915$$

14. *These calculations indicate that retention of lycopene is 91.5% during processing of tomato concentrate in a pouch.*

PROBLEMS

5.1 Determine the D value for the microorganism when given the following thermal resistance data for a spore suspension:

Time (min)	Number of survivors
0	10^6
15	2.9×10^5
30	8.4×10^4
45	2.4×10^4
60	6.9×10^3

Plot the survivor curve on regular and semilogarithmic coordinates.

5.2 The results of a thermal resistance experiment gave a D value of 7.5 minutes at 110°C. If there were 4.9×10^4 survivors at 10 minutes determine the ratio (N/N_0) for 5, 15, and 20 minutes.

5.3 Determine the z value for a microorganism that has the following decimal reduction times: $D_{110} = 6$ minutes, $D_{116} = 1.5$ minutes, $D_{121} = 0.35$ minutes, and $D_{127} = 0.09$ minutes.

5.4 An F_0 value of 7 minutes provides an acceptable economic spoilage for a given product. Determine the process time at 115°C.

5.5 If the z value of a microorganism is 16.5°C and D_{121} is 0.35 minute, what is D_{110}?

5.6 A thermal process is being used to reduce the microbial population on product contact surfaces of the packages needed for aseptic packaging of the liquid product. The spoilage microorganism has a $D_{121} = 10$ min and a $z = 8$°C. The microbial population on the surface is 100 per cm^2 and the product contact surface area is 8 cm^2 for each package. A 9D process has been recommended.

a. Determine the process time for the containers at 140°C.

b. Estimate the number of containers with spoilage based on survivors on the package surface when 10 million packages have been filled.

5.7 The following temperature–time data have been collected during UHT processing of a liquid food:

Time (s)	Temperature (°C)	Time (s)	Temperature (°C)
0	15	35	80
5	50	40	75
10	60	45	70
15	70	50	65
20	80	55	60
25	80	60	50
30	80	65	35

Estimate the lethality (F_{75}) for a microbial pathogen with $z = 5$°C. Determine the number of survivors of this process if the initial population is 3.5×10^5 and the $D_{75} = 0.1$ min.

5.8 A liquid food in a can is being processed in a retort at 125°C. The can dimensions are 4 cm in diameter and 5 cm in height. The product has a density of 1000 kg/m^3 and a specific heat of 3.9 kJ/kg°C. The overall heat transfer coefficient from steam to product is 500 W/m^2°C.

a. If the initial temperature of the product is 60°C, estimate the product temperature at 30 s intervals until the product reaches 120°C.

b. If the D_{121} for the spoilage microorganism is 0.7 min and the z is 10°C, estimate the lethality (F_{121}) for the heating portion of the process.

5.9 The product temperatures at the center of the particles carried through the holding tube of an aseptic processing system have been predicted as follows:

Time (s)	Temperature (°C)
0	85
5	91
10	103
15	110
20	116
25	120
30	123

The purpose of the thermal process is to reduce the population of a thermally resistant microbial spore with a D_{121} of 0.4 min and z of 9°C.

a. If the initial population is 100 per particle, estimate the probability of spore survivors from the aseptic process.

b. Estimate the retention of a quality attribute with D_{121} of 2 min and z of 20°C.

***5.10** A thermal process for a conduction heating product in a 307×407 can is being established. The initial product temperature is 20°C and the heating medium (retort) temperature is 125°C. The spoilage microorganism has a $D_{121} = 4$ min and $z = 10$°C. The retort come-up-time is 10 min, and the $f_h = 40$ min, $j_h = 1.7$, and $j_c = 1.2$.

a. If the initial microbial population is 2×10^5 per container, determine the target lethality (F) needed to ensure a spoilage rate of 1 can per 10 million processed.

b. Compute the operator time for the retort, and determine the temperature at the end of product heating.

***5.11** D values for *Salmonella* and *Listeria* in chicken breast meat are given in the following table.

Temperature (°C)	55.0	57.5	60.0	62.5	65.0	67.5	70.0
Salmonella D (min)	30.1	12.9	5.88	2.51	1.16	0.358	0.238
Listeria D (min)	50.8	27.2	5.02	2.42	1.71	0.400	0.187

Reference: *Murphy, R.Y. Marks, B.P., Johnson, E.R., and Johnson, M.G. (2000). Thermal inactivation kinetics of Salmonella and Listeria in ground chicken breast meat and liquid medium.* J. Food Science **65**(4): 706–710.

*Indicates an advanced level in solving.

5.11 a. Write a MATLAB® script to plot the log10(D) versus temperature. Using the basic curve fitting feature, determine the z values. Compare your results to Murphy et al. of $z = 6.53\,°C$ and $6.29\,°C$ for *Salmonella* and *Listeria*.

b. Calculate the k values for the D values for *Listeria* determined in part (a). Determine the activation energy, E_a, and Arrhenius constant, A. Compare your results to Murphy's of $E_a = 352.2$ (kJ/mol) and $A = 5.06 \times 10^{54}$ (1/min).

LIST OF SYMBOLS

α	thermal diffusivity (m^2/s)
B	Arrhenius constant
d_c	characteristic dimension (m)
D	decimal reduction time (s)
D'	time constant in Equation (5.8)
E_a	activation energy (kJ/kg)
f_h	heating rate factor (s)
F	thermal death time (s)
F_o	standard thermal death time at 121 °C and $z = 10\,°C$
F_R	thermal death time at reference temperature (s)
g	difference between temperature of heating medium and product at end of heating portion of process (°C)
h	convective heat transfer coefficient (W/[m^2 °C])
j_c	temperature lag factor for cooling
j_h	temperature lag factor for heating
k	reaction rate constant (s^{-1})
L	lethality (s)
n	order of reaction
N	microbial population
N_0	microbial population at time zero
N_{Re}	Reynolds number
N_{Pr}	Prandtl number
N_{Nu}	Nusselt number
P	pressure (kPa)
P_R	reference pressure (kPa)
r	number of containers processed
R_g	gas constant (kJ/[kg K])
t	time (s)
t_{cut}	come-up-time for retort (s)
t_B	operator time from mathematical method (s)
t_P	operator time (s)

T_B	temperature at end of heating portion of process (°C)
T_{cm}	temperature of cooling medium (°C)
T_i	initial temperature (°C)
T_m	medium temperature (°C)
T_R	reference temperature (°C)
T	temperature (°C)
T_A	absolute temperature (K)
$\bar{u}$	mean velocity (m/s)
V	activation volume constant
z	thermal resistance factor (°C)

BIBLIOGRAPHY

Ball, C. O. (1923). Thermal process time for canned foods. *Bull. Natl. Res. Council,* **7**: Part 1 (37) 76.

Ball, C. O. (1936). Apparatus for a method of canning. U.S. Patent 2,020,303.

Ball, C. O. and Olson, F. C. W. (1957). *Sterilization in Food Technology*. McGraw-Hill, New York.

Barbosa-Canovas, G. V., Pothakumury, U. R., Palou, E., and Swanson, B. G. (1998). *Nonthermal Preservation of Foods*. Marcel Dekker, New York.

Bigelow, W. D., Bohart, G. S., Richardson, A. C., and Ball, C. O. (1920). Heat penetration in processing canned foods. *National Canners Association Bulletin*, No. 16L.

Chandarana, D. and Gavin, A., III (1989). Establishing thermal processes for heterogeneous foods to be processed aseptically: A theoretical comparison of process development methods. *J. Food Sci.* **54**(1): 198–204.

Chandarana, D., Gavin, A., III, and Wheaton, F. W. (1989). Simulation of parameters for modeling aseptic processing of foods containing particulates. *Food Technol.* **43**(3): 137–143.

Chandarana, D. I., Gavin, A., III, and Wheaton, F. W. (1990). Particle/fluid interface heat transfer under UHT conditions at low particle/fluid relative velocities. *J. Food Process Eng.* **13**: 191–206.

Danckwerts, P. V. (1953). Continuous flow systems. *Chem. Eng. Sci.* **2**(1).

deRuyter, P. W. and Burnet, R. (1973). Estimation of process conditions for continuous sterilization of food containing particulates. *Food Technol.* **27**(7): 44.

Dignan, D. M., Barry, M. R., Pflug, I. J., and Gardine, T. D. (1989). Safety considerations in establishing aseptic processes for low-acid foods containing particulates. *Food Technol.* **43**(3): 118–121.

Dixon, M. S., Warshall, R. B., Crerar, J. B. (1963). Food processing method and apparatus. U.S. Patent No. 3,096,161.

Heppell, N. J. (1985). Measurement of the liquid–solid heat transfer coefficient during continuous sterilization of foodstuffs containing particles. *Proceedings of Symposium of Aseptic Processing and Packing of Foods*. Tylosand, Sweden, Sept. 9–12.

Lopez, A. (1969). *A Complete Course in Canning*, 9th ed. The Canning Trade, Baltimore, Maryland.

Manson, J. E. and Cullen, J. F. (1974). Thermal process simulation for aseptic processing of foods containing discrete particulate matter. *J. Food Sci.* **39**: 1084.

Martin, W. M. (1948). Flash process, aseptic fill, are used in new canning unit. *Food Ind.* **20**: 832–836.

Mertens, B. and Desplace, G. (1993). Engineering aspects of high pressure technology in the food industry. *Food Technol.* **47**(6): 164–169.

McCoy, S. C., Zuritz, C. A., Sastry, S. K. (1987). Residence time distribution of simulated food particles in a holding tube. ASAE Paper No. 87-6536. American Society of Agricultural Engineers, St. Joseph, Michigan.

Mitchell, E. L. (1989). A review of aseptic processing. *Adv. Food Res.* **32**: 1–37.

Murphy, R. Y., Marks, B. P., Johnson, E. R., and Johnson, M. G. (2000). Thermal inactivation kinetics of Salmonella and Listeria in ground chicken breast meat and liquid medium. *J. Food Sci.* **65**(4): 706–710.

Mussa, D. M., Ramaswamy, H. S., and Smith, J. P. (1999). High pressure destruction kinetics of *Listeria Monocytogenes* in pork. *J. Food Protection*. **62**(1):40–45.

NFPA (1980). *Laboratory Manual for Food Canners and Processors*. AVI Publishing Co, Westport, Connecticut.

Palmer, J. and Jones, V. (1976). Prediction of holding times for continuous thermal processing of power law fluids. *J. Food Sci.* **41**(5): 1233.

Ranz, W. E. and Marshall, W. R. Jr. (1952). Evaporation from drops. *Chem. Eng. Prog.* **48**: 141–180.

Rumsey, T. R. (1982). Energy use in food blanching. In *Energy in Food Processing*, R. P. Singh, ed. Elsevier Science Publishers, Amsterdam.

Sastry, S. (1986). Mathematical evaluation of process schedules for aseptic processing of low-acid foods containing discrete particulates. *J. Food Sci.* **51**: 1323.

Sastry, S. K., Heskitt, B. F., and Blaisdell, J. L. (1989). Experimental and modeling studies on convective heat transfer at the particle–liquid interface in aseptic processing systems. *Food Technol.* 43(3): 132–136.

Sizer, C. (1982). Aseptic system and European experience. *Proc. Annu. Short Course Food Ind., 22nd,* 93–100. University of Florida, Gainesville.

Stumbo, C. R. (1973). *Thermobacteriology in Food Processing*. Academic Press, New York.

Teixiera, A. A. (2007). Thermal processing of canned foods. Ch. 11. In *Handbook of Food Engineering*, D. R. Heldman and D. B. Lund, eds. CRC Press, Taylor & Francis Group, Boca Raton, Florida.

Torpfl, S., Heinz, V., and Knorr, D. (2005). Overview of pulsed electric field processing for food. In *Emerging Technologies for Food Processing*, Da-Wen Sun, ed. Elsevier Academic Press, London.

Weisser, H. (1972). *Untersuchungen zum Warmeubergang im Kratzkuhler*. Ph.D. thesis, Karlsruhe Universitat, Germany.

Zuritz, C. A., McCoy, S., and Sastry, S. K. (1987). Convective heat transfer coefficients for non-Newtonian flow past food-shaped particulates. ASAE Paper No. 87-6538. American Society of Agricultural Engineers, St. Joseph, Michigan.

Chapter 6

Refrigeration

Temperature plays an important role in maintaining the quality of stored food products. Lowering the temperature retards the rates of reactions that cause quality deterioration. It is generally agreed that the reaction rate is reduced by half by lowering the temperature by 10°C.

In earlier days, a lower temperature was obtained by the use of ice. Ice was allowed to melt in an insulated chamber that contained food products (Fig. 6.1). During melting, ice requires latent heat (333.2 kJ/kg) to be converted from the solid phase to liquid water. This heat was extracted from the product that was kept next to ice in an insulated chamber.

Figure 6.1 An ice box.

Today, the cooling process is achieved by the use of a mechanical refrigeration system. Refrigeration systems allow transfer of heat from the cooling chamber to a location where the heat can easily be discarded. The transfer of heat is accomplished by using a refrigerant, which like water changes state—from liquid to vapor. Unlike water, a refrigerant has a much lower boiling point. For example, ammonia, a commonly used refrigerant in industrial plants, has a boiling point of −33.3°C. This is a much lower temperature compared with 100°C, the boiling point of water at atmospheric pressure. Similar to water, ammonia needs latent heat to change its phase from liquid to gas at its boiling point. The boiling point of a refrigerant can be varied by changing the pressure. Thus, to increase the boiling point of ammonia to 0°C, its pressure must be raised to 430.43 kPa.

A very simple refrigeration system that utilizes a refrigerant is shown in Figure 6.2. The only drawback in this illustration is the onetime use of the refrigerant. Because refrigerants are expensive, they must

 Professor R. Paul Singh, Department of Biological and Agricultural Engineering, University of California, Davis, CA 95616, USA. Email: rps@rpaulsingh.com.

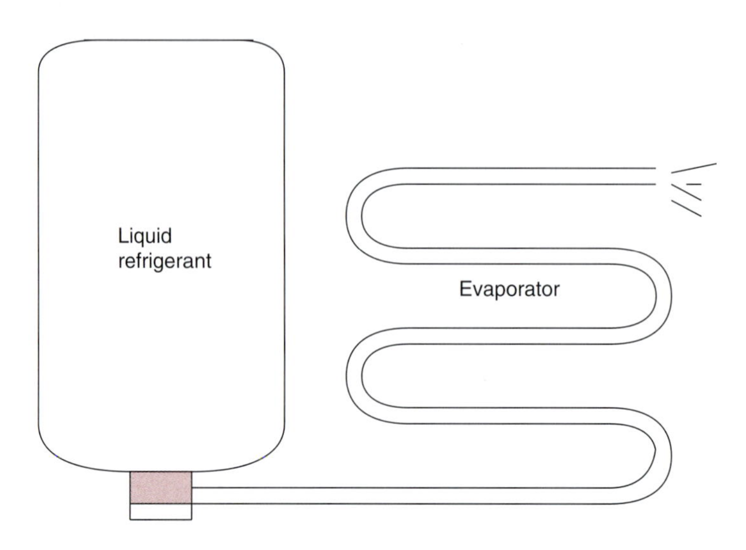

■ **Figure 6.2** Use of liquid refrigerant to accomplish refrigeration.

be reused. Thus, the system must be modified to allow collection of the refrigerant vapors and their conversion to liquid state so that the same refrigerant can be used repetitively. This is accomplished with the use of a mechanical vapor-compression system.

Before we discuss a mechanical vapor-compression system, it is necessary to examine the properties of refrigerants. The selection of a refrigerant is based on suitable properties at a desired temperature. After examining the properties of refrigerants, we will consider various components of a mechanical vapor-compression system. We will use a thermodynamic chart for a refrigerant to examine a mechanical vapor-compression refrigeration system. The concluding section in this chapter will include several application problems where mathematical expressions will be used to design simple refrigeration systems.

6.1 SELECTION OF A REFRIGERANT

A wide variety of refrigerants are commercially available for use in vapor-compression systems. Selection of a refrigerant is based on several performance characteristics that assist in determining the refrigerant's suitability for a given system. The following is a list of important characteristics that are usually considered:

1. *Latent heat of vaporization.* A high latent heat of vaporization is preferred. For a given capacity, a high value of latent heat

of vaporization indicates that a smaller amount of refrigerant will be circulated per unit of time.

2. *Condensing pressure*. Excessively high condensing pressure requires considerable expenditure on heavy construction of condenser and piping.
3. *Freezing temperature*. The freezing temperature of the refrigerant should be below the evaporator temperature.
4. *Critical temperature*. The refrigerant should have sufficiently high critical temperature. At temperatures above the critical temperature, the refrigerant vapor cannot be liquefied. Particularly in the case of air-cooled condensers, the critical temperature should be above the highest ambient temperature expected.
5. *Toxicity*. In many applications, including air conditioning systems, the refrigerant must be nontoxic.
6. *Flammability*. The refrigerant should be nonflammable.
7. *Corrosiveness*. The refrigerant should not be corrosive to the materials used in the construction of the refrigeration system.
8. *Chemical stability*. The refrigerant must be chemically stable.
9. *Detection of leaks*. If a leak develops in the refrigeration system, the detection of such a leak should be easy.
10. *Cost*. Low-cost refrigerant is preferred in industrial applications.
11. *Environmental impact*. The refrigerant released from the refrigeration systems due to leaks should not cause environmental damage.

Table 6.1 includes properties and performance characteristics of some commonly used refrigerants. The performance characteristics are given at $-15°C$ (5°F) and 30°C (86°F), the evaporator and condenser temperatures, respectively. These temperatures are used as standard conditions to make comparisons of refrigerants by the American Society of Heating, Refrigeration, and Air Conditioning Engineers (ASHRAE).

Ammonia offers an exceptionally high latent heat of vaporization among all other refrigerants. It is noncorrosive to iron and steel but corrodes copper, brass, and bronze. It is irritating to mucous membranes and eyes. It can be toxic at concentrations of 0.5% by volume in air. A leak in the refrigeration system that uses ammonia as a refrigerant can easily be detected either by smell or by burning sulfur candles and noting white smoke created by the ammonia vapors.

Standard designations for refrigerants are based on ANSI/ ASHRAE Standard 34-1978. Some of the commonly used refrigerants and their standard designations are listed in Table 6.2.

Table 6.1 Comparison between Commonly Used Refrigerants (Performance Based on −15°C Evaporator Temperature and 30°C Condenser Temperature)

Chemical formula	**Freon 12 (dichlorodifluoro-methane, CCl_2F_2)**	**Freon 22 (monochlorodi-fluoro-methane, $CHClF_2$)**	**HFC 134a (CH_2FCF_3)**	**Ammonia (NH_3)**
Molecular weight	120.9	86.5	102.3	17.0
Boiling point (°C) at 101.3 kPa	−29.8	−40.8	−26.16	−33.3
Evaporator pressure at −15°C (kPa)	182.7	296.4	164.0	236.5
Condensing pressure at 30°C (kPa)	744.6	1203.0	770.1	1166.5
Freezing point (°C) at 101.3 kPa	−157.8	−160.0	−96.6	−77.8
Critical temperature (°C)	112.2	96.1	101.1	132.8
Critical pressure (kPa)	4115.7	4936.1	4060	11423.4
Compressor discharge temperature (°C)	37.8	55.0	43	98.9
Compression ratio (30°C/−15°C)	4.07	5.06	4.81	4.94
Latent heat of vaporization at −15°C (kJ/kg)	161.7	217.7	209.5	1314.2
Horsepower/ton refrigerant, ideal	1.002	1.011	1.03	0.989
Refrigerant circulated/ton refrigeration (kg/s), ideal	2.8×10^{-2}	2.1×10^{-2}	2.38×10^{-2}	0.31×10^{-2}
Compressor displacement/ton refrigeration (m^3/s)	2.7×10^{-3}	1.7×10^{-3}	2.2×10^{-3}	1.6×10^{-3}
Stability (toxic decomposition products)	Yes	Yes	No (at high temperatures)	No
Flammability	None	None	None	Yes
Odor	Ethereal	Ethereal	Ethereal	Acrid

A number of refrigerants used in commercial practice are halocarbons, although their use is being severely curtailed as described later in this section. Refrigerant-12, also called Freon 12, is a dichlorodifluoromethane. Its latent heat of vaporization is low compared with ammonia (R-717); therefore, considerably more weight of the refrigerant must be circulated to achieve the same refrigeration capacity.

Table 6.2 Standard Designations of Refrigerants

Refrigerant number	Chemical name	Chemical formula
Halocarbons		
12	Dichlorodifluoromethane	CCl_2F_2
22	Chlorodifluoromethane	$CHClF_2$
30	Methylene Chloride	CH_2Cl_2
114	Dichlorotetrafluoroethane	$CClF_2CClF_2$
134a	1,1,1,2-tetrafluoroethane	CH_2FCF_3
Azeotropes	R-22/R-115	$CHClF_2/CClF_2CF_3$
502		
Inorganic compound	Ammonia	NH_3
717		

Refrigerant-22 (monochlorodifluoromethane) is useful with low-temperature applications (−40 to −87°C). Use of Refrigerant-22, which has a low specific volume, can result in greater heat removal than with Refrigerant-12 for a compressor with the same size piston.

During the mid-1970s, it was first postulated that chlorofluorocarbons (CFCs), because of their extremely stable characteristics, have a long life in the lower atmosphere and they migrate to the upper atmosphere over a period of time. In the upper atmosphere, the chlorine portion of the CFC molecule is split off by the sun's ultraviolet radiation, and it reacts with ozone, resulting in the depletion of ozone concentration. Ozone depletion in the upper atmosphere permits more of the sun's harmful ultraviolet radiation to reach the earth's surface. In the early 1990s, increasing concern was directed at the role of CFCs in damaging the protective ozone layer surrounding the planet. Many of the commonly used refrigerants have been fully halogenated chlorofluorocarbons (CFCs) containing chlorine. Alternatives to the CFCs are hydrofluorocarbons (HFCs) and hydrochlorofluorocarbons (HCFCs). Hydrogen-containing fluorocarbons have weak carbon–hydrogen bonds, which are more susceptible to cleavage; thus they are postulated to have shorter lifetimes.

The Montreal Protocol (agreed upon on September 16, 1987) with its amendments has provided the framework to control substances that lead to stratospheric ozone depletion. A major result of this international agreement has been the phase out of CFCs, HCFCs, Halons, and methyl bromide. Developing countries are still allowed production of these substances but they are also required to phase out CFC production and import by 2010. In examples in this chapter, we will use the thermodynamic data for R-12 (Freon), R-717 (ammonia), and HFC 134a to solve typical problems involving design and operation of refrigerated facilities.

6.2 COMPONENTS OF A REFRIGERATION SYSTEM

Major components of a simple mechanical vapor compression refrigeration system are shown in Figure 6.3. As the refrigerant flows through these components its phase changes from liquid to gas and then back to liquid. The flow of refrigerant can be examined by tracing the path of the refrigerant in Figure 6.3.

At location D on Figure 6.3, just prior to the entrance to the expansion valve, the refrigerant is in a saturated liquid state. It is at or below its condensation temperature. The expansion valve separates the high-pressure region from the low-pressure region. After passing through the expansion valve, the refrigerant experiences a drop in pressure accompanied by a drop in temperature. Due to the drop in pressure,

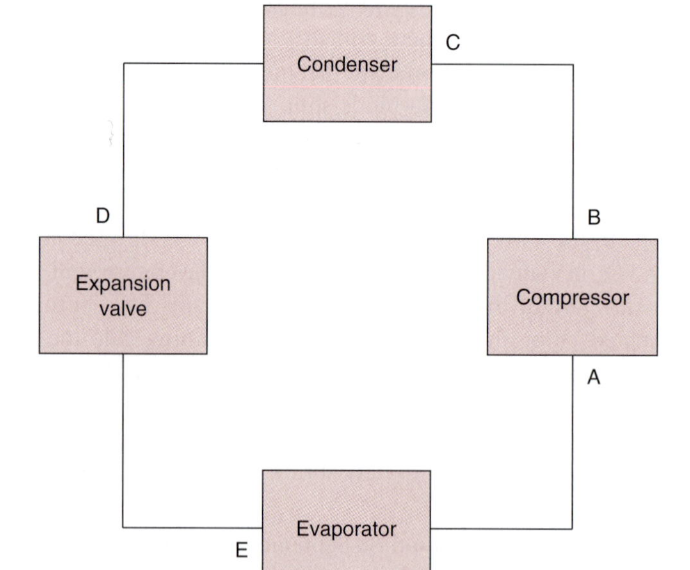

W

Figure 6.3 A mechanical vapor-compression refrigeration system.

some of the liquid refrigerant changes to gas. The liquid/gas mixture leaving the expansion valve is termed "flash gas."

The liquid/gas mixture enters the evaporator coils at location E. In the evaporator, the refrigerant completely vaporizes to gas by accepting heat from the media surrounding the evaporator coils. The saturated vapors may reach a superheated stage due to gain of additional heat from the surroundings.

The saturated or superheated vapors enter the compressor at location A, where the refrigerant is compressed to a high pressure. This high pressure must be below the critical pressure of the refrigerant and high enough to allow condensation of the refrigerant at a temperature slightly higher than that of commonly available heat sinks, such as ambient air or well water. Inside the compressor, the compression process of the vapors occurs at constant entropy (called an isentropic process). As the pressure of the refrigerant increases, the temperature increases, and the refrigerant becomes superheated as shown by location B.

The superheated vapors are then conveyed to a condenser. Using either an air-cooled or a water-cooled condenser, the refrigerant discharges heat to the surrounding media. The refrigerant condenses back to the liquid state in the condenser as shown by location D. After the entire amount of refrigerant has been converted to saturated liquid, the temperature of the refrigerant may decrease below that of its condensation temperature due to additional heat discharged to the surrounding media; in other words, it may be subcooled. The subcooled or saturated liquid then enters the expansion valve and the cycle continues.

6.2.1 Evaporator

Inside the evaporator, the liquid refrigerant vaporizes to a gaseous state. The change of state requires latent heat, which is extracted from the surroundings.

Based on their use, evaporators can be classified into two categories. *Direct-expansion* evaporators allow the refrigerant to vaporize inside the evaporator coils; the coils are in direct contact with the object or fluid being refrigerated. *Indirect-expansion* evaporators involve the use of a carrier medium, such as water or brine, which is cooled by the refrigerant vaporizing in the evaporator coils. The cooled carrier medium is then pumped to the object that is being refrigerated. The indirect-expansion evaporators require additional equipment. They are useful when cooling is desired at several locations in the system. Water may be used as a carrier medium if the temperature stays above freezing. For lower temperatures, brine (a proper concentration of $CaCl_2$) or glycols, such as ethylene or propylene glycol, are commonly used.

The evaporators are either bare-pipe, finned-tube, or plate type, as shown in Figure 6.4. Bare-pipe evaporators are most simple, easy to defrost and clean. The fins added to the finned-tube evaporators allow increase in surface area, thus increasing the rate of heat transfer. The plate evaporators allow an indirect contact between the product (e.g., a liquid food) to be cooled and the refrigerant.

Evaporators can also be classified as direct-expansion and flooded types. In the direct-expansion type of evaporators, there is no recirculation of the refrigerant within the evaporator. The liquid refrigerant changes to gas as it is conveyed through a continuous tube. In contrast, the flooded evaporator allows recirculation of liquid refrigerant. The liquid refrigerant, after going through the metering device, enters a surge chamber. As shown in Figure 6.5, the liquid refrigerant boils

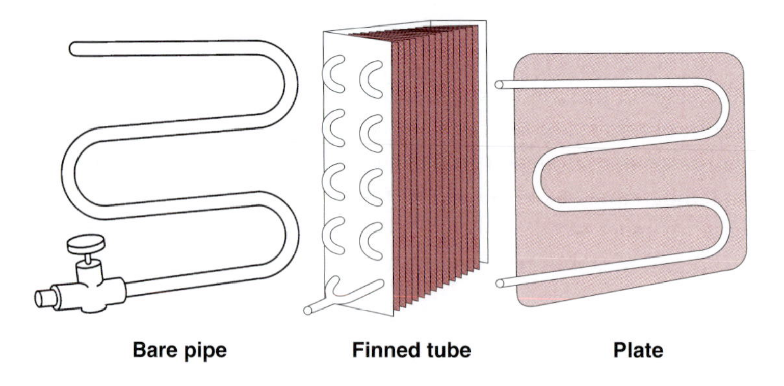

■ **Figure 6.4** Different types of evaporator coils. (Courtesy of Carrier Co.)

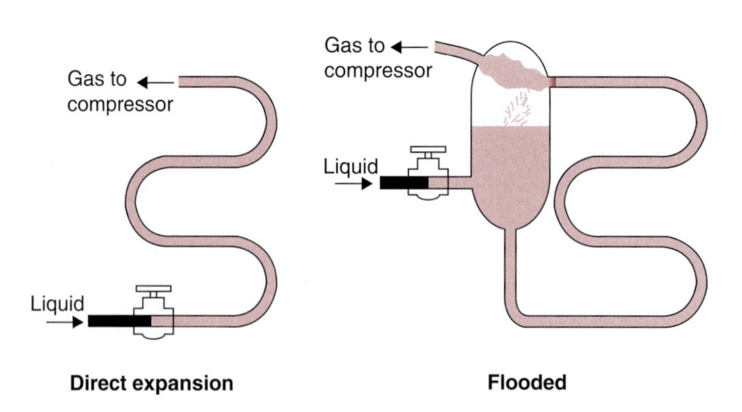

■ **Figure 6.5** A direct-expansion evaporator and a flooded-type evaporator. (Courtesy of Carrier Co.)

in the evaporator coil and extracts heat from the surroundings. The liquid refrigerant is recirculated through the surge tank and the evaporator coil. The refrigerant gas leaves the surge tank for the compressor.

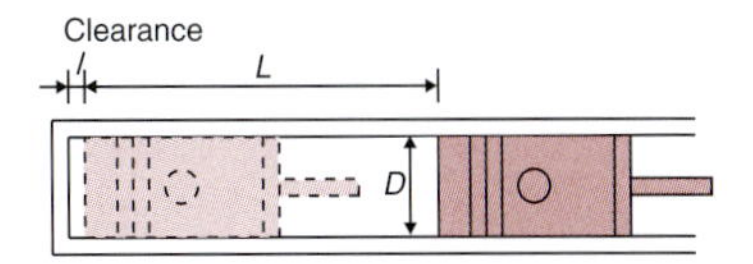

■ **Figure 6.6** Operation of a cylinder.

6.2.2 Compressor

The refrigerant enters the compressor in a vapor state at low pressure and temperature. The compressor raises the pressure and temperature of the refrigerant. It is due to this action of the compressor that heat can be discharged by the refrigerant in the condenser. The compression processes raise the temperature of the refrigerant sufficiently above the ambient temperature surrounding the condenser, so that the temperature gradient between the refrigerant and the ambient promotes the heat flow from the refrigerant to the ambient.

The three common types of compressors are reciprocating, centrifugal, and rotary. As is evident from the name, the reciprocating compressor contains a piston that travels back and forth in a cylinder (as shown in Fig. 6.6). Reciprocating compressors are most commonly used and vary in capacity from a fraction of a ton to 100 tons of refrigeration per unit (for definition of ton of refrigeration, see Section 6.4.1). The centrifugal compressor contains an impeller with several blades that turn at high speed. The rotary compressor involves a vane that rotates inside a cylinder.

The compressor may be operated with an electric motor or an internal combustion engine. Figure 6.7 shows a typical installation of a reciprocating compressor operated with an electric motor.

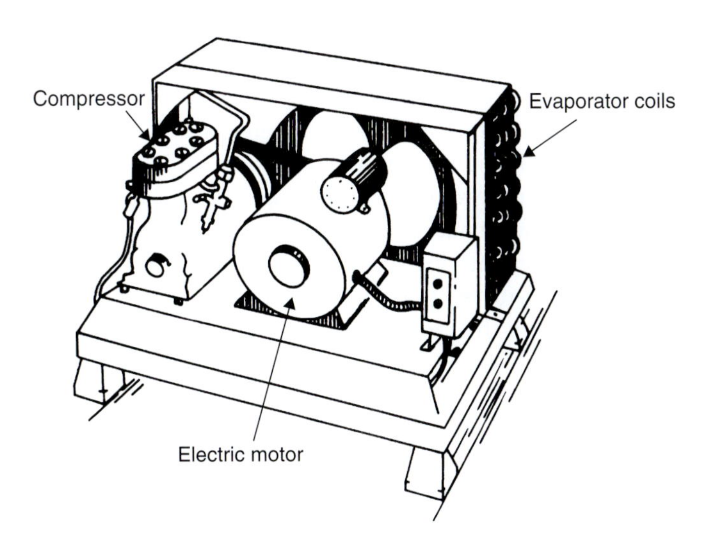

■ **Figure 6.7** A typical compression refrigeration system of a two-cylinder, air-cooled condenser driven by an electric motor.

An important parameter that influences the performance of a compressor is the compressor capacity. The compressor capacity is affected by several factors. Factors that are inherent to the design of equipment include (a) piston displacement, (b) clearance between the piston head and the end of the cylinder when the piston is at the top of its stroke, and (c) size of the suction and discharge valves. Other factors that influence the compressor capacity are associated with the operating conditions. These factors include (a) revolutions per minute, (b) type of refrigerant, (c) suction pressure, and (d) discharge pressure.

The piston displacement can be computed from the following equation:

$$\text{Piston displacement} = \frac{\pi D^2 L N}{4}$$

where D is the diameter of cylinder bore (cm), L is the length of stroke (cm), and N is the number of cylinders.

The compressor displacement can be computed from the following equation:

$$\text{compressor displacement} = (\text{piston displacement})(\text{revolutions per minute})$$

The piston displacement can be examined by considering Figure 6.8. Point A refers to the cylinder volume of 100% with the vapors at the suction pressure. The vapors are compressed to about 15% of cylinder volume, shown by point B. During the compression cycle, both the suction valve and discharge valve remain closed. During the discharge cycle BC, the discharge valve opens and the gas is released. The cylinder volume is decreased to 5%. This value represents the volume between the head of the piston and the end of the cylinder. As the piston starts traveling in the opposite direction, during the expansion cycle CD, the high-pressure gas remaining in the clearance expands. Attempts are always made to keep the clearance as small as practically possible, since the clearance space represents a loss in capacity. Process DA represents suction; the suction valve remains open during this process and the vapors are admitted to the cylinder. The piston retracts to the location indicated by point A. The cycle is then repeated.

In actual practice, the suction and discharge pressure lines (BC and DA) are not straight lines, due to the pressure drop across the valves. The discharge line lies above, and the suction line lies below the

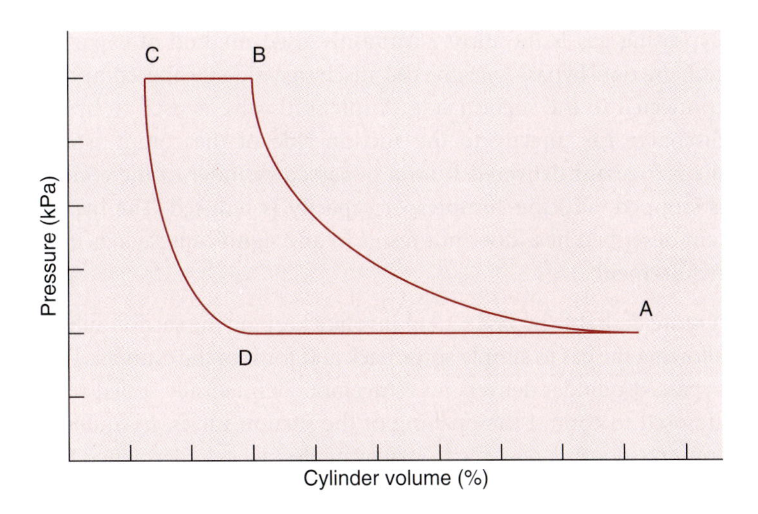

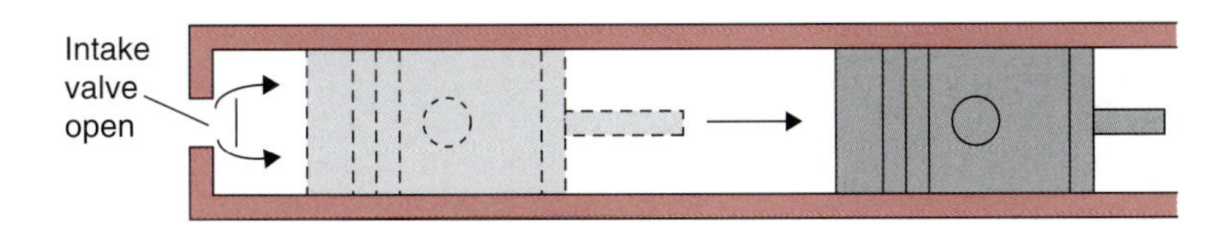

Figure 6.8 A complete cycle of a reciprocating compressor. (Courtesy of Carrier Co.)

theoretical straight lines indicated in Figure 6.8. This difference can be explained by the fact that at the beginning of discharge, the cylinder pressure must be higher than the discharge pressure to force the valve open against the spring pressure. Similarly, at the beginning of the suction process, the pressure must be lower than the suction pressure to open the suction valve.

It is often necessary to control the compressor capacity, since the refrigeration loads are seldom constant. Thus, the compressor is operated mostly at partial loads compared with the refrigeration load used in the design of a compressor. The compressor capacity can be controlled by (a) controlling the speed (revolutions per minute), (b) bypassing gas from the high-pressure side to the low-pressure side of the compressor, and (c) internal bypassing of gas in a compressor, by keeping the suction valve open.

The speed can be controlled by the use of a variable-speed electric motor.

Bypassing gas is the most commonly used method of capacity control. In one bypass system, the discharge side of the compressor is connected to the suction side. A solenoid valve is used to bypass the discharge gas directly to the suction side of the compressor. Thus, the refrigerant delivered from a bypassed cylinder of the compressor is stopped, and the compressor capacity is reduced. The bypass system described here does not result in any significant savings in power requirement.

A more desirable bypass system involves keeping the suction valve open, allowing the gas to simply surge back and forth in the cylinder. Thus, the bypassed cylinder delivers no refrigerant. Hermetically sealed solenoids are used to control the opening of the suction valves. In multicylinder compressors, it is possible to inactivate desired cylinders when the load is low. For a four-cylinder compressor with three cylinders inoperative, the compressor capacity can be reduced by as much as 75%.

6.2.3 Condenser

The function of the condenser in a refrigeration system is to transfer heat from the refrigerant to another medium, such as air and/or water. By rejecting heat, the gaseous refrigerant condenses to liquid inside the condenser.

The major types of condensers used are (1) water-cooled, (2) air-cooled, and (3) evaporative. In evaporative condensers, both air and water are used.

Three common types of water-cooled condensers are (1) double pipe, (2) shell and tube (as shown in Fig. 6.9), and (3) shell and coil.

In a double-pipe condenser, water is pumped in the inner pipe and the refrigerant flows in the outer pipe. Countercurrent flows are maintained to obtain high heat-transfer efficiencies. Although double-pipe condensers commonly have been used in the past, the large number of gaskets and flanges used in these heat exchangers leads to maintenance problems.

In a shell-and-tube condenser, water is pumped through the pipes while refrigerant flows in the shell. Installations of fins in pipes allows better heat transfer. The shell-and-tube condensers are generally low in cost and easy to maintain.

In a shell-and-coil condenser, a welded shell contains a coil of finned water tubing. It is generally most compact and low in cost.

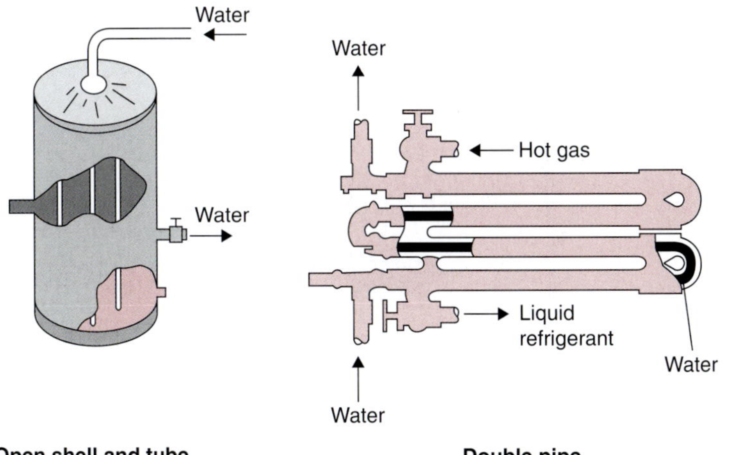

■ **Figure 6.9** An open shell-and-tube condenser and double-pipe condenser. (Courtesy of Carrier Co.)

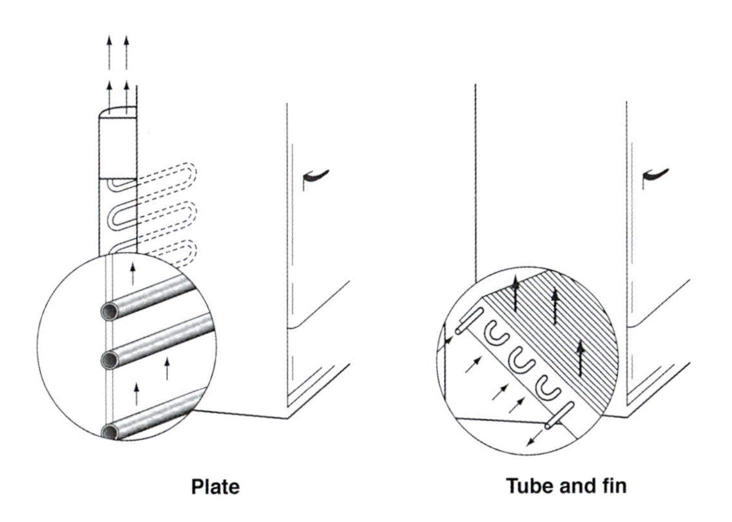

■ **Figure 6.10** A plate and tube-and-fin condenser. (Courtesy of Carrier Co.)

Air-cooled condensers can be either tube-and-fin type or plate type, as shown in Figure 6.10. Fins on tubes allow a large heat transfer area in a compact case. The plate condensers have no fins, so they require considerably larger surface areas. However, they are cheaper to construct and require little maintenance. Both these types of condensers can be found in household refrigerators.

Air-cooled condensers can also employ artificial movement of air by using a fan. The fan helps in obtaining higher convective heat-transfer coefficients at the surface of the condenser.

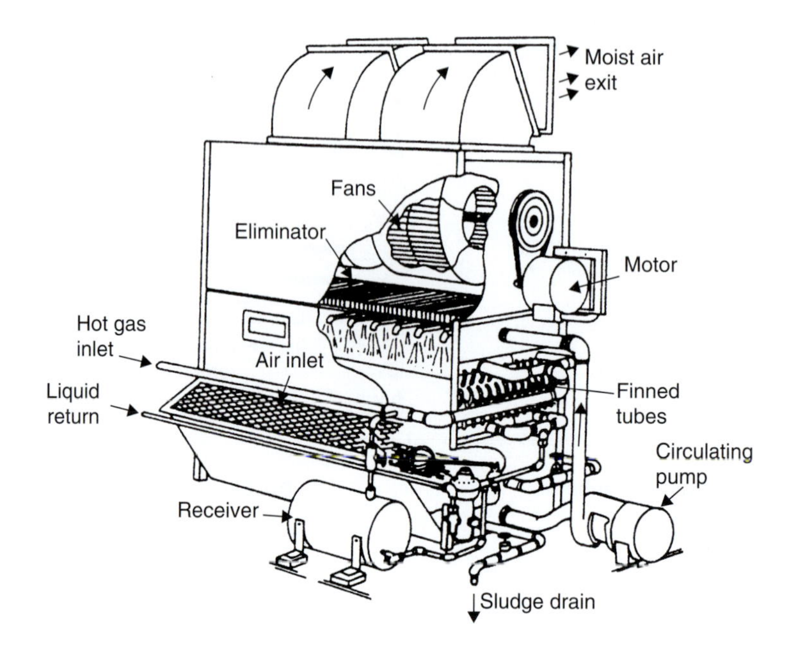

■ **Figure 6.11** An evaporative condenser. (From Jennings, 1970. Copyright © 1939, 1944, 1949, 1956, 1958, 1970 by Harper and Row, Publishers, Inc. Reprinted with permission of the publisher.)

In evaporative condensers, a circulating water pump draws water from a pan at the base of the condenser and sprays the water onto the coils. In addition, a large amount of air is drawn over the condenser coils. Evaporation of water requires latent heat, which is extracted from the refrigerant. Figure 6.11 shows an evaporative condenser. These units can be quite large.

6.2.4 Expansion Valve

An expansion valve is essentially a metering device that controls the flow of liquid refrigerant to an evaporator. The valve can be operated either manually or by sensing pressure or temperature at another desired location in the refrigeration system.

The common type of metering devices used in the refrigeration system include (1) manually operated expansion valve, (2) automatic low-side float valve, (3) automatic high-side float valve, (4) automatic expansion valve, and (5) thermostatic expansion valve.

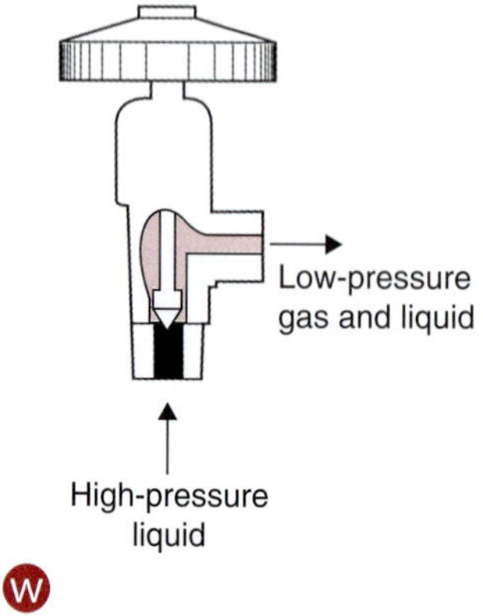

■ **Figure 6.12** A manually operated expansion valve. (Courtesy of Carrier Co.)

A simple, manually operated expansion valve is shown in Figure 6.12. The valve, manually adjusted, allows a desired amount of flow of refrigerant from the high-pressure liquid side to the low-pressure gas/liquid side. The refrigerant cools as it passes through the valve. The heat given up by the liquid refrigerant is absorbed to convert some of

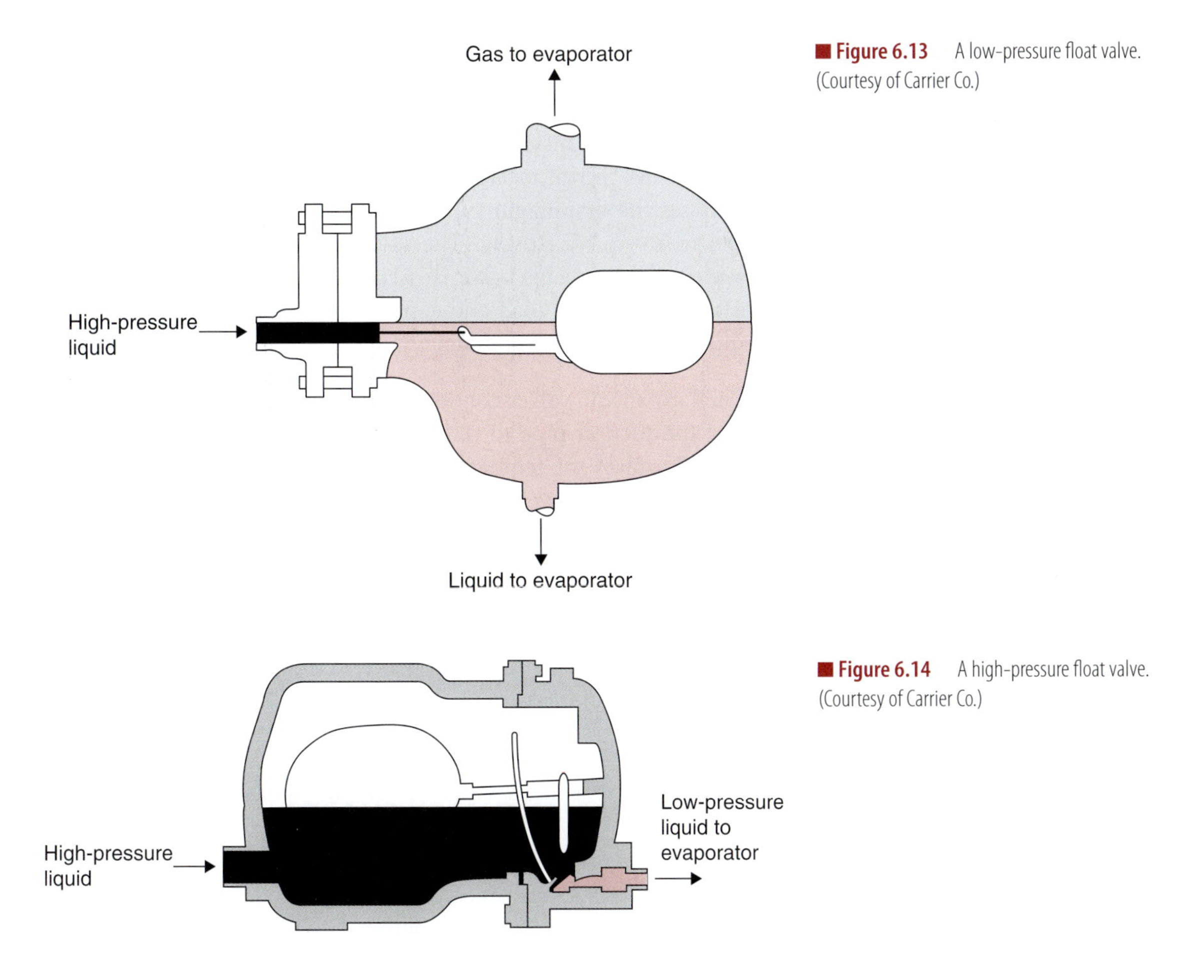

Figure 6.13 A low-pressure float valve. (Courtesy of Carrier Co.)

Figure 6.14 A high-pressure float valve. (Courtesy of Carrier Co.)

the liquid into vapor. This partial conversion of the liquid refrigerant to gas as it passes through the expansion valve is called *flashing*.

The automatic low-pressure float valve is used in a flooded evaporator. The float ball is located on the low-pressure side of the system, as shown in Figure 6.13. As more liquid is boiled away in the evaporator, the float ball drops and opens the orifice to admit more liquid from the high-pressure side. The orifice closes as the float rises. This type of expansion valve is simple, almost trouble-free, and provides excellent control.

In an automatic high-pressure float valve, the float is immersed in high-pressure liquid (Fig. 6.14). As the heated gas is condensed in the condenser, the liquid-refrigerant level rises inside the chamber. The

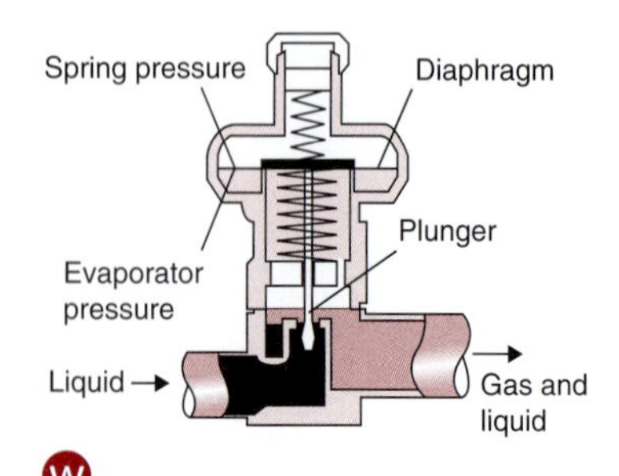

W

■ Figure 6.15 An automatic expansion valve. (Courtesy of Carrier Co.)

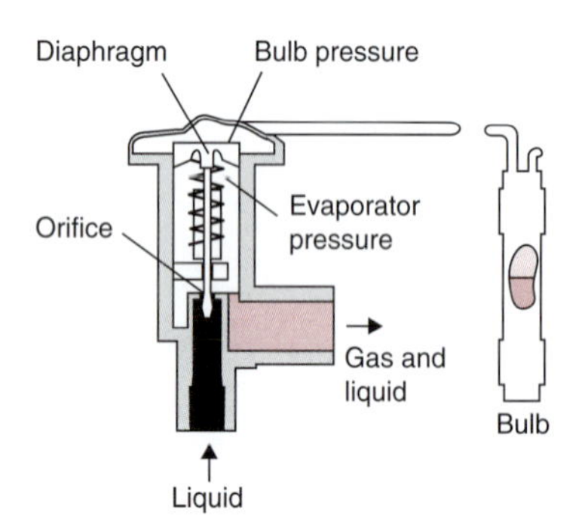

■ Figure 6.16 A thermostatic expansion valve. (Courtesy of Carrier Co.)

float consequently rises and opens the orifice, allowing the refrigerant to flow to the evaporator.

The automatic expansion valve maintains a constant pressure in the evaporator. As shown in Figure 6.15, an increase in evaporator pressure causes the diaphragm to rise against the spring pressure, which results in the valve closing. The valve opens when the evaporator pressure decreases. This valve is used in applications that require a constant refrigeration load and constant evaporator temperature—for example, in a household refrigerator.

Thermal expansion valves contain a thermostatic bulb clamped to the side of the suction pipe to the compressor (Fig. 6.16). The thermostatic bulb senses the temperature of the superheated gas leaving the evaporator. The relatively high temperature of the thermostatic bulb causes the fluid in the bulb (usually the same refrigerant) to increase in pressure. The increased pressure is transmitted via the thermostatic tube to the bellows and the diaphragm chamber. The valve consequently opens to allow more liquid refrigerant to flow through. Thermostatic valves are the most widely used of all metering devices in the refrigeration industry.

6.3 PRESSURE–ENTHALPY CHARTS

Both pressure and enthalpy of the refrigerant change as the refrigerant is conveyed through various components of a refrigeration system. In both the evaporator and the condenser, the enthalpy of the refrigerant changes and the pressure remains constant. During the compression step, work is done by the compressor, resulting in an increase in the enthalpy of the refrigerant along with an increase in pressure. The expansion valve is a constant-enthalpy process that allows the liquid refrigerant under high pressure to pass at a controlled rate into the low-pressure section of the refrigeration system.

Charts or diagrams have been used extensively in the literature to present thermodynamic properties of refrigerants. These charts are particularly useful during the early, conceptual stages of a refrigeration system design. Looking at a chart, we can easily comprehend a standard process, as well as any deviations from the standard. Most commonly used charts depict enthalpy and pressure values on the x and y axes, respectively. Another type of chart involves entropy and temperature values plotted along x and y axes, respectively. The entire refrigeration cycle comprising evaporator, compressor, condenser, and expansion valve can be conveniently depicted on the pressure–enthalpy charts. Figure A.6.1 (in the appendix) is a pressure–enthalpy chart for Freon R-12 refrigerant. In this chart, which conforms to the specifications of the International Institute of Refrigeration (IIR), the value of the enthalpy of saturated liquid is assumed to be 200 kJ/kg at a chosen datum temperature of 0°C. Similar charts can be obtained for other refrigerants from their manufacturers. Charts conforming to the specifications of the American Society of Heating and Refrigerating and Air Conditioning Engineers (ASHRAE) use different reference enthalpy values (ASHRAE, 2005).

A skeleton description of the pressure–enthalpy chart is given in Figure 6.17. Pressure (kPa) is plotted on a logarithmic scale on the vertical axis. The horizontal axis gives enthalpy (kJ/kg).

The pressure–enthalpy chart may be divided into different regions, based on saturated liquid and saturated vapor curves. In the sketch shown in Figure 6.17, the area enclosed by the bell-shaped curve

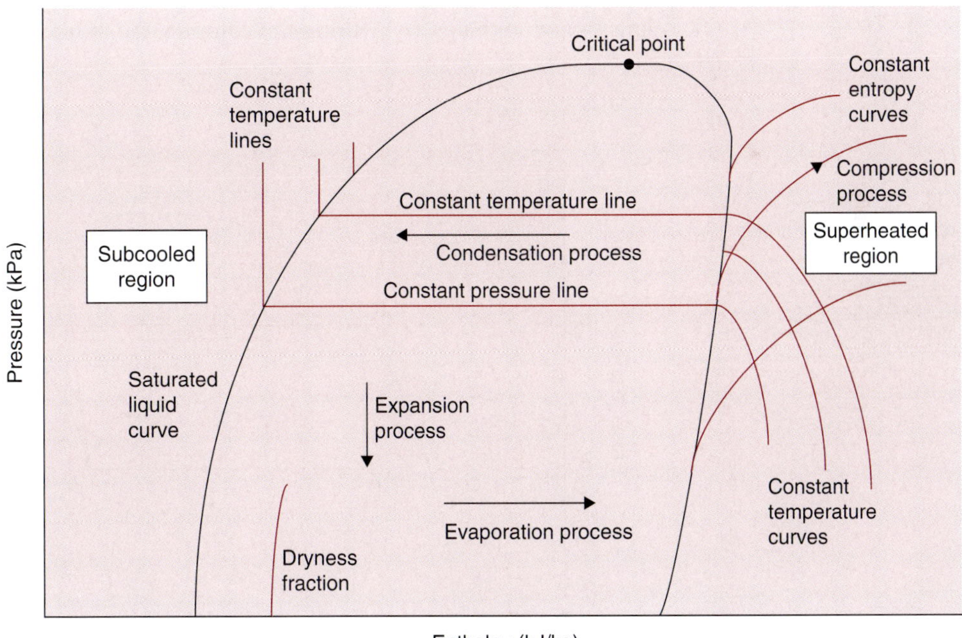

Figure 6.17 A pressure–enthalpy diagram.

represents a two-phase region containing a mixture of both liquid and vapor refrigerant. The horizontal lines extending across the chart are constant-pressure lines. The temperature lines are horizontal within the bell-shaped area, vertical in the subcooled liquid region, and skewed downward in the superheated region. The area on the left-hand side of the saturated liquid curve denotes subcooled liquid refrigerant with temperatures below the saturation temperature for a corresponding pressure. The area to the right-hand side of the dry saturated vapor curve depicts the region where the refrigerant vapors are at superheated temperatures above the saturation temperature of vapor at the corresponding pressure. Within the bell-shaped curve, the dryness fraction curves are useful in determining the liquid and vapor content of the refrigerant.

Let us consider a simple vapor-compression refrigeration system, where the refrigerant enters the expansion valve as saturated liquid and leaves the evaporator as saturated vapor. Such a system is shown on a pressure–enthalpy diagram in Figure 6.18.

As dry saturated vapors enter the compressor, the condition of refrigerant is represented by location A. The refrigerant vapors are at pressure P_1 and enthalpy H_2. During the compression stroke, the vapors

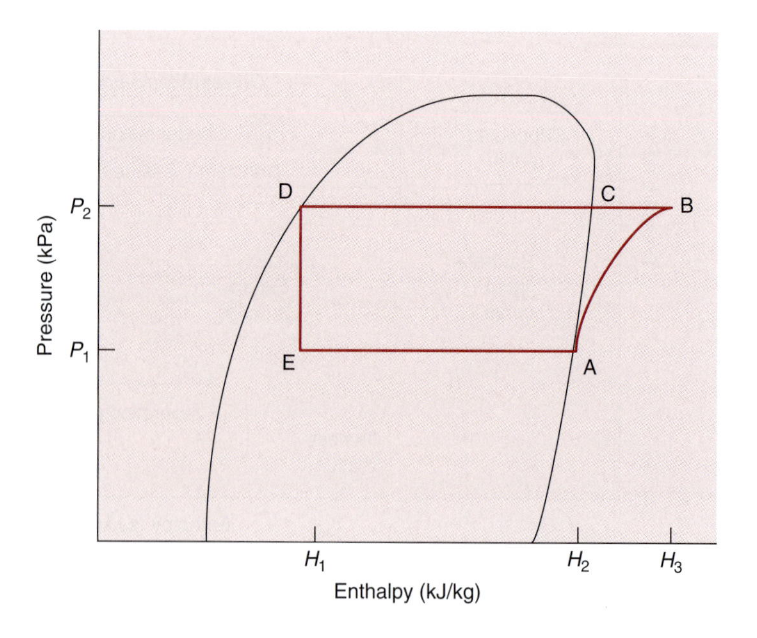

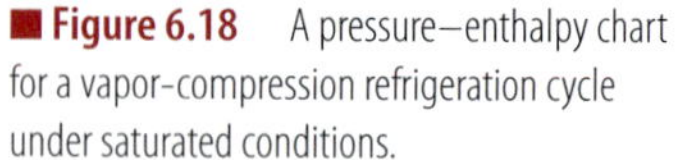

Figure 6.18 A pressure–enthalpy chart for a vapor-compression refrigeration cycle under saturated conditions.

are compressed isoentropically (or at constant entropy) to pressure P_2. Location B is in the superheated vapor region. The enthalpy of the refrigerant increases from H_2 to H_3 during the compression process. In the condenser, first the superheat is removed in the desuperheater section of the condenser, and then the latent heat of condensation is removed from C to D. The saturated liquid enters the expansion valve at location D. As refrigerant moves through the expansion valve, the pressure drops to P_1 while the enthalpy remains constant at H_1. Some flashing of the refrigerant occurs within the expansion valve; as a result, location E indicates refrigerant containing liquid as well as vapor. The liquid–vapor mix of refrigerant accepts heat in the evaporator and converts completely to the vapor phase. The evaporator section is represented by the horizontal line from location E to A; the pressure remains constant at P_1 and the enthalpy of the refrigerant increases from H_1 to H_2.

In actual practice, deviations from the cycle just discussed may be observed. For example, it is common to encounter a refrigeration cycle as shown in Figure 6.19. To prevent a refrigerant in liquid state from entering the compressor, the refrigerant is allowed to convert completely into the saturated vapor state inside the evaporator coils before reaching the exit location. Once the refrigerant

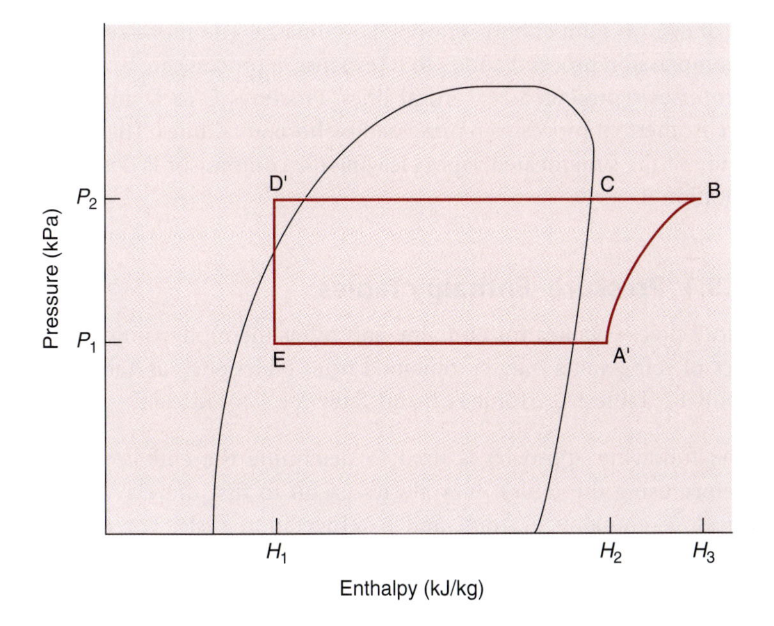

W

Figure 6.19 A pressure–enthalpy chart for a vapor-compression refrigeration cycle with deviations.

is converted into vapors, and if those vapors are still inside the evaporator coils, the refrigerant will gain additional heat from its surroundings due to the temperature gradient. Thus, by the time the refrigerant vapors enter the compressor suction, they are superheated, while pressure P_1 remains the same as in the evaporator as shown by location A′.

Another deviation from the ideal cycle involves subcooled refrigerant. The refrigerant may be subcooled in a receiver tank between the condenser and the expansion valve. Subcooled refrigerant remains at pressure P_2, the same as in the condenser. Another factor causing subcooling is the heat loss from the refrigerant that is completely converted to saturated liquid but still inside the condenser coils. The subcooled refrigerant is represented by location D′ on the pressure–enthalpy chart.

These pressure–enthalpy charts, such as those shown in Figure A.6.1, are useful in obtaining values of H_1, H_2, and H_3. Since the superheated region is crowded with a number of curves, an expanded portion for that region is more convenient to use to draw process A to B, and obtain a value for H_3 (Figure A.6.2 for R-12 and Figure A.6.3 for R-717).

As noted previously, temperature–entropy charts can also be used to represent the refrigeration cycle. In Figure 6.20 a refrigeration cycle is drawn on temperature–entropy coordinates. The processes A to B (compression process) and D to E (expansion process) are isoentropic processes represented by vertical lines. Processes C to D and E to A are isothermal processes represented by horizontal lines. The temperature of the superheated vapors leaving the compressor is denoted by location B.

6.3.1 Pressure–Enthalpy Tables

More precise values for enthalpy and other thermodynamic properties of refrigerants may be obtained from tables such as Table A.6.1 for R-12, Table A.6.2 for R-717, and Table A.6.3 for R-134a.

The following approach is used to determine the enthalpy values. Before using the tables, it is always useful to first draw a sketch of pressure–enthalpy diagram and a refrigeration cycle. For example, in Figure 6.21 a refrigeration cycle depicting an evaporator temperature of $-20°C$ and a condenser temperature of $+30°C$ is shown.

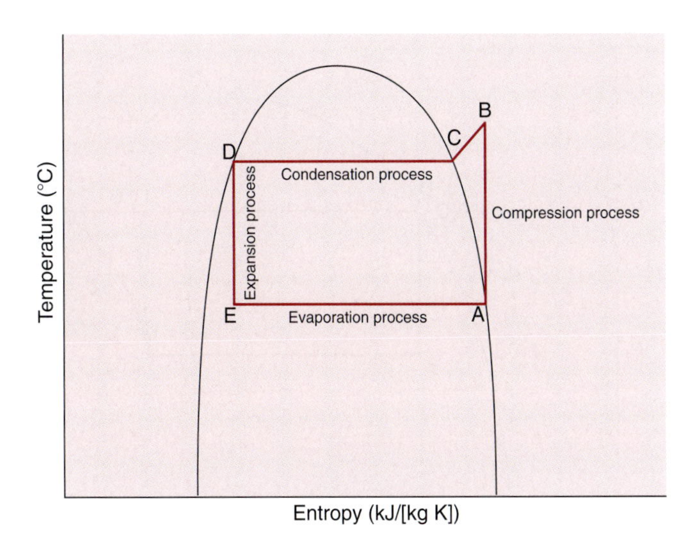

Figure 6.20 A temperature–entropy diagram.

Assume that the refrigerant being used is ammonia. Since location A represents saturated vapor conditions, from Table A.6.2 we can find at −20°C the enthalpy of the refrigerant in saturated vapor state is 1437.23 kJ/kg. Thus, H_2 is 1437.23 kJ/kg. At location D, the refrigerant is in saturated liquid state at the condenser temperature. Thus, from Table A.6.2, at 30°C, the enthalpy of ammonia refrigerant under saturated liquid conditions is 341.76 kJ/kg. Thus, H_1 is 341.76 kJ/kg. To determine the enthalpy value H_3, although tables for superheated properties at various conditions are available, they are more cumbersome to use as they require several interpolations. It is more convenient to use the expanded chart for the superheated region: for example, Figure A.6.3 is useful to determine the value for H_3 as 1710 kJ/kg.

6.3.2 Use of Computer-Aided Procedures to Determine Thermodynamic Properties of Refrigerants

Another approach to determine the thermodynamic properties of refrigerants is to use computer-aided methods with empirical correlations. Cleland (1986) has provided such empirical correlations

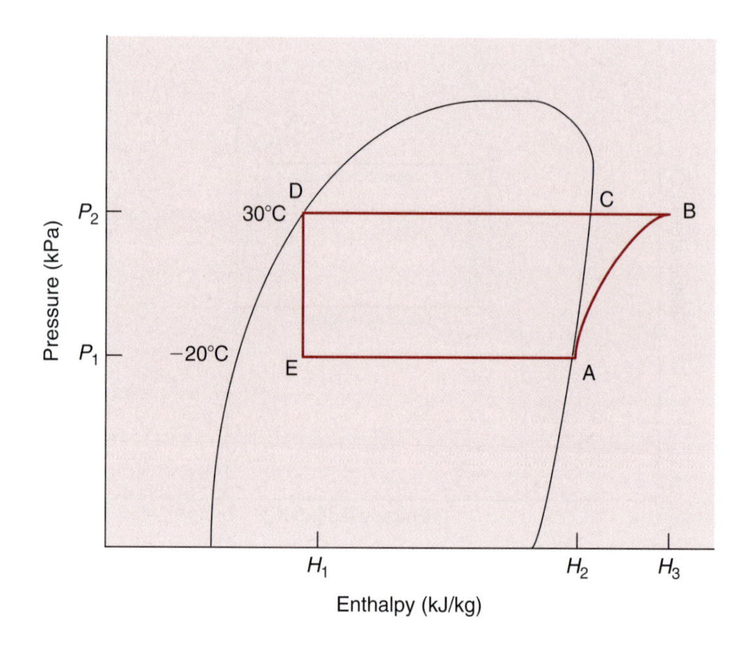

Figure 6.21 A pressure–enthalpy chart for evaporator temperature of −20°C and condenser temperature of 30°C.

for several commonly used refrigerants. Some selected equations and their coefficients for refrigerants R-12, R-22, R-134a, and R-717 follow.

Vapor pressure

$$P_{sat} = \exp(a_1 + a_2/(T_{sat} + a_3)) \quad (6.1)$$

where P_{sat} is saturation pressure and T_{sat} is saturation temperature.

Saturation temperature

$$T_{sat} = a_2/[\ln(P_{sat}) - a_1] - a_3 \quad (6.2)$$

Liquid enthalpy

$$\Delta T_b = T_{sat} - T_L \quad (6.3)$$

$$H_L = a_4 + a_5 T_L + a_6 T_L^2 + a_7 T_L^3 \quad (6.4)$$

where T_b is the temperature of subcooled liquid, T_L is the temperature of liquid refrigerant, and H_L is the enthalpy of liquid refrigerant.

Saturated vapor enthalpy

$$H_{i1} = a_8 + a_9 T_{sat} + a_{10} T_{sat}^2 + a_{11} T_{sat}^3 \tag{6.5}$$

$$H_v = H_{i1} + a_{12} \tag{6.6}$$

where H_{i1} is an intermediate enthalpy value and H_v is the enthalpy of saturated vapor.

Superheated vapor enthalpy

$$\Delta T_s = T_s - T_{sat} \tag{6.7}$$

$$\begin{aligned} H_{i2} = H_{i1}(1 &+ a_{13}\Delta T_s + a_{14}(\Delta T_s)^2 + a_{15}(\Delta T_s)(T_{sat}) \\ &+ a_{16}(\Delta T_s)^2(T_{sat}) + a_{17}(\Delta T_s)(T_{sat})^2 \\ &+ a_{18}(\Delta T_s)^2(T_{sat})^2) \end{aligned} \tag{6.8}$$

$$H_s = H_{i2} + a_{12} \tag{6.9}$$

where T_s is the temperature of superheated vapors, H_{i2} is an intermediate enthalpy value, and H_s is the enthalpy of superheated vapors.

Saturated vapor specific volume

$$v_v = \exp[a_{19} + a_{20}/(T_{sat} + 273.15)](a_{21} + a_{22}T_{sat} + a_{23}T_{sat}^2 + a_{24}T_{sat}^3) \tag{6.10}$$

where v_v is the specific volume of saturated vapors.

Superheated vapor specific volume

$$\begin{aligned} v_s = v_v(1.0 &+ a_{25}(\Delta T_s) + a_{26}(\Delta T_s)^2 + a_{27}(\Delta T_s)(T_{sat}) \\ &+ a_{28}(\Delta T_s)^2(T_{sat}) + a_{29}(\Delta T_s)(T_{sat})^2 + a_{30}(\Delta T_s)^2(T_{sat})^2) \end{aligned} \tag{6.11}$$

where v_s is the specific volume of superheated vapors.

Enthalpy change in isentropic compression with no vapor superheat at suction

$$\Delta h = \frac{c}{c-1} P_1 v_1 \left[\left(\frac{P_2}{P_1} \right)^{(c-1)/c} - 1 \right] \tag{6.12}$$

where c is an empirical coefficient, P is absolute pressure, v is specific volume, and subscripts 1 and 2 represent conditions at suction and discharge of the compressor.

$$\Delta T_c = T_{sat2} - T_{sat1} \tag{6.13}$$

where ΔT_c is the change in saturation temperature due to compression.

$$\begin{aligned} c_{i1} &= a_{31} + a_{32}(T_{sat1}) + a_{33}(T_{sat1})^2 + a_{34}(T_{sat1})(\Delta T_c) \\ &\quad + a_{35}(T_{sat1})^2(\Delta T_c) + a_{36}(T_{sat1})(\Delta T_c)^2 \\ &\quad + a_{37}(T_{sat1})^2(\Delta T_c) + a_{38}(\Delta T_c) \end{aligned} \tag{6.14}$$

$$c = c_{i1} \tag{6.15}$$

The coefficients for these equations are given in Table 6.3. More details about the applicable range and accuracy of these equations are given in Cleland (1986).

6.4 MATHEMATICAL EXPRESSIONS USEFUL IN ANALYSIS OF VAPOR-COMPRESSION REFRIGERATION

6.4.1 Cooling Load

The cooling load is the rate of heat energy removal from a given space (or object) in order to lower the temperature of that space (or object) to a desired level. Before the era of mechanical refrigeration, ice was the cooling medium most widely used. Cooling capacity was often related to melting of ice. A typical unit for cooling load still commonly used in commercial practice is *ton of refrigeration*. One ton of refrigeration is equivalent to the latent heat of fusion of one ton of ice (2000 pounds $\times$ 144 Btu/pound)/24 h = 288,000 Btu/24 h = 303,852 kJ/24 h = 3.5168 kW. Thus, a mechanical refrigeration system that has the capacity to absorb heat from the refrigerated space at the rate of 3.5168 kW is rated at *one ton of refrigeration*.

Several factors should be considered in calculating the cooling load from a given space. If the product stored in the space is a fresh fruit or vegetable, it gives out heat due to respiration. This heat of respiration must be removed to keep the product and the space at a low temperature. Heat of respiration of fresh fruits and vegetables are tabulated in Table A.2.6. Other factors affecting the cooling load calculations include heat infiltration through walls, floor, and ceiling; heat gain

Table 6.3 Coefficients for Empirical Equations to Calculate Thermodynamic Properties of Refrigerants

Coefficient	R-12	R-22	R-717	R-134a
a_1	20.82963	21.25384	22.11874	21.51297
a_2	−2033.5646	−2025.4518	−2233.8226	−2200.981
a_3	248.3	248.94	244.2	246.61
a_4	200,000	200,000	200,000	100,000
a_5	923.88	1170.36	4751.63	1335.29
a_6	0.83716	1.68674	2.04493	1.7065
a_7 ($\times 10^{-3}$)	5.3772	5.2703	−37.875	7.6741
a_8	187,565	250,027	1,441,467	249,455
a_9	428.992	367.265	920.154	606.163
a_{10}	−0.75152	−1.84133	−10.20556	−1.50644
a_{11} ($\times 10^{-3}$)	5.6695	−11.4556	−26.5126	−18.2426
a_{12}	163,994	155,482	15689	299,048
a_{13} ($\times 10^{-3}$)	3.43263	2.85446	1.68973	3.48186
a_{14} ($\times 10^{-7}$)	7.27473	4.0129	−3.47675	16.886
a_{15} ($\times 10^{-6}$)	7.27759	13.3612	8.55525	9.2642
a_{16} ($\times 10^{-8}$)	−6.63650	−8.11617	−3.04755	−7.698
a_{17} ($\times 10^{-8}$)	6.95693	14.1194	9.79201	17.07
a_{18} ($\times 10^{-10}$)	−4.17264	−9.53294	−3.62549	−12.13
a_{19}	−11.58643	−11.82344	−11.09867	−12.4539
a_{20}	2372.495	2390.321	2691.680	2669
a_{21}	1.00755	1.01859	0.99675	1.01357
a_{22} ($\times 10^{-4}$)	4.94025	5.09433	4.02288	10.6736
a_{23} ($\times 10^{-6}$)	−6.04777	−14.8464	2.64170	-9.2532
a_{24} ($\times 10^{-7}$)	−2.29472	-2.49547	-1.75152	-3.2192
a_{25} ($\times 10^{-3}$)	4.99659	5.23275	4.77321	4.7881
a_{26} ($\times 10^{-6}$)	−5.11093	−5.59394	−3.11142	−3.965
a_{27} ($\times 10^{-5}$)	2.04917	3.45555	1.58632	2.5817
a_{28} ($\times 10^{-7}$)	−1.51970	−2.31649	−0.91676	−1.8506
a_{29} ($\times 10^{-7}$)	3.64536	5.80303	2.97255	8.5739
a_{30} ($\times 10^{-9}$)	−1.67593	−3.20189	−0.86668	−5.401
a_{31}	1.086089	1.137423	1.325798	1.06469
a_{32} ($\times 10^{-3}$)	−1.81486	−1.50914	0.24520	-1.6907
a_{33} ($\times 10^{-6}$)	−14.8704	−5.59643	3.10683	-8.56
a_{34} ($\times 10^{-6}$)	2.20685	-8.74677	−11.3335	−21.35
a_{35} ($\times 10^{-7}$)	1.97069	-1.49547	−1.42736	−6.173
a_{36} ($\times 10^{-8}$)	−7.86500	5.97029	6.35817	20.74
a_{37} ($\times 10^{-9}$)	−1.96889	1.41458	0.95979	7.72
a_{38} ($\times 10^{-4}$)	−5.62656	−4.52580	−3.82295	−6.103

through doors; heat given by lights, people, and use of fork lifts for material handling.

Example 6.1

Calculate the cooling load present in a walk-in chamber, caused by the heat evolution of 2000 kg of cabbage stored at 5°C.

Given

Amount of cabbage stored = 2000 kg
Storage temperature = 5°C

Approach

We will use Table A.2.6 to obtain the value of heat evolution (due to respiration) of cabbage.

Solution

1. *From Table A.2.6, heat evolution for cabbage stored at 5°C is 28–63 W/Mg.*
2. *Choose the larger value of 63 W/Mg for design purposes.*
3. *Total heat evolution for 2000 kg of cabbage is*

$$(2000\ kg)(63\ W/Mg) \times \left(\frac{1\ Mg}{1000\ kg} \right) = 126\ W$$

4. *The cooling load rate due to 2000 kg of cabbage stored at 5°C is 126 W.*

6.4.2 Compressor

The work done on the refrigerant during the isoentropic compression step can be calculated from the enthalpy rise of the refrigerant and the refrigerant flow rate.

$$q_w = \dot{m}\,(H_3 - H_2) \tag{6.16}$$

where $\dot{m}$ is refrigerant mass flow rate (kg/s), H_3 is enthalpy of refrigerant at the end of compression stroke (kJ/kg refrigerant), H_2 is enthalpy of refrigerant at the beginning of compression stroke (kJ/kg refrigerant), and q_w is rate of work done on the refrigerant (kW).

6.4.3 Condenser

Within the condenser, the refrigerant is cooled at constant pressure. The heat rejected to the environment can be expressed as

$$q_c = \dot{m}(H_3 - H_1) \tag{6.17}$$

where q_c is rate of heat exchanged in the condenser (kW) and H_1 is enthalpy of refrigerant at exit from the condenser (kJ/kg refrigerant).

6.4.4 Evaporator

Within the evaporator the refrigerant changes phase from liquid to vapor and accepts heat from the surroundings at a constant pressure. The enthalpy difference of the refrigerant between the inlet and the outlet locations of an evaporator is called the *refrigeration effect*. The rate of heat accepted by the refrigerant as it undergoes evaporation process in the evaporator is given by

$$q_c = \dot{m}(H_2 - H_1) \tag{6.18}$$

where q_e is the rate of heat exchanged in the evaporator (kW), and the refrigeration effect is $H_2 - H_1$.

6.4.5 Coefficient of Performance

The purpose of a mechanical refrigeration system is to transfer heat from a low-temperature environment to one that is at a higher temperature. The refrigeration effect or the amount of heat absorbed from the low-temperature environment is much greater than the heat equivalence of the work required to produce this effect. Therefore, the performance of a refrigeration system is measured, like that of an engine, by the ratio of the useful refrigeration effect obtained from the system to the work expended on it to produce that effect. This ratio is called the coefficient of performance. It is used to indicate the efficiency of the system.

The coefficient of performance (C.O.P.) is defined as a ratio between the heat absorbed by the refrigerant as it flows through the evaporator to the heat equivalence of the energy supplied to the compressor.

$$\text{C.O.P.} = \frac{H_2 - H_1}{H_3 - H_2} \tag{6.19}$$

6.4.6 Refrigerant Flow Rate

The refrigerant flow rate depends on the total cooling load imposed on the system and the refrigeration effect. The total cooling load on the system is computed from the heat to be removed from the space or object that is intended to be refrigerated (see Section 6.4.1). The following expression is used to determine the refrigerant flow rate:

$$\dot{m} = \frac{q}{(H_2 - H_1)} \tag{6.20}$$

where $\dot{m}$ is the refrigerant flow rate (kg/s), and q is the total cooling load rate (kW).

Example 6.2

A cold storage room is being maintained at 2°C using a vapor-compression refrigeration system that uses R-134a. The evaporator and condenser temperatures are −5 and 40°C, respectively. The refrigeration load is 20 tons. Calculate the mass flow rate of refrigerant, the compressor power requirement, and the C.O.P. Assume the unit operates under saturated conditions and the compressor efficiency is 85%.

Given

Room temperature = 2°C
Evaporator temperature = −5°C
Condenser temperature = 40°C
Refrigeration load = 20 tons
Compressor efficiency = 85%

Approach

We will draw the refrigeration cycle on a pressure–enthalpy diagram for R-134a. From the diagram we will obtain the required enthalpy values for use in Equations (6.16) through (6.20).

Solution

1. *On a pressure–enthalpy chart for R-134a, draw lines EA and DC, representing the evaporator and condenser conditions, as shown in Figure E6.1. Follow the constant-entropy curve (may require interpolation) from A to intersect horizontal line DC extended to B. From point D, draw a vertical line to intersect line EA at point E. Thus, ABCDE represents the refrigeration cycle under saturated conditions for the given data.*
2. *From the chart, read the following:*
   Evaporator pressure = 243 kPa
   Condenser pressure = 1,015 kPa
   $H_1 = 156$ kJ/kg
   $H_2 = 296$ kJ/kg
   $H_3 = 327$ kJ/kg
3. *From Equation (6.20), mass flow rate of refrigerant (noting that 1 ton of refrigeration = 303,852 kJ/24 h) is*

$$\dot{m} = \frac{(20\text{ tons})(303852\text{ kJ/ton})}{(24\text{ h})(3600\text{ s/h})(296\text{ kJ/kg} - 156\text{ kJ/kg})}$$
$$= 0.502\text{ kg/s}$$

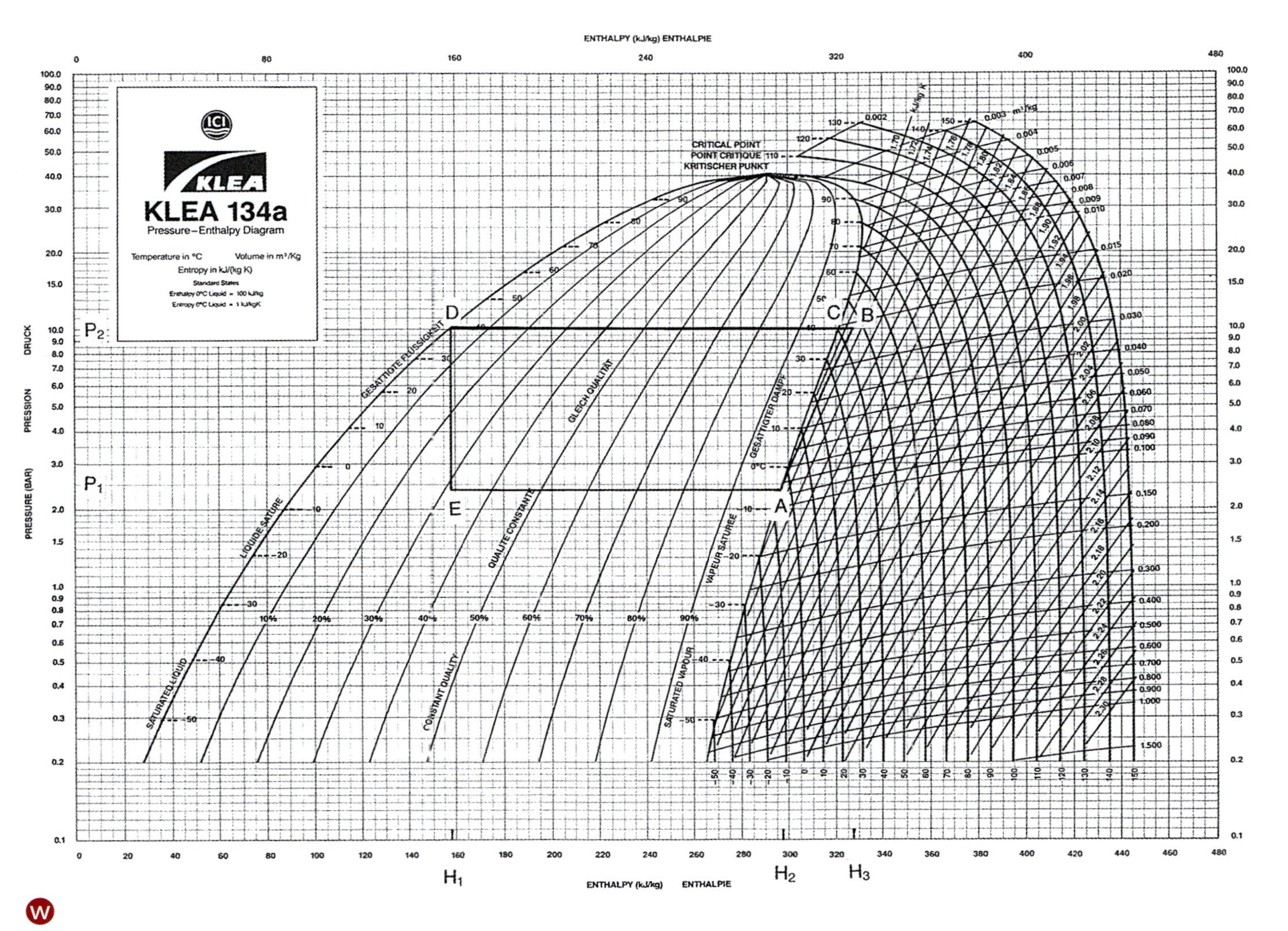

■ Figure E6.1 A pressure–enthalpy chart for a vapor-compression refrigeration cycle for conditions given in Example 6.2.

4. *From Equation (6.16), compressor power requirement, assuming compressor efficiency of 85%, is*

$$q_w = \frac{(0.502\,kg/s)(327\,kJ/kg - 296\,kJ/kg)}{0.85}$$
$$= 18.31\,kW$$

5. *From Equation (6.19), coefficient of performance is*

$$C.O.P. = \frac{(296\,kJ/kg - 156\,kJ/kg)}{(327\,kJ/kg - 296\,kJ/kg)}$$
$$= 4.52$$

Example 6.3 Redo Example 6.2 using pressure–enthalpy tables.

Given

Room temperature = 2°C
Evaporator temperature = −5°C
Condenser temperature = 40°C
Refrigeration load = 20 tons
Compressor efficiency = 85%

Approach

We will use pressure–enthalpy Table A.6.3 for R-134a given in the appendix. In addition, we will use Figure A.6.5 for the expanded portion of the superheated gas region.

Solution

1. *It is convenient to draw a skeleton pressure–enthalpy diagram (similar to Fig. 6.18) and represent values determined from the tables.*
2. *In using pressure–enthalpy tables, it is important to know the condition of the refrigerant at locations A, B, D, and E. We know that at A the refrigerant exists as saturated vapor, and at D it exists as saturated liquid. The temperature of the refrigerant at A is −5°C, and at D it is 40°C.*
3. *At −5°C, the enthalpy for saturated vapor is 295.59 kJ/kg. Therefore,* $H_2 = 295.59$ *kJ/kg.*
4. *At 40°C, the enthalpy for saturated liquid is 156.49 kJ/kg. Therefore,* $H_1 = 156.49$ *kJ/kg.*
5. *The superheated portion of the cycle is drawn on Figure A.6.5, and* H_3 *is obtained as 327 kJ/kg.*
6. *The remaining calculations are done in the same manner as in Example 6.2.*

Example 6.4

Repeat Example 6.2 assuming that the vapors leaving the evaporator are heated by an additional 10°C before entering the compressor, and the liquid refrigerant discharging from the condenser is subcooled an additional 15°C before entering the expansion valve.

Given

See Example 6.2.

Superheat = 10°C

Subcooling = 15°C

Approach

On a pressure–enthalpy chart we will account for the superheating and subcooling of the refrigerant.

Solution

1. *Since the additional superheat is 10°C, the temperature of vapors entering the compressor is 5°C, and the temperature of liquid leaving the condenser is 25°C.*
2. *Draw line EA, representing the evaporator temperature of −5°C.*
3. *Extend line EA, on Figure E6.2, to EA_1. Point A_1 is found by following the 5°C isotherm in the superheated region.*
4. *Draw A_1B, by following a constant-entropy curve originating from A_1.*
5. *Draw BD, corresponding to a condenser temperature of 40°C.*
6. *Extend BD to D_1, where point D_1 is located by drawing a vertical line from the point representing 25°C on the saturated liquid curve.*
7. *Draw D_1E, a vertical line for the adiabatic process in the expansion valve.*
8. *From the refrigeration cycle EA_1BCD_1, determine the following enthalpy values:*

$$H_1 = 137 \text{ kJ/kg}$$

$$H_2 = 305 \text{ kJ/kg}$$

$$H_3 = 338 \text{ kJ/kg}$$

9. *Thus,*

   Mass flow rate of refrigerant

$$= \frac{(20 \text{ ton})(303852 \text{ kJ/ton})}{(24 \text{ h})(3600 \text{ s/h})(305 \text{ kJ/kg} - 137 \text{ kJ/kg})}$$

$$= 0.42 \text{ kg/s}$$

10. *From Equation (6.16), compressor power requirement, assuming compressor efficiency of 85%, is*

$$q_w = \frac{(0.42 \text{ kg/s})(338 \text{ kJ/kg} - 305 \text{ kJ/kg})}{0.85}$$

$$= 16.3 \text{ kW}$$

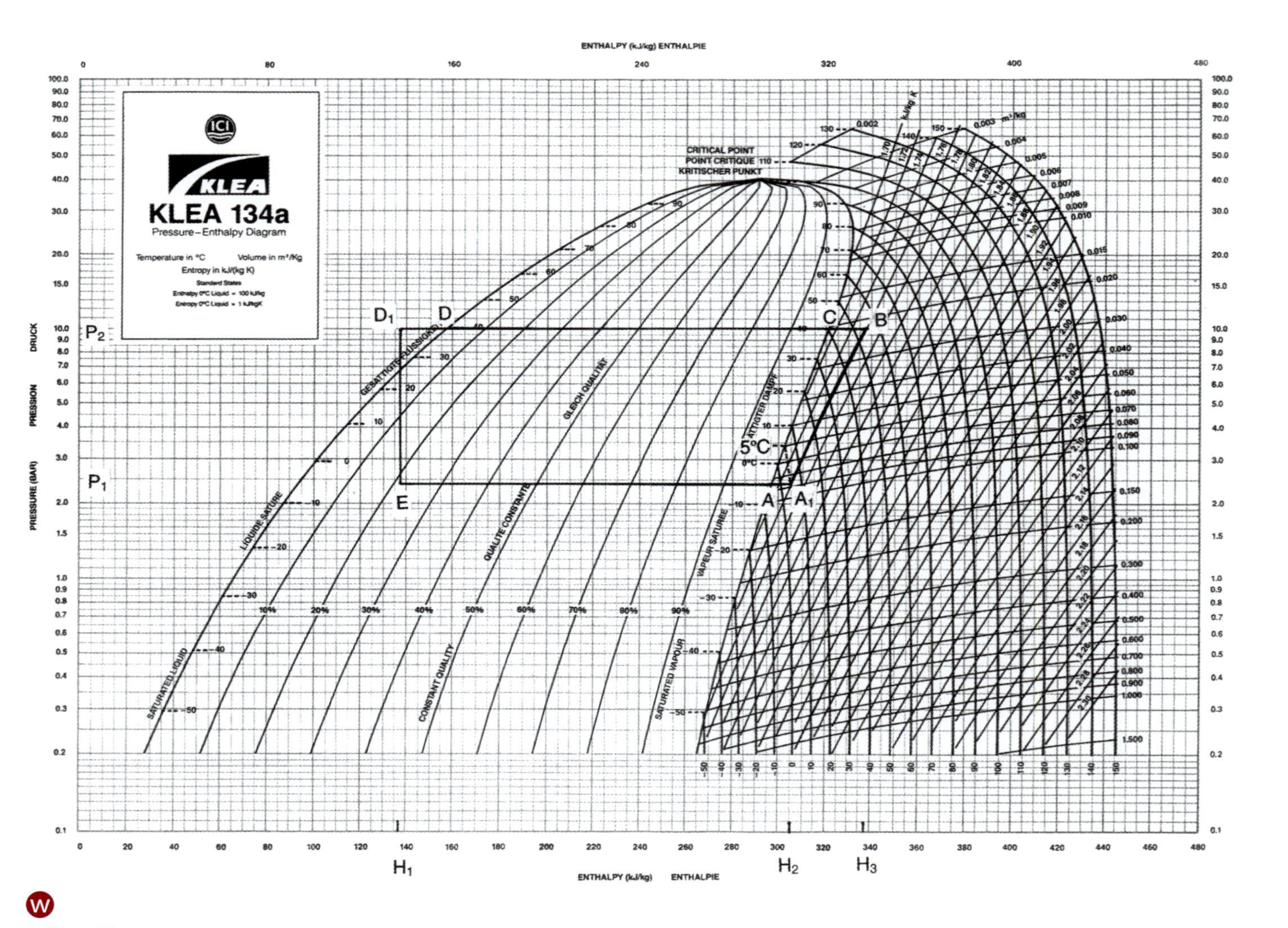

■ Figure E6.2 A pressure–enthalpy chart for a vapor-compression refrigeration cycle for conditions given in Example 6.4.

11. *From Equation (6.19), coefficient of performance is*

$$\text{C.O.P.} = \frac{(305\ \text{kJ/kg} - 137\ \text{kJ/kg})}{(338\ \text{kJ/kg} - 305\ \text{kJ/kg})}$$

$$= 5.1$$

12. *The example shows the influence of superheating and subcooling on the refrigerant flow rate and the compressor horsepower.*

Example 6.5

Redo Example 6.2, using ammonia as a refrigerant instead of R.134a.

Given

Same as in Example 6.2

Approach

We will use pressure–enthalpy Table A.6.2, and the expanded diagram in Figure A.6.3 for ammonia.

Solution

1. *Draw a skeleton pressure–enthalpy diagram and identify points A, B, C, D, and E (similar to Fig. 6.18).*
2. *At A, the refrigerant is in saturated vapor state. From Table A.6.2, the enthalpy of saturated vapor at* $-5°C$ *is 1456.15 kJ/kg. Thus,* $H_2 = 1456.15$ *kJ/kg.*
3. *At D, the refrigerant is in saturated liquid state. From Table A.6.2, the enthalpy of saturated liquid at 40°C is 390.59 kJ/kg. Therefore,* $H_1 = 390.59$ *kJ/kg.*
4. *From Figure A.6.3, point B is identified and* H_3 *is obtained as 1680 kJ/kg.*
5. *The mass flow rate of refrigerant, from Equation (6.20) is*

$$\dot{m} = \frac{(20\ \text{tons})(303{,}852\ \text{kJ/ton})}{(24\ \text{h})(3600\ \text{s/h})(1456.15\ \text{kJ/kg} - 390.59\ \text{kJ/kg})}$$

$$= 0.066\ \text{kg/s}$$

6. *From Equation (6.16), compressor power requirement, assuming compressor efficiency of 85%, is*

$$q_w = \frac{(0.66\ \text{kg/s})(1680\ \text{kJ/kg} - 1456.15\ \text{kJ/kg})}{0.85}$$

$$= 17.38\ \text{kW}$$

7. *From Equation (6.19), coefficient of performance is*

$$\text{C.O.P.} = \frac{(1456.15\ \text{kJ/kg} - 390.59\ \text{kJ/kg})}{(1680\ \text{kJ/kg} - 1456.15\ \text{kJ/kg})}$$

$$= 4.76$$

8. *In contrast to R-134a, the use of ammonia reduces the refrigerant flow rate by 84%.*

Example 6.6

Using a spreadsheet, develop a computer-aided procedure to determine the following refrigerant properties for R-12, R-134a, and R-717:

- **a.** Enthalpy of saturated liquid at condenser temperature.
- **b.** Enthalpy of saturated vapor at evaporator temperature.
- **c.** Specific volume of saturated vapors at evaporator temperature.
- **d.** Determine the enthalpy values H_1, H_2, and H_3 for Example 6.2 using the spreadsheets for R-12, R-134a, and R-717.

Given

Same as in Example 6.2.

Approach

We will use the empirical correlations given by Cleland (1986, 1994) for developing a spreadsheet using Excel™. The spreadsheets will be developed for R-12, R-134a, and R-717.

Solution

1. *Using empirical relations given by Cleland (1986, 1994), the following spreadsheets are prepared (Figs. E6.3, E6.4, and E6.5). All empirical correlations are presented in D6:H11.*

	A	B	C	D	E	F	G	H	I	J	K
1	T_evaporator (°C)	−5									
2	T_condenser (°C)	40									
3	T_condenser-T_evaporator (°C)	45	← =B2−B1	Coefficients from Cleland (1986)							
4				R-12 (Freon)							
5											
6				20.82963	−2033.546	248.3	200000	923.88			
7				0.83716	5.38E−03	187565	428.992	−0.75152			
8				5.67E−03	163994	−11.58643	2372.495	1.00755			
9				4.94E−04	−6.05E−06	−2.29E−07	1.09E+00	−1.81E−03			
10				−1.49E−05	2.21E−06	1.97E−07	−7.87E−08	−1.97E−09			
11				−5.63E−04							
12											
13	P_suction	260.76	← =EXP(D6+E6/(B1+F6))/1000								
14	P_discharge	961.25	← =EXP(D6+E6/(B2+F6))/1000								
15											
16	H_1	238.64	← =(G6+H6*B2+D7*B2^2+E7*B2^3)/1000								
17	H_2	349.39	← =(F7+G7*B1+H7*B1^2+D8*B1^3+E8)/1000								
18	v_saturated	0.06	← =EXP(F8+G8/(B1+273.15))*(H8+D9*B1+E9*B1^2+F9*B1^3)								
19	c_constant	1.07	← =G9+H9*B1+D10*B1^2+E10*B1*B3+F10*B1^2*B3+G10*B1*B3^2+H10*B1^2*B3+D11*B3								
20	delta_H (kJ/kg)	23.07	← =((B19/(B19−1))*B13*1000*B18*((B14/B13)^((B19−1)/B19)−1))/1000								
21	H_3 (kJ/kg)	372.47	← =B17+B20								

■ **Figure E6.3** Spreadsheet for determining various refrigerant properties of Freon (R-12).

	A	B	C	D	E	F	G	H	I	J	K
1	T_evaporator (°C)	−40									
2	T_condenser (°C)	25									
3	T_condenser-T_evaporator (°C)	65	← =B2−B1		Coefficients from Cleland (1994)						
4					R-134a						
5											
6				21.51297	−2200.981	246.61	100000	1335.29			
7				1.7065	0.007674	249.455	606.163	−1.50644			
8				−0.018243	299048	−12.4539	2669	1.01357			
9				1.07E−03	−9.25E−06	−3.22E−07	1.06469	−0.001691			
10				−8.56E−06	−2.14E−05	−6.17E−07	2.074E−07	7.72E−09			
11				−0.00061							
12											
13	P_suction	52.06	← =EXP(D6+E6/(B1+F6))/1000								
14	P_discharge	666.31	← =EXP(D6+E6/(B2+F6))/1000								
15											
16	H_1	134.57	← =(G6+H6*B2+D7*B2^2+E7*B2^3)/1000								
17	H_2	273.81	← =(F7+G7*B1+H7*B1^2+D8*B1^3+E8)/1000								
18	v_saturated	0.36	← =EXP(F8+G8/(B1+273.15))*(H8+D9*B1+E9*B1^2+F9*B1^3)								
19	c_constant	1.04	← =G9+H9*B1+D10*B1^2+E10*B1*B3+F10*B1^2*B3+G10*B1*B3^2+H10*B1^2*B3+D11*B3								
20	delta_H (kJ/kg)	49.55	← =((B19/(B19−1))*B13*1000*B18*((B14/B13)^((B19−1)/B19)−1))/1000								
21	H_3 (kJ/kg)	323.36	← =B17+B20								

■ **Figure E6.4** Spreadsheet for determining various refrigerant properties of R-134a.

	A	B	C	D	E	F	G	H	I	J	K
1	T_evaporator (°C)	−5									
2	T_condenser (°C)	40									
3	T_condenser-T_evaporator (°C)	45	← =B2−B1		Coefficients from Cleland (1986)						
4					R-717 (Ammonia)						
5											
6				22.11874	−2233.823	244.2	200000	4751.63			
7				2.04493	−0.37875	1441467	920.154	−10.20556			
8				−0.026513	15689	−11.09867	2691.68	0.99675			
9				0.000402	2.64−E06	−1.75E−07	1.325798	000245			
10				3.11E−06	1.13E−05	−1.43E−07	6.36E−08	9.60E−10			
11				−0.000382							
12											
13	P_suction	355.05	← =EXP(D6+E6/(B1+F6))/1000								
14	P_discharge	1557.67	← =EXP(D6+E6/(B2+F6))/1000								
15											
16	H_1	390.91	← =(G6+H6*B2+D7*B2^2+E7*B2^3)/1000								
17	H_2	1452.30	← =(F7+G7*B1+H7*B1^2+D8*B1^3+E8)/1000								
18	v_saturated	0.34	← =EXP(F8+G8/(B1+273.15))*(H8+D9*B1+E9*B1^2+F9*B1^3)								
19	c_constant	1.30	← =G9+H9*B1+D10*B1^2+E10*B1*B3+F10*B1^2*B3+G10*B1*B3^2+H10*B1^2*B3+D11*B3								
20	delta_H (kJ/kg)	215.91	← =((B19/(B19−1))*B13*1000*B18*((B14/B13)^((B19−1)/B19)−1))/1000								
21	H_3 (kJ/kg)	1668.22	← =B17+B20								

■ **Figure E6.5** Spreadsheet for determining various refrigerant properties of ammonia (R-717).

2. *The following input parameters may be used: Temperature at suction (saturated vapor); temperature of saturated liquid (at the entrance to expansion valve).*
3. *The calculations include the following:*
   a. *Pressure (saturated) calculated at the suction temperature of compressor*
   b. *Pressure (saturated) calculated for the temperature of refrigerant in the condenser*
   c. H_1
   d. H_2
   e. *Specific volume of saturated vapors at inlet to the compressor suction*
   f. *An interim constant "c" (see Cleland, 1986)*
   g. *For an isoentropic compression process, for Example 6.2, the value of ΔH can be calculated and then added to H_2 to determine H_3*
   h. *Figures E6.3, E6.4, and E6.5 provide a listing of the pressure–enthalpy spreadsheets for Freon R-12, R-134a and ammonia R-717, respectively*
4. *For conditions given in Example 6.2,*

$$T_{evaporator} = -5°C$$

$$T_{condenser} = 40°C$$

5. *From the spreadsheet for R-12,*

$$H_1 = 238.64 \text{ kJ/kg}$$

$$H_2 = 349.4 \text{ kJ/kg}$$

$$H_3 = 372.46 \text{ kJ/kg}$$

6. *From the spreadsheet for R-717,*

$$H_1 = 390.91 \text{ kJ/kg}$$

$$H_2 = 1452.30 \text{ kJ/kg}$$

$$H_3 = 1668.21 \text{ kJ/kg}$$

7. *From the spreadsheet for R-134a,*

$$H_1 = 134.57 \text{ kJ/kg}$$

$$H_2 = 273.81 \text{ kJ/kg}$$

$$H_3 = 323.36 \text{ kJ/kg}$$

6.5 USE OF MULTISTAGE SYSTEMS

Multistage systems involve using more than one compressor, often with an objective of reducing the total power requirements. Although adding additional compressors requires an increase in capital spending, the total operating costs must be reduced for the multistage systems to be justifiable. The following discussion involves a commonly used approach for using a dual-stage refrigeration system—a flash gas removal system.

6.5.1 Flash Gas Removal System

As seen previously in Figure 6.18, a refrigerant leaves the condenser in a saturated liquid state, and in the expansion valve there is a pressure drop from a high condenser pressure to the low evaporator pressure. The drop in pressure of the refrigerant with a partial conversion to vapor state, commonly called "flashing," is accompanied by the conversion of some of the refrigerant from liquid to vapor state. If we consider the state of the refrigerant at some intermediate pressure between P_1 and P_2, such as at location K on Figure 6.22, the refrigerant is existing partially as a vapor, but mostly in liquid state. The refrigerant already converted to vapors in the expansion valve can no longer provide any useful purpose in the evaporator. It therefore may be desirable to take the vapors at that intermediate pressure in the expansion valve, and compress them with another small compressor to the condensing pressure.

The liquid refrigerant, with only a small fraction of vapors (due to further flashings that occur with lowering of the pressure to the evaporator pressure), then enters the evaporator.

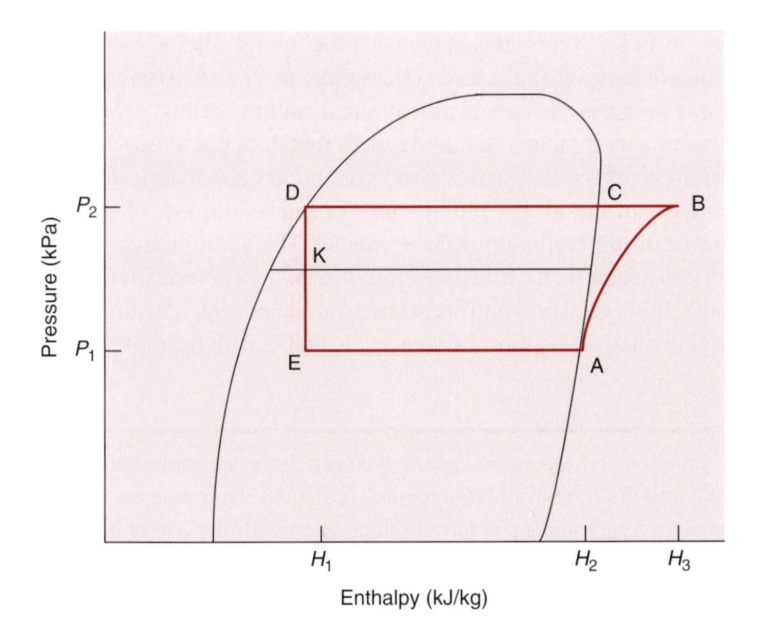

■ **Figure 6.22** A pressure–enthalpy diagram for a flash-gas removal system.

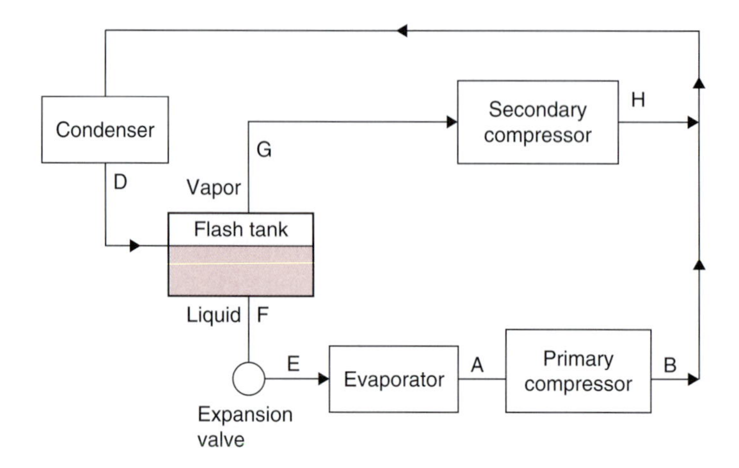

■ **Figure 6.23** A flash-gas removal system.

A schematic of the flash gas removal system is shown in Figure 6.23 The refrigerant leaving the condenser in liquid state is allowed to go through a throttling valve that is controlled by the liquid level in a flash tank. The separation of the vapors from the liquid is done in the flash tank. The liquid refrigerant is then conveyed to an expansion valve prior to entering the evaporator, while the vapors are conveyed to a secondary compressor. The use of the flash gas system results in an overall decrease in power requirements.

The reduction in power requirements becomes significant when the temperature difference between the condenser and the evaporator is large; in other words, this system is most useful for low-temperature chilling or freezing applications. The secondary compressor required for compressing the flash gas is usually small, and the primary compressor is also smaller than in a standard system that does not remove flash gas. Since most of the flash gas is removed, and gas has significantly larger volume than liquid, the piping used for intake and exit of the refrigerant from the evaporator can be smaller. The additional costs of the flash gas system include the need for a secondary compressor (although small), flash gas tank, and associated valves, piping, and fittings. The use of the flash gas removal system is illustrated with Example 6.7.

Example 6.7

An ammonia refrigeration system involves an evaporator operating at −20°C and the condenser temperature set at 40°C. Determine the reduction in power requirements per ton of refrigeration with the use of a flash gas removal system at an intermediate pressure of 519 kPa (Fig. 6.23).

Given

Evaporator temperature = −20°C

Condenser temperature = 40°C

Intermediate pressure for the flash gas removal system = 519 kPa

Approach

We will determine the enthalpy values of ammonia for the ideal cycle as well as when a flash gas removal system is considered. These values will be obtained using the spreadsheet developed in Example 6.6.

Solution

1. *Using the spreadsheet program for ammonia, the enthalpy values for the ideal and modified system are as follows (Fig. E6.6):*

   Ideal System

$$H_1 = 391\,\text{kJ/kg}$$

$$H_2 = 1435\,\text{kJ/kg}$$

$$H_3 = 1750\,\text{kJ/kg}$$

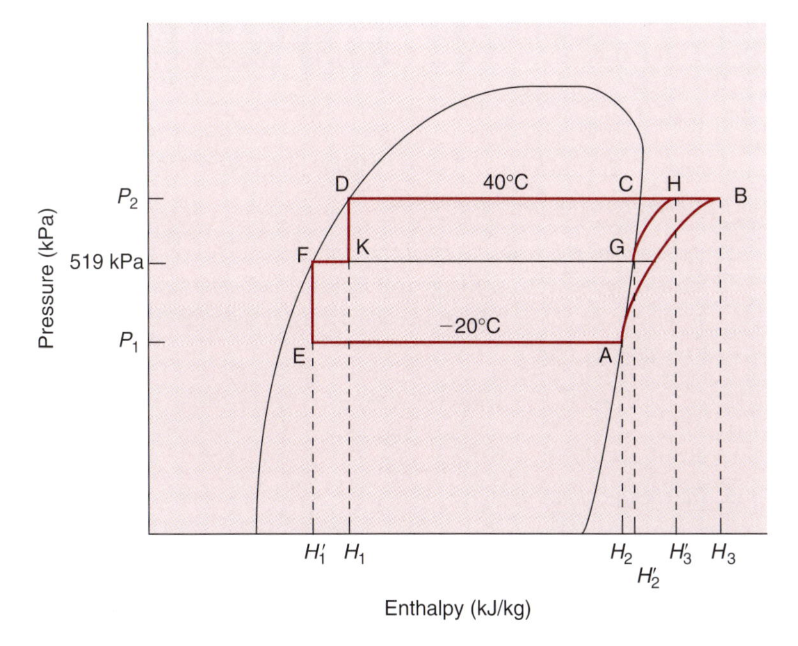

Figure E6.6 A pressure–enthalpy diagram for a flash-gas removal system as described in Example 6.7.

Modified System

$$H_1' = 200 \text{ kJ/kg}$$

$$H_2' = 1457 \text{ kJ/kg}$$

$$H_3' = 1642 \text{ kJ/kg}$$

Ideal System

2. *The refrigerant flow rate may be calculated per ton of refrigeration load (noting that one ton of refrigeration is equivalent to 3.517 kW) as*

$$\dot{m} = \frac{3.517 \text{ kJ/s}}{1435 \text{ kJ/kg} - 391 \text{ kJ/kg}}$$
$$= 0.00337 \text{ kg/s}$$

3. *Then compressor power*

$$q_w = (0.00337 \text{ kg/s})(1750 \text{ kJ/kg} - 1435 \text{ kJ/kg})$$
$$= 1.062 \text{ kW/ton of refrigeration}$$

Modified System

4. *To determine the power requirements of the compressor for the modified system, it is necessary to determine the refrigerant flow rates. The refrigerant flow rate through the evaporator and the primary compressor is*

$$\dot{m}_A = \dot{m}_B = \dot{m}_F = \dot{m}_E = \frac{3.517 \text{ kJ/s}}{1435 \text{ kJ/kg} - 200 \text{ kJ/kg}}$$
$$= 0.00285 \text{ kg/s}$$

where $\dot{m}$ is the refrigerant flow rate and the subscripts A, B, F, E correspond to locations as indicated in Figure E6.6.

5. *To determine the power requirement of the secondary compressor in the modified system, it is necessary to calculate the refrigerant flow rate through it. This can be obtained by conducting a mass and energy balance on the flash gas tank. Thus,*

$$\dot{m}_D = \dot{m}_E + \dot{m}_G$$
$$\dot{m}_D = 0.00285 + \dot{m}_G$$

6. *The energy balance gives*

$$(\dot{m}_D)(H_1) = (0.00285)(H_1') + (\dot{m}_G)(H_2')$$

$$(0.00285 + \dot{m}_G)(391) = (0.00285 \times 200) + (\dot{m}_G)(1457)$$

7. *Solving the mass and energy balances we get*

$$\dot{m}_G = 0.000511\,kg/s$$

8. *Then the power required by the primary compressor is*

$$q_{w_1} = 0.00285(1750 - 1435)$$
$$= 0.8977\ kW$$

and the power required by the secondary compressor is

$$q_{w_2} = 0.000511(1642 - 1457)$$
$$= 0.0945\ kW$$

9. *Then the total power required by the flash gas removal system is*

$$q_w = 0.8977 + 0.0945$$
$$= 0.992\ kW$$

10. *Thus the use of the flash gas removal system results in a 7% reduction in power requirements.*

PROBLEMS

6.1 Determine the C.O.P. of a simple saturated ammonia (R-717) compression refrigeration cycle. The evaporation temperature is −20°C and the condenser temperature is 30°C.

6.2 For a 10-ton-capacity refrigeration system, the pressure of refrigerant in the evaporator is 210 kPa, whereas in the condenser it is 750 kPa. If ammonia (R-717) is used under saturated conditions, calculate the theoretical power required to operate the compressor.

6.3 A food storage chamber requires a refrigeration system of 15-ton capacity operating at an evaporator temperature of −8°C and a condenser temperature of 30°C. Refrigerant ammonia (R-717) is used and the system operates under saturated conditions. Determine:

a. C.O.P.

b. Refrigerant flow rate

c. Rate of heat removed by condenser

***6.4** Repeat problem 6.3, with the refrigerant subcooled by 5°C before entering the expansion valve and vapor superheated by 6°C[1].

* Indicates an advanced level of difficulty in solving

***6.5** A vapor-compression refrigeration system, using ammonia (R-717) as the refrigerant, is operating with an evaporator temperature of −5°C and condenser temperature of 40°C. It is desired to determine the influence of raising the evaporator temperature to 5°C while holding condenser temperature at 40°C. Calculate the percent changes in

a. Refrigeration effect per kg refrigerant flow rate
b. C.O.P.
c. Heat of compression
d. Theoretical power requirements
e. Rate of heat rejected at the condenser, anticipated due to raising the evaporator temperature

6.6 A vapor-compression refrigeration system, using ammonia (R-717) as the refrigerant, is operating with an evaporator temperature of −20°C and condenser temperature of 30°C. It is desired to determine the influence of raising the condenser temperature to 35°C. Calculate the percent changes in

a. Refrigeration effect per kg of refrigerant flow rate
b. C.O.P.
c. Heat of compression
d. Theoretical power requirement
e. Rate of heat rejected at the condenser, anticipated due to raising the evaporator temperature

6.7 A refrigerated room, used for product cooling, is being maintained at the desired temperature using an evaporator temperature of 0°C and condenser pressure of 900 kPa with an R-134a vapor-compression refrigeration system. The condenser is a countercurrent tubular heat exchanger with water entering at 25°C and leaving at 35°C. Evaluate the following for a refrigeration load of 5 tons:

a. The rate of heat exchange at the condenser
b. The compressor power requirement at 80% efficiency
c. The coefficient of performance for the system
d. The heat transfer surface area in the condenser when the overall heat transfer coefficient is 500 W/m^2 °C
e. The flow rate of water through the condenser

6.8 An ammonia (R-717) vapor-compression refrigeration system is operating under saturated conditions. The following data

* Indicates an advanced level of difficulty in solving

were obtained: condenser pressure, 900 kPa; state of vapors entering the evaporator, 70% liquid. This refrigeration system is being used to maintain a controlled-temperature chamber at $-5\,^\circ$C. The walls of the chamber provide an overall resistance to conductive heat transfer equivalent to $0.5\,\text{m}^2\,^\circ\text{C/W}$. The convective heat-transfer coefficient on the outside wall/ceiling is $2\,\text{W/m}^2\,^\circ\text{C}$ and on the inside wall/ceiling is $10\,\text{W/m}^2\,^\circ\text{C}$. The outside ambient temperature is $38\,^\circ$C. The total wall and ceiling area is $100\,\text{m}^2$. Ignore any heat gain from the floor.

a. Calculate the flow rate of refrigerant for this system.
b. Calculate the power requirements (kW) of the compressor operating to maintain these conditions.

6.9 A vapor compression refrigeration system, using R-134a, is operating under ideal conditions. The system is being used to provide cold air for a walk-in freezer. The evaporator temperature is $-35\,^\circ$C and the condenser temperature is $40\,^\circ$C. You hire a consultant to find different ways to conserve energy. The consultant suggests that the work done by the compressor can be reduced by half if the condenser temperature is lowered to $0\,^\circ$C. Check the consultant's numerical results. Prepare a reply to either accept or deny the consultant's conclusions and justify your opinion.

6.10 A mass of 100 kg of a liquid food contained in a large vessel is to be cooled from $40\,^\circ$C to $5\,^\circ$C in 10 min. The specific heat of the liquid food is $3600\,\text{J/(kg}\,^\circ\text{C)}$. Cooling is done by immersing an evaporator coil inside the liquid. The evaporator coil temperature is $1\,^\circ$C, and condenser temperature is $41\,^\circ$C. The refrigerant being used is R-134a. Draw the refrigeration cycle on a pressure–enthalpy diagram and use the refrigerant property tables to solve the following parts of this problem:

a. Show the refrigeration cycle on a pressure–enthalpy diagram for R-134a.
b. Determine the flow rate of R-134a.
c. Determine the C.O.P. for the system.
d. If the condenser is a countercurrent tubular heat exchanger, and water is used as a cooling medium, and the temperature of water increases from $10\,^\circ$C to $30\,^\circ$C, determine the length of the heat exchanger pipe needed for the condenser. The overall heat transfer coefficient is $U_0 = 1000\,\text{W/(m}^2\,^\circ\text{C)}$; for the inner pipe, the outside diameter is 2.2 cm and the inner diameter is 2 cm.

LIST OF SYMBOLS

c	empirical coefficient in Equation (6.12)
c_{i1}	empirical coefficient in Equation (6.14)
D	diameter of cylinder bore (cm)
Δh	enthalpy change in isoentropic compression (kJ/kg)
H_1	enthalpy of refrigerant at the exit from the condenser (kJ/kg refrigerant)
H_2	enthalpy of refrigerant at the beginning of compression stroke (kJ/kg refrigerant)
H_3	enthalpy of refrigerant at the end of compression stroke (kJ/kg refrigerant)
H_{i1}, H_{i2}	intermediate enthalpy values in Equations (6.5) and (6.8)
H_L	enthalpy of liquid refrigerant (same as H_1) (kJ/kg refrigerant)
H_s	enthalpy of superheated vapor (same as H_3) (kJ/kg refrigerant)
H_v	enthalpy of saturated vapor (same as H_2) (kJ/kg refrigerant)
L	length of stroke (cm)
$\dot{m}$	refrigerant mass flow rate (kg/s)
N	number of cylinders
P	pressure (kPa)
p	vapor pressure (kPa)
p_c	vapor pressure at critical point (kPa)
P_{sat}	saturation pressure (kPa)
q	rate of cooling load (kW)
q_c	rate of heat exchanged in the condenser (kW)
q_e	rate of heat absorbed by the evaporators (kW)
q_w	rate of work done on the refrigerant (kW)
T_b	temperature of subcooled liquid (°C)
T_c	temperature of critical point (°C)
T_L	temperature of liquid refrigerant (°C)
T_s	temperature of superheated vapor (°C)
T_{sat}	saturation temperature (°C)
U	overall heat transfer coefficient (W/[m^2 K])
v	specific volume (m^3/kg)
v_s	specific volume of superheated vapor (m^3/kg)
v_v	specific volume of saturated vapor (m^3/kg)

BIBLIOGRAPHY

American Society of Heating, Refrigerating and Air-Conditioning Engineers, Inc (2005). *ASHRAE Handbook of 2005 Fundamentals*. ASHRAE, Atlanta, Georgia.

Cleland, A. C. (1986). Computer subroutines for rapid evaluation of refrigerant thermodynamic properties. *Int. J. Refrigeration* **9**(Nov.): 346–351.

Cleland, A. C. (1994). Polynomial curve-fits for refrigerant thermodynamic properties: extension to include R134a. *Int. J. Refrigeration* **17**(4): 245–249.

Jennings, B. H. (1970). *Environmental Engineering, Analysis and Practice*. International Textbook Company, New York.

McLinden, M. O. (1990). Thermodynamic properties of CFC alternatives: A survey of the available data. *Int. J. Refrigeration* **13**(3): 149–162.

Stoecker, W. F. and Jones, J. W. (1982). *Refrigeration and Air Conditioning*. McGraw-Hill, New York.

Chapter 7

Food Freezing

The preservation of food by freezing has become a major industry in the United States as well as in other parts of the world. For example, from 1970 to 2005, the annual per capita consumption of frozen vegetables in the United States increased from 20 to 34 kg.

Preservation of a food by freezing occurs by several mechanisms. At temperatures below 0°C there is a significant reduction in growth rates for microorganisms and in the corresponding deterioration of the product due to microbial activity. The same temperature influence applies to most other reactions that might normally occur in the product, such as enzymatic and oxidative reactions. In addition, the formation of ice crystals within the product changes the availability of water to participate in reactions. As the temperature is reduced and more water is converted to a solid state, less water is available to support deteriorative reactions.

Although freezing as a preservation process generally results in a high-quality product for consumption, the quality is influenced by the freezing process and frozen-storage conditions. The freezing rate or time allowed for the product temperature to be reduced from above to below the initial freezing temperature will influence product quality, but in a variable manner depending on the food commodity. For some products, rapid freezing (short freezing time) is required to ensure formation of small ice crystals within the product structure and minimal damage to the product texture. Other products are not influenced by structural changes and do not justify the added costs associated with rapid freezing. Still other products have geometric configurations and sizes that do not allow rapid freezing. The storage-temperature conditions influence frozen food quality in a significant

All icons in this chapter refer to the author's web site, which is independently owned and operated. Academic Press is not responsible for the content or operation of the author's web site. Please direct your web site comments and questions to the author: Professor R. Paul Singh, Department of Biological and Agricultural Engineering, University of California, Davis, CA 95616, USA. Email: rps@rpaulsingh.com.

manner. Any elevation in storage temperature tends to reduce the preservation quality of the process, and fluctuations in storage temperature tend to be even more detrimental to product quality.

From these brief introductory comments, it is obvious that the optimum freezing process depends on the product characteristics. Numerous freezing systems are available, each designed to achieve freezing of a particular product in the most efficient manner and with the maximum product quality. The importance of residence time in the freezing system must be emphasized, as well as the need for accurate freezing time prediction.

7.1 FREEZING SYSTEMS

To achieve freezing of a food product, the product must be exposed to a low-temperature medium for sufficient time to remove sensible heat and latent heat of fusion from the product. Removal of the sensible and latent heat results in a reduction in the product temperature as well as a conversion of the water from liquid to solid state (ice). In most cases, approximately 10% of the water remains in the liquid state at the storage temperature of the frozen food. To accomplish the freezing process in desired short times, the low-temperature medium is at much lower temperature than the desired final temperature of the product, and large convective heat-transfer coefficients are created.

The freezing process can be accomplished by using either indirect or direct contact systems. Most often, the type of system used will depend on the product characteristics, both before and after freezing is completed. There are a variety of circumstances where direct contact between the product and refrigerant is not possible.

7.1.1 Indirect Contact Systems

In numerous food-product freezing systems, the product and refrigerant are separated by a barrier throughout the freezing process. This type of system is illustrated schematically in Figure 7.1. Although many systems use a nonpermeable barrier between product and refrigerant, indirect freezing systems include any system without direct contact, including those where the package material becomes the barrier.

7.1.1.1 *Plate Freezers*

The most easily recognized type of indirect freezing system is the plate freezer, illustrated in Figure 7.2. As indicated, the product is frozen while held between two refrigerated plates. In most cases, the barrier

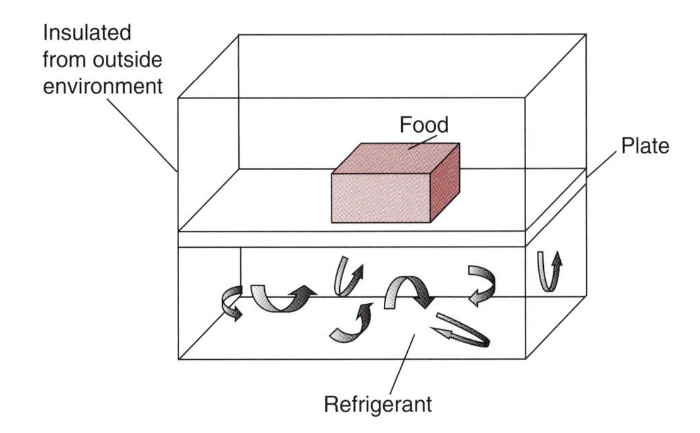

Figure 7.1 Schematic diagram of an indirect-contact freezing system.

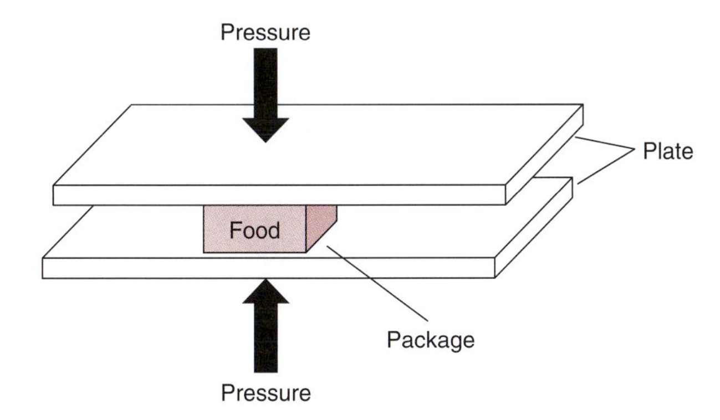

Figure 7.2 Schematic illustration of a plate freezing system.

between product and refrigerant will include both the plate and package material. The heat transfer through the barrier (plate and package) can be enhanced by using pressure to reduce resistance to heat transfer across the barrier as illustrated (Fig. 7.2). In some cases, plate systems may use single plates in contact with the product and accomplish freezing with heat transfer across a single package surface. As would be expected, these systems are less efficient, and they are costly to acquire and operate.

Plate-freezing systems can be operated as a batch system with the product placed on the plates for a specified residence time before being removed. In this situation, the freezing time is the residence time and represents the total time required to reduce the product from the initial temperature to some desired final temperature. In general, the batch plate-freezing system has significant flexibility in terms of handling diverse product types and product sizes.

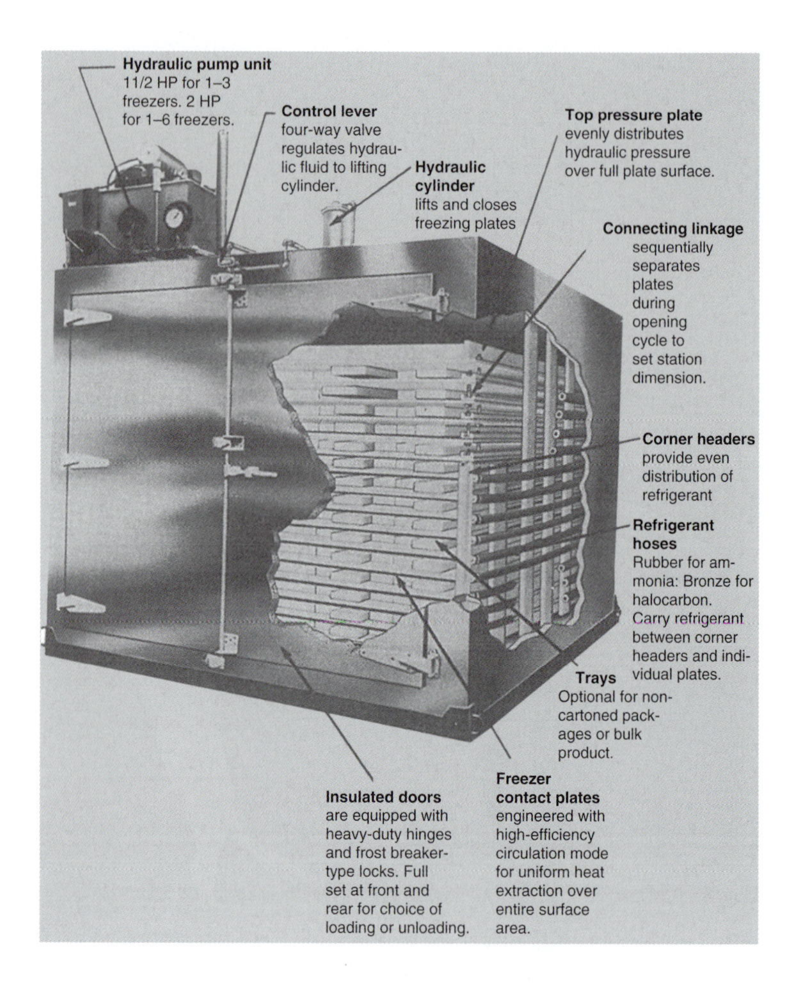

W

Figure 7.3 Plate contact freezing system. (Courtesy of CREPACO, Inc.)

The plate-freezing system operates in a continuous mode by moving the plates holding the product through an enclosure in some prescribed manner. In Figure 7.3, the product is held between two refrigerated plates throughout the freezing process. The movement of plates (and product) occurs as the plates index upward or across within the compartment. At the entrance and exit to the freezing system, the plates are opened to allow the product to be conveyed to or from the system. In a continuous plate-freezing system, the freezing time is the total time required for the product to move from entrance to exit. During the residence time, the desired amounts of sensible and latent heat are removed to achieve the desired frozen product temperature.

7.1.1.2 *Air-Blast Freezers*

In many situations, the product size and/or shape may not accommodate plate freezing. For these situations, air-blast freezing systems become the best alternative. In some cases, the package film is the barrier for the indirect freezing, with cold air being the source of refrigeration.

Air-blast freezers can be a simple design, as in the case of a refrigerated room. In this situation, the product is placed in the room, and the low-temperature air is allowed to circulate around the product for the desire residence of freezing time. This approach represents a batch mode, and the refrigerated room may act as a storage space in addition to the freezing compartment. In most cases, freezing times will be long because of lower air speeds over the product, inability to achieve intimate contact between product and cold air, and the smaller temperature gradients between product and air.

Most air-blast freezers are continuous, such as those illustrated in Figure 7.4. In these systems, the product is carried on a conveyor that moves through a stream of high-velocity air. The length and speed of the conveyor establish the residence of freezing time. These times can be relatively small based on the use of very low-temperature air, high

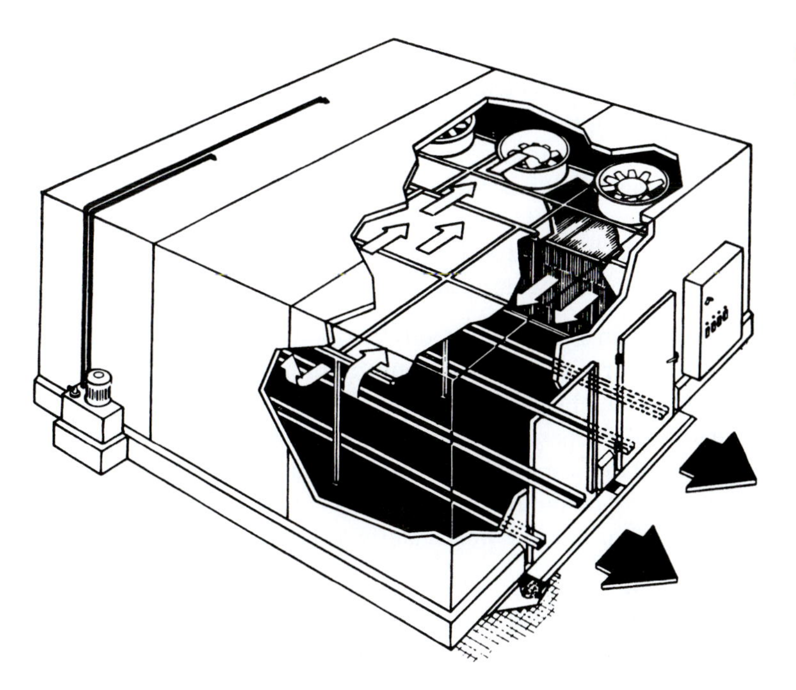

W

Figure 7.4 Continuous air-blast freezing system. (Courtesy of Frigoscandia Contracting, Inc.)

air velocities, and good contact between individual product packages and the cold air.

Continuous air-blast freezing systems use a variety of different conveying arrangements for movement of product through the refrigerated air. Alternate arrangements to Figure 7.4 include tray conveyors, spiral conveyors, and roller conveyors. Most often, the system used will depend on product characteristics.

7.1.1.3 *Freezers for Liquid Foods*

The third general type of indirect freezing systems includes those designed primarily for liquid foods. In many situations, the most efficient removal of thermal energy from a liquid food can be accomplished before the product is placed in a package. Although any indirect heat exchanger designed for a liquid food would be acceptable, the most common type is a scraped-surface system, as described in Section 4.1.3. The heat exchangers for freezing liquid foods are designed specifically for freezing, with the heat-exchange shell surrounding the product compartment becoming an evaporator for a vapor-compression refrigeration system. This approach provides precise control of the heat-exchange surface by adjustment of pressure on the low-pressure side of the refrigeration system.

For freezing liquid foods, the residence time in the freezing compartment is sufficient to decrease the product temperature by several degrees below the temperature of initial ice-crystal formation. At these temperatures, between 60 and 80% of the latent heat has been removed from the product, and the product is in the form of a frozen slurry. In this condition, the product flows quite readily and can be placed in a package for final freezing in a low-temperature refrigerated space. The scraped-surface heat exchanger ensures efficient heat exchange between the slurry and the cold surface.

Freezing systems for liquid foods can be batch or continuous. The batch system places a given amount of unfrozen liquid in the compartment and allows the freezing process to continue until the desired final temperature is reached. The product compartment is a scraped-surface heat exchanger but is operated as a batch system. In the case of ice-cream freezing, the system is designed with facility for injection of air into the frozen slurry to achieve the desired product consistency.

A continuous freezing system for liquid foods is illustrated in Figure 7.5. As indicated, the basic system is a scraped-surface heat

Figure 7.5 Continuous freezing system for liquid foods. (Courtesy of Cherry-Burrell Corporation)

exchanger using refrigerant during phase change as the cooling medium. The rotor acts as a mixing device, and the scraper blades enhance heat transfer at the heat-exchange surface. For these systems, the residence time for product in the compartment will be sufficient to reduce the product temperature by the desired amount and to provide time for any other changes in the product before it is placed in the package for final freezing.

7.1.2 Direct-Contact Systems

Several freezing systems for food operate with direct contact between the refrigerant and the product, as illustrated in Figure 7.6. In most situations, these systems will operate more efficiently since there are no barriers to heat transfer between the refrigerant and the product. The refrigerants used in these systems may be low-temperature air at high speeds or liquid refrigerants with phase change while in contact with the product surface. In all cases, the systems are designed

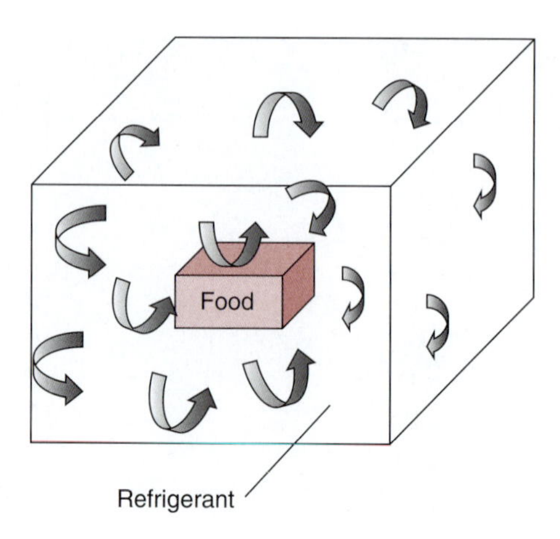

Figure 7.6 Schematic diagram of a direct-contact freezing system.

to achieve rapid freezing, and the term *individual quick freezing* (IQF) will apply.

7.1.2.1 *Air Blast*

The use of low-temperature air at high speeds in direct contact with small product objects is a form of IQF. The combination of low-temperature air, high convective heat-transfer coefficient (high air speed), and small product shape leads to short freezing time or rapid freezing. In these systems, the product is moved through the high-speed-air region on a conveyor in a manner that controls the residence time. The types of product that can be frozen in these systems are limited to those that have the appropriate geometries and that require rapid freezing for maximum quality.

A modification of the regular air-blast IQF system is the fluidized-bed IQF freezing system, as illustrated in Figure 7.7. In these systems, the high-speed air is directed vertically upward through the mesh conveyor carrying product through the system. By careful adjustment of the air speed in relation to the product size, the product is lifted from the conveyor surface and remains suspended in the low-temperature air. Although the air flow is not sufficient to maintain the product in suspension at all times, the fluidized action results in the highest possible convective heat-transfer coefficients for the freezing process. This type of freezing process results in rapid freezing of product shapes and sizes that can be fluidized in the manner described. The use of the process is limited by the size of product that can be fluidized at air speeds of reasonable magnitude.

Figure 7.7 A fluidized-bed freezing system. (Courtesy of Frigoscandia Contracting, Inc.)

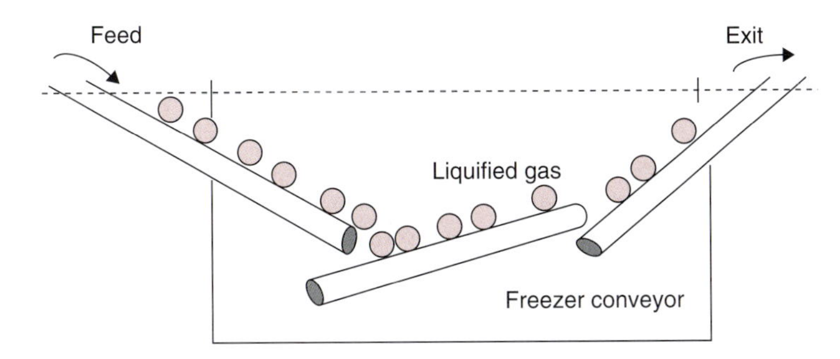

Figure 7.8 Schematic illustration of an immersion freezing system.

7.1.2.2 *Immersion*

By immersion of the food product in liquid refrigerant, the product surface is reduced to a very low temperature. Assuming the product objects are relatively small, the freezing process is accomplished very rapidly or under IQF conditions. For typical products, the freezing time is shorter than for the air-blast or fluidized-bed systems. As illustrated in Figure 7.8, the product is carried into a bath of liquid refrigerant and is conveyed through the liquid while the refrigerant changes from liquid to vapor and absorbs heat from the product. The most common refrigerants for this purpose are nitrogen and carbon dioxide.

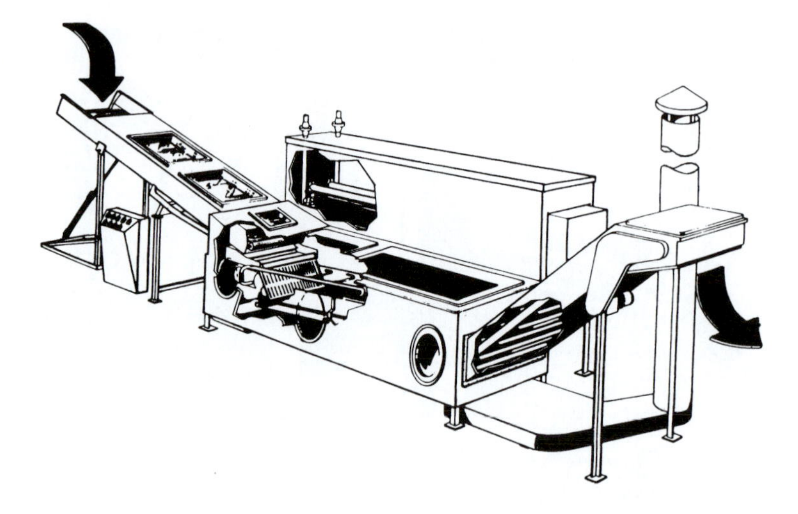

■ **Figure 7.9** Individual quick freezing (IQF) using liquid refrigerant. (Courtesy of Frigoscandia Contracting, Inc.)

A commercial version of the immersion IQF freezing system is shown in Figure 7.9. In this illustration, the product freezing compartment is filled with refrigerant vapors while the product is conveyed through the system. In addition, the product is exposed to a spray of liquid refrigerant, which absorbs thermal energy from the product while changing phase from liquid to vapor. One of the major disadvantages of immersion-type freezing systems is the cost of the refrigerant. Since the refrigerant changes from liquid to vapor while the product freezes, it becomes difficult to recover the vapors leaving the freezing compartment. These refrigerants are expensive, and the overall efficiency of the freezing system is a function of the ability to recover and reuse the vapors produced in the freezing compartments.

7.2 FROZEN-FOOD PROPERTIES

The freezing process has a dramatic influence on the thermal properties of the food product. Because of the significant amount of water in most foods and the influence of phase change on properties of water, the properties of the food product change in a proportional manner. As the water within the product changes from liquid to solid, the density, thermal conductivity, heat content (enthalpy), and apparent specific heat of the product change gradually as the temperature decreases below the initial freezing point for water in the food.

7.2.1 Density

The density of solid water (ice) is less than the density of liquid water. Similarly, the density of a frozen food will be less than the unfrozen

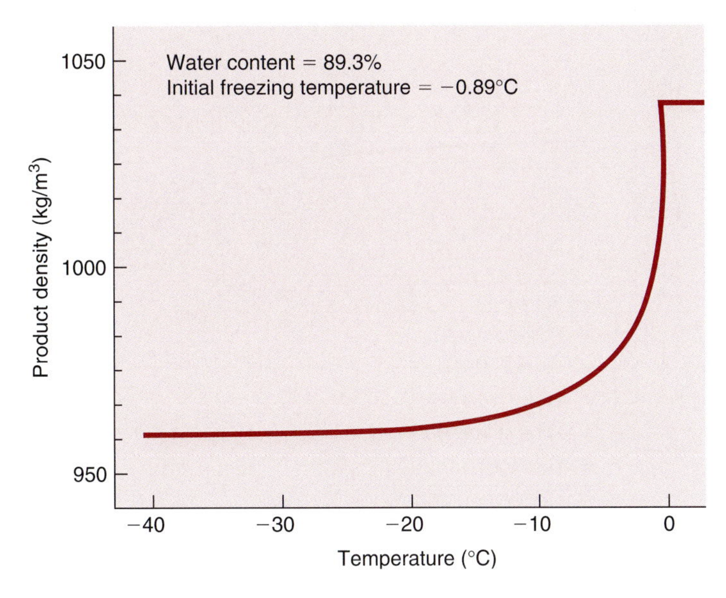

■ Figure 7.10 Influence of freezing on the predicted density of strawberries. (From Heldman and Lund, 2007)

product. Figure 7.10 illustrates the influence of temperature on density. The gradual change in density is due to the gradual change in the proportion of water frozen as a function of temperature. The magnitude of change in density is proportional to the moisture content of the product.

7.2.2 Thermal Conductivity

The thermal conductivity of ice is approximately a factor of four larger than the thermal conductivity of liquid water. This relationship has a similar influence on the thermal conductivity of a frozen food. Since the change of phase for water in the product is gradual as temperature decreases, the thermal conductivity of the product changes in a manner illustrated in Figure 7.11. The majority of the thermal conductivity increase occurs within 10 °C below the initial freezing temperature for the product. If the product happens to contain a fibrous structure, the thermal conductivity will be less when measured in a direction perpendicular to the fibers.

7.2.3 Enthalpy

The heat content or enthalpy of a frozen food is an important property in computations of refrigeration requirements for freezing the

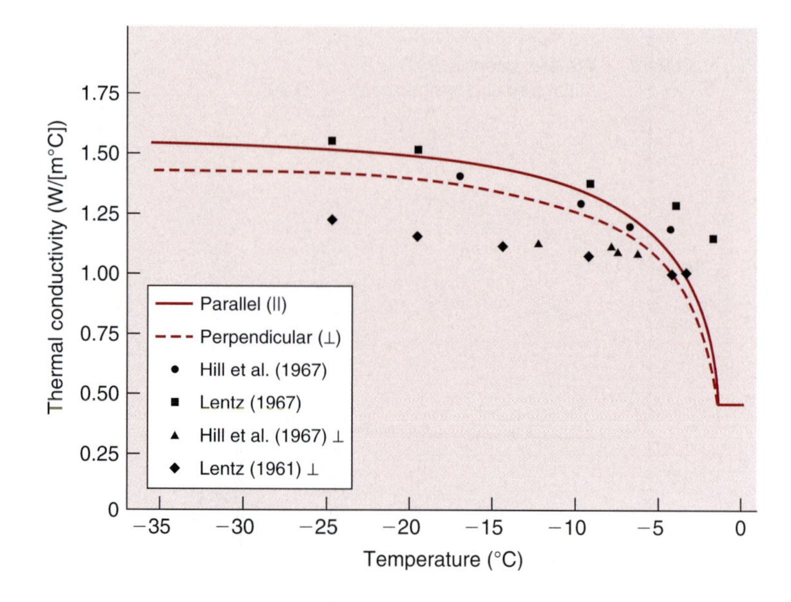

■ **Figure 7.11** Thermal conductivity of frozen lean beef as a function of temperature. (From Heldman and Lund, 2007)

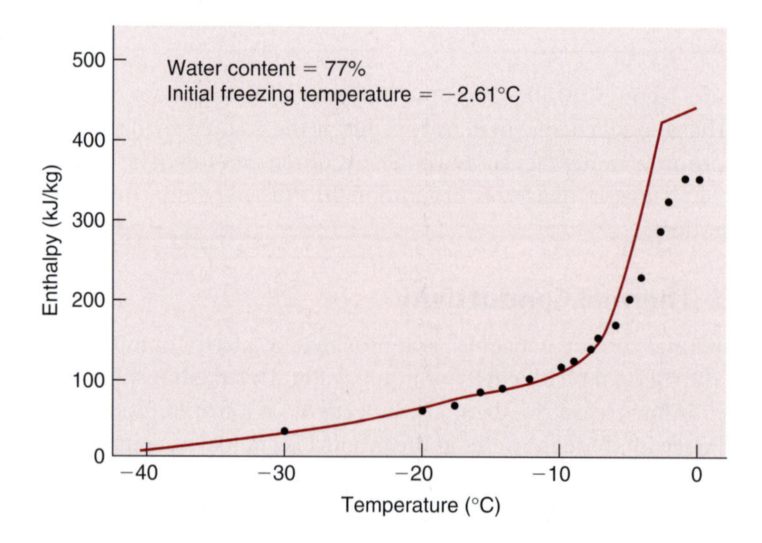

■ **Figure 7.12** Enthalpy of sweet cherries as a function of temperature. (From Heldman and Lund, 2007)

product. The heat content is normally zero at −40°C and increases with increasing temperature in a manner illustrated in Figure 7.12. Significant changes in enthalpy occur at 10°C just below the initial freezing temperature, when most of the phase change in product water occurs.

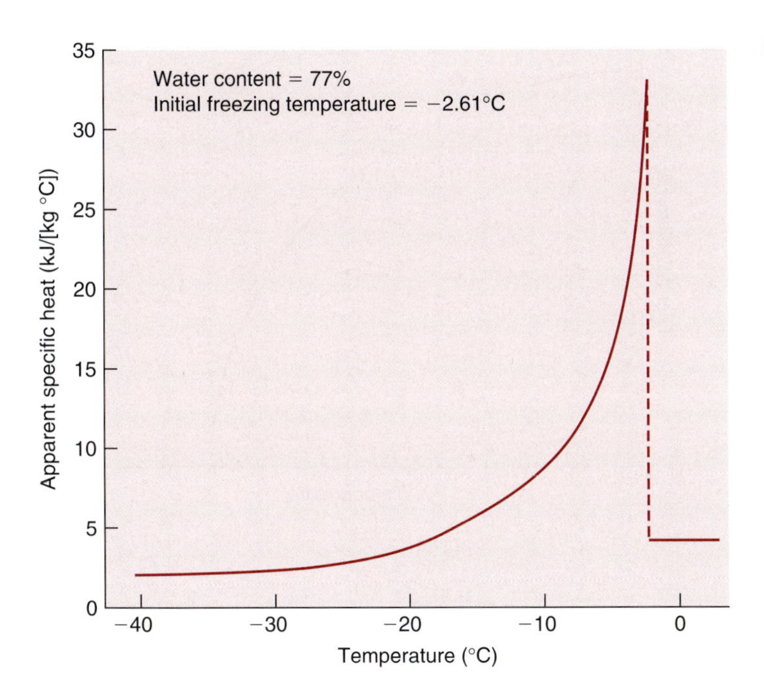

Figure 7.13 Predicted apparent specific heat of frozen sweet cherries as a function of temperature. (From Heldman and Lund, 2007)

7.2.4 Apparent Specific Heat

Based on the thermodynamic definition of specific heat, the profile of apparent specific heat for a food product as a function of temperature would appear as in Figure 7.13. This illustration reveals that the specific heat of a frozen food at a temperature greater than 20°C below the initial freezing point is not significantly different from the specific heat of the unfrozen product. The apparent specific heat profile clearly illustrates the range of temperature where most of the phase changes for water in the product occur.

7.2.5 Apparent Thermal Diffusivity

When the density, thermal conductivity, and apparent specific heat of a frozen food are combined to compute apparent thermal diffusivity, the profile in Figure 7.14 is obtained. The relationship illustrates that apparent thermal diffusivity increases gradually as the temperature decreases below the initial freezing point. The magnitudes of the property for the frozen product are significantly larger than for the unfrozen food.

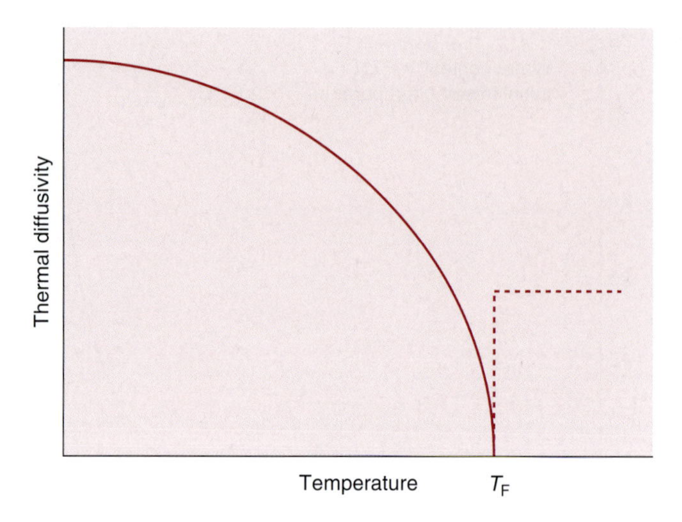

■ **Figure 7.14** Relationship between thermal diffusivity and temperature during freezing of food product as predicted from initial freezing temperature assumption. (From Heldman and Lund, 2007)

7.3 FREEZING TIME

A key calculation in the design of a freezing process is the determination of freezing time. Three distinct periods are noticeable at any location within a food undergoing freezing: prefreezing, phase change, and postfreezing. Consider a simple experiment that illustrates these three periods. First, we measure temperature change in freezing pure water into ice by placing an ice-cube tray filled with water in the freezer section of a home freezer, with a thermocouple located inside the tray. In the second part of the experiment, we measure temperature in a small stick of potato (e.g., a french fry) placed in a freezer with a thermocouple embedded inside the potato stick. The temperature–time plots for water and potato obtained in these experiments will be similar to those shown in Figure 7.15. During the precooling period, the temperature of water decreases to the freezing point as sensible heat is removed. The temperature plot shows a small amount of supercooling (below 0°C); once nucleation occurs and ice crystals begin to form, the freezing point increases to 0°C. The temperature remains at the freezing point until a complete phase change occurs as latent heat of fusion is removed from liquid water to convert it into solid ice. When all the liquid water has changed into solid ice, the temperature of the ice decreases rapidly as sensible heat is removed during the postfreezing period.

For the potato, the temperature plot obtained during freezing is similar to that of water but with important differences. Like water, the

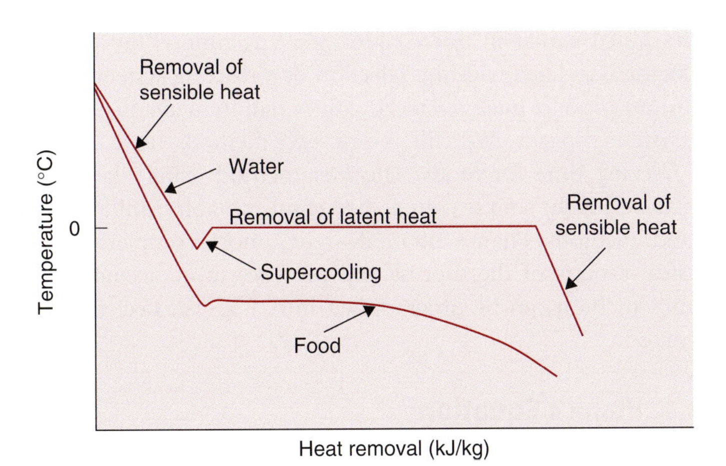

Figure 7.15 Freezing diagram of water and a food material.

temperature decreases during precooling as sensible heat is removed. However, the temperature at which initial nucleation occurs and ice crystals begin to form is lower than that of water due to the presence of solutes in the food. After a brief period of supercooling, latent heat is gradually removed with decreasing temperature. This deviation in temperature profile from that of pure water is the result of the concentration effect during freezing of foods. As water in the food converts into ice, the remaining water becomes more concentrated with solutes and depresses the freezing point. This gradual change in temperature with the additional removal of latent heat continues until the food is largely a mix of the initial solid food components and ice. After this time, mostly sensible heat is removed until some final, preselected, endpoint temperature is reached. Typically, fruits and vegetables are frozen to a temperature of −18 °C, and foods with higher fat content such as ice cream and fatty fish are frozen to lower temperatures of around −25 °C. We can draw several conclusions from these simple freezing experiments:

1. Freezing involves removal of both sensible and latent heat.
2. Freezing of pure water exhibits sharp transitions between the different freezing periods, whereas with foods, the transitions are more gradual.
3. At the endpoint temperature for freezing foods, the frozen food may still have some water present as a liquid; in fact, up to 10 percent water may be in liquid state for foods frozen to −18 °C.

This highly concentrated unfrozen water may play an important role in determining the storage of frozen foods.

As indicated earlier in this chapter, freezing time is the most critical factor associated with the selection of a freezing system to ensure optimum product quality. Freezing-time requirements help establish the system capacity. We will review two methods used in predicting freezing time for foods. The first method using Plank's equation is relatively simple, but it has some notable limitations. The second method—Pham's method—relies more completely on the physical aspects of the process and provides more accurate results. Pham's method can be programmed into a spreadsheet for ease of calculation.

7.3.1 Plank's Equation

The first and most popular equation for predicting freezing time was proposed by Plank (1913) and adapted to food by Ede (1949). This equation describes only the phase change period of the freezing process. Consider an infinite slab (Fig 7.16) of thickness a. We assume that the material constituting the slab is pure water. Because this method ignores the prefreezing step, the initial temperature of the slab is the same as the initial freezing point of the material, T_F. With water, the initial freezing point is 0 °C. The slab is exposed to a freezing medium (e.g., low-temperature air in a blast freezer) at temperature T_a. The heat transfer is one-dimensional. After some duration of time, there will be three layers: two frozen layers each of thickness x and a middle unfrozen layer. Consider the right half of the slab.

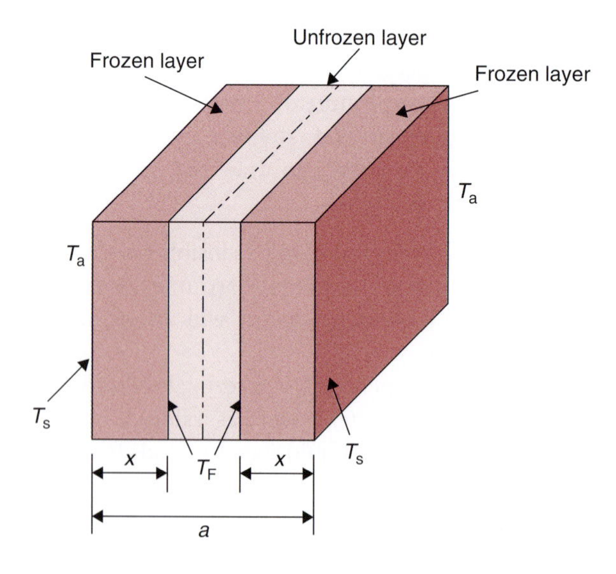

Figure 7.16 Use of Plank's equation in determining freezing time.

A moving front inside the slab separates the frozen from the unfrozen region. As water is converted into ice at the moving front, latent heat of fusion, L, is generated. This latent heat of fusion from the moving front must be transferred through the frozen layer to the outside freezing medium. The convective heat transfer coefficient at the surface of the slab is h. The temperature of the unfrozen region remains at TF until the freezing front moves all the way to the center plane of the slab. Next, we consider the rate of heat transfer, q, from the moving front to the surrounding freezing medium. There are two layers, a conductive frozen layer and a convective boundary layer. Thus, we can write the following expression:

$$q = \frac{A(T_F - T_a)}{\frac{1}{h} + \frac{x}{k_f}} \tag{7.1}$$

where the denominator is the sum of thermal resistances for the outer convective and internal conductive frozen layer. The moving front advances with a velocity of dx/dt, and the heat generated is the latent heat of fusion, L. Thus,

$$q = AL\rho_f \frac{dx}{dt} \tag{7.2}$$

Since all the heat generated at the freezing front must be transferred out to the surrounding medium, equating Equations (7.1) and (7.2), we obtain

$$L\rho_f \frac{dx}{dt} = \frac{(T_F - T_a)}{\frac{1}{h} + \frac{x}{k_f}} \tag{7.3}$$

Separating variables, rearranging the terms and setting up integrals, noting that the freezing process is completed when the moving front advances to the center of the slab, $a/2$, we get,

$$\int_0^{t_f} dt = \frac{L\rho_f}{(T_F - T_a)} \int_0^{a/2} \left[\frac{1}{h} + \frac{x}{k_f}\right] dx \tag{7.4}$$

Integrating, we obtain the freezing time, t_f,

$$t_f = \frac{L\rho_f}{(T_F - T_a)} \left[\frac{a}{2h} + \frac{a^2}{8k_f}\right] \tag{7.5}$$

We derived Equation (7.5) for an infinite slab. However, using the same steps we can obtain similar expressions for either an infinite cylinder or a sphere, but with different geometrical constants. Furthermore, to apply Equation (7.5) to a food material with a moisture content, m_m, we must replace latent heat of fusion of water, L, with L_f, the latent heat of fusion for the food material, or,

$$L_f = m_m L \tag{7.6}$$

where m_m is the moisture content of the food (fraction) and L is the latent heat of fusion of water, 333.3 kJ/(kg K).

Therefore, a general expression appropriate for a food material for calculating freezing time, known as Plank's equation, is

$$t_F = \frac{\rho_f L_f}{T_F - T_a}\left(\frac{P'a}{h} + \frac{R'a^2}{k_f}\right) \tag{7.7}$$

where ρ_f is the density of the frozen material, L_f is the change in the latent heat of the food (kJ/kg), T_F is the freezing temperature (°C), T_a is the freezing air temperature (°C), h is the convective heat transfer coefficient at the surface of the material (W/[m^2 °C]), a is the thickness/diameter of the object (m), k is the thermal conductivity of the frozen material (W/[m °C]), and the constants P' and R' are used to account for the influence of product shape, with $P' = \frac{1}{2}$, $R' = \frac{1}{8}$ for infinite plate; $P' = \frac{1}{4}$, $R' = \frac{1}{16}$ for infinite cylinder; and $P' = \frac{1}{6}$, $R' = \frac{1}{24}$ for sphere.

From Equation (7.7) it is evident that the freezing time t_F will increase with increasing density ρ_f, the latent heat of freezing L_f, and increasing size a. With an increase in the temperature gradient, the convective heat-transfer coefficient h, and the thermal conductivity k of frozen product, the freezing time will decrease. The dimension a is the product thickness for an infinite slab, and the diameter for an infinite cylinder or a sphere.

The limitations to Plank's equation are related primarily to assignment of quantitative values to the components of the equation. Density values for frozen foods are difficult to locate or measure. Although the initial freezing temperature is tabulated for many foods, the initial and final product temperatures are not accounted for in the equation for computation of freezing time. The thermal conductivity k should be for the frozen product, and accurate values are not readily available for most foods.

Even with these limitations, the ease of using Plank's equation has made it the most popular method for predicting freezing time. Most other available analytical methods are modifications of Plank's equation, with an emphasis on developments to overcome limitations to the original equation.

Example 7.1

A spherical food product is being frozen in an air-blast freezer. The initial product temperature is 10°C and the cold air −40°C. The product has a 7 cm diameter with density of 1000 kg/m^3, the initial freezing temperature is −1.25°C, the thermal conductivity of the frozen product is 1.2 W/(m K), and the latent heat of fusion is 250 kJ/kg. Compute the freezing time.

Given

Initial product temperature $T_i = 10°C$
Air temperature $T_\infty = -40°C$
Initial freezing temperature $T_F = -1.25°C$
Product diameter $a = 7\,cm = 0.07\,m$
Product density $\rho_f = 1000\,kg/m^3$
Thermal conductivity of frozen product $k = 1.2\,W/(m\,K)$
Latent heat $H_L = 250\,kJ/kg$
Shape constants for spheres:

$$P' = \frac{1}{6}$$

$$R' = \frac{1}{24}$$

Convective heat-transfer coefficient $h_c = 50\,W/(m^2\,K)$

Approach

Use Plank's equation (Eq. 7.7) and insert parameters as given to compute freezing time.

Solution

1. *Using Equation (7.7),*

$$t_F = \frac{(1000\,kg/m^3)(250\,kJ/kg)}{[-1.25°C - (-40°C)]} \times \left[\frac{0.07\,m}{6(50\,W/[m^2\,K])} + \frac{(0.07\,m)^2}{24(1.2\,W/[m\,K])}\right]$$

$$= (6.452 \times 10^3\,kJ/[m^3\,°C]) \times \left(2.33 \times 10^{-4}\,\frac{m^3\,K}{W} + 1.7014 \times 10^{-4}\,\frac{m^3\,K}{W}\right)$$

$$= 2.6\,kJ/W$$

2. *Since 1000 J = 1 kJ and 1 W = 1 J/s,*

$$t_F = 2.6 \times 10^3 \text{ s}$$
$$= 0.72 \text{ h}$$

7.3.2 Other Freezing-Time Prediction Methods

Numerous attempts have been made to improve freezing-time capabilities using analytical equations. These equations or approaches include those of Nagaoka et al. (1955), Charm and Slavin (1962), Tao (1967), Joshi and Tao (1974), Tien and Geiger (1967, 1968), Tien and Koumo (1968, 1969), and Mott (1964). In general, all these approaches have been satisfactory for conditions closely related to defined experimental conditions. In addition to analytical methods, numerical procedures have been developed to predict freezing times as reviewed by Cleland (1990) and Singh and Mannapperuma (1990). Mannapperuma and Singh (1989) found that prediction of freezing/thawing times of foods using a numerical method based on enthalpy formulation of heat conduction with gradual phase change provided good agreement with experimental results.

7.3.3 Pham's Method to Predict Freezing Time

Pham (1986) proposed a method for predicting food freezing and thawing times. His method can be used for finite-size objects of irregular shapes by approximating them to be similar to an ellipsoid. Another advantage of this method is that it is easy to use, yet it provides answers with reasonable accuracy. In the following sections, we will use this method to determine the freezing time for a one-dimensional infinite slab and then consider objects of other shapes. The following assumptions are used in developing this method:

- The environmental conditions are constant.
- The initial temperature, T_i, is constant.
- The value for the final temperature, T_c, is fixed.
- The convective heat transfer at the surface of an object is described by Newton's law of cooling.

Consider a freezing diagram, as shown in Figure 7.17. We will use a "mean freezing temperature," T_{fm}, to separate the diagram into two parts: one mostly for the precooling period with some phase change component, and the other comprising largely the phase change and

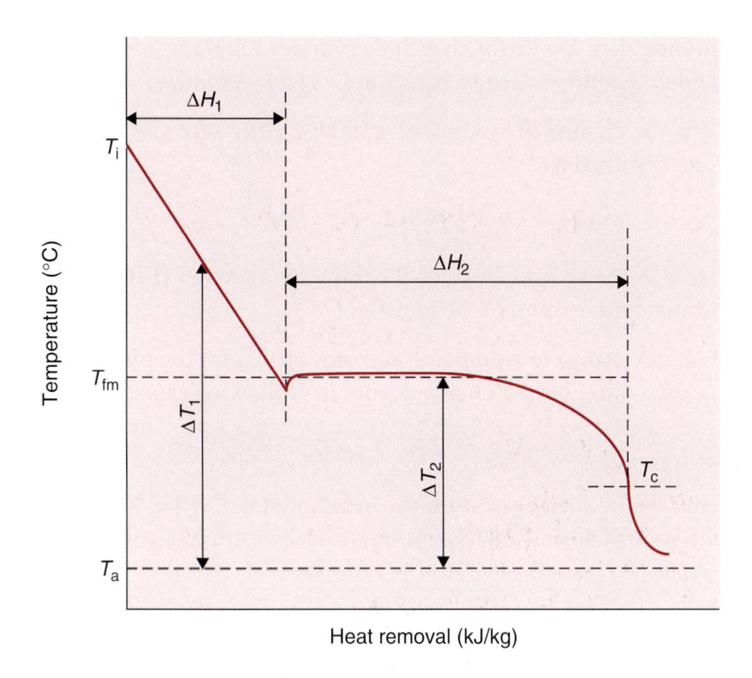

Figure 7.17 Freezing diagram of food, divided into sections for Pham's method.

the postcooling periods. Using experimental data obtained from freezing a wide variety of foods, Pham (1986) determined the following equation for T_{fm},

$$T_{fm} = 1.8 + 0.263T_c + 0.105T_a \tag{7.8}$$

where T_c is final center temperature (°C), and T_a is freezing medium temperature. Equation (7.8) is an empirically derived equation that is valid for most water-rich biological materials. This equation is the only empirical expression used in Pham's method.

The time for freezing of any simple-shaped object is calculated from the following equation:

$$t = \frac{d_c}{E_f h}\left[\frac{\Delta H_1}{\Delta T_1} + \frac{\Delta H_2}{\Delta T_2}\right]\left(1 + \frac{N_{Bi}}{2}\right) \tag{7.9}$$

where d_c is a characteristic dimension, either shortest distance to the center, or radius (m), h is the convective heat transfer coefficient (W/[m^2 K]), E_f is the shape factor, an equivalent heat transfer dimension. $E_f = 1$ for an infinite slab, $E_f = 2$ for an infinite cylinder, and $E_f = 3$ for a sphere. The other quantities used in Equation (7.9) are as follows.

ΔH_1 is the change in volumetric enthalpy (J/m^3) for the precooling period, obtained as

$$\Delta H_1 = \rho_u c_u (T_i - T_{fm}) \tag{7.10}$$

where c_u is the specific heat for the unfrozen material (J/[kg K]), Ti is the initial temperature of the material (°C).

ΔH_2 is the change in volumetric enthalpy (J/m^3) for the phase change and postcooling period obtained from the following expression:

$$\Delta H_2 = \rho_F [L_f + c_f (T_{fm} - T_c)] \tag{7.11}$$

where c_f is the specific heat for the frozen material (J/[kg K]), L_f is the latent heat of fusion of food (J/kg), and ρ_f is the density of frozen material.

The temperature gradients ΔT_1 and ΔT_2 are obtained from following equations:

$$\Delta T_1 = \left(\frac{T_i + T_{fm}}{2}\right) - T_a \tag{7.12}$$

$$\Delta T_2 = T_{fm} - T_a \tag{7.13}$$

Pham's procedure involves first calculating various factors given in Equation (7.8) and Equations (7.10) to (7.13) and then substituting in Equation (7.9) to obtain freezing time. Note that, depending upon the factor E_f, the equation is useful in determining freezing time for an infinite slab, infinite cylinder, or a sphere shape.

Example 7.2

Recalculate the freezing time in Example 7.1, using Pham's method with the following additional information. Final center temperature is −18°C, density of unfrozen product is 1000 kg/m^3, density of frozen product is 950 kg/m^3, moisture content of the product is 75%, specific heat of unfrozen product is 3.6 kJ/(kgK), and specific heat of frozen product is 1.8 kJ/(kgK).

Given

Initial product temperature = 10°C
Air temperature = −40°C
Product diameter = 0.07 m
Product density, unfrozen = 1000 kg/m^3

Product density, frozen = 950 kg/m³
Thermal conductivity of frozen product = 1.2 W/(m K)
Product specific heat, unfrozen = 3.6 kJ/(kg K)
Product specific heat, frozen = 1.8 kJ/(kg K)
Final center temperature = −18°C
Moisture content = 0.75

Approach

We will use Pham's method to calculate freezing time and compare the results with the answer obtained in Example 7.1 using Plank's equation.

Solution

1. Using Equation (7.8) calculate T_{fm}

$$T_{fm} = 1.8 + [0.263 \times (-18)] + [0.105 \times (-40)]$$
$$T_{fm} = -7.134°C$$

2. Using Equation (7.10) calculate ΔH_1

$$\Delta H_1 = 1000[kg/m^3] \times 3.6[kJ/(kg\ K)] \times 1000\ [J/kJ] \times (10 - (-7.134))[°C]$$
$$\Delta H_1 = 61{,}682{,}400\ J/m^3$$

3. Using Equation (7.11) calculate ΔH_2

$$\Delta H_2 = 950[kg/m^3] \times \left(\begin{array}{l} 0.75 \times 333.2[kJ/kg] \times 1000[J/kJ] + \{1.8[kJ/(kg\ K)] \\ \times 1000[J/kJ] \times (-7.134 - (-18))[°C]\} \end{array} \right)$$
$$\Delta H_2 = 255{,}985{,}860\ J/m^3$$

4. Using Equation (7.12) calculate ΔT_1

$$\Delta T_1 = \left(\frac{10 + (-7.134)}{2} \right) - (-40)$$
$$\Delta T_1 = 41.43°C$$

5. Using Equation (7.13) calculate ΔT_2

$$\Delta T_2 = (-7.134 - (-40))$$
$$\Delta T_2 = 32.87°C$$

6. Biot number is calculated as

$$N_{Bi} = \frac{50[W/(m^2\ K)] \times 0.035[m]}{1.2[W/(m\ K)]}$$
$$N_{Bi} = 1.46$$

7. *Substituting results of steps (1) through (6) in Equation (7.9), noting that for a sphere, $E_f = 3$,*

$$t = \frac{0.035[m]}{3 \times 50[W/(m^2\ K)]} \times \left(\frac{61{,}682{,}400[J/m^3]}{41.43[°C]} + \frac{255{,}985{,}860[J/m^3]}{32.87[°C]} \right) \left(1 + \frac{1.46}{2}\right)$$

$$time = 3745.06\ s = 1.04\ h$$

8. *As expected, freezing time predicted with Plank's equation is shorter (0.72 h) when compared with prediction using Pham's method (1.04 h). The main reason for this difference is that Plank's equation does not account for the time required for sensible heat removal during prefreezing and postfreezing periods.*

7.3.4 Prediction of Freezing Time of Finite-Shaped Objects

Pham's method can also be used to calculate freezing times of other shapes such as a finite cylinder, infinite rectangular rod, and rectangular brick, which are commonly encountered in freezing foods. Pham's equation, Equation (7.9), may be used for this purpose using an appropriate value of the shape factor, E_f. In order to calculate E_f, two-dimensional ratios are required, β_1 and β_2. Referring to Figure 7.18 for the finite object shown,

$$\beta_1 = \frac{\text{second shortest dimension of object}}{\text{shortest dimension}} \qquad (7.14)$$

■ **Figure 7.18** Determining shape factors of finite objects.

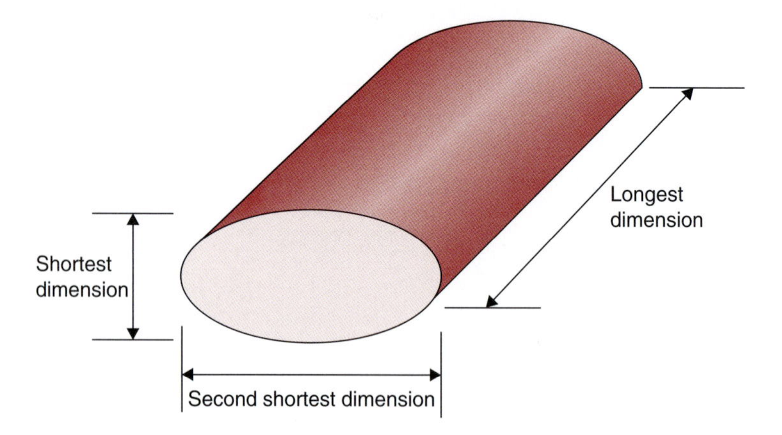

and

$$\beta_2 = \frac{\text{longest dimension of object}}{\text{shortest dimension}} \tag{7.15}$$

The equivalent dimension E_f is obtained as follows:

$$E_f = G_1 + G_2E_1 + G_3E_2 \tag{7.16}$$

where values of G_1, G_2, and G_3 are obtained from Table 7.1, and E_1 and E_2 are obtained from the following equations:

$$E_1 = \frac{X_1}{\beta_1} + [1 - X_1]\frac{0.73}{\beta_1^{2.5}} \tag{7.17}$$

Table 7.1 G Values for Different Shapes

	G_1	G_2	G_3
Finite cylinder, height < diameter	1	2	0
Finite cylinder, height > diameter	2	0	1
Rectangular rod	1	1	0
Rectangular brick	1	1	1

and

$$E_2 = \frac{X_2}{\beta_2} + [1 - X_2]\frac{0.73}{\beta_2^{2.5}} \tag{7.18}$$

where factors X_1 and X_2 are obtained from

$$X_1 = \frac{2.32\beta_1^{-1.77}}{(2N_{Bi})^{1.34} + 2.32\beta_1^{-1.77}} \tag{7.19}$$

and

$$X_2 = \frac{2.32\beta_2^{-1.77}}{(2N_{Bi})^{1.34} + 2.32\beta_2^{-1.77}} \tag{7.20}$$

Example 7.3

Lean beef in the shape of a large slab with 1 m length, 0.6 m width and 0.25 m thickness is to be frozen in an air-blast freezer with a Biot Number of 2.5. Calculate the shape factor from the given dimensions.

Given

Length = 1 m

Width = 0.6 m

Thickness = 0.25 m

$N_{Bi} = 2.5$

Approach

We will use Equations (7.14) through (7.20) to determine the shape factor E_f for this finite slab.

Solution

1. *The longest dimension is 1 m, the shortest dimension is 0.25 m. Therefore*

$$\beta_1 = \frac{0.6}{0.25} = 2.4$$

and

$$\beta_2 = \frac{1}{0.25} = 4$$

2. *From Equation (7.19)*

$$X_1 = \frac{2.32 \times 2.4^{-1.77}}{(2 \times 2.5)^{1.34} + (2.32 \times 2.4^{-1.77})}$$

$$X_1 = 0.05392$$

and from Equation (7.20)

$$X_2 = \frac{2.32 \times 4^{-1.77}}{(2 \times 2.5)^{1.34} + (2.32 \times 4^{-1.77})}$$

$$X_2 = 0.02256$$

3. *From Equations (7.17) and (7.18) we calculate E_1 and E_2*

$$E_1 = \frac{0.05393}{2.4} + [1 - 0.05393]\frac{0.73}{2.4^{2.5}}$$

$$E_1 = 0.09987$$

$$E_2 = \frac{0.02256}{4} + [1 - 0.02256]\frac{0.73}{4^{2.5}}$$

$$E_2 = 0.027938$$

4. *From Table 7.1 we obtain G_1, G_2 and G_3, as 1, 1, 1, respectively.*

5. *From Equation (7.16)*

$$E_f = 1 + 0.09987 + 0.02794$$
$$E_f = 1.128$$

6. *The shape factor for the finite-slab-shaped beef is 1.128. This is expected, as the value should be greater than 1 but less than 2 (E_f for an infinite cylinder).*

Example 7.4

Lean beef with 74.5% moisture content and 1 m length, 0.6 m width, and 0.25 m thickness is being frozen in an air-blast freezer with $h_c = 30\,\text{W/(m}^2\,\text{K)}$ and air temperature of $-30°C$. If the initial product temperature is 5°C, estimate the time required to reduce the product temperature to $-10°C$. An initial freezing temperature of $-1.75°C$ has been measured for the product. The thermal conductivity of frozen beef is 1.5 W/(m K), and the specific heat of unfrozen beef is 3.5 kJ/(kg K). A product density of 1050 kg/m^3 can be assumed, and a specific heat of 1.8 kJ/(kg K) for frozen beef can be estimated from properties of ice.

Given

Product length $d_2 = 1$ m
Product width $d_1 = 0.6$ m
Product thickness $a = 0.25$ m
Convective heat-transfer coefficient $h_c = 30\,\text{W/(m}^2\,\text{K)}$
Air temperature $T_\infty = -30°C$
Initial product temperature $T_i = 5°C$
Initial freezing temperature $T_F = -1.75°C$
Product density $\rho = 1050\,\text{kg/m}^3$
Enthalpy change (ΔH) = 0.745(333.22 kJ/kg) = 248.25 kJ/kg (estimated for moisture content of product)
Thermal conductivity k of frozen product = 1.5 W/(m K)
Specific heat of product (c_{pu}) = 3.5 kJ/(kg K)
Specific heat of frozen product (c_{pf}) = 1.8 kJ/(kg K)

Approach

The freezing time will be computed using Pham's model. The problem will be solved using a spreadsheet.

Solution

Figure E.7.1 shows the spreadsheet solution for the given data. It will require 26.6 hours to accomplish freezing.

	A	B	C	D	E	F	G	H
1	Given							
2	T_i	5						
3	T_a	−30						
4	T_c	−10						
5	c_{pu}	3.5						
6	rho	1050						
7	mc	0.745						
8	cpf	1.8						
9	rhou	1050						
10	rhof	1050						
11	h3	0						
12	kf	1.5						
13	dc	0.125						
14								
15	beta1	2.4	← =0.6/0.25					
16	beta2	4	← =1/0.25					
17	X1	0.0539285	← =(2.32*B15^−1.77)/((2*B28)^1.34+2.32*B15^−1.77)					
18	X2	0.0225586	← =(2.32*B16^−1.77)/((2*B28)^1.34+2.32*B16^−1.77)					
19	E1	0.0998662	← =B17/B15+(1−B17)*0.73/(B15^2.5)					
20	E2	0.0279375	← =B18/B16+(1−B18)*0.73/(B16^2.5)					
21	E	1.1278038	← =1+1*B19+1*B20					
22								
23	Tfm	−3.98	← =1.8+0.263*B4+0.105*B3					
24	DH1	33001500	← =B9*B5*1000*(B2−B23)					
25	DH2	272023500	← =B10*(B7*333.2*1000+B8*1000*(B23−B4))					
26	DT1	30.51	← =(B2+B23)/2−B3					
27	DT2	26.02	← =B23−B3					
28	Biot No	2.5	← =B11*B13/B12					
29								
30	Time (s)	95894.858	← =(B13/(B21*B11))*(B24/B26+B25/B27)*(1+B28/2)					
31	Time (hr)	26.637461	← =B30/3600					

Figure E7.1 Spreadsheet solution of Example 7.4.

7.3.5 Experimental Measurement of Freezing Time

Experimental methods are needed to evaluate situations where freezing times must be verified or where predicting the freezing time is difficult. Methods should be designed to simulate the actual conditions as closely as possible and provide for measurement of temperature history at a minimum of one location as the freezing process is completed. If only one location is used, the temperature sensor should be located at the slowest-cooling point in the product or at a well-defined location near the slowest-cooling point. The boundary conditions should be identical to the actual conditions in terms of medium temperature and factors influencing the magnitude of the convective heat-transfer coefficient. Under some conditions, the temperature history required to evaluate freezing time should be measured for different variables influencing the magnitude of freezing time.

7.3.6 Factors Influencing Freezing Time

As indicated by Plank's equation, several factors influence freezing time and will influence the design of equipment used for food freezing. One of the first factors is the freezing-medium temperature, where lower magnitudes will decrease freezing time in a significant manner. The product size influences freezing time, and the effect can be quite significant.

The factor with the most significant influence on freezing time is the convective heat-transfer coefficient, h. This parameter can be used to influence freezing time through equipment design and should be analyzed carefully. At low magnitudes of the convective heat-transfer coefficient, small changes will influence the freezing time in a significant manner. Both initial and final product temperatures will influence freezing times slightly but are not accounted for in the original Plank's equation. Product properties (T_F, ρ, k) will influence freezing-time predictions, as indicated in Plank's equation. Although these factors are not variables to be used in equipment design, the selection of appropriate values is important in accurate freezing-time prediction. A detailed analysis of all factors influencing freezing-time prediction has been presented by Heldman (1983).

7.3.7 Freezing Rate

The freezing rate (°C/h) for a product or a package is defined as the difference between the initial and the final temperature divided by the freezing time (IIR, 1986). Since the temperature at different locations of a product may vary during freezing, a *local freezing rate* is defined for a given location in a product as the difference between the initial temperature and the desired temperature, divided by the time elapsed until the moment at which the desired temperature is achieved at that location.

7.3.8 Thawing Time

In industrial processes, frozen foods may be thawed for additional processing. Although freezing and thawing processes have some similarities in that both involve phase change processes, there are also a number of differences. For example, surface boundary conditions during thawing become complicated due to the formation and later melting of frost on the surface (Mannapperuma and Singh, 1989). The following equation proposed by Cleland is relatively easy to use for predicting thawing times:

$$t_t = \frac{d_c}{E_f h} \frac{\Delta H_{10}}{(T_a - T_F)} (P_1 + P_2 N_{Bi}) \tag{7.21}$$

where

$$\text{Biot number, } N_{Bi} = \frac{hd_c}{k_u} \quad (7.22)$$

$$\text{Stefan number, } N_{Ste} = \rho_u c_u \frac{(T_a - T_F)}{\Delta H_{10}} \quad (7.23)$$

$$\text{Plank number, } N_{Pk} = \rho_f c_f \frac{(T_F - T_i)}{\Delta H_{10}} \quad (7.24)$$

$$P_1 = 0.7754 + 2.2828 N_{Ste} \times N_{Pk} \quad (7.25)$$

$$P_2 = 0.5(0.4271 + 2.122 N_{Ste} - 1.4847 N_{Ste}^2) \quad (7.26)$$

ΔH_{10} is the volumetric enthalpy change of the product from 0 to −10°C.

7.4 FROZEN-FOOD STORAGE

Although the efficiency of food freezing is influenced most directly by the freezing process, the quality of a frozen food is influenced significantly by the storage conditions. Since the influence of factors causing quality loss is reduced at lower temperatures, the storage temperature for the frozen food is very important. Major consideration must be given to using the lowest storage temperatures feasible in terms of extending storage life without using refrigeration energy inefficiently.

The most detrimental factor influencing frozen food quality is a fluctuation in storage temperature. If frozen foods are exposed to temperature cycles resulting in changes in product temperature, the storage life is reduced significantly. The quantitative aspects of frozen-food storage life have been investigated by Schwimmer et al. (1955), Van Arsdel and Guadagni (1959), and summarized by Van Arsdel et al. (1969). More recently, Singh and Wang (1977) and Heldman and Lai (1983) have described numerical and computer predictions of frozen-food storage time.

Based on experimental data for storage of frozen foods, recommendations on frozen-food storage have been developed (IIR, 1986).

7.4.1 Quality Changes in Foods during Frozen Storage

A commonly used descriptor for the storage life of frozen foods is the practical storage life (PSL). The practical storage life of a product is the period of frozen storage after freezing during which the product retains its characteristic properties and remains suitable for consumption or other intended process (IIR, 1986).

Table 7.2 lists the practical storage life of a variety of frozen foods. The shelf life of frozen fish is considerably less than that of most other commodities. The typical temperature used for storing frozen foods in the commercial food chain is −18 °C. However, for seafood, lower storage temperatures may be more desirable for retaining quality.

Another term used to describe the storage life of frozen foods is the high-quality life (HQL). As defined by IIR (1986), the high-quality life is the time elapsed from freezing of an initially high-quality product and the moment when, by sensory assessment, a statistically significant difference ($p < 0.01$) from the initial high quality (immediately after freezing) can be established. The observed difference is termed the *just noticeable difference* (JND). When a triangular test is used for sensory assessment of the quality, a just noticeable difference can be postulated when 70% of experienced panelists successfully distinguish the product from the control sample stored under such conditions as have been proven to produce no detectable degradation during the time period under consideration (IIR, 1986). A typical temperature used for control experiments is −35 °C.

The quality loss in frozen foods can be predicted using the experimentally obtained data on acceptable time for storage. Singh and Wang (1977) and Heldman and Lai (1983) have described numerical and computer predictions of storage time based on a kinetic analysis of changes occurring in foods during frozen storage.

Jul (1984) has provided values for typical exposure times that a product may encounter in different components of the frozen food chain. Figure 7.19 presents data on acceptable times for storage of frozen strawberries (Jul, 1984). The acceptable shelf life in this figure is based on experimental data. Looking at Table 7.3, column 1 describes a frozen chain, starting with the producer's warehouse and finishing at the consumer's freezer. Columns 2 and 3 denote the expected time and temperature at various locations in the chain. Column 4 is the acceptable days corresponding to the temperature as obtained from Figure 7.19. Values in column 5 are the inverse of those in column 4, multiplied by 100 to obtain percent loss per day. Column 6 gives the calculated values of loss obtained by multiplying values in column 2 with those in column 5. Thus, in the example shown in Table 7.3, the frozen strawberries have lost 77.3% of their acceptable quality after

Table 7.2 The Practical Storage Life (PSL) of Frozen Foods at Several Storage Temperatures

Product	Storage time (months)		
	−12°C	−18°C	−24°C
Fruits			
Raspberries/Strawberries (raw)	5	24	>24
Raspberries/Strawberries in sugar	3	24	>24
Peaches, Apricots, Cherries (raw)	4	18	>24
Peaches, Apricots, Cherries in sugar	3	18	>24
Fruit juice concentrate	–	24	>24
Vegetables			
Asparagus (with green spears)	3	12	>24
Beans, green	4	15	>24
Beans, lima	–	18	>24
Broccoli	–	15	24
Brussels sprouts	6	15	>24
Carrots	10	18	>24
Cauliflower	4	12	24
Corn-on-the-cob	–	12	18
Cut corn	4	15	>24
Mushrooms (cultivated)	2	8	>24
Peas, green	6	24	>24
Peppers, red and green	–	6	12
Potatoes, French fried	9	24	>24
Spinach (chopped)	4	18	>24
Onions	–	10	15
Leeks (blanched)	–	18	–
Meats and poultry			
Beef carcass (unpackaged)[a]	8	15	24
Beef steaks/cuts	8	18	24
Ground beef	6	10	15
Veal carcass (unpackaged)a	6	12	15
Veal steaks/cuts	6	12	15
Lamb carcass, grass fed (unpackaged)[a]	18	24	>24
Lamb steaks	12	18	24
Pork carcass (unpackaged)[a]	6	10	15
Pork steaks/cuts	6	10	15
Sliced bacon (vacuum packed)	12	12	12
Chicken, whole	9	18	>24
Chicken, parts/cuts	9	18	>24
Turkey, whole	8	15	>24
Ducks, geese, whole	6	12	18
Liver	4	12	18

(Continued)

Table 7.2 (*Continued*)

Product	Storage time (months)		
	−12°C	−18°C	−24°C
Seafood			
Fatty fish, glazed	3	5	>9
Lean fish[b]	4	9	>12
Lobster, crab, shrimps in shell (cooked)	4	6	>12
Clams and oysters	4	6	>9
Shrimps (cooked/peeled)	2	5	>9
Eggs			
Whole egg magma	–	12	>24
Milk and milk products			
Butter, lactic, unsalted pH 4.7	15	18	20
Butter, lactic, salted pH 4.7	8	12	14
Butter, sweet-cream, unsalted pH 6.6	–	>24	>24
Butter, sweet-cream, salt (2%) pH 6.6	20	>24	>24
Cream	–	12	15
Ice cream	1	6	24
Bakery and confectionary products			
Cakes (cheese, sponge, chocolate, fruit, etc.)	–	15	24
Breads	–	3	–
Raw dough	–	12	18

Source: IIR (1986)
[a]*Carcass may be wrapped in stockinette.*
[b]*The PSL for single fillets of lean fish would be 6, 9, and 12 months at −18°C, −24°C, and −30°C, respectively.*

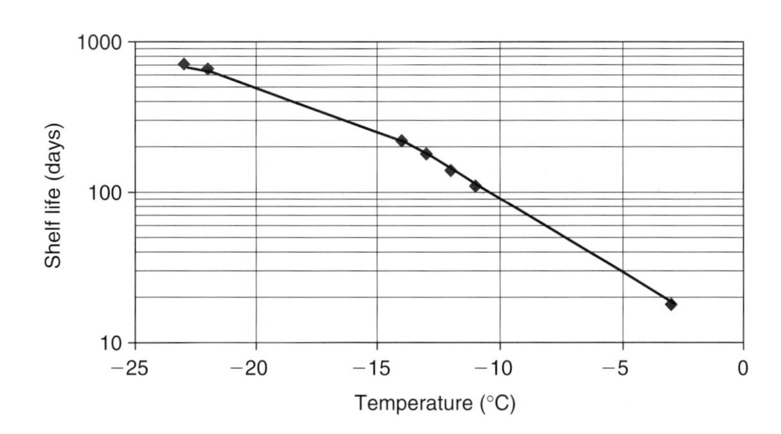

Figure 7.19 Semilogarithmic plot of shelf life versus temperature for frozen strawberries.

Table 7.3 An Example of the Conditions of Storage and Quality Loss in Strawberries at Different Stages in a Frozen Food Chain

Stage	Time (days)	Temperature (°C)	Acceptability (days)	Loss per day (percent/day)	Loss percent
Producer	250	−22	660	0.15152	37.88
Transport	2	−14	220	0.45455	0.91
Wholesale	50	−23	710	0.14085	7.04
Transport	1	−12	140	0.71429	0.71
Retail	21	−11	110	0.90909	19.09
Transport	0.1	−3	18	5.55556	0.56
Home freezer	20	−13	180	0.55556	11.11
Total storage (days) = 344.1			Total quality loss (percent) = 77.30		

Source: Jul (1984)

344.1 days in the frozen food chain. This analysis helps to identify those components of the chain where major losses are occurring. For the example shown in Table 7.3, major losses in quality occur in the producer's and retailer's warehouses. A reduction of either storage temperature or time in the identified components can help reduce quality loss for a given food.

PROBLEMS

7.1 Estimate the latent heat of fusion for a food product with 68% moisture content.

7.2 A food product with 82% moisture content is being frozen. Estimate the specific heat of the product at −10 °C when 80% of the water is in a frozen state. The specific heat of dry product solid is 2.0 kJ/(kg °C). Assume specific heat of water at −10 °C is similar to specific heat of water at 0 °C.

7.3 A 5 cm-thick beef steak is being frozen in a −30 °C room. The product has 73% moisture content, density of 970 kg/m^3, and thermal conductivity (frozen) of 1.1 W/(m K). Estimate the freezing time using Plank's equation. The product has an initial freezing temperature of −1.75 °C, and the movement of air in the freezing room provides a convective heat-transfer coefficient of 5 W/(m^2 K).

***7.4** Partially frozen ice cream is being placed in a package before completion of the freezing process. The package has dimensions of 8 cm by 10 cm by 20 cm and is placed in air-blast freezing with convective heat coefficient of 50 W/(m^2 K) for freezing. The product temperature is −5°C when placed in the package, and the air temperature is −25°C. The product density is 700 kg/m^3, the thermal conductivity (frozen) is 1.2 W/(m K), and the specific heat of the frozen product is 1.9 kJ/(kg K). If the latent heat to be removed during blast freezing is 100 kJ/kg, estimate the freezing time.

***7.5** A food product with 80% moisture content is being frozen in a 6 cm-diameter can. The product density is 1000 kg/m^3, the thermal conductivity is 1.0 W/(m K), and the initial freezing temperature is −2°C. After 10 h in the −15°C freezing medium, the product temperature is −10°C. Estimate the convective heat-transfer coefficient for the freezing medium. Assume the can has infinite height.

7.6 Develop a spreadsheet to solve Example 7.2. Determine time for freezing for the following h_c values: 30, 50, 80, and 100 W/(m^2 °C).

7.7 Using the spreadsheet developed in Example 7.4, determine the time for freezing if the shape of lean beef is assumed to be

a. a finite cylinder, with a diameter of 0.5 m and 1 m long.
b. an infinite cylinder with a diameter of 0.5 m.
c. a sphere of diameter 0.5 m.

7.8 A fabricated food, in the form of small spherical pellets, is to be frozen in an air-blast freezer. The air-blast freezer is operating with air at −40°C. The initial product temperature is 25°C. The pellets have a diameter of 1 cm, and the density of the product is 980 kg/m^3. The initial freezing temperature is −2.5°C. The latent heat of fusion for this product is 280 kJ/kg. The thermal conductivity of the frozen product is 1.9 W/(m°C). The convective heat transfer coefficient is 50 W/(m^2°C). Calculate the freezing time.

---

*Indicates an advanced level of difficulty in solving.

7.9 Plank's equation is being used to estimate the convective heat transfer coefficient in a food being frozen in an air-blast freezer. It took a product 20 minutes to freeze in the freezer. The product is in the shape of an infinite cylinder with a diameter of 2 cm. The properties of the product are as follows: thermal conductivity of frozen material = 1.8 W/(m°C), density = 890 kg/m^3, and latent heat of fusion = 260 kJ/kg, initial freezing point = −1.9°C. The initial temperature of the product is 25°C and the air temperature is −35°C.

7.10 Using Plank's equation, determine the freezing time for a potato sphere with a moisture content of 88%. The potato will be frozen in a blast freezer where air is available at −40°C and the convective heat transfer coefficient is 40 W/(m^2°C). The thermal conductivity of frozen potato is estimated to be 1.3 W/(m°C) and its density is 950 kg/m^3. The initial freezing temperature of potato is −2°C. The diameter of the potato sphere is 2 cm.

7.11 Calculate the shape factor E_f for the following objects; assume that Biot number is 1.33 in each case:

a. A meat loaf has the following dimensions: length = 25 cm, width = 12 cm, height = 10 cm.
b. A finite cylinder has the following dimensions: length = 25 cm, diameter = 12 cm.
c. A sphere of diameter = 12 cm.
d. What conclusions can you draw from the results obtained for the three shapes?

7.12 Using Pham's method, calculate the time for freezing a meatloaf (moisture content = 85%) with the following dimensions: length = 25 cm, width = 12 cm, height = 10 cm. The convective heat transfer coefficient is 40 W/(m^2 °C). The air temperature is −40°C. The initial temperature of the product is 10°C and the required final temperature of the frozen meatloaf is −18°C. The initial freezing temperature is −1.8°C. The thermal conductivity of the frozen meat is 1.5 W/(m °C) and the specific heat of the unfrozen meat is 3.4 kJ/(kg °C). The specific heat of frozen meat is 1.9 kJ/(kg °C). The density of the meatloaf is 1020 kg/m^3.

7.13 A frozen dessert is to be marketed in a rectangular package that has dimensions of 15 cm long, 10 cm wide, and 7 cm high. In the manufacturing process, the dessert mix is prepared and filled into the package at 1°C. The moisture content of the mix is 90%. The specific heat of unfrozen dessert mix is 3.5 kJ/(kg °C). The package is then placed in a blast freezer where the convective heat transfer coefficient is 35 W/(m^2 °C), and the air temperature is −40°C. The final temperature of frozen dessert is −25°C. The properties of the mix are as follows: unfrozen dessert density = 750 kg/m^3, thermal conductivity of frozen dessert is 1.3 W/(m °C), specific heat of the frozen dessert is 1.85 kJ/(kg °C). The latent heat removed during the blast freezing is 120 kJ/kg. Estimate the freezing time using Pham's method.

7.14 Use MATLAB® to plot the enthalpy versus temperature for peas over the range of temperature from −30°C to 20°C using the data in Table A.2.7. On the same graph, plot the temperature versus enthalpy for peas using the empirical equation given by Pham et al. (1994).

$$H = \begin{cases} A + c_f T + \dfrac{B}{T}; & T \leq T_i \\ H_0 + c_u T; & T > T_i \end{cases}$$

Pham et al. reported parameter values for the enthalpy of peas to be

$A = 90.3$ kJ/kg
$c_f = 2.45$ kJ/kg − °C
$B = -199$ kJ °C/kg
$T_i = -0.75$°C
$H_0 = 357$ kJ/kg
$c_u = 4.17$ kJ/kg °C

Discuss possible reasons for differences in the results.

7.15 Use MATLAB® to plot the enthalpy versus temperature for beef over the range of temperature from −30°C to 20°C using the data in Table A.2.7. On the same graph, plot the temperature versus enthalpy for lean beef and ground hamburger patties using the empirical equation given by Pham et al. (1994).

$$H = \begin{cases} A + c_f T + \dfrac{B}{T}; & T \leq T_i \\ H_0 + c_u T; & T > T_i \end{cases}$$

Pham et al. reported parameter values for the enthalpy of lean beef to be

$A = 83.0$ kJ/kg, $c_f = 2.26$ kJ/kg °C, $B = -163$ kJ °C/kg,
$T_i = -0.70$°C

$H_0 = 317\,\text{kJ/kg}$, $c_u = 3.64\,\text{kJ/kg °C}$

Pham et al. reported parameter values for the enthalpy of hamburger patties to be

$A = 53.3\,\text{kJ/kg}$, $c_f = 1.59\,\text{kJ/kg °C}$, $B = -444\,\text{kJ °C/kg}$, $T_i = -1.94\,°\text{C}$
$H_0 = 285\,\text{kJ/kg}$, $c_u = 3.53\,\text{kJ/kg °C}$

Discuss possible reasons for differences in the results.

LIST OF SYMBOLS

A	area (m^2)
a	thickness of plate (m)
c_f	specific heat of frozen material (J/[kg °C])
c_u	specific heat of unfrozen material (J/[kg °C])
d_c	characteristic dimension (m)
E_1, E_2	shape constants
E_f	shape factor
G_1, G_2, G_3	constants
h	convective heat transfer coefficient (W/[m^2 °C])
k_f	thermal conductivity of frozen material (W/[m °C])
L	latent heat of fusion of water, 333.2 (kJ/kg)
L_f	latent heat of fusion for food material (kJ/kg)
m_m	moisture content (fraction)
N_{Bi}	Biot number, dimensionless
N_{Pk}	Plank number, dimensionless
N_{Pr}	Prandtl number, dimensionless
N_{Ste}	Stefan number, dimensionless
P'	constant
P_1, P_2	constants
q	rate of heat transfer (W)
R'	constant
T_a	freezing medium temperature (°C)
T_c	final center temperature (°C)
T_F	initial freezing temperature (°C)
T_{fm}	mean freezing temperature (°C)
T_i	initial temperature (°C)
T_s	surface temperature (°C)
t_t	thawing time (s)
x	space coordinate
ΔH_1	change in volumetric enthalpy for precooling (J/m^3)
ΔH_{10}	volumetric enthalpy change from 0°C to −10°C

ΔH_2	change in volumetric enthalpy for phase change and postfreezing (J/m^3)
β_1, β_2	dimensionless ratio
ρ	density (kg/m^3)
ρ_f	density of frozen material (kg/m^3)
ρ_u	density of unfrozen material (kg/m^3)

BIBLIOGRAPHY

Chandra, P. K. and Singh, R. P. (1994). *Applied Numerical Methods for Agricultural Engineers*. CRC Press, Boca Raton, Florida.

Charm, S. E. and Slavin, J. (1962). A method for calculating freezing time of rectangular packages of food. *Annexe. Bull. Inst. Int. Froid.* 567–578.

Cleland, A. C. (1990). *Food Refrigeration Processes, Analysis, Design and Simulation*. Elsevier Applied Science, New York.

Cleland, A. C. and Earle, R. L. (1976). A new method for prediction of surface heat-transfer coefficients in freezing. *Annexe. Bull. Inst. Int. Froid.* **1**: 361.

Cleland, A. C. and Earle, R. L. (1977). A comparison of analytical and numerical methods of predicting the freezing times of foods. *J. Food Sci.* **42**: 1390–1395.

Cleland, A. C. and Earle, R. L. (1979a). A comparison of methods for predicting the freezing times of cylindrical and spherical foodstuffs. *J. Food Sci.* **44**: 958–963.

Cleland, A. C. and Earle, R. L. (1979b). Prediction of freezing times for foods in rectangular packages. *J. Food Sci.* **44**: 964–970.

Cleland D. J. (1991). A generally applicable simple method for prediction of food freezing and thawing times. Paper presented at the 18th International Congress of Refrigeration. Montreal, Quebec.

Cleland, D. J., Cleland, A. C. and Earle, R. L. (1987). Prediction of freezing and thawing times for multi-dimensional shapes by simple formulae: I—Regular shapes. *Int. J. Refrig.* **10**: 156–164.

Ede, A. J. (1949). The calculation of the freezing and thawing of foodstuffs. *Mod. Refrig.* **52**, 52.

Heldman, D. R. (1983). Factors influencing food freezing rates. *Food Technol.* **37**(4): 103–109.

Heldman, D. R. and Gorby, D. P. (1975). Prediction of thermal conductivity of frozen foods. *Trans. ASAE* **18**: 740.

Heldman, D. R. and Lai, D. J. (1983). A model for prediction of shelf-life for frozen foods. *Proc. Int. Congr. Refrig. 16th Commission* C2, 427–433.

Heldman, D. R. and Lund, D. B. (2007). *Handbook of Food Engineering*, 2nd ed. Taylor and Francis Group, Boca Raton, Florida.

Hill, J. E., Litman, J. E. and Sutherland, J. E. (1967). Thermal conductivity of various meats. *Food Technol.* **21**: 1143.

IIR (1986). *Recommendations for the Processing and Handling of Frozen Foods*, 3rd ed. International Institute of Refrigeration, Paris, France.

Joshi, C. and Tao, L. C. (1974). A numerical method of simulating the axisymmetrical freezing of food systems. *J. Food Sci.* **39**: 623.

Jul, M. (1984). *The Quality of Frozen Foods*. Academic Press, Orlando.

Lentz, C. P. (1961). Thermal conductivity of meats, fats, gelatin gels and ice. *Food Technol.* **15**: 243.

Mannapperuma, J. D. and Singh, R. P. (1989). A computer-aided method for the prediction of properties and freezing/thawing times of foods. *J. Food Eng.* **9**: 275–304.

Mott, L. F. (1964). The prediction of product freezing time. *Aust. Refrig., Air Cond. Heat.* **18**: 16.

Nagaoka, J., Takagi, S. and Hotani, S. (1955). Experiments on the freezing of fish in an air-blast freezer. *Proc. Int. Congr. Refrig.*, 9th **2**: 4.

Pham, Q. T. (1986). Simplified equation for predicting the freezing time of foodstuffs. *J. Food Technol.* **21**: 209–219.

Pham, Q. T., Wee, H. K. Kemp, R. M. and Lindsay, D. T. (1994). Determination of the enthalpy of foods by an adiabatic calorimeter. *J. Food Engr.* **21**: 137–156.

Plank, R. Z. (1913). *Z. Gesamte Kalte-Ind.* **20**: 109, (cited by Ede, 1949).

Schwimmer, S., Ingraham, L. L. and Hughes, H. M. (1955). Temperature tolerance in frozen food processing. *Ind. Eng. Chem.* **47**(6): 1149–1151.

Singh, R. P. and Sarkar, A. (2005). Thermal properties of frozen foods. In *Engineering Properties of Foods*, M. A. Rao, S. S. H. Rizvi and A. K. Datta, eds., 175–207. CRC Press, Taylor and Francis Group, Boca Raton, Florida.

Singh, R. P. (2000). Scientific principles of shelf life evaluation. In *Shelf-Life Evaluation of Foods*, 2nd ed. C. M. D. Man and A. A. Jones, eds., 3–22. Aspen Publishers Inc., Gaithesberg, Maryland.

Singh, R. P. (1995). Principles of heat transfer. In *Frozen and Refrigerated Doughs and Batters*, K. Kulp, K. Lorenz, and J. Brümmer, eds. American Association of Cereal Chemists, Inc, Chapter 11.

Singh, R. P. and Mannapperuma, J. D. (1990). Developments in food freezing. In *Biotechnology and Food Process Engineering*, H. Schwartzberg and A. Rao, eds. Marcel Dekker, New York.

Singh, R. P. and Wang, C. Y. (1977). Quality of frozen foods—A review. *J. Food Process Eng.* **1**(2): 97.

Singh, R. P. and Wirakartakusumah, M. A. (1992). *Advances in Food Engineering*. CRC Press, Boca Raton, Florida.

Tao, L. C. (1967). Generalized numerical solutions of freezing a saturated liquid in cylinders and spheres. *AIChE J.* **13**: 165.

Taub, I. A. and Singh, R. P. (1998). *Food Storage Stability*. CRC Press, Boca Raton, Florida.

Tien, R. H. and Geiger, G. E. (1967). A heat transfer analysis of the solidification of a binary eutectic system. *J. Heat Transfer* **9**: 230.

Tien, R. H. and Geiger, G. E. (1968). The unidimensional solidification of a binary eutectic system with a time-dependent surface temperature. *J. Heat Transfer* **9C**(1): 27.

Tien, R. H. and Koumo, V. (1968). Unidimensional solidification of a slab variable surface temperature. *Trans. Metal Soc. AIME* **242**: 283.

Tien, R. H. and Koumo, V. (1969). Effect of density change on the solidification of alloys. *Am. Soc. Mech. Eng.*, [Pap.] 69-HT-45.

Van Arsdel, W. B. and Guadagni, D. G. (1959). Time-temperature tolerance of frozen foods. XV. Method of using temperature histories to estimate changes in frozen food quality. *Food Technol.* **13**(1): 14–19.

Van Arsdel, W. B., Copley, M. J. and Olson, R. D. (1969). *Quality and Stability of Frozen Foods*. Wiley, New York.

Chapter 8

Evaporation

Evaporation is an important unit operation commonly employed to remove water from dilute liquid foods to obtain concentrated liquid products. Removal of water from foods provides microbiological stability and assists in reducing transportation and storage costs. A typical example of the evaporation process is in the manufacture of tomato paste, usually around 35% to 37% total solids, obtained by evaporating water from tomato juice, which has an initial concentration of 5% to 6% total solids. Evaporation differs from dehydration, since the final product of the evaporation process remains in liquid state. It also differs from distillation, since the vapors produced in the evaporator are not further divided into fractions.

In Figure 8.1 a simplified schematic of an evaporator is shown. Essentially, an evaporator consists of a heat exchanger enclosed in a large chamber; a noncontact heat exchanger provides the means to transfer heat from low-pressure steam to the product. The product inside the evaporation chamber is kept under vacuum. The presence of vacuum causes the temperature difference between steam and the product to increase, and the product boils at relatively low temperatures, thus minimizing heat damage. The vapors produced are conveyed through a condenser to a vacuum system. The steam condenses inside the heat exchanger and the condensate is discarded.

In the evaporator shown in Figure 8.1, the vapors produced are discarded without further utilizing their inherent heat, therefore this type of evaporator is called a single-effect evaporator, since the vapors produced are discarded. If the vapors are reused as the heating medium in another evaporator chamber, as shown in Figure 8.2, the evaporator system is called a multiple-effect evaporator. More specifically, the

Professor R. Paul Singh, Department of Biological and Agricultural Engineering, University of California, Davis, CA 95616, USA.
Email: rps@rpaulsingh.com.

■ Figure 8.1 Schematic diagram of a single-effect evaporator.

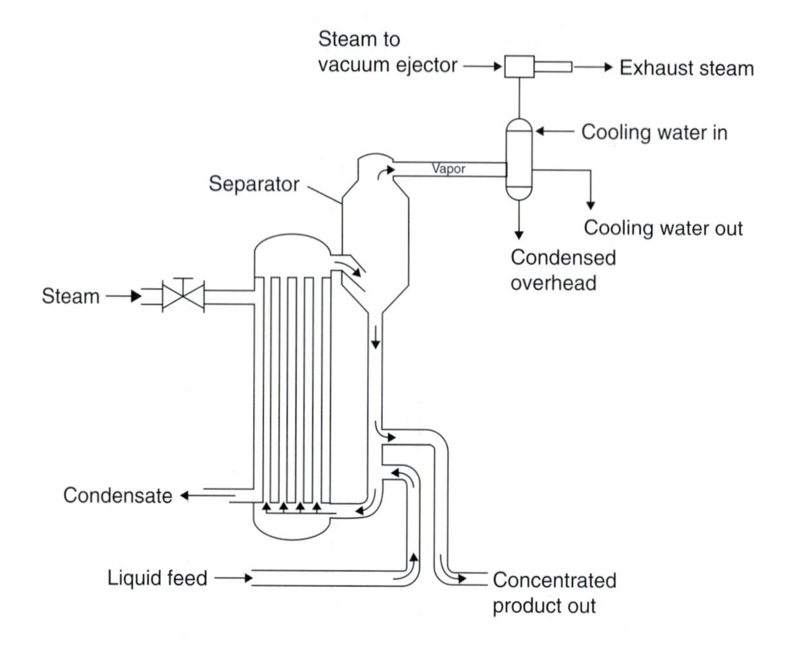

■ Figure 8.2 Schematic diagram of a triple-effect evaporator.

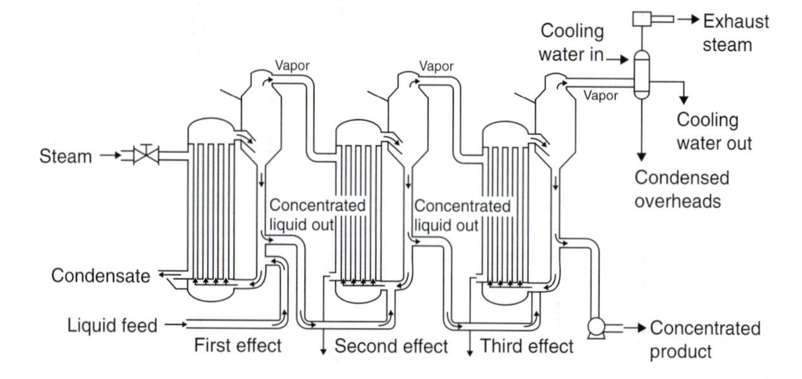

evaporator shown in Figure 8.2 is a triple-effect evaporator, as vapors produced from first and second effects (or evaporation chambers) are used again as the heating medium in second and third effects, respectively.

Note that in a multi-effect evaporator, steam is used only in the first effect. The use of vapors as a heating medium in additional effects results in obtaining higher energy-use efficiency from the system. The partially concentrated product leaving the first effect is introduced as feed into the second effect. After additional concentration, product from the second effect becomes feed for the third effect. The product from the third effect leaves at the desired concentration. This particular arrangement is called a forward feed system. Other flow arrangements used in industrial practice include backward feed systems and parallel feed systems.

The characteristics of the liquid food have a profound effect on the performance of the evaporation process. As water is removed, the liquid becomes increasingly concentrated, resulting in reduced heat transfer. The boiling point rises as the liquid concentrates, resulting in a smaller differential of temperature between the heating medium and the product. This causes reduced rate of heat transfer.

Food products are noted for their heat sensitivity. Evaporation processes must involve reducing the temperature for boiling as well as the time of heating, to avoid excessive product degradation.

In addition, fouling of the heat-exchange surface can seriously reduce the rate of heat transfer. Frequent cleaning of heat-exchange surfaces requires shutdown of the equipment, thus decreasing the processing capacity. Liquid foods that foam during vaporization cause product losses as a result of escape through vapor outlets. In designing evaporation systems, it is important to keep in perspective the preceding characteristics of liquid food.

In this chapter we will consider boiling-point elevation in liquid foods during concentration, describe various types of evaporators based on the method of heat exchange from steam to product, and then we will design single- and multiple-effect evaporators.

8.1 BOILING-POINT ELEVATION

Boiling-point elevation of a solution (liquid food) is defined as the increase in boiling point over that of pure water, at a given pressure.

A simple method to estimate boiling-point elevation is the use of Dühring's rule. The Dühring rule states that a linear relationship exists between the boiling-point temperature of the solution and the boiling-point temperature of water at the same pressure. The linear relationship does not hold over a wide range of temperatures, but over moderate temperature ranges, it is quite acceptable. Dühring lines for a sodium chloride–water system are shown in Figure 8.3. Example 8.1 illustrates the use of the figure to estimate boiling-point elevation.

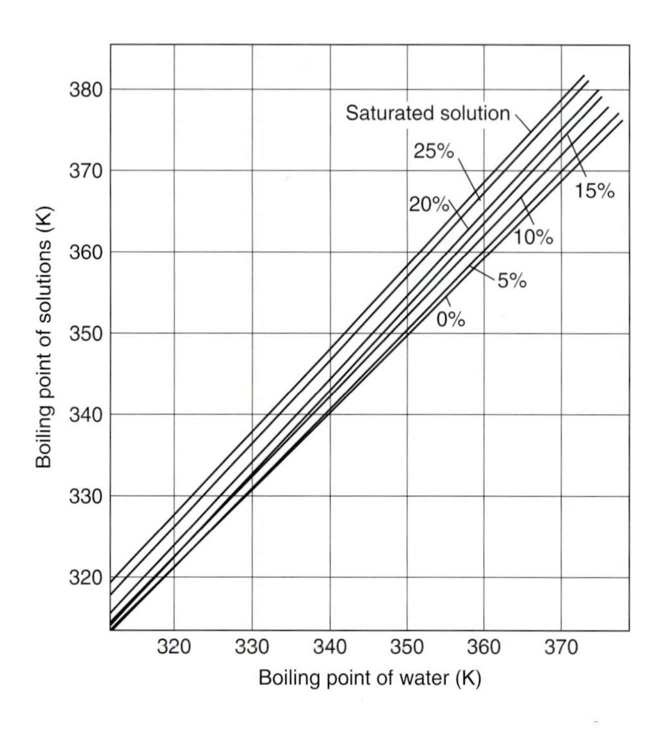

Figure 8.3 Dühring lines illustrating the influence of solute concentrations on boiling point elevation of NaCl. (From Coulson and Richardson, 1978)

Example 8.1

Use Dühring's chart to determine the initial and final boiling point of a liquid food with a composition that exerts vapor pressure similar to that of sodium chloride solution. The pressure in the evaporator is 20 kPa. The product is being concentrated from 5% to 25% total solids concentration.

Given

Initial concentration = 5% total solids

Final concentration = 25% total solids

Pressure = 20 kPa

Approach

To use Dühring's chart, given in Figure 8.3, we need the boiling point of water. This value is obtained from the steam tables. The boiling point of the liquid food can then be read directly from Figure 8.3.

Solution

1. *From a steam table (Appendix A.4.2) at 20 kPa, the boiling point of water is 60°C or 333 K.*
2. *From Figure 8.3,*

Boiling point at initial concentration of 5% total solids is 333 K = 60°C

Boiling point at final concentration of 25% total solids is 337 K = 64°C

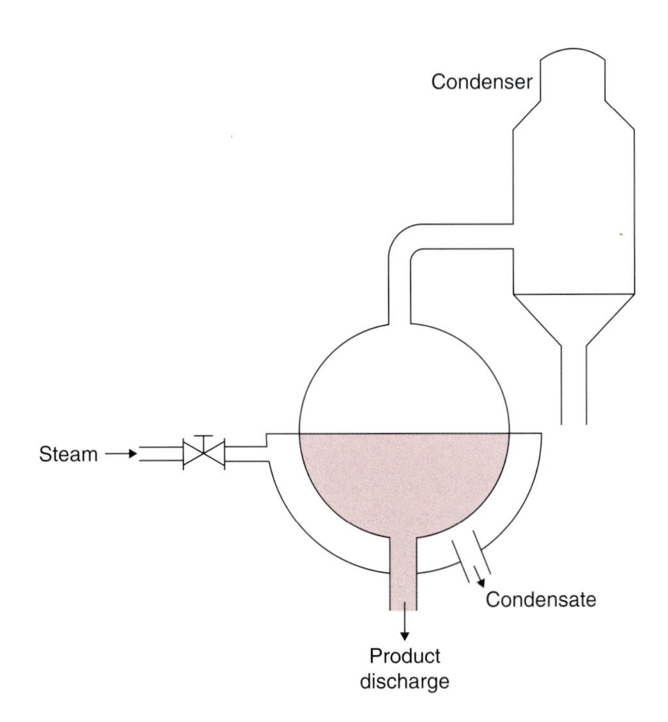

■ **Figure 8.4** A batch-type pan evaporator. (Courtesy of APV Equipment, Inc.)

The boiling-point elevation merits consideration since the temperature difference between steam and product decreases as the boiling point of the liquid increases due to concentration. The reduced temperature differential causes a reduction in rate of heat transfer between steam and product.

8.2 TYPES OF EVAPORATORS

Several types of evaporators are used in the food industry. In this section, a brief discussion of the more common types is given.

8.2.1 Batch-Type Pan Evaporator

One of the simplest and perhaps oldest types of evaporators used in the food industry is the batch-type pan evaporator, shown in Figure 8.4. The product is heated in a steam jacketed spherical vessel. The heating vessel may be open to the atmosphere or connected to a condenser and vacuum. Vacuum permits boiling the product at temperatures lower than the boiling point at atmospheric pressure, thus reducing the thermal damage to heat-sensitive products.

The heat-transfer area per unit volume in a pan evaporator is small. Thus, the residence time of the product is usually very long, up to

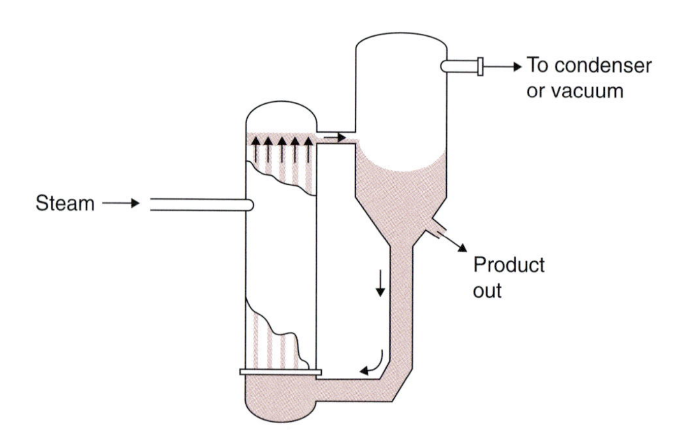

■ **Figure 8.5** A natural-circulation evaporator. (Courtesy of APV Equipment, Inc.)

several hours. Heating of the product occurs mainly due to natural convection, resulting in smaller convective heat-transfer coefficients. The poor heat-transfer characteristics substantially reduce the processing capacities of the batch-type pan evaporators.

8.2.2 Natural Circulation Evaporators

In natural circulation evaporators, short vertical tubes, typically 1–2 m long and 50–100 mm in diameter, are arranged inside the steam chest. The whole calandria (tubes and steam chest) is located in the bottom of the vessel. The product, when heated, rises through these tubes by natural circulation while steam condenses outside the tubes. Evaporation takes place inside the tubes, and the product is concentrated. The concentrated liquid falls back to the base of the vessel through a central annular section. A natural-circulation evaporator is shown in Figure 8.5. A shell-and-tube heat exchanger can be provided outside the main evaporation vessel to preheat the liquid feed.

8.2.3 Rising-Film Evaporator

In a rising-film evaporator (Fig. 8.6), a low-viscosity liquid food is allowed to boil inside 10–15 m-long vertical tubes. The tubes are heated from the outside with steam. The liquid rises inside these tubes by vapors formed near the bottom of the heating tubes. The upward movement of vapors causes a thin liquid film to move rapidly upward. A temperature differential of at least 14 °C between the product and the heating medium is necessary to obtain a well-developed film. High convective heat-transfer coefficients are achieved in these evaporators. Although the operation is mostly once-through, liquid can be recirculated if necessary to obtain the required solid concentration.

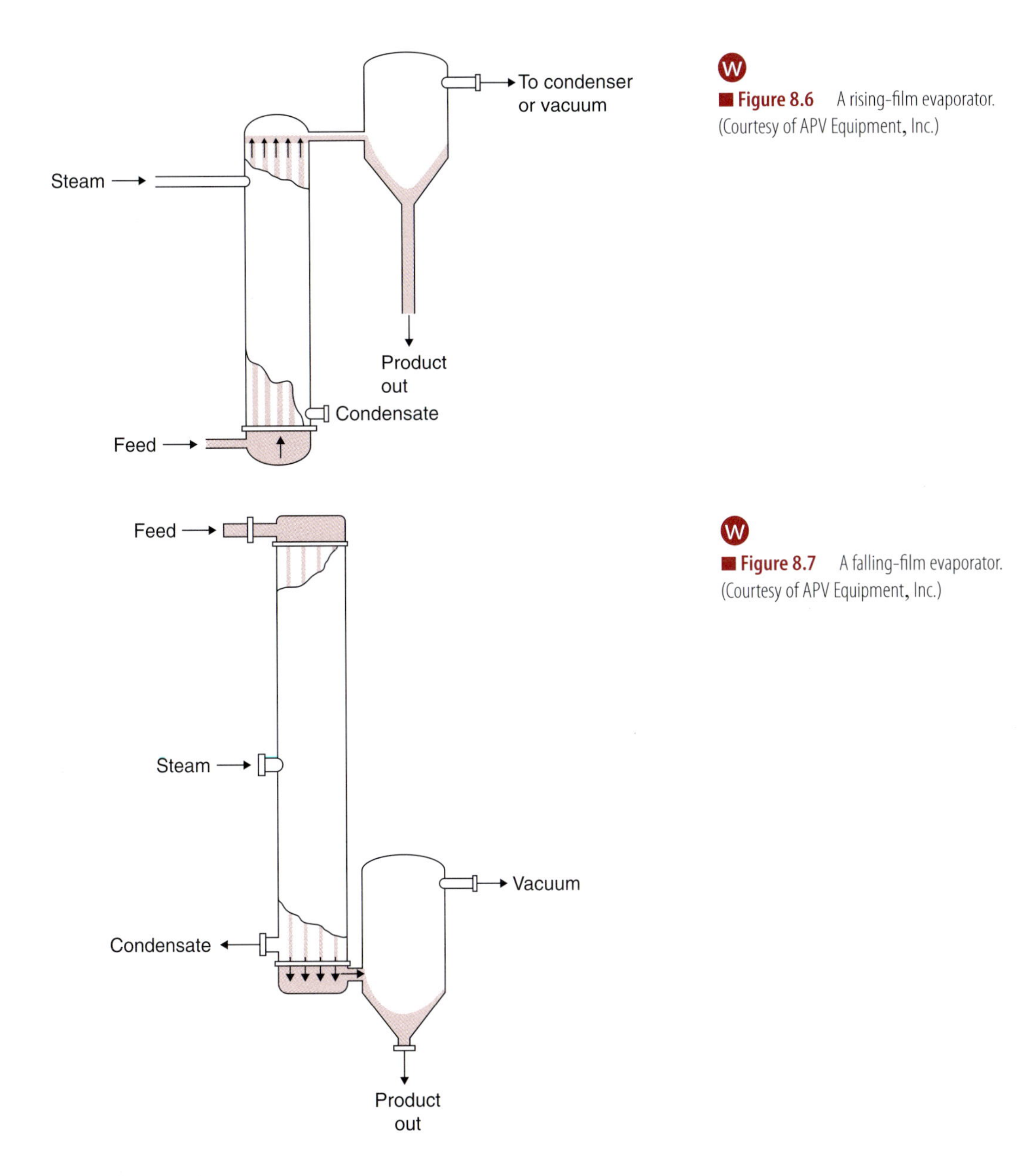

W

■ **Figure 8.6** A rising-film evaporator. (Courtesy of APV Equipment, Inc.)

W

■ **Figure 8.7** A falling-film evaporator. (Courtesy of APV Equipment, Inc.)

8.2.4 Falling-Film Evaporator

In contrast to the rising-film evaporator, the falling-film evaporator has a thin liquid film moving downward under gravity on the inside of the vertical tubes (Fig. 8.7). The design of such evaporators is complicated by the fact that distribution of liquid in a uniform film flowing downward in a tube is more difficult to obtain than an upward-flow system such as in a rising-film evaporator. This is accomplished by the use of specially designed distributors or spray nozzles.

The falling-film evaporator allows a greater number of effects than the rising-film evaporator. For example, if steam is available at 110°C and the boiling temperature in the last effect is 50°C, then the total available temperature differential is 60°C. Since rising-film evaporators require 14°C temperature differential across the heating surface, only four effects are feasible. However, as many as 10 or more effects may be possible using a falling-film evaporator. The falling-film evaporator can handle more viscous liquids than the rising-film type. This type of evaporator is best suited for highly heat-sensitive products such as orange juice. Typical residence time in a falling-film evaporator is 20 to 30 seconds, compared with a residence time of 3 to 4 minutes in a rising-film evaporator.

8.2.5 Rising/Falling-Film Evaporator

In the rising/falling-film evaporator, the product is concentrated by circulation through a rising-film section followed by a falling-film section of the evaporator. As shown in Figure 8.8, the product is first concentrated as it ascends through a rising tube section, followed by

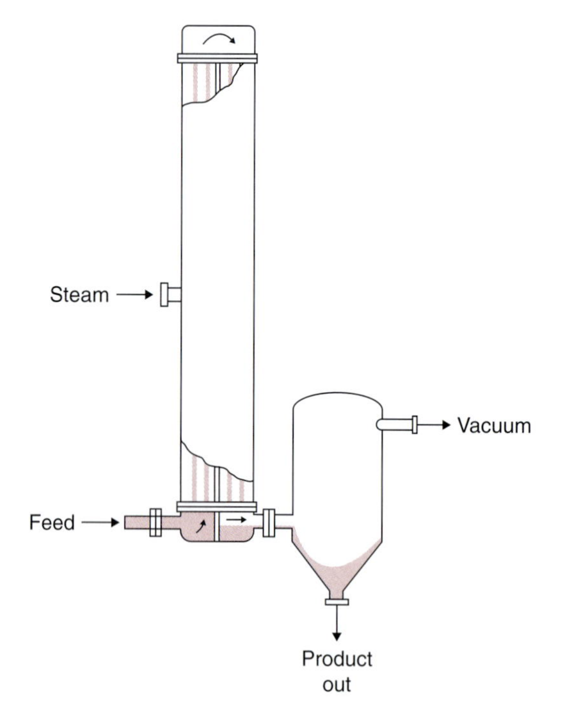

W

Figure 8.8 A rising/falling-film evaporator. (Courtesy of APV Equipment, Inc.)

the preconcentrated product descending through a falling-film section; there it attains its final concentration.

8.2.6 Forced-Circulation Evaporator

The forced-circulation evaporator involves a noncontact heat exchanger where liquid food is circulated at high rates (Fig. 8.9). A hydrostatic head, above the top of the tubes, eliminates any boiling of the liquid. Inside the separator, absolute pressure is kept slightly lower than that in the tube bundle. Thus, the liquid entering the separator flashes to form a vapor. The temperature difference across the heating surface in the heat exchanger is usually 3–5 °C. Axial flow pumps are generally used to maintain high circulation rates with linear velocities of 2–6 m/s, compared with a linear velocity of 0.3–1 m/s in natural-circulation evaporators. Both capital and operating costs of these evaporators are very low in comparison with other types of evaporators.

8.2.7 Agitated Thin-Film Evaporator

For very viscous fluid foods, feed is spread on the inside of the cylindrical heating surface by wiper blades, as shown in Figure 8.10. Due to high agitation, considerably higher rates of heat transfer are obtained. The cylindrical configuration results in low heat-transfer

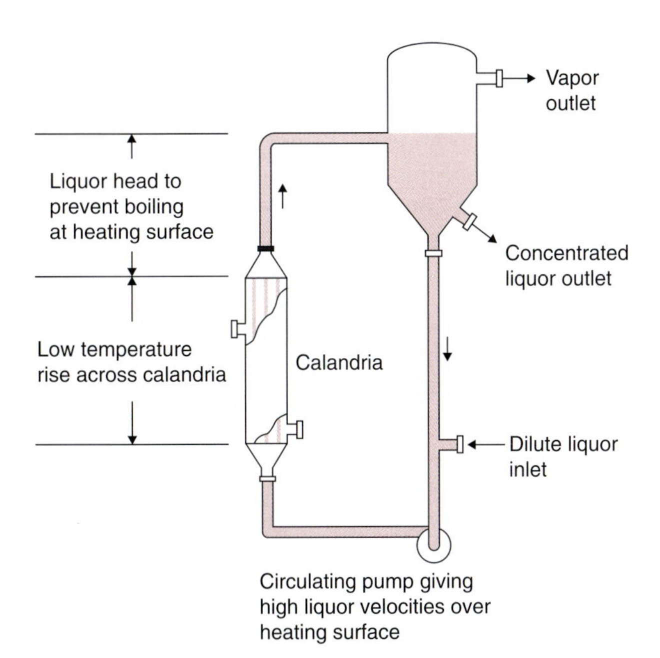

W

Figure 8.9 A forced-circulation evaporator. (Courtesy of APV Equipment, Inc.)

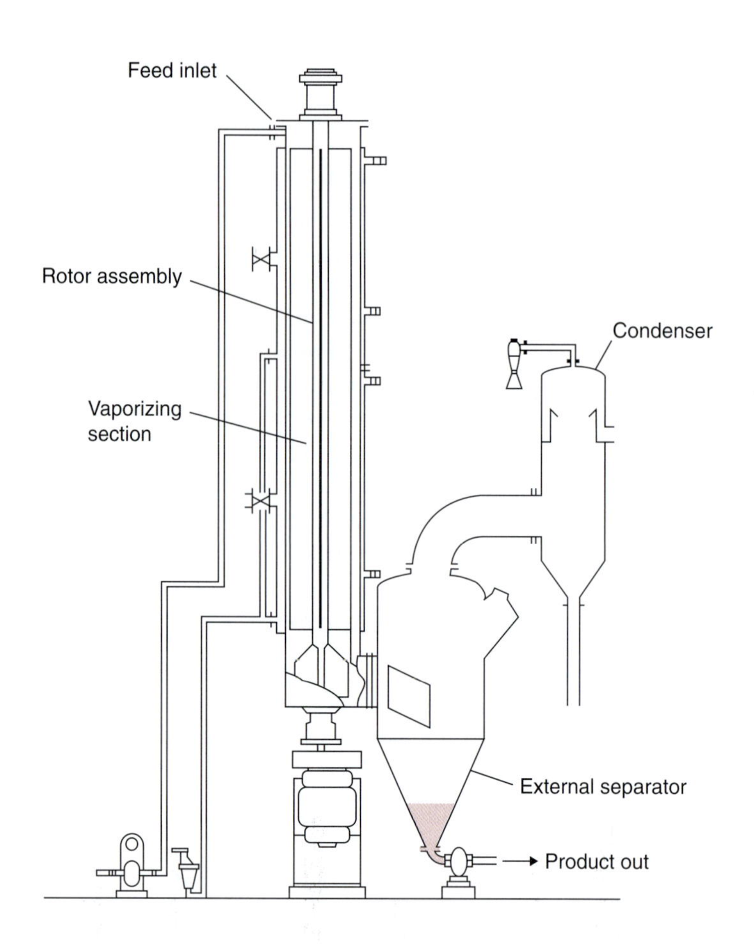

■ **Figure 8.10** An agitated thin-film evaporator. (Courtesy of APV Equipment, Inc.)

area per unit volume of the product. High-pressure steam is used as the heating medium to obtain high wall temperatures for reasonable evaporation rates. The major disadvantages are the high capital and maintenance costs and low processing capacity.

In addition to the tubular shape, plate evaporators are also used in the industry. Plate evaporators use the principles of rising/falling-film, falling-film, wiped-film, and forced-circulation evaporators. The plate configuration often provides features that make it more acceptable. A rising/falling-film plate evaporator is more compact, thus requiring less floor area than a tubular unit. The heat-transfer areas can easily be inspected. A falling-film plate evaporator with a capacitor of 25,000 to 30,000 kg water removed per hour is not uncommon.

Table 8.1 summarizes the various comparative features of different types of evaporators used in concentrating liquid foods. The characteristics

Table 8.1 Types of Evaporators Employed in Concentrating Liquid Foods[a]

Evaporator type	Tube Depth	Circulation	Viscosity capability ($\times 10^{-3}$ Pa s)	Able to handle suspended solids	Applicable to multiple effects	Applicable to mechanical vapor decompression	Heat transfer rate	Residence time	Capital cost	Remarks
Vertical tubular	Long	Natural	Up to 50	Yes	Yes	No	Medium	High	Low	
Vertical tubular	Long	Assisted	Up to 150	Yes	Yes	No	Good	High	Low-medium	
Vertical tubular	Short	Natural	Up to 20	Yes	Yes	No	Medium	High	High	Calandria usually internal to separator
Vertical tubular	Short	Assisted	Up to 2000	Yes	No	No	Low	Very high	High	Calandria usually internal to separator
Vertical tubular	Long	Forced-suppressed boiling	Up to 500	Yes	Limited	No	Medium	Very high	Very high	Used on scaling duties
Plate heat exchanger	N/A	Forced-suppressed boiling	Up to 500	Limited	Yes	No	Good	Medium	Medium	Used on scaling duties
Vertical tubular rising film	Long	None or limited	Up to 1000	Not desirable	Yes	No	Good	Low	Medium	
Vertical tubular rising/falling film	Medium	None or limited	Up to 2000	Not desirable	Yes	No	Good	Low	Medium	
Plate rising/falling film	N/A	None or limited	Up to 2000	Very limited	Yes	No	Good	Low	Low-medium	
Vertical tubular falling film	Long	None or limited	Up to 3000	Not desirable	Yes	Yes	Excellent	Very low	Medium	Used on heat-sensitive products
Vertical tubular falling film	Long	Medium	Up to 1000	Yes	Yes	Yes	Good	Low	Medium	
Plate falling film	N/A	None or limited	Up to 3000	No	Yes	Yes	Excellent	Very low	Medium	Used on heat-sensitive products
Swept surface	N/A	None	Up to 10,000	Yes	No	No	Excellent	Low	Very high	

[a] *Courtesy of APV Equipment, Inc.*

are presented in a general way. Custom-designed modifications of these types can significantly alter the specific duties.

8.3 DESIGN OF A SINGLE-EFFECT EVAPORATOR

In a single-effect evaporator, as shown in Figure 8.11, dilute liquid feed is pumped into the heating chamber, where it is heated indirectly with steam. Steam is introduced into the heat exchanger, where it condenses to give up its heat of vaporization to the feed, and exits the system as condensate.

The temperature of evaporation, T_1, is controlled by maintaining vacuum inside the heating chamber. The vapors leaving the product are conveyed through a condenser to a vacuum system, usually a steam ejector or a vacuum pump. In a batch system, the feed is heated until the desired concentration is obtained. The concentrated product is then pumped out of the evaporator system.

Heat and mass balances conducted on the evaporator system allow determination of various design and operating variables. Such variables may include mass flow rates, final concentration of product, and heat-exchanger area.

The following expressions can be obtained by conducting a mass balance on flow streams and product solids, respectively.

$$\dot{m}_f = \dot{m}_v + \dot{m}_p \tag{8.1}$$

■ **Figure 8.11** Schematic diagram of a single-effect evaporator.

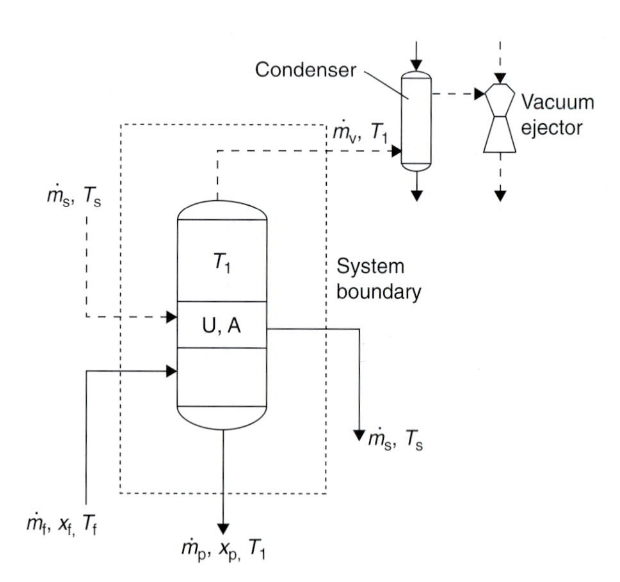

where $\dot{m}_f$ is the mass flow rate of dilute liquid feed (kg/s), $\dot{m}_v$ is the mass flow rate of vapor (kg/s), and $\dot{m}_p$ is the mass flow rate of concentrated product (kg/s),

$$x_f \dot{m}_f = x_p \dot{m}_p \tag{8.2}$$

where x_f is the solid fraction in the feed stream (dimensionless) and x_p is the solid fraction in the product stream (dimensionless).

An enthalpy balance conducted on the evaporator system gives the following expression:

$$\dot{m}_f H_f + \dot{m}_s H_{vs} = \dot{m}_v H_{v1} + \dot{m}_p H_{p1} + \dot{m}_s H_{cs} \tag{8.3}$$

where $\dot{m}_s$ is the mass flow rate of steam (kg/s); H_f is enthalpy of dilute liquid feed (kJ/kg); H_{p1} is enthalpy of concentrated product (kJ/kg); H_{vs} is enthalpy of saturated vapor at temperature T_s (kJ/kg); H_{v1} is enthalpy of saturated vapor at temperature T_1 (kJ/kg); H_{cs} is enthalpy of condensate (kJ/kg); T_s is temperature of steam (°C); T_1 is the boiling temperature maintained inside the evaporator chamber (°C); and T_f is the temperature of dilute liquid feed (°C).

The first term in Equation (8.3), $\dot{m}_f H_f$, represents the total enthalpy associated with the incoming dilute liquid feed, where H_f is a function of T_f and x_f. The enthalpy content H_f can be computed from

$$H_f = c_{pf}(T_f - 0°C) \tag{8.4}$$

The specific heat may be obtained either from Table A.2.1 or by using Equation (4.3) or Equation (4.4).

The second term, $\dot{m}_s H_{vs}$, gives the total heat content of steam. It is assumed that saturated steam is being used. The enthalpy, H_{vs}, is obtained from the steam table (Table A.4.2) as enthalpy of saturated vapors evaluated at the steam temperature T_s.

On the right-hand side of Equation (8.3), the first term, $\dot{m}_v H_{v1}$, represents total enthalpy content of the vapors leaving the system. The enthalpy H_{v1} is obtained from the steam table (Table A.4.2) as the enthalpy of saturated vapors evaluated at temperature T_1.

The second term, $\dot{m}_p H_{p1}$, is the total enthalpy associated with the concentrated product stream leaving the evaporator. The enthalpy content H_{p1} is obtained using the following equation:

$$H_{p1} = c_{pp}(T_1 - 0°C) \tag{8.5}$$

where c_{pp} is the specific heat content of concentrated product (kJ/[kg °C]).

Again, c_{pp} is obtained from Table A.2.1 or by using Equation (4.3) or Equation (4.4).

The last term, $\dot{m}_s H_{cs}$, represents the total enthalpy associated with the condensate leaving the evaporator. Since an indirect type of heat exchanger is used in evaporator systems, the rate of mass flow of incoming steam is the same as the rate of mass flow of condensate leaving the evaporator. The enthalpy H_{cs} is obtained from the steam table (Table A.4.2) as enthalpy of saturated liquid evaluated at temperature T_s. If the condensate leaves at a temperature lower than T_s, then the lower temperature should be used to determine the enthalpy of the saturated liquid.

In addition to the mass and enthalpy balances given previously, the following two equations are also used in computing design and operating variables of an evaporator system.

For the heat exchanger, the following expression gives the rate of heat transfer:

$$q = UA(T_s - T_1) = \dot{m}_s H_{vs} - \dot{m}_s H_{cs} \tag{8.6}$$

where q is the rate of heat transfer (W), U is the overall heat transfer coefficient (W/[m^2 K]), and A is the area of the heat exchanger (m^2).

The overall heat-transfer coefficient decreases as the product becomes concentrated, due to increased resistance of heat transfer on the product side of the heat exchanger. In addition, the boiling point of the product rises as the product becomes concentrated. In Equation (8.6), a constant value of the overall heat-transfer coefficient is used and would result in some "overdesign" of the equipment.

Steam economy is a term often used in expressing the operating performance of an evaporator system. This term is a ratio of rate of mass of water vapor produced from the liquid feed per unit rate of steam consumed.

$$\text{Steam economy} = \dot{m}_v / \dot{m}_s \tag{8.7}$$

A typical value for steam economy of a single-effect evaporator system is close to 1.

Example 8.2

Apple juice is being concentrated in a natural-circulation single-effect evaporator. At steady-state conditions, dilute juice is the feed introduced at a rate of 0.67 kg/s. The concentration of the dilute juice is 11% total solids. The juice is concentrated to 75% total solids. The specific heats of dilute apple juice and concentrate are 3.9 and 2.3 kJ/(kg °C), respectively. The steam pressure is measured to be 304.42 kPa. The inlet feed temperature is 43.3°C. The product inside the evaporator boils at 62.2°C. The overall heat-transfer coefficient is assumed to be 943 W/(m^2 °C). Assume negligible boiling-point elevation. Calculate the mass flow rate of concentrated product, steam requirements, steam economy, and the heat-transfer area. The system is sketched in Figure E8.1.

Given

Mass flow rate of feed $\dot{m}_f = 0.67$ *kg/s*
Concentration of food $x_f = 0.11$
Concentration of product $x_p = 0.75$
Steam pressure = *304.42 kPa*
Feed temperature $T_f = 43.3$*°C*
Boiling temperature T_1 *in evaporator* = *62.2°C*
Overall heat transfer coefficient $U = 943$ *W/(m^2 K)*
Specific heat of dilute feed $c_{pf} = 3.9$ *kJ/(kg °C)*
Specific heat of concentrated product $c_{pp} = 2.3$ *kJ/(kg °C)*

Approach

We will use the heat and mass balances given in Equations (8.1), (8.2), and (8.3) to determine the unknowns. Appropriate values of enthalpy for steam and vapors will be obtained from steam tables.

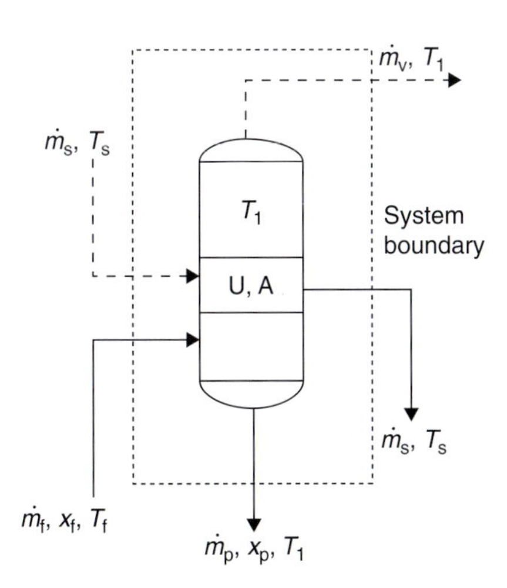

Figure E8.1 Schematic diagram of a single-effect evaporator.

Solution

1. *From Equation (8.2),*

$$(0.11)(0.67\ kg/s) = (0.75)\dot{m}_p$$
$$\dot{m}_p = 0.098\ kg/s$$

Thus, mass flow rate of concentrated product is 0.098 kg/s.

2. *From Equation (8.1),*

$$\dot{m}_v = (0.67\ kg/s) - (0.098\ kg/s)$$
$$\dot{m}_v = 0.57\ kg/s$$

Thus, mass flow rate of vapors is 0.57 kg/s.

3. *To use the enthalpy balance of Equation (8.3), the following quantities are first determined:*

From Equation (8.4),

$$H_f = (3.9\ kJ/[kg\ °C])(43.3°C - 0°C) = 168.9\ kJ/kg$$

From Equation (8.5),

$$H_{p1} = (2.3\ kJ/[kg\ °C])(62.2°C - 0°C) = 143.1\ kJ/kg$$

From the steam table (Table A.4.2),
Temperature of steam at 304.42 kPa = 134°C
Enthalpy for saturated vapor H_{vs} (at T_s = 134°C) = 2725.9 kJ/kg
Enthalpy for saturated liquid Hcs (at Ts = 134°C) = 563.41 kJ/kg
Enthalpy for saturated vapor H_{v1} (at T_1 = 62.2°C) = 2613.4 kJ/kg

$$(0.67\ kg/s)(168.9\ kJ/kg) + (\dot{m}_s\ kg/s)(2725.9\ kJ/kg)$$
$$= (0.57\ kg/s)(2613.4\ kJ/kg) + (0.098\ kg/s)(143.1\ kJ/kg)$$
$$+ (\dot{m}_s\ kg/s)(563.41\ kJ/kg)$$
$$2162.49\ \dot{m}_s = 1390.5$$
$$\dot{m}_s = 0.64\ kg/s$$

4. *To calculate steam economy, we use Equation (8.7).*

$$\text{Steam economy} = \frac{\dot{m}_v}{\dot{m}_s} = \frac{0.57}{0.67} = 0.85\ kg\ \text{water evaporated/kg steam}$$

5. *To compute area of heat transfer, we use Equation (8.6).*

$$A(943\ W/[m^2\ °C])(134°C - 62.2°C)$$
$$= (0.64\ kg/s)(2725.9 - 563.14\ kJ/kg)(1000\ J/kJ)$$
$$A = 20.4\ m^2$$

8.4 DESIGN OF A MULTIPLE-EFFECT EVAPORATOR

In a triple-effect evaporator, shown in Figure 8.12, dilute liquid feed is pumped into the evaporator chamber of the first effect. Steam enters the heat exchanger and condenses, thus discharging its heat to the product. The condensate is discarded. The vapors produced from the first effect are used as the heating medium in the second effect, where the feed is the partially concentrated product from the first effect. The vapors produced from the second effect are used in the third effect as heating medium, and the final product with the desired final concentration is pumped out of the evaporator chamber of the third effect. The vapors produced in the third effect are conveyed to a condenser and a vacuum system. In the forward feed system shown, partially concentrated product from the first effect is fed to the second effect. After additional concentration, product leaving the second effect is introduced into the third effect. Finally, product with the desired concentration leaves the third effect.

Design expressions for multiple-effect evaporators can be obtained in the same manner as for a single-effect evaporator, discussed in Section 8.3.

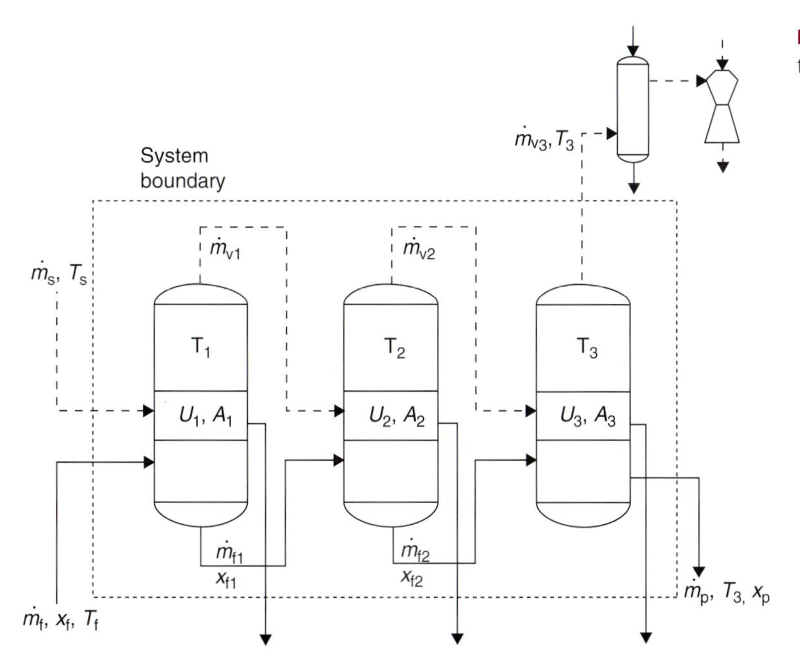

Figure 8.12 Schematic diagram of a triple-effect evaporator.

Conducting mass balance analysis on the flow streams,

$$\dot{m}_f = \dot{m}_{v1} + \dot{m}_{v2} + \dot{m}_{v3} + \dot{m}_p \tag{8.8}$$

where $\dot{m}_f$ is the mass flow rate of dilute liquid feed to the first effect (kg/s); $\dot{m}_{v1}$, $\dot{m}_{v2}$, and $\dot{m}_{v3}$ are the mass flow rates of vapor from the first, second, and third effect, respectively (kg/s); and $\dot{m}_p$ is the mass flow rate of concentrated product from the third effect (kg/s).

Using mass balance on the solids fraction in the flow streams,

$$x_f \dot{m} = x_p \dot{m}_p \tag{8.9}$$

where x_f is the solid fraction in the feed stream to be consistent with the first effect (dimensionless) and x_p is the solid fraction in the product stream from the third effect (dimensionless).

We write enthalpy balances around each effect separately.

$$\dot{m}_f H_f + \dot{m}_s H_{vs} = \dot{m}_{v1} H_{v1} + \dot{m}_{f1} H_{f1} + \dot{m}_s H_{cs} \tag{8.10}$$

$$\dot{m}_{f1} H_{f1} + \dot{m}_{v1} H_{v1} = \dot{m}_{v2} H_{v2} + \dot{m}_{f2} H_{f2} + \dot{m}_{v1} H_{c1} \tag{8.11}$$

$$\dot{m}_{f2} H_{f2} + \dot{m}_{v2} H_{v2} = \dot{m}_{v3} H_{v3} + \dot{m}_p H_{p3} + \dot{m}_{v2} H_{c2} \tag{8.12}$$

where the subscripts 1, 2, and 3 refer to the first, second, and third effect, respectively. The other symbols are the same as defined previously for a single-effect evaporator.

The heat transfer across heat exchangers of various effects can be expressed by the following three expressions:

$$q_1 = U_1 A_1 (T_s - T_1) = \dot{m}_s H_{vs} - \dot{m}_s H_{cs} \tag{8.13}$$

$$q_2 = U_2 A_2 (T_1 - T_2) = \dot{m}_{v1} H_{v1} - \dot{m}_{v1} H_{c1} \tag{8.14}$$

$$q_3 = U_3 A_3 (T_2 - T_3) = \dot{m}_{v2} H_{v2} - \dot{m}_{v2} H_{c2} \tag{8.15}$$

The steam economy for a triple-effect evaporator as shown in Figure 8.12 is given by

$$\text{Steam economy} = \frac{\dot{m}_{v1} + \dot{m}_{v2} + \dot{m}_{v3}}{\dot{m}_s} \tag{8.16}$$

Example 8.3 illustrates the use of these expressions in evaluating the performance of multiple-effect evaporators.

Example 8.3

Calculate the steam requirements of a double-effect forward-feed evaporator (Fig. E8.2) to concentrate a liquid food from 11% total solids to 50% total solids concentrate. The feed rate is 10,000 kg/h at 20°C. The boiling of liquid inside the second effect takes place under vacuum at 70°C. The steam is being supplied to the first effect at 198.5 kPa. The condensate from the first effect is discarded at 95°C and from the second effect at 70°C. The overall heat-transfer coefficient in the first effect is 1000 W/(m^2 °C); in the second effect it is 800 W/(m^2 °C). The specific heats of the liquid food are 3.8, 3.0, and 2.5 kJ/(kg °C) at initial, intermediate, and final concentrations. Assume the areas and temperature gradients are equal in each effect.

Given

Mass flow rate of feed $= \dot{m}f = 10{,}000$ *kg/h* $= 2.78$ *kg/s*
Concentration of feed $x_f = 0.11$
Concentration of product $x_p = 0.5$
Steam pressure $= 198.5$ *kPa*
Feed temperature $= 20°C$
Boiling temperature T_2 *in second effect* $= 70°C$
Overall heat-transfer coefficient U_1 *in first effect* $= 1000$ *W/(m^2 °C)*
Overall heat-transfer coefficient U_2 *in second effect* $= 800$ *W/(m^2 °C)*
Specific heat of dilute feed $c_{pf} = 3.8$ *kJ/(kg °C)*
Specific heat of feed at intermediate concentration $c'_{pf} = 3.0$ *kJ/(kg °C)*
Specific heat of concentrated food product $c_{pp} = 2.5$ *kJ/(kg °C)*

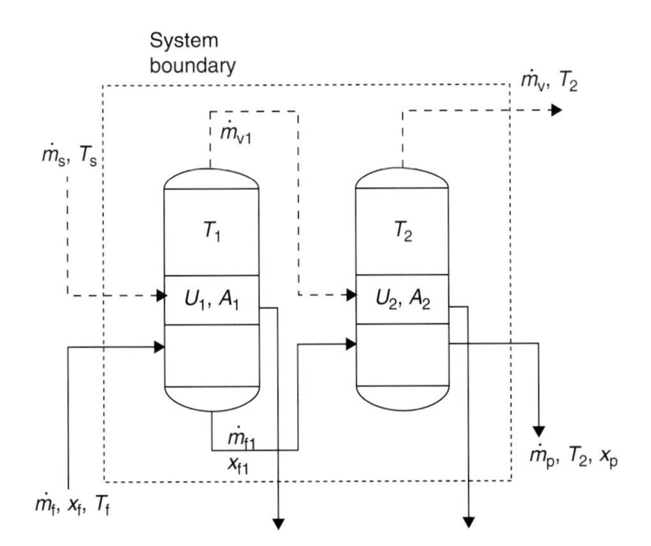

Figure E8.2 Schematic diagram of a double-effect evaporator.

Approach

Since this is a double-effect evaporator, we will use modified forms of Equations (8.8), (8.9), (8.10), (8.11), (8.13), and (8.14). Enthalpy values of steam and vapors will be obtained from steam tables.

Solution

1. *From Equation (8.9),*

$$(0.11)(2.78\,kg/s) = (0.75)\dot{m}_p$$
$$\dot{m}_p = 0.61\,kg/s$$

2. *From Equation (8.8),*

$$2.78 = \dot{m}_{v1} + \dot{m}_{v2} + 0.61$$

Thus, the total amount of water evaporating is

$$\dot{m}_{v1} + \dot{m}_{v2} = 2.17\,kg/s$$

3. *Steam is being supplied at 198.5 kPa or 120°C, the temperature in the second effect is 70°C, and thus the total temperature gradient is 50°C.*

$$\Delta T_1 + \Delta T_2 = 50°C$$

Assuming equal temperature gradient in each evaporator effect,

$$\Delta T_1 = \Delta T_2 = 25°C$$

4. *The area of heat transfer in the first and second effects are the same. Thus, from Equations (8.13) and (8.14),*

$$\frac{q_1}{U_1(T_s - T_1)} = \frac{q_2}{U_2(T_1 - T_2)}$$

or

$$\frac{\dot{m}_s H_{vs} - \dot{m}_s H_{cs}}{U_1(T_s - T_1)} = \frac{\dot{m}_{v1} H_{v1} - \dot{m}_{v1} H_{c1}}{U_2(T_1 - T_2)}$$

5. *To use Equations (8.10) and (8.11), we need values for enthalpy of product.*

$$H_f = c_{pf}(T_f - 0) = (3.8\,kJ/[kg°C])(20°C - 0°C) = 76\,kJ/kg$$
$$H_{f1} = c'_{pf}(T_1 - 0) = (3.0\,kJ/[kg°C])(95°C - 0°C) = 285\,kJ/kg$$
$$H_{f2} = c_{pp}(T_2 - 0) = (2.5\,kJ/[kg°C])(70°C - 0°C) = 175\,kJ/kg$$

In addition, from steam tables,

$$\begin{aligned} &\text{At } T_s = 120°C \quad H_{vs} = 2706.3 \text{ kJ/kg} \\ &\qquad\qquad\qquad\; H_{cs} = 503.71 \text{ kJ/kg} \\ &\text{At } T_1 = 95°C \quad H_{v1} = 2668.1 \text{ kJ/kg} \\ &\qquad\qquad\qquad\; H_{c1} = 397.96 \text{ kJ/kg} \\ &\text{At } T_2 = 70°C \quad H_{v2} = 2626.8 \text{ kJ/kg} \\ &\qquad\qquad\qquad\; H_{c2} = 292.98 \text{ kJ/kg} \end{aligned}$$

6. Thus, substituting enthalpy values from step (5) in the equation given in step (4),

$$\frac{[(\dot{m}_s \text{ kg/s}(2706.3 \text{ kJ/kg}) - (\dot{m}_s \text{ kg/s})(503.71 \text{ kJ/kg})](1000 \text{ J/kJ})}{(1000 \text{ W/[m}^2 °C])(120°C - 95°C)}$$

$$= \frac{[(\dot{m}_{v1} \text{ kg/s}(2668.1 \text{ kJ/kg}) - (\dot{m}_{v1} \text{ kg/s})(397.96 \text{ kJ/kg})](1000 \text{ J/kJ})}{(800 \text{ W/[m}^2 °C])(95°C - 70°C)}$$

or

$$\frac{2205.59\dot{m}_s}{25{,}000} = \frac{2270.14\dot{m}_{v1}}{20{,}000}$$

7. Using Equations (8.10) and (8.11),

$$\begin{aligned} &(2.78)(76) + (\dot{m}_s)(2706.3) \\ &\quad = (\dot{m}_{v1})(2668.1) + (\dot{m}_{f1})(285) + (\dot{m}_s)(503.71) \\ &(\dot{m}_{f1})(285) + (\dot{m}_{v1})(2668.1) \\ &\quad = (\dot{m}_{v2})(2626.8) + (\dot{m}_p)(175) + (\dot{m}_{v1})(397.96) \end{aligned}$$

8. Let us assemble all equations representing mass flow rates of product, feed, vapor, and steam.

From step (1): $\dot{m}_p = 0.61$

From step (2): $\dot{m}_{v1} + \dot{m}_{v2} = 2.17$

From step (6): $0.088\dot{m}_s = 0.114\dot{m}_{v1}$

From step (7):

$$\begin{aligned} 2202.59\dot{m}_s &= 2668.1\dot{m}_{v1} + 285\dot{m}_{f1} - 211.28 \\ 2270.14\dot{m}_{v1} &= 2626.8\dot{m}_{v2} + 175\dot{m}_p - 285\dot{m}_{f1} \end{aligned}$$

9. In step (8), we have five equations with five unknowns, namely, $\dot{m}_p$, $\dot{m}_{v1}$, $\dot{m}_{v2}$, $\dot{m}_s$, and $\dot{m}_{f1}$. We will solve these equations using a spreadsheet procedure to solve simultaneous equations. The method described in the following was executed on Excel™.

10. The simultaneous equations are rewritten so that all unknown variables are collected on the right-hand side. The equations are rewritten so that the

	A	B	C	D	E	F	G	H
1								
2		1.000	0.000	0.000	0.000	0.000		0.61
3		0.000	0.000	1.000	1.000	0.000		2.17
4		0.000	0.088	−0.114	0.000	0.000		0
5		0.000	2202.590	−2668.100	0.000	−285.000		−211.28
6		−175.000	0.000	2270.140	−2626.800	285.000		0
7	=+MINVERSE(B2:F6)						=+MMULT(B9:F13,H2:H6)	
8								
9		1.000	0.000	0.000	0.000	0.000		0.61
10		0.045	0.670	4.984	0.000	0.000		1.43
11		0.034	0.517	−4.925	0.000	0.000		1.10
12		−0.034	0.483	4.925	0.000	0.000		1.07
13		0.022	0.336	84.621	−0.003	0.000		1.46
14								

■ Figure E8.3 A spreadsheet to solve simultaneous equations.

coefficients can easily be arranged in a matrix. The spreadsheet method will use a matrix inversion process to solve the simultaneous equations.

$$\dot{m}_p + 0\dot{m}_s + 0\dot{m}_{v1} + 0\dot{m}_{v2} + 0\dot{m}_{f1} = 0.61$$

$$0\dot{m}_p + 0\dot{m}_s + \dot{m}_{v1} + \dot{m}_{v2} + 0\dot{m}_{f1} = 2.17$$

$$0\dot{m}_p + 0.088\dot{m}_s - 0.114\dot{m}_{v1} + 0\dot{m}_{v2} + 0\dot{m}_{f1} = 0$$

$$0\dot{m}_p + 2202.59\dot{m}_s - 2668.1\dot{m}_{v1} + 0\dot{m}_{v2} - 285\dot{m}_{f1} = -211.28$$

$$-175\dot{m}_p + 0\dot{m}_s + 2270.14\dot{m}_{v1} - 2626.8\dot{m}_{v2} + 285\dot{m}_{f1} = 0$$

11. *As shown in Figure E8.3, enter the coefficients of the left-hand side of the preceding equations in array B2:F6; enter the right-hand side coefficients in a column vector H2:H6.*

12. *Select another array B9:F13 (by dragging the cursor starting from cell B9). Type + MINVERSE(B2:F6) in cell B9 and press the CTRL, SHIFT, and ENTER keys simultaneously. This procedure will invert the matrix B2:F6 and give the coefficients of the inverted matrix in array B9:F13.*

13. *Highlight cells H9:H13 by dragging the cursor starting from cell H9. Type + MMULT(B9:F13,H2:H6) into cell H9; press the CTRL, SHIFT, and ENTER keys simultaneously. The answers are displayed in the column vector H9:H13. Thus,*

$$\dot{m}_p = 0.61\,\text{kg/s}$$

$$\dot{m}_s = 1.43\,\text{kg/s}$$

$$\dot{m}_{v1} = 1.10\,\text{kg/s}$$

$$\dot{m}_{v2} = 1.07\,\text{kg/s}$$

$$\dot{m}_{f1} = 1.46\,\text{kg/s}$$

14. *The steam requirements are computed to be 1.43 kg/s.*

15. *The steam economy can be computed as*

$$\frac{\dot{m}_{v1} + \dot{m}_{v2}}{\dot{m}_s} = \frac{1.10 + 1.07}{1.43} = 1.5\ \text{kg water vapor/kg steam}$$

8.5 VAPOR RECOMPRESSION SYSTEMS

The preceding discussion on multiple-effect evaporators has shown how energy requirements of the total system are decreased by using exit vapors as the heating medium in subsequent effects. Two additional systems that employ vapor recompression assist in reduction of energy requirements. These systems are thermal recompression and mechanical vapor recompression. A brief introduction to these two systems follows.

8.5.1 Thermal Recompression

Thermal recompression involves the use of a steam jet booster to recompress part of the exit vapors, as shown in Figure 8.13. Through recompression, the pressure and temperature of exit vapors are increased. These systems are usually applied to single-effect evaporators or to the first effect of multiple-effect evaporators. Application of this system requires that steam be available at high pressure, and low-pressure steam is needed for the evaporation process.

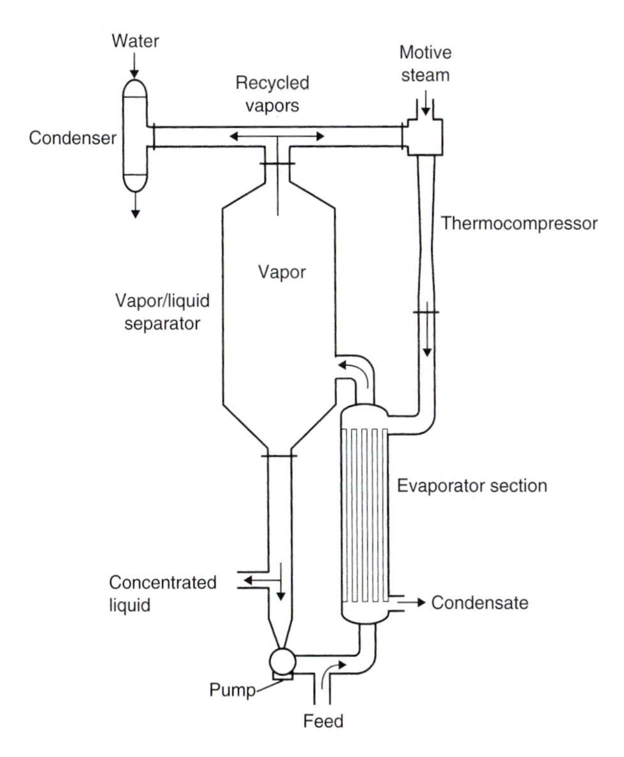

■ **Figure 8.13** Schematic diagram of a thermal recompression system.

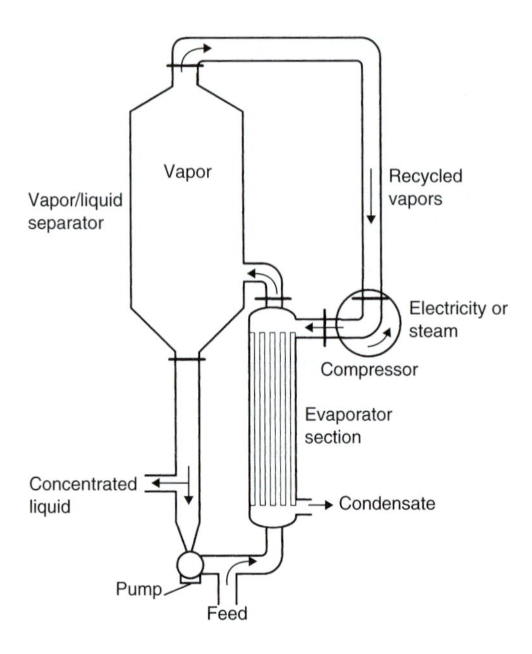

■ **Figure 8.14** Schematic diagram of a mechanical vapor recompression system.

8.5.2 Mechanical Vapor Recompression

Mechanical vapor recompression involves compression of all vapors leaving the evaporator, as shown in Figure 8.14. Vapor compression is accomplished mechanically, using a compressor driven by an electric motor, a steam turbine, or a gas engine. A steam-turbine-driven compressor is most suitable for mechanical recompression if high-pressure steam is available. Availability of electricity at low cost would favor the use of an electric motor.

Mechanical vapor recompression systems are very effective in reducing energy demands. Under optimum conditions, these systems can lower the energy requirements by an amount equivalent to adding 15 effects. These systems can be very noisy to operate due to the use of large compressors.

Mathematical procedures useful in designing vapor recompression systems are beyond the scope of this text. Students could consult Heldman and Singh (1981) for more information on these procedures.

PROBLEMS

8.1 A fruit juice at 20°C with 5% total solids is being concentrated in a single-effect evaporator. The evaporator is being operated at a sufficient vacuum to allow the product moisture to evaporate at 80 °C, and steam with 85% quality is being supplied at 169.06 kPa. The desired concentration of the final product is 40% total solids. The concentrated product exits the evaporator at a rate of 3000 kg/h. Calculate the (a) steam requirements and (b) steam economy for the process, when condensate is released at 90 °C. The specific heat of liquid feed is 4.05 kJ/(kg °C), and of concentrated product is 3.175 kJ/(kg °C).

8.2 A single-effect evaporator is being used to concentrate 10,000 kg/h of tomato juice from 5% total solids to 30% total solids. The juice enters the evaporator at 15 °C. The evaporator is operated with steam (80% quality) at 143.27 kPa. The vacuum inside the evaporator allows the juice to boil at 75 °C. Calculate (a) the steam requirements and (b) steam economy for the process. Assume the condensate is discharged at 75 °C. The specific heat of the liquid feed is 4.1 kJ/(kg °C) and the concentrated product is 3.1 kJ/(kg °C).[1]

***8.3** A four-effect evaporator is being considered for concentrating a fruit juice that has no appreciable boiling-point elevation. Steam is available at 143.27 kPa, and the boiling point of the product in the fourth effect is 45 °C. The overall heat-transfer coefficients are 3000 W/(m^2 °C) in the first effect, 2500 W/(m^2 °C) in the second effect, 2100 W/(m^2 °C) in the third effect, and 1800 W/(m^2 °C) in the fourth effect. Calculate the boiling-point temperatures of the product in the first, second, and third effects. Assume the heating areas in all the effects are equal to 50 m^2 each. The mass flow rate of steam to the first effect is 2400 kg/h, the feed rate to the first effect of 5% total solids fluid is 15,000 kg/h, the concentrated product from the first effect leaves at 6.25% total solids, and the concentration of product leaving the second effect is 8.82% total solids.

***8.4** A double-effect evaporator is being used to concentrate fruit juice at 25,000 kg/h. The juice is 10% total solid at 80 °C. The juice must be concentrated to 50% total solids. Saturated steam at 1.668 atm is available. The condensing temperature of the vapor in the second effect is 40 °C. The overall heat-transfer coefficient in the first effect is 1000 W/(m^2 °C), and in the second effect it is 800 W/(m^2 °C). Calculate the steam

---

*Indicates an advanced level of difficulty.

economy and area required in each effect, assuming the areas are equal in each effect. (Hint: Assume $(\Delta T)_2 = 1.3 \times (\Delta T)_1$.)

***8.5** A double-effect evaporator is being used to concentrate a liquid food from 5 to 35% total solids. The concentrated product leaves the second effect at a rate of 1000 kg/h. The juice enters the first effect at 60°C. Saturated steam at 169.06 kPa is available. Assume the areas in each effect are equal and the evaporation temperature inside the second effect is 40°C. The overall heat-transfer coefficient inside the first effect is 850 W/(m^2 °C) and inside the second effect is 600 W/(m^2 °C). Calculate the steam economy and the area required in each effect. (Hint: Assume $(\Delta T)_{1\text{st effect}} = (\Delta T)_{2\text{nd effect}}$ and do at least one iteration.)

***8.6** Solve Problem 8.5 using MATLAB®. For a tutorial on using MATLAB® see www.rpaulsingh.com.

***8.7** A single effect evaporator is being used to concentrate tomato paste. The outlet concentration, x_1, is controlled by regulating the product flow out, $\dot{m}_p$, and the feed flow in, $\dot{m}_f$, is used to maintain the level of the product in the evaporator. Measurements show that:

$$T_S = 130 \text{ C} \qquad T_1 = 48 \text{ C} \qquad T_f = 25 \text{ C}$$
$$x_f = 0.05 \qquad x_1 = 0.30 \qquad A = 28\,\text{m}^2$$
$$U = 1705\,\text{W/(m}^2 \text{ C)}$$

a. Using MATLAB® calculate the feed rate, $\dot{m}_f$, steam consumption, $\dot{m}_s$, and the steam economy.

b. If the concentration of the feed is increased to $x_f = 0.10$ by a preprocessing step, calculate the new values for $\dot{m}_f$, $\dot{m}_s$ and steam economy.
Assume no boiling point elevation and steady state. Use the Irvine and Liley (1984) equations to obtain steam properties and the ASHRAE model to calculate the enthalpy of the juice.

8.8 The overall heat transfer coefficient in a single effect tomato paste evaporator is related to product temperature and concentration by the empirical equation:

$$U = (4.086 * T_1 + 72.6)/x_1 \qquad (\text{W/m}^2\,°\text{C})$$

---

*Indicates an advanced level of difficulty.

Given the following information:

$$\dot{m}_f = 2.58\,\text{kg/s} \quad x_f = 0.05 \quad T_f = 71.1°\text{C}$$
$$A = 46.5\,\text{m}^2 \quad T_1 = 93.3°\text{C} \quad T_s = 126.7°\text{C}$$

Find $\dot{m}_p$, $\dot{m}_v$, $\dot{m}_s$, x_1 and the economy using MATLAB® function called *fsolve* nonlinear equation solver. Assume steady state, and there is no boiling point elevation. Use the Irvine and Liley (1984) equations to obtain steam properties and the ASHRAE model to calculate the enthalpy of the juice.

LIST OF SYMBOLS

A	area of heat exchanger (m^2)
c_{pf}	specific heat content of dilute liquid feed (kJ/[kg °C])
c_{pp}	specific heat content of concentrated product (kJ/[kg °C])
ΔT	temperature gradient inside an evaporator; temperature of steam—temperature of boiling liquid inside the evaporator chamber (°C)
H_{cs}	enthalpy of condensate at temperature T_s (kJ/kg)
H_f	enthalpy of liquid feed (kJ/kg)
H_p	enthalpy of concentrated product (kJ/kg)
H_{v1}	enthalpy of saturated vapor at temperature T_1 (kJ/kg)
H_{vs}	enthalpy of saturated vapor at temperature T_s (kJ/kg)
$\dot{m}_f$	mass flow rate of dilute liquid feed (kg/s)
$\dot{m}_p$	mass flow rate of concentrated product (kg/s)
$\dot{m}_s$	mass flow rate of steam or condensate (kg/s)
$\dot{m}_v$	mass flow rate of vapor (kg/s)
q	rate of heat transfer (W)
T_1	boiling temperature maintained inside the evaporator chamber (°C)
T_f	temperature of dilute liquid feed (°C)
T_s	temperature of steam (°C)
T	temperature (°C)
U	overall heat transfer coefficient (W/[m^2 K])
x_f	solid fraction in feed stream, dimensionless
x_p	solid fraction in product stream, dimensionless

BIBLIOGRAPHY

American Society of Heating, Refrigerating and Air Conditioning Engineers, Inc (2005). *2005 ASHRAE Handbook of Fundamentals.* ASHRAE, Atlanta, Georgia.

Anonymous (1977). *Upgrading Existing Evaporators to Reduce Energy Consumption*. Technical Information Center, Department of Energy, Oak Ridge, Tennessee.

Blakebrough, N. (1968). *Biochemical and Biological Engineering Science*. Academic Press, New York.

Charm, S. E. (1978). *The Fundamentals of Food Engineering*, 3rd ed. AVI Publ. Co, Westport, Connecticut.

Coulson, J. M. and Richardson, J. F. (1978). *Chemical Engineering*, 3rd ed., Vol. II. Pergamon Press, New York.

Geankoplis, C. J. (2003). *Transport Processes and Separation Process Principles*, 4th ed. Pearson Education Inc, New Jersey.

Heldman, D. R. and Singh, R. P. (1981). *Food Process Engineering*, 2nd ed. AVI Publ. Co, Westport, Connecticut.

Irvine, T. F. and Liley, P. E. (1984). *Steam and Gas Tables with Computer Equations*, Appendix I, 23. Academic Press, Inc, Orlando.

Kern, D. W. (1950). *Process Heat Transfer*. McGraw-Hill, New York.

McCabe, W. L., Smith, J. C., and Harriott, P. (1985). *Unit Operations of Chemical Engineering*, 4th ed. McGraw-Hill, New York.

Chapter 9

Psychrometrics

The subject of psychrometrics involves determination of thermodynamic properties of gas–vapor mixtures. The most common applications are associated with the air–water vapor system.

An understanding of the procedures used in computations involving psychrometric properties is useful in design and analysis of various food processing and storage systems. Knowledge of properties of air–water vapor mixture is imperative in design of systems such as air-conditioning equipment for storage of fresh produce, dryers for drying cereal grains, and cooling towers in food processing plants.

In this chapter, important thermodynamic properties used in psychrometric computations are defined. Psychrometric charts useful in determining such properties are presented. In addition, procedures to evaluate certain air-conditioning processes are discussed.

9.1 PROPERTIES OF DRY AIR

9.1.1 Composition of Air

Air is a mixture of several constituent gases. The composition of air varies slightly, depending on the geographical location and altitude. For scientific purposes, the commonly accepted composition is referred to as standard air. The composition of standard air is given in Table 9.1.

The apparent molecular weight of standard dry air is 28.9645. The gas constant for dry air, R_a, is computed as

$$\frac{8314.41}{28.9645} = 287.055\,(\mathrm{m^3\ Pa})/(\mathrm{kg\ K})$$

All icons in this chapter refer to the author's web site, which is independently owned and operated. Academic Press is not responsible for the content or operation of the author's web site. Please direct your web site comments and questions to the author: Professor R. Paul Singh, Department of Biological and Agricultural Engineering, University of California, Davis, CA 95616, USA. Email: rps@rpaulsingh.com

Table 9.1 Composition of Standard Air

Constituent	Percentage by volume
Nitrogen	78.084000
Oxygen	20.947600
Argon	0.934000
Carbon dioxide	0.031400
Neon	0.001818
Helium	0.000524
Other gases (traces of methane, sulfur dioxide, hydrogen, krypton, and xenon)	0.000658
	100.0000000

9.1.2 Specific Volume of Dry Air

Ideal gas laws can be used to determine the specific volume of dry air. Therefore,

$$V_a' = \frac{R_a T_A}{p_a} \tag{9.1}$$

where V_a' is the specific volume of dry air (m^3/kg); T_a is the absolute temperature (K); p_a is partial pressure of dry air (kPa); and R_a is the gas constant ([m^3 Pa]/[kg K]).

9.1.3 Specific Heat of Dry Air

At 1 atm (101.325 kPa), the specific heat of dry air c_{pa} in a temperature range of −40 to 60°C varying from 0.997 to 1.022 kJ/(kg K) can be used. For most calculations, an average value of 1.005 kJ/[kg K] may be used.

9.1.4 Enthalpy of Dry Air

Enthalpy, the heat content of dry air, is a relative term and requires selection of a reference point. In psychrometric calculations the reference pressure is selected as the atmospheric pressure and the reference temperature is 0°C. Use of atmospheric pressure as the reference allows the use of the following equation to determine the specific enthalpy:

$$H_a = 1.005(T_a - T_0) \tag{9.2}$$

where H_a is enthalpy of dry air (kJ/kg); T_a is the dry bulb temperature (°C); and T_0 is the reference temperature, usually selected as 0°C.

9.1.5 Dry Bulb Temperature

Dry bulb temperature is the temperature indicated by an unmodified temperature sensor. This is in contrast to the wet bulb temperature (described in Section 9.3.8) where the sensor is kept covered with a layer of water. Whenever the term temperature is used without any prefix in this book, dry bulb temperature is implied.

9.2 PROPERTIES OF WATER VAPOR

In Section 9.1, constituents of standard dry air were given. However, atmospheric air always contains some moisture. Moist air is a binary mixture of dry air and vapor. The vapor in the air is essentially superheated steam at low partial pressure and temperature. Air containing superheated vapor is clear; however, under certain conditions the air may contain suspended water droplets leading to the condition commonly referred to as "foggy."

The molecular weight of water is 18.01534. The gas constant for water vapor can be determined as

$$R_w = \frac{8314.41}{18.01534} = 461.52\,(\text{m}^3\ \text{Pa})/(\text{kg mol K})$$

9.2.1 Specific Volume of Water Vapor

Below temperatures of 66°C, the saturated or superheated vapor follows ideal gas laws. Thus, the characteristic state equation can be used to determine its properties.

$$V'_w = \frac{R_w T_A}{p_w} \tag{9.3}$$

where p_w is the partial pressure of water vapor (kPa), V'_w is the specific volume of water vapor (m^3/kg), R_w is the gas constant for water vapor ([m^3 Pa]/[kg K]), and T_A is the absolute temperature (K).

9.2.2 Specific Heat of Water Vapor

Experiments indicate that within a temperature range of −71 to 124°C, the specific heat of both saturated and superheated vapor changes only slightly. For convenience, a specific-heat value of 1.88 kJ/(kg K) can be selected.

9.2.3 Enthalpy of Water Vapor

The following expression can be used to determine the enthalpy of water vapor:

$$H_w = 2501.4 + 1.88(T_a - T_0) \tag{9.4}$$

where H_w is enthalpy of saturated or superheated water vapor (kJ/kg); T_a is the dry bulk temperature (°C); and T_0 is the reference temperature (°C).

9.3 PROPERTIES OF AIR–VAPOR MIXTURES

Similar to gas molecules, the water molecules present in an air–vapor mixture exert pressure on the surroundings. Air–vapor mixtures do not exactly follow the perfect gas laws, but for total pressures up to about 3 atm these laws can be used with sufficient accuracy.

9.3.1 Gibbs–Dalton Law

In atmospheric air–steam mixtures, the Gibbs–Dalton law is followed closely. Thus, the total pressure exerted by a mixture of perfect gases is the same as that exerted by the constituent gases independently. Atmospheric air exists at a total pressure equal to the barometric pressure. From the Gibbs–Dalton law,

$$p_B = p_a + p_w \tag{9.5}$$

where p_B is the barometric or total pressure of moist air (kPa), p_a is the partial pressure exerted by dry air (kPa), and p_w is the partial pressure exerted by water vapor (kPa).

9.3.2 Dew-Point Temperature

Water vapors present in the air can be considered steam at low pressure. The water vapor in the air will be saturated when air is at a temperature equal to the saturation temperature corresponding to the partial pressure exerted by the water vapor. This temperature of air is called the dew-point temperature. The dew-point temperature can be obtained from the steam table; for example, if the partial pressure of water vapor is 2.064 kPa, then the dew-point temperature can be directly obtained as the corresponding saturation temperature, 18°C.

A conceptual description of the dew-point temperature is as follows. When an air–vapor mixture is cooled at constant pressure and constant humidity ratio, a temperature is reached when the mixture becomes saturated. Further lowering of temperature results in condensation of moisture. The temperature at which this condensation process begins is called the dew-point temperature.

9.3.3 Humidity Ratio (or Moisture Content)

The humidity ratio W (sometimes called moisture content or specific humidity) is defined as the mass of water vapor per unit mass of dry air. The common unit for the humidity ratio is kg water/kg dry air. Thus,

$$W = \frac{m_w}{m_a} \tag{9.6}$$

or

$$W = \left(\frac{18.01534}{28.9645}\right)\frac{x_w}{x_a} = 0.622\frac{x_w}{x_a} \tag{9.7}$$

where x_w is the mole fraction for water vapor and x_a is the mole fraction for dry air.

The mole fractions x_w and x_a can be expressed in terms of partial pressure as follows. From the perfect gas equations for dry air, water vapor, and mixture, respectively,

$$p_a V = n_a RT \tag{9.8}$$

$$p_w V = n_w RT \tag{9.9}$$

$$pV = nRT \tag{9.10}$$

Equation (9.10) can be written as

$$(p_a + p_w)V = (n_a + n_w)RT \tag{9.11}$$

Dividing Equation (9.8) by Equation (9.11),

$$\frac{p_a}{p_a + p_w} = \frac{n_a}{n_a + n_w} = x_a \tag{9.12}$$

and, dividing Equation (9.9) by Equation (9.11),

$$\frac{p_w}{p_a + p_w} = \frac{n_w}{n_a + n_w} = x_w \tag{9.13}$$

Thus, from Equations (9.7), (9.12), and (9.13),

$$W = 0.622\frac{p_w}{p_a}$$

Since $p_a = p_B - p_w$,

$$W = 0.622 \frac{p_w}{p_B - p_w} \tag{9.14}$$

9.3.4 Relative Humidity

Relative humidity ϕ is the ratio of mole fraction of water vapor in a given moist air sample to the mole fraction in an air sample saturated at the same temperature and pressure. Thus,

$$\phi = \frac{x_w}{x_{ws}} \times 100 \tag{9.15}$$

From Equation (9.13),

$$\phi = \frac{p_w}{p_{ws}} \times 100$$

where p_{ws} is the saturation pressure of water vapor.

For conditions where the perfect gas laws hold, the relative humidity can also be expressed as a ratio of the density of water vapor in the air to the density of saturated water vapor at the dry bulb temperature of air. Thus,

$$\phi = \frac{\rho_w}{\rho_s} \times 100 \tag{9.16}$$

where ρ_w is the density of water vapor in the air (kg/m^3) and ρ_s is the density of saturated water vapor at the dry bulb temperature of air (kg/m^3). As the name suggests, relative humidity is not a measure of the absolute amount of moisture in the air. Instead, it provides a measure of the amount of moisture in the air relative to the maximum amount of moisture in air saturated at the dry bulb temperature. Since the maximum amount of moisture in the air increases as the temperature increases, it is important to express the temperature of the air whenever relative humidity is expressed.

9.3.5 Humid Heat of an Air–Water Vapor Mixture

The humid heat c_s is defined as the amount of heat (kJ) required to raise the temperature of 1 kg dry air plus the water vapor present by 1 K. Since the specific heat of dry air is 1.005 kJ/(kg dry air K) and 1.88 kJ/(kg water K) for water vapor, humid heat of the air–vapor mixture is given by

$$c_s = 1.005 + 1.88\,W \tag{9.17}$$

where c_s is the humid heat of moist air (kJ/[kg dry air K]) and W is the humidity ratio (kg water/kg dry air).

9.3.6 Specific Volume

The volume of 1 kg dry air plus the water vapor in the air is called specific volume. The commonly used units are cubic meter per kilogram (m^3/kg) of dry air.

$$V'_m = \left(\frac{22.4\ m^3}{1\ kg\ mol}\right)\left(\frac{1\ kg\ mol\ air}{29\ kg\ air}\right)\left(\frac{T_a + 273}{0 + 273}\right) + \left(\frac{22.4\ m^3}{1\ kg\ mol}\right)\left(\frac{1\ kg\ mol\ water}{18\ kg\ water}\right)\left(\frac{T_a + 273}{0 + 273}\right)\frac{W\ kg\ water}{kg\ air} \quad (9.18)$$

$$V'_m = (0.082T_a + 22.4)\left(\frac{1}{29} + \frac{W}{18}\right) \quad (9.19)$$

Example 9.1

Calculate the specific volume of air at 92°C and a humidity ratio of 0.01 kg water/kg dry air.

Given

Dry bulb temperature = 92°C
Humidity ratio = 0.01 kg water/kg dry air

Solution

Using Equation (9.19),

$$V'_m = (0.082 \times 92 + 22.4)\left(\frac{1}{29} + \frac{0.01}{18}\right)$$
$$= 1.049\ m^3/kg\ dry\ air$$

9.3.7 Adiabatic Saturation of Air

The phenomenon of adiabatic saturation of air is applicable to the convective drying of food materials.

The adiabatic saturation process can be visualized by the following experiment. In a well-insulated chamber as shown in Figure 9.1, air is allowed to contact a large surface area of water. The insulated chamber assures no gain or loss of heat to the surroundings (adiabatic conditions). In this process, part of the sensible heat of entering air is transformed into latent heat.

For the conditions just described, the process of evaporating water into the air results in saturation by converting part of the sensible heat of the entering air into latent heat and is defined as adiabatic saturation.

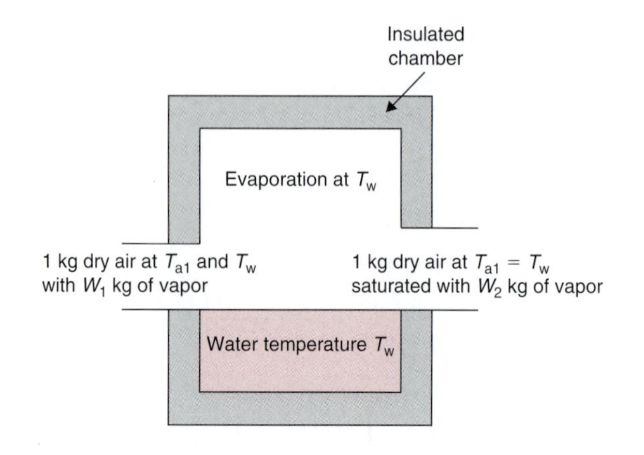

Figure 9.1 Adiabatic saturation of air in an insulated chamber. (From Jennings, 1970. Copyright © 1970 by Harper and Row, Publishers. Reprinted with permission of the publisher.)

The equation for adiabatic saturation is

$$T_{a1} = H_L \frac{(W_2 - W_1)}{(1.005 + 1.88W_1)} + T_{a2} \tag{9.20}$$

Equation (9.20) can be written as

$$\frac{W_2 - W_1}{T_{a1} - T_{a2}} = \frac{\bar{c}_s}{H_L} \tag{9.21}$$

where $\bar{c}_s = 1.005 + 1.88(W_1 + W_2)/2$.

Example 9.2

Air at 60°C dry bulb temperature and 27.5°C wet bulb temperature, and a humidity ratio of 0.01 kg water/kg dry air is mixed with water adiabatically and is cooled and humidified to a humidity ratio of 0.02 kg water/kg dry air. What is the final temperature of the conditioned air?

Given

Inlet: dry bulb temperature = 60°C
wet bulb temperature = 27.5°C
Initial humidity ratio W_1 = 0.01 kg water/kg dry air
Final humidity ratio W_2 = 0.02 kg water/kg dry air

Solution

1. *From Table A.4.2, latent heat of vaporization at 27.5°C = 2436.37 kJ/kg*
2. *Using Equation (9.20),*

$$T_{exit} = 60 - \frac{2436.37(0.02 - 0.01)}{[1.005 + (1.88)(0.015)]}$$
$$= 36.4°C$$

9.3.8 Wet Bulb Temperature

In describing air–vapor mixtures, two wet bulb temperatures are commonly used: the psychrometric wet bulb temperature and the thermodynamic wet bulb temperature. For moist air, the numerical values of these two temperatures are approximately the same. However, in other gas–vapor systems, the difference between the two temperatures can be substantial.

The psychrometric wet bulb temperature is obtained when the bulb of a mercury thermometer is covered with a wet wick and exposed to unsaturated air flowing past the bulb at high velocity (about 5 m/s). Alternatively, the bulb covered with a wet wick can be moved through unsaturated air. When the wick is exposed to unsaturated air, moisture evaporates due to the vapor pressure of saturated wet wick being higher than that of the unsaturated air.

The evaporation process requires latent heat from the wick and causes the temperature of the covered bulb to decrease. As the temperature of the wick decreases below the dry bulb temperature of air, the sensible heat flows from the air to the wick and tends to raise its temperature. A steady state is achieved when the heat flow from air to wick is equal to the latent heat of vaporization required to evaporate the moisture from the wick. This equilibrium temperature indicated by a wet bulb thermometer or similarly modified temperature sensor is called the wet bulb temperature.

As mentioned previously, the movement of air past the wet wick is essential, otherwise the wick will attain an equilibrium temperature between T_a and T_w.

In contrast to the psychrometric wet bulb temperature, the thermodynamic wet bulb temperature is reached by moist air when it is adiabatically saturated by the evaporating water. The thermodynamic wet bulb temperature is nearly equal to the psychrometric wet bulb temperature for moist air.

A mathematical equation that relates partial pressures and temperatures of air–vapor mixtures, developed by Carrier, has been used widely in calculations to determine psychrometric properties. The equation is

$$p_w = p_{wb} - \frac{(p_B - p_{wb})(T_a - T_w)}{1555.56 - 0.722T_w} \quad (9.22)$$

where p_w is the partial pressure of water vapor at dew-point temperature (kPa); p_B is the barometric pressure (kPa); p_{wb} is the saturation pressure of water vapor at the wet bulb temperature (kPa); T_a is the dry bulb temperature (°C); and T_w is the wet bulb temperature (°C).

Example 9.3

Find the dew-point temperature, humidity ratio, humid volume, and relative humidity of air having a dry bulb temperature of 40°C and a wet bulb temperature of 30°C.

Given

Dry bulb temperature = 40°C
Wet bulb temperature = 30°C

Solution

1. *From Table A.4.2,*
   Vapor pressure at 40°C = 7.384 kPa
   Vapor pressure at 30°C = 4.246 kPa
2. *From Equation (9.22),*

$$p_w = 4.246 - \frac{(101.325 - 4.246)(40 - 30)}{1555.56 - (0.722 \times 30)}$$
$$= 3.613 \text{ kPa}$$

   From Table A.4.2, the corresponding temperature for 3.613 kPa vapor pressure is 27.2°C. Thus, the dew-point temperature = 27.2°C.

3. *Humidity ratio, from Equation (9.14),*

$$W = \frac{(0.622)(3.613)}{(101.325 - 3.613)} = 0.023 \text{ kg water/kg dry air}$$

4. *Humid volume, from Equation (9.19),*

$$V'_m = (0.082 \times 40 + 22.4)\left(\frac{1}{29} + \frac{0.023}{18}\right)$$
$$= 0.918 \text{ m}^3/\text{kg dry air}$$

5. *Relative humidity: Based on Equation (9.15), the relative humidity is the ratio of the partial pressure of water vapor in the air (3.613 kPa) to the vapor pressure at dry bulb temperature (7.384 kPa), or*

$$\phi = \frac{3.613}{7.384} \times 100 = 48.9\%$$

Example 9.4

Develop a spreadsheet that can be used to determine psychrometric properties such as dew-point temperature, humidity ratio, humid volume, and relative humidity for air with a dry bulb temperature of 35°C and a wet bulb temperature of 25°C.

Given

Dry bulb temperature = 35°C
Wet bulb temperature = 25°C

Solution

1. *The spreadsheet is developed using Excel™. To determine, under saturation conditions, temperature when pressure is known or pressure when temperature is known, empirical expressions developed for a steam table by Martin (1961) and Steltz et al. (1958) are used. The equations used for this spreadsheet are valid between 10° and 93°C, and 0.029 and 65.26 kPa.*
2. *The spreadsheet formulas and results are shown in Figure E9.1. The procedure for calculation of psychrometric properties is the same as that used in Example 9.3.*

	A	B	C	D	E
1	Dry bulb temperature	35	C	7.46908269	−7.50675994E−03
2	Wet bulb temperature	25	C	−4.6203229E−09	−1.215470111E−03
3	x1	−610.398			
4	x2	−628.398			35.15789
5	vapor pressure (at dbt)	5.622	kPa		24.592588
6	vapor pressure (at wbt)	3.167	kPa		2.1182069
7	pw (SI units)	2.529	kPa		−0.3414474
8	pw (English units)	0.367	psia		0.15741642
9	pw_inter	1.299			−0.031329585
10	Dew-point temperature	21.27	C		0.003865828
11	Humidity ratio	0.016	kg water/kg dry air		−2.49018E−05
12	Specific volume	0.904	m^3/kg dry air		6.8401559E−06
13	Relative humidity	44.98	%		
14					
15	B3 =(B1*1.8+32)−705.398				
16					
17	B4 =(B2*1.8+32)−705.398				
18					
19	B5 =6.895*EXP(8.0728362+(B3*(D1+E1*B3+D2*B3^3)/((1+E2*B3)*((B1*1.8+32)+459.688))))				
20					
21	B6 =6.895*EXP(8.0728362+(B4*(D1+E1*B4+D2*B4^3)/((1+E2*B4)*((B2*1.8+32)+459.688))))				
22	B7 =B6−((101.325−B6)*(B1−B2)/(1555.56−(0.722*B2)))				
23					
24	B8 =B7/6.895				
25					
26	B9 =LN(10*B8)				
27					
28	B10 =((E4+E5*B9+E6*(B9)^2+E7*(B9)^3+E8*(B9)^4+E9*(B9)^5+ E10*(B9)^6+E11*(B9)^7+E12*(B9)^7+E12*(B9)^8)−32)/1.8				
29					
30					
31	B11 =0.622*B7/(101.325−B7)				
32					
33	B12 =(0.082*B1+22.4)*(1/29+0.023/18)				
34	B13 =B7/B5*100				
35					
36					

Steps:
1) Enter equations for cells B3 to B13 as shown
2) Enter coefficients in cells D1, D2, E1, E2, E4 to E12 as shown above. These coefficients are from Martin (1961)
3) Enter temperature values in cells B1 and B2, results are calculated in B10 to B13

Figure E9.1 Calculation of the psychrometric properties for Example 9.4.

3. *Enter empirical coefficients in cells D1:E2 and E4:E12; these coefficients are obtained from Martin (1961) and Steltz et al. (1958).*
4. *Enter mathematical expressions for calculating various psychrometric properties in cells B3:B13.*
5. *Enter 35 in cell B1 and 25 in cell B2. The results are calculated in the spreadsheet.*

9.4 THE PSYCHROMETRIC CHART

9.4.1 Construction of the Chart

From the preceding sections, it should be clear that the various properties of air–vapor mixtures are interrelated, and such properties can be computed using appropriate mathematical expressions. Another method to determine such properties is the use of a psychrometric chart drawn for a given barometric pressure. If two independent property values are known, the chart allows rapid determination of all psychrometric properties.

The construction of a psychrometric chart can be understood from Figure 9.2. The basic coordinates of the chart are dry bulb temperature plotted as the abscissa and humidity ratio (or specific humidity) as the ordinate. The wet bulb and dew-point temperatures are plotted

Figure 9.2 A skeleton psychrometric chart.

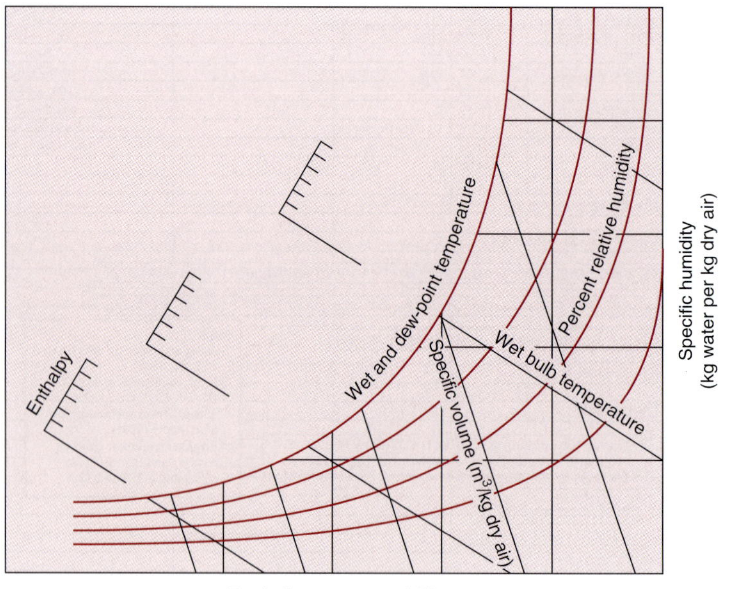

on the curve that swings upward to the right. The constant wet bulb temperature lines drawn obliquely are shown on Figure 9.2. The constant enthalpy lines coincide with the wet bulb temperature lines. The relative humidity curves also swing upward to the right. Note that the saturation curve represents 100% relative humidity. The constant specific volume lines are drawn obliquely; however, they have a different slope than the wet bulb temperature lines.

The psychrometric chart with all the thermodynamic data is shown in Appendix A.5. To use this chart, any two independent psychrometric properties are required. This allows location of a point on the psychrometric chart. The remaining property values can then be read from the chart. As an example, in Figure E9.2, a point A is located for known dry bulb and wet bulb temperatures. The various property values such as relative humidity, humidity ratio, specific volume, and enthalpy can then be read from the chart. It may be necessary to interpolate a property value, depending on the location of the point.

It should be noted that the psychrometric chart given in Appendix A.5 is for a barometric pressure of 101.325 kPa. All example problems discussed in this book assume a barometric pressure of 101.325 kPa. For other pressure values, charts drawn specifically for those pressures would be required.

Example 9.5

An air–vapor mixture is at 60°C dry bulb temperature and 35°C wet bulb temperature. Using the psychrometric chart (Appendix A.5), determine the relative humidity, humidity ratio, specific volume, enthalpy, and dew-point temperature.

Solution

1. *From the two given independent property values, identify a point on the psychrometric chart. As shown in the skeleton chart (Fig. E9.2), the following steps illustrate the procedure.*
2. *Location of point A: Move up on the 60°C dry bulb line until it intersects with the 35°C wet bulb temperature line.*
3. *Relative humidity: Read the relative humidity curve passing through A; $\phi = 20\%$.*
4. *Specific humidity: Move horizontally to the right of the ordinate to read $W = 0.026$ kg water/kg dry air.*
5. *Enthalpy: Move left on the oblique line for constant enthalpy (same as constant wet bulb temperature) to read $H_w = 129$ kJ/kg dry air.*
6. *Specific volume: By interpolation between specific volume lines, read $V'_m = 0.98\ m^3/kg$ dry air.*

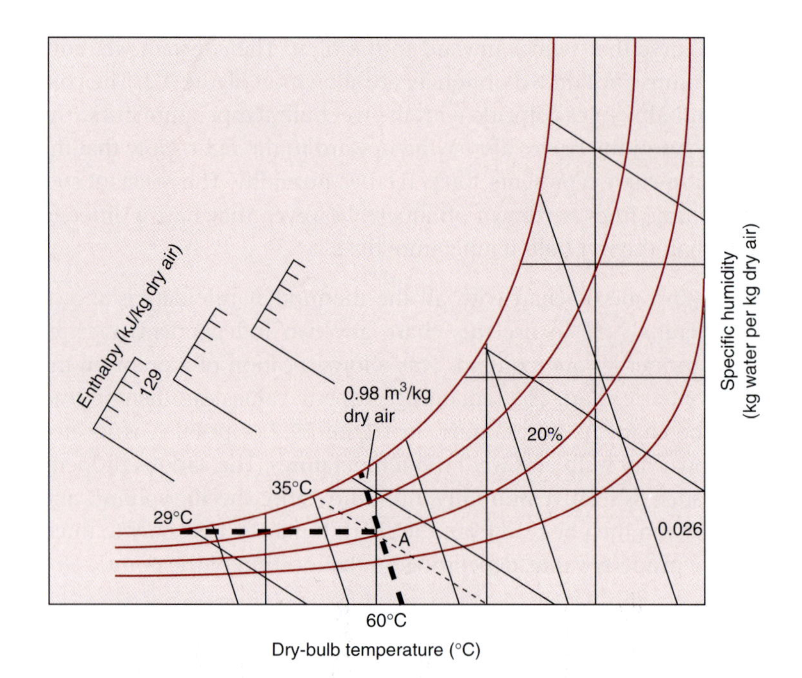

■ **Figure E9.2** A psychrometric chart with conditions of air given in Example 9.5.

7. *Dew-point temperature: Move horizontally to the left to intersect the 100% relative humidity (saturation curve). The temperature at the intersection is the dew-point temperature, or 29°C.*

9.4.2 Use of Psychrometric Chart to Evaluate Complex Air-Conditioning Processes

Several air-conditioning processes can be evaluated using the psychrometric chart. Usually, it is possible to describe an entire process by locating certain points as well as drawing lines on the chart that describe the psychrometric changes occurring during the given process. The value of such analysis is in relatively quick estimation of information useful in the design of equipment used in several food storage and processing plants, including air-conditioning, heating, drying, evaporative cooling, and humidification, as well as dehumidification of air. The following are some processes with important applications to food processing.

9.4.2.1 *Heating (or Cooling) of Air*

Heating (or cooling) of air is accomplished without addition or removal of moisture. Thus, the humidity ratio remains constant. Consequently, a straight horizontal line on the psychrometric chart exhibits a heating (or cooling) process.

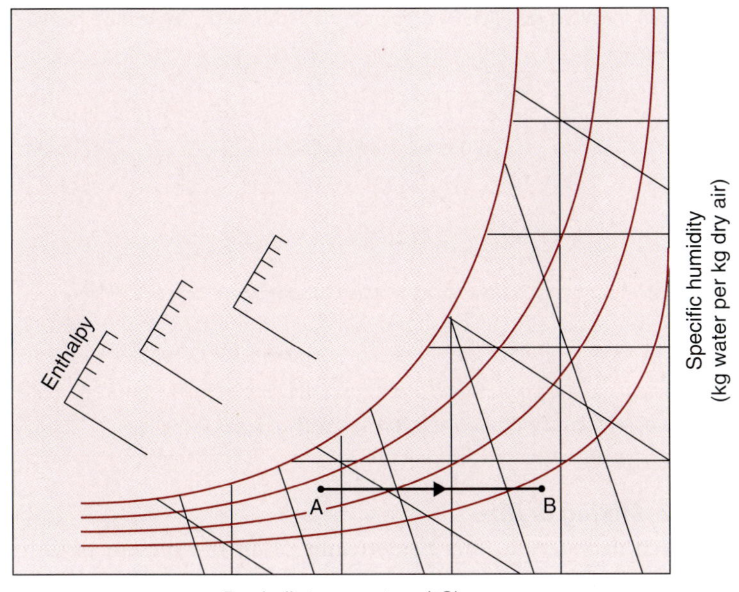

W

Figure 9.3 A heating process A–B shown on a psychrometric chart.

As shown in Figure 9.3, the process identified by line AB indicates a heating/cooling process. It should be obvious that if the air–vapor mixture is heated, the dry bulb temperature would increase; thus the process conditions will change from A to B. Conversely, the cooling process will change from B to A.

To calculate the amount of thermal energy necessary to heat moist air from state A to state B, the following equation can be used:

$$q = \dot{m}(H_B - H_A) \tag{9.23}$$

where H_B and H_A are enthalpy values read from the chart.

Example 9.6

Calculate the rate of thermal energy required to heat 10 m^3/s of outside air at 30°C dry bulb temperature and 80% relative humidity to a dry bulb temperature of 80°C.

Solution

1. *Using the psychrometric chart, we find at 30°C dry bulb temperature and 80% relative humidity, the enthalpy $H_1 = 85.2$ kJ/kg dry air, humidity ratio $W_1 = 0.0215$ kg water/kg dry air, and specific volume $V'_1 = 0.89$ m^3/kg dry air. At the end of the heating process, the dry bulb temperature is 80°C with a humidity ratio of 0.0215 kg water/kg dry air. The remaining values are read*

from the chart as follows: enthalpy $H_2 = 140\,kJ/kg$ dry air; relative humidity $\phi_2 = 7\%$.

2. *Using Equation (9.23),*

$$q = \frac{10}{0.89}(140 - 85.2)$$
$$= 615.7\,kJ/s$$
$$= 615.7\,kW$$

3. *The rate of heat required to accomplish the given process is 615.7 kW*
4. *In these calculations, it is assumed that during the heating process there is no gain of moisture. This will not be true if a directly fired gas or oil combustion system is used, since in such processes small amounts of water are produced as part of the combustion reaction (see Section 3.2.2).*

9.4.2.2 *Mixing of Air*

It is often necessary to mix two streams of air of different psychrometric properties. Again, the psychrometric chart can easily be used to determine the state of the mixed air.

The procedure involves first locating the conditions of the two air masses on the chart, as shown in Figure 9.4, points A and B. Next, the two points are joined with a straight line. This straight line is then

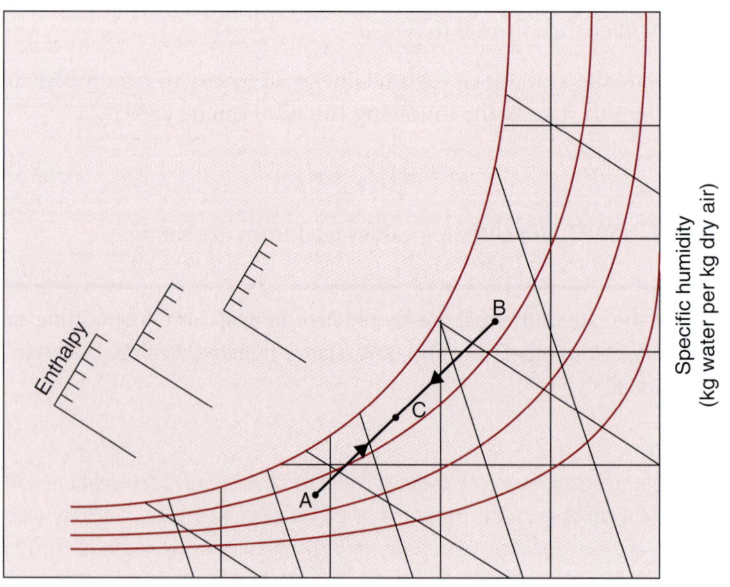

Figure 9.4 Mixing of air in equal parts shown on a psychrometric chart.

divided in inverse proportion to the weights of the individual air quantities. If the two air quantities are equal in weight, the air mixture will be denoted by point C (midpoint of line AB), as shown in Figure 9.4.

Example 9.7

In efforts to conserve energy, a food dryer is being modified to reuse part of the exhaust air along with ambient air. The exhaust airflow of 10 m^3/s at 70°C and 30% relative humidity is mixed with 20 m^3/s of ambient air at 30°C and 60% relative humidity. Using the psychrometric chart (Appendix A.5), determine the dry bulb temperature and humidity ratio of the mixed air.

Solution

1. *From the given data, locate the state points A and B, identifying the exit and ambient air as shown on the skeleton chart (Fig. E9.3).*
2. *Join points A and B with a straight line.*
3. *The division of line AB is done according to the relative influence of the particular air mass. Since the mixed air contains 2 parts ambient air and 1 part exhaust air, line AB is divided in 1:2 proportion to locate point C. Thus, the shorter length of line AC corresponds to larger air mass.*
4. *The mixed air, represented by point C, will have a dry bulb temperature of 44°C and a humidity ratio of 0.032 kg water/kg dry air.*

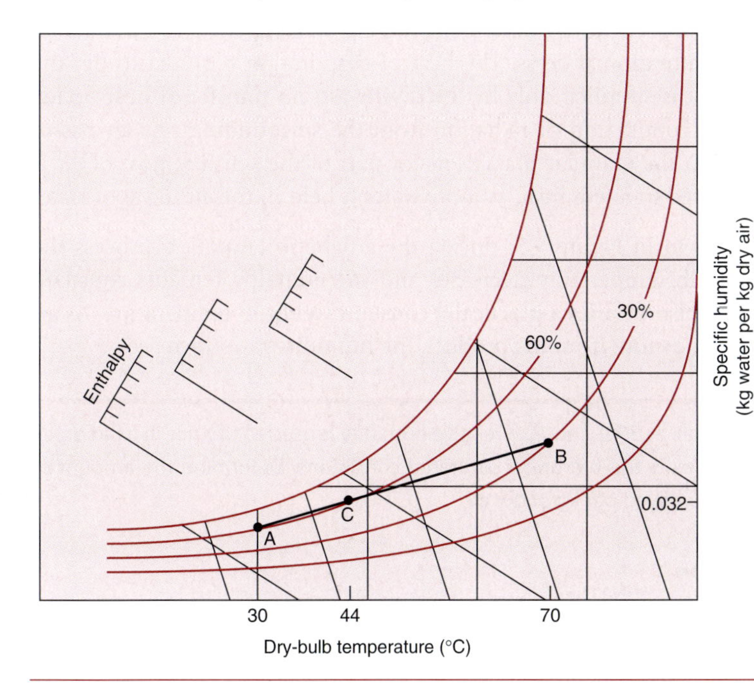

Figure E9.3 Mixing of air in unequal parts for data given in Example 9.7.

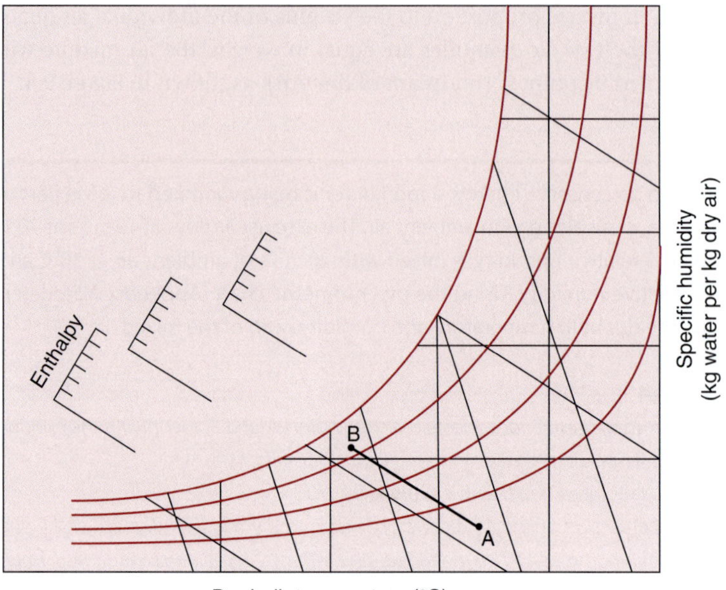

Figure 9.5 Drying (or adiabatic saturation) process shown on a psychrometric chart.

9.4.2.3 *Drying*

When heated air is forced through a bed of moist granular food, the drying process can be described on the psychrometric chart as an adiabatic saturation process. The heat of evaporation required to dry the product is supplied only by the drying air; no transfer of heat occurs due to conduction or radiation from the surroundings. As air passes through the granular mass, a major part of the sensible heat of air is converted to latent heat, as more water is held in the air in vapor state.

As shown in Figure 9.5, during the adiabatic saturation process the dry bulb temperature decreases and the enthalpy remains constant, which also implies a practically constant wet bulb temperature. As air gains moisture from the product, the humidity ratio increases.

Example 9.8

Heated air at 50°C and 10% relative humidity is used to dry rice in a bin dryer. The air exits the bin under saturated conditions. Determine the amount of water removed per kg of dry air.

Solution

1. *Locate point A on the psychrometric chart, as shown in Figure E9.4. Read humidity ratio = 0.0078 kg water/kg dry air.*
2. *Follow the constant enthalpy line to the saturation curve, point B.*

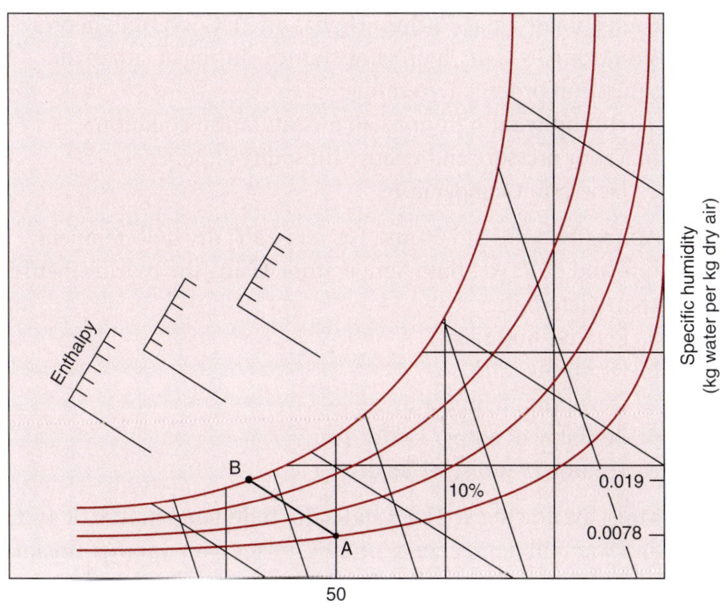

■ **Figure E9.4** Drying process for conditions given in Example 9.8.

3. At point B, read the humidity ratio = 0.019 kg water/kg dry air.

4. The amount of moisture removed from rice = 0.019 − 0.0078 = 0.0112 kg water/kg dry air.

PROBLEMS

9.1 The barometer for atmospheric air reads 750 mm Hg; the dry bulb temperature is 30°C; wet bulb temperature is 20°C. Determine:

a. The relative humidity.
b. The humidity ratio.
c. The dew-point temperature.

9.2 The humidity ratio of moist air at atmospheric pressure and at 27°C is 0.015 kg water/kg dry air. Determine

a. The partial pressure of water vapor.
b. The relative humidity.
c. The dew-point temperature.

9.3 Calculate (a) specific volume, (b) enthalpy, and (c) humidity ratio for moist air at 21°C and relative humidity of 30%, at a barometric pressure of 755 mm Hg.

***9.4** Atmospheric air at 750 mm Hg has an 11°C wet bulb depression from 36°C dry bulb temperature, during an adiabatic saturation process. Determine:

a. Humidity ratio from adiabatic saturation equation.
b. Vapor pressure and relative humidity at 36°C.
c. Dew-point temperature.

9.5 Atmospheric air at 760 mm Hg is at 22°C dry bulb temperature and 20°C wet bulb temperature. Using the psychrometric chart, determine:

a. Relative humidity.
b. Humidity ratio.
c. Dew-point temperature.
d. Enthalpy of air per kg dry air.
e. Volume of moist air/kg dry air.

9.6 Moist air flowing at 2 kg/s and a dry bulb temperature of 46°C and wet bulb temperature of 20°C mixes with another stream of moist air flowing at 3 kg/s at 25°C and relative humidity of 60%. Using a psychrometric chart, determine the (a) humidity ratio, (b) enthalpy, and (c) dry bulb temperature of the two streams mixed together.

***9.7** Air at a dry bulb temperature of 20°C and relative humidity of 80% is to be heated and humidified to 40°C and 40% relative humidity. The following options are available for this objective: (a) by passing air through a heated water-spray air washer; (b) by preheating sensibly, and then passing through a water-spray washer with recirculated water until relative humidity rises to 95% and then again heating sensibly to the final required state. Determine for (a) and (b) the total heating required, the make-up water required in water-spray air washer, and the humidifying efficiency of the recirculated spray water.

9.8 Moist air at 35°C and 55% relative humidity is heated using a common furnace to 70°C. From the psychrometric chart, determine how much heat is added per m^3 initial moist air and what the final dew-point temperature is.

***9.9** A water-cooling tower is to be designed with a blower capacity of 75 m^3/s. The moist air enters at 25°C and wet bulb temperature of 20°C. The exit air leaves at 30°C and relative humidity of 80%. Determine the flow rate of water, in kg/s, that can be

---

* Indicates an advanced level of difficulty in solving.

cooled if the cooled water is not recycled. The water enters the tower at 40°C and leaves the tower at 25°C.

***9.10** Ambient air with a dew point of 1°C and a relative humidity of 60% is conveyed at a rate of 1.5 m^3/s through an electric heater. The air is heated to a dry bulb temperature of 50°C. The heated air is then allowed to pass through a tray drier that contains 200 kg of apple slices with an initial moisture content of 80% wet basis. The air exits the dryer with a dew-point temperature of 21.2°C.

a. If the electrical energy costs 5¢/(kW h), calculate the electrical costs for heating the air per hour of operation.
b. Calculate the amount of water removed by air from apple slices per hour of operation.
c. If the dryer is operated for 2 h, what will be the final moisture content of the apple slices (wet basis)?

9.11 Air is at a dry bulb temperature of 20°C and a wet bulb temperature of 15°C. Determine the following properties from a psychrometric chart.

a. Moisture content
b. Relative humidity
c. Enthalpy
d. Dew point
e. Specific volume

9.12 In a humidifier, air at a dry bulb temperature of 40°C and relative humidity of 10% is humidified to a relative humidity of 40%. Determine the amount of moisture added in the humidifier per kg of dry air.

9.13 Air at a dry bulb temperature of 100°C and 4% relative humidity is passed over cooling coils to cool it to a dry bulb temperature of 40°C. How much heat is removed from the air in the process?

9.14 Air at a dry bulb temperature of 40°C and a wet bulb temperature of 20°C is first heated in a heater to a dry bulb temperature of 90°C. Then it is passed through a bed of apricot slices to dry them. The air exiting from the top of the apricot bed is at a dry bulb temperature of 60°C. It is then passed through a dehumidifier to reduce its relative humidity to 10%. Clearly show the various paths of the process on a psychrometric chart.

---

* Indicates an advanced level of difficulty in solving.

The air velocity through the bed and the dehumidifier is 4 m/s and the cross-sectional diameter of the bed is 0.5 m.

a. Determine the amount of moisture removed, in grams of water/second from the apricot bed.

b. Determine the amount of moisture removed, in grams of water/second in the dehumidifier.

9.15 On a copy of a psychrometric chart, show the heating and humidification process from the following data. Initially the air is at a dry bulb temperature (dbt) of 40°C and a relative humidity of 30%. The air is heated to a dbt of 80°C. The heated air is then conveyed through a humidifier to raise its relative humidity to 25%. For this process, calculate

a. The change in moisture content per kg of dry air.

b. The change in enthalpy from the initial unheated air to final humidified air.

9.16 Air at a dbt of 30°C and a relative humidity of 30% is conveyed through a heated dryer where it is heated to a dbt of 80°C. Then it is conveyed through a bed of granular pet food to dry it. The air exits the dryer at a dbt of 60°C. The exit air is again heated to 80°C and conveyed through another dryer containing another batch of pet food. The exit air from the second dryer leaves at saturation. Clearly show the paths of air, starting from the ambient air to the saturated air exiting the second dryer on a copy of a psychrometric chart. Determine the amount of water removed in the first and second dryer per kg of dry air.

LIST OF SYMBOLS

c_{pa} specific heat of dry air (kJ/[kg K])
c_{pw} specific heat of water vapor (kJ/[kg K])
c_s humid heat of moist air (kJ/kg dry air K)
H_a enthalpy of dry air (kJ/kg)
H_L latent heat of vaporization (kJ/kg)
H_w enthalpy of saturated or superheated water vapor (kJ/kg)
m_a mass of dry air (kg)
$\dot{m}$ mass flow rate of moist air (kg/s)
M_w molecular weight of water
m_w mass of water vapor (kg)
n number of moles
n_a number of moles of air
n_w number of moles of water vapor

p	partial pressure (kPa)
p_a	partial pressure of dry air (kPa)
p_B	barometric or total pressure of moist air (kPa)
p_w	partial pressure of water vapor (kPa)
p_{wb}	partial pressure of water vapor at wet bulb temperature (kPa)
p_{ws}	saturation pressure of water vapor (kPa)
ϕ	relative humidity (%)
q	rate of heat transfer (kW)
R	gas constant ([m^3 Pa]/[kg K])
R_a	gas constant dry air ([m^3 Pa]/[kg K])
R_0	universal gas constant (8314.41 [m^3 Pa]/[kg mol K])
R_w	gas constant water vapor ([m^3 Pa]/[kg K])
ρ_s	density of saturated water vapor at the dry bulb temperature (kg/m^3)
ρ_w	density of water vapor in the air (kg/m^3)
T	temperature (°C)
T_A	absolute temperature (K)
T_a	dry bulb temperature (°C)
T_0	reference temperature (°C)
T_w	wet bulb temperature (°C)
V	volume (m^3)
V'_a	specific volume of dry air (m^3/kg dry air)
V'_m	specific volume of moist air (m^3/kg)
V'_w	specific volume of water vapor (m^3/kg)
W	humidity ratio (kg water/kg dry air)
x_a	mole fraction for dry air
x_w	mole fraction for water vapor
x_{ws}	mole fraction for saturated air

BIBLIOGRAPHY

American Society of Heating, Refrigerating and Air-Conditioning Engineers, Inc. (1997). *ASHRAE Handbook of 1997 Fundamentals*, ASHRAE, Atlanta, Georgia.

Geankoplis, C. J. (1978). *Transport Processes and Unit Operations*, Allyn & Bacon, Boston.

Jennings, B. H. (1970). *Environmental Engineering, Analysis and Practice*, International Textbook Company, New York.

Martin, T. W. (1961). Improved computer oriented methods for calculation of steam properties. *J. Heat Transfer*, **83**: 515–516.

Steltz, W. G. and Silvestri, G. J. (1958). The formulation of steam properties for digital computer application. *Trans. ASME*. **80**: 967–973.

Chapter 10

Mass Transfer

In food processing, we often create conditions to encourage chemical reactions that produce desirable end-products in the most efficient manner. Frequently, in addition to desirable products, several by-products may be produced. These by-products may be undesirable from the process standpoint, but may have considerable economic value. In order to recover these secondary products, a separation step must be used to isolate the primary product of interest. In designing separation processes, an understanding of the mass transfer processes becomes important.

Mass transfer plays a key role in the creation of favorable conditions for reactants to physically come together, allowing a particular reaction to occur. Once the reactants are in proximity to a particular site, the reaction will proceed at an optimum rate. Under these circumstances, we may find that the reaction is limited by the movement of the reactants to the reaction site, or movement of end-products away from the reaction site. In other words, the reaction is mass-transfer limited, instead of being limited by the kinetics of the reaction.

To study mass transfer in food systems, it is important that we understand the term *mass transfer* as used throughout this textbook. In situations where we have a bulk flow of a fluid from one location to another, there is a movement of the fluid (of a certain mass), but the process is *not* mass transfer, according to the context being used. Our use of the term *mass transfer* is restricted to the migration of a constituent of a fluid or a component of a mixture. The migration occurs because of changes in the physical equilibrium of the system caused by the concentration differences. Such transfer may occur within one phase or may involve transfer from one phase to another.

All icons in this chapter refer to the author's web site, which is independently owned and operated. Academic Press is not responsible for the content or operation of the author's web site. Please direct your web site comments and questions to the author: Professor R. Paul Singh, Department of Biological and Agricultural Engineering, University of California, Davis, CA 95616, USA.
Email: rps@rpaulsingh.com.

Consider this example: If we carefully allow a droplet of ink to fall into a stagnant pool of water, the ink will migrate in various directions from the point where the ink made contact with the water. Initially, the concentration of ink in the droplet is very high, and the concentration of ink in the water is zero, thus establishing a concentration gradient. As the ink migration continues, the concentration gradient will decrease. When the ink becomes fully dissipated in the water, the concentration gradient becomes zero, and the mass transfer process will cease. The concentration gradient is considered the "driving force" for movement of a given component within a defined environment. For example, if you open a bottle of highly volatile material such as nail polish remover in a room, the component (acetone) will migrate to various parts of the room because of the concentration gradients of acetone. If the air is stationary, the transfer occurs as a result of random motion of the acetone molecules. If a fan or any other external means are used to cause air turbulence, the eddy currents will enhance the transfer of acetone molecules to distant regions in the room.

As we will find in this chapter, a number of similarities exist between mass transfer and heat transfer. In mass transfer, we will encounter terms that are also used in heat transfer, such as flux, gradient, resistance, transfer coefficient, and boundary layer.

According to the second law of thermodynamics discussed in Chapter 1, systems that are not in equilibrium tend to move toward equilibrium with time. For chemical reactions, any difference in the chemical potential of a species in one region of a space as compared to another region of the same space, is a departure from an equilibrium state. Over time, there will be a shift toward equilibrium, such that the chemical potential of that species is uniform throughout the region. Differences in chemical potential may occur because of the different concentration of the species from one point to another, differences in temperature and/or pressure, or differences caused by other external fields, such as gravitational force.

10.1 THE DIFFUSION PROCESS

Mass transfer involves both mass diffusion occurring at a molecular scale and bulk transport of mass due to convection flow. The diffusion process can be described mathematically using Fick's law of diffusion, which states that the mass flux per unit area of a component is proportional to its concentration gradient. Thus, for a component B,

$$\frac{\dot{m}_B}{A} = -D\frac{\partial c}{\partial x} \tag{10.1}$$

where $\dot{m}_B$ is mass flux of component B (kg/s); c is the concentration of component B, mass per unit volume (kg/m^3); D is the mass diffusivity (m^2/s); and A is area (m^2). Mass flux may also be expressed as kg-mole/s, and the concentration of component B will be kg-mole/m^3.

We note that Fick's law is similar to Fourier's law of heat conduction,

$$\frac{q}{A} = -k\frac{\partial T}{\partial x}$$

and Newton's equation for shear-stress–strain relationship,

$$\sigma = -\mu\frac{\partial u}{\partial y}$$

These similarities between the three transport equations suggest additional analogies among mass transfer, heat transfer, and momentum transfer. We will examine these similarities later in Section 10.1.2.

Consider two gases B and E in a chamber, initially separated by a partition (Figure 10.1a). At some instant in time, the partition is removed, and B and E diffuse in opposite directions as a result of the concentration gradients. The following derivation is developed to express the mass diffusion of gas B into gas E, and gas E into gas B. Figure 10.1b shows the gas concentrations at some time after the partition is removed. The concentrations are expressed as molecules per unit volume. In our simplistic diagram, circles represent molecules of a gas, and the molecules move in random directions. However, since the initial concentration of gas B is high on the right-hand side of the partition, there is a greater likelihood for molecules of B crossing the partition from right to left—a net transport of B from right to left. Similarly there is a net transport of E from left to right.

Using the ideal gas law,

$$p_B = \rho_B R_B T \tag{10.2}$$

where p_B is the partial pressure of the gas B (kPa); R_B is the gas constant for gas B; T is absolute temperature (K); and ρ_B is mass concentration of B (kg/m^3).

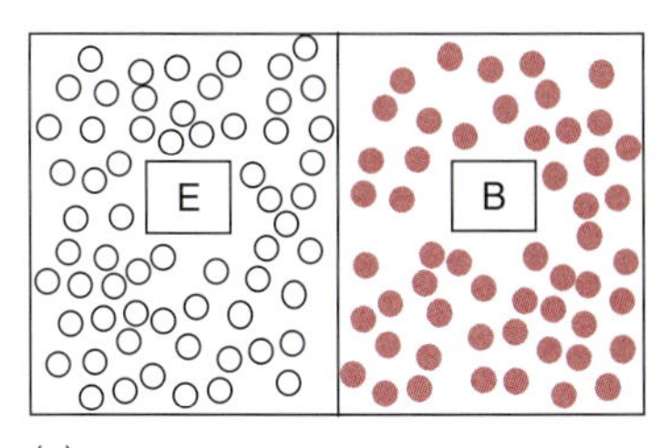

(a)

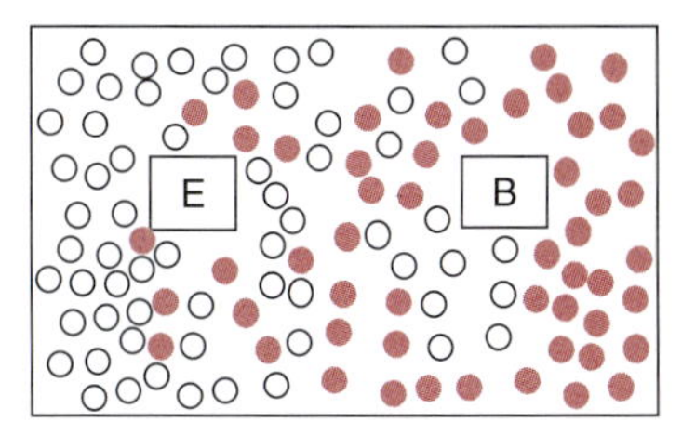

(b)

Figure 10.1 Diffusion of gases in an enclosed chamber.

The gas constant R_B for gas B can be written in terms of the universal gas constant R_u as follows:

$$R_B = \frac{R_u}{M_B} \tag{10.3}$$

where R_u is the universal gas constant 8314.41 (m^3 Pa)/(kg-mol K) or 8.314 (m^3 Pa)/(g-mol K) and M_B is molecular weight of gas B.

Thus, from Equation (10.2):

$$\rho_B = \frac{p_B}{R_B T} \tag{10.4}$$

or

$$\rho_B = \frac{p_B M_B}{R_u T} \tag{10.5}$$

Since ρ_B is mass concentration, we can substitute Equation (10.5) in Equation (10.1). Thus,

$$\frac{\dot{m}_B}{A} = -D_{BE} \frac{d}{dx}\left(\frac{p_B M_B}{R_u T}\right) \tag{10.6}$$

or

$$\frac{\dot{m}_B}{A} = -\frac{D_{BE} M_B}{R_u T} \frac{dp_B}{dx} \tag{10.7}$$

The mass diffusivity D_{BE} refers to diffusivity of gas B in gas E.

Equation (10.7) expresses the diffusion of gas B in gas E. Similarly, we can obtain Equation (10.8) to express diffusion of gas E in gas B.

$$\frac{\dot{m}_E}{A} = -\frac{D_{EB} M_E}{R_u T} \frac{dp_E}{dx} \tag{10.8}$$

The magnitude of mass diffusivities for liquids or gases in solids *are* less than the mass diffusivities for gases in liquids. These differences are due to the mobility of the molecules. Mass diffusivity values are expressed as centimeters squared per second (cm^2/s). In solids, the mass diffusivities range from 10^{-9} to 10^{-1} cm^2/s; in liquids the range of mass diffusivities is from 10^{-6} to 10^{-5} cm^2/s; and for gases, the range is from 5×10^{-1} to 10^{-1} cm^2/s. The mass diffusivity magnitudes are a function of temperature and concentration; in the case of gases, the mass diffusivity is substantially influenced by pressure.

Some representative values of mass diffusivities of gases in air and in water are presented in Tables 10.1a and 10.1b.

Table 10.1a Diffusion Coefficients of Selected Gases in Water at 20°C

Gas	D ($\times 10^{-9}$ m/s^2)
Ammonia	1.8
Carbon dioxide	1.8
Chlorine	1.6
Hydrogen	5.3
Nitrogen	1.9
Oxygen	2.1

For other temperatures $D_T = D_{20}[1 + 0.02(T - 20)]$

Table 10.1b Diffusion Coefficients of Selected Gases and Vapors in Air (under Standard Conditions)

Gas	D ($\times 10^{-6}$ m/s^2)
Ammonia	17.0
Benzene	7.7
Carbon dioxide	13.8
Ethyl alcohol	10.2
Hydrogen	61.1
Methyl alcohol	13.3
Nitrogen	13.2
Oxygen	17.8
Sulfur dioxide	10.3
Sulfur trioxide	9.4
Water vapor	21.9

10.1.1 Steady-State Diffusion of Gases (and Liquids) through Solids

Assuming the mass diffusivity does not depend on concentration, from Equation (10.1) we obtain

$$\frac{\dot{m}_A}{A} = -D_{AB}\frac{dc_A}{dx} \tag{10.9}$$

where D_{AB} is mass diffusivity for gas A (or liquid A) in a solid B. Subscript A for $\dot{m}$ and c represents a gas or liquid diffusing through a solid. In reality, D_{AB} represents an effective diffusivity through solids.

By separating variables and integrating Equation (10.9):

$$\frac{\dot{m}_A}{A}\int_{x_1}^{x_2} dx = -D_{AB}\int_{c_{A1}}^{c_{A2}} dc_A \qquad (10.10)$$

$$\frac{\dot{m}_A}{A} = \frac{D_{AB}(c_{A1} - c_{A2})}{(x_2 - x_1)} \qquad (10.11)$$

Equation (10.11) applies to one-dimensional diffusion when the concentration gradient is $[c_{A1} - c_{A2}]$ and constant with time at locations x_2 and x_1. In addition, the expression applies to rectangular coordinates. For a cylindrical shape, radial coordinates would apply and the following equation is obtained:

$$\dot{m}_A = \frac{D_{AB} 2\pi L(c_{A1} - c_{A2})}{\ln \frac{r_2}{r_1}} \qquad (10.12)$$

Equation (10.12) applies to a situation when diffusion is occurring in the radial direction of a cylinder; from the center to the surface or from the surface to the center. In order for the mass transfer to be steady-state the concentrations at the surface and the center must be constant with time.

The conditions of steady-state diffusion need to be emphasized. The concentrations at the boundaries must be constant with time, and diffusion is limited to molecular motion within the solid being described. In addition, the mass diffusivities, D, are not influenced by magnitude of concentration, and no temperature gradients exist within the solid. The magnitudes for mass diffusivity, D, depend on both the solid and the gas or liquid diffusing in the solid.

10.1.2 Convective Mass Transfer

When the transport of a component due to a concentration gradient is enhanced by convection, the mass flux of the component will be higher than would occur by molecular diffusion. Convective mass transfer will occur in liquids and gases, and within the structure of a porous solid. The relative contributions of molecular diffusion and convective mass transfer will depend on the magnitude of convective currents within the liquid or gas.

The convective mass transfer coefficient k_m is defined as the rate of mass transfer per unit area per unit concentration difference. Thus,

$$k_m = \frac{\dot{m}_B}{A(c_{B1} - c_{B2})} \tag{10.13}$$

where $\dot{m}_B$ is the mass flux (kg/s); c is concentration of component B, mass per unit volume (kg/m^3); A is area (m^2). The units of k_m are m^3/m^2 s or m/s. The coefficient represents the volume (m^3) of component B transported across a boundary of one square meter per second.

By using the relationship presented in Equation (10.5), the mass transport due to convection becomes:

$$\dot{m}_B = \frac{k_m A M_B}{R_u T_A}(p_{B1} - p_{B2}) \tag{10.14}$$

This expression is used to estimate the mass flux based on the vapor pressure gradient in the region of mass transport.

When the specific application of mass transport is water vapor in air, Equation (9.14) can be incorporated in Equation (10.14) to obtain:

$$\dot{m}_B = \frac{k_m A M_B p}{0.622 R_u T_A}(W_1 - W_2) \tag{10.15}$$

When computing the convective transport of water vapor in air, Equation (10.15) is used, and the gradient is in the form of a humidity ratio gradient in the region of convective mass transport.

Convective mass transfer coefficients can be predicted using dimensional analysis, analogous to the methods described in Chapter 4 for convective heat transfer coefficients. In this section, we will consider some of the important dimensionless numbers involved in mass transfer.

In situations that involve molecular diffusion and mass transfer due to forced convection, the following variables are important: mass diffusivity D_{AB}, for component A in fluid B; the velocity of the fluid, u; the density of the fluid, ρ; the viscosity of the fluid, μ; the characteristic dimension d_c; and the convective mass transfer coefficient k_m. In the case of natural convection, additional important variables include the acceleration due to gravity, g, and the mass density difference $\Delta\rho$. The variables are grouped in the following dimensionless numbers:

$$N_{Sh} = \frac{k_m d_c}{D_{AB}} \tag{10.16}$$

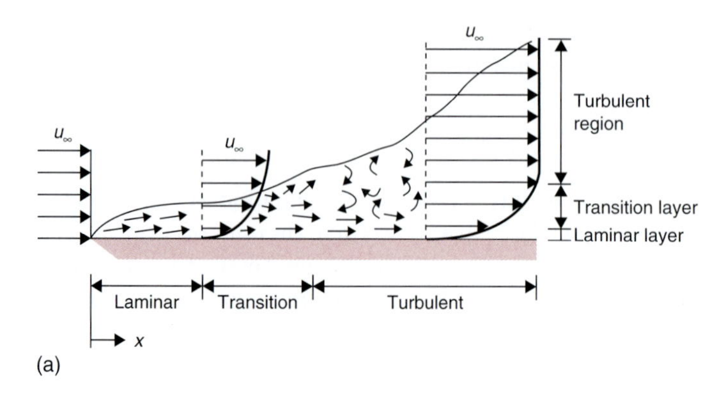

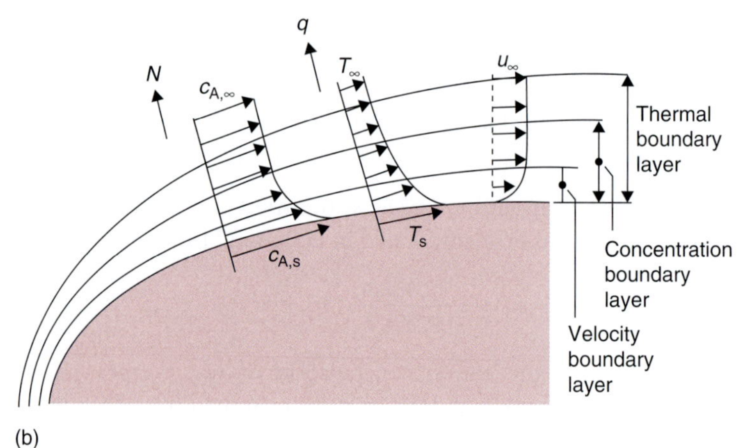

Figure 10.2 (a) The development of a boundary layer on a flat plate. (b) The development of the thermal, concentration, and velocity boundary layers on a surface. (From Incropera, Dewitt, Bergman, and Lavine, 2007. Copyright © 2007 by John Wiley & Sons.)

$$N_{\text{Sc}} = \frac{\mu}{\rho D_{\text{AB}}} \tag{10.17}$$

$$N_{\text{Re}} = \frac{\rho u d_{\text{c}}}{\mu} \tag{10.18}$$

$$N_{\text{Le}} = \frac{k}{\rho c_{\text{p}} D_{\text{AB}}} \tag{10.19}$$

Consider a fluid flowing over a flat plate as shown in Figure 10.2. For the boundary layer from the leading edge of the plate, we can write the following equations for momentum, energy, and concentration, respectively.

$$u_x \frac{\partial u_x}{\partial x} + u_y \frac{\partial u_x}{\partial y} = \frac{\mu}{\rho} \frac{\partial^2 u_x}{\partial y^2} \tag{10.20}$$

$$u_x \frac{\partial T}{\partial x} + u_y \frac{\partial T}{\partial y} = \alpha \frac{\partial^2 T}{\partial y^2} \tag{10.21}$$

$$u_x \frac{\partial c_A}{\partial x} + u_y \frac{\partial c_A}{\partial y} = D_{AB} \frac{\partial^2 c_A}{\partial y^2} \tag{10.22}$$

In Equation (10.22), c_A represents concentrations of component A at locations within the boundary layer.

Note that

$$\frac{\mu}{\rho\alpha} = \frac{\mu c_p}{k} = N_{Pr} = \textit{Prandtl number} \tag{10.23}$$

Thus, the Prandtl number provides the link between velocity and temperature profiles.

From Equations (10.20) and (10.22), if

$$\frac{\mu}{\rho D_{AB}} = 1 \tag{10.24}$$

then velocity and concentration profiles have the same shape. The ratio

$$\frac{\mu}{\rho D_{AB}} = N_{Sc} = \textit{Schmidt number} \tag{10.25}$$

The concentration and temperature profiles will have the same shape if

$$\frac{\alpha}{D_{AB}} = 1 \tag{10.26}$$

The ratio

$$\frac{\alpha}{D_{AB}} = N_{Le} = \textit{Lewis number} \tag{10.27}$$

The functional relationships that correlated these dimensional numbers for forced convection are:

$$N_{Sh} = f(N_{Re}, N_{Sc}) \tag{10.28}$$

If we compare the correlations for mass transfer with those presented for heat transfer in Chapter 4, the analogies are evident. If the dimensionless profiles of velocity, temperature, and concentration are assumed to be similar, the Nusselt and Prandtl numbers for heat transfer can be replaced by the Sherwood and Schmidt numbers, respectively, in mass transfer. Thus, it may be deduced that

$$N_{\mathrm{Sh}} = \frac{\text{total mass transferred}}{\text{total mass transferred by molecular diffusion}} \tag{10.29}$$

$$N_{\mathrm{Sc}} = \frac{\text{molecular diffusion of momentum}}{\text{molecular diffusion of mass}} \tag{10.30}$$

Next, we will consider a number of dimensionless correlations used in evaluating the convective mass transfer coefficient (k_{m}). These correlations are based on the following assumptions:

- Constant physical properties
- No chemical reactions in the fluid
- Small bulk flow at the interface
- No viscous dissipation
- No interchange of radiant energy
- No pressure, thermal, or forced diffusion.

10.1.3 Laminar Flow Over a Flat Plate

Laminar flow over a flat plate exists when $N_{\mathrm{Re}} < 5 \times 10^5$, and the correlation is:

$$N_{\mathrm{Sh}_x} = \frac{k_{\mathrm{m},x} x}{D_{\mathrm{AB}}} = 0.322 N_{\mathrm{Re_L}}^{1/2} N_{\mathrm{Sc}}^{1/3} \qquad N_{\mathrm{Sc}} \geq 0.6 \tag{10.31}$$

In Equation (10.31), the convective mass transfer coefficient $k_{\mathrm{m},x}$ in the Sherwood number is at a fixed location; therefore, the $N_{\mathrm{Sh},x}$ is termed the local Sherwood number. The characteristic dimension used in the Sherwood and Reynolds numbers is the distance from the leading edge of the plate.

When flow is laminar over the entire length of the plate, we can obtain an average Sherwood number from the following relationship:

$$N_{\mathrm{Sh_L}} = \frac{k_{\mathrm{m,L}} L}{D_{\mathrm{AB}}} = 0.664 N_{\mathrm{Re_L}}^{1/2} N_{\mathrm{Sc}}^{1/3} \qquad N_{\mathrm{Sc}} \geq 0.6 \tag{10.32}$$

In Equation (10.32), the characteristic dimension is the total length of the plate, L; and the convective mass transfer coefficient $k_{\mathrm{m,L}}$, obtained from the Sherwood number, is the average value for the entire plate.

Example 10.1

Determine the rate of water evaporated from a tray full of water. Air at a velocity of 2 m/s is flowing over the tray. The temperature of water and air is 25°C. The width of the tray is 45 cm and its length along the direction of air flow is 20 cm. The diffusivity of water vapor in air is $D = 0.26 \times 10^{-4}\ m^2/s$. The relative humidity of air is 50%.

Given

Velocity = 2 m/s
Temperature of water and air = 25°C
Width of the tray = 45 cm
Length of the tray = 20 cm
Diffusivity = $0.26 \times 10^{-4}\ m^2/s$
Kinematic viscosity of air at 25°C = $16.14 \times 10^{-6}\ m^2/s$

Approach

We will first determine the Reynolds number and then use an appropriate dimensionless correlation to obtain the mass transfer coefficient and the water evaporation rate.

Solution

1. *Reynolds number for the 20 cm long tray is*

$$N_{Re} = \frac{2 \times 0.2}{16.14 \times 10^{-6}} = 24{,}783$$

Since $N_{Re} < 5 \times 10^5$, the flow is laminar.

2. *We use Equation (10.32):*

$$N_{Sh} = \frac{k_m L}{D_{AB}} = 0.664 (N_{Re})^{1/2} (\dot{N}_{Sc})^{1/3}$$

where

$$N_{Sc} = \frac{\nu}{D_{AB}} = \frac{16.14 \times 10^{-6}}{0.26 \times 10^{-4}} = 0.62$$

3. *Thus,*

$$\frac{k_m \times 0.2}{0.26 \times 10^{-4}} = 0.664 (24{,}783)^{1/2} (0.62)^{1/3}$$

$$k_m = 1.1587 \times 10^{-2}\ m/s$$

4. *The evaporation rate for the tray is*

$$\dot{m}_A = k_m A (c_{A,s} - c_{A,\infty})$$

where $c_{A,s}$ is the concentration under saturated conditions,

$$c_{A,S} = \rho_{A,S} = 0.02298\ kg/m^3$$

and where $c_{A,\infty}$ is the concentration of water in the free stream; since relative humidity is 50%, then

$$\rho_{A,\infty} = (0.5)(0.02298) = 0.01149\ kg/m^3$$

5. *Therefore,*

$$\dot{m}_A = (1.1587 \times 10^{-2}\ m/s) \times (0.45\ m \times 0.2\ m) \times (0.02298\ kg/m^3 - 0.01149\ kg/m^3)$$

$$\dot{m}_A = 1.1982 \times 10^{-5}\ kg/s$$

6. *The water evaporation rate from the tray is 0.043 kg/h.*

Example 10.2

Determine the rate of water evaporated from a tray of water described in Example 10.1 by using the partial pressures of water vapor in the air and at the water surface. Relative humidity of air is 50%.

Given

Air velocity $= 2\ m/s$
Temperature (air and water) $= 25°C$
Tray width $= 0.45\ m$
Tray length $= 0.2\ m$
Diffusivity of water vapor in air $= 0.26 \times 10^{-4}\ m^2/s$
Kinematic viscosity of air (25°C) $= 16.14 \times 10^{-6}\ m^2/s$
Relative humidity of air $= 50\%$
Vapor pressure of water at saturation $= 3.179\ kPa$ *(from Table A.4.2 at 25°C)*
Molecular weight of water $= 18\ kg/(kg\ mol)$
Gas constant, $R = 8.314\ m^3\ kPa/(kg\ mol\ K)$

Approach

We will use the same approach as in Example 10.1 to obtain the mass transfer coefficient. The partial pressure gradient will be used to compute the water evaporation rate.

Solution

1. *Based on computation from Example 10.1, the mass transfer coefficient,*

$$k_m = 1.16 \times 10^{-2}\ m/s$$

2. *Using the definition of relative humidity, the partial pressure of 50% RH air is*

$$p_{B2} = \left(\frac{\%RH}{100}\right)(p_{B1}) = \left(\frac{50}{100}\right)(3.179) = 1.5895\ kPa$$

3. *Using Equation (10.14):*

$$\dot{m}_B = \frac{1.16\times10^{-2}[m/s]\times(0.2\times0.45)[m^2]\times18[kg/(kg\,mol)]}{8.314[m^3\,kPa/(kg\,mol\,K)]\times(25+273)[K]}\times(3.179-1.5895)[kPa]$$

4. *Then*

$$\dot{m}_B = 1.2\times10^{-5}\text{ kg/s} = 0.043\text{ kg water/h}$$

Example 10.3

Determine the rate of water evaporated from a tray of water described in Example 10.1 by using the humidity ratios for water vapor in the air and at the water surface.

Given

Air velocity = 2 m/s
Temperature (air and water) = 25°C
Tray width = 0.45 m
Tray length = 0.2 m
Diffusivity of water vapor in air = $0.26\times10^{-4}\ m^2/s$
Kinematic viscosity of air (25°C) = $16.14\times10^{-6}\ m^2/s$
Relative humidity of air = 50%
Molecular weight of water = 18 kg/(kg mol)
Gas constant, R = $8.314\ m^3\ kPa/(kg\ mol\ K)$
Atmospheric pressure = 101.325 kPa

Approach

The steps used in Example 10.1 are followed to obtain the mass transfer coefficient. The gradient of humidity ratios is used to compute the water evaporation rate.

Solution

1. *From Example 10.1, the mass transfer coefficient $k_m = 1.16\times10^{-2}$ m/s*
2. *From the psychrometric chart (Fig. A.5), the humidity ratio for saturated air (25°C) at the water surface is determined.*

$$W_1 = 0.0202\text{ kg water/kg dry air}$$

3. *From the psychrometric chart (Fig. A.5) the humidity ratio for 25°C air at 50% relative humidity is*

$$W_2 = 0.0101\text{ kg water/kg dry air}$$

4. *Using Equation (10.15), the mass flux of water from the surface to air is determined:*

$$\dot{m}_B = \frac{1.16\times10^{-2}[\text{m/s}]\times(0.2\times0.45)[\text{m}^2]\times18[\text{kg/(kg mol)}]\times101.325[\text{kPa}]}{0.622\times8.314[\text{m}^3\text{ kPa/(kg mol K)}]\times(25+273)[\text{K}]}\times(0.0202-0.0101)[\text{kg water/kg dry air}]$$

5. *Then*

$$\dot{m}_B = 1.25\times10^{-5}\text{ kg/s} = 0.045\text{ kg water/h}$$

10.1.4 Turbulent Flow Past a Flat Plate

The dimensionless relationship for dimensionless groups during turbulent flow ($N_{Re} > 5 \times 10^5$) past a flat plate is as follows:

$$N_{Sh_x} = \frac{k_{m,x}x}{D_{AB}} = 0.0296 N_{Re_x}^{4/5} N_{Sc}^{1/3} \quad 0.6 < N_{Sc} < 3000 \tag{10.33}$$

In Equation (10.33), the characteristic dimension is the distance from the leading edge of the plate, and the convective mass transfer coefficient is the local coefficient at the characteristic dimension, x.

The correlation to be used to determine the average convective mass transfer coefficient during turbulent flow is:

$$N_{Sh_L} = \frac{k_{mL}L}{D_{AB}} = 0.036 N_{Re}^{0.8} N_{Pr}^{0.33} \tag{10.34}$$

In Equation (10.34), the characteristic dimension is the total length of the flat plate.

10.1.5 Laminar Flow in a Pipe

For laminar flow in a pipe, the following equation is suggested:

$$\bar{N}_{Sh_d} = \frac{k_m d_c}{D_{AB}} = 1.86\left(\frac{N_{Re_d} N_{Sc}}{L/d_c}\right)^{1/3} \quad N_{Re} < 10{,}000 \tag{10.35}$$

where the characteristic dimension, d_c, is the diameter of the pipe.

10.1.6 Turbulent Flow in a Pipe

For turbulent flow in a pipe,

$$\bar{N}_{Sh_d} = \frac{k_m d_c}{D_{AB}} = 0.023 N_{Re_d}^{0.8} N_{Sc}^{1/3} \qquad N_{Re} > 10{,}000 \qquad (10.36)$$

where d_c is the characteristic dimension and diameter of the pipe.

10.1.7 Mass Transfer for Flow over Spherical Objects

Mass transfer to or from a spherical object is obtained from an expression similar to the Froessling correlation presented as Equation (4.69) for heat transfer.

$$\bar{N}_{Sh_d} = 2.0 + (0.4 N_{Re_d}^{1/2} + 0.06 N_{Re_d}^{2/3}) N_{Sc}^{0.4} \qquad (10.37)$$

For mass transfer from a freely falling liquid droplet, the following expression is recommended.

$$\bar{N}_{Sh_d} = 2.0 + 0.6 N_{Re_d}^{1/2} N_{Sc}^{1/3} \qquad (10.38)$$

Example 10.4

A 0.3175 cm sphere of glucose is placed in a water stream flowing at a rate of 0.15 m/s. The temperature of water is 25°C. The diffusivity of glucose in water is 0.69×10^{-5} cm²/s. Determine the mass transfer of coefficient.

Given

Diameter of sphere = 0.3175 cm = 0.003175 m
Velocity of water = 0.15 m/s
Temperature of water = 25°C
Diffusivity of glucose in water = 0.69×10^{-5} cm²/s
From Table A.4.1 @ 25°C

$$\text{Density} = 997.1\ \text{kg/m}^3$$

$$\text{Viscosity} = 880.637 \times 10^{-6}\ \text{Pa s}$$

Approach

We will first determine the Reynolds number and Schmidt number. Since the glucose sphere is submerged in a stream of water, we will use Equation (10.38) to determine the Sherwood number. The mass transfer coefficient will be obtained from the Sherwood number.

Solution

1. *The Reynolds number is*

$$N_{Re} = \frac{997.1\,kg/m^3 \times 0.15\,m/s \times 0.003175\,m}{880.637 \times 10^{-6}\,Pa\,s}$$
$$= 539$$

2. *The Schmidt number is*

$$N_{Sc} = \frac{880.637 \times 10^{-6}\,Pa\,s \times 10{,}000\,cm^2/m^2}{997.1\,kg/m^3 \times 0.69 \times 10^{-5}\,cm^2/s}$$
$$= 1279$$

3. *The Sherwood number can be obtained from Equation (10.38)*

$$N_{Sh} = 2.0 + 0.6(1279)^{1/3} \times (539)^{1/2}$$
$$= 153$$

4. *The mass transfer coefficient*

$$k_m = \frac{153 \times 0.69 \times 10^{-5}\,cm^2/s}{0.003175\,m \times 10{,}000\,cm^2/m^2}$$
$$= 3.32 \times 10^{-5}\,m/s$$

5. *The mass transfer coefficient will be 3.32×10^{-5} m/s, assuming that by dissolving glucose in water we will not alter the physical properties of water to any significant magnitude.*

10.2 UNSTEADY-STATE MASS TRANSFER

In many applications, the changes in concentration of a component within a food will occur under conditions where the rate of concentration change may increase or decrease with time. Examples would include the diffusion of salt within a solid food matrix, the diffusion of a volatile flavor within a dry food or the diffusion of an antimicrobial substance within a food. Under some conditions, the diffusion of liquid phase water may occur within a food under isothermal conditions. Finally, the uptake of moisture by a dry food during storage will occur due to the diffusion of water vapor within the dry food structure.

10.2.1 Transient-State Diffusion

The diffusion of the food component with a product mass would be described by:

$$\frac{\partial c}{\partial t} = D\left(\frac{\partial^2 c}{\partial x^2}\right) \tag{10.39}$$

where c is the concentration of the component diffusing within the solid food structure, as a function of time, t. The mass diffusivity, D, is the same property of the product and the diffusing component as described for steady-state diffusion. The analytical solutions to Equation (10.39) have been presented in many references, with Crank (1975) having the most complete array of geometries and boundary conditions. The key factors influencing the type of solution obtained are the geometry of the solid food object, and the boundary conditions needed to describe the conditions at the surface of the object. The series solutions are similar to the solutions referenced in unsteady-state heat transfer.

Unsteady-state mass transfer charts have been developed, such as Figure 10.3 from Treybal (1968). The chart presents concentration ratio versus dimensionless ratios Dt/d_c^2 for three standard geometries: infinite plate, infinite cylinder, and sphere. When using the chart in Figure 10.3, the concentration ratio contains the mass average concentration, c_{ma}, at any time, t; the concentration of the diffusing component in the medium surrounding the food object, c_m; and the initial concentration of the diffusing component within the food, c_i.

As introduced for unsteady-state heat transfer in Chapter 4, the characteristic dimension, d_c, changes depending on the geometry: one-half thickness for the infinite plate, radius of the infinite cylinder, and radius of the sphere. In addition, the chart in Figure 10.3 assumes that the boundary condition would represent negligible resistance to mass transfer at the surface of the object, as compared with diffusion within the food. This is a reasonable assumption in most food applications since the mass diffusivities, D, for liquids or gases within solid food structures have small magnitudes as compared with mass transfer of the gases or liquids at the boundary. Any convection at the surface will enhance mass transfer within the boundary layer at the object surface. It should be noted that the mass average concentration of the food as a function of time may not provide sufficient information, and the concentration distribution history within the food should be considered.

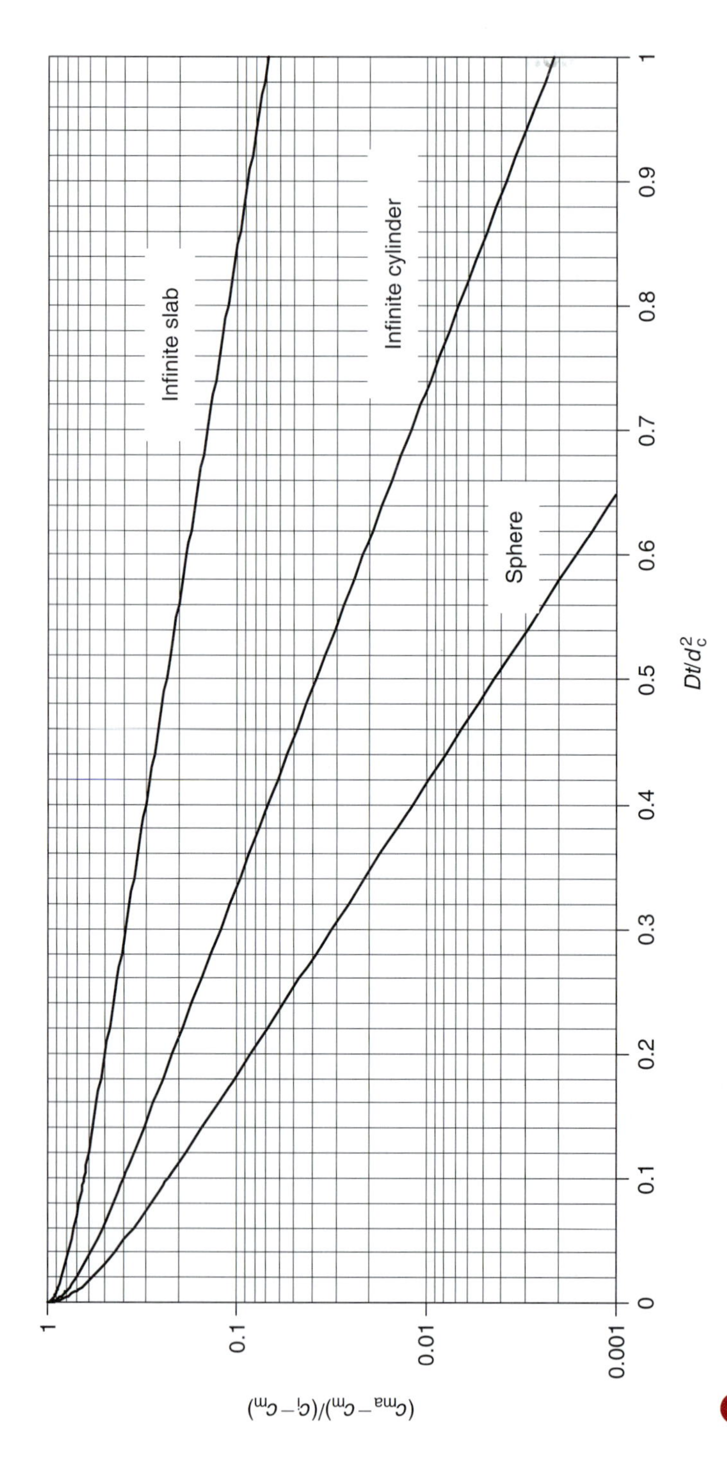

W **Figure 10.3** Unsteady-state mass transfer chart for mass average concentration in three standard geometries. (From Treybal, 1968)

Example 10.5

Salt is being used to preserve a 4.8 mm slice of salmon muscle. The concentration of salt at the surface is 0.533 kg/kg salt free salmon (SFS), and the initial concentration is 0.012 kg/kg SFS. If the mass diffusivity, D, of salt in salmon muscle is 8.78×10^{-11} m^2/s, determine the time required for the mass average concentration to reach 0.4 kg/kg SFS.

Given

Characteristic dimension for an infinite slab, $d_c = 2.4\,\text{mm} = 2.4 \times 10^{-3}\,\text{m}$
Salt concentration at the surface, $c_m = 0.533$ *kg/kg SFS*
Initial salt concentration, $c_i = 0.012$ *kg/kg SFS*
Mass average concentration, $c_{ma} = 0.4$ *kg/kg SFS*
$D = 8.78 \times 10^{-11}\,\text{m}^2/\text{s}$

Approach

The unsteady-state mass transfer chart (Fig. 10.3) will be used to estimate the dimensionless quantity, Dt/d_c^2*, from the concentration ratio.*

Solution

1. *The concentration ratio*

$$\frac{c_{ma} - c_m}{c_i - c_m} = \frac{0.4 - 0.533}{0.012 - 0.533} = 0.255$$

2. *From Figure 10.3,*

$$\frac{Dt}{d_c^2} = 0.46$$

3. *Then*

$$\text{Time} = \frac{0.46 d_c^2}{D} = \frac{0.46 \times (2.4 \times 10^{-3})^2 [\text{m}^2]}{8.78 \times 10^{-11} [\text{m}^2/\text{s}]} = 3.018 \times 10^4\,\text{s}$$

$$\text{Time} = 8.38\,\text{h}$$

4. *It takes 8.38 h for mass average salt concentration to reach 0.4 kg/kg SFS.*

A more useful relationship for applications to unsteady-state mass transfer in foods would be:

$$\frac{c - c_m}{c_i - c_m} = f(\bar{N}_{Bi}, \bar{N}_{Fo}) \qquad (10.40)$$

where

N_{Bi} = mass transfer Biot number;
$\bar{N}_{Fo}$ = mass transfer Fourier number.

The more complete use of the solutions to Equation (10.40) would predict concentration distribution histories within a food, based on knowledge of the convective mass transfer coefficient at the boundary of the object, and the mass diffusivity of the gas or liquid within the food structure.

An alternative approach to charts, such as Figure 10.3, is based on an analogy to heat transfer, as described in Chapter 4. When applied to mass transfer, the basic expression is the **diffusion rate equation**, as follows:

$$\log(c_m - c) = -\frac{t}{\bar{f}} + \log[\bar{j}(c_m - c_i)] \tag{10.41}$$

where the diffusion rate constant, $\bar{f}$, represents the time required for a one log-cycle change in the concentration gradient, and the lag coefficient, $\bar{j}$, describes the region of nonlinearity in the relationship between the concentration gradient and time during the initial stages of diffusion.

By adapting the charts developed by Pflug et al. (1965) and presented in Chapter 4, the coefficients $(\bar{f}, \bar{j})$ needed for the diffusion rate equation can be determined. The diffusion rate constant, $\bar{f}$, is predicted by using Figure 4.40, where the dimensionless number $\bar{f}D/d_c^2$ is presented as a function of the mass transfer Biot number. Note that when using Figure 4.40 for mass transfer we use the symbols on the chart appropriately. As is evident, the influence of the mass transfer Biot number is most dramatic between magnitudes of 0.1 and 100. At values less than 0.1, the internal resistance to mass transfer is negligible, and changes in the concentration within the food would be controlled by the magnitude of the convective mass transfer coefficient at the product surface. An application of this situation might be the transport of a gas or vapor through a packaging film to a porous food during storage. At mass transfer Biot numbers greater than 100, the external resistance to mass transfer is negligible, and changes in concentration within the food as a function of time are controlled by the magnitude of the mass diffusivity, D. Since Equation (10.41) was obtained by using the first term of the series solutions, as discussed in Chapter 4, it is valid only for Fourier numbers greater than 0.2.

The magnitude of the lag coefficient, $\bar{j}$, is influenced by the mass transfer Biot number, as illustrated in Figures. 4.41 and 4.42. In Figure 4.41, the relationships describe the influence of the mass transfer Biot number on the lag coefficient $\bar{j}_c$ at the geometric center of the object. The lag coefficient $\bar{j}_m$ at the location defining the mass average concentration of the object is influenced by the mass transfer Biot number as illustrated in Figure 4.42. For both coefficients, the influence of the mass transfer Biot number is most dramatic between 0.1 and 100.

The approach presented can be used to predict the time required for the mass average concentration of a food, or for the concentration at the center of the product, to reach some defined magnitude. After the magnitude of the mass transfer Biot number is established, the appropriate values are obtained from Figures. 4.40, 4.41, and 4.42. The magnitude of the diffusion rate constant is computed, based on magnitudes of the mass diffusivity, *D*, and the characteristic dimension for the product. These coefficients are used to compute the time from the diffusion rate equation, when given the concentration of diffusing component in the medium surrounding the product, as well as the initial concentration in the food.

Example 10.6

The diffusion of salt in salmon muscle described in Example 10.5 can be described by the diffusion rate equation. Determine the time required to increase the mass average concentration to 0.4 kg salt per kg SFS.

Given

$d_c = 2.4\,mm = 2.4 \times 10^{-3}\,m$

$c_m = 0.533\,kg/kg\,SFS$

$c_i = 0.012\,kg/kg\,SFS$

$D = 8.78 \times 10^{-11}\,m^2/s$

$c_{mc} = 0.4\,kg/kg\,SFS$

Approach

The diffusion rate equation will be used to determine the time required, after estimating the diffusion rate constant ($\bar{f}$) and lag coefficient ($\bar{j}_m$) from the charts (Figs. 4.40 and 4.42).

Solution

1. *Estimate the diffusion rate constant.*
   Since the slab concentration has been measured at the surface of the muscle, the resistance to mass transport is negligible at the surface and $\bar{N}_{Bi} \gg 40$
   From Figure 4.40,

$$\frac{\bar{f}D}{d_c^2} = 0.97$$

for an infinite slab at $\bar{N}_{Bi} \gg 40$.

2. *Then*

$$\bar{f} = \frac{0.97 \times (2.4 \times 10^{-3})^2 [m^2]}{8.78 \times 10^{-11} [m^2/s]} = 6.36 \times 10^4 \text{ s} = 17.68 \text{ h}$$

3. *Using Figure 4.42 at* $\bar{N}_{Bi} \gg 40$,

$$\bar{j}_m = 0.82$$

4. *Using the diffusion rate equation:*

$$\log(0.533 - 0.4) = -\frac{t}{17.68} + \log[0.82(0.533 - 0.012)]$$

$$-0.876 = -\frac{t}{17.68} + (-0.369)$$

$$t = 17.68(0.876 - 0.369)$$

$$t = 8.96 \text{ h}$$

10.2.2 Diffusion of Gases

The specific application of the diffusion rate equation to unsteady-state mass transfer of a gas can be accomplished by recognizing that the concentration is directly related to the partial pressure, as indicated in Equation (10.5). Given this relationship, the diffusion rate equation can be expressed as follows:

$$\log(p_m - p) = -\frac{t}{\bar{f}} + \log[\bar{j}(p_m - p_i)] \quad (10.42)$$

and the changes in partial pressure of a diffusing gas within a food product structure can be predicted in terms of partial pressures of that gas. This form of the diffusion rate equation would have specific application to diffusion of oxygen and similar gases within food products.

By considering the definition of water activity in terms of partial pressures of water vapor, the diffusion rate equation can be presented as:

$$\log(a_{wm} - a_w) = -\frac{t}{\bar{f}} + \log[\bar{j}(a_{wm} - a_{wi})] \quad (10.43)$$

and the changes in water activity of a dry food can be predicted, based on exposure to an environment with a water activity (relative humidity) different for the product. This form of the equation can be used to predict water activity within a food after a defined storage period in a defined environment, or to predict the time required for the product to reach a water activity limit during storage. These applications are closely associated with shelf-life predictions for dry and intermediate moisture content foods.

Example 10.7

Individual pieces of dry pasta are exposed to an environment at 15°C and 50% relative humidity. The mass diffusivity for water vapor within the pasta is 12×10^{-12} m^2/s, and the mass transfer coefficient in the environment around the pasta has been estimated to be1.2×10^{-4} m/s. The pieces of pasta have a diameter of 1 cm. If the initial water activity is 0.05, estimate the water activity of the pasta after one week.

Given

Characteristic dimension, infinite cylinder, $d_c = 0.005$ m

$k_m = 1.2 \times 10^{-4}$ m/s

$D = 12 \times 10^{-12}$ m^2/s

$a_{wm} = 0.5$ *(from relative humidity = 50%)*

$a_{wi} = 0.05$

Approach

The first step in the solution is computation of the mass transfer Biot number, followed by determination of the appropriate coefficients to be used in the diffusion rate equation.

Solution

1. *The mass transfer Biot number for the individual pieces of pasta is*

$$\bar{N}_{Bi} = 5 \times 10^4$$

2. *Using Figure 4.40,*

$$\frac{\bar{f}D}{d_c^2} = 0.4$$

$$\bar{f} = \frac{0.4 \times (0.005)^2 [\text{m}^2]}{12 \times 10^{-12} [\text{m}^2/\text{s}]} = 8.3 \times 10^5 \text{ s} = 231.5 \text{ h}$$

3. *Using Figure 4.42,*

$$\bar{j}_m = 0.7$$

4. *Using the diffusion rate equation,*

$$\log(0.5 - a_w) = -\frac{168}{231.5} + \log[0.7(0.5 - 0.05)]$$

and $a_w = 0.44$

5. *Based on the steps used, the mass average water activity of the pasta after one week is 0.44.*

The application of the diffusion rate equation to finite geometries is accomplished in the same manner as previously described for heat transfer. The key expressions for a finite cylinder and a finite slab, respectively, are:

$$\frac{1}{\bar{f}} = \frac{1}{\bar{f}_{IS}} + \frac{1}{\bar{f}_{IC}} \tag{10.44}$$

$$\frac{1}{\bar{f}} = \frac{1}{\bar{f}_{IS1}} + \frac{1}{\bar{f}_{IS2}} + \frac{1}{\bar{f}_{IS3}} \tag{10.45}$$

Similarly, coefficient $\bar{j}$ for finite cylinder and finite slab are, respectively:

$$\bar{j} = \bar{j}_{IS} \times \bar{j}_{IC} \tag{10.46}$$

$$\bar{j} = \bar{j}_{IS1} \times \bar{j}_{IS2} \times \bar{j}_{IS3} \tag{10.47}$$

By using the appropriate expressions, the coefficients $(\bar{f}, \bar{j})$ are obtained and the diffusion rate equation is used to predict concentrations, partial pressures, or water activities, as a function of time.

Example 10.8

Determine the time required for the center of pieces of the pasta in Example 10.7 to reach a water activity of 0.3. The pieces have a length of 2 cm and diameter of 1 cm.

Given

Characteristic dimension, infinite cylinder, $d_c = 0.005$ m
Characteristic dimension, infinite slab, $d_c = 0.01$ m
$k_m = 1.2 \times 10^{-4}$ m/s
$D = 12 \times 10^{-12}$ m²/s
$a_{wm} = 0.5$ (from relative humidity = 50%)
$a_{wi} = 0.05$
$a_w = 0.3$

Approach

After determination of the mass transfer Biot number, the charts (Figs. 4.40 and 4.41) are used to determine the coefficients for the diffusion rate equation.

Solution

1. *Both mass transfer Biot numbers (based on infinite slab and infinite cylinder) exceed 5×10^4.*

2. *Using computations from Example 10.7,*

$$\bar{f}_{IC} = 231.5\ h$$

3. *Using Figure 4.40 (for infinite slab),*

$$\frac{\bar{f}D}{d_c^2} = 0.97$$

4. *Then*

$$\bar{f}_{IS} = \frac{0.97 \times (0.01)^2}{12 \times 10^{-12}} = 8.08 \times 10^6\ s = 2245.4\ h$$

5. *From Figure 4.41,*

$$\bar{j}_{cs} = 1.27 \text{ (for infinite slab)}$$

$$\bar{j}_{ci} = 1.60 \text{ (for infinite cylinder)}$$

6. *Using Equation (10.44),*

$$\frac{1}{\bar{f}} = \frac{1}{2245.4} + \frac{1}{231.5}$$

$$\bar{f} = 209.86\ h$$

7. *Using Equation (10.46),*

$$\bar{j}_c = 1.27 \times 1.60 = 2.04$$

8. *Based on the diffusion rate equation,*

$$\log(0.5 - 0.3) = -\frac{t}{209.86} + \log[2.04(0.5 - 0.05)]$$

$$t = 138.89\ h$$

9. *Time for water activity to reach 0.3 at the center of the pasta pieces is 138.89 h or 5.8 days.*

PROBLEMS

10.1 A droplet of water is falling through 20°C air at the terminal velocity. The relative humidity of the air is 10% and the droplet is at the wet bulb temperature. The diffusivity of water vapor in air is $0.2 \times 10^{-4}\ m^2/s$. Estimate the convective mass transfer coefficient for a 100 μm diameter droplet.

10.2 The Sherwood Number for vapor transport from the surface of a high moisture food product to the surrounding air is 2.78.

Compute the convective mass transfer coefficient, when the dimension of the product in the direction of air movement is 15 cm, and the mass diffusivity for water vapor in air is 1.8×10^{-5} m^2/s.

10.3 A flavor compound is held within a 5 mm diameter time-release sphere. The sphere is placed in a liquid food and will release the flavor compound after one month of storage at 20°C. The concentration of flavor within the sphere is 100% and the mass diffusivity for the flavor compound within the liquid food is 7.8×10^{-9} m^2/s. Estimate the steady-state mass flux of flavor into the liquid food from the surface of the time-release sphere. The convective mass transfer coefficient is 50 m/s.

10.4 A desiccant is being used to remove water vapor from a stream of air with 90% RH at 50°C. The desiccant is a flat plate with 25 cm length and 10 cm width, and maintains a water activity of 0.04 at the surface exposed to the air stream. The stream of air has velocity of 5 m/s moving over the surface in the direction of the 25 cm length. The mass diffusivity for water vapor in air is 0.18×10^{-3} m^2/s. Compute the mass flux of water vapor from the air to the desiccant surface.

*__10.5__ Cucumbers are preserved by storage in a salt brine with a concentration of 20% NaCl. The initial NaCl concentration in the cucumber is 0.6% and the moisture content is 96.1% (wb). The convective mass transfer coefficient at the surface of the cucumbers is sufficiently high to cause the mass transfer Biot number to be greater than 100. The mass diffusivity for NaCl in water is 1.5×10^{-9} m^2/s. Estimate the time required for the center of a 2 cm cucumber to reach 15%. Note that the concentration percentages in the cucumber are kg NaCl per kg of cucumber, whereas the brine concentration is kg NaCl per kg water.

10.6 An apple slice with 1 cm thickness is exposed to 80% RH for one week. After one week, the water activity of the apple has increased to 0.6, from an initial value of 0.1. The convective mass transfer coefficient at the product surface is 8×10^{-3} m/s. Estimate the mass diffusivity for water vapor in the apple.

---

*Indicates an advanced level in solving.

LIST OF SYMBOLS

$\dot{m}$	mass flow rate (kg/s)
$\bar{j}$	lag coefficient for mass transfer (dimensionless)
$\bar{N}_{\text{Bi}}$	mass transfer Biot number (dimensionless)
$\bar{N}_{\text{Fo}}$	mass transfer Fourier number (dimensionless)
$\bar{f}$	time required for a one log-cycle change in concentration gradient (s)
A	area (m^2)
a_{w}	water activity
c	concentration (kg/m^3 or kg mol/m^3)
c_{p}	specific heat (kJ/[kg °C])
D	mass diffusivity (m^2/s)
d_{c}	characteristic dimension (m)
E_{p}	activation energy for permeability (kcal/mol)
k	thermal conductivity (W/[m °C])
k_{m}	mass transfer coefficient (m/s)
L	length (m)
$\dot{m}$	mass flow rate (kg/s)
M	molecular weight
N_{Le}	Lewis number (dimensionless)
N_{Re}	Reynolds number (dimensionless)
N_{Sc}	Schmidt number (dimensionless)
N_{Sh}	Sherwood number (dimensionless)
p	partial pressure of gas (kPa)
P	permeability coefficient
q	rate of heat transfer (W)
R	gas constant (m^3 Pa/[kg K])
r	radial coordinate (m)
R_{u}	universal gas constant (m^3 Pa/[kg mol K])
σ	shear stress (Pa)
S	solubility (mol/[cm^3 atm])
T	temperature (K)
t	time (s)
ν	kinematic viscosity (m^2/s)
u	fluid velocity (m/s)
W	humidity ratio (kg water/kg dry air)
x	distance coordinate (m)
α	thermal diffusivity (m^2/s)
μ	viscosity (Pa s)
ρ	mass concentration (kg/m^3)

Subscripts: A, component A; B, component B; E, component E; i, initial; IC, infinite cylinder; IS, infinite slab; m, medium; ma, mass average; S, surface location; x, variable distance; 1, location 1; 2, location 2.

BIBLIOGRAPHY

Crank, J. (1975). *The Mathematics of Diffusion*, 2nd ed. Oxford University Press, London.

Incropera, F. P., Dewitt, D. P., Bergman, T., and Lavine, A. (2007). *Fundamentals of Heat and Mass Transfer*, 6th ed. John Wiley & Sons Inc., New York.

McCabe, W. L., Smith, J. C., and Harriott, P. (1985). *Unit Operations of Chemical Engineering*. McGraw-Hill, New York.

Pflug, I. J., Blaisdell, J. L., and Kopelman, J. (1965). Developing temperature–time curves for objects that can be approximated by a sphere, infinite plate, or infinite cylinder. *ASHRAE Trans.* **71**(1): 238–248.

Rotstein, E., Singh, R. P., and Valentas, K. (1997). *Handbook of Food Engineering Practice*. CRC Press, Boca Raton, Florida.

Treybal, R. E. (1968). *Mass Transfer Operations*, 2nd ed. McGraw-Hill, New York.

Chapter 11

Membrane Separation

Membrane separation systems have been used extensively in the chemical process industry. Their use in the food industry is now becoming more common. Some of the typical food-related applications include purification of water, and the concentration and clarification of fruit juices, milk products, alcoholic beverages, and wastewater.

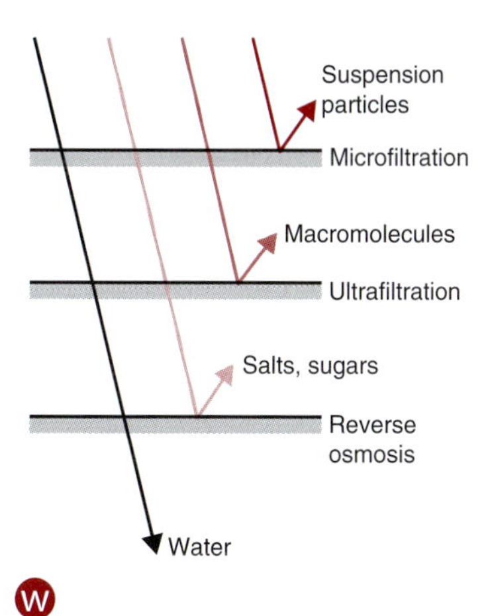

W

■ **Figure 11.1** Use of membrane systems to separate substances of different-sized molecules. (From Cheryan, 1989)

The most popular method for concentration of a liquid food is the evaporation process. During evaporation, a sufficient amount of thermal energy, equivalent to the latent heat of vaporization, must be added to the product to initiate phase change for water within the product (see Chapter 8). In an evaporator, the latent heat of vaporization represents a substantial part of the energy requirements or operating costs. In membrane separation systems, water is removed from the liquid product without phase change.

In a membrane separation system, a fluid containing two or more components is in contact with a membrane that permits selected components (for example, water in the fluid) to permeate more readily than other components. The physical and chemical nature of the membrane—for example, pore size and pore-size distribution—affect the separation of liquid streams. As shown in Figure 11.1, the membrane in a reverse-osmosis system allows water to permeate whereas salts and sugars are rejected. Ultrafiltration membranes are useful in fractionating components by rejecting macromolecules. In microfiltration, the membranes separate suspended particulates. The application of each of the membranes for separating different materials is shown in Figure 11.2.

Permeation of the selected component(s) is the result of a "driving force." For dialysis, the concentration difference across the membrane

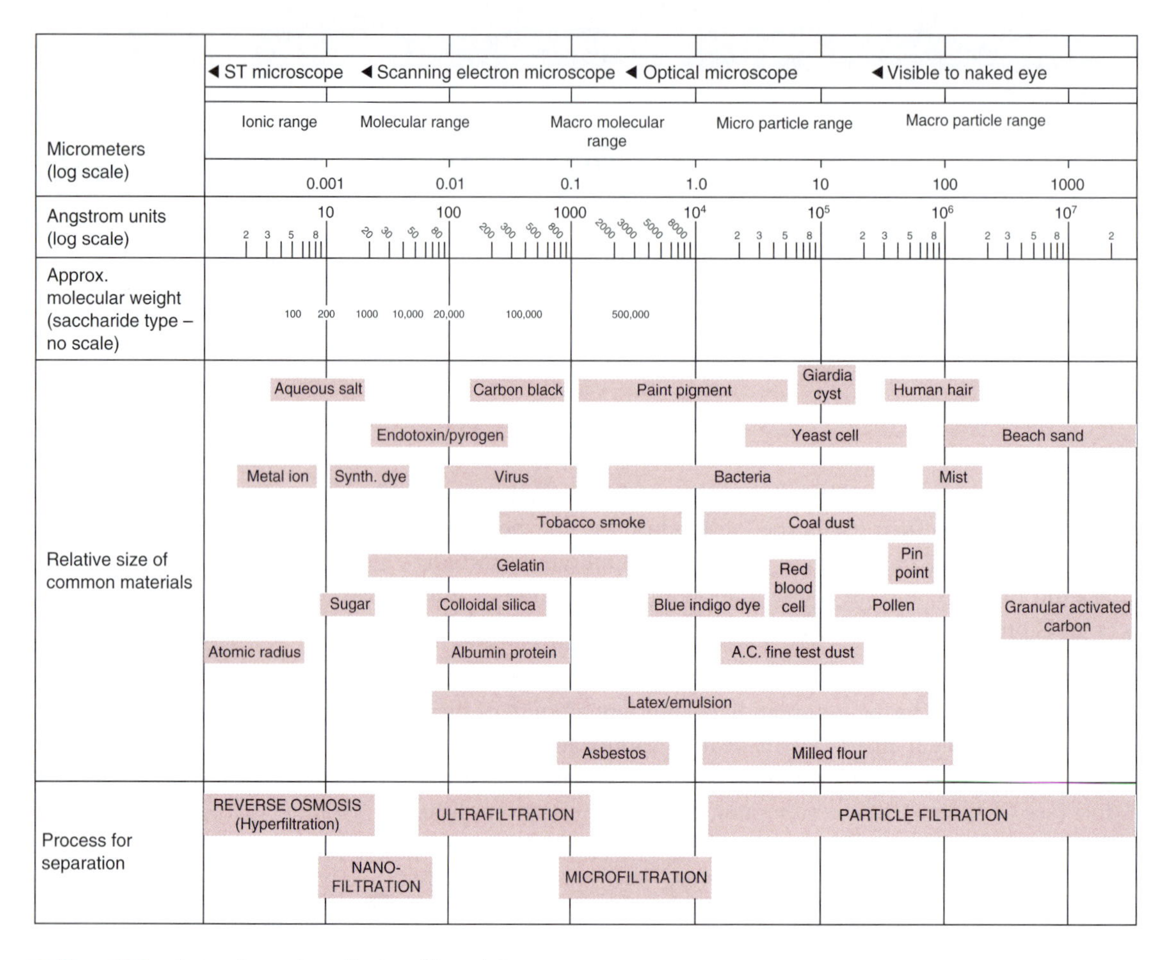

Figure 11.2 A separation spectrum. (Courtesy of Osmonics)

is the driving force, whereas in the case of reverse osmosis, ultrafiltration, and microfiltration systems, hydrostatic pressure is the key driving force. Microfiltration membrane systems require the lowest amount of hydraulic pressure; about 1 to 2 bar (or 15 to 30 psig). Ultrafiltration membrane systems operate at higher pressures; on the order of 1 to 7 bar (or 15 to 100 psig). These pressure levels are required to overcome the hydraulic resistance caused by a macromolecular layer at the membrane surface, as discussed in Section 11.5. In a reverse-osmosis system, considerably higher hydraulic pressures, in the range of 20 to 50 bar (300 to 750 psig), are necessary to overcome the osmotic pressures.

We can best understand the process of selective permeation by examining the structure of a membrane. Figure 11.3 is a visualization of a

Figure 11.3 The structure of an ultrafiltration membrane. (From Lacey, 1972)

membrane represented as a composite material consisting of polymeric chain interconnected by cross-linking. Any material being transported through a membrane must move through the interstitial spaces. When the interstitial openings are small, the transporting material makes its way through a membrane by pushing aside the neighboring polymeric chains. The resistance to the movement of a given material through a membrane depends on how "tight" or "loose" a membrane is. A polymeric membrane with a considerable degree of cross-linkages and crystallinity is considered tight, and it will offer considerable resistance to the permeation of a transporting material.

We will now consider three types of membrane systems: electrodialysis, reverse osmosis, and ultrafiltration.

11.1 ELECTRODIALYSIS SYSTEMS

The separation process in an electrodialysis system is based on the selective movement of ions in solution. The membranes in these systems derive their selectivity from the ions (anions or cations) that are allowed to permeate through them. Selected ions are removed from water as they pass through the semipermeable membranes. These membranes do not permit permeation of water.

If a polymeric chain in a membrane has a fixed negative charge, it will repulse any anion that tries to enter the membrane. This is shown schematically in Figure 11.4. For example, a negatively charged chain attracts cations and allows them to move through. In this type of membrane,

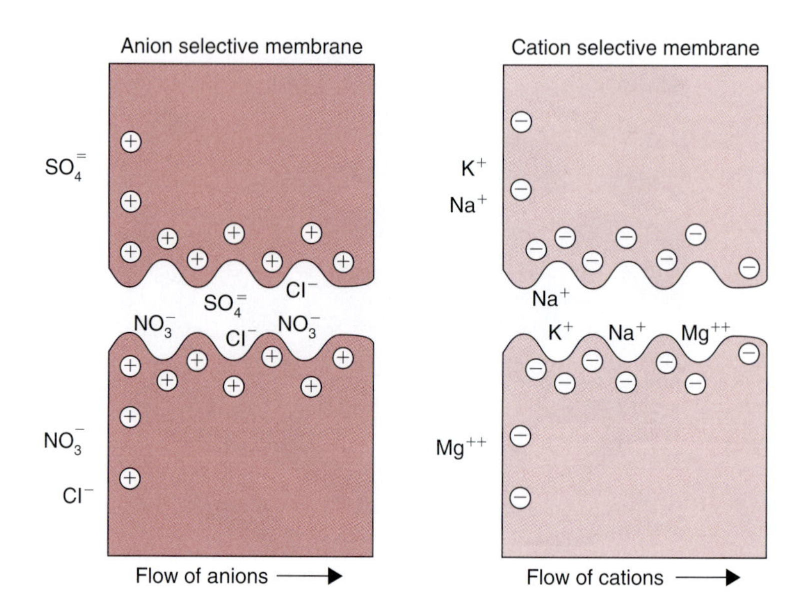

Figure 11.4 The movement of ions in ion-selective membranes. (From Applegate, 1984)

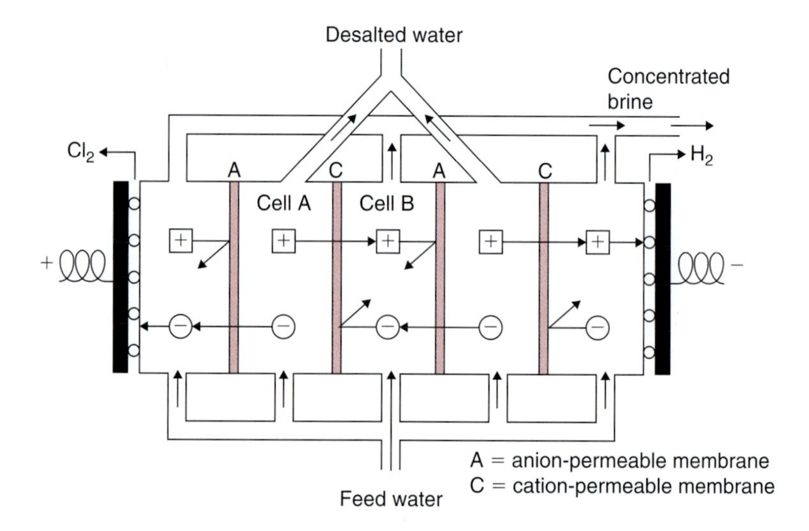

Figure 11.5 Desalting of water with an electrodialysis system. (From Lacey, 1972)

the distance between the cross-linkages of the polymeric chain should be large enough to minimize the resistance offered to the transporting ion. At the same time, the distance between the cross-linkages must not be too large, or the repulsive forces will be insufficient to provide the desired selectivity. This functional property of an ion membrane system is used in the electrodialysis system, as shown in Figure 11.5.

The electrodialysis system uses an electric current to transfer ions through a membrane. The membranes have fixed ionic groups that are chemically bound to the structure of the membrane. As shown in Figure 11.5, the electrodialysis system involves an array of membranes, alternating between anion-exchange and cation-exchange membranes. In between the membranes are small compartments (0.5 to 1.0 mm thick) that contain the solution. Electrodes are used to apply an electric charge to both sides of the membrane assembly. Depending on the fixed charge of the polymer chain in the membrane, only the compatible ions are able to move through a given membrane. Thus, cations will readily move through cation-exchange membranes but will be repulsed by anion-exchange membranes.

Let us follow the path of cations (such as Na^+ in a salt solution) shown in Figure 11.5. The cations in cell A are attracted toward the anode, and they readily move through the cation-exchange membrane. However, the anions are repulsed back into the solution. After the cations move to cell B, they cannot move farther to the right because they are repulsed toward the left by the anion-exchange membrane. Thus, either ion-concentrated or ion-depleted solution streams are obtained from alternating chambers.

Electrodialysis has been used extensively in desalting processes employing membranes that are permeable to ions, but impervious to water. Water leaving from the ion-depleted cells is the desalted product, and brine is obtained from the ion-concentrated cells. For the desalting applications, anion-selective membranes are made from cross-linked polystyrene with quaternary ammonia groups; cation-selective membranes are made from cross-linked polystyrene that is sulfonated, such that sulfonate groups are attached to the polymer. The sulfonate (SO_4) and ammonium (NH_4) groups provide the electronegative and electropositive charges, respectively. Ions attach with opposite electric charges on the membrane and easily migrate from one charge to another. This migration of ions causes the flow of electric current. The pores in these membranes are too small to allow water to transport through them. The electrodialysis process does not remove colloidal material, bacteria, or un-ionized matter.

The energy consumption for the electrodialysis process is given by

$$E = I^2 n R_c t \tag{11.1}$$

where E is energy consumption (J), I is electric current through the stack (A), n is the number of cells in the stack, R_c is the resistance of the cell (Ω), and t is time (s).

The electric current I can be calculated from the following equation:

$$I = \frac{zF\dot{m}\Delta c}{U} \tag{11.2}$$

where z is electrochemical valence; F is Faraday's constant, 96,500 A/s equivalent; $\dot{m}$ is feed solution flow rate (L/s); Δc is concentration difference between feed and product; U is current utilization factor; and I is direct current (A).

From Equations (11.1) and (11.2),

$$E = (nR_c)\left(\frac{zF\dot{m}\Delta c}{U}\right)^2 \tag{11.3}$$

For applications involving desalination of water, it is evident from Equation (11.3) that the energy required to desalt water is directly proportional to the concentration of salt in the feed. When the salt concentration is very high, the energy consumption will be correspondingly high. Application of electrodialysis to feedwater of less than 10,000 ppm total dissolved solids is usually limited by economics. Typically, for commercial applications, the most favorable economic application of electrodialysis requires a feed with a total dissolved solids content (TDS) of 1000 to 5000 mg/L to obtain a product with a TDS content of 500 mg/L. The U.S. Public Health Service Drinking Water Standards require that potable water should not contain more than 500 ppm TDS (although up to 1000 ppm may be considered acceptable). Table 11.1 presents the terms used to express the solids content of different levels of saline water.

Table 11.1 Total Dissolved Solids Content of Saline Water

Term	Total dissolved solids (ppm)
Fresh	$<$1000
Brackish	
Mildly brackish	1000–5000
Moderately brackish	5000–15,000
Heavily brackish	15,000–35,000
Sea water	35,000 (approximately)

In Japan, the electrodialysis process has been used extensively to obtain table salt from sea water. Other food applications of the electrodialysis process include removing salts from whey and orange juice.

11.2 REVERSE OSMOSIS MEMBRANE SYSTEMS

It is well known that when a plant or an animal membrane is used to separate two solutions of different solute concentrations, pure water passes through the membrane. The movement of water occurs from a solution with high concentration of water to a solution with low concentration of water, thus tending to equalize the water concentration on the two sides of the membrane. This movement of water is generally referred to as *osmosis*. Plant root hairs absorb water from the soil according to this phenomenon. The water is usually present in high concentration in soil surrounding the root hairs, whereas the water concentration inside the root cell is low due to dissolved sugars, salts, and other substances. Osmotic diffusion moves water from the soil into the root hairs.

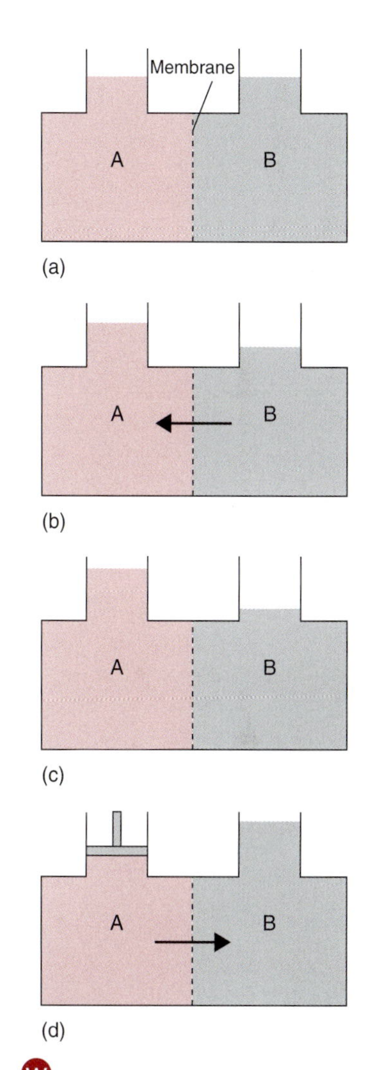

Figure 11.6 The reverse-osmosis process: (a) solute concentration same in cells A and B; (b) water movement from cell B to cell A; (c) osmosis equilibrium; (d) reverse osmosis—water movement from cell A to cell B.

Consider a solution of water containing a solute. In Figure 11.6a, a semipermeable membrane separates the solutions of the same solute concentration contained in chambers A and B. Since the chemical potential of the solvent (water) is the same on both sides of the membrane, no net flow of water occurs through the membrane. In Figure 11.6b, chamber A contains a solution with a higher solute concentration than chamber B; that is, chamber A has lower water concentration than chamber B. This also means that the chemical potential of the solvent (water) in chamber A will be lower compared with that of chamber B. As a result, water will flow from chamber B to chamber A. As seen in Figure 11.6c, this movement of water will cause an increase in the volume of water in chamber A. Once equilibrium is reached, the increased volume represents a change in head, or pressure, which will be equal to the osmotic pressure. If an external pressure greater than the *osmotic pressure* is then applied to chamber A, as shown in Figure 11.6d, the chemical potential of water in chamber A will increase, resulting in water flow from chamber A to chamber B. The reversal in the direction of water flow, obtained by application of an external pressure that exceeds the osmotic pressure, is termed *reverse osmosis*.

A reverse-osmosis membrane system is used to remove water from a water–solute mixture by application of external pressure. In contrast

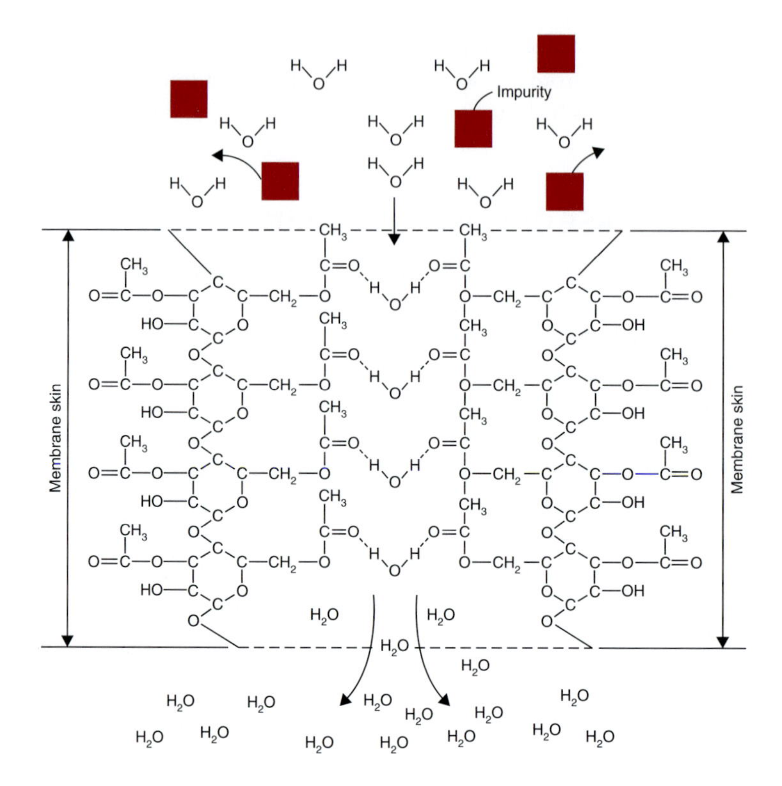

■ Figure 11.7 The movement of water through a cellulose acetate membrane. (From Lacey, 1972)

to electrodialysis, the membrane used in the reverse-osmosis system must be permeable to water.

In the 1950s, it was discovered that cellulose acetate, a highly organized polymer, has groups that can hydrogen-bond with water as well as with other solvents such as ammonia or alcohol. Figure 11.7 shows the chemical structure of cellulose acetate polymer. The hydrogen in the water molecule bonds to the carbonyl group of cellulose acetate. The water molecules hydrogen-bond on one side of the membrane, then move through the membrane by bonding to neighboring carbonyl groups. This process continues as water permeates the membrane to the other side. As illustrated in Figure 11.7, the polymer must be highly organized with carbonyl groups occurring at fixed locations, or water molecules wili be unable to permeate the membrane. The driving force for the water molecules to move from one set of hydrogen-bonding sites to the next is the pressure difference across the membrane.

Structurally, a polymeric membrane can be viewed as strands of the polymer and interstitial spaces. Since the polymer used in a reverse-osmosis membrane is highly organized, its structure must be tight, such that the interstitial spaces are small.

To obtain high flux rates throughout the membrane, the thickness of the membrane must be small. In the late 1950s, Loeb and Sourirajan invented a method to fabricate extremely thin films of anisotropic cellulose acetate attached to a supporting matrix with an open structure. Since their initial discovery, many developments have taken place in the selection of membrane materials.

In a reverse-osmosis system, water is the permeating material referred to as "permeate," and the remaining solution concentrated with the solutes is called "retentate."

The osmotic pressure Π of a dilute solution can be obtained by Van't Hoff's equation, which uses colligative properties of dilute solutions.

$$\Pi = \frac{cRT}{M} \tag{11.4}$$

where Π is osmotic pressure (Pa), c is solute concentration (kg/m^3) of solution, T is absolute temperature (K), R is gas constant, and M is molecular weight.

Example 11.1

Estimate the osmotic pressure of orange juice with 11% total solids at 20°C.

Given

Concentration of solids = 11% = 0.11 kg solids/kg product
Temperature = 20°C = 293 K

Approach

The Van't Hoff equation (Eq. (11.4)) will be used for computation, while assuming that glucose is the predominant component influencing the osmotic pressure.

Solution

1. *The density of orange juice is estimated based on density of carbohydrates at 1593 kg/m^2.*

$$\begin{aligned}\rho &= 0.11(1593) + 0.89(998.2)\\ &= 1063.6\ \text{kg/m}^3\end{aligned}$$

2. *The concentration, c, for Equation (11.4) becomes*

$$c = 0.11\ [\text{kg solids/kg product}] \times 1063.6\ [\text{kg product/m}^3\ \text{product}] = 117\ \text{kg solids/m}^3\ \text{product}$$

3. *Using Equation (11.4),*

$$\Pi = \frac{117\,[\text{kg solids / m}^3\ \text{product}] \times 8.314\,[\text{m}^3\ \text{kPa / (kg mol K)}] \times 293\,[\text{K}]}{180\,[\text{kg/(kg mol)}]}$$

4. $\Pi = 1583.5\ \text{kPa}$

From Equation (11.4), we observe that the presence of small molecules in a solution results in a high osmotic pressure. Another equation found to be more accurate over a wider range of solute concentrations uses Gibb's relationship, given as

$$\Pi = -\frac{RT \ln X_A}{V_m} \quad (11.5)$$

where V_m is the molar volume of pure liquid, and X_A is mole fraction of pure liquid.

Osmotic pressures of some food materials are given in Table 11.2. Foods with smaller molecular-weight constituents have higher

Table 11.2 Osmotic Pressure of Foods and Food Constituents at Room Temperature

Food	Concentration	Osmotic pressure (kPa)
Milk	9% solids-not-fat	690
Whey	6% total solids	690
Orange juice	11% total solids	1587
Apple juice	15% total solids	2070
Grape juice	16% total solids	2070
Coffee extract	28% total solids	3450
Lactose	5% w/v	380
Sodium chloride	1% w/v	862
Lactic acid	1% w/v	552

Source: Cheryan (1998)

osmotic pressure. Data on osmotic pressures of foods or food components is very limited. These data are important in membrane processing. For example, in order to achieve separation in a reverse-osmosis membrane, the pressure applied to the feed solution must exceed the osmotic pressure.

Example 11.2

Estimate the osmotic pressure of orange juice with 11% total solids at 20°C, using the Gibb's relationship.

Given

Concentration of solids = 11%

Temperature = 20°C

Approach

The Gibb's relationship is Equation (11.5) and requires computation of the mole fraction and the molar volume of pure liquid.

Solution

1. *Based on the total solids of 0.11 kg solids/kg product, the mole fraction is:*

$$X_A = \frac{\dfrac{0.89}{18}}{\dfrac{0.89}{18} + \dfrac{0.11}{180}} = 0.9878$$

2. *The molar volume for water:*

$$\frac{1}{V_m} = \frac{0.89[\text{kg water/kg product}] \times 1063.6[\text{kg product/m}^3 \text{ product}]}{18[\text{kg/kg mol}]}$$

$$V_m = 0.019\,\text{m}^3/\text{kg mol}$$

where product density is obtained from Example 11.1.

3. *Using Equation (11.5),*

$$\Pi = -\frac{8.314[\text{m}^3\,\text{kPa/(kg mol K)}] \times 293[\text{K}]}{0.019[\text{m}^3/\text{kg mol}]} \ln(0.9878)$$

$$\Pi = 1573.8\,\text{kPa}$$

The Hagen–Poiseuille law is useful in developing a relationship between the flux through a membrane and the pressure differential across it. Thus,

$$N = K_p(\Delta P - \Delta \Pi) \quad (11.6)$$

where N is flux of solvent permeation, K_p is the permeability coefficient of the membrane, ΔP is the difference in transmembrane hydrostatic pressure, and $\Delta \Pi$ is the difference in the osmotic pressure between the feed solution and the permeate.

The more specific expression for water flux through a reverse osmosis membrane has been proposed by Matsuura et al. (1973), as follows:

$$N = K_p[\Delta P - \pi(X_{c2}) + \pi(X_{c3})] \quad (11.7)$$

where X_c is the weight fraction of carbon in the solution being separated. The weight fraction (X_{c2}) is the carbon content in the concentrated boundary solution at the membrane surface, and the X_{c3} is the carbon weight fraction in the water passing through the membrane. The magnitude of X_{c2} will exceed the magnitude in the feed stream and the concentrated product leaving the system. Matsuura et al. (1973) have proposed the following expression for the permeability coefficient:

$$K_p = \frac{N_w}{3600 A_e \Delta P} \quad (11.8)$$

where N_w is a pure water permeation rate for an effective area of membrane surface, and A_e is the effective membrane area. The magnitude of K_p is a function of membrane properties such as porosity, pore-size distribution, and membrane thickness. In addition, the viscosity of the solvent will influence the permeability coefficient. For membranes that provide high rejection—that is, they do not allow most of the impurities in water to pass through—the osmotic pressure of the permeate is negligible. Matsuura et al. (1973) have measured the permeability coefficient for the cellulose acetate membrane and found a value of 3.379×10^{-6} kg water/(m^2 s kPa).

Matsuura et al. (1973) have proposed alternative expressions for water flux in a reverse-osmosis membrane, as follows:

$$N = S_p\left[\frac{1 - X_{c3}}{X_{c3}}\right][c_2 X_{c2} - c_3 X_{c3}] \quad (11.9)$$

$$N = k_m c_1 (1 - X_{c3}) \ln \left[\frac{X_{c2} - X_{c3}}{X_{c1} - X_{c3}} \right] \tag{11.10}$$

where S_p is a solute transport parameter, which is a function of the solute and the membrane characteristics. Typical magnitudes for the parameter have been measured, as presented in Table 11.3. The concentrations (c_1, c_2, c_3) in Equations (11.9) and (11.10) are the weight concentrations (kg water/m^3) in the feed stream, at the membrane boundary, and in the water passing though the membrane pore, respectively. The magnitude of the mass transfer coefficient (k_m) is a

Table 11.3 Effect of Feed Concentration on $D/K\delta$ for Fruit Juice Solutes at 4137 kPa (600 psig)[a]

Film number	Feed solution	Carbon content in feed solution (ppm)	Soluble transport parameter S_p ($\times 10^5$ cm/s)
J7	Apple juice	29,900	0.81
		43,800	0.84
		61,900	0.66
		84,800	0.36
J8	Pineapple juice	29,800	0.64
		47,300	0.43
		62,200	0.24
		80,400	0.35
J9	Orange juice	30,800	1.32
		45,000	0.97
		80,200	1.18
J10	Grapefruit juice	31,700	0.66
		45,900	0.35
		58,500	0.77
		86,900	0.43
J11	Grape juice	33,300	1.12
		48,100	0.63
		62,700	0.39
		81,500	0.69

Source: Matsuura et al. (1973)
[a] *Experiments carried out in nonflow-type cell.*

function of product flow over the membrane surface, and can be evaluated by the dimensionless expressions presented in Chapter 10.

The osmotic pressure of a liquid feed is useful in selecting membranes, since the membrane must be able to physically withstand transmembrane pressure higher than the osmotic pressure.

11.3 MEMBRANE PERFORMANCE

The flow of water through a membrane is described by

$$\dot{m}_w = \frac{K_w A(\Delta P - \Delta \Pi)}{t} \tag{11.11}$$

where $\dot{m}_w$ is water flow rate (kg/s), ΔP is the hydraulic pressure differential across the membrane (kPa), $\Delta \Pi$ is the osmotic pressure difference across the membrane (kPa), t is time (s), A is area (m^2), and K_w is the coefficient of water permeability through the membrane (kg/[m^2 kPa]).

The flow of a solute through a membrane is given by

$$\dot{m}_s = \frac{K_s A \Delta c}{t} \tag{11.12}$$

where $\dot{m}_s$ is the solute flow rate, Δc is the differential of solute concentration across the membrane (kg/m^3), and K_s is the coefficient of solute permeability through the membrane (L/m).

From Equations (11.11) and (11.12), it is evident that water flow rate through the membrane is increased by increasing the hydraulic pressure gradient across the membrane. The hydraulic pressure gradient has no effect on the solute flow rate. The solute flow is influenced by the concentration gradient across the membrane.

The performance of a membrane system is often described by the "retention factor," R_f.

$$R_f = \frac{(c_f - c_p)}{c_f} \tag{11.13}$$

where c_f is the concentration of a solute in the feed stream (kg/m^3) and c_p is the concentration of a solute in the permeate stream (kg/m^3).

Another factor used to describe the performance of a membrane system is the "rejection factor," R_j.

$$R_j = \frac{(c_f - c_p)}{c_p} \tag{11.14}$$

Membrane performance may be expressed as "molecular weight cutoff," or the maximum molecular weight for the solute to pass through the membrane.

Another term used to denote membrane performance is conversion percentage, Z.

$$Z = \frac{\dot{m}_p \times 100}{\dot{m}_f} \tag{11.15}$$

where $\dot{m}_p$ is product flow rate and $\dot{m}_f$ is the feed flow rate. Thus, operating a membrane at a conversion percentage of 70% means that a feed of 100 kg/h will yield 70 kg/h of product (permeate) and 30 kg/h of retentate.

11.4 ULTRAFILTRATION MEMBRANE SYSTEMS

Ultrafiltration membranes have pore sizes much larger than the reverse-osmosis membrane. Ultrafiltration membranes are used primarily for fractionating purposes: that is, to separate high-molecular-weight solutes from those with low molecular weight. Since the ultrafiltration membranes have larger pore sizes, the hydraulic pressures required as a driving force are much smaller when compared with the reverse osmosis membrane systems. Typically, pressures in the range of 70 to 700 kPa are needed for ultrafiltration membrane systems. As shown in Figure 11.2, the pore size of ultrafiltration membrane ranges from 0.001 to 0.02 m, with molecular weight cutoffs from 1000 to 80,000.

The flux rate through an ultrafiltration membrane can be obtained from the following equation:

$$N = KA\Delta P \tag{11.16}$$

where ΔP is pressure difference across the membrane, K is membrane permeability constant (kg/[m^2 kPa s]), and A is membrane surface area (m^2).

Example 11.3

The concentration of whey is being accomplished by using an ultrafiltration membrane to separate water. The 10 kg/min feed stream has 6% total solids and is being increased to 20% total solids. The membrane tube has a 5 cm inside diameter, and the pressure difference applied is 2000 kPa. Estimate the flux of water through the membrane and the length of the membrane tube when the permeability constant is 4×10^{-5} kg water/(m^2 kPa s).

Given

Feed concentration = 6% total solids = 0.06 kg solids/kg product
Final concentration = 20% total solids = 0.2 kg solids/kg product
Tube diameter = 5 cm = 0.05 m
Operating pressure = 2000 kPa
Membrane permeability constant = 4×10^{-5} kg water/(m^2 kPa s)

Approach

Use Equation (11.16) and a mass balance to determine the mass flux and the membrane tube length.

Solution

1. *Using a mass balance on the membrane system, feed stream = water flux + concentrated product*

$$10 = N + N_p$$

and

$$10(0.06) = N_p(0.2)$$

$$N_p = 3 \text{ kg/min of concentrated product}$$

Then

$$N = 7 \text{ kg/min of water through membrane}$$

2. *Using Equation (11.16)*

$$A = \frac{7[\text{kg water/min}]}{4 \times 10^{-5} \text{ kg water/(m}^2 \text{ kPa s)} \times 2000[\text{kPa}] \times 60[\text{s/min}]}$$

$$A = 1.46 \text{ m}^2$$

3. *Since d = 0.05 m*

$$L = \frac{1.46[\text{m}^2]}{\pi \times 0.05[\text{m}]} = 9.28 \text{ m}$$

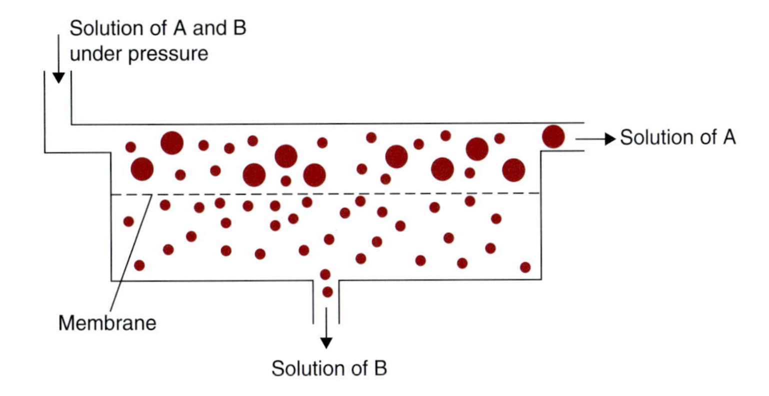

Figure 11.8 Separation process in a pressure-driven membrane system. (From Lacey, 1972)

11.5 CONCENTRATION POLARIZATION

In membrane separation processes, when a liquid solution containing salts and particulates is brought next to a semipermeable membrane, some of the molecules accumulate in the boundary layer next to the membrane surface (Fig. 11.8). Thus, the concentration of a retained species will be higher in the boundary layer adjacent to the membrane than in bulk. This phenomenon is called concentration polarization, and it has a major effect on the performance of a membrane system.

Concentration polarization occurs in both reverse-osmosis and ultrafiltration systems. In addition, the causes of concentration polarization are the same, but the consequences are different. During reverse-osmosis, the low-molecular-weight material is retained on the membrane surface and increases the solute concentration and the osmotic pressure (Eq. (11.4)). For a given transmembrane pressure, increasing the osmotic pressure will decrease the flux through the membrane (Eq. (11.6)). In ultrafiltration membranes, the influence of larger molecules on osmotic pressure is small since the molecules are retained on the membrane surface. However, the retained molecules can lead to precipitation and formation of a solid layer at the membrane surface. This phenomenon of gel formation will be explained later in this section.

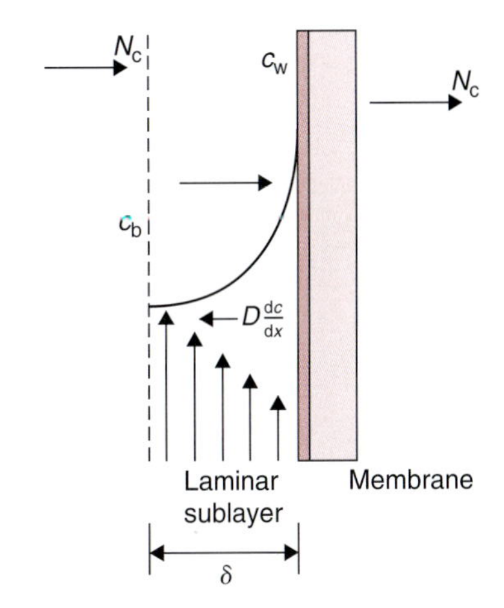

Figure 11.9 A profile of solute concentration during ultrafiltration, showing concentration polarization. (From Schweitzer, 1979)

The concentration profiles of a solute adjacent to the membrane surface can be described using Figure 11.9. As permeation of a solute proceeds through the membrane, the solute concentration at the membrane surface, c_w, increases compared with the solute concentration in the bulk fluid, c_b. This is attributed to convective transport of the solute toward the membrane. Due to the increased concentration of the solute at the membrane surface, there will be a concentration gradient set up between the concentration at the wall and the concentration in the bulk, resulting in back diffusion of the solute. At steady state, the back diffusion must equal the convective flux. The rate of convective transport of the solute can be written as

$$\text{convective transport of a solute} = N_c c \tag{11.17}$$

where N_c is the permeate flux rate, $m^3/(m^2\,s)$; and c is the concentration of a solute (kg/m^3).

The solute rejected at the wall moves back into the bulk liquid. The back-transport flux rate of the solute is expressed as

$$\text{flux rate of a solute due to back transport} = D\frac{dc}{dx} \tag{11.18}$$

where D is the diffusion coefficient of the solute (m^2/s).

Under steady-state conditions, the convective transport of a solute equals the back transport due to concentration gradient; thus,

$$D\frac{dc}{dx} = N_c c \tag{11.19}$$

Separating the variables and integrating with boundary conditions, $c = c_w$ at $x = 0$ and $c = c_b$ at $x = \delta$, we obtain

$$\frac{N_c \delta}{D} = \ln\frac{c_w}{c_b} \tag{11.20}$$

$$N_c = \frac{D}{\delta}\ln\frac{c_w}{c_b} \tag{11.21}$$

The preceding equation may be rearranged as

$$\frac{c_w}{c_b} = \exp\left(\frac{N_c \delta}{D}\right) \tag{11.22}$$

According to Equation (11.22), c_w/c_b, also referred to as concentration modulus, increases exponentially with transmembrane flux and with the thickness of the boundary layer, and it decreases exponentially with increasing value of solute diffusivity. Thus, the influence of concentration polarization is particularly severe with membranes that have high permeability, such as ultrafiltration membranes, and solutes that have high molecular weights. The thickness of the boundary layer is the result of the flow conditions next to the membrane surface.

This derivation is valid for both reverse osmosis and ultrafiltration membranes. In the case of ultrafiltration membranes, as observed in the preceding paragraph, the solute concentration will increase rapidly and the solute may precipitate, forming a gel layer (Fig. 11.10). The resistance to permeation by this gel layer may become more prominent than the membrane resistance. Under these conditions, the solute concentration at the surface of the gel layer, referred to as c_g, becomes a constant and it is no longer influenced by the solute concentration in the bulk, the membrane characteristics, operating pressures, or the fluid flow conditions.

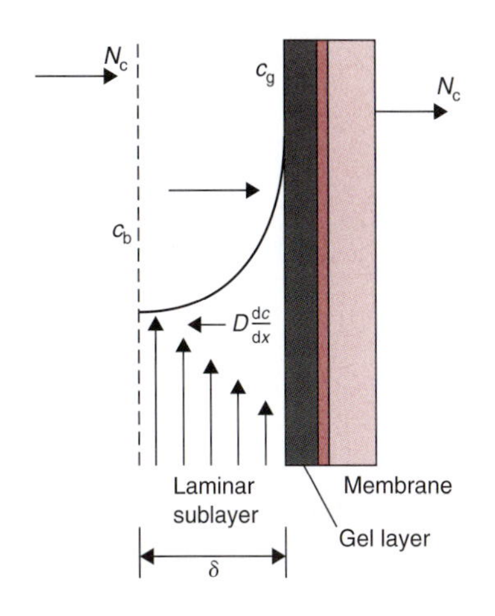

Figure 11.10 A concentration profile showing the formation of a gel layer on an ultrafiltration membrane. (From Schweitzer, 1979)

For situations when the pressure has no more influence on the flux, we can rewrite Equation (11.22) as

$$N_c = k_m \ln\left(\frac{c_w}{c_b}\right) \tag{11.23}$$

where k_m is the mass transfer coefficient (L/[m^{2}h]).

Equation (11.23) indicates that the ultrafiltration rate N_c is influenced mainly by the solute concentration in the bulk, c_b, and the mass transfer coefficient, k_m.

The mass transfer coefficient can be evaluated using dimensional analysis.

$$N_{sh} = (N_{Re})^a (N_{Sc})^b \tag{11.24}$$

where

$$\text{Sherwood number, } N_{Sh} = \frac{k_m d_c}{D}$$

$$\text{Reynolds number, } N_{Re} = \frac{\rho \bar{u} d_c}{\mu}$$

$$\text{Schmidt number, } N_{Sc} = \frac{\mu}{\rho D}$$

$$d_c = 4\left(\frac{\text{cross section available for flow}}{\text{wetted perimeter}}\right)$$

For turbulent flow,

$$N_{Sh} = 0.023(N_{Re})^{0.8}(N_{Sc})^{0.33} \tag{11.25}$$

Examples 11.4 and 11.5 illustrate the use of dimensional analysis to determine the mass-transfer coefficient (Cheryan, 1998).

Typical materials used in ultrafiltration membranes include cellulose acetate, polyvinyl chloride, polysulfones, polycarbonates, and polyacrylonitriles.

Example 11.4

A reverse osmosis (RO) system is being used to concentrate apple juice at 20°C with an initial total solids content of 10.75%. The system contains 10 tubes with 1.5 cm diameter and the feed rate is 150 kg/min. The density of the feed stream is 1050 kg/m^3 and the viscosity is 1×10^{-3} Pa s. Estimate the flux of water through the RO membrane, when the solute diffusivity is 8×10^{-8} m^2/s, and operating pressure is 6895 kPa.

Given

Feed concentration = 0.1075 kg solids/kg product
Feed rate = 150 kg/min or 15 kg/min per tube
Product density = 1050 kg/m^3
Product viscosity = 1×10^{-3} Pa s

Approach

The molar concentration at the membrane boundary will be estimated using Equations (11.7) and (11.10). After the concentration at the boundary is determined, the flux of water will be estimated from Equation (11.7).

Solution

1. *Since Equation (11.10) requires the convective mass transfer coefficient k_m as input, the dimensionless relationship (Eq. (11.24)) will be used.*
2. *Based on the mass flow rate of 15 kg/min:*

$$\bar{u} = \frac{15[\text{kg/min}]}{\left(\pi \frac{(1.5/100)^2}{4}\right)[\text{m}]^2 \times 60[\text{s/min}] \times 1050[\text{kg/m}^3]}$$

$$= 1.35 \text{ m/s}$$

3. *Then the Reynolds number is:*

$$N_{Re} = \frac{1050[\text{kg/m}^3] \times 1.35[\text{m/s}] \times 0.015[\text{m}]}{1.0 \times 10^{-3}[\text{Pa s}]} = 21{,}263$$

4. Then the Schmidt number is:

$$N_{Sc} = \frac{1\times10^{-3}[Pa\,s]}{1050[kg/m^3]\times(8\times10^{-8})[m^2/s]} = 11.9$$

5. Using Equation (11.25),

$$N_{Sh} = 0.023(21,263)^{0.8}(11.9)^{0.33} = 150.9$$

and

$$k_m = \frac{150.9\times(8\times10^{-8})}{0.015} = 8.05\times10^{-4}\,m^2/s$$

6. Using Equation (11.7), with $K_p = 3.379\times10^{-6}$ kg water/(m^2 s kPa) (for cellulose acetate membrane) and using Equation (11.4),

$$\Pi = \frac{0.1075\times1050[kg/m^3]\times8.314[m^3\,kPa/(kg\,mol\,K)]\times293[K]}{180[kg/kg\,mol]}$$

$$= 1528\,kPa$$

$$X_{c2} = \text{unknown}$$

$$X_{c3} = \text{assumed to be zero}$$

$$\Delta P = 6895\,kPa$$

7. Using Equation (11.10), where c_1 = (0.1075 kg solids/kg product) (1050 kg product/m^3 product)

$$c_1 = 112.875\ \text{kg solids/m}^3\ \text{product}$$

$$k_m = 8.05\times10^{-4}\,m^2/s$$

$$X_{c3} = \text{assumed to be zero}$$

$$X_{c1} = 0.0438\ \text{(carbon weight fraction from Table 11.3)}$$

then

$$N = 112.875[\text{kg solids/m}^3\ \text{product}]\times(8.05\times10^{-4})[m^2/s] \times \ln\left[\frac{X_{c2}}{0.0438}\right]$$

8. Using Equations (11.7) and (11.10):

$$X_{c2} = 0.0565$$

becomes the weight fraction of carbon at the membrane boundary

9. Using the magnitude of X_{c2}

$$N = 23.1\times10^{-3}\ \text{kg water/(m}^2\ \text{s)}$$
$$= 1.388\ \text{kg water/(m}^2\ \text{min)}$$
$$= 83.3\ \text{kg water/(m}^2\ \text{h)}$$

Example 11.5

Determine the flux rate expected in a tubular ultrafiltration system being used to concentrate milk. The following conditions apply: density of milk = $1.03\,g/cm^3$, viscosity = 0.8 cP, diffusivity = $7 \times 10^{-7}\,cm^2/s$, $c_B = 3.1\%$ weight per unit volume. Diameter of tube = 1.1 cm, length = 220 cm, number of tubes = 15, and fluid velocity = 1.5 m/s.

Given

Density of milk = $1.03\,g/cm^3 = 1030\,kg/m^3$
Viscosity of milk = $0.8\,cP = 0.8 \times 10^{-3}\,Pa\,s$
Mass diffusivity = $7 \times 10^{-7}\,cm^2/s = 7 \times 10^{-11}\,m^2/s$
Bulk concentration = $0.031\,kg/m^3$
Gel concentration = $0.22\,kg/m^3$
Tube diameter = 0.011 m
Length of tube = 220 cm = 2.2 m
Number of tubes = 15
Fluid velocity = 1.5 m/s

Approach

The convective mass transfer coefficient will be estimated by the dimensionless equation, and the flux of water will be determined by Equation (11.23).

Solution

1. *Computation of the Reynolds number:*

$$N_{Re} = \frac{1030[kg/m^3] \times 1.5[m/s] \times 0.011[m]}{0.8 \times 10^{-3}[Pa\,s]} = 21,244$$

2. *Computation of Schmidt number:*

$$N_{Sc} = \frac{0.8 \times 10^{-3}[Pa\,s]}{1030[kg/m^3] \times (7 \times 10^{-11})[m^2/s]} = 11.1 \times 10^3$$

3. *Since flow is turbulent*

$$N_{Sh} = 0.023(21,244)^{0.8}(11.1 \times 10^3)^{0.33}$$

$$N_{Sh} = 1440$$

4. *Then*

$$k_m = \frac{1440 \times (7 \times 10^{-11})[m^2/s]}{0.011[m]} = 9.16 \times 10^{-6}\,m/s$$

5. *Using Equation (11.23),*

$$N = (9.16 \times 10^{-6})[m/s] \times 998.2[kg/m^3] \times \ln\left(\frac{0.22}{0.031}\right)$$

$$= 0.018\,kg/(m^2\,s)$$

where density of water at 20°C is used

$$N = 64.8 \text{ kg water}/(\text{m}^2 \text{ h})$$

6. *By using the membrane surface area*

$$A = \pi(0.011)(2.2) = 0.076 \text{ m}^2/\text{tube}$$

$$\text{Total area} = 0.076\ (15 \text{ tubes}) = 1.14 \text{ m}^2$$

7. *Total flux of water through membrane:*

$$\begin{aligned} \text{flux} &= (64.8 \text{ kg water}/[\text{m}^2 \text{ h}])\,(1.14 \text{ m}^2) \\ &= 73.87 \text{ kg water/h} \end{aligned}$$

11.6 TYPES OF REVERSE-OSMOSIS AND ULTRAFILTRATION SYSTEMS

Four major types of membrane devices are used for reverse-osmosis and ultrafiltration systems: plate-and-frame, tubular, spiral-wound, and hollow-fiber. Table 11.4 provides a general comparison among

Table 11.4 Comparison of Process-Related Characteristics for Membrane Module Configurations

	Module type			
Characteristic	**Plate-and-frame**	**Spiral-wound**	**Tube-in-shell**	**Hollow-fiber**
Packing density (m^2/m^3)	200–400	300–900	150–300	9000–30,000
Permeate flux ($m^3/[m^2$ day])	0.3–1.0	0.3–1.0	0.3–1.0	0.004–0.08
Flux density ($m^3/[m^3$ day])	60–400	90–900	45–300	36–2400
Feed channel diameter (mm)	5	1.3	13	0.1
Method of replacement	As sheets	As module assembly	As tubes	As entire module
Replacement labor	High	Medium	High	Medium
Pressure drop				
Product side	Medium	Medium	Low	High
Feed side	Medium	Medium	High	Low
Concentration polarization	High	Medium	High	Low
Suspended solids buildup	Low/medium	Medium/high	Low	High

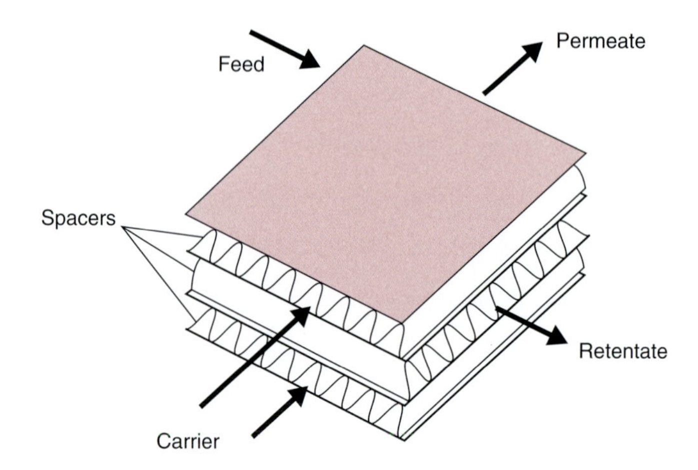

■ Figure 11.11 A module used in a plate-and-frame membrane system.

these four types. A brief description of each type of membrane is given in the following.

11.6.1 Plate-and-Frame

The plate-and-frame membrane systems involve a large number of flat membranes that are sandwiched together with the use of spacers. As shown in Figure 11.11, the spacers provide the channels for flow. The membranes (usually 50 to 500 μm thick) are bonded on a porous, inert matrix that offers little resistance to the fluid flow. The flow of feed and retentate occur in alternate channels. This arrangement of membranes is very similar to the plate heat-exchanger described in Chapter 4.

11.6.2 Tubular

The tubular design was the first commercial design of a reverse-osmosis system. It consists of a porous tube coated with the membrane material such as cellulose acetate. Typically, feed solution is pumped into the tube through one end and forced in the radial direction through the porous pipe and the membrane (Fig. 11.12). Water drips from the outer surface of the membrane while the concentrated stream "retenate" leaves through the outer end of the tube. This type of reverse-osmosis device is expensive to use for high volumetric flow rates, since the membrane area is relatively small.

11.6.3 Spiral-Wound

To increase the membrane surface area per unit volume, a spiral-wound configuration was a key commercial development following

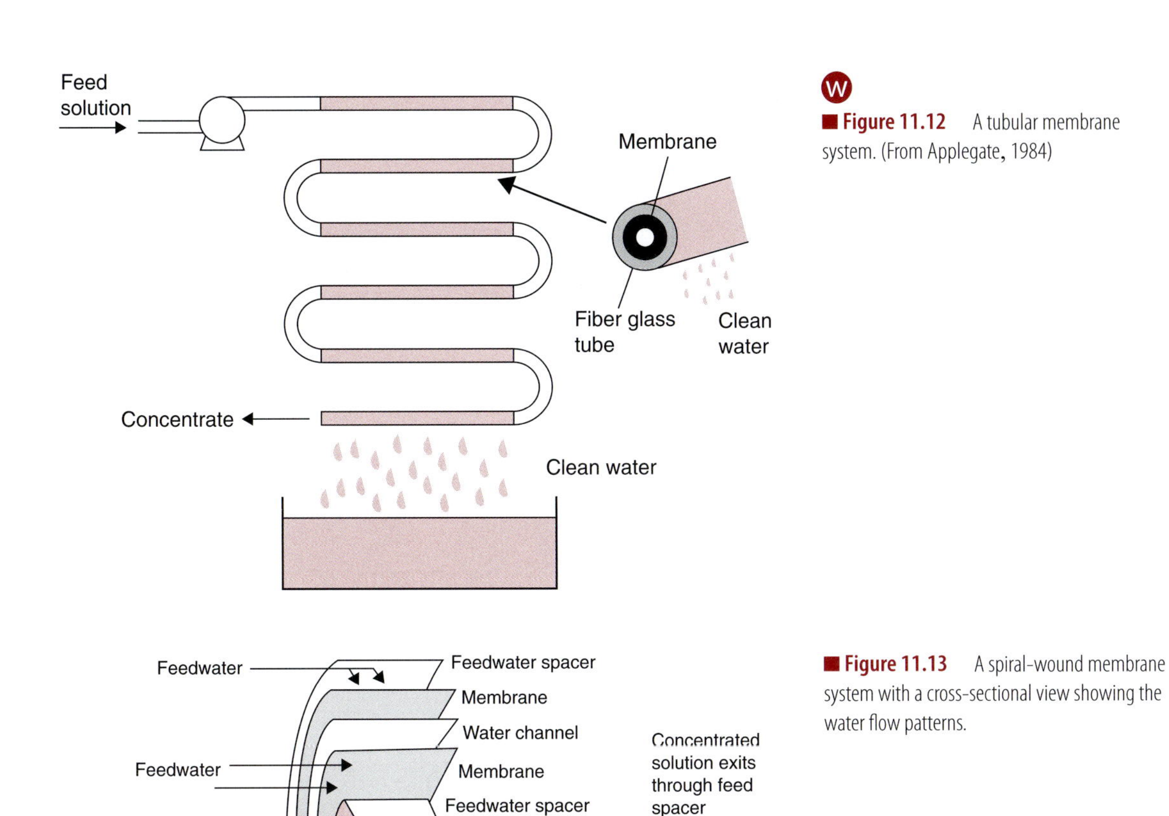

W

Figure 11.12 A tubular membrane system. (From Applegate, 1984)

Figure 11.13 A spiral-wound membrane system with a cross-sectional view showing the water flow patterns.

the tubular design. This design, shown in Figure 11.13, can be visualized as a composite of multilayers. The two layers of membrane are separated by a plastic mesh, and on either side of the membrane is a porous sheet. These five layers are then spirally wound around a perforated tube. The ends of the rolled layers are sealed to prevent mixing of feed and product streams. The whole spiral assembly is housed in a tubular metal jacket that can withstand applied pressures. Feed is pumped through the perforated tube on one side of the spiral-wound roll. Feed enters inside the plastic mesh, which aids in creating

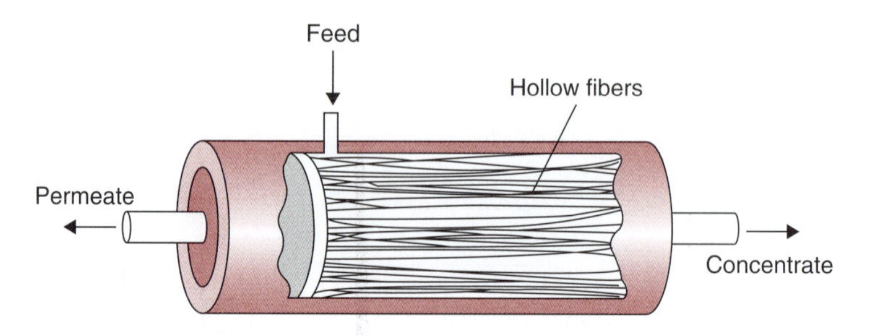

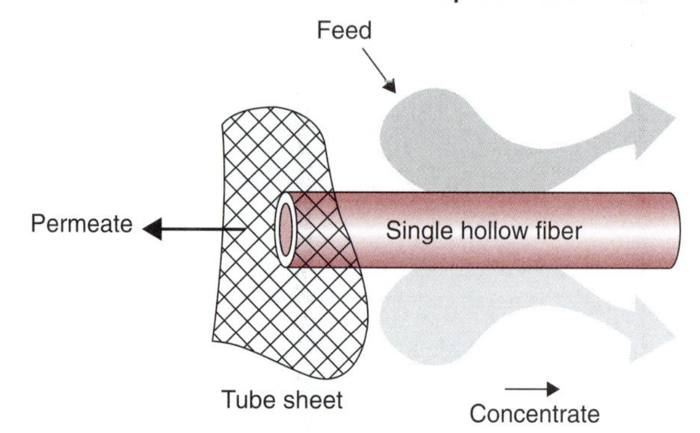

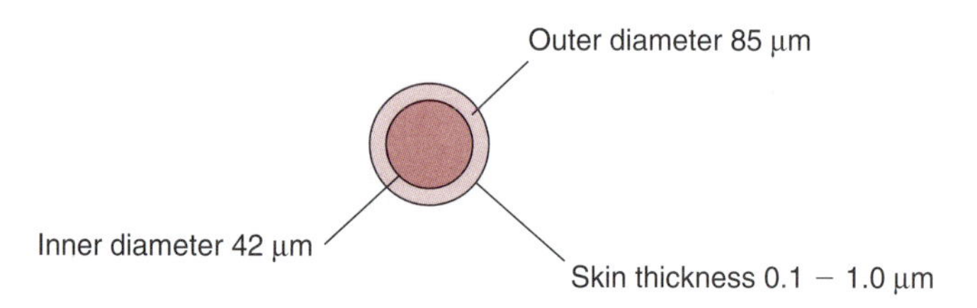

Figure 11.14 A hollow-fiber membrane system.

turbulence and minimizing fouling. The feed then permeates the membrane in a radial direction and exits the membrane into porous layers. The permeate (water) transports through the porous sheet in a spiral manner and leaves the assembly through the exit tube, whereas the retentate leaves through the other end of the spiral-wound roll. Typical dimensions of spiral cartridges are 11 cm in diameter, 84 cm in length, with 0.7 mm membrane spacing, and 5 m^2 area. The spiral-wound systems are similar for both reverse-osmosis and ultrafiltration applications.

11.6.4 Hollow-Fiber

A hollow-fiber, made out of aramid, was first introduced by DuPont in 1970. Hollow-fibers, finer than human hair, have an internal diameter of about 40 μm, and an external diameter of about 85 μm. A large number of hollow-fibers (several millions) are arranged in a bundle around a perforated distributor tube (Fig. 11.14). In the reverse-osmosis system, the fibers are glued with epoxy to either end. These fibers provide extremely large surface areas; thus, hollow fiber membrane systems can be made very compact. Feed water is introduced through the distributor pipe; the permeate flows through the annular space of the fibers into the hollow bore of the fibers and moves to the tube sheet end, discharging from the exit port. The retentate or brine stays on the outside of the fibers and leaves the device from the brine port. Hollow fibers are used mainly to purify water. Liquid foods are difficult to handle in hollow fiber systems because of problems associated with fouling of the fibers.

The hollow-fibers used for ultrafiltration membranes are quite different from those used for reverse-osmosis systems. For ultrafiltration applications, hollow fibers are made from acrylic copolymers.

Most membrane systems in the food industry have been in dairy and fruit juice applications. Other commercial applications include processing of coffee, tea, alcohol, gelatins, eggs, and blood, and corn refining and soybean processing.

PROBLEMS

11.1 An ultrafiltration system is being used to concentrate gelatin. The following data were obtained: A flux rate was 1630 L/m^2 per day at 5% solids by weight concentration, and a flux rate was 700 L/m^2 per day at 10% solids by weight. Determine the concentration of the gel layer and the flux rate at 7% solids.

11.2 Estimate the osmotic pressure of a 20% sucrose solution at a temperature of 10°C.

11.3 Determine the transmembrane pressure required to maintain a flux of 220 kg/(m^2 h) in a reverse-osmosis system when the feed solution is 6% total solids whey. The permeability coefficient of the membrane is 0.02 kg/(m^2 h kPa).

11.4 An ultrafiltration system is being used to concentrate orange juice at 30°C from an initial solids content of 10%

to 35% total solids. The ultrafiltration system contains six tubes with 1.5 cm diameter. The product properties include density of 1100 kg/m^3, viscosity of 1.3×10^{-3} Pa s, and solute diffusivity of 2×10^{-8} m^2/s. The concentration of solute at the membrane surface is 25%. Estimate the length of ultrafiltration tubes required to achieve the desired concentration increase.

***11.5** A reverse osmosis (RO) system, with 100 tubes of 10 m length and 1.0 cm diameter, is used to concentrate orange juice from 11% to 40% total solids. The permeability coefficient for the RO membrane is 0.2 kg water/m^2 hr kPa and the product feed rate is 200 kg/min.

a. Determine the flux of water (kg water/hr) through the membrane needed to accomplish the magnitude of concentration indicated.

b. Estimate the difference in transmembrane hydrostatic pressure (ΔP) needed for the system to operate.

LIST OF SYMBOLS

A	area (m^2)
A_e	effective membrane area (m^2)
c	concentration of a solute (kg/m^3)
c_b	concentration in bulk stream (kg/m^3)
c_f	concentration of a solute in the feed stream (kg/m^3)
c_g	concentration at gel layer surface (kg/m^3)
c_p	concentration of a solute in permeate stream (kg/m^3)
c_w	concentration of water (kg/m^3)
d_c	characteristic dimension (m)
D	diffusion coefficient of the solute (m^2/s)
Δc	differential of solute concentration (kg/m^3)
δ	laminar sublayer
E	energy consumption (J)
F	Faraday's constant (96,500 A/s equivalent)
I	electric current through the stack (A)
K	membrane permeability constant (kg/[m^2 kPa s])
k_m	mass transfer coefficient (L/[m^{2}h])
K_p	permeability coefficient of the membrane

* Indicates an advanced level in solving.

K_s	coefficient of solute permeability through the membrane (L/m)
K_w	coefficient of water permeability through the membrane (kg/[m^2 kPa])
$\dot{m}$	feed solution flow rate (L/s)
$\dot{m}_p$	product flow rate (kg/s)
$\dot{m}_s$	solute flow rate (kg/s)
$\dot{m}_w$	water flow rate (kg/s)
M	molecular weight
μ	viscosity (Pa s)
n	number of cells in the stack
N	permeate flux rate (m^3/[m^2 s])
N_c	convective permeate flux rate (m^3/[m^2 s])
N_{Re}	Reynolds Number, dimensionless
N_{Sc}	Schmidt Number, dimensionless
N_{Sh}	Sherwood Number, dimensionless
N_w	pure water permeation rate for an effective area of membrane surface (m^3/[m^2 s])
Π	osmotic pressure (Pa)
$\Delta\Pi$	difference in the osmotic pressure between the feed solution and the permeate
ΔP	difference in transmembrane hydrostatic pressure (Pa)
ρ	density (kg/m^3)
R	universal gas constant (m^{3}Pa/[kg mol K])
R_c	resistance of the cell (Ω)
R_f	retention factor
R_j	rejection factor
S_p	solute transport parameter
T	temperature (absolute)
t	time (s)
U	current utilization factor
$\bar{u}$	mean velocity (m/s)
V_m	molar volume of pure liquid
X_A	mole fraction of pure liquid
X_c	weight fraction of carbon in the solution being separated
Z	membrane performance conversion percentage
z	electrochemical valence

BIBLIOGRAPHY

Applegate, L. (1984). Membrane separation processes. *Chem. Eng.* (June 11), 64–89.

Cheryan, M. (1998). *Ultrafiltration Handbook*. Technomic Publishing Co., Lancaster, Pennsylvania.

Cheryan, M. (1989). Membrane separations: Mechanisms and models. In *Food Properties and Computer-Aided Engineering of Food Processing Systems*, R. P. Singh and A. Medina, eds. Kluwer Academic Publishers, Amsterdam.

Lacey, R. E. (1972). Membrane separation processes. *Chem. Eng.* (Sept. 4), 56–74.

Matsuura, T., Baxter, A. G., and Sourirajan, S. (1973). Concentration of juices by reverse osmosis using porous cellulose acetate membranes. *Acta Aliment.* **2**: 109–150.

McCabe, W. L., Smith, J. C., and Harriott, P. (1985). *Unit Operations of Chemical Engineering*. McGraw-Hill, New York.

Schweitzer, P. A. (1979). *Handbook of Separation Processes*. McGraw-Hill, New York.

Chapter 12

Dehydration

The removal of moisture from a food product is one of the oldest preservation methods. By reducing the water content of a food product to very low levels, the opportunity for microbial deterioration is eliminated and the rates of other deteriorative reactions are reduced significantly. In addition to preservation, dehydration reduces product mass and volume by significant amounts and improves the efficiency of product transportation and storage. Often, the dehydration of a food product results in a product that is more convenient for consumer use.

The preservation of fruits, vegetables, and similar food products by dehydration offers a unique challenge. Due to the structural configuration of these types of products, the removal of moisture must be accomplished in a manner that will be the least detrimental to product quality. This requires that the process produce a dry product that can be returned to approximately the original quality after rehydration. To achieve the desired results for dehydrated foods, with a defined physical structure, the process must provide the optimum heat and mass transfer within the product. The design of these processes requires careful analysis of the heat and mass transfer occurring within the product structure. Only through analysis and understanding of the transfer processes can the maximum efficiency and optimum quality be achieved.

12.1 BASIC DRYING PROCESSES

To achieve moisture removal from a food product in the most efficient manner, the design of the dehydration system must account for the various processes and mechanisms occurring within the product.

All icons in this chapter refer to the author's website which is independently owned and operated. Academic Press is not responsible for the content or operation of the author's website. Please direct your website comments and questions to the author: Professor R. Paul Singh, Department of Biological and Agricultural Engineering, University of California, Davis, CA 95616, USA.
Email: rps@rpaulsingh.com

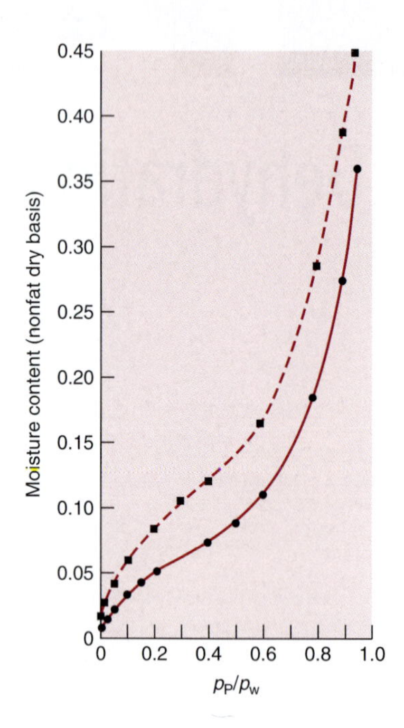

■ Figure 12.1 Equilibrium moisture content isotherm for a freeze-dried food product, illustrating hysteresis. ■, Desorption; ●, adsorption. (Adapted from Heldman and Singh, 1981)

These processes and mechanisms are of particular importance for foods with a defined physical structure, due to the influence of product structure on the movement of moisture from the product.

12.1.1 Water Activity

One of the important parameters in food dehydration is the equilibrium condition that establishes a limit to the process. Although the parameter represents an important portion of the gradient for moisture movement, water activity is important in analysis of storage stability for dry foods.

By definition, water activity is the equilibrium relative humidity of the product divided by 100. For most food products, the relationship between moisture content and water activity is as illustrated in Figure 12.1. The sygmoid isotherm is typical for dry foods, as is the difference between the adsorption and desorption isotherms for the same product. In addition to the water activity values establishing the storage stability of the product against various deterioration reactions (Fig. 12.2), the equilibrium moisture contents are the lower limit of the gradient for moisture removal from the product. As might be expected, higher temperatures result in lower equilibrium moisture content and a large moisture gradient for moisture movement.

One of the most widely used models for description of equilibrium moisture isotherms is the GAB model (named after Guggenheim–Anderson–DeBoer). This model is used to fit and draw the sorption data obtained for a given food. The GAB model is expressed as follows:

$$\frac{w}{w_m} = \frac{Cka_w}{(1 - ka_w)(1 - ka_w + Cka_w)} \qquad (12.1)$$

where:

w = the equilibrium moisture content, fraction dry basis
w_m = the monolayer moisture content, fraction dry basis
C = the Guggenheim constant = $C' \exp(H_1 - H_m)/RT$
H_1 = heat of condensation of pure water vapor
H_m = total heat of sorption of the first layer on primary sites
k = a factor correcting properties of multiplayer with respect to the bulk liquid,
= $k' \exp(H_1 - H_q)/RT$
H_q = total heat of sorption of the multilayers.

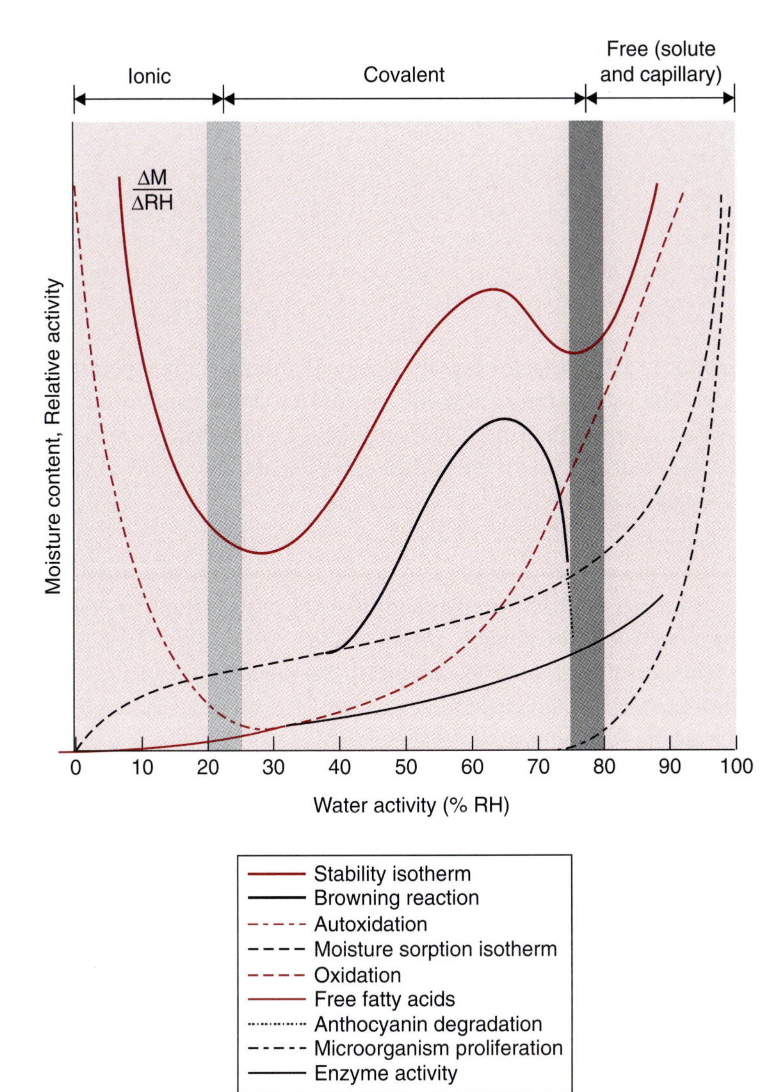

Figure 12.2 Influence of water activity on rates of various deterioration reactions in foods. (From Rockland and Nishi, 1980. Copyright © by Institute of Food Technologists. Reprinted with permission of *Food Technology*.)

The GAB model can be used to a maximum water activity of 0.9. The following procedure is suggested by Bizot (1983) to fit data on water activities and equilibrium moisture content.

Equation (12.1) can be transformed as follows:

$$\frac{a_w}{w} = \alpha a_w^2 + \beta a_w + \gamma \tag{12.2}$$

where

$$\alpha = \frac{k}{w_m}\left[\frac{1}{C} - 1\right]$$

$$\beta = \frac{1}{w_m}\left[1 - \frac{2}{C}\right]$$

$$\gamma = \frac{1}{w_m C k}$$

Equation (12.2) indicates that the GAB equation is a three-parameter model. The water activity and equilibrium moisture content data are regressed using Equation (12.2), and values of the three coefficients α, β, and γ are obtained. From these coefficients, the values of k, w_m, and C can be obtained.

Example 12.1

A dry food product has been exposed to a 30% relative-humidity environment at 15°C for 5 h without a weight change. The moisture content has been measured and is at 7.5% (wet basis). The product is moved to a 50% relative-humidity environment, and a weight increase of 0.1 kg/kg product occurs before equilibrium is achieved.

a. Determine the water activity of the product in the first and second environments.

b. Compute the moisture contents of the product on a dry basis in both environments.

Given

Equilibrium relative humidity = 30% in first environment

Product moisture content = 7.5% wet basis in first environment

For 30% relative-humidity environment, moisture content will be 0.075 kg H_2O/kg product.

Approach

Water activities of product are determined by dividing equilibrium relative humidity by 100. Dry-basis moisture contents for product are computed by expressing mass of water in product on a per unit dry solids basis.

Solution

1. *The water activity of the food product is equilibrium relative humidity divided by 100; the water activities are 0.3 in the first environment and 0.5 in the second.*

2. *The dry basis moisture of the product at equilibrium in 30% RH is*

$$7.5\% = \frac{7.5 \text{ kg } H_2O}{100 \text{ kg product}} = 0.075 \text{ kg } H_2O/\text{kg product}$$

$$\frac{0.075 \text{ kg } H_2O/\text{product}}{0.925 \text{ kg solids/kg product}} = 0.08108 \text{ kg } H_2O/\text{kg solids}$$

$$= 8.11\% \text{ MC (dry basis)}$$

3. *Based on weight gain at 50% RH,*

$$0.075 \text{ kg } H_2O/\text{kg product} + 0.1 \text{ kg } H_2O/\text{kg product}$$

$$= 0.175 \text{ kg } H_2O/\text{kg product} = 17.5\% \text{ MC (wet basis)}$$

or

$$0.175 \text{ kg } H_2O/\text{kg product} = 0.212 \frac{\text{kg } H_2O}{\text{kg solids}}$$

$$= 21.2\% \text{ MC (dry basis)}$$

12.1.2 Moisture Diffusion

Significant amounts of moisture removal from a food product will occur due to diffusion of liquid water and/or water vapor through the product structure. This phase of moisture movement will follow the evaporation of water at some location within the product. The rate of moisture diffusion can be estimated by the expressions for molecular diffusion, as introduced in Chapter 10. The mass flux for moisture movement is a function of the vapor pressure gradient, as well as the mass diffusivity for water vapor in air, the distance for water vapor movement within the product structure, and temperature. Since thermal energy is required for moisture evaporation, the process becomes simultaneous heat and mass transfer.

The moisture removal from the product will depend, in part, on convective mass transfer at the product surface. Although this transport process may not be rate-limiting, the importance of maintaining the optimum boundary conditions for moisture transport cannot be overlooked.

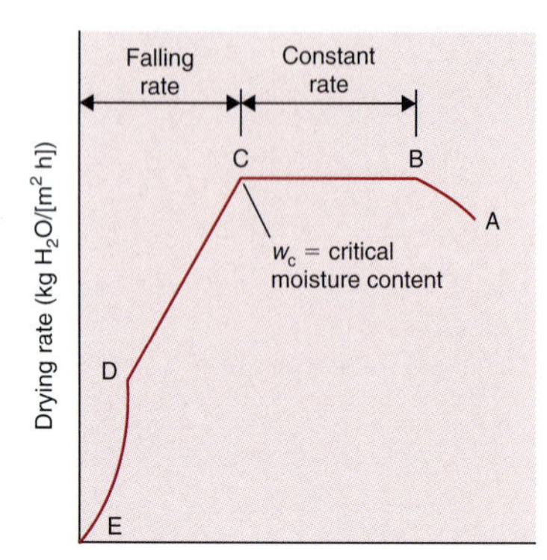

Figure 12.3 Illustration of constant-rate and falling-rate drying periods. (Adapted from Heldman and Singh, 1981)

12.1.3 Drying-Rate Curves

The removal of moisture from a typical food product will follow a series of drying rates, as illustrated in Figure 12.3. The initial removal of moisture (AB) occurs as the product and the water within the product experience a slight temperature increase. Following the initial stages of drying, significant reductions in moisture content will occur at a constant rate (BC) and at a constant product temperature. The constant-rate drying period occurs with the product at the wet bulb temperature of the air. In most situations, the constant-rate drying period will continue until the moisture content is reduced to the critical moisture content. At moisture contents below the critical moisture content, the rate of moisture removal decreases with time. One or more falling-rate drying periods (CE) will follow. The critical moisture content is well defined due to the abrupt change in the rate of moisture removal.

12.1.4 Heat and Mass Transfer

As indicated previously, the removal of moisture from a food product involves simultaneous heat and mass transfer. Heat transfer occurs within the product structure and is related to the temperature gradient between the product surface and the water surface at some location within the product. As sufficient thermal energy is added to the water to cause evaporation, the vapors are transported from the water surface within the product to the product surface. The gradient causing moisture-vapor diffusion is vapor pressure at the liquid water surface, as compared with the vapor pressure of air at the product surface. The heat and the mass transfer within the product structure occurs at the molecular level, with heat transfer being limited by thermal conductivity of the product structure, whereas mass transfer is proportional to the molecular diffusion of water vapor in air.

At the product surface, simultaneous heat and mass transfer occurs but is controlled by convective processes. The transport of vapor from the product surface to the air and the transfer of heat from the air to the product surface is a function of the existing vapor pressure and temperature gradients, respectively, and the magnitude of the convective coefficients at the product surface.

Since the drying rate is directly proportional to the slowest of the four processes, it is important to account for all processes. In most products, the heat and mass transfer within the product structure will be rate-limiting processes.

Example 12.2

The initial moisture content of a food product is 77% (wet basis), and the critical moisture content is 30% (wet basis). If the constant drying rate is 0.1 kg $H_2O/(m^2\ s)$, compute the time required for the product to begin the falling-rate drying period. The product has a cube shape with 5-cm sides, and the initial product density is 950 kg/m^3.

Given

Initial moisture content = 77% wet basis
Critical moisture content = 30% wet basis
Drying rate for constant rate period = 0.1 kg $H_2O/(m^2\ s)$
Product size = cube with 5-cm sides
Initial product density = 950 kg/m^3

Approach

The time for constant-rate drying will depend on mass of water removed and the rate of water removal. Mass of water removed must be expressed on dry basis, and rate of water removal must account for product surface area.

Solution

1. *The initial moisture content is*

$$0.77\ \text{kg}\ H_2O/\text{kg product} = 3.35\ \text{kg}\ H_2O/\text{kg solids}$$

2. *The critical moisture content is*

$$0.3\ \text{kg}\ H_2O/\text{kg product} = 0.43\ \text{kg}\ H_2O/\text{kg solids}$$

3. *The amount of moisture to be removed from product during constant-rate drying will be*

$$3.35 - 0.43 = 2.92\ \text{kg}\ H_2O/\text{kg solids}$$

4. *The surface area of the product during drying will be*

$$0.05\ \text{m} \times 0.05\ \text{m} = 2.5 \times 10^{-3}\ \text{m}^2/\text{side}$$

$$2.5 \times 10^{-3} \times 6\ \text{sides} = 0.015\ \text{m}^2$$

5. *The drying rate becomes*

$$0.1\ \text{kg}\ H_2O/(\text{m}^2\ \text{s}) \times 0.015\ \text{m}^2 = 1.5 \times 10^{-3}\ \text{kg}\ H_2O/\text{s}$$

6. *Using the product density, the initial product mass can be established.*

$$950\ \text{kg/m}^3 \times (0.05)^3\ \text{m}^3 = 0.11875\ \text{kg product}$$

$$0.11875\ \text{kg product} \times 0.23\ \text{kg solid/kg product} = 0.0273\ \text{kg solid}$$

7. *The total amount of water to be removed becomes*

$$2.92\ kg\ H_2O/kg\ solids \times 0.0273\ kg\ solids = 0.07975\ kg\ H_2O$$

8. *Using the drying rate, the time for constant-rate drying becomes*

$$\frac{0.07975\ kg\ H_2O}{1.5 \times 10^{-3}\ kg\ H_2O/s} = 53.2\ s$$

12.2 DEHYDRATION SYSTEMS

Based on the analysis of heat and mass transfer, the most efficient dehydration systems will maintain the maximum vapor-pressure gradient and maximum temperature gradient between the air and the interior parts of the product. These conditions, along with high convective coefficients at the product surface, can be maintained in several different designs of dehydration systems. The systems we will describe are representative of the systems used for dehydration of foods.

12.2.1 Tray or Cabinet Dryers

These types of drying systems use trays or similar product holders to expose the product to heated air in an enclosed space. The trays holding the product inside a cabinet or similar enclosure (Fig. 12.4) are exposed to heated air so that dehydration will proceed. Air movement over the product surface is at relatively high velocities to ensure that heat and mass transfer will proceed in an efficient manner.

A slight variation of the cabinet dryer incorporates a vacuum within the chamber (Fig. 12.5). This type of dehydration system uses a

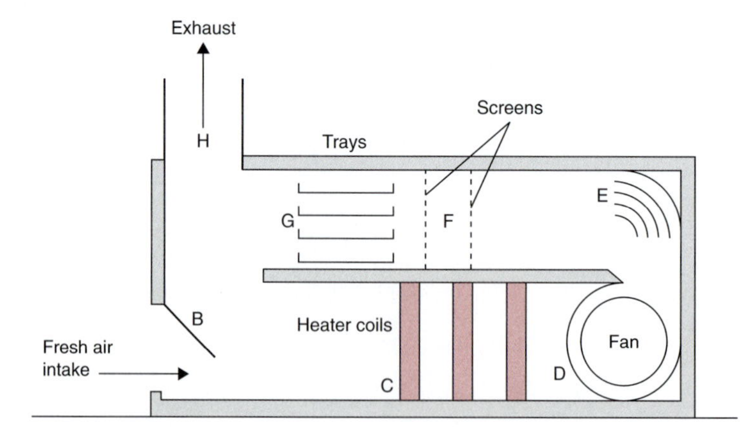

■ **Figure 12.4** Schematic illustration of a cabinet-type tray drier. (From Van Arsdel et al., 1973)

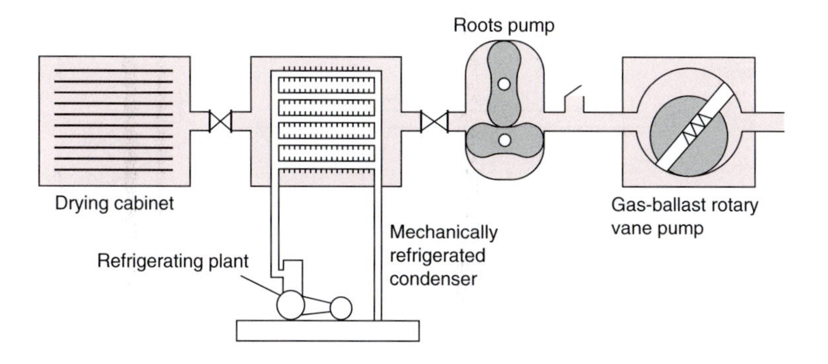

Figure 12.5 Cabinet dryer with vacuum. (From Potter, 1978)

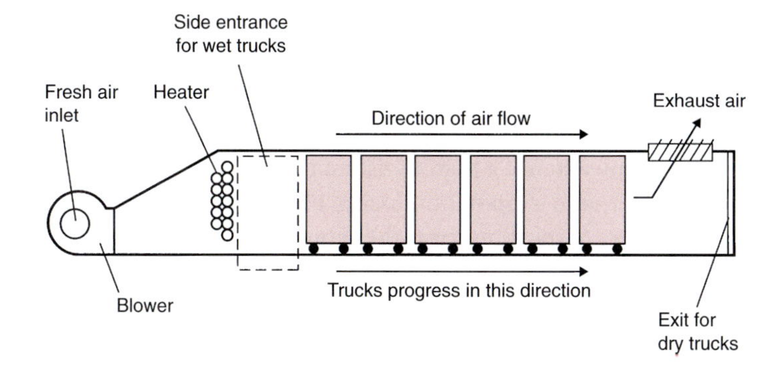

Figure 12.6 Schematic illustration of a concurrent-flow tunnel dryer. (From Van Arsdel, 1951)

vacuum to maintain the lowest possible vapor pressure in the space around the product. The reduction in pressure also reduces the temperature at which product moisture evaporates, resulting in improvements in product quality.

In most cases, cabinet dryers are operated as batch systems and have the disadvantage of nonuniform drying of a product at different locations within the system. Normally, the product trays must be rotated to improve uniformity of drying.

12.2.2 Tunnel Dryers

Figures 12.6 and 12.7 show examples of tunnel dryers. As illustrated, the heated drying air is introduced at one end of the tunnel and moves at an established velocity through trays of products being carried on trucks. The product trucks are moved through the tunnel at a rate required to maintain the residence time needed for dehydration. The product can be moved in the same direction as the air flow to provide concurrent dehydration (Fig. 12.6), or the tunnel can

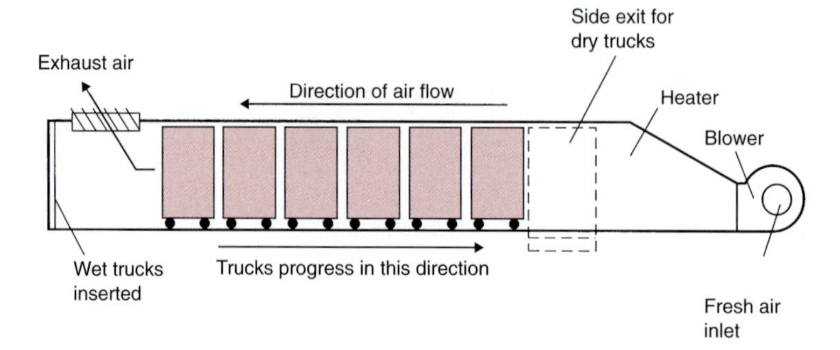

■ **Figure 12.7** Schematic illustration of a countercurrent-flow tunnel dryer. (From Van Arsdel, 1951)

be operated in countercurrent manner (Fig. 12.7), with the product moving in the direction opposite to air flow. The arrangement used will depend on the product and the sensitivity of quality characteristics to temperature.

With concurrent systems, a high-moisture product is exposed to high-temperature air, and evaporation assists in maintaining lower product temperature. At locations near the tunnel exit, the lower-moisture product is exposed to lower-temperature air. In countercurrent systems, a lower-moisture product is exposed to high-temperature air, and a smaller temperature gradient exists near the product entrance to the tunnel. Although the overall efficiency of the countercurrent system may be higher than the concurrent, product quality considerations may not allow its use. The concept of air recirculation is used whenever possible to conserve energy.

12.2.3 Puff-Drying

A relatively new process that has been applied successfully to several different fruits and vegetables is explosion puff-drying. This process is accomplished by exposing a relatively small piece of product to high pressure and high temperature for a short time, after which the product is moved to atmospheric pressure. This results in flash evaporation of water and allows vapors from the interior parts of the product to escape. Products produced by puff-drying have very high porosity with rapid rehydration characteristics. Puff-drying is particularly effective for products with significant falling-rate drying periods. The rapid moisture evaporation and resulting product porosity contribute to rapid moisture removal during the final stages of drying.

The puff-drying process is accomplished most efficiently by using 2-cm cube shapes. These pieces will dry rapidly and uniformly and

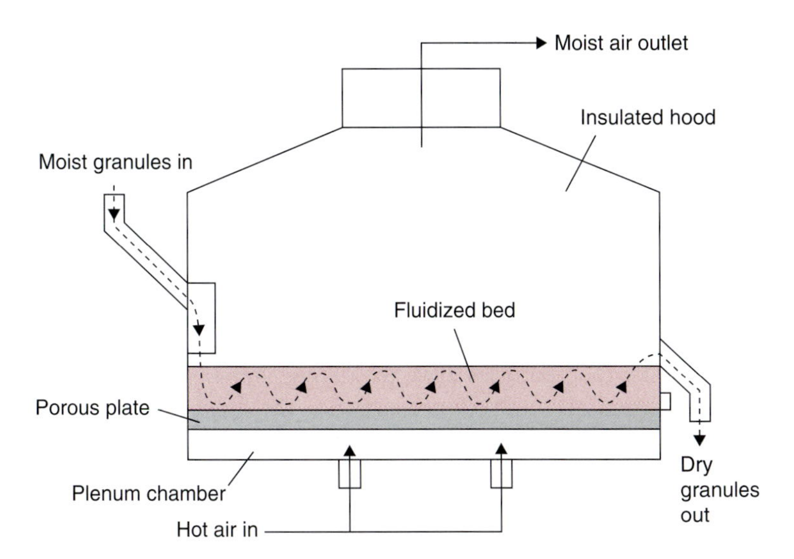

Figure 12.8 Schematic illustration of a fluidized-bed dryer. (From Joslyn, 1963)

will rehydrate within 15 minutes. Although the process may not have applications for all foods, the superior quality encourages additional investigation of the process.

12.2.4 Fluidized-Bed Drying

A second relatively new design for drying solid-particle foods incorporates the concept of the fluidized bed. In this system, the product pieces are suspended in the heated air throughout the time required for drying. As illustrated in Figure 12.8, the movement of product through the system is enhanced by the change in mass of individual particles as moisture is evaporated. The movement of the product created by fluidized particles results in equal drying from all product surfaces. The primary limitation to the fluidized-bed process is the size of particles that will allow efficient drying. As would be expected, smaller particles can be maintained in suspension with lower air velocities and will dry more rapidly. Although these are desirable characteristics, not all products can be adapted to the process.

12.2.5 Spray Drying

The drying of liquid food products is often accomplished in a spray dryer. Moisture removal from a liquid food occurs after the liquid is atomized or sprayed into heated air within a drying chamber. Although various configurations of the chamber are used, the arrangement

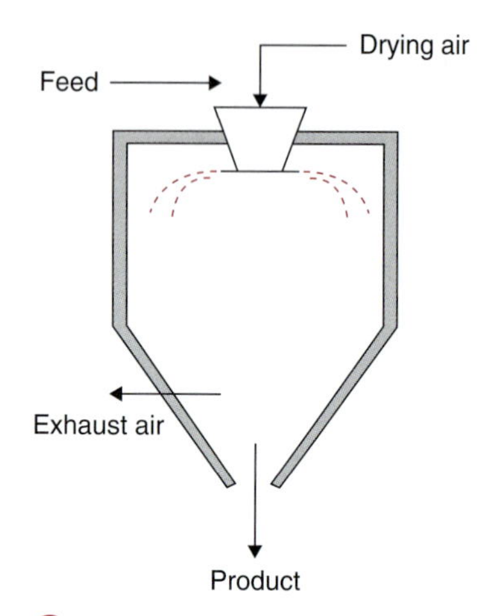

■ **Figure 12.9** Schematic illustration of a spray-drying system.

shown in Figure 12.9 illustrates the introduction of liquid droplets into a heated air stream.

While liquid food droplets are moving with the heated air, the water evaporates and is carried away by the air. Much of the drying occurs during a constant-rate period and is limited by mass transfer at the droplet surface. After reaching the critical moisture content, the dry food particle structure influences the falling-rate drying period. During this portion of the process, moisture diffusion within the particle becomes the rate-limiting parameter.

After the dry food particles leave the drying chamber, the product is separated from air in a cyclone separator. The dried product is then placed in a sealed container at moisture contents that are usually below 5%. Product quality is considered excellent due to the protection of product solids by evaporative cooling in the spray dryer. The small particle size of dried solids promotes easy reconstitution when mixed with water.

Air molecules and other noncondensables evacuated
Low-temperature condenser
Product vapor collected as ice
Water vapors
Ice interface
Dry product
Product ice
Heat input
Vacuum chamber

■ **Figure 12.10** Schematic illustration of a freeze-drying system. (Courtesy of the Vitris Company, Inc.)

12.2.6 Freeze-Drying

Freeze-drying is accomplished by reducing the product temperature so that most of the product moisture is in a solid state, and by decreasing the pressure around the product, sublimation of ice can be achieved. When product quality is an important factor for consumer acceptance, freeze-drying provides an alternative approach for moisture removal.

The heat- and mass-transfer processes during freeze-drying are unique. Depending on the configuration of the drying system (Fig. 12.10), heat transfer can occur through a frozen product layer or through a dry product layer. Obviously, heat transfer through the frozen layer will be rapid and not rate-limiting. Heat transfer through the dry product layer will be at a slow rate due to the low thermal conductivity of the highly porous structure in a vacuum. In both situations, the mass transfer will occur in the dry product layer. The diffusion of water vapor would be expected to be the rate-limiting process because of the low rates of molecular diffusion in a vacuum.

The advantages of the freeze-drying process are superior product quality resulting from low temperature during sublimation and the maintenance of product structure. These advantages are balanced against the energy-intensive aspects of the product freezing and vacuum requirements.

Figure 12.11 Mass and energy balance for a dehydration process.

12.3 DEHYDRATION SYSTEM DESIGN

The design of a dehydration system involves several considerations. The factors that have a direct influence on the capacity of a drying system include the quantity and characteristics of air available for drying, along with the drying time required for individual components of the product being dried. These factors require somewhat different approaches to analysis.

12.3.1 Mass and Energy Balance

By application of a mass and energy balance to the entire dehydration system, as illustrated in Figure 12.11, several parameters influencing design of the system are considered. Although the analysis being illustrated is for a countercurrent system, a similar approach could be applied to a concurrent system or a batch system.

An overall balance on water entering and leaving the system gives

$$\dot{m}_a W_2 + \dot{m}_p w_1 = \dot{m}_a W_1 + \dot{m}_p w_2 \tag{12.3}$$

where

$\dot{m}_a$ = air flow rate (kg dry air/h)
$\dot{m}_p$ = product flow rate (kg dry solids/h)
W = absolute humidity (kg water/kg dry air)
w = product moisture content, dry basis (kg water/kg dry solids)

It is important to note that the balance presented in Equation (12.3) is conducted on the basis of dry air and moisture-free product solids.

Example 12.3

A cabinet dryer is being used to dry a food product from 68% moisture content (wet basis) to 5.5% moisture content (wet basis). The drying air enters the system at 54°C and 10% RH and leaves at 30°C and 70% RH. The product temperature is 25°C throughout drying. Compute the quantity of air required for drying on the basis of 1 kg of product solids.

Given

Initial product moisture content $w_1 = 0.68/0.32 = 2.125$ kg H_2O/kg solids
Final product moisture content $w_2 = 0.055/0.945 = 0.0582$ kg H_2O/kg solids
Air entering dryer = 54°C and 10% RH
Air leaving dryer = 30°C and 70% RH
Product temperature = 25°C
Computations based on 1 kg product solids

Approach

The air requirements for drying can be determined by using Equation (12.3) with the following modifications:

$$(\dot{m}_a/\dot{m}_p)W_2 + w_1 = (\dot{m}_a/\dot{m}_p)W_1 + w_2$$

Solution

1. *Using the psychrometric chart from Appendix A.5:*

   $W_1 = 0.0186$ *kg H_2O/kg dry air (using 30°C and 70% RH)*

   $W_2 = 0.0094$ *kg H_2O/kg dry air (using 54°C and 10% RH)*

2. *Using the modification of Equation (12.3):*

$$\begin{aligned}
\dot{m}_a/\dot{m}_p&(0.0094 \text{ kg } H_2O/\text{kg dry air}) \\
&+ 2.125 \text{ kg } H_2O/\text{kg solids} \\
&= \dot{m}_a/\dot{m}_p(0.0186 \text{ kg } H_2O/\text{kg dry air}) \\
&+ 0.0582 \text{ kg } H_2O/\text{kg solids} \\
0.0092\dot{m}_a/\dot{m}_p &= 2.067 \\
\dot{m}_a/\dot{m}_p &= 224.65 \text{ kg dry air/kg solids}
\end{aligned}$$

An energy balance on the dehydration system is described by the following relationship:

$$\dot{m}_a H_{a2} + \dot{m}_p H_{p1} = \dot{m}_a H_{a1} + \dot{m}_p H_{p2} + q \qquad (12.4)$$

where

q = thermal energy loss from the dehydration system
H_a = the thermal energy content of air (kJ/kg dry air)
H_p = the thermal energy content of product (kJ/kg dry solids)

The expressions for thermal energy content of air and product include:

$$H_a = c_s(T_a - T_0) + WH_L \qquad (12.5)$$

where

c_s = humid heat of air (kJ/[kg dry air K]) = 1.005 + 1.88 W (see Eq. (9.17))
T_a = air temperature (°C)
T_0 = reference temperature = 0°C
H_L = latent heat of vaporization for water (kJ/kg water).

In addition:

$$H_p = c_{pp}(T_p - T_0) + wc_{pw}(T_p - T_0) \quad (12.6)$$

where

c_{pp} = specific heat of product solids (kJ/[kg solids K])
T_p = product temperature (°C)
c_{pw} = specific heat of water (kJ/[kg water K])

Using these equations, we can determine the quantity of air required for drying a defined amount of product for a known quantity of air at the inlet and establish the moisture characteristics of air at the system outlet.

Example 12.4

A fluidized-bed dryer is being used to dry diced carrots. The product enters the dryer with 60% moisture content (wet basis) at 25°C. The air used for drying enters the dryer at 120°C after being heated from ambient air with 60% RH at 20°C. Estimate the production rate when air is entering the dryer at 700 kg dry air/h and product leaving the dryer is at 10% moisture content (wet basis). Assume product leaves the dryer at the wet bulb temperature of air and the specific heat of product solid is 2.0 kJ/(kg °C). Air leaves the dryer 10°C above the product temperature.

Given

Initial product moisture content w_1 = 0.6/0.4 = 1.5 kg H_2O/kg solids
Initial air condition = 20°C and 60% RH (W_2 = 0.009 kg H_2O/kg dry air)
Air temperature entering dryer = 120°C
Air flow rate $\dot{m}_a$ = 700 kg dry air/h
Final product moisture content (w_2) = 0.1/0.9 = 0.111 kg H_2O/kg solids
Specific heat of product solids (c_{pp}) = 2.0 kJ/(kg K)
Final product temperature = wet bulb temperature of air
Final air temperature (T_{a1}) = T_{p2} + 10°C

Approach

The product production rate will be determined using the material balance, Eq. (12.3), and the energy balance, Eq. (12.4), along with parameters obtained from the psychrometric chart (Appendix A.5).

Solution

1. *Using Equation (12.3):*

$$(700 \text{ kg dry air/h})(0.009 \text{ kg water/kg dry air}) + \dot{m}_p\,(1.5 \text{ kg water/kg solids}) = (700 \text{ kg dry air/h})W_1 + \dot{m}_p(0.111 \text{ kg water/kg solids})$$

2. *To use Equation (12.4), the following computations are required from Equations (12.5) and (12.6):*

For the air:

$$H_{a2} = c_{s2}(120 - 0) + 0.009H_{L2}$$

where

$$c_{s2} = 1.005 + 1.88(0.009) = 1.0219 \text{ kJ/(kg dry air K)}$$

$$H_{L2} = 2202.59 \text{ kJ/kg water at } 120°\text{C (from Table A.4.2)}$$

Therefore,

$$H_{a2} = (1.0219 \text{ kJ/[kg dry air K]})(120°\text{C}) + (0.009 \text{ kg water/kg dry air})(2202.59 \text{ kJ/kg water})$$

$$H_{a2} = 142.45 \text{ kJ/kg dry air}$$

since $T_{p2} = 38°\text{C}$ *[wet bulb temperature of air (Appendix A.5)]*

$$T_{a1} = 38 + 10 = 48°\text{C}$$

Then,

$$H_{a1} = c_{s1}(T_{a1} - 0) + W_1(H_{L1})$$

where

$$c_{s1} = 1.005 + 1.88W_1$$

and

$$H_{L1} = 2387.56 \text{ kJ/kg water at } 48°\text{C (from Table A.4.2)}$$

Therefore,

$$H_{a1} = (1.005 + 1.88W_1)(48°\text{C}) + W_1(2387.56 \text{ kJ/kg water})$$

For the product:

$$H_{p1} = (2.0\ \text{kJ/[kg solids K]})(25^\circ\text{C} - 0) + (1.5\ \text{kg water/kg solids})(4.178\ \text{kJ/[kg water K]}) \times (25^\circ\text{C} - 0)$$

Then $H_{p1} = 206.75$ *kJ/kg solids*

$$H_{p2} = (2.0\ \text{kJ/[kg solids K]})(T_{p2}\,^\circ\text{C} - 0) + (0.111\ \text{kg water/kg solids})(4.175\ \text{kJ/[kg water K]}) \times (T_{p2}\,^\circ\text{C} - 0)$$

Then from Equation (12.4),

$$\begin{aligned}(700\ \text{kg dry air/h})(142.45\ \text{kJ/kg dry air}) + \dot{m}_p(206.75\ \text{kJ/kg solids})\\ = (700\ \text{kg dry air/h})[(1.005 + 1.88W_1)(48^\circ\text{C})\\ + W_1(2387.56/\text{kg water})]\\ + \dot{m}_p\{(2.0\ \text{kJ/[kg solids K]})(38^\circ\text{C})\\ + (0.111\ \text{kg water/kg solids})(4.175\ \text{kJ/[kg water K]})(38^\circ\text{C})\}\\ + 0\end{aligned}$$

where $q = 0$, *indicating negligible heat loss from surface of dryer.*

3. *The material-balance (step 1) and energy-balance (step 2) equations can be solved simultaneously:*

a. $700(0.009) + 1.5\,\dot{m}_p = 700W_1 + 0.111\,\dot{m}_p$

b. $700(142.45) + \dot{m}_p(206.75) = 700\{(1.005 + 1.88W_1)(48) + 2387.56W_1\} + \dot{m}_p\{(2.0)(38) + (0.111)(4.175)(38)\}$

a. $6.3 + 1.5\dot{m}_p = 700W_1 + 0.111\dot{m}_p$

b. $99{,}715 + 206.75\dot{m}_p = 700(48.24 + 2477.8W_1) + 93.61\dot{m}_p$

a. $W_1 = (1.389\dot{m}_p + 6.3)/700$

b. $65{,}947 + 113.14\dot{m}_p = 1{,}734{,}460W_1$

Then

$$\begin{aligned}65{,}947 + 113.14\dot{m}_p &= 1{,}734{,}460(1.389\dot{m}_p + 6.3)/700\\ \dot{m}_p &= 15.12\ \text{kg solids/h}\end{aligned}$$

4. *The absolute humidity of air leaving the dryer is*

$$\begin{aligned}W_1 &= (1.389 \times 15.12 + 6.3)/700\\ &= 0.039\ \text{kg water/kg dry air}\end{aligned}$$

indicating that air leaving the dryer will be 48°C and 55% RH.

12.3.2 Drying-Time Prediction

To determine the time required to achieve the desired reduction in product moisture content, the rate of moisture removal from the product must be predicted. For the constant-rate drying period, the moisture removal rate would be described by:

$$\dot{m}_c = \frac{w_o - w_c}{t_c} \tag{12.7}$$

where:

$\dot{m}_c$ = moisture removal rate during constant rate drying (s^{-1})
w_c = critical moisture content (kg water/kg dry solids)
w_0 = initial moisture content (kg water/kg dry solids)
t_c = time for constant-rate drying (s)

During the constant-rate drying period, thermal energy is being transferred from the hot air to the product surface, while water vapor is transferred from the product surface to the heated air. The magnitude of thermal energy transfer is described by:

$$q = hA(T_a - T_s) \tag{12.8}$$

where:

q = rate of heat transfer (W)
h = convective heat transfer coefficient (W/[m^2 K])
A = surface area of product exposed to heated air (m^2)
T_a = heated air temperature (°C)
T_s = product surface temperature (°C)

It should be noted that during constant-rate drying, the product surface temperature will remain at the wet bulb temperature of the heated air.

The magnitude of water vapor transfer during constant-rate drying is described by the following mass transfer expression:

$$\dot{m}_c = \frac{k_m A M_w P}{0.622 R T_A}(W_s - W_a) \tag{12.9}$$

where:

k_m = convective mass transfer coefficient (m/s)
A = product surface area (m^2)
M_w = molecular weight of water

P = atmospheric pressure (kPa)
R = universal gas constant (8314.41 m^{3}Pa/[kg mol K])
T_A = absolute temperature (K)
W_a = humidity ratio for air (kg water/kg dry air)
W_s = humidity ratio at product surface (kg water/kg dry air).

When using Equation (12.9), the magnitude of the humidity ratio at the product surface is the value at saturation for air moving over the surface, and would be determined by using a psychrometric chart.

If the rate of constant-rate drying is based on thermal energy transfer, where the thermal energy is used to cause phase change of water at the product surface, the following expression will apply:

$$q = \dot{m}_c H_L \tag{12.10}$$

where H_L is the latent heat of vaporization for water at the wet bulb temperature of heated air (kJ/kg water).

By combining Equations (12.10) and (12.8), the following expression for rate of water vapor transfer is obtained:

$$\dot{m}_c = \frac{w_o - w_c}{t_c} = \frac{hA}{H_L}(T_a - T_s) \tag{12.11}$$

Then, by solving for time, the following equation for prediction of the constant-rate drying time is obtained:

$$t_c = \frac{H_L(w_o - w_c)}{hA(T_a - T_s)} \tag{12.12}$$

As indicated by Equation (12.12), the time for constant-rate drying is directly proportional to the difference between the initial moisture content and the critical moisture content, and is inversely proportional to the temperature gradient between the product surface and the heated air.

The time for constant-rate drying can be based on mass transfer by using Equations (12.7) and (12.9), and solving for drying time, as follows:

$$t_c = \frac{0.622RT_A(w_o - w_c)}{k_m A M_w P(W_s - W_a)} \tag{12.13}$$

This expression indicates that the constant-rate drying time is inversely proportional to the gradient of humidity ratios at the product surface and in the heated air.

Example 12.5

Air at 90°C is being used to dry a solid food in a tunnel dryer. The product, with 1 cm thickness and a 5 cm by 10 cm surface, is exposed to the heated air with convective mass transfer coefficient of 0.1 m/s. Estimate the constant-rate drying time, when the initial moisture content is 85% and the critical moisture content is 42%. The air has been heated from 25°C and 50% RH. The product density is 875 kg/m^3.

Given

Initial moisture content, $w_0 = 0.85/0.15 = 5.67$ kg water/kg solids
Critical moisture content, $w_c = 0.42/0.58 = 0.724$ kg water/kg solids
Air temperature, $T_a = 90°C$
Convective mass transfer coefficient, $k_m = 0.1$ m/s
Product surface area, $A = 0.05 \times 0.1 = 0.005\ m^2$
Molecular weight of water, $M_w = 18$ kg/mol
Atmospheric pressure, $P = 101.325$ kPa

Approach

During the constant-rate period of drying, the rate of water removal from the product surface will be described by mass transfer from the product surface to the air flowing over the surface.

Solution

1. *In order to use Equation (12.13), the humidity ratio for the heated air and at the product surface must be evaluated.*
2. *Since the heated air was heated from 25°C and 50% RH, the humidity ratio (W_a) is 0.01 kg water/kg dry air, as determined from the psychrometric chart (Fig. A.5).*
3. *The humidity ratio (W_s) at the product surface is evaluated at the saturation condition for the heated air. From the psychrometric chart, W_s is 0.034 kg water/kg dry air at a wet bulb temperature of 33.7°C.*
4. *The temperature in Equation (12.13) is the mean of heated air temperature (90°C) and the product surface (33.7°C), so:*

$$T_A = 62.85 + 273 = 335.85\ K$$

5. *Then Equation (12.13) becomes*

$$t_c = \frac{0.622[\text{kg water/kg dry air}] \times 8.314\ [\text{m}^3\text{kPa/(kg mol K)}] \times 335.85\ [\text{K}] \times (5.67 - 0.724)[\text{kg water/kg solids}]}{0.1\ [\text{m/s}] \times 0.005[\text{m}^2] \times 18\ [\text{kg water/mol}] \times 101.325\ [\text{kPa}] \times (0.034 - 0.01)[\text{kg water/kg dry air}]}$$

$$t_c = 3.925 \times 10^5\ \text{s/kg solids}$$

6. *Since each piece of product has*

$$0.05 \times 0.1 \times 0.01 = 5 \times 10^{-5}\ m^3$$

and

$$5 \times 10^{-5}\,[m^3] \times 875\ [kg/m^3] = 0.04375\ kg$$

or

$$0.04375\ [kg] \times 0.15\ [kg\ solids/kg\ product] = 6.5625 \times 10^{-3}\ kg\ solids$$

7. *Based on the solids content of each product piece*

$$t_c = 3.925 \times 10^5\,[s/kg\ solids] \times 6.5625 \times 10^{-3}\,[kg\ solids]$$
$$t_c = 2575.85\ s = 42.9\ min$$

8. *The result indicates that the constant rate drying time is 42.9 min for each product piece.*

As indicated in Section 12.1.3, a portion of the moisture removal from a food occurs during a falling-rate drying period. This period of drying begins at the critical moisture content, w_c, and continues until the moisture content decreases to the equilibrium moisture content, w_e. During this period of drying, the product temperature begins to increase to magnitudes above the wet bulb temperature, and the diffusion of moisture from the internal product structure becomes a rate-controlling factor. In addition, the expressions used to describe the moisture diffusion process are dependent on product shape. For an infinite-plate geometry, the falling-rate drying period is described as follows:

$$\frac{w - w_e}{w_c - w_e} = \frac{8}{\pi^2} \exp\left[-\frac{\pi^2 D t}{4 d_c^2}\right] \tag{12.14}$$

where:

d_c = characteristic dimension, half-thickness of the slab (m)
D = effective mass diffusivity (m^2/s)
t = drying time (s)

The expression for drying time will be:

$$t_F = \frac{4 d_c^2}{\pi^2 D} \ln\left[\frac{8}{\pi^2}\left(\frac{w_c - w_e}{w - w_e}\right)\right] \tag{12.15}$$

An expression for the drying time of a product with an infinite cylinder geometry will be:

$$t_F = \frac{d_c^2}{\beta^2 D} \ln \left[\frac{4}{\beta^2} \left(\frac{w_c - w_e}{w - w_e} \right) \right] \tag{12.16}$$

where d_c is the characteristic dimension, the radius of the cylinder (m), and β is the first root of the zero-order Bessel function equation, with a magnitude of 2.4048. Finally, the falling-rate drying time for a spherical product will be predicted by the following expression:

$$t_F = \frac{d_c^2}{\pi^2 D} \ln \left[\frac{6}{\pi^2} \left(\frac{w_c - w_e}{w - w_e} \right) \right] \tag{12.17}$$

where d_c is the characteristic dimension, the radius of the sphere (m).

The key parameter in all three of the prediction equations for the falling-rate drying period is the effective mass diffusivity for moisture movement within the product structure. The magnitude of this property will approach a value for diffusion of water vapor in air, but will be influenced by the structure of the food product.

Example 12.6

The drying of a noodle occurs during the falling-rate period between the critical moisture content of 0.58 kg water/kg solids and a final moisture content of 0.22 kg water/kg solids. The mass diffusivity for water vapor within the noodle is 2×10^{-7} cm^2/s and the noodle thickness is 3 mm. The equilibrium moisture content is 0.2 kg water/kg solids. Estimate the falling-rate drying time.

Given

Characteristic dimension (half-thickness), $d_c = 0.0015$ m
Mass diffusivity, $D = 2 \times 10^{-7}$ cm^2/s $= 2 \times 10^{-11}$ m^2/s
Critical moisture content, $w_c = 0.58$ kg water/kg solids
Equilibrium moisture content, $w_e = 0.22$ kg water/kg solids
Final moisture content, $w = 0.2$ kg water/kg solids

Approach

The falling-rate drying time can be estimated by assuming the noodle geometry is described as an infinite slab.

Solution

Using Equation (12.15),

$$t_F = \frac{4 \times 0.0015^2 [m^2]}{\pi^2 \times 2 \times 10^{-11} [m^2/s]} ln\left(\frac{8}{\pi^2}\frac{(0.58-0.20)}{(0.22-0.20)}\right)$$

$$= 1.24675 \times 10^5 \ s$$

$$t_F = 34.6 \ h$$

An example of a drying process where constant-rate drying is limited by evaporation from a free water surface is spray drying. Within the drying chamber, the droplets of liquid food are moving through heated air, while water at the droplet surface changes to a vapor phase. The water vapor is removed from the droplet surface by heated air. During the constant-rate drying period, the rate is limited by heat and mass transfer at the droplet surface.

During the constant-rate period of drying, a spray drying process may be described by heat transfer from the heated air to the droplet surface or by mass transfer from the droplet surface to the heated air. When the process is described in terms of heat transfer, the convective heat transfer coefficient may be estimated by Equation (4.69). For descriptions in terms of mass transfer, the mass transfer coefficient is estimated by Equation (10.38). The similarity between the expressions used to estimate the surface coefficients is obvious.

During the falling-rate drying periods, the rate of spray drying is controlled by heat and mass transfer within the product particle structure. For a heat transfer limited process, the heat transfer is described by conduction through the product structure and will be proportional to the thermal conductivity of the product solids and parameters describing the porous structure of the product. Mass transfer occurs as a result of diffusion of water vapor through the gas phase in the porous structure of the product particle.

A specific expression for drying time during spray drying is based on Equations (12.12)and (12.17). By recognizing that the surface area of a sphere is $4\pi R^2$, Equation (12.12) is modified to become:

$$t_c = \frac{H_L(w_o - w_c)}{4\pi R_d^2 h(T_a - T_s)} \quad (12.18)$$

where R_d is the radius of the liquid food droplet.

Ranz and Marshall (1952) have demonstrated that the convective heat transfer coefficient at the surface of a droplet during spray drying can be estimated by the ratio of the thermal conductivity of air to the droplet radius. Using this relationship, the time for constant-rate drying of a liquid droplet during spray drying is

$$t_c = \frac{H_L(w_0 - w_c)}{4\pi R_d k_a (T_a - T_s)} \tag{12.19}$$

A similar relationship for constant-rate drying time, based on mass transfer, can be developed.

Using these developments, the prediction expression for total drying time during spray drying of a liquid food droplet will be

$$t = \frac{H_L(w_0 - w_c)}{4\pi R_d k_a (T_a - T_s)} + \frac{R_p^2}{\pi^2 D} \ln\left[\frac{6}{\pi^2}\left(\frac{w_c - w_e}{w - w_e}\right)\right] \tag{12.20}$$

where R_p is the radius of the product particle at the critical moisture content. In Equation (12.20), the first term on the right side of the equation is the time for the constant-rate drying period. The second term represents the time for falling-rate drying. One of the critical parameters in Equation (12.20) is the droplet/particle radius (R_p) at the critical moisture content (w_c). Although these two parameters are most often established through experimental measurements, most other inputs required for Equation (12.20) can be obtained from handbooks or tables of properties.

Example 12.7

Skim milk with 5% total solids is being spray dried to a final moisture content of 4% using 120°C air with 7% RH. The density of the skim milk product is 1000 kg/m^3, and the largest droplet diameter is 120 μm. The critical moisture content is 45%, and the diameter of the particle at the critical moisture content is 25.5 μm. The equilibrium moisture content is 3.85%, and the mass diffusivity for water vapor within the particle is 5×10^{-11} m^2/s. Estimate the drying time for product within the spray drier.

Given

Initial moisture content, $w_0 = 0.95$ *kg water/kg product* $= 19$ *kg water/kg solids*
Critical moisture content, $w_c = 0.45$ *kg water/kg product* $= 0.818$ *kg water/kg solids*
Final moisture content, $w = 0.04$ *kg water/kg product* $= 0.042$ *kg water/kg solids*
Equilibrium moisture content, $w_e = 0.0385$ *kg water/kg product* $= 0.04$ *kg water/kg solids*
Heated air temperature $= T_a = 120°C$
Heated air relative humidity $= RH = 7\%$
Droplet radius, $R_d = 60 \times 10^{-6}$ *m*

Particle radius, $R_p = 12.75 \times 10^{-6}$ m
Mass diffusivity, $D = 5 \times 10^{-11}$ m^2/s
Liquid product density, $\rho = 1000$ kg/m^3

Approach

The total time for drying the droplets of skim milk will include time for constant-rate drying and time for the falling-rate period.

Solution

1. *In order to use Equation (12.19) to predict the constant-rate drying time, the temperature at the droplet surface must be estimated, and the latent heat of the vaporization must be determined.*
2. *Using the psychrometric chart (Fig. A.5), the wet bulb temperature for the heated air (120°C, 7% RH) is 57.1°C. Since the droplet temperature will not be above the wet bulb temperature of air during the constant-rate period of drying, the droplet surface temperature can be estimated at 57.1°C.*
3. *Using the properties of saturated steam (Table A.4.2) at 57.1°C, the latent heat of vaporization is determined as 2354 kJ/kg.*
4. *The thermal conductivity of the heated air at 120°C is* $k_a = 0.032$ W/(m °C) *from Table A.4.4.*
5. *Using Equation (12.19),*

$$t_c = \frac{2354[\text{kJ/kg water}] \times (19 - 0.818)[\text{kg water/kg solids}] \times 1000[\text{J/kJ}]}{4\pi \times 60 \times 10^{-6}[\text{m}] \times 0.032[\text{W/(m °C)}] \times (120 - 57.1)[\text{°C}]}$$
$$= 2.82 \times 10^{10} \text{ s per kg solids}$$

6. *Since Equation (12.19) presents the results per unit of product solids, the mass of solids per droplet must be estimated.*
   Using the volume, density, and fraction of solids in each droplet,

$$\text{mass [kg solids/droplet]} = \rho V (0.05)[\text{kg solids/kg product}]$$

$$\text{mass} = 1000[\text{kg product/m}^3] \times \left(\frac{4}{3}(60 \times 10^{-6})^3 \pi [\text{m}^3]\right) \times 0.05\ [\text{kg solids/kg product}]$$

$$\text{mass} = 4.52 \times 10^{-11} \text{ kg solids/droplet}$$

7. *Constant rate drying time* = 1.276 s.
8. *In order to estimate the falling-rate drying time, the particle is assumed to be spherical, and Equation (12.17) will be used as follows:*

$$t_F = \frac{(12.75 \times 10^{-6})^2[\text{m}^2]}{\pi^2 \times 5 \times 10^{-11}[\text{m}^2/\text{s}]} \ln\left\{\frac{6}{\pi^2}\left[\frac{(0.818 - 0.04)}{(0.042 - 0.04)}\right]\right\}$$

$$t_F = 1.801 \text{ s}$$

9. *Total drying time:*

$$t = 1.276 + 1.801 = 3.077 \text{ s}$$

A drying process that illustrates the type of drying that occurs when the drying rate is limited by internal mass transfer is freeze-drying. An analysis of the heat and mass transfer for the system in Figure 12.10 is simplified, since heat transfer occurs from the hot plate to the drying front through the frozen product layer of dry product, and moisture diffusion becomes the rate-limiting process.

The heat and mass transfer during freeze-drying as illustrated in Figure 12.10 has been analyzed by King (1970, 1973). His analysis provided an equation for estimating the drying time, as follows:

$$t = \frac{RT_A L^2}{8DMV_w(P_i - P_a)}\left(1 + \frac{4D}{k_m L}\right) \quad (12.21)$$

where L is the thickness of product layer (m), T_A is the absolute temperature (K), M is the molecular weight (kg/kg mol), V_w is the specific volume of water (m^3/kg water), P_i is the vapor pressure of ice (Pa), P_a is the vapor pressure of air at the condenser surface (Pa), k_m is the mass transfer coefficient (kg mol/[s m^2 Pa]), R is universal gas constant (8314.41 m^3 Pa/[kg mol K]), and D is the diffusion coefficient (m^2/s).

Equation (12.21) is limited to situations where the drying time is based on a process where rate is limited by moisture diffusion within the structure of the dry product layer. The calculation of drying time requires knowledge of the moisture diffusivity, D, and mass transfer coefficient, k_m, and the magnitude of both is likely to be product-dependent. Often, these property values must be measured for individual situations.

Example 12.8

A concentrated liquid coffee is being freeze-dried by placing a 2 cm thick frozen layer of the product over a heated platen. The product is frozen to –75°C initially and before placing over the 30°C platen. The freeze-drying is accomplished in a chamber at a pressure of 38.11 Pa with a condenser temperature of –65°C. Properties needed to describe the process have been measured in an experimental system; mass diffusivity = 2×10^{-3} m^2/s and mass transfer coefficient = 1.5 kg mole/s m^2 Pa. The initial moisture content of the concentrate is 40% and the density of dry product solids is 1400 kg/m^3. Compute the drying time for the product.

Given

Thickness of product layer = 2 cm = 0.02 m
Mass diffusivity = 2×10^{-3} m^2/s
Mass transfer coefficient = 1.5 kg mole/s m^2 Pa

Approach

We will compute the drying time for the coffee concentrate by using Equation (12.21).

Solution

1. *To use Equation (12.21), several parameters must be determined from the thermodynamic tables.*

   Universal gas constant $= 8314.41\ m^3\ Pa/kg\ mol\ K$
   Absolute temperature $(T_A) = 243\ K$ *(based on ice temperature at pressure 38.11 Pa)*
   Vapor pressure of ice $(P_I) = 38.11\ Pa$
   Vapor pressure of condenser surface $(P_a) = 0.5\ Pa$
   Molecular weight of water $(M_w) = 18$

2. *The specific volume of water is computed from initial moisture content of product and density of product solids.*

   Moisture content (dry basis) $= 0.4/0.6 = 0.667$ *kg water/kg solids*

   Then

$$V_w = \frac{1}{(0.667)(1400)} = 0.00107 \frac{m^3\ solid}{kg\ water}$$

3. *Using Equation (12.21),*

$$t = \frac{(8314.41)(243)(0.02)^2}{8(2 \times 10^{-3})(18)(0.00107)(38.11 - 0.5)} \left(1 + \frac{4(2 \times 10^{-3})}{(1.5)(0.02)} \right)$$

$$t = 88324\ s = 1472\ min = 24.5\ hr$$

The dehydration process for foods results in products with reduced mass and possible reductions in volume. These reductions provide for increased efficiency in transportation and storage, in addition to effective preservation. The selection of the most appropriate process for a given food product will depend on two primary considerations: cost of drying and product quality.

Of the processes discussed in this chapter, air drying by cabinet or tunnel will have the lowest cost when expressed as cost per unit of water removed. The use of vacuum increases the cost of water removal somewhat, and the value will be comparable to fluidized-bed drying. The explosion puff-drying process will be more costly than previously mentioned processes, but it appears to be more efficient than freeze-drying.

The product quality resulting from the various processes will be opposite to the cost comparisons. The freeze-drying process produces the highest-quality product and can be used for products where the quality characteristics allow product pricing to recover the extra drying cost. Based on available literature, the puff-drying process produces quality that is comparable to that of freeze-drying, with somewhat lower cost. The product quality resulting from fluidized-bed drying and vacuum drying should be somewhat similar to, but lower than, that for puff-drying and freeze-drying processes. The lowest-quality dehydrated products are produced by the lowest-cost processes—tunnel and cabinet drying using heated air. Although these comparisons suggest a direct trade-off between dehydration cost and dry product quality, we need to recognize that the selection of the process will depend on the adaptability of the product to a given process. The physical characteristics of the product before and/or after drying may dictate which process is used for dehydration.

PROBLEMS

12.1 The following equilibrium moisture-content data have been collected for a dry food:

Water activity	Equilibrium moisture content (g H_2O/g product)
0.1	0.060
0.2	0.085
0.3	0.110
0.4	0.122
0.5	0.125
0.6	0.148
0.7	0.173
0.8	0.232

Develop a plot of the equilibrium moisture isotherm for moisture content on a dry weight basis.

12.2 A product enters a tunnel dryer with 56% moisture content (wet basis) at a rate of 10 kg/h. The tunnel is supplied with 1500 kg dry air/h at 50°C and 10% RH, and the air leaves at 25°C in equilibrium with the product at 50% RH. Determine the moisture content of product leaving the dryer and the final water activity.

***12.3** A countercurrent tunnel dryer is being used to dry apple slices from an initial moisture content (wet basis) of 70% to 5%. The heated air enters at 100°C with 1% RH and leaves at 50°C. If the product temperature is 20°C throughout the dryer and the specific heat of product solids is 2.2 kJ/(kg °C), determine the quantity of heated air required for drying the product at a rate of 100 kg/h. Determine the relative humidity of outlet air.

12.4 A cabinet dryer is to be used for drying of a new food product. The product has an initial moisture content of 75% (wet basis) and requires 10 minutes to reduce the moisture content to a critical level of 30% (wet basis). Determine the final moisture of the product if a total drying time of 15 minutes is used.

12.5 The following data were obtained by Labuza et al. (1985) on fish flour at 25°C. Determine the GAB model for these data and the values of k, w_m, and C.

a_w	g water/ 100 g solids
0.115	2.12
0.234	3.83
0.329	5.53
0.443	6.82
0.536	7.65
0.654	10.29
0.765	13.40
0.848	17.50

12.6 Pistachios are to be dried using a countercurrent dryer operating at steady state. The nuts are dried from 80% (wet basis) to 12% (wet basis) at 25°C. Air enters the heater at 25 °C (dry bulb temperature) and 80% relative humidity. The heater supplies 84 kJ/kg dry air. The air exits the dryer at 90% relative humidity. For this given information solve the following parts:

a. What is the relative humidity of the air leaving the heater section of the dryer?
b. What is the temperature (dry bulb temperature) of air leaving the dryer?
c. What is the flow rate (m^3/s) of air required to dry 50 kg/h of pistachio nuts?

*Indicates an advanced level of difficulty in solving.

12.7 A sample of a food material weighing 20 kg is initially at 450% moisture content dry basis. It is dried to 25% moisture content wet basis. How much water is removed from the sample per kg of dry solids?

12.8 Air enters a counterflow drier at 60°C dry bulb temperature and 25°C dew point temperature. Air leaves the drier at 40°C and 60% relative humidity. The initial moisture content of the product is 72% (wet basis). The amount of air moving through the drier is 200 kg of dry air/h. The mass flow rate of the product is 1000 kg dry solid per hour. What is the final moisture content of the dried product (in wet basis)?

12.9 The constant rate portion of drying for a new food product must be accomplished within 5 min; a reduction from an initial moisture content of 75% to the critical moisture content of 40%. Heated air at 95°C and 10% RH is used for drying. The surface of product exposed to the air is 10 cm wide and 20 cm in the direction of air movement over the surface. The mass diffusivity for water vapor in air is $1.3 \times 10^{-3}\ m^2/s$. The product thickness is 5 cm and density is $900\ kg/m^3$.

a. Estimate the mass transfer coefficient needed at the product surface.

b. Compute the air velocity required.

12.10 A spray drier with a 5 m diameter and 10 m height is used to dry skim milk to a final moisture content of 5%. The air entering the drier is 120°C, is heated from ambient air at 25°C and 75% RH, and flows concurrently with the product droplets (and particles) through the system. The skim milk is atomized into the hot air at 45°C with a maximum droplet size of 120 micron diameter. The critical moisture content for the product is 30%, and the particle diameter is 25 micron at the critical moisture content. The equilibrium moisture content of the dry skim milk is 3.5% at the exit temperature of 55°C from the drier. The mass diffusivity for water vapor within the product particles is $7.4 \times 10^{-7}\ m^2/s$ and the specific heat of product solids is 2.0 kJ/kgK. The heated air leaves the system at 5°C above the product temperature. The initial moisture content of skim milk is 90.5% wet basis. Determine the production rate for the spray drier; kg of 5% moisture content product per unit time.

12.11 A spray drier with a 5 m diameter and 10 m height is used to dry skim milk to a final moisture content of 5%. The air entering the drier is 120°C, is heated from ambient air at 25°C and 75% RH, and flows concurrently with the product droplets (and particles) through the system. The air flow rate is 1000 m^3/min. The skim milk enters the dryer at 45°C with 90.5% moisture content and the exit temperature is 55°C. The specific heat of product solids is 2.0 kJ/kgK. The heated air leaves the system at 5°C above the product temperature. Determine

a. The production rate for the system; kg of 5% moisture content product per unit time.
b. The temperature and % RH of air leaving the dryer.
c. The thermal energy for heating the air entering the dryer.

12.12 A tunnel dryer is used to reduce the moisture content of a food from an initial magnitude of 85%. The produce is conveyed through the dryer on a 1 m wide conveyor with a 10 m length. The volumetric flow rate of heated air is 240 m^3/min at 100°C, heated from 25°C and 40% RH. The cross-sectional area for air flow through the dryer is 4 m^2, and the thickness of product on the conveyor is 1.5 cm.

a. Estimate the temperature and % RH of air leaving the dryer, when all moisture removal from the product occurs during a constant-rate drying period.
b. Determine the moisture content of product leaving the dryer.

12.13 The falling rate portion of the drying time for a particle of skim milk begins at the critical moisture content of 25%, and the final moisture content is 4%. The air used for drying is 120°C and was heated from ambient air of 20°C and 40% RH. The particle size at the critical moisture content is 20 microns, and the specific heat of the product solids is 2.0 kJ/kg K. The mass diffusivity for water within the product particle is 3.7×10^{-12} m^2/s, and the density of the product particle is 1150 kg/m^3.

a. If the equilibrium moisture content for the product is 3.5%, estimate the time for the falling rate portion of drying.
b. If 5000 m^3/min of heated air is needed for product drying, determine the thermal energy needed to heat the air to 120°C.

12.14 A concurrent tunnel dryer is used to dry a new food product carried in 10 cm by 10 cm by 1 cm trays. Moisture removal from the product occurs from the upper surface only. The heated air is 100°C, and has been heated from ambient conditions at 25°C and 40% RH at a rate of 1000 m^3/min.

The product leaves the system at a temperature 5°C above the wet bulb temperature, and the air leaves 10°C above the product temperature. The initial moisture content of the product is 85%, the density of the product entering the dryer is 800 kg/m^3, and the initial temperature is 20°C. The critical moisture content of the product is 30% and the equilibrium moisture content is 3.0%. The product volume does not change during drying. The convective heat transfer coefficient at the product surface in the dryer is 500 W/m^{2}K and the mass diffusivity for vapor within the product is 1.7×10^{-7} m^2/s. The specific heat of product solids ($c_{psolids}$) is 2.0 kJ/kgK. Estimate the following:

a. The amount of final product leaving the dryer at 4.0% moisture.

b. The temperature and RH of air leaving the dryer.

c. The thermal energy for heating the air entering the dryer.

d. The length of the dryer tunnel, if the conveyor carrying the product moves at a rate of 1 m/min.

12.15 Use MATLAB® to evaluate and plot equilibrium desorption and adsorption moistures over the range of humidities ($0.01 < a_w < 0.95$) for rough rice at 25°C predicted by the empirical equations given as follows.

Desorption (Basunia and Abe, 2001):

$$M = C_1 - C_3 \ln[-(T + C_2)\ln(a_w)]; \quad C_1 = 31.652,\ C_2 = 19.498,\ C_3 = 5.274$$

Adsorption (Basunia and Abe, 1999):

$$M = \frac{-1}{b_3}\ln\left[-\left(\frac{T + b_2}{b_1}\right)\ln(a_w)\right]; \quad b_1 = 594.85,\ b_2 = 49.71,\ b_3 = 0.2045$$

M = Moisture content (% dry basis)
T = temperature (°C)
a_W = water activity (decimal)

12.16 Use the MATLAB® nonlinear regression function, *nlinfit*, to find the Guggenheim-Anderson-DeBoer (GAB) parameters (m_0, A, B) for rough rice (desorption) given in the following table.

$$m = \frac{m_0 A B a_w}{(1 - B a_w)(1 - B a_w + A B a_w)}$$

Water activity	Sample 1(%db)	Sample 2 (%db)	Sample 3 (%db)
0.05	6.5	6.2	6.3
0.1	7.5	7.3	7.7
0.2	8.2	8.5	8.4
0.3	10.5	10.5	10.7
0.4	12.2	12.0	12.0
0.5	13.5	13.4	13.7
0.6	15.0	15.1	15.3
0.7	17.0	17.2	17.0
0.8	19.8	19.7	19.4
0.9	23.5	23.3	23.8
0.95	27.4	27.6	27.9

LIST OF SYMBOLS

a_w water activity (dimensionless)
A surface area of product exposed to heated air (m^2)
β first root of the zero-order Bessel function equation
C the Guggenheim constant $= C' \exp(H_1 - H_m)/RT$
c_{pp} specific heat of product solids (kJ/[kg solids K])
c_{pw} specific heat of water (kJ/[kg water K])
c_s humid heat of air (kJ/[kg dry air K]) $= 1.005 + 1.88\,W$ (Eq. 9.17)
D diffusion coefficient (m^2/s)
d_c characteristic dimension (m)
h convective heat transfer coefficient (W/[m^2K])
H_1 heat of condensation of pure water vapor
H_a thermal energy content of air (kJ/[kg dry air])
H_L latent heat of vaporization for water (kJ/kg water)
H_m total heat of sorption of the first layer on primary sites
H_p thermal energy content of product (kJ/kg dry solids)
H_q total heat of sorption of the multilayers
K factor correcting properties of multilayer with respect to the bulk liquid, $= k' \exp(H_1 - H_q)/RT$
k_m convective mass transfer coefficient (m/s) or (kg mol/[s m^2 Pa])
L thickness of product layer (m)
M molecular weight (kg/[kg mol])
$\dot{m}_a$ air flow rate (kg dry air/h)
$\dot{m}_c$ moisture removal rate during constant-rate drying (s^{-1})
$\dot{m}_p$ product flow rate (kg dry solids/h)
P atmospheric pressure (kPa)
P_a vapor pressure of air at the condenser surface (Pa)

P_i	vapor pressure of ice (Pa)
q	rate of heat transfer (W)
R	universal gas constant (8314.41 m^3 Pa/[kg mol K])
R_d	radius of the liquid food droplet
R_p	radius of the product particle at the critical moisture content
t	drying time
T_0	reference temperature at 0°C
T_a	air temperature (°C)
T_A	absolute temperature (K)
t_c	time for constant-rate drying (s)
T_p	product temperature (°C)
T_s	product surface temperature (°C)
V_w	specific volume of water (m^3/kg water)
w	product moisture content, dry basis (kg water/kg dry solids)
w_c	critical moisture content (kg water/kg dry solids)
w_e	equilibrium moisture content, fraction dry basis
w_m	monolayer moisture content, fraction dry basis
w_0	initial moisture content (kg water/kg dry solids)
W	absolute humidity (kg water/kg dry air)
W_a	humidity ratio for air (kg water/kg dry air)
W_s	humidity ratio at product surface (kg water/kg dry air)

Subscript: w, water

BIBLIOGRAPHY

Basunia, M. A. and Abe, T. (1999). Moisture adsorption isotherms of rough rice. *J. Food Engr.* **42**: 235–242.

Basunia, M. A. and Abe, T. (2001). Moisture desorption isotherms of medium grain rough rice. *J. Stored Prod. Res.* **37**: 205–219.

Bizot, H. (1983). Using the GAB model to construct sorption isotherms. In *Physical Properties of Foods*, R. Jowitt, F. Escher, B. Hallstrom, H. Th. Meffert, W. E. L. Spiess, and G. Vos, eds., 43–54. Applied Science Publishers, London.

Charm, S. E. (1978). *The Fundamentals of Food Engineering*, 3rd ed. AVI Publ. Co., Westport, Connecticut.

Eisenhardt, N. H., Cording, J., Jr., Eskew, R. K., and Sullivan, J. F. (1962). Quick-cooking dehydrated vegetable pieces. *Food Technol.* **16**(5): 143–146.

Fish, B. P. (1958). Diffusion and thermodynamics of water in potato starch gel. *Fundamental Aspects of Dehydration of Foodstuffs*, 143–157. Macmillan, New York.

Flink, J. M. (1977). Energy analysis in dehydration processes. *Food Technol.* **31**(3): 77.

Forrest, J. C. (1968). Drying processes. In *Biochemical and Biological Engineering Science*, N. Blakebrough, ed., 97–135. Academic Press, New York.

Gorling, P. (1958). Physical phenomena during the drying of foodstuffs. In *Fundamental Aspects of the Dehydration of Foodstuffs*, 42–53. Macmillan, New York.

Heldman, D. R. and Hohner, G. A. (1974). Atmospheric freeze-drying processes of food. *J. Food Sci.* **39**: 147.

Jason, A. C. (1958). A study of evaporation and diffusion processes in the drying of fish muscle. In *Fundamental Aspects of the Dehydration of Foodstuffs*, 103–135. Macmillan, New York.

Joslyn, M. A. (1963). Food processing by drying and dehydration. In *Food Processing Operations*, M. A. Joslyn and J. L. Heid, eds., Vol. 2, 545–584. AVI Publ. Co., Westport, Connecticut.

King, C. J. (1970). Freeze-drying of foodstuffs. *CRC Crit. Rev. Food Technol.* **1**: 379.

King, C. J. (1973). Freeze-drying. In *Food Dehydration,* W. B. Van Arsdel, M. J. Copley, and A. I. Morgan, Jr, eds., *2nd ed.*, Vol. 1, 161–200. AVI Publ. Co, Westport, Connecticut.

Labuza, T. P. (1968). Sorption phenomena in foods. *CRC Crit. Rev. Food Technol.* **2**: 355.

Labuza, T. P., Kaanane, A., and Chen, J. Y. (1985). Effect of temperature on the moisture sorption isotherms and water activity shift of two dehydrated foods. *J. Food. Sci.* **50**: 385.

Potter, N. N. (1978). *Food Science*, 3rd ed. AVI Publ. Co., Westport, Connecticut.

Ranz, W. E. and Marshall, W. R., Jr (1952). Evaporation from drops. *Chem. Eng. Prog.* **48**: 141–180.

Rockland, L. B. and Nishi, S. K. (1980). Influence of water activity on food product quality and stability. *Food Technol.* **34**(4): 42–51.

Sherwood, T. K. (1929). Drying of solids. *Ind. Eng. Chem.* **21**: 12.

Sherwood, T. K. (1931). Application of theoretical diffusion equations to the drying of solids. *Trans. Am. Inst. Chem. Eng.* **27**: 190.

Van Arsdel, W. B. (1951). Tunnel-and-truck dehydrators, as used for dehydrating vegetables. *USDA Agric. Res. Admin. Pub.* **AIC-308**., Washington, D.C.

Van Arsdel, W. B., Copley, M. J., and Morgan, A. I., Jr., (eds.) (1973). *Food Dehydration, 2nd ed.*, Vol. 1. AVI Publ. Co., Westport, Connecticut.

Chapter 13

Supplemental Processes

In this chapter, various supplemental processes used in the food industry are presented. Some of these processes are ubiquitous in food processing such as mixing. Others such as filtration, sedimentation, and centrifugation are more uniquely applied to specific applications. For each process, we will consider mathematical relationships that are useful in their design and operation.

13.1 FILTRATION

Solid particles are removed from liquids by somewhat different mechanisms than used for removing solid particles from air. The removal process can be described by standard equations. In this section, we will examine some of these equations and discuss specific applications of filtration techniques.

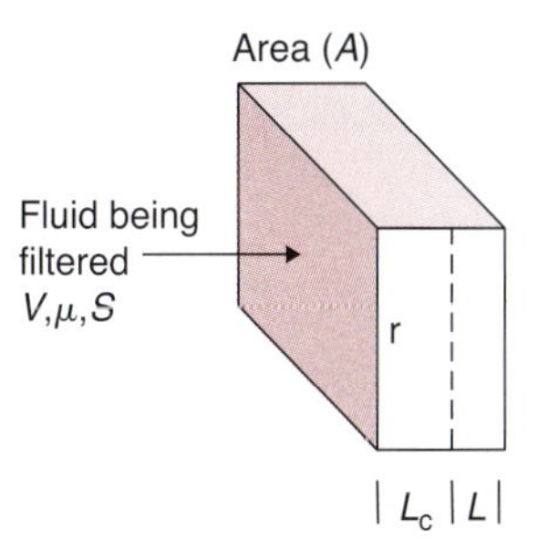

Figure 13.1 Schematic illustration of filtration process.

13.1.1 Operating Equations

In general, the filtration process is described by the manner in which the fluid being filtered flows through the filter medium where the solids are deposited (Fig 13.1). As the solids are removed from the fluid, they accumulate in the filter medium, resulting in an increased resistance to flow as the filtration process continues. All these factors result in a description of filtration rate. In addition, this filtration rate depends on several other factors, including

- The pressure drop across the filter medium
- The area of the filtering surface
- The viscosity of the filtrate
- The resistance of the filter cake as determined by the solids removed from the fluid
- The resistance of the filter medium.

The rate of filtration can be written as follows:

$$\text{Rate of filtration} = \frac{\text{Driving force}}{\text{Resistance}} \tag{13.1}$$

where the driving force is the pressure required to move the fluid through the filter medium and the resistance is dependent on several factors. The overall resistance can be described by the following expression:

$$R = \mu r'(L_c + L) \tag{13.2}$$

where L_c represents the thickness of accumulated solids in the filter cake, μ is the fluid viscosity, and L is a fictitious thickness of the filter material or medium. The parameter r' in Equation (13.2) represents the specific resistance of the filter cake and it is a property of the particles forming the filter cake. Earle (1983) describes L_c by the following expression:

$$L_c = \frac{SV}{A} \tag{13.3}$$

where S is the solids content of the fluid being filtered, and V is the volume that has passed through the filter with cross-sectional area (A).

The thickness of the filter cake represents a fictitious value to describe the total thickness of all solids accumulated. In some filtration processes, this may approach the real situation. Utilizing Equations (13.2) and (13.3), the total resistance can be written in the following manner:

$$R = \mu r' \left(\frac{SV}{A} + L \right) \tag{13.4}$$

Combining Equations (13.1) and (13.4), an expression for the rate of filtration is obtained as follows:

$$\frac{dV}{dt} = \frac{A\Delta P}{\mu r' \left(\frac{SV}{A} + L \right)} \tag{13.5}$$

Equation (13.5) is an expression used to describe the filtration process and can be used for scale-up if converted to appropriate forms.

The filtration process may occur in two phases: (1) constant-rate filtration, normally occurring during the early stages of the process, and (2) constant-pressure filtration, occurring during the final stages of the process.

13.1.1.1 Constant-Rate Filtration

Constant-rate filtration is described by the following integrated forms of Equation (13.5), which can be used to determine the required pressure drop as a function of the filtration rate:

$$\frac{V}{t} = \frac{A\Delta P}{\mu r'\left(\frac{SV}{A} + L\right)} \tag{13.6}$$

or

$$\Delta P = \frac{\mu r'}{A^2 t}\left[SV^2 + LAV\right] \tag{13.7}$$

Equation (13.6) can be expressed in a different form if the thickness (L) of the filter medium is considered negligible. The following equation for pressure drop as a function of time is obtained

$$\Delta P = \frac{\mu r' S V^2}{A^2 t} \tag{13.8}$$

In many situations, Equation (13.8) can be used to predict pressure drop requirements for a filter during the early stages of the process.

Example 13.1

An air filter is used to remove small particles from an air supply to a quality control laboratory. The air is supplied at a rate of 0.5 m^3/s through an air filter with a 0.5 m^2 cross-section. If the pressure drop across the filter is 0.25 cm water after one hour of use, determine the life of the filter if a filter change is required when the pressure drop is 2.5 cm of water.

Given

Volumetric flow rate of air = 0.5 m^3/s
Area of filter = 0.5 m^2
Pressure drop = 0.25 cm water
Time of use = 1 hr
Pressure drop before filter change = 2.5 cm of water

Approach

We will use Equation (13.8) to determine the product of specific resistance and particle content of air. Using this calculated value and Equation (13.8) with a pressure drop of 2.5 cm of water we will obtain the time when the filter should be changed.

Solution

1. *Due to the nature of air filtration, constant-rate filtration can be assumed and the filter medium thickness is negligible.*
2. *Based on the information given and Equation (13.8), the product of the specific resistance (r′) and the particle content of air (S) can be established*

$$r'S = \frac{\Delta P A^2 t}{\mu V^2}$$

3. *The pressure drop of 0.25 cm of water can be expressed in consistent units (Pa) by*

$$0.25 \text{ cm of water} = 24.52 \text{ Pa}$$

4. *Using Equation (13.8) (for 1 hr of filtration)*

$$r'S = \frac{(24.52\,\text{Pa})(0.5\text{m}^2)^2(3600\,\text{s})}{(1.7\times10^{-5}\,\text{kg/ms})(0.5\times3600)^2}$$

where $\mu = 1.7 \times 10^{-5}$ (viscosity of air at 0°C from Table A.4.4)

$$r'S = 396/\text{m}^2$$

5. *Using $r'S = 396$ and Equation (13.8), the time required for ΔP to increase to 2.5 cm water can be computed.*
6. *Converting ΔP to consistent units*

$$2.5 \text{ cm of water} = 245.16 \text{ Pa}$$

then

$$\frac{t}{V^2} = \frac{(1.7\times10^{-5})(396)}{(0.5)^2(245.16)} = 1.11\times10^{-4}\,\text{s/m}^6$$

7. *Since the volumetric flow rate through the filter is $0.5\,\text{m}^3/\text{s}$*

$$\frac{V}{t} = 0.5\text{m}^3/\text{s}$$

Therefore

$$V^2 = 0.25t^2$$

or

$$t = 1.11 \times 10^{-4} \times 0.25t^2 = 2.778 \times 10^{-5} t^2$$

and

$$t = \frac{1}{2.775 \times 10^{-5}} = 35{,}997 s$$

$$t = 10\ hr$$

8. *Thus the filter should be changed after 10 hours of operation.*

13.1.1.2 *Constant-Pressure Filtration*

An expression for describing constant-pressure filtration can be obtained from the following form of Equation (13.5):

$$\frac{\mu r' S}{A} \int_0^V V dV + \mu r' L \int_0^V dV = A \Delta P \int_0^t dt \qquad (13.9)$$

Integration leads to the following design equation:

$$\frac{tA}{V} = \frac{\mu r' S V}{2 \Delta P A} + \frac{\mu r' L}{\Delta P} \qquad (13.10)$$

or the following equation if filter medium thickness (L) can be assumed to be negligible:

$$t = \frac{\mu r' S V^2}{2 A^2 \Delta P} \qquad (13.11)$$

Essentially, Equation (13.11) indicates the time required to filter a given volume of fluid when a constant pressure is maintained. Various procedures are followed in using this equation to obtain information that is not readily available. For example, the specific resistance of the filter cake (r') may not be known for some types of solids and must be determined experimentally. Earle (1983) presented procedures for determining these parameters, and Charm (1978) discussed determination of filtration constants that are more complex than those proposed in the previous presentation.

Example 13.2

A liquid is filtered at a pressure of 200 kPa through a 0.2-m^2 filter. Initial results indicate that 5 min is required to filter 0.3 m^3 of liquid. Determine the time that will elapse until the rate of filtration drops to 5×10^{-5} m^3/s.

Given

Pressure drop = 200 kPa
Filter area = 0.2 m^2
Time = 5 min
Volume of liquid = 0.3 m^3
Rate of filtration = 5×10^{-5} m^3/s

Approach

We will use Equation (13.11) to determine $\mu r'S$. Next we will use Equation (13.5) to obtain the volume followed by calculation of time using Equation (13.11).

Solution

1. *Since the filtration is assumed to be in the constant-pressure regime, based on data obtained at 5 min, Equation (13.11) will apply as follows*

$$\mu r'S = \frac{2A^2 \Delta P t}{V^2} = \frac{2(0.2)^2(200{,}000)(5\times 60)}{0.3^2}$$

$$\mu r'S = 53.33\times 10^6 \text{ kg/m}^3\text{s}$$

2. *Using Equation (13.5) (where L is assumed to be negligible)*

$$5\times 10^{-5} = \frac{(0.2)^2(200{,}000)}{(53.33\times 10^6)V}$$

$$V = 3 \text{ m}^3$$

3. *Using Equation (13.11):*

$$t = \frac{(53.33\times 10^6)(3)^2}{2(0.2)^2(200{,}000)} = 29{,}998 \text{ s}$$

t = 499.9 min; indicates that 500 min of filtration at a constant pressure would occur before the rate dropped to 5×10^{-5} m^3/s.

Reference to Equation (13.5) reveals that the filtration process is directly dependent on two factors: the filter medium and the fluid being filtered. In Equation (13.5) a filter medium is described in terms of area (A) and the specific resistance (r'). The filter medium will depend considerably on the type of fluid being filtered. In the case of liquid filtration, the filter medium, to a large extent, will contain the solids removed from the liquid. This filter cake must be supported by some type of structure that plays only a limited role in the filtration process. In some cases, these supporting materials may be woven (wool, cotton, linen) or they may be granular materials for particular types of liquids. In any case, the primary role of the material is to support the collected solids so that the solids can act as a filter medium for the liquid.

The design of filters for the removal of particles from air is significantly different. In this application, the entire filter medium is designed into the filter and the collected solids play a very minor role in the filtration process. The filter medium is a porous collection of filter fibers of the same magnitude in size as the particles to be removed from the air. This results in filtration processes that are nearly constant-flow rate in all situations.

The second factor of Equation (13.5) that influences filtration rate is the fluid being filtered as described by the fluid viscosity (μ). The rate of filtration and the viscosity of the fluid being filtered are inversely related; as the viscosity of the fluid increases, the rate of filtration must decrease. Fluid viscosity plays a very important role in the filtration process and must be accounted for in all design computations.

13.1.2 Mechanisms of Filtration

The mechanisms involved in removing small particles from air are relatively well-defined. Whitby and Lundgren (1965) have listed four mechanisms as follows: (1) Brownian diffusion, (2) interception, (3) inertial impaction, and (4) electrical attraction. Decker et al. (1962) indicated that deposition according to Stoke's law could be considered as an additional mechanism. The mechanism of Brownian diffusion will have very little influence on removal of particles larger than 0.5 micron. This particular mechanism contributes to the particle collection by causing particles to deviate from the streamline of air flow around the filter fibers and bringing them into contact with the fiber. Somewhat larger particles may be removed by interception even when the particles do not deviate from the air streamline.

When these particles follow the air streamline and the streamline brings them sufficiently close to the fiber, the particles will be removed by direct interception. The larger particles will not follow the air streamline and will be removed by inertial impaction. Particles larger than 1 micron will normally be removed by this mechanism. When the particles and fiber have different or opposite electrical charges, the particles will deviate from the air streamline and deposit on the fiber as a result of this mechanism. Very large particles may be influenced by gravity and deviate from the air streamline due to gravitational force, resulting in deposition on the filter fiber. The contribution of this mechanism to small-particle removal is probably very small.

In the case of removing solids from liquids, the mechanisms of removal are not well-defined and may be considerably different, depending on the mode of filtration considered. During the initial stages when the liquid is moving through the filter medium at a rapid rate, the mechanisms are most likely to be direct interception or inertial impaction. After the filter cake is established and constant-pressure filtration occurs, the liquid will flow through the filter cake in a streamline fashion.

13.1.3 Design of a Filtration System

Although the expressions presented for describing filtration rate by a given system are relatively straightforward, use of these equations requires knowledge of several system parameters. In most cases, the viscosity and specific resistance values may not be known. To obtain some indication of the relationships involved, the approach normally followed is to make a small-scale filtration experiment followed by a scale-up of the entire process. Expressions of the type presented as Equations (13.7) and (13.11) are ideal for this purpose. A small or pilot-scale operation usually results in determination of the filtrate volume after given periods of time at a constant pressure on the small-scale filter.

Earle (1983) showed that this information can be used to determine a plot of the type given in Figure E13.1. To obtain the filtration graph, we use Equation (13.10), in which L is not assumed to be negligible. By using pilot-scale data to obtain a relationship for Figure E13.1, we can evaluate appropriate constants for use in the scale-up of the filtration process. In some situations, it is desirable to collect pilot-scale data at different pressures and/or different filter areas.

Example 13.3

A filtration system is designed to filter 4 m^3 of a slurry in 2 hr using a constant pressure of 400 kPa. The necessary design conditions were established on a laboratory scale using a filter with 0.1 m^2 surface area and 140 kPa constant pressure. The following results were obtained on a laboratory scale:

Time (min)	Filtration Volume ($\times 10^{-2}$ m^3)
10	2.3
20	3.7
30	4.9
40	6.1
50	6.8

Determine the filter area required to provide the desired conditions for the design situation.

Given

Volume = 4 m^3
Time = 2 hour
Pressure = 400 kPa
Laboratory filter area = 1 m^2
Pressure in laboratory scale experiment = 140 kPa

Approach

We will first plot tA/V vs. V/A and obtain slope and intercept. This will provide us with a design equation that we will use to determine the required area.

Solution

1. *Using Equation (13.10), a plot of tA/V versus V/A should provide a linear relationship with the slope equal to $\mu r' S/\Delta P$ and a vertical axis intercept of $\mu r' L/\Delta P$.*
2. *From the experimental data provided*

$V/A \times 10^{-2}$	tA/V
23	43.48
37	54.05
49	61.22
61	65.57
68	73.53

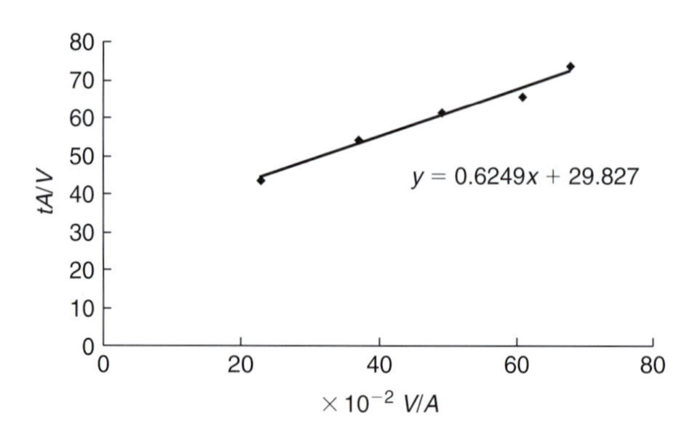

Figure E13.1 A graph of data obtained from operating pilot-scale filtration equipment.

3. From the results presented in Figure E13.1, a slope of 62.5 and a vertical intercept of 29.83 provide the design equation

$$\frac{tA}{V} = 62.5\frac{V}{A} + 29.83$$

4. Since pressure and area are the two variables existing between the laboratory and design situation, appropriate changes must be incorporated

$$\text{Slope} = 62.5 = \frac{\mu r' S}{2(140{,}000)}; \quad \mu r' S = 1.75 \times 10^7$$

$$\text{Intercept} = 29.83 = \frac{\mu r' L}{140{,}000}; \quad \mu r' L = 4{,}176{,}200$$

Thus the design equation becomes

$$\frac{tA}{V} = \frac{1.75 \times 10^7}{\Delta P}\left(\frac{V}{A}\right) + \frac{4{,}176{,}200}{\Delta P}$$

5. For the filtration system being designed: $t = 2\,\text{hr} = 120$ min.; $V = 4\ m^3$; $\Delta P = 400$ kPa.

Then

$$\frac{120A}{4} = \frac{1.75 \times 10^7}{400{,}000}\left(\frac{4}{A}\right) + \frac{4{,}176{,}200}{400{,}000}$$

$$30A = \frac{175}{A} + 10.44$$

$$A = 2.6\ m^2$$

6. An area of $2.6\,m^2$ is obtained when the positive solution of the quadratic equation is selected.

13.2 SEDIMENTATION

Sedimentation is the separation of solids from fluid streams by gravitational or centrifugal force. In the food industry, most processes involving sedimentation are used to remove particle solids from either liquid or gas. As will become obvious, the use of gravity in particular for removing solids from fluids has considerable application.

13.2.1 Sedimentation Velocities for Low-Concentration Suspensions

Particles in a low-concentration suspension will settle at a rate representing the terminal velocity of the particle in the suspension fluid.

The terminal velocity of each particle is established as though each individual particle is the only particle in the suspension. This type of sedimentation is usually referred to as free settling, since there is no interaction between particles and the suspension. The terminal velocity of the particle can be predicted and established by examining the forces acting on the particle. The force that resists the gravitational force is referred to as the drag force and can be described by the following expression:

$$F_D = 3\pi\mu d u \tag{13.12}$$

where u represents the relative velocity between the particle and the fluid, d is particle diameter, and μ is fluid viscosity. Equation (13.12) applies as long as the particle Reynolds number is less than 0.2, which will be the case for most applications in food processing. Expressions to use when the particle Reynolds number is greater than 0.2 have been developed and are presented by Coulson and Richardson (1978). The gravitational force (F_G) is a function of particle volume and density difference along with gravitational acceleration, as illustrated by the following equation:

$$F_G = \frac{1}{6}\pi d^3(\rho_p - \rho_f)g \tag{13.13}$$

where d is the particle diameter, assuming a spherical shape. By setting Equation (13.12) equal to Equation (13.13) (the conditions that must exist at the terminal velocity of the particle), we can solve and obtain an equation representing the terminal velocity as follows:

$$u_t = \frac{d^2 g}{18\mu}(\rho_p - \rho_f) \tag{13.14}$$

From Equation (13.14), we can see that the terminal velocity is related directly to the square of the particle diameter. In addition, the terminal velocity is dependent on the density of the particle and properties of the fluid. Equation (13.14) is the most common form of Stoke's law and should be applied only for streamline flow and spherical particles. When the particles are not spherical, the usual approach is to introduce a shape factor to account for the irregular shape of the particle and the corresponding influence of this factor on terminal velocity.

When streamline or laminar flow does not exist at Reynolds numbers above 1000, the conditions must be modified to account for the drag forces. For these conditions, the equation for terminal velocity becomes:

$$u_t = \sqrt{\left(\frac{4dg\,(\rho_p - \rho_g)}{3C_d\,\rho_f}\right)} \tag{13.15}$$

where the drag coefficient C_d is a function of the Reynolds Number, as illustrated in Figure 13.2.

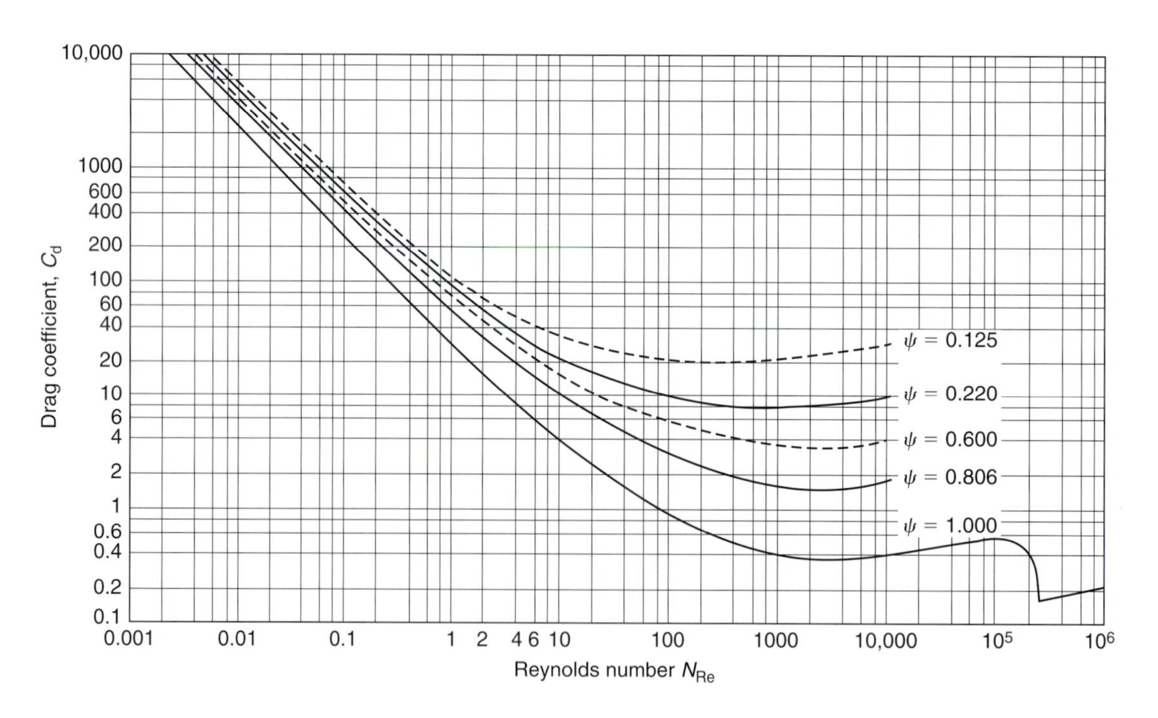

Figure 13.2 Influence of the Reynolds number on drag coefficient for spheres. ψ represents sphericity. (Adapted from Foust et al., 1960)

Example 13.4

The damage to blueberries and other fruits during handling immediately after harvest is closely related to the terminal velocity in air. Compute the terminal velocity of a blueberry with a diameter of 0.60 cm and density of 1120 kg/m^3 in air at 21°C and atmospheric pressure.

Given

Diameter of blueberries = 0.6 cm

Density = 1120 kg/m^3

Air temperature = 21°C

Approach

We will use Equation (13.14) to determine the terminal velocity. Next, we will check the validity of the equation by calculating the Reynolds number. If the equation is not valid then we will use Figure 13.2 and Equation (13.15) to obtain drag coefficient and recalculate terminal velocity.

Solution

1. *Equation (13.14) can be used with the following parameter values:* $d = 0.60\,cm$, $g = 9.806\,m/s^2$, $\mu = 1.828 \times 10^{-5}\,kg/ms$, $\rho_p = 1120\,kg/m^3$, $\rho_f = 1.2\,kg/m^3$.

2. *Using Equation (13.14)*

$$u_t = \frac{(0.006)^2(9.806)}{(18)(1.828 \times 10^{-5})}[1120 - 1.2] = 1200\ m/s$$

3. *A check of the particle Reynolds number*

$$N_{Re} = \frac{(1.2)(0.006)(1200)}{1.828 \times 10^{-5}} = 4.73 \times 10^5$$

indicates that streamline flow does not exist and conditions for using Equation (13.14) do not apply.

4. *By using Equation (13.15), and the* $C_d = 0.2$ *from Figure 13.2, obtained at* $N_{Re} = 4.73 \times 10^5$

$$u_t = \left[\frac{4}{3}\frac{(0.006)(9.806)}{(0.2)(1.2)}(1120 - 1.2)\right]^{1/2} = 19.1\ m/s$$

5. *A check of the Reynolds number*

$$N_{Re} = \frac{(1.2)(0.006)(19.1)}{1.828 \times 10^{-5}} = 7523$$

indicates that $C_d = 0.4$ *and*

$$u_t = \left[\frac{4}{3}\frac{(0.006)(9.806)}{(0.4)(1.2)}(1120 - 1.20)\right]^{1/2} = 13.5 \text{ m/s}$$

6. *A final check of the Reynolds number*

$$N_{Re} = \frac{(1.2)(0.006)(13.5)}{1.828 \times 10^{-5}} = 5317$$

indicates that $C_d = 0.4$ *is appropriate and the terminal velocity is 13.5 m/s when the spherical geometry is assumed.*

13.2.2 Sedimentation in High-Concentration Suspensions

When the concentration of solids in the suspension becomes sufficiently high, Stoke's law no longer describes the velocity of settling. In situations where the range of particle sizes is between 6 and 10 microns, the particles making up the solid suspension will fall at the same rate. This rate will correspond to the velocity predicted by Stoke's law, using a particle size that is the mean of the smallest and the largest particle size in the suspension. Physically it is apparent that the rate at which the larger particles settle is reduced by the influence of the smaller particles on the properties of the fluid. The rate at which the smaller particles fall is accelerated by the movement of the larger particles.

The settling of particles results in very well-defined zones of solids concentrations. The movement of the solids at a constant rate results in a clear liquid at the top of the column suspension. Directly below the clear liquid is the concentration of solids, which is moving at a constant rate. Below the solids that are settling is a zone of variable concentration within which the solids are collecting. At the bottom of the suspension column is the collected sediment, which contains the largest particles in the suspension. The definition of these zones will be influenced by the size of particles involved and the range of these sizes.

The rates at which solids settle in high-concentration suspensions have been investigated on an empirical basis and have been reviewed by Coulson and Richardson (1978). One approach is to modify Stoke's law by introducing the density and viscosity of the suspension in place of the density and viscosity of the fluid. This may result in an equation for the settling rate of the suspension as follows:

$$u_s = \frac{KD^2(\rho_p - \rho_s)g}{\mu_s} \tag{13.16}$$

where K is a constant to be evaluated experimentally. In most cases, attempts are made to predict the density and viscosity of the suspension on the basis of the composition. Another approach is to account for the void between suspension particles, which allows movement of fluid to the upper part of the suspension column. The expression for the particle settling velocity was obtained as follows:

$$u_p = \frac{D^2(\rho_p - \rho_s)g}{18\mu} f(e) \tag{13.17}$$

where $f(e)$ is a function of the void in a suspension. Equation (13.17) is still a modification of Stoke's law, where the density of the suspension and the viscosity of the fluid are used. The function of the void space in the suspension must be determined experimentally for each situation in which the expression is utilized. Although there are definite limitations to Equations (13.16) and (13.17), they will provide practical results for high-concentration suspensions of large particles. Description of settling rates for small-particle suspensions is subject to considerable error and acceptable approaches need to be developed.

Example 13.5

A sedimentation tank is used to remove larger-particle solids from wastewater leaving a food-processing plant. The ratio of liquid mass to solids in the inlet to the tank is 9 kg liquid/kg solids; the inlet flow rate is 0.1 kg/s. The sediment leaving the tank bottom should have 1 kg liquid/kg solids. Density of water is 993 kg/m^3. If the sedimentation rate for the solids in water is 0.0001 m/s, determine the area of sedimentation necessary.

Given

Ratio of liquid mass to solids in the inlet = 9 kg liquid/kg solids
Inlet flow rate = 0.1 kg/s
Sediment composition = 1 kg liquid/kg solids
Sedimentation rate of solids in water = 0.0001 m/s

Approach

We will first obtain an upward velocity of the liquid in the tank. Then using the given information we will obtain the area when the upward liquid velocity is equal to the sedimentation velocity for solids.

Solution

1. *Using the mass flow equation and the difference between inlet and outlet conditions on the wastewater, the following equation for upward velocity of liquid in the tank is obtained*

$$u_f = \frac{(c_i - c_o)w}{A\rho}$$

where c_i and c_o are mass ratios for liquid to solids in waste water.

2. *Based on information given*

$$c_i = 9 \qquad w = 0.1$$
$$c_o = 1 \qquad \rho = 993$$

3. *Since the upward liquid velocity should equal sedimentation velocity for the solids, the area becomes*

$$A = \frac{(9-1)(0.1)}{(0.0001)(993)} = 8\ m^2$$

In addition to the properties of the solid and fluid in the suspension, other factors influence the sedimentation process. The height of the suspension generally will not affect the rate of sedimentation or the consistency of the sediment produced. By experimentally determining the height of the liquid or fluid-sediment interface as a function of time for a given initial height, the results for any other initial height can be predicted. The diameter of the sedimentation column may influence the rate of sedimentation if the ratio of the column diameter to the particle diameter is less than 100. In this particular situation, the walls may have a retarding influence on sedimentation rate. In general, the concentration of the suspension will influence the rate of sedimentation, with higher concentrations tending to reduce the rate at which the sediment settles.

It is possible to use sedimentation to remove solid particles from air, such as after spray-drying processes. By introducing the particle suspension into a static air column, the solid particles tend to settle to the floor surface. Relatively simple, straightforward computations using expressions presented previously in this chapter will illustrate that such sedimentation processes are quite slow for particles of the type normally encountered. Because of this, the sedimentation is not used often in this regard. Procedures using other forces on the particle to accelerate removal are preferred and will be discussed more thoroughly in the next section.

13.3 CENTRIFUGATION

In many processes, the use of sedimentation to separate two liquids or a liquid and a solid does not progress rapidly enough to accomplish separation efficiently. In these types of applications the separation can be accelerated through the use of centrifugal force.

13.3.1 Basic Equations

The first basic equation describes the force acting on a particle moving in a circular path as follows:

$$F_c = \frac{mr\omega^2}{g_c} \tag{13.18}$$

where ω represents the angular velocity of the particle. Since ω can be expressed as the tangential velocity of the particle and its radial distance to the center of rotation (r), Equation (13.18) can be presented as:

$$F_c = \left(\frac{m}{g_c}\right)\frac{u^2}{r} \tag{13.19}$$

Earle (1983) illustrated that if the rotational speed is expressed in revolutions per minute, Equation (13.19) can be written as:

$$F_c = 0.011\frac{mrN^2}{g_c} \tag{13.20}$$

where N represents the rotational speed in revolutions per minute.

Equation (13.20) indicates that the centrifugal force acting on a particle is directly related to and dependent on the distance of the particle from the center of rotation r, the centrifugal speed of rotation N, and the mass of the particle considered m. For example, if a fluid containing particles of different densities is placed in a rotating bowl, the higher-density particles will move to the outside of the bowl as a result of the greater centrifugal force acting upon them. This will result in a movement of the lower-density particles toward the interior portion of the bowl, as illustrated in Figure 13.3. This principle is used in the separation of liquid food products that contain components of different densities.

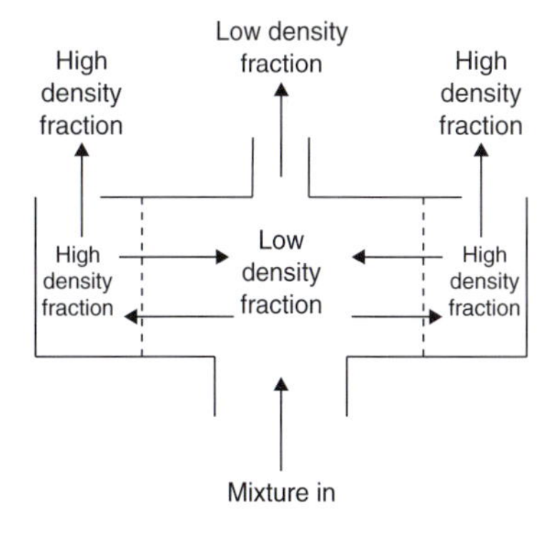

Figure 13.3 Separation of fluids with different densities using centrifugal force.

13.3.2 Rate of Separation

The rate at which the separation of materials of different densities can occur is usually expressed in terms of the relative velocity between the two phases. An expression for this velocity is the same as the equation for terminal velocity given as Equation (13.14), in which the gravitational acceleration (g) is replaced by an acceleration parameter describing the influence of centrifugal force. This acceleration can be expressed in the following way:

$$a = r\left(\frac{2\pi N}{60}\right)^2 \tag{13.21}$$

where N is the rotational speed of the centrifuge in revolutions per minute. Substitution of Equation (13.21) into Equation (13.14) results in the following expression to describe the velocity of spherical particles in a centrifugal force field:

$$u_c = \frac{D^2 N^2 r(\rho_p - \rho_s)}{1640\mu} \tag{13.22}$$

Equation (13.22) can be applied to any situation in which there are two phases with different densities. The expression clearly illustrates that the rate of separation as expressed by velocity (u_c) is directly related to the density between phases, the distance from the center of rotation, the speed of rotation, and the diameter of the particles in the higher density phase.

Example 13.6

The solid particles in a liquid-solid suspension are to be separated by centrifugal force. The particles are 100 microns in diameter with a density of 800 kg/m^3. The liquid is water with a density of 993 kg/m^3, and the effective radius for separation is 7.5 cm.If the required velocity for separation is 0.03 m/s, determine the required rotation speed for the centrifuge.

Given

Particle diameter = 100 microns
Density = 800 kg m^3
Density of water = 993 kg/m^3
Effective radius for separation = 7.5 cm
Required velocity for separation = 0.03 m/s

Approach

We will use Equation (13.22) to obtain the rotational speed of the cartridge.

Solution

1. Using Equation (13.22) with the viscosity of water at 5.95×10^{-4} kg/ms

$$N^2 = \frac{1640\mu u_c}{D^2 r(\rho_p - \rho_s)} = \frac{1640(5.95 \times 10^{-4})}{(0.075)(100 \times 10^{-6})^2(993 - 800)}(0.03)$$

$$N^2 = 2.02 \times 10^5\ (1/s^2)$$

$$N = 26{,}940\ \text{rpm}$$

13.3.3 Liquid-Liquid Separation

In the case of separation involving two liquid phases, it is usually easier to describe the process in terms of the surface that separates the two phases during separation. The differential centrifugal force acting on an annulus of the liquid in the separation cylinder can be written as:

$$dF_c = \frac{dm}{g_c} r\omega^2 \tag{13.23}$$

where (dm) represents the mass in the annulus of liquid. Equation (13.23) can be rewritten as:

$$\frac{dF_c}{2\pi rb} = dP = \frac{\rho\omega^2 r dr}{g_c} a \tag{13.24}$$

where (dP) represents the differential pressure across the annulus of liquid and b represents the height of the separation bowl. By integration of Equation (13.24) between two different radii in the separation cylinder, the difference in pressure between these two locations can be computed from the following expression:

$$P_2 - P_1 = \frac{\rho\omega^2(r_2^2 - r_1^2)}{2g_c} \tag{13.25}$$

At some point in the cylinder, the pressure of one phase must equal that of the other phase so that expressions of the type given by Equation (13.25) can be written for each phase and represent the radius of equal pressure in the following manner:

$$\frac{\rho_A\omega^2(r_n^2 - r_1^2)}{2g_c} = \frac{\rho_B\omega^2(r_n^2 - r_2^2)}{2g_c} \tag{13.26}$$

By solving for the radius of equal pressures for the two phases, the following expression is obtained:

$$r_n^2 = \left[\frac{(\rho_A r_1^2 - \rho_B r_2^2)}{\rho_A - \rho_B}\right] \tag{13.27}$$

where ρ_A equals the density of the heavy liquid phase and ρ_B is the low-density liquid phase. Equation (13.27) is a basic expression for use in the design of the separation cylinder. The radius of equal pressure, which represents the radius at which the two phases may be or are separated is dependent on two radii (r_1 and r_2). These two values can be varied independently to provide optimum separation of the two phases involved and will account for the density of each phase as illustrated.

Example 13.7

Design the inlet and discharge for a centrifugal separation of cream from whole milk. The density of the skim milk is 1025 kg/m^3. Illustrate the discharge conditions necessary if the cream outlet has a 2.5-cm radius and the skim milk outlet has a 5-cm radius. Suggest a desirable radius for the inlet. Cream density is 865 kg/m^3.

Given

Density of milk = 1025 kg/m^3
Cream outlet radius = 2.5 cm
Skim milk outlet radius = 5 cm
Cream density = 865 kg/m^3

Approach

We will use Equation (13.27) to obtain the radius of the neutral zone and then we will use this information to develop a diagram to illustrate the design.

Solution

1. *Using Equation (13.27), the radius of the neutral zone can be computed:*

$$r_n^2 = \frac{1025(0.05)^2 - 865(0.025)^2}{1025 - 865}$$

$$r_n^2 = 0.0126$$

$$r_n = 0.112 \text{ m} = 11.2 \text{ cm}$$

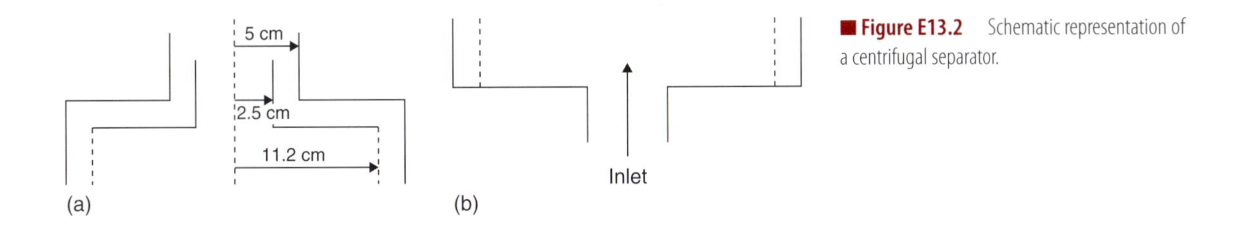

Figure E13.2 Schematic representation of a centrifugal separator.

2. *Based on information given and the computation of neutral zone radius, the separator discharge should be designed as shown in Figure E13.2a.*
3. *Similarly the inlet should be designed to allow the product to enter with least disturbance possible to the neutral zone (Fig. E13.2b).*

13.3.4 Particle-Gas Separation

The separation of solids from a gas phase is a common operation in many food-processing operations. Probably the most common is the separation of a spray-dried product from the air stream after the drying operation is completed. This is usually accomplished in what is known as a cyclone separator. The basic equations presented earlier in this section will apply, and Equation (13.22) will provide some indication of the rate at which separation of solid particles from an air stream can be accomplished.

It is obvious from this expression that the diameter of the particles must be known, along with the density of the solid and the density of the air stream.

13.4 MIXING

Mixing is a common operation in the food industry. A mixing device is required whenever an ingredient is added to a food. The type of device used is usually based on the properties of the food undergoing mixing. Because the quality of a mixed product is judged by the even distribution of its ingredients, the design of the mixing device and its operation must be chosen carefully to accomplish the desired results.

In industrial applications, agitation is used for mixing. Agitation is defined as the application of mechanical force to move a material in a circulatory or similar manner in a vessel. Mixing may involve two or more materials of the same or different phases. Mechanical force is used in mixing to cause one material to randomly distribute in another. Examples of mixing include mixing a solid in a fluid, a solid in a solid, or a fluid in a fluid, where a fluid may be a liquid or a gas.

The difference between the terms "mixing" and "agitation" can be understood from the following. Mixing always requires two or more materials or two or more phases of the same material. For example, we can *agitate* a vessel full of cold water, but in *mixing* we would add hot water to cold water to raise the temperature of the mixture, or we would add a material of a different phase such as sugar into water.

When two or more materials are mixed, the goal of mixing is to obtain a homogenous final mixture. However, "homogeneity" depends on the sample size taken from a mixture. Mixing two liquids such as corn syrup and water may yield a greater homogeneity than mixing two solids of different particulate size, such as raisins mixed in flour. Obviously, the sample size taken from a mixture containing raisins and flour must be larger than the size of a single raisin or the mixture will be devoid of raisins.

In food processing, we agitate liquid foods for a variety of reasons such as to blend one liquid into another miscible liquid (corn syrup and water), disperse gas in a liquid such as in carbonation, or create emulsions such as mayonnaise where one liquid is immiscible in another. Many fragile foods require mixing when ingredients are added to them while taking care to prevent any physical damage to the product (e.g., potato salad, macaroni salad).

Mixing operations in the food industry vary in their complexity. Simple systems involve mixing two miscible liquids of low viscosity, whereas mixing gum into a liquid can become complicated as the viscosity of the mix changes during mixing. When small quantities of one material are mixed into a bulk, such as spices into flour, the mixer must ensure that the entire bulk contents are well mixed. In mixing of many free-flowing particulate materials such as breakfast cereals, there is a tendency of the materials being mixed to actually segregate, counteracting the desired objective.

Mixing processes in the industry are carried out either in bulk or continuous mode. Although continuous systems are smaller in size and desirable for minimizing variations between runs, they require materials that have appropriate flow characteristics. In batch systems, there is more variation between runs during mixing. Furthermore, batch units require more labor but they are easy to modify simply by changing impellers.

13.4.1 Agitation Equipment

A typical vessel used for agitation is shown in Figure 13.4. The vessel may be open or closed and at times may be operated under vacuum. Inside the vessel, a shaft along the middle axis of the vessel is connected at the top to a motor. When heating or cooling is desired during agitation, the vessel is surrounded with a jacket containing a circulating heat transfer medium. An impeller is installed at the bottom of the shaft. In some applications, the shaft may be located off the central axis. In other applications, more than one impeller may be installed in the same vessel. The tank bottom is rounded because sharp corners restrict fluid movement. Typical geometric ratios of the impeller size and location with respect to the tank size are shown in Table 13.1.

There are three types of commonly used impellers: propellers, paddles, and turbines.

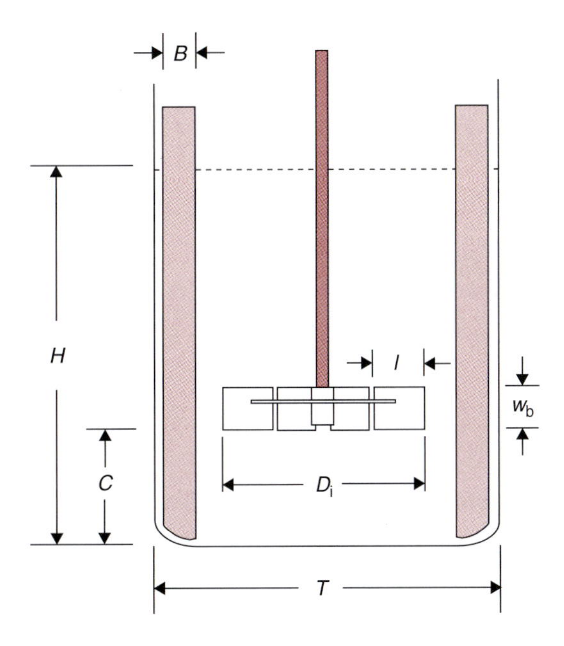

Figure 13.4 Schematic illustration of an agitation vessel with an impeller and baffles.

Table 13.1 Typical Geometric Ratios of Commonly Used Impellers

Ratio	Typical Range	Ratios for a "standard" agitating system
H/T	1–3	1
D_i/T	$^1/_4$–$^2/_3$	$^1/_3$
C/T	$^1/_4$–$^1/_2$	$^1/_3$
C/D_i	~1	1
B/T	$^1/_{12}$–$^1/_{10}$	$^1/_{10}$
w_b/D_i	$^1/_8$–$^1/_5$	$^1/_5$

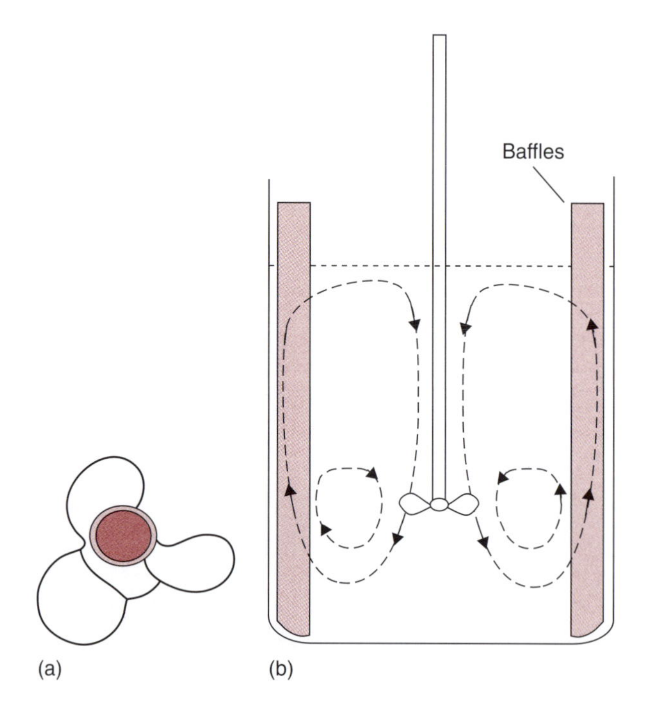

Figure 13.5 Amarine type propeller and flow behavior inside an agitation vessel.

13.4.1.1 *Marine-type Propeller Impellers*

As shown in Figure 13.5, a propeller-type impeller typically has three blades, similar to propellers used in boats for propulsion in water. Propeller-type impellers are used largely for low-viscosity fluids and

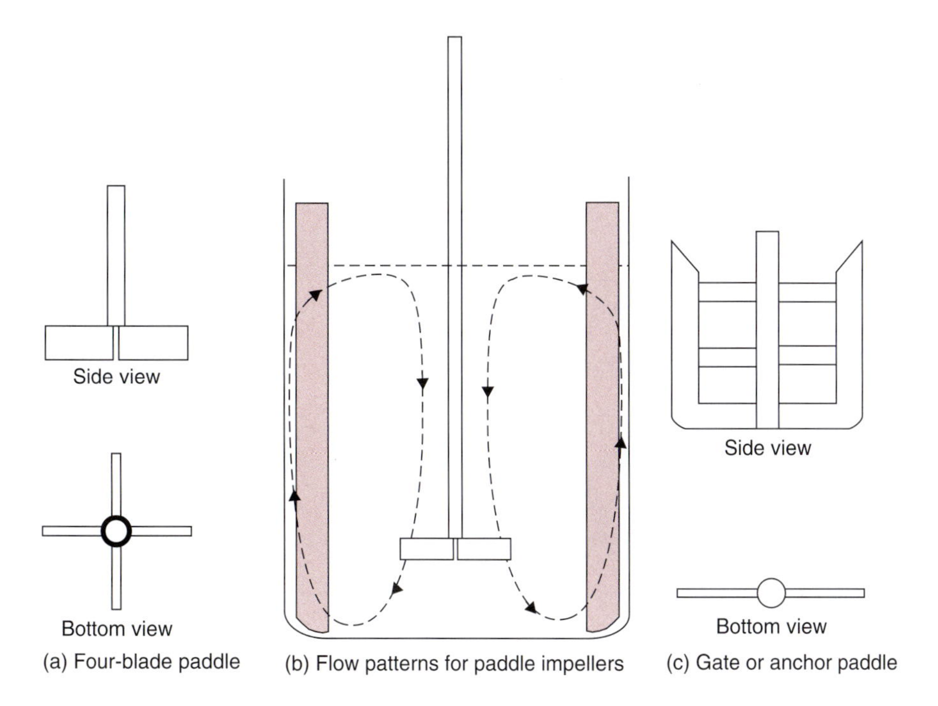

Figure 13.6 Paddle impellers and flow behavior inside an agitation vessel.

are operated at high speeds. The discharge flow in a propeller is parallel to the axis. This type of flow pattern is called axial flow. As shown in Figure 13.5, the fluid moves up along the sides and down along the central axis.

13.4.1.2 *Paddle Impellers*

Paddle impellers usually have either two or four blades (Fig. 13.6). The blades may be flat or pitched at an angle. As the paddles turn, the liquid is pushed in the radial and tangential direction. There is no motion in the vertical direction. These impellers are effective for agitating fluid at low speeds (20 to150 rpm). When paddle impellers are operated at high speeds, the mixing vessel must be equipped with a baffle to prevent the material from moving in a plug-flow pattern. The ratio of impeller-to-tank diameter is in the range of 0.5 to 0.9. In situations requiring scraping of the inside surface of the vessel to minimize fouling when the product is heated, an anchor-type design for the impeller is used, as shown in Figure 13.6.

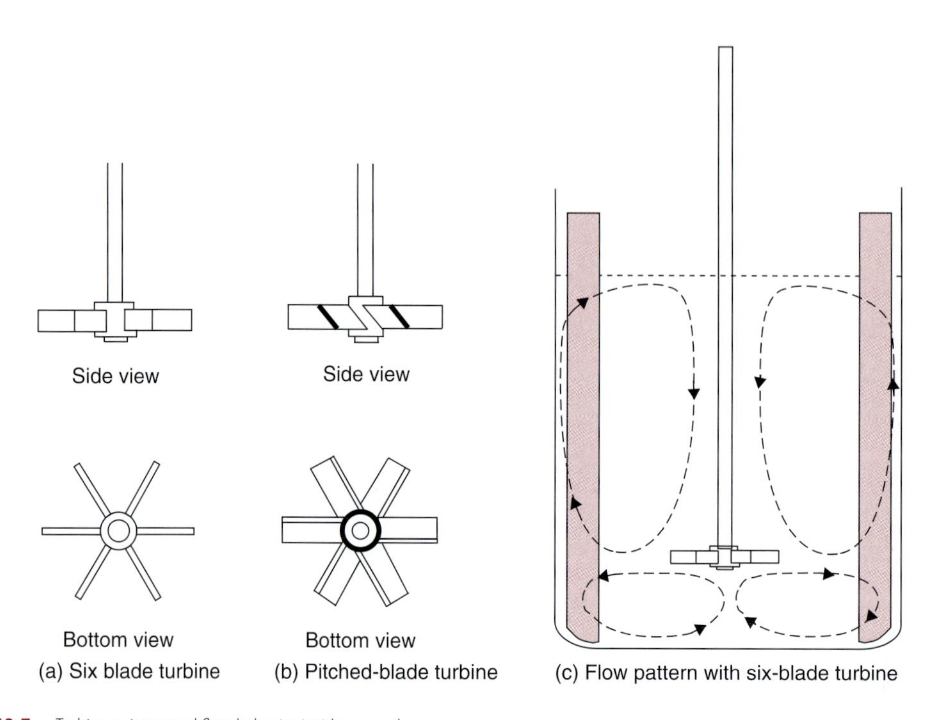

■ **Figure 13.7** Turbine agitator and flow behavior inside a vessel.

13.4.1.3 *Turbine Agitators*

Turbine agitators are similar to paddle impellers equipped with short multiple blades. The diameter of the turbine is usually less than half of the diameter of the vessel. Axial flow is induced when pitched blades are used, as shown in Figure 13.7. Flat-bladed turbines discharge in the radial direction. In curved-bladed turbines, the blade curves away from the direction of rotation. This modification causes less mechanical shear of the product and is more suitable for products with friable solids.

Table 13.2 shows various types of agitators and the recommended range of product viscosities.

13.4.2 Power Requirements of Impellers

In designing agitation systems, the power needed to drive an impeller must be known. The power requirement of an impeller is influenced by several product and equipment variables. Because the large number of variables does not permit development of an analytical

Table 13.2 Suitability of Agitators for Different Ranges of Viscosity

Agitator type	Viscosity range of product (Pas)
Propellor	<3
Turbine	<100
Anchor	50–500
Helical and ribbon type	500–1000

relationship useful for design purposes, we must use empirical computations to estimate the required power for driving the impeller.

In the empirical approach, we use dimensional analysis to develop relationships between various measurable quantities. The power required by the impeller is a function of the size and shape of the impeller and tank, the rotational speed of the impeller, gravitational forces, and properties of the fluid such as viscosity and density. Using Buckingham π theorem (Appendix A.9), the following functional equation is obtained for various quantities important in an agitation system:

$$f\left(\frac{\rho N D_i^2}{\mu}, \frac{N^2 D_i}{g}, \frac{P_r}{\rho N^3 D_i^5}, \frac{D_i}{T}, \frac{D_i}{H}, \frac{D_i}{C}, \frac{D_i}{p}, \frac{D_i}{w_b}, \frac{D_i}{l}, \frac{n_2}{n_1}\right) = 0 \tag{13.28}$$

where ρ is the fluid density, kg/m^3; D_i is the diameter of the impeller, m; and μ is the viscosity of the fluid, kg/ms; N is the rotational speed of the impeller, revolutions/s; g is the acceleration due to gravity, m/s^2; P_r is the power consumption by the impeller, J/s; H is the liquid depth in the tank, m; C is clearance of the impeller off the vessel bottom, m; l is the blade length, m; w_b is the blade width, m; p is the pitch of blades; n is the number of blades.

In Equation (13.28), the last seven terms on the left side are related to the geometry of the impeller and the vessel. The last term is the ratio of number of blades.

The first three dimensionless numbers in Equation (13.28) are as follows. The first term is the Reynolds number,

$$N_{Re} = \frac{\rho N D_i^2}{\mu} \tag{13.29}$$

The second number is the Froude number that is a ratio of the inertial to gravitational forces.

$$N_{Fr} = \frac{N^2 D_i}{g} \tag{13.30}$$

The Froude number plays an important role in many situations involving agitation. Consider a vessel open to the atmosphere and with a free liquid surface. If the tank does not contain baffles, then upon the action of the impeller, a vortex is formed and the role of gravitational forces becomes important in defining the shape of the vortex. When vessels contain baffles, the role of Froude number becomes negligible.

The third number is called the Power number, N_P. It is defined as follows:

$$N_P = \frac{P_r}{\rho N^3 D_i^5} \tag{13.31}$$

N_p is a dimensionless number. The dimensions cancel when the units of power are written in base units as $kg\,m^2/s^3$.

Relationships between the Power number and the Reynolds number have been reported for a variety of impeller designs and geometrical considerations by several investigators (Rushton et al., 1950; Bates et al., 1966). For agitated tanks, a Reynolds number less than 10 indicates laminar flow, greater than 100,000 is turbulent flow, and between 10 and 100,000 the flow field is considered to be transitional.

Experimental data obtained from trials with different type of agitation systems are plotted on a log-log graph, with the Power number as the y-ordinate and the Reynolds number as the abscissa as shown in Figure 13.8. We can use Figure 13.8 for calculating power requirements of an agitation system, but it is important to note that this figure is applicable only for the impeller design and geometrical considerations for which this figure was developed. Therefore, we must first ensure that the conditions of a given problem are similar to the ones shown in the figure. Similar graphical relationships are available in the literature for other types of impeller designs and geometrical considerations.

From Figure 13.8, we observe that for paddle and turbine impellers operating in the laminar region, there is a linear decrease in the Power number with Reynolds number. In the turbulent region, the Power number remains fairly constant.

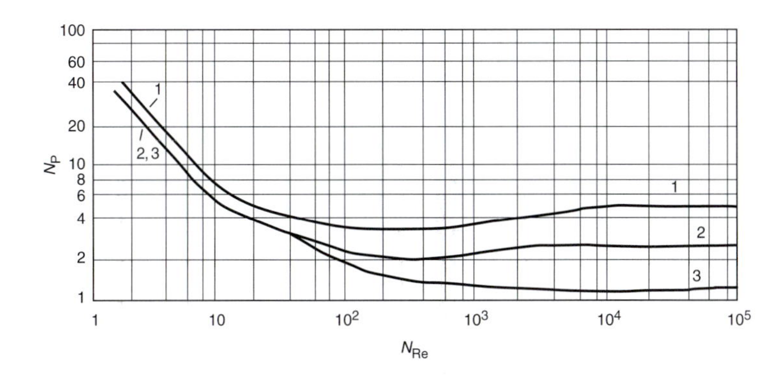

Figure 13.8 Correlation between the Power number and the Reynolds number for various impellers. (Curve 1 is for a flat six-blade turbine with disk and four baffles; $w_b/D_i = 1/5$, $B/T = 1/12$; Curve 2 is for a flat six-blade open turbine with four baffles, $w_b/D_i = 1/8$, $B/T = 1/12$; Curve 3 is for a six blade open turbine with blades pitched at 45° and four baffles, $w_b/D_i = 1/8$, $B/T = 1/12$). (Adapted from Bates et al., 1963)

Example 13.8

A concentrated fruit juice with a viscosity of 0.03 Pa.s and a density of 1100 kg/m^3 is being agitated in an agitation system containing a turbine impeller. The impeller has a disk with six blades. The tank diameter is 1.8 m. The height of the liquid in the tank is the same as the tank diameter. The impeller diameter is 0.5 m. The width of the blade is 0.1 m. The tank is equipped with four baffles, each with a width of 0.15 m. If the turbine is operated at 100 rpm, determine the required power.

Given

Diameter of the impeller, $D_i = 0.5\,m$
Diameter of the tank, $T = 1.8\,m$
Width of the blade, $w_b = 0.1\,m$
Width of the baffle, $B = 0.15\,m$
Impeller speed, $N = 100\,rpm$
Density of fluid, $\rho = 1100\,kg/m^3$
Viscosity of fluid, $\mu = 0.03\,Pa.s$

Approach

We will first calculate the Reynolds number. We observe that the turbine type, size and number of blades, and size and number of baffles in this example are the same as shown for curve 1 in Figure 13.8. Therefore, we will use Figure 13.8 to determine the Power number using the calculated Reynolds number. Power requirements will be obtained from the Power number.

Solution

1. *From Equation (13.29),*

$$N_{Re} = \frac{D_i^2 N\rho}{\mu} = \frac{(0.5m)^2(1.667rev/s)(1100kg/m^3)}{0.03(kg/ms)}$$

$$N_{Re} = 15,280$$

2. *In Figure 13.8, curve 1 is for a flat six-blade turbine with disk, for $D_i/w_b = 5$ and the tank containing four baffles with $B/T = 1/12$. These conditions match the given dimensions therefore we will use curve 1 to obtain the Power number. From Figure 13.8, for a Reynolds number of 1.5×10^4, the power number*

$$N_p = 5$$

3. *From Equation (13.31)*

$$P_r = N_p \rho N^3 D_i^5 = 5.0(1100kg/m^3)(1.667rev/s)^3(0.5m)^5$$

$$P_r = 796.2\ J/s = 0.796\ kW$$

4. *The impeller requires power input of 0.796 kW. Therefore the motor selected for the mixer must be larger than 0.8 kW.*

PROBLEMS

13.1 An air filter normally requires change after 100 hours of use when the pressure drop is 5 cm. The filter is 1 m × 3 m and is designed for an air flow rate of $1.5\ m^3/s$. Compute the influence of increasing the air flow rate to $2.5\ m^3/s$ on the life of the filter.

13.2 The effective diameter of strawberry fruit is being estimated from terminal velocity measurements. Compute the effective diameter from a terminal velocity of 15 m/s and fruit density of $1150\ kg/m^3$.

13.3 The 50-micron particles in a water suspension are to be removed using a centrifugal separator operating at 70,000 rpm. Compute the effective radius for separation when the velocity of separation is 0.05 m/s and the particle density is $850\ kg/m^3$.

***13.4** A $0.5\text{-}m^2$ filter is used to filter a liquid at a pressure of 100 kPa. A $1\text{-}cm^2$ laboratory-scale version of the filter media was used to filter $0.01\ m^3$ of the liquid in 1 minute at the same pressure. Compute the rate of filtration after 2 hours through the commercial filter.

* Indicates an advanced level in solving.

13.5 Cream is separated from skim milk using a centrifugal separator. The separator inlet has a 25-cm radius, and the skim milk outlet is 10 cm radius. Estimate the cream outlet radius when the cream density is 865 kg/m^3 and skim milk density is 1025 kg/m^3.

LIST OF SYMBOLS

A	area, m^2
a	acceleration, m/s^2
b	height of centrifugal separation bowl, m
B	width of baffles, m
c	solids concentration in solution expressed as mass of liquid per unit mass solids, kg liquid/kg solid
C	clearance of the impeller off the vessel bottom, m
d	particle diameter, micron, 10^{-6} m
D_i	Diameter of impeller, m
F	force, N
$f(e)$	function used in Equation (13.17)
g	acceleration due to gravity, m/s^2
g_c	constant in Equations (13.18) and (13.19)
H	Liquid height in the tank, m
K	experimental constant
L	thickness of filter material in liquid filtration system, m
L_c	thickness of filter cake in liquid filtration system, m
l	blade length, m
m	particle mass in centrifugal force field, kg
n	number of blades
N	rotational speed, revolutions/min
N_p	Power number
N_{Re}	Reynolds number
N_{Fr}	Froude number
p	pitch of blades, degrees
P	total pressure at some location in the separation, Pa
P_r	Power, kg m^2/s^3
R	resistance in filtration separation system
N_{Re}	Reynolds number
r	radius, m
r'	specific resistance of filter cake
S	solids content of liquid being filtered, kg solid/m^3 liquid
t	time, s
u	velocity, m/s

V	volume of liquid being filtered, m^3
w	mass flow rate, kg/s
w_b	blade width, m
μ	viscosity, kg/ms
ω	angular velocity
ρ	density, kg/m^3
ψ	sphericity

Subscripts: A, high-density fraction in Equation (13.26); B, low-density fraction in Equation (13.26); c, centrifugal separation system reference; D, drag in particle; f, fluid fraction reference; G, gravitation reference in separation system; i, input condition reference; n, location of equal pressure in centrifugal separation system; o, outlet condition reference; p, solid or particle fraction reference; s, suspension during separation; t, terminal condition; 1,2, general location or time references

BIBLIOGRAPHY

Bates, R. L., Fondy, P. L., and Corpstein, R. R. (1963). An examination of some geometric parameters of impeller power. *Ind. Eng. Chem. Proc. Des. Dev.* **2**: 310.

Charm, S. E. (1978). *The Fundamentals of Food Engineering*, 3rd ed. AVI Publishing Co, Westport, Conn.

Coulson, J. M. and Richardson, J. F. (1978). *Chemical Engineering*, 3rd ed., Volume 11. Pergamon Press, Elmsford, N.Y.

Decker, H. M., Buchanan, L. M., Hall, L. B., and Goddard, K. R. (1962). *Air Filtration of Microbial Particles.* Public Health Service, Publ. 593, U.S. Govt. Printing Office, Washington, D.C.

Earle, R. L. (1983). *Unit Operations in Food Processing*, 2nd ed. Pergamon Press, Elmsford, N.Y.

Foust, A. S., Wenzel, L. A., Clump, C. W., Maus, L., and Anderson, L. B. (1960). *Principles of Unit Operations*. John Wiley & Sons, New York.

Geankoplis, C. J. (2003). *Transport Processes and Separation Principles*, 4th ed., Pearson Education Inc, New Jersey.

Rushton, J. H., Costich, E. W., and Everett, H. J. (1950). Power characteristics of mixing impellers. *Chem. Eng. Progr.* **46**: 395, 467.

Whitby, K. T. and Lundgren, D. A. (1965). The mechanics of air cleaning. *Trans. Am. Soc. Chem. Engrs.* **8**(3): 342.

Chapter 14

Extrusion Processes for Foods

Extrusion is a process that converts raw material into a product with desired shape and form by forcing the material through a small opening using pressure. The process involves a series of unit operations such as mixing, kneading, shearing, heating, cooling, shaping and forming. Many food products are manufactured by extrusion cooking—a process that uses both thermal energy and pressure to convert raw food ingredients into popular products such as breakfast cereals, pastas, pet foods, snacks and meat products.

14.1 INTRODUCTION AND BACKGROUND

The origins of the extrusion process are closely associated with polymer science and technology. In the mid-1850s, extrusion was used to produce the first seamless lead pipe. The first man-made thermoplastic, celluloid, was manufactured in the 1860s based on a reaction between cellulose and nitric acid. The manufacturing of Bakelite in 1907, and the protective coating resin, glyptal, in 1912, was dependent on extrusion processing. Formal applications of extrusion processes to foods began in the 1930s and evolved over the following 50 years, as equipment for extrusion processing increased in capabilities and complexity.

In general, all extrusion systems contain five key components. These components are:

- Primary feed system consisting of a container and delivery system for the primary ingredients involved in the process
- Pump to move all ingredients through the steps associated with the extrusion process
- Reaction vessel where key actions such as mixing, kneading, shearing, heating, and cooling occur

 Please direct your web site comments and questions to the author: Professor R. Paul Singh, Department of Biological and Agricultural Engineering, University of California, Davis, CA 95616, USA.
Email: rps@rpaulsingh.com.

- Secondary feed system for adding secondary ingredients or energy as needed to achieve the desired product characteristics
- Exit assembly designed to restrict flow and contribute to the shaping and forming of the final product (usually referred to as the die)

These components appear in different ways and may be identified differently in various extrusion systems, depending on the specific equipment used and the product being manufactured.

Early examples of extrusion processing include pasta products (macaroni, spaghetti, etc.) and pellets for conversion into ready-to-eat cereals. Current applications include commercial production of cereal-based products (cornflakes, puffed rice, crispbreads, snacks), fruit-based products (fruit gums, licorices, hard candies), protein-based products (textured vegetable proteins), animal feeds (pet foods), and spice-based products (flavors).

The application of the extrusion process to food manufacturing is very complex and cannot be described and discussed in detail in this text. Numerous references provide in-depth information on the design of the process, the equipment used, and the products created by the process. Some key references include Harper (1981); Mercier, Linko, and Harper (1989); Kokini, Ho and Karwe (1992); and Levine and Miller (2006). The goal of the information presented here is to provide undergraduate food science students with an introduction to the quantitative and qualitative aspects of the extrusion process.

14.2 BASIC PRINCIPLES OF EXTRUSION

Extrusion involves a combination of transport processes, including flow of materials within the system, thermal energy transfer to and within the material, and mass transfer to and within the material during extrusion. The flow of materials within the channels of the system occurs in all types of extrusion. Food ingredients of various types may be processed by extrusion and are referred to as *extrudates*. All ingredients involved in the extrusion process flow through a channel with a defined geometry. The power requirements for the process are directly dependent on the flow characteristics through the channel. As indicated in Chapter 2, these requirements are also dependent on properties of the fluid used. In general, these properties are part of extrudate rheology.

The relationships used in extrudate rheology include many of the basic expressions presented in Chapter 2, beginning with basic relationships presented to define viscosity. The relationship between shear stress (σ) and rate of shear (γ) was presented as Equation (2.10):

$$\sigma = \mu \left[\frac{du}{dy} \right]$$

The primary property of the extrudate in this relationship is viscosity (μ). Since most food extrudates are highly non-Newtonian, the apparent viscosity decreases with increasing rate of shear. As illustrated in Chapter 2, these materials are described by the Herschel-Bulkley model or Equation (2.161):

$$\sigma = K \left[\frac{du}{dy} \right]^n + \sigma_o$$

When applied to extrudates, this three-parameter model is usually reduced to the following two-parameter model:

$$\sigma = K \left[\frac{du}{dy} \right]^n \tag{14.1}$$

where the consistency coefficient (K) and the flow behavior index (n) are parameters describing the rheological properties of the extrudate. Based on investigations of the flow behavior of food extrudates, the flow behavior index is normally less than 1.0.

Two additional parameters with significant influence on flow of food extrudates are moisture content and temperature. In order to account for the additional factors, the following relationship has been proposed:

$$\sigma = K_o e^{\left(\frac{A}{T}\right)} e^{(BM)} \left[\frac{du}{dy} \right]^n \tag{14.2}$$

This relationship contains two additional parameters: an activation energy constant (A) to account for the influence of temperature (T), and a similar exponential constant (B) to account for the influence of dry basis moisture content (M). Typical rheological properties and constants for food extrudates are presented in Table 14.1. These properties and constants are likely to change during the extrusion process, but the magnitudes presented are acceptable for preliminary estimates of extruder performance.

Although the flow of an extrudate within the extrusion system may occur in a tube or pipe geometry, the channels include several flow

Table 14.1 Reported Power Law Models (Eq. 14.2) for Food Extrudates

Material	K_o	n	Temperature range (°C)	Moisture range (%)	A (K)	B (1/%M_{DB})	Reference
Cooked cereal dough (80% corn grits, 20% oat flour)	78.5	0.51	67–100	25–30	2500	−7.9[a]	Harper et al., 1971
Pregelatinized corn flour	36.0	0.36	90–150	22–35	4390	−14	Cervone and Harper, 1978
Soy grits	0.79	0.34	35–60	32	3670	–	Remsen and Clark, 1978
Hard wheat dough	1885	0.41	35–52	27.5–32.5	1800	−6.8	Levine, 1982
Corn grits	28,000	~0.5	177	13	–	–	van Zuilichem et al., 1974
	17,000	~0.5	193	13	–	–	
	7600	~0.5	207	13	–	–	
Full-fat soybeans	3440	0.3	120	15–30	–	–	Fricke et al., 1977
Moist food products	223	0.78	95	35	–	–	Tsao et al., 1978
Pregelatinized corn flour	17,200	0.34	88	32	–	–	Hermann and Harper, 1974
Sausage emulsion	430	0.21	15	63	–	–	Toledo et al., 1977
Semolina flour	20,000	0.5	45	30	–	–	Nazarov et al., 1971
Defatted soy	110,600	0.05	100	25	–	–	Jao et al., 1978
	15,900	0.40	130	25	–	–	
	671	0.75	160	25	–	–	
	78,400	0.13	100	28	–	–	
	23,100	0.34	130	28	–	–	
	299	0.65	160	28	–	–	
	28,800	0.19	100	35	–	–	
	28,600	0.18	130	35	–	–	
	17,800	0.16	160	35	–	–	
Wheat flour	4450	0.35	33	43	–	–	Launay and Bure, 1973
Defatted soy flour	1210	0.49	54	25	–	–	Luxenburg et al., 1985
	868	0.045	54	50	–	–	
	700	0.43	54	75	–	–	
	1580	0.37	54	85	–	–	
	2360	0.31	54	100	–	–	
	2270	0.31	54	110	–	–	

[a] *Wet basis moisture content*

geometries. In many situations, the flow geometry may be described as a rectangular cross-section within the barrel of the extruder (Harper, 1981). To model the flow behavior within the extruder, it is assumed that the flow is incompressible, steady, laminar and fully developed. In addition, the rectangular cross-section channel formed between the screw flights and the barrel surface is assumed as an infinite plate sliding across the channel (Harper, 1981). Based on these assumptions, for an axial flow in rectangular Cartesian coordinates, the differential equation for the momentum flux is

$$\frac{d\sigma}{dy} = \frac{\Delta P}{L} \tag{14.3}$$

where L represents the length of the channel (distance in the downstream direction), and y is the element distance from the barrel surface to the surface of the screw (distance in the vertical direction). Upon integration, we obtain an expression for the shear stress (σ) for a laminar flow in the channel, as:

$$\sigma = \frac{\Delta P}{L} y + C \tag{14.4}$$

where C is a constant of integration. It must be noted that the preceding equation holds for both Newtonian and non-Newtonian fluids. In the case of a Newtonian fluid this expression is combined with Equation (2.28) to obtain the following expression.

$$\mu \frac{du}{dy} = \frac{\Delta P}{L} y + C \tag{14.5}$$

The fluid velocity distribution within the channel can be obtained by integration of the previous expression in the following manner,

$$\int du = \int \left[\left(\frac{\Delta P}{\mu L} \right) y + C_1 \right] dy \tag{14.6}$$

Following integration, the velocity profile is described as:

$$u = \frac{\Delta P}{2\mu L} y^2 + C_1 y + C_2 \tag{14.7}$$

where the constants of integration (C_1 and C_2) are determined by considering that the fluid velocity distribution ranges from zero (at the barrel surface, $y = 0$) to u_{wall} at the surface of the screw ($y = H$)

$$u = \frac{\Delta P H^2}{2\mu L} \left[\frac{y}{H} - \frac{y^2}{H^2} \right] + \frac{u_{wall}}{H} y \tag{14.8}$$

where ΔP is the absolute value of the pressure drop across the channel length (L).

In order to evaluate the volumetric flow rate for this fluid, the following general expression applies

$$dV = u(y)W dy \tag{14.9}$$

By integration of Equation (14.9) over the rectangular cross-section we obtain

$$\int_0^V dV = \int_0^H \left(\frac{\Delta P H^2}{2\mu L} \left[\frac{y}{H} - \frac{y^2}{H^2} \right] + \frac{u_{wall}}{H} y \right) W dy \tag{14.10}$$

Upon evaluating the integrals, the following expression for volumetric flow rate of the Newtonian fluid in the channel cross-section is obtained.

$$V = \frac{\Delta P W H^3}{12\mu L} + \frac{u_{wall} H W}{2} \tag{14.11}$$

The mean velocity through the channel may be calculated from the following equation

$$u_{mean} = \frac{V}{WH} = \frac{\Delta P H^2}{12\mu L} + \frac{u_{wall}}{2} \tag{14.12}$$

Example 14.1

Corn meal with a moisture content of 18% (wb) is being extruded through a metering zone of an extruder with the following dimensions of the channel: width 5 cm, height 2 cm, length 50 cm. The wall velocity is estimated to be 0.3 m/s. The rheological properties of the extrudate can be estimated by a viscosity of 66,700 Pa s and a density of 1200 kg/m^3. If the pressure drop is maintained at 3000 kPa, estimate the mass flow rate of extrudate through the die.

Given

Moisture content	*18% wet basis*
Channel cross-section	*5 cm × 2 cm*
Channel length	*50 cm*
Viscosity	*66,700 Pa.s*
Density	*1200 kg/m^3*
Pressure	*3000 kPa*

Approach

We will use Equation (14.11) to obtain the volumetric flow rate and convert it into mass flow rate using the given density.

Solution

Using Equation (14.11),

$$V = \frac{(3\times10^{6})(0.05)(0.02)^{3}}{(12)(66,700)(0.5)} + \frac{(0.3)(0.02)(0.05)}{2}$$

$$V = 1.53\times10^{-4}\ m^{3}/s$$

Using the density of the extrudate of 1200 kg/m³, the mass flow rate may be computed as follows:

$$\dot{m} = (1.53\times10^{-4})(1200) = 0.1835\ kg/s = 660\ kg/hr$$

As indicated earlier, most food materials involved in extrusion are non-Newtonian and require appropriate relationships to describe the flow characteristics. The flow characteristics of these types of fluids in cylindrical pipes were described in Section 2.9. In an extrusion system, the flow would most likely occur in a rectangular cross-section channel (generally assumed to be a plane narrow slit) and the expression for shear stress is given by Equation (14.4).

In the case of a power-law fluid the shear stress is given by

$$\sigma = K\left(\frac{du}{dy}\right)^{n} \qquad (14.13)$$

where the consistency coefficient (K) and the flow behavior index (n) are properties of the power-law fluid.

By combining Equations (14.4) and (14.13), the following relationship is obtained.

$$K\left(\frac{du}{dy}\right)^{n} = \frac{\Delta P}{L}y + C \qquad (14.14)$$

By assuming that du/dy is positive inside the channel (where the velocity increases from zero at $y = 0$ to u_{wall} at $y = H$), the previous equation can be arranged as,

$$\frac{du}{dy} = \left(\frac{\Delta P}{KL}y + C_1\right)^{\frac{1}{n}} \qquad (14.15)$$

The previous equation is integrated in the following manner

$$\int du = \int \left(\frac{\Delta P}{KL} y + C_1 \right)^{\frac{1}{n}} dy \qquad (14.16)$$

Following integration the following expression is obtained.

$$u(y) = \frac{nKL}{(n+1)\Delta P} \left(\frac{\Delta P}{KL} y + C_1 \right)^{\frac{n+1}{n}} + C_2 \qquad (14.17)$$

The integration constants (C_1 and C_2) are determined by considering the boundary conditions for the velocity field within the channel (i.e., $u = 0$ at $y = 0$ and $u = u_{wall}$ at $y = H$). However, unlike the case of a Newtonian fluid, the explicit expressions for C_1 and C_2 can not be obtained. Instead, the following relationships are obtained.

$$\left[\left(\frac{\Delta P}{KL} H + C_1 \right)^{\frac{n+1}{n}} - C_1^{\frac{n+1}{n}} \right] = \frac{u_{wall}(n+1)\Delta P}{nKL} \qquad (14.18)$$

$$C_2 = -\frac{nKL}{(n+1)\Delta P} C_1^{\frac{n+1}{n}} \qquad (14.19)$$

In order to obtain the velocity profile of a power-law fluid, Equation (14.17) must be combined with Equation (14.18) and (14.19). Note that in Equation (14.17), (14.18) and (14.19) ΔP corresponds to the actual pressure drop across a channel of length "L" (i.e., ΔP is a negative value).

The analytical solutions of this system of equations are usually not possible and numerical or approximate solutions are generally pursued.

Rauwendaal (1986) approximated the volumetric flow rate at the extruder outlet for non-Newtonian fluids by using the following equation.

$$V = \frac{(4+n)}{10} WHu_{wall} - \frac{1}{(1+2n)} \frac{WH^3}{4K} \left(\frac{u_{wall}}{H} \right)^{1-n} \frac{\Delta P}{L} \qquad (14.20)$$

This approximation is valid for screw pitch angles between 15 and 25 degrees and flow indices between 0.2 and 1.0.

Example 14.2

A non-Newtonian (power-law) soy flour extrudate with 25% moisture content (wb) is being pumped through an extruder. The channel in the metering section has the following dimensions: width 5 cm, height 2 cm, length 50 cm. The properties of the extrudate are described by a consistency coefficient of 1210 Pa s^n, flow behavior index of 0.49, and density of 1100 kg/m^3. Estimate the pressure drop if the mass flow rate of 600 kg/hr is to be maintained.

Given

Moisture content	*25% wet basis*
Channel cross-section	*5 cm × 2 cm*
Channel length	*50 cm*
Consistency coefficient	*1210 Pa.s^n*
Flow behavior index	*0.49*
Density	*1100 kg/m^3*
Mass flow rate	*600 kg/hr*

Approach

We will first obtain volumetric flow rate and then use Equation (14.20) to calculate pressure drop.

Solution

For a mass flow rate of 600 kg/hr, the volumetric flow rate is

$$V = \frac{600}{1100} = 0.545 \frac{m^3}{hr} = 1.39 \times 10^{-4} m^3/s$$

Using Equation (14.20):

$$\Delta P = \left[1.39 \times 10^{-4} - \frac{(4+0.49)0.05 \times 0.02 \times 0.3}{10}\right] \times \frac{(1+2 \times 0.49) \times 4 \times 1210 \times 0.5}{0.05 \times 0.02^3}\left[\frac{0.02}{0.3}\right]^{(0.49-1)}$$

$$\Delta P = 204{,}972\ Pa = 205\ kPa$$

14.3 EXTRUSION SYSTEMS

Extrusion systems can be divided into four different categories. These four categories include two different methods of operations—cold extrusion or extrusion cooking—and two different barrel configurations—single or twin screw. Both barrel configurations may be used for either method of operation.

14.3.1 Cold Extrusion

Cold extrusion is used most often to form specific shapes of extrudate at locations downstream from the die. In this process, the extrudate is pumped through a die without the addition of external thermal energy. Systems as simple as kneading steps in the preparation of dough before baking could be referred to as cold extrusion. Alternatively, more complex systems used to create coextruded products (Fig. 14.1) can also be cold extrusion. As illustrated, one component of the final product is pumped through an opening of defined shape to create a continuous tube of the first component. At the same time, the second component of the final product is introduced just before the die and becomes a filler for the interior space within the outer tube, as illustrated in Figure 14.2. During a final step in the process, the tubular product is cut into appropriate lengths.

In general, cold extrusion is used to mix, knead, disperse, texturize, dissolve, and form a food product or product ingredient. Typical food products include pastry dough, individual pieces of candy or confections, pasta pieces, hot dogs, and selected pet foods. These types of extruders would be considered low-shear systems and would create relatively low pressures upstream from the die.

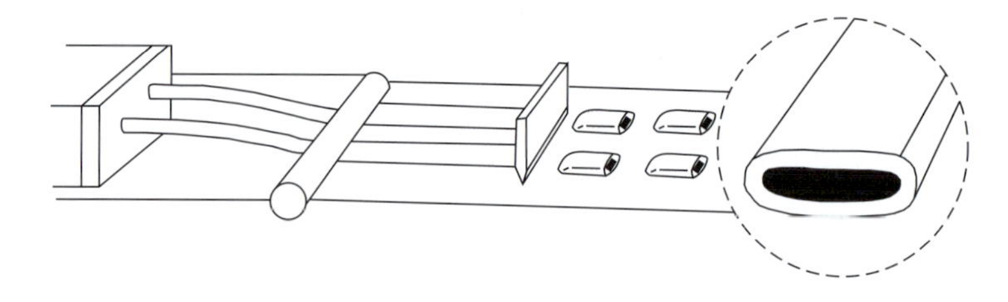

Figure 14.1 A cold extrusion operation. (Adapted from Moore, 1994)

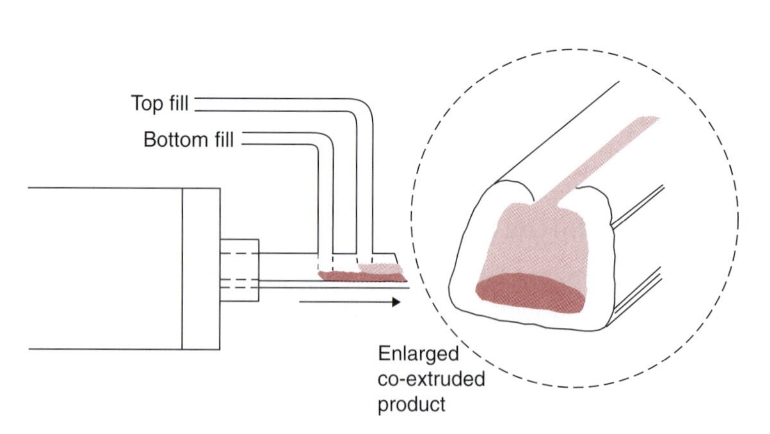

Figure 14.2 Filled products using co-extrusion. (Adapted from Moore, 1994)

14.3.2 Extrusion Cooking

When thermal energy becomes a part of the extrusion process, the process is referred to as extrusion cooking. Thermal energy may be added to the extrudate during the process from an external source or may be generated by friction at internal surfaces of the extruder in contact with the extrudate. As illustrated in Figure 14.3, the addition of thermal energy occurs at the surface of the barrel of the extrusion system. Thermal energy may be transferred through the walls and surfaces of the barrel to ingredients used to create the extrudate. In addition, mechanical energy created by friction between surfaces and ingredients within the barrel is dissipated as thermal energy in the extrudate.

The "cooking" process during extrusion is unique from most other thermal processes. As the ingredients are introduced to form the extrudate, they are exposed to elevated pressures as well as temperatures. The geometry of the extrusion barrel is designed to increase the pressure on the ingredients as movement from entrance to exit proceeds. The exit from the barrel is the "die"—an opening with much smaller cross-sectional area than that of the barrel. A portion of the cooking process occurs

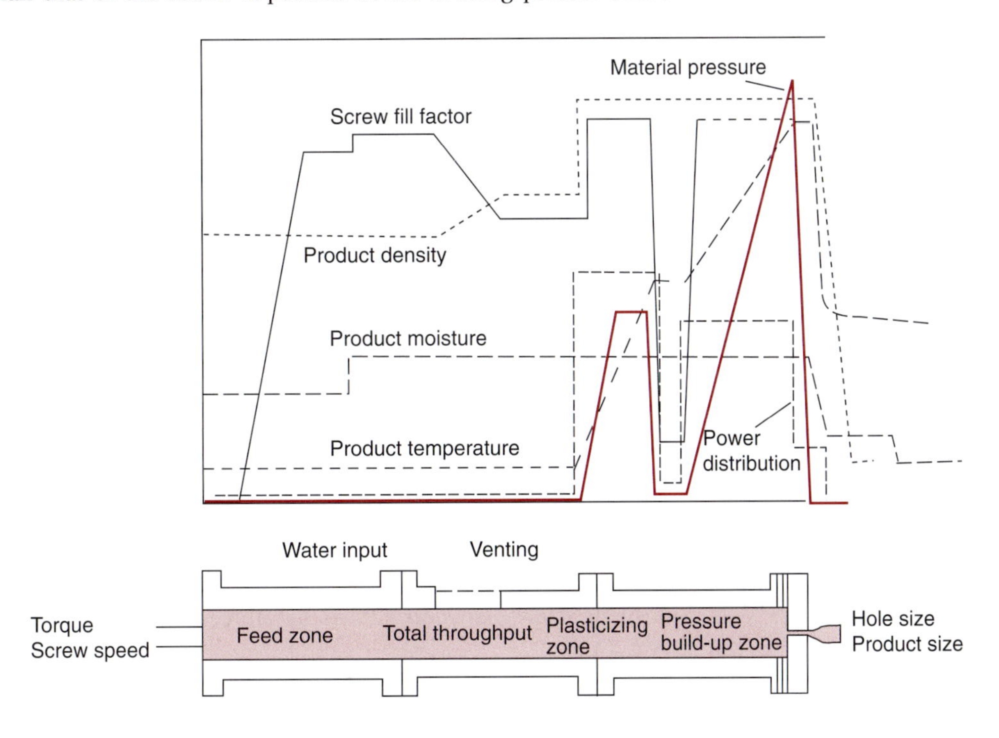

Figure 14.3 Single-screw ex truder components. (Adapted from Werner and Pfleiderer, Ltd.)

■ **Figure 14.4** Process variables during extrusion cooking. (From Miller and Mulvaney, 2000)

downstream from the die due to the rapid change in pressure. The pressure change results in a rapid reduction in temperature and a release of moisture from the extrudate. These changes are illustrated in Figure 14.4. The combinations of temperatures, pressures, and moisture contents may be used to create an unlimited range of product characteristics. For example, the density of individual pieces of extrudate will be a function of the pressure difference across the die.

14.3.3 Single Screw Extruders

In a single-screw extrusion system, the barrel of the extruder contains a single screw or auger that moves the extrudate through the barrel. As shown in the photograph in Figure 14.5, the extrudate is carried through the barrel in the space between the core of the screw and the barrel. The flow rate of the extrudate through the system will be proportional to the rotation speed (rpm) of the screw.

A single-screw extrusion system has three components or sections (Fig. 14.6):

- Feed section, where the various ingredients are introduced and initial mixing occurs. The rotating action of the screw moves the ingredients to the transition or compression section.

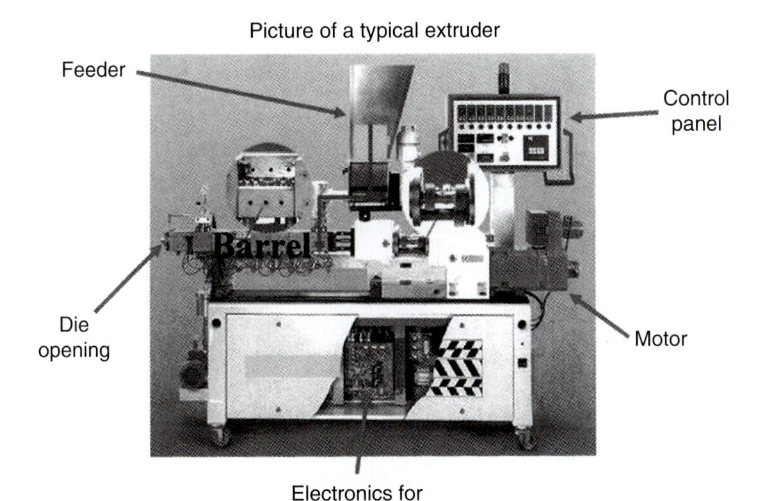

Figure 14.5 Photograph of single-screw extruder. (Courtesy of Leistritz Company)

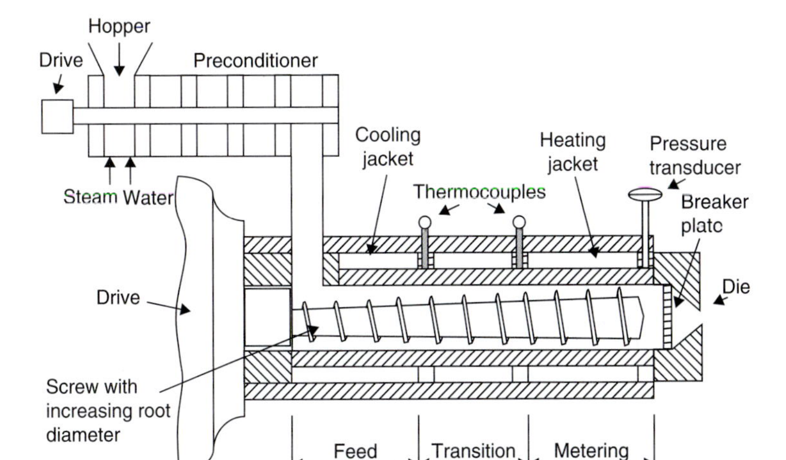

Figure 14.6 Sections within the barrel of a single screw extruder. (Adapted from Harper, 1989)

- Compression or transition section, where the ingredients begin the transition to the extrudate as pressure and temperature begin to increase. As the dimensions of the flow channel decrease, the material is compressed and mechanical energy is dissipated as temperature increases. This section may be referred to as a kneading section, with significant changes in the physical and chemical characteristics of the ingredients occurring.
- Metering (or cooking) section, where additional compression of the extrudate occurs as a result of additional reductions in

the dimensions of the flow channel and increased shearing action. In some designs, the overall dimensions of the barrel are reduced as well.

Single-screw extruders may be classified according to shearing action as well. Generally, low shear extruders will include smooth barrel surfaces, relatively large flow channels, and low screw speeds (3–4 rpm). Characteristics of moderate shear extruders include grooved barrel surfaces, reduced flow channel cross-sections and moderate screw speeds (10–25 rpm). High shear extrusion systems operate at high screw speeds (30–45 rpm), variable pitch and flight depth screws, and grooved barrel surfaces. Each of these types of extrusion systems will create extruded products with different properties and characteristics.

An additional dimension of operation within the single-screw extrusion system is associated with flow characteristics within the barrel and around the flights of the screw. Although the forward flow is caused by action of the screw, backward flow occurs between the flights and the barrel surface. The backward flow is the result of increased pressure as the extrudate moves from one section of the extruder to the next. The second component of backward flow is the leakage between the flights and the barrel; this type of flow may be reduced by grooves in the barrel surface.

14.3.4 Twin-Screw Extruders

Twin-screw extrusion systems incorporate two parallel screws into the extruder barrel. The screws may be co-rotating or counter-rotating, as illustrated in Figure 14.7. Various configurations of the screw or auger have been developed, including the fully intermeshing, self-wiping, co-rotating twin-screw system in Figure 14.8. This particular system has been used in many food applications due to the self-cleaning, better mixing, moderate shear force and higher capacity characteristics.

Twin-screw extrusion systems have numerous advantages. The throughput of these systems can be independent of feed rate and screw speed. Process variables include degree of fill, temperature-shear history, and heat transfer, all of which may influence the properties of the extruded product. Twin-screw systems provide increased flexibility in terms of higher moisture content extrudates, as well as higher concentrations of ingredients (lipids, carbohydrates, etc.). These systems usually have less wear due to shorter sections of the barrel being exposed to the high pressures required for product extrusion. Finally, twin-screw extruders will accommodate a wider range of particle sizes in the ingredients.

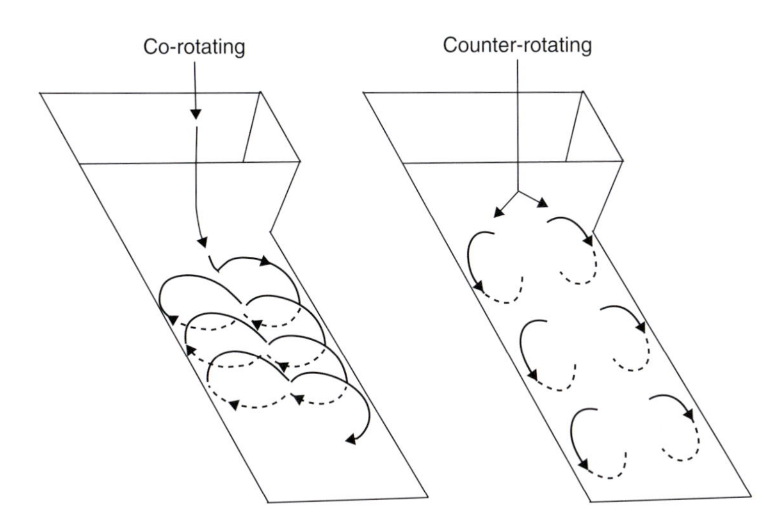

■ **Figure 14.7** Screw rotation in a twin-screw extruder. (Adapted from Weidman, 1992)

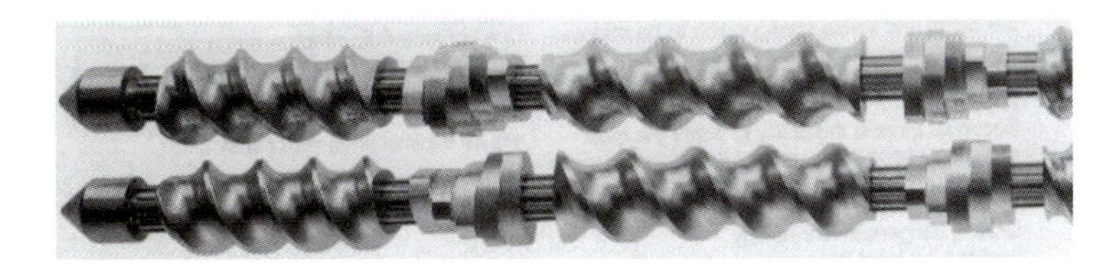

■ **Figure 14.8** Configuration of screws in a twin-screw extruder.

Other configurations of twin-screw extrusion systems include counter-rotating with non-intermeshing systems, counter-rotating with inter-meshing systems, and co-rotating with non-intermeshing systems. Each of these configurations may be used in specific applications and require unique and complex analysis for design and scale-up.

14.4 EXTRUSION SYSTEM DESIGN

The power requirement for operating an extrusion system is a key design factor. Power consumption is a complex function of properties of the material being extruded, extruder design, extruder motor type and extrusion conditions. Although there are unique aspects associated with estimating power consumption for single- versus twin-screw systems, a general approach for estimating total power consumption (p_t) is:

$$p_t = p_s + V_d \Delta P \tag{14.21}$$

where p_s is the portion of the power consumption for viscous dissipation associated with shear of the feed ingredients, and the second term of the expression is the power needed to maintain flow through the barrel and die of the extrusion system.

The power needed for viscous dissipation has been expressed in terms of the Screw Power number (N_p), as follows:

$$N_p = \frac{p_s}{\rho N^3 D^4 L} \qquad (14.22)$$

where the screw speed (N), screw diameter (D), screw length (L), as well as density of extrudate (ρ) are considered. The magnitude of the Screw Power number is dependent on the Screw Rotational Reynolds number (N_{Res}), as illustrated by Figure 14.9. For extrudate with rheological properties described by the power-law model, the Screw Rotational Reynolds number is defined as follows:

$$N_{Res} = \frac{\left[\rho (DN)^{2-n} H^n\right]}{K\pi^{2+n}} \qquad (14.23)$$

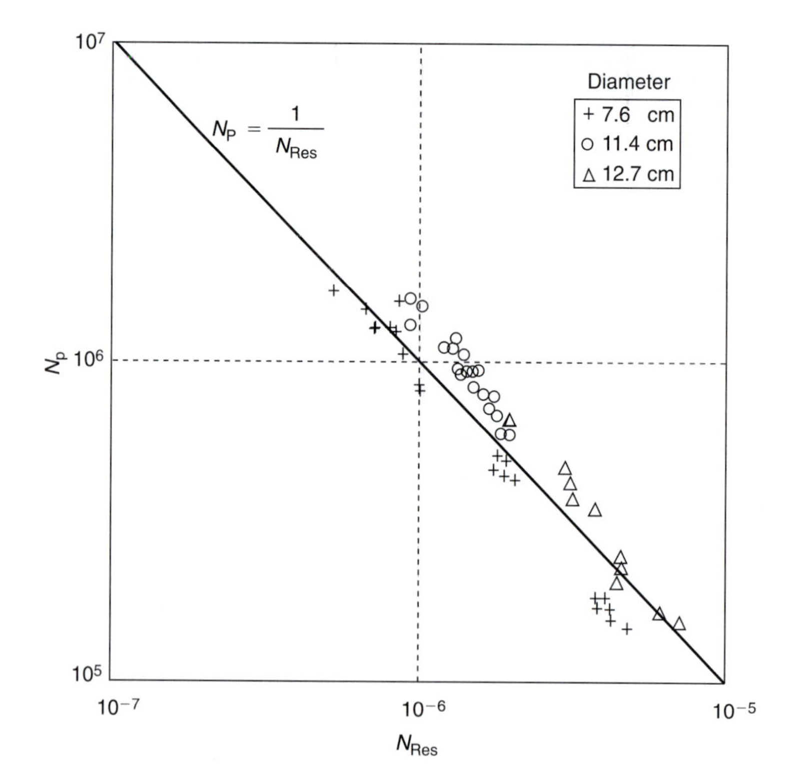

■ **Figure 14.9** Dimensionless correlation for extruder power consumption. (Adapted from Levine, 1982)

The drag flow rate (V_d) for the extruder screw can be estimated as follows:

$$V_d = \frac{(\pi NDWH)}{2} \qquad (14.24)$$

and is a key expression in approximating the overall power requirements for the extrusion system.

Example 14.3

Estimate the mechanical power requirements for a single-screw extrusion system during extrusion of 30% moisture content (wb) corn meal. The apparent viscosity of the extrudate is 1765 Pa.s at 135°C, and the density is 1200 kg/m^3. The diameter of the screw is 7.5 cm, the length of the screw is 50 cm. The channel depth within the barrel is 2 cm and width is 2.5 cm. The system is operating at a flow rate of 300 kg/hr with a screw speed of 75 rpm. The estimated wall velocity is 0.3 m/s.

Given

Moisture content	*30% wet basis*
Apparent viscosity	*1765 Pa.s*
Temperature	*135°C*
Density	*1200 kg/m^3*
Diameter of screw	*7.5 cm*
Length of screw	*50 cm*
Channel depth	*2 cm*
Channel width	*2.5 cm*
Mass flow rate	*300 kg/hr*
Screw speed	*75 rpm*

Approach

We will use Equation (14.23) to calculate the Screw Rotational Reynolds Number, and Figure 14.9 to obtain the Screw Power number. This result will be used to calculate power consumption due to viscous dissipation. We will calculate the total power requirement from the pressure drop determined from volumetric flow rate and the power consumption due to viscous dissipation.

Solution

1. *The total mechanical power for the extrusion system is estimated by using Equation (14.21). The power for viscous dissipation is obtained by computing*

the Screw Rotational Reynolds Number (Eq. (14.23)). Note that n = 1

$$N_{Res} = \frac{(1200)(0.075)\left(\frac{75}{60}\right)(0.02)}{(1765)(\pi)^3}$$

$$N_{Res} = 4.11 \times 10^{-5}$$

2. *Using Figure 14.9, the Screw Power Number is determined as follows*

$$N_p = 2.4323 \times 10^4$$

3. *Based on Equation (14.22), the power consumption for viscous dissipation is computed as follows*

$$p_s = (1200)\left(\frac{75}{60}\right)^3 (0.075)^4 (0.5)(2.4323 \times 10^4) = 901.8\ W$$

4. *The pressure change across the die is estimated using Equation (14.11), after converting the mass flow rate to volumetric flow rate*

$$V = \frac{(300\ kg/hr)}{(3600\,s/hr)(1200\ kg/m^3)} = 6.944 \times 10^{-5}\,m^3/s$$

Then

$$\Delta P = \left(6.944 \times 10^{-5} - \frac{0.3 \times 0.02 \times 0.025}{2}\right)\frac{(12)(1765)(0.5)}{(0.025)(0.02)^3}$$

$$\Delta P = -2.94 \times 10^5\ Pa$$

Note that we will use the absolute value of ΔP in the following calculations.

5. *The power requirement using Equation (14.21), becomes*

$$p_t = 901.8 + (2.94 \times 10^5)(6.944 \times 10^{-5})$$
$$p_t = 922\ W = 0.92\ kW$$

Example 14.4

A wheat flour dough is being extruded in a single-screw extruder. The screw length is 50 cm and its diameter is 6 cm. The channel dimensions are as follows, width 2 cm and height 1 cm. The properties of the dough include a consistency coefficient of 4450 Pa s^n, flow behavior index of 0.35 and density of 1200 kg/m^3. Estimate the power requirements for the system when the flow rate is 270 kg/hr and the screw rotation speed is 200 rpm with an estimated wall velocity of 0.6 m/s.

Given

Screw diameter	6 cm
Screw length	50 cm
Channel depth	1 cm
Channel width	2 cm
Consistency coefficient	4450 Pa s^n
Flow behavior index	0.35
Density	1200 kg/m^3
Flow rate	270 kg/hr
Screw rotation speed	200 rpm
Wall velocity	0.6 m/s

Approach

We will first calculate the Screw Rotation Reynolds Number and use Figure 14.9 to obtain the Screw Power Number. Then using Equation (14.22), we will obtain the power requirement due to viscous dissipation. Next, we will determine pressure change across the die using Equation (14.20). Then Equation (14.21) will be used to calculate total power requirement.

Solution

1. The first step in estimating the power for viscous dissipation within the shear within the barrel is obtained by using Equation (14.23). The Screw Rotation Reynolds Number is computed as follows

$$N_{Res} = \frac{(1200)\left[\left(\frac{6}{100}\right)\left(\frac{200}{60}\right)\right]^{(2-0.35)}\left(\frac{1}{100}\right)^{0.35}}{(4450)(\pi)^{2.35}}$$

$$= 2.566 \times 10^{-4}$$

2. Then, using Figure 14.9

$$N_p = \frac{1}{N_{Res}} = \frac{1}{2.566 \times 10^{-4}}$$

$$N_p = 3.897 \times 10^3$$

3. Based on Equation (14.22), the power for viscous dissipation is

$$p_s = (1200)\left(\frac{200}{60}\right)^3\left(\frac{6}{100}\right)^4\left(\frac{50}{100}\right)(3.897 \times 10^3)$$

$$p_s = 1122\ W = 1.12\ kW$$

4. *The second step in estimating mechanical power is for maintaining flow through the die. The pressure change across the die is obtained by using Equation (14.20) with its terms rearranged to solve for* ΔP

$$\Delta P = \left(6.25 \times 10^{-5} - \frac{4.35 \times 0.02 \times 0.01 \times 0.6}{10}\right) \times \left(\frac{(1+0.7) \times 4 \times 4450}{0.02 \times (0.01)^3}\right)(0.5)\left(\frac{0.6}{0.01}\right)^{(0.35-1)}$$

with

$$V = \left(\frac{270}{1200 \times 3600}\right) = 6.25 \times 10^{-5}\, m^3/s$$

then

$$\Delta P = 5.443 \times 10^5\, Pa$$

5. *Using Equation (14.21)*

$$p_t = 1122 + (5.443 \times 10^5)(6.25 \times 10^{-5})$$
$$p_t = 1122 + 34.02$$
$$p_t = 1156\, W = 1.16\, kW$$

14.5 DESIGN OF MORE COMPLEX SYSTEMS

The concepts presented in this chapter provide the basic steps involved in the design of an extrusion system. However, numerous additional factors should be considered when designing a typical extrusion system for a food product. The analysis and examples presented have distinguished between materials with Newtonian flow characteristics as compared with materials with non-Newtonian flow characteristics.

Many of the complex design considerations are associated with the extrusion system barrel and screw configuration. Adjustments in the basic design expressions are needed in order to account for the influence of barrel configuration, including channel depth and width, screw diameter, downstream flow geometry, and so forth. These adjustments are a function of flow characteristics such as the flow behavior index. Additional adjustments are required for leakage flow within the extruder barrel, usually backward flow through gaps between the screw and barrel. The die configuration also has a significant impact on the operating characteristics of the extrusion system. The dimensions and shape of the die as well as the relationship to barrel dimensions and shape can influence many operating parameters. Temperature is not normally considered in detail.

The properties and characteristics of the material being extruded will change dramatically as the material moves from the feed zone to the exit from the die. Typical design expressions require specific input properties and do not account for changes in properties imposed by the process. The configuration of the screw will impact the operating characteristics of the system. The diameter and channel dimensions often change with distance along the barrel length. When considering twin-screw systems, the configuration of the two screws within the barrel requires special consideration. Adjustment factors for the expressions presented have been developed and are available in appropriate references.

Since control of temperature during the extrusion process is important, heat transfer to and from the material being extruded must be considered. Although limited information is available, a few key expressions are available for use in estimating the impacts. An additional factor impacting the characteristics of the extruded material is residence time distribution. All reactions and changes occurring within the system are a function of time, and the characteristics of the extruded material will be influenced by residence time.

PROBLEMS

14.1 A cereal dough is being extruded in a single-screw extruder with the following dimensions, screw length 25 cm, channel height 0.5 cm, and channel width 3 cm. The apparent viscosity of the dough is 1700 Pa s. The density is 1200 kg/m^3. The wall velocity is estimated to be 0.3 m/s. If a mass flow rate of 108 kg/hr is desired, calculate the pressure drop.

14.2 A single screw extruder is being used to process pregelatinized starch. The screw dimensions are as follows: channel width 3 cm, channel height 0.5 cm and screw length 50 cm. The wall velocity is estimated to be 0.31 m/s. The consistency coefficient is 3300 Pa s^n and the flow behavior index is 0.5. A mass flow rate of 91 kg/hr is desired. The density is 1200 kg/m^3. Calculate the pressure drop.

***14.3** A cereal dough is being processed in a single-screw extruder. Estimate the power requirements if the following information is known. The screw length is 50 cm and its diameter is 10 cm.

*Indicates an advanced level in solving.

The channel width is 3 cm, and the channel height is 1 cm. The estimated wall velocity is 0.4 m/s. The density of the dough is 1200 kg/m^3 and the apparent viscosity is 1765 Pa s. The mass flow rate is 254 kg/hr. The rotational speed is 75 rpm.

LIST OF SYMBOLS

A	activation energy constant (J/kg)
B	constant to account for moisture content (% db)
C_1	integration constant (see Eq. (14.16))
D	screw diameter (m)
γ	rate of shear (1/s)
H	height; channel (m)
K	consistency coefficient (Pa s^n)
K_o	consistency coefficient (see Eq. (14.2))
L	length; channel or screw (m)
μ	viscosity (Pa s)
$\dot{m}$	mass flow rate (kg/s)
M	moisture content (% db)
n	flow behavior index
N	screw speed (rps)
N_p	Screw Power number
N_{Res}	Screw Rotation Reynolds number
ΔP	Pressure drop (Pa)
p_t	extruder power consumption (W)
p_s	power consumption for viscous dissipation (W)
ρ	density (kg/m^3)
σ	shear stress (Pa)
T	temperature (C) or (K)
u	velocity (m/s)
u_{max}	maximum velocity (m/s)
u_{mean}	mean velocity (m/s)
V	volumetric flow rate (m^3/s)
V_d	drag flow rate (m^3/s)
W	width of channel (m)
z	vertical dimension
ΔP	pressure difference or change (Pa)

BIBLIOGRAPHY

Cervone, N. W. and Harper, J. M. (1978). Viscosity of an intermediate moisture dough. *J. Food Process Eng.* **2**: 83–95.

Fang, Q., Hanna, M. A., and Lan, Y. (2003). Extrusion system components. *Encyclopedia of Agricultural, Food and Biological Engineering*, 301–305. Marcel Dekker, Inc, New York.

Fang, Q., Hanna, M. A., and Lan, Y. (2003). Extrusion system design. *Encyclopedia of Agricultural, Food and Biological Engineering*, 306–309. Marcel Dekker, Inc., New York.

Fellows, P. J. (1988). Chapter 13: Extrusion. In *Food Processing Technology*. Ellis Horwood, Ltd., London.

Fricke, A. L., Clark, J. P., and Mason, T. F. (1977). Cooking and drying of fortified cereal foods: extruder design. *AIChE Symp. Ser.* **73**: 134–141.

Harper, J. M. (1981). *Extrusion of Foods*, Vol. 1, Vol. 2. CRC Press, Inc., Boca Raton, Florida.

Harper, J. M. (1989). Food extruders and their applications. In *Extrusion Cooking*, C. Mercier, P. Linko, and J. M. Harper, eds. American Association of Cereal Chemists, St. Paul, Minnesota.

Harper, J. M., Rhodes, T. P., and Wanninger, L. A. (1971). Viscosity model for cooked cereal dough. *AIChE Symp. Ser.* **67**: 40–43.

Hermann, D. V. and Harper, J. M. (1974). Modeling a forming foods extruder. *J. Food Sci.* **39**: 1039–1044.

Hsieh, Fu-hung. (2003). Extrusion power requirements. *Encyclopedia of Agricultural, Food and Biological Engineering*, 298–300. Marcel Dekker, Inc., New York.

Jao, Y. C., Chen, A. H., Leandowski, D., and Irwin, W. E. (1978). Engineering analysis of soy dough. *J. Food Process Eng.* **2**: 97–112.

Kokini, J. L., Ho, C-T., and Karwe, M. V. (eds.) (1992). *Food Extrusion Science and Technology*. Marcel Dekker, Inc., New York.

Launay, B. and Bure, J. (1973). Application of a viscometric method to the study of wheat flour doughs. *J. Texture Stud.* **4**: 82–101.

Levine, L. and Miller, R. C. (2006). Chapter 12: Extrusion Processes. *Handbook of Food Engineering*. Marcel Dekker, Inc., New York.

Levine, L. (1982). Estimating output and power of food extruders. *J. Food Process Eng.* **6**: 1–13.

Luxenburg, L. A., Baird, D. O., and Joseph, E. O. (1985). Background studies in the modeling of extrusion cooking processes for soy flour doughs. *Biotechnol. Prog.* **1**: 33–38.

Mercier, C., Linko, P., and Harper, J. M. (eds.). *Cooking Extrusion*. American Association of Cereal Chemists, St. Paul, Minnesota.

Miller, R. C. and Mulvaney, S. J. (2000). Unit operations and equipment IV. Extrusion and Extruders, Chapter 6. In *Breakfast Cereals and How They Are Made*, R. B. Fast and E. F. Caldwell, eds. American Association of Cereal Chemists, St. Paul, Minnesota.

Moore, G. (1994). Snack food extrusion, Chapter 4. In *The Technology of Extrusion Cooking*, N. D. Frame, ed. Blackie Academic & Professional, London.

Nazarov, N. I., Azarov, B. M., and Chaplin, M. A. (1971). Capillary viscometry of macaroni dough. *Izv. Vyssh. Uchebn. Zaaved. Pishch. Teknol.* **1971**: 149.

Rauwendaal, C. (1986). *Polymer Extrusion*. Carl Hanser, New York.

Remson, C. H. and Clark, P. J. (1978). Viscosity model for a cooking dough. *J. Food Process Eng.* **2**: 39–64.

Toledo, R., Cabot, J., and Brown, D. (1977). Relationship between composition, stability and rheological properties of rat comminuted meat batters. *J. Food Sci.* **42**: 726.

Tsao, T. F., Harper, J. M., and Repholz, K. M. (1978). The effects of screw geometry on the extruder operational characteristics. *AIChE Symp. Ser.* **74**: 142–147.

van Zuilichem, D. J., Buisman, G., and Stolp, W. (1974). Shear behavior of extruded maize. Paper presented at the 4th Int. Cong. Food Sci. Technol., International Union of Food Science and Technology, Madrid.

Wiedmann, W. (1992). Improved product quality through twin-screw extrusion and closed-loop quality control, Chapter 35. In *Food Extrusion Science and Technology*, J. L. Kokini, C. T. Ho, and M. V. Karwe, eds. Marcel Dekker, Inc, New York.

Chapter 15

Packaging Concepts

Developments in food packaging have evolved in response to the need for protection of the food product from both external and internal environments and in response to consumer expectations for convenience and product safety. A recent survey indicated that 72% of consumers in the United States are willing to pay extra for guaranteed product freshness delivered by the type of packaging. Many new packaging developments have focused on extending the shelf life of the product and on delivering a higher quality product to the consumer. These developments would not be possible without significant advances in the materials used in packaging and the incorporation of various types of sensors into food packaging.

15.1 INTRODUCTION

Historically, the packaging of foods has evolved in response to a variety of expectations. The functions of packaging for food have been documented by Yam et al. (1992), March (2001), Robertson (2006), and Krochta (2007). The four basic functions of a food package are:

- Containment
- Protection
- Communication
- Convenience

Containment is defined by the food product, with different types of packages required for liquids as compared with solids or dry powders. Product protection is a key function for most packages in order to maintain the quality and safety of the food. The communication function is most obvious in the various types of information presented on the outside surface of packages, including simple product

All icons in this chapter refer to the author's web site which is independently owned and operated. Academic Press is not responsible for the content or operation of the author's web site. Please direct your web site comments and questions to the author: Professor R. Paul Singh, Department of Biological and Agricultural Engineering, University of California, Davis, CA 95616, USA.
Email: rps@rpaulsingh.com

descriptions and details of product composition. The convenience of many foods is highly dependent on package design. These four basic functions of food packages take on different levels of importance with different food products. Additional factors that can impact packaging include efficiency in package manufacturing, the impact of the package on the environment, and level of food safety provided by the package.

15.2 FOOD PROTECTION

The degree of protection required by a food product is a key factor in selecting the packaging material and design. Figure 15.1 summarizes the role of the food package in this regard. In general, protection is defined in terms of a variety of factors that can impact the quality attributes of the food product from the time the product is placed in the container to the time of consumption. Environmental parameters such as oxygen, nitrogen, carbon dioxide, water vapor or aromas, in direct or indirect contact with the product, are influenced by the package properties. Many foods are sensitive to the oxygen concentration

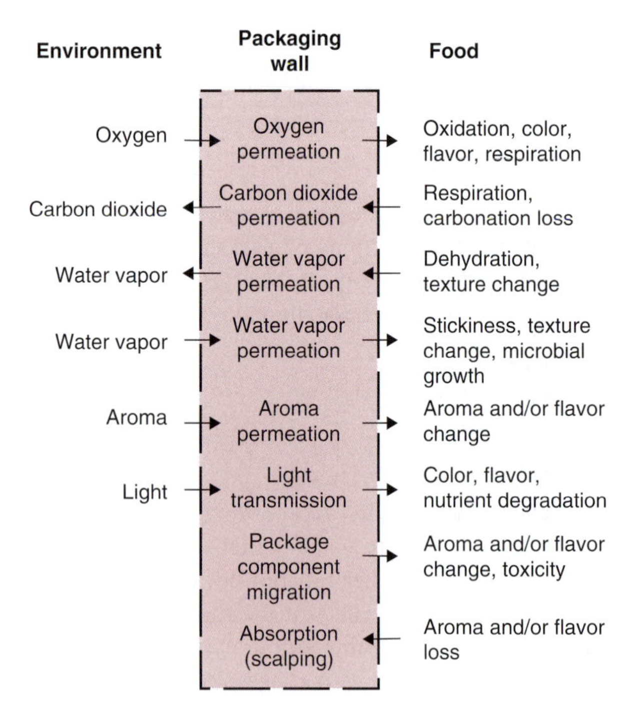

Figure 15.1 Interactions among the food, the package, and the environment. (Adapted from Linssen and Roozen, 1994)

in the immediate environment due to the deterioration associated with oxidation. The shelf life of fresh food commodities is impacted by concentrations of carbon dioxide in direct contact with the product. In a similar manner, the water vapor concentration in the environment in direct contact with a dry and/or intermediate moisture food must be controlled. In all situations, the properties of the packaging material have a significant role in establishing the shelf life and quality of the product reaching the consumer.

15.3 PRODUCT CONTAINMENT

The function of the package in containing the food product is directly related to the packaging material. A variety of materials are used for food packages including glass, metals, plastics, and paper. Each material has unique properties and applications for food packaging.

Glass containers for foods provide an absolute barrier for gases, water vapors, and aromas but do not protect products with sensitivities to light in the environment surrounding the product package. A major disadvantage of glass is weight compared with other packaging materials.

Metal containers are used for a significant number of shelf-stable food products such as fruits and vegetables, and offer an excellent alternative to glass containers. The metals used include steel, tin, and aluminum, with each representing a unique application for foods or beverages. Due to structural integrity, metal containers have been used for thermally processed foods, with many applications for foods processed in retorts at high temperatures and pressures. Metal containers are relatively heavy, and the container manufacturing process is complex.

Plastic packaging materials are used for an increasing number and variety of food products. Most plastic packaging materials are either thermoplastic or thermoset polymers. Thermoplastic polymers are the basic material used for a large number of food products, and provide significant flexibility in package design based on the specific needs of the food. Plastic films are lightweight and provide an unobscured view of the product within the package. The permeability of polymers to oxygen, carbon dioxide, nitrogen, water vapor, and aromas provides both challenges and opportunities in the design of packaging materials for specific food requirements.

Due to its broad use in all levels of packaging, paper is used for food packaging more than any other material. It is the most versatile and flexible type of material. Its key disadvantage is the lack of a barrier to oxygen, water vapor, and similar agents that cause deterioration of product quality.

15.4 PRODUCT COMMUNICATION

The package of a food product is used to communicate information about the product to consumers. This information is presented on the label and includes both legally required information about the ingredients and information needed to market the product.

15.5 PRODUCT CONVENIENCE

A variety of designs have been incorporated into food packages in an effort to increase convenience. These designs include innovation in opening the packages, dispensing the product, resealing the package, and the ultimate preparation of the product before consumption. Convenience will continue to provide innovation in the future.

15.6 MASS TRANSFER IN PACKAGING MATERIALS

An important requirement in selecting packaging systems for foods is the barrier property of the packaging material. To keep a food product crisp and fresh, the package must provide a barrier to moisture. Rancidity can be minimized by keeping a food protected from light. To reduce oxidation of food constituents, the packaging material must provide a good barrier to oxygen. The original aroma and flavor of a food can be maintained by using a packaging material that offers a barrier to a particular aroma. Thus, properly selected packaging materials are beneficial in extending the shelf life of foods. The barrier properties of a packaging material can be expressed in terms of permeability.

The permeability of a packaging material provides a measure of how well a certain gas or vapor can penetrate the packaging material. In quantitative terms, permeability is the mass of gas or vapor transferred per unit of time, area, and a "driving force." In the case of diffusional mass transfer, the driving force is a difference in concentration or in partial pressures. If the driving force is a difference in total pressure, the mass transfer occurs due to bulk flow of a gas or vapor. A polymeric membrane may be thought of as an aggregate of wriggling

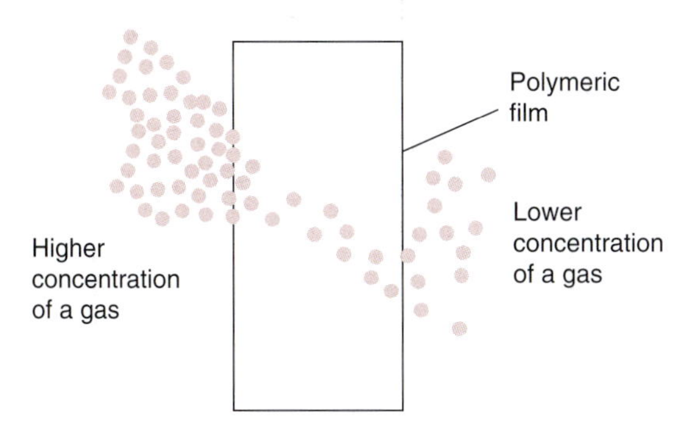

Figure 15.2 Mass transfer of a gas through a polymeric material.

worms, with worms representing the long chains of polymers. The space between the worms is like the interstitial space through which a species passes. The wriggling of worms is representative of the thermal motion of polymeric chains.

Mass transport through polymeric materials can be described as a step process. Referring to Figure 15.2, in step 1, the gas vapor or liquid molecules dissolve in the polymeric material on the side of the film exposed to the higher concentration. In step 2, the gas or vapor molecules diffuse through the polymeric material moving toward the side of the film exposed to the lower concentration. The movement of molecules depends on the availability of "holes" in the polymeric material. The "holes" are formed as large chain segments of the polymer slide over each other due to thermal agitation. Finally, step 3 involves the desorption of the gas or vapor molecules and evaporation from the surface of the film.

We can again use Fick's law of diffusion to develop an expression for the transport process of a gas through a polymeric material. From Equation (10.11):

$$\frac{\dot{m}_B}{A} = \frac{D_B(c_{B1} - c_{B2})}{(x_2 - x_1)} \tag{15.1}$$

This equation would be sufficient to determine the rate of flux, $\dot{m}_B/A$, but the concentrations of a gas at the film surfaces are more difficult to measure than partial pressures. The concentrations can be converted to partial pressures by using Henry's law,

$$c = Sp \tag{15.2}$$

where S is solubility (moles/[cm^3 atm]) and p is partial pressure of gas (atm). Thus, we have

$$\dot{m}_B = \frac{D_B SA(p_{B1} - p_{B2})}{(x_2 - x_1)} \quad (15.3)$$

The quantity $D_B S$ is known as the permeability coefficient, P_B.

$$P_B = \frac{\text{(amount of gas vapor)(thickness of film)}}{\text{(area)(time)(pressure difference across the film)}} \quad (15.4)$$

A wide variety of units are used to report the permeability coefficient (Table 15.1).

Another parameter used by some authors is *permeance*, which is not corrected to a unit thickness. Sometimes permeance to water vapor is

Table 15.1 Conversion Factors for Various Units of Permeability Coefficient

	$\frac{\text{cm}^3\ \text{cm}}{\text{s cm}^2\ \text{(cmHg)}}$	$\frac{\text{cm}^3\ \text{cm}}{\text{s cm}^2\ \text{(Pa)}}$	$\frac{\text{cm}^3\ \text{cm}}{\text{day m}^2\ \text{(atm)}}$
$\frac{\text{cm}^3\ \text{cm}}{\text{s cm}^2\ \text{(cmHg)}}$	1	7.5×10^{-4}	6.57×10^{10}
$\frac{\text{cm}^3\ \text{mm}}{\text{s cm}^2\ \text{(cmHg)}}$	10^{-1}	7.5×10^{-5}	6.57×10^{9}
$\frac{\text{cm}^3\ \text{mm}}{\text{s cm}^2\ \text{(atm)}}$	1.32×10^{-2}	9.9×10^{-6}	8.64×10^{8}
$\frac{\text{cm}^3\ \text{mil}}{\text{day m}^2\ \text{(atm)}}$	3.87×10^{-14}	2.9×10^{-17}	2.54×10^{-3}
$\frac{\text{in}^3\ \text{mil}}{\text{day 100 in}^2\ \text{(atm)}}$	9.82×10^{-12}	7.37×10^{-15}	6.46×10^{-1}
$\frac{\text{cm}^3\ \text{cm}}{\text{day m}^2\ \text{(atm)}}$	1.52×10^{-11}	1.14×10^{-14}	1

Source: Yasuda and Stannett (1989)

reported in units that are neither corrected to unit thickness nor to unit pressure, but this value must always be reported with specified thickness, humidity, and temperature. For example, water vapor permeability is defined as grams of water per day per 100 cm^2 of package surface for a specified thickness and temperature and/or a relative humidity on one side of approximately 0% and on the other side of 95%.

15.6.1 Permeability of Packaging Material to "Fixed" Gases

Gases such as oxygen, nitrogen, hydrogen, and carbon dioxide, which have low boiling points, are known as "fixed" gases. They show similar ideal behavior with respect to permeability through packaging materials. The permeability of O_2, CO_2, and N_2 for several polymeric materials is shown in Table 15.2. It is evident that for any given gas

Table 15.2 Permeability Coefficients, Diffusion Constants, and Solubility Coefficients of Polymers[a]

Polymer	Permeant	T[°C]	$P \times 10^{10}$	$D \times 10^6$	$S \times 10^2$
Poly(ethylene) (density 0.914)	O_2	25	2.88	0.46	4.78
	CO_2	25	12.6	0.37	25.8
	N_2	25	0.969	0.32	2.31
	H_2O	25	90		
Poly(ethylene) (density 0.964)	O_2	25	0.403	0.170	1.81
	CO_2	25	1.69	0.116	11.1
	CO	25	0.193	0.096	1.53
	N_2	25	0.143	0.093	1.17
	H_2O	25	12.0		
Poly(propylene)	H_2	20	41	2.12	
	N_2	30	0.44		
	O_2	30	2.3		
	CO_2	30	9.2		
	H_2O	25	51		
Poly(oxyethyleneoxyterephthaloyl) (Poly(ethylene terephthalate)) crystalline	O_2	25	0.035	0.0035	7.5
	N_2	25	0.0065	0.0014	5.0
	CO_2	25	0.17	0.0006	200
	H_2O	25	130		
Cellulose acetate	N_2	30	0.28		
	O_2	30	0.78		
	CO_2	30	22.7		
	H_2O	25	5500		

(Continued)

Table 15.2 *(Continued)*

Polymer	Permeant	T[°C]	$P \times 10^{10}$	$D \times 10^6$	$S \times 10^2$
Cellulose (Cellophane)	N_2	25	0.0032		
	O_2	25	0.0021		
	CO_2	25	0.0047		
	H_2O	25	1900		
Poly(vinyl acetate)	O_2	30	0.50	0.055	6.3
Poly(vinyl alcohol)	H_2	25	0.009		
	N_2	14[b]	<0.001		
		14[c]	0.33	0.045	5.32
	O_2	25	0.0089		
	CO_2	25	0.012		
		23[b]	0.001		190
		23[d]	11.9	0.0476	
	ethylene oxide	0	0.002		
Poly(vinyl chloride)	H_2	25	1.70	0.500	2.58
	N_2	25	0.0118	0.00378	2.37
	O_2	25	0.0453	0.0118	2.92
	CO_2	25	0.157	0.00250	47.7
	H_2O	25	275	0.0238	8780.0
Poly(vinylidene chloride) (Saran)	N_2	30	0.00094		
	O_2	30	0.0053		
	CO_2	30	0.03		
	H_2O	25	0.5		
Poly[imino (1-oxohexamethylene)] (Nylon 6)	N_2	30	0.0095		
	O_2	30	0.038		
	CO_2	20	0.088		
		30[b]	0.10		
		30[e]	0.29		
	H_2O	25	177		
Poly[imino (1-oxoundecamethylene)] (Nylon 11)	CO_2	40	1.00	0.019	40

Notes: See overleaf.
Source: Yasuda and Stannett (1989)
[a] *Units used are as follows: P in [cm^3 (STP) cm cm^{-2} s^{-1} (cm Hg)$^{-1}$], D in [cm^2 s^{-1}], and S in [cm^3 (STP) cm^{-3} atm^{-1}]. To obtain corresponding coefficients in SI units, the following factors should be used: $P \times 7.5 \times 10^{-4}$ = [cm^3 (STP) cm cm^{-2} s^{-1} Pa^{-1}]; $S \times 0.987 \times 10^{-5}$ = [cm^3 (STP) cm^{-3} Pa^{-1}]*
[b] *Relative humidity 0%.*
[c] *Relative humidity 90%*
[d] *Relative humidity 94%*
[e] *Relative humidity 95%*

there exist materials with widely differing permeabilities. For example, Saran is 100,000 times less permeable to oxygen than silicone rubber. Moreover, there are certain regularities in the transmission of different gases through the same material. For example, carbon dioxide permeates four to six times faster than oxygen, and oxygen four to six times faster than nitrogen. Since carbon dioxide is the largest of the three gas molecules, we would expect its diffusion coefficient to be low, and it is. Its permeability coefficient is high because its solubility S in polymers is much greater than that for other gases.

Fixed gases also show ideal behaviors:

1. Permeabilities can be considered independent of concentration.
2. The permeabilities change with temperature in accordance with the following relation:

$$P = P_o e^{-E_p/RT} \tag{15.5}$$

where E_p is the activation energy for permeability (k cal/mol).

For some materials there is a break in the permeability temperature curve, and above a critical temperature the material is much more permeable. For polyvinyl acetate, that temperature is around 30°C and for polystyrene it is around 80°C. Breaks are due to a glass transition temperature T_g', below which the material is glassy, and above which it is rubbery.

Example 15.1

The permeability coefficient for a 0.1-mm polyethylene film is being measured by maintaining a moisture vapor gradient across the film in a sealed test apparatus. The high moisture vapor side of the film is maintained at 90% RH and a salt ($ZnCl \cdot \frac{1}{2} H_2O$) maintains the opposite side at 10% RH. The area of film exposed to vapor transfer is 10 cm by 10 cm. When the test is conducted at 30°C, a weight gain of 50 g in the desiccant salt is recorded after 24 h. From these given data, calculate the permeability coefficient of the film.

Given

Film thickness $= 0.1\ mm = 1 \times 10^{-4}\ m$
High relative humidity $= 90\%$
Low relative humidity $= 10\%$
Temperature $= 30°C$
Film area $= 10\ cm \times 10\ cm = 100\ cm^2 = 0.01\ m^2$
Moisture rate of flow $= 50\ g/24\ h = 5.787 \times 10^{-4}$ *g water/s*

Approach

We will use Equation (15.4) to calculate the permeability coefficient (P_B) after vapor pressures are expressed in terms of moisture contents of air.

Solution

1. *By using Equation (9.16) modified with vapor pressures,*

$$\phi = \frac{p_w}{p_{ws}} \times 100$$

2. *From Table A.4.2,*

$$p_{ws} = 4.246 \text{ kPa at } 30°\text{C}$$

3. *From steps 1 and 2, at 10% relative humidity,*

$$p_w = 4.246 \times 10/100$$
$$= 0.4246 \text{ kPa}$$

At 90% relative humidity,

$$p_w = 4.246 \times 90/100$$
$$= 3.821 \text{ kPa}$$

Using Equation (15.4) and solving for permeability coefficient,

$$P_B = \frac{(5.787 \times 10^{-4} \text{ g water/s})(1 \times 10^{-4} \text{ m})}{(0.01 \text{ m}^2)(3.821 \text{ kPa} - 0.4246 \text{ kPa})(1000 \text{ Pa/kPa})}$$

$$P_B = 1.7 \times 10^{-9} \text{ [(g water m)/(m}^2 \text{ Pa s)]}$$

4. *The permeability coefficient of the film is calculated to be 1.7×10^{-9} [(g water m)/(m^2 Pa s)]; the units may be converted to any other form desired using Table 15.1.*

15.7 INNOVATIONS IN FOOD PACKAGING

Innovations in food packaging have created an array of new terms associated with the role of packaging in the improvement of safety, shelf life, and convenience of the food product. There are three broad categories of packaging for foods: passive, active, and intelligent. The two types of active packaging systems include simple and advanced. Intelligent packaging systems are simple and interactive.

15.7.1 Passive Packaging

A passive packaging system is a system that serves as a physical barrier between the product and the environment surrounding the package.

Most conventional packaging used for food products would be described as passive packaging systems. Metal cans, glass bottles, and many of the flexible packaging materials provide a physical barrier between the product and the environment. These packaging systems ensure that most properties of the environment and the agents contained in the environment are prevented from making contact with the food. In general, these packages are expected to provide maximum protection of the product but are not responsive to any of the changes that might occur within the container.

Innovations in passive packaging systems continue with the development of new barrier coatings for polymer containers and films. These new materials reduce or control permeability of agents that could impact the safety or shelf-life of the food product within the container.

15.7.2 Active Packaging

An active packaging system is a system that detects or senses changes within the package environment, followed by modification of package properties in response to the detected change.

15.7.2.1 *Simple Active*

A packaging system that does not incorporate an active ingredient and/or actively functional polymer is referred to as a simple active system. An active packaging system responds in some manner to changes occurring within the package. One of the earliest active packaging systems was Modified Atmosphere Packaging (MAP). The package materials for MAP have properties that attempt to control the atmosphere within the package in contact with the food product. A simple example is packaging films that help maintain desired concentration of oxygen and carbon dioxide in a package containing respiring fruits and vegetables.

15.7.2.2 *Advanced Active*

A packaging system that contains an active ingredient and/or an actively functional polymer is an advanced active system. Advanced active packaging systems can be divided into two categories. The incorporation of an oxygen scavenger into the package is an example of a system that absorbs an unwanted agent within the environment of the package and in contact with the product. Usually, scavengers are included as small sachets inserted into the package to reduce the level of oxygen within the package. An alternative to the sachets is the integration of an active scavenging system into packaging materials, such as films or closures.

A second example of the "absorbing" type of system is an ethylene scavenger. Ethylene triggers ripening, accelerates senescence, and reduces the shelf life of climacteric fruits and vegetables. By incorporating an ethylene-absorbing material into the packaging system, the shelf life of the product can be extended.

Moisture content or water activity becomes a shelf life limiting factor for many foods. For these products, the incorporation of water-absorbing material into the package can be beneficial. More sophisticated control may be required within packages for products requiring regulation of humidity levels.

Another example of an advanced active packaging system would be MAP with barrier properties in combination with gas flushing to achieve a desirable steady-state atmosphere composition around the product. Other active absorbing packaging systems are available to control concentrations of undesirable flavor constituents or carbon dioxide.

Advanced active packaging systems are continuing to evolve with the development of systems designed to release agents in response to undesirable activity within the package. For example, in response to microbial growth, the release of antimicrobial agents can be used to extend the shelf-life of refrigerated foods. Similarly, antioxidants can be incorporated into packaging films to protect the film and the product within the package from degradation. The potential for incorporation of flavor compounds into a package represents a significant opportunity. The release of these compounds during the shelf life of a food product could mask off-odors or improve the sensory attributes of the product.

15.7.3 Intelligent Packaging

A packaging system that senses changes in the environment and responds with corrective action is defined as an intelligent packaging system. Four objectives for intelligent packaging systems have been identified:

1. Improved product quality and product value
2. Increased convenience

3. Changes in gas permeability properties
4. Protection against theft, counterfeiting, and tampering

15.7.3.1 *Simple Intelligent*

A packaging system that incorporates a sensor, an environmental reactive component, and/or a computer-communicable device is a simple intelligent system. Quality or freshness indicators are examples of simple intelligent packaging systems and are used to communicate changes in a product quality attribute. These internal or external indicators/sensors may indicate elapsed time, temperature, humidity, time-temperature, shock abuse, or gas concentration. The primary function of the sensor is to indicate quality losses during storage and distribution. Other sensors are available for changes in color, physical condition (damage), and microbial growth.

A second type of simple intelligent package system is the time-temperature indicator (Taoukis et al., 1991). Temperature-indicating labels are attached to the external surface of the package and report the maximum temperature to which the surface has been exposed. Time-temperature integrators provide more refined information by integrating the impact of time and temperature at the package surface (Wells and Singh, 1988a, 1988b; Taoukis and Labuza, 1989). The disadvantage of both types of indicators is that the output may not reveal the actual impact on the product.

Internal gas-level indicator sensors are components of a simple intelligent packaging system. These sensors are placed in the package to monitor gas concentrations and provide rapid visual monitoring based on color changes. In addition to monitoring food product quality, the sensor can detect package damage. Indicators are available for monitoring oxygen (O_2) and carbon dioxide (CO_2). A variation on these types of indicators may be used for detecting microbial growth by monitoring for increased CO_2 levels.

Simple intelligent package systems for supply chain management and traceability use automatic data capture during distribution and with connection to the Internet. Examples include systems to trace containers of fruits from the fields to the point of delivery to the customer. Currently, Radio Frequency Identification (RFID) tags are attached to bulk containers to provide information on content, weight, location, and times throughout the distribution channels. It is possible that in the future RFID labels will be placed on individual food packages.

Another convenience-type of simple intelligent packaging system may be incorporated into intelligent cooking appliance systems. These systems carry essential information about the food product and the package on a bar code and assist appliances by providing preparation instructions, maintaining food product inventories, and identifying expiration dates.

15.7.3.2 *Interactive Intelligent*

Interactive intelligent packaging systems incorporate mechanisms to respond to a signal (sensor, indicator, or integrator). For example, an interactive intelligent packaging system can change permeability properties to accommodate changes in freshness and quality of the packaged food product during storage and distribution. New and improved intelligent breathable films change permeability in response to different fresh produce or changes in temperature of the produce. These packaging films have been developed based on knowledge of respiration rates as a function of temperature. The systems are ideal for high-respiratory-rate fresh and fresh-cut produce.

An evolving expectation for interactive intelligent packaging systems includes detection of theft, counterfeiting, and tampering. To accomplish this, simple intelligent packaging systems could include labels or tapes that are invisible before tampering but change color permanently if tampering occurs. More complex devices would respond to counterfeiting and theft by using holograms, special inks and dyes, laser labels, bar codes, and electronic data tags.

15.8 FOOD PACKAGING AND PRODUCT SHELF-LIFE

The packaging provided for a food product may have a significant impact on its shelf-life. These impacts are closely related to the types of protection provided by the package, as mentioned earlier. The impact of the package can be quantified using information on the deterioration of the product in combination with the type of protection provided by the package. We will first develop mathematical relationships that are useful in describing the deterioration of food products during storage.

15.8.1 Scientific Basis for Evaluating Shelf Life

The scientific basis for evaluating shelf life of a food product relies on principles of chemical kinetics (Labuza, 1982; Singh, 2000). In this section, we will examine changes in a quality attribute, Q, measured over time. The general rate expression for the quality attribute may be written as

$$\pm\frac{d[Q]}{dt} = k[Q]^n \quad (15.6)$$

where $\pm$ indicates that the quality attribute may increase or decrease during storage, k is the pseudo forward rate constant, and n is the order of reaction. We will first assume that the environmental factors that influence shelf life such as storage temperature, moisture, and light are kept constant.

If the quality attribute decreases with time, then we may write Equation (15.6) as

$$-\frac{d[Q]}{dt} = k[Q]^n \quad (15.7)$$

15.8.1.1 *Zero-Order Reaction*

In Figure 15.3, the measured amount of a quality attribute remaining at different storage times is plotted. The plot is linear, suggesting that the rate of loss of quality attribute remains constant over the entire period. This type of behavior is observed in many shelf-life studies, including quality changes due to enzymatic degradation and non-enzymatic browning. Similar behavior is observed for lipid oxidation, which often causes the development of rancid flavors.

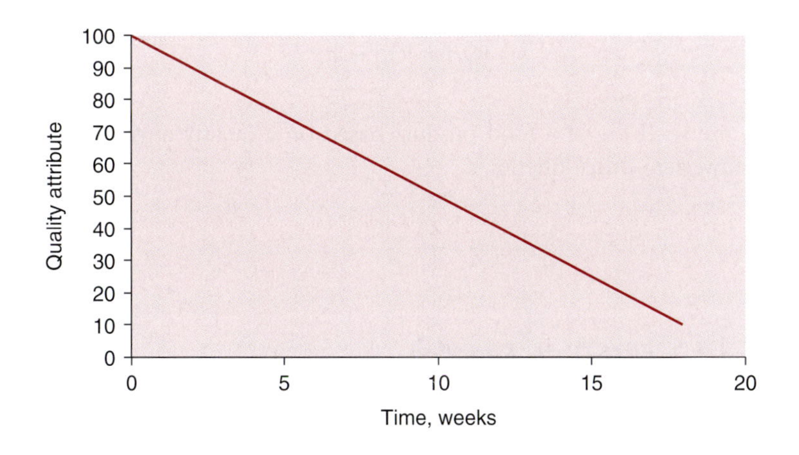

Figure 15.3 A linear decrease in the remaining quality attribute plotted against time.

The linear plot in Figure 15.3 represents a zero-order reaction. Therefore, if we substitute $n = 0$ in Equation (15.7), then we get

$$-\frac{dQ}{dt} = k \tag{15.8}$$

We can solve this equation using the procedure of separation of variables to obtain an algebraic solution. We will consider that initially the amount of quality attribute is represented by Q_i and after some storage time, t, it is Q.

Then,

$$-\int_{[Q_i]}^{[Q]} d[Q] = k\int_0^t dt \tag{15.9}$$

Integrating, we get,

$$-\left|[Q]\right|_{[Q_i]}^{[Q]} = kt \tag{15.10}$$

or

$$[Q_i] - [Q] = kt \tag{15.11}$$

In Equation (15.11), the left-hand side denotes the extent of reaction, ξ, for a reaction that follows a zero-order kinetics. Thus,

$$\xi = kt \tag{15.12}$$

if we specify that the end of shelf life of a food product, t_s, is reached when the quality attribute, Q, reaches Q_f. Then, from Equation (15.11),

$$[Q_f] = [Q_i] - kt_s \tag{15.13}$$

or, the shelf life of a food product based on a quality attribute that follows zero-order kinetics is

$$t_s = \frac{[Q_i] - [Q_f]}{k} \tag{15.14}$$

15.8.1.2 *First-Order Reaction*

Let us consider another shelf life study where the measured quality attribute follows a profile as shown in Figure 15.4. This plot shows an

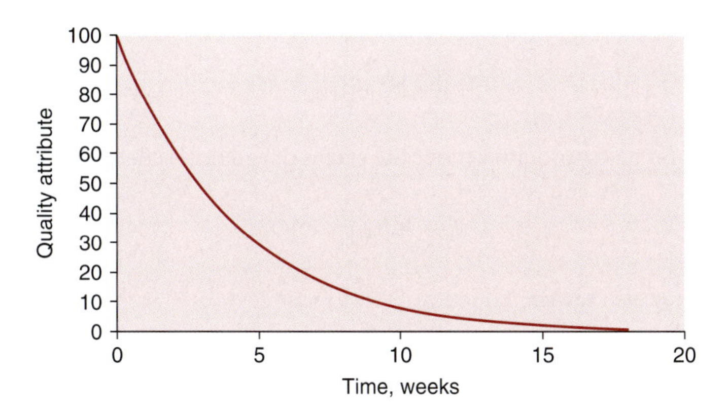

Figure 15.4 An exponential decrease in the remaining quality attribute plotted against time.

exponential decrease of the quality attribute. In this study, the rate of loss of quality attribute depends upon the amount of remaining quality attribute. Food deterioration reactions that show an exponential decrease include loss of vitamins and protein, and microbial growth as we learned in Chapter 5. The exponential decrease in a quality attribute is described by the first-order kinetics, where the rate of reaction, $n = 1$. Thus we may rewrite Equation (15.7) as

$$-\frac{\mathrm{d}[Q]}{\mathrm{d}t} = k[Q] \tag{15.15}$$

We again use the method of separation of variables to integrate this equation as

$$-\frac{\mathrm{d}[Q]}{[Q]} = k\mathrm{d}t \tag{15.16}$$

Again, we assume that initially the quality attribute is $[Q_i]$ and after time t it decreases to $[Q]$, then

$$-\int_{[Q_i]}^{[Q]} \frac{\mathrm{d}[Q]}{[Q]} = k\int_0^t \mathrm{d}t \tag{15.17}$$

Integrating, we get,

$$-\left|\ln[Q]\right|_{[Q_i]}^{[Q]} = k\left|t\right|_0^t \tag{15.18}$$

or

$$\ln[Q_i] - \ln[Q] = kt \tag{15.19}$$

Thus for a first-order reaction, the extent of reaction is determined as

$$\xi = \ln[Q_i] - \ln[Q] \tag{15.20}$$

We may also rewrite Equation (15.19) as

$$\ln\frac{[Q]}{[Q_i]} = -kt \tag{15.21}$$

To determine the fraction of quality change after a certain storage period, Equation (15.21) may be rewritten as

$$\frac{[Q]}{[Q_i]} = e^{-kt} \tag{15.22}$$

In Equation (15.21), $[Q]$ is the amount of quality attribute remaining after a storage time t.

If $[Q_f]$ represents the amount of quality attribute at the end of shelf life, then from Equation (15.21) we obtain

$$t_s = \frac{\ln\frac{[Q_i]}{[Q_f]}}{k} \tag{15.23}$$

Equation (15.23) can easily be modified if we want to know time for half-life of a quality attribute. We substitute $Q_f = 0.5Q_i$ to obtain

$$t_{1/2} = \frac{\ln 2}{k} = \frac{0.693}{k} \tag{15.24}$$

Equations (15.14) and (15.24) have been used to predict shelf life of a food product, or the time that would elapse from when the product is placed in storage until the quality attribute deteriorates to an unacceptable level $[Q_f]$. The rate constant (k) has been established for key deterioration reactions in foods, or it can be determined during shelf life evaluations.

The impact of environmental conditions during product storage is expressed through the magnitude of the rate constant (k). The most evident environmental variable is temperature, and the influence of temperature on the rate constant is expressed by the Arrhenius equation (Eq. (5.9)):

$$k = Be^{\left(-\frac{E_A}{RT_A}\right)} \tag{15.25}$$

where the activation energy constant (E_A) quantifies the influence of temperature on the rate of product quality deterioration during storage. This expression is used primarily during accelerated shelf life testing, when elevated temperatures are used under experimental conditions to accelerate the deterioration of the product. These approaches allow us to quantify shelf life in a relatively short period of time (days or weeks) at elevated temperatures, whereas the actual shelf life may be much longer (1 to 2 years).

In most situations, the impact of the package on product shelf life is a function of environmental parameters other than temperature. As indicated earlier, the concentrations of oxygen, water vapor, nitrogen, carbon dioxide, and other environmental parameters may influence the deterioration reactions in foods. In order to incorporate the impact of these parameters into the prediction of shelf life, we will need to know the influence of these variables on rate constants. These additional relationships are then used in combination with expressions describing the transport of the agent across the package barrier. These expressions, as based on Equation (15.1) and appropriate permeability coefficients, were presented earlier in this chapter.

Example 15.2

A dry breakfast cereal has been fortified with ascorbic acid (Vitamin C). The degradation of ascorbic acid during storage of the cereal is a function of water activity of the product. The package used for the cereal must ensure that the water activity is maintained at a sufficiently low level to preserve the desired level of Vitamin C.

The initial water activity of the cereal is 0.1, and the moisture content is 3.0% (d.b.). The relationship between product moisture content and water activity was established as

$$w = 0.175\, a_w + 0.0075$$

The rate constant (k) for degradation of ascorbic acid is 1.701×10^{-5}/min at a water activity of 0.1, and the rate constant increases with increasing water activity according to the following relationship:

$$k = 2.733 \times 10^{-5} a_w + 1.428 \times 10^{-5}$$

The package dimensions are 20 cm × 30 cm × 5 cm, and there is 0.5 kg of product in the package. Select the package film permeability, with 1 mm thickness, to ensure that 60% of the original ascorbic acid is retained in the product after 14 days of storage in an environment with a relative humidity of 60% and 30°C.

Given

Product initial moisture content (w) = 3.0% (d.b.)
Product initial water activity (a_w) = 0.1
Storage environment relative humidity = 60%
Ascorbic acid degradation rate constant (k) = 1.701×10^{-5}/min
Shelf-life expectation = 14 days
Package surface area = 0.17 m^2

Approach

The key expression used in the solution is Equation (15.3), with the permeability (P_B) = $D_B S$. The degradation rate of ascorbic acid is used to establish the maximum product water activity allowed to prevent degradation beyond 60% within the 14-day shelf life. Given the maximum allowable water activity, we can determine the total amount of water that can be added to the product over the 14 days of storage, and the permeability of the package film with a given thickness.

Solution

1. *Use Equation (15.22) to determine the average rate constant for the time period of product shelf life:*

$$0.6 = \exp[-(k_{ave})(14\ \text{days})(24\ \text{hr/day})(60\ \text{min/hr})]$$

$$k_{ave} = 2.53 \times 10^{-5}/\text{min}$$

2. *Given the relationship between water activity and the rate constant:*

$$2.53 \times 10^{-5} = 2.733 \times 10^{-5} a_w + 1.428 \times 10^{-5}$$

$$a_w = 0.4$$

3. *In order to determine the amount of moisture change needed to increase the water activity to 0.4, the water activity must be converted to moisture content. The typical relationship is given by Equation (12.1), but the relationship for this product for the range of water activities between 0.1 and 0.4 is*

$$w = 0.175a_w + 0.0075$$

Given this relationship, the product moisture content at a water activity of 0.4 is

$$w = 0.175(0.4) + 0.0075 = 0.0775 \quad \text{or} \quad 7.75\% \text{ (d.b)}$$

4. *During the 14 days of storage, the moisture content can increase from 3% to 7.75% or*

$$0.0775 - 0.03 = 0.0475 \text{ kg water/kg product solids}$$

$$\text{and } 0.02375 \text{ kg water per package}$$

is the amount of moisture transfer through the package film over a period of 14 days. Then the transfer rate becomes

$$= \frac{0.02375 \text{ kg water}}{(14 \text{ days})(24 \text{ hr/day})(60 \text{ min/hr})}$$

$$= 1.178 \times 10^{-6} \text{ kg water/min}$$

5. *Using Equation (15.3):*

$$P_B = \frac{(1.178 \times 10^{-6} \text{ kg water/min})(1 \times 10^{-3} \text{ m})}{(60 \text{ s/min})(0.17 \text{ m}^2)(2.5476 - 0.4246 \text{ kPa})(1000 \text{ Pa/kPa})}$$

where: $p_{ws} = 4.246(0.6) = 2.5476$

$$p_w = 4.246(0.1) = 0.4426$$

$$P_B = 5.44 \times 10^{-14} \text{ kg m/(m}^2 \text{ Pa s)}$$

6. *By converting to units consistent with Table 15.1:*

$$P_B = 6.44 \times 10^{-10} \text{ cm}^3 \text{ cm/(cm}^2 \text{ s cm Hg)}$$

7. *Based on information presented in Table 15.2, polyvinylidene chloride would be the appropriate package film to ensure the desired shelf life.*

15.9 SUMMARY

Recent innovations in packaging for foods and food products have resulted in an array of developments that improve the safety, convenience, shelf life, and overall quality of the products. New packaging provides opportunities for detection of changes in the product during storage and distribution and the potential for corrective action based on package design. Future developments will provide more sophisticated packaging to extend shelf life and improve quality attributes of the food product.

The evolution of "nano-scale science" has potential impacts for future packaging innovations. The outcomes from nano-scale research will lead to nano-materials with unique properties for protection of foods from all types of harmful agents. These materials will provide such opportunities as antimicrobial surfaces and sensing of microbiological and biochemical activities for all packaging systems. The application of nano-scale science will lead to packaging materials that will adjust to changes in pH, pressure, temperature, and light as well as to other byproducts of reactions occurring within the package.

PROBLEMS

15.1 A dry food product is contained in a 1 cm × 4 cm × 3 cm box using a polymer film to protect the oxygen sensitivity of the product. The concentration gradient across the film is defined by the oxygen concentration in air and 1% within the package. The oxygen diffusivity for the polymer film is $3 \times 10^{-16}\, m^2/s$. Estimate the film thickness needed to ensure a product shelf life of 10 months. The shelf life of the product is established as the time when oxidation reactions within the product have used 0.5 mol of oxygen.

***15.2** A dry food product is being stored in a package with a 0.75 mm polypropylene film. The dimensions of the package are 15 cm × 15 cm × 5 cm, and the amount of product in the package is 0.75 kg. The initial water activity of the product is 0.05, and the initial moisture content is 2% (d.b.). A key component of the product is sensitive to water activity, and the rate constant for degradation of the component is described by

$$k = 5 \times 10^{-5}\, a_w + 1.5$$

* Indicates an advanced level in solving.

The relationship of the product water activity to moisture content is described by the following GAB constants

$$K = 1.05$$

$$C = 5.0$$

$$w_o = 1.1\%$$

The product is stored in a 25°C environment with 50% relative humidity. Predict the shelf life of the product when the shelf life is established by the time period for the intensity of the key component to decrease to 50% of the original amount.

LIST OF SYMBOLS

A	area (m^2)
a_w	water activity
B	Arrhenius constant
C	GAB constant
c	concentration (kg/m^3 or kg mole/m^3)
D	mass diffusivity (m^2/s)
E_A	activation energy for temperature (kJ/kg)
E_p	activation energy for permeability (kcal/mole)
K	GAB constant
k	rate constant for quality change (1/s)
m	mass flux (kg/s)
n	order of reaction
p	partial pressure of gas (kPa)
P	package film permeability
Q	amount of quality attribute
R	gas constant (m^3 Pa/kg mol K)
S	solubility (moles/cm^3 atm)
ξ	extent of reaction
T	temperature (C)
T_A	absolute temperature (K)
t	time (s)
t_s	shelf life (s)
$t_{1/2}$	half life (s)
w	moisture content (%, d.b.)
w_o	monolayer moisture content (%, d.b.)
x	distance coordinate (m)

Subscripts: B, component B; i, initial condition; f, final condition; o, standard condition; w, water vapor; ws, water vapor at saturation; 1, location 1; 2, location 2.

BIBLIOGRAPHY

Brody, A. L. (2003). Modified atmosphere packaging. In *Encyclopedia of Agricultural, Food and Biological Engineering*, D. R. Heldman, ed., 666–670. Marcel Dekker Inc, New York.

De Kruijf, N. and van Beest, M. D. (2003). Active Packaging. In *Encyclopedia of Agricultural, Food and Biological Engineering*, D. R. Heldman, ed., 5–9. Marcel Dekker Inc., New York.

Krochta, J. M. (2003). Package Permeability. In *Encyclopedia of Agricultural, Food and Biological Engineering*, D. R. Heldman, ed., 720–726. Marcel Dekker Inc., New York.

Krochta, J. M. (2007). Food Packaging. In *Handbook of Food Engineering*, D. R. Heldman and D. B. Lund, eds. CRC Press, Taylor & Francis Group, Boca Raton, Florida.

Labuza, T. P. (1982). *Shelf-Life Dating of Foods*. Food and Nutrition Press, Westport, Connecticut.

Linssen, J. P. H. and Roozen, J. P. (1994). Food flavour and packaging interactions. In *Food Packaging and Preservation*, M. Mathlouthi, ed., 48–61. Blackie Academic and Professional, New York.

March, K. S. (2001). Looking at packaging in a new way to reduce food losses. *Food Technology* **55**: 48–52.

Robertson, G. L. (2006). *Food Packaging—Principles and Practice*. CRC Press, Taylor & Francis Group, Boca Raton, Florida.

Rodrigues, E. T. and Han, J. H. (2003). Intelligent Packaging. In *Encyclopedia of Agricultural, Food and Biological Engineering*, D. R. Heldman, ed., 528–535. Marcel Dekker Inc., New York.

Shellhammer, T. H. (2003). Flexible Packaging. In *Encyclopedia of Agricultural, Food and Biological Engineering*, D. R. Heldman, ed., 333–336. Marcel Dekker Inc, New York.

Singh, R. P. (2000). Scientific Principles of Shelf-Life Evaluation. In *Shelf-Life Evaluation of Foods*, C. M. D. Man and A. A. Jones, eds., 3–22. Aspen Publication, Maryland.

Steven, M. D. and Hotchkiss, J. H. (2003). Package Functions. In *Encyclopedia of Agricultural, Food and Biological Engineering*, D. R. Heldman, ed., 716–719. Marcel Dekker Inc., New York.

Taoukis, P. S. and Labuza, T. P. (1989). Applicability of time-temperature indicators as food quality monitors under non-isothermal conditions. *J. Food Sci.* **54**: 783.

Taoukis, P. S., Fu, B., and Labuza, T. P. (1991). Time temperature indicators. *Food Tech.* **45**(10): 70–82.

Wells, J. H. and Singh, R. P. (1988a). A kinetic approach to food quality prediction using full-history time-temperature indicators. *J. Food Sci.* **53**(6): 1866–1871, 1893.

Wells, J. H. and Singh, R. P. (1988b). Application of time-temperature indicators in monitoring changes in quality attributes of perishable and semiperishable foods. *J. Food Sci.* **53**(1): 148–156.

Yam, K. L., Paik, J. S., and Lai, C. C. (1992). Food Packaging, Part 1. General Considerations. In *Encyclopedia of Food Science & Technology*, Y. H. Hui, ed. John Wiley & Sons, Inc., New York.

Yasuda, H. and Stannett, V. (1989). Permeability coefficients. In *Polymer Handbook, 3rd*, J. Brandrup and E. H. Immergut, eds. Wiley, New York.

Appendices

A.1 SI SYSTEM OF UNITS AND CONVERSION FACTORS

A.1.1 Rules for Using SI Units

The following rules for SI usage are based on recommendations from several international conferences, the International Organization for Standardization, and the American Society of Agricultural Engineers.

A.1.1.1 SI Prefixes

The prefixes along with the SI symbols are given in Table A.1.1. The prefix symbols are printed in roman (upright) type without spacing between the prefix symbol and the unit symbol. The prefixes provide an order of magnitude, thus eliminating insignificant digits and decimals. For example,

$$19{,}200 \text{ m or } 19.2 \times 10^3 \text{ m} \quad \text{becomes} \quad 19.2 \text{ km}$$

Table A.1.1 SI Prefixes

Factor	Prefix	Symbol	Factor	Prefix	Symbol
10^{18}	exa	E	10^{-1}	deci	d
10^{15}	peta	P	10^{-2}	centi	c
10^{12}	tera	T	10^{-3}	milli	m
10^{9}	giga	G	10^{-6}	micro	μ
10^{6}	mega	M	10^{-9}	nano	n
10^{3}	kilo	k	10^{-12}	pico	p
10^{2}	hecto	h	10^{-15}	femto	f
10^{1}	deka	da	10^{-18}	atto	a

An exponent attached to a symbol containing a prefix indicates that the multiple or submultiple of the unit is raised to the power expressed by the exponent. For example,

$$1\,\mathrm{mm}^3 = (10^{-3}\,\mathrm{m})^3 = 10^{-9}\,\mathrm{m}^3$$

$$1\,\mathrm{cm}^{-1} = (10^{-2}\,\mathrm{m})^{-1} = 10^2\,\mathrm{m}^{-1}$$

Compound prefixes, formed by the juxtaposition of two or more SI prefixes, are not to be used. For example,

1 nm but not 1 mμm

Among the base units, the unit of mass is the only one whose name, for historical reasons, contains a prefix. To obtain names of decimal multiples and submultiples of the unit mass, attach prefixes to the word "gram."

Attach prefixes to the numerator of compound units, except when using "kilogram" in the denominator. For example, use

2.5 kJ/s not 2.5 J/ms

but

550 J/kg not 5.5 dJ/g

In selecting prefixes, a prefix should be chosen so that the numerical value preferably lies between 0.1 and 1000. However, double prefixes and hyphenated prefixes should not be used. For example, use

GJ not kMJ

A.1.1.2 Capitalization

The general principle governing the writing of unit symbols is as follows: roman (upright) type, in general lowercase, is used for symbols of units; however, if the symbols are derived from proper names, capital roman type is used (for the first letter), for example, K, N. These symbols are not followed by a full stop (period).

If the units are written in an unabbreviated form, the first letter is not capitalized (even for those derived from proper nouns): for example, kelvin, newton. The numerical prefixes are not capitalized except for symbols E (exa), P (peta), T (tera), G (giga), and M (mega).

A.1.1.3 Plurals

The unit symbols remain the same in the plural form. In unabbreviated form the plural units are written in the usual manner. For example:

45 newtons or 45 N

22 centimeters or 25 cm

A.1.1.4 Punctuation

For a numerical value less than one, a zero should precede the decimal point. The SI symbols should not be followed by a period, except at the end of a sentence. English-speaking countries use a centered dot for a decimal point; others use a comma. Large numbers should be grouped into threes (thousands) by using spaces instead of commas. For example,

3 456 789.291 22

not

3,456,789.291,22

A.1.1.5 Derived Units

The product of two or more units may be written in either of the following ways:

$$\mathrm{N \cdot m} \qquad \mathrm{Nm}$$

A solidus (oblique stroke, /), a horizontal line, or negative powers may be used to express a derived unit formed from two others by division. For example:

$$\mathrm{m/s} \qquad \frac{\mathrm{m}}{\mathrm{s}} \qquad \mathrm{m\,s^{-1}}$$

A solidus must not be repeated on the same line. In complicated cases, parentheses or negative powers should be used. For example:

$$\mathrm{m/s^2} \text{ or } \mathrm{m\,s^{-2}} \quad \text{but not} \quad \mathrm{m/s/s}$$

$$\mathrm{J/(s\,m\,K)} \text{ or } \mathrm{J\,s^{-1}\,m^{-1}\,K^{-1}} \quad \text{but not} \quad \mathrm{J/s/m/K}$$

Table A.1.2 Useful Conversion Factors

Useful Conversion Factors
Acceleration of gravity $g = 9.80665\ m/s^2$ $g = 980.665\ cm/s^2$ $g = 32.174\ ft/s^2$ $1\ ft/s^2 = 0.304799\ m/s^2$
Area 1 acre $= 4.046856 \times 10^3\ m^2$ $1\ ft^2 = 0.0929\ m^2$ $1\ in^2 = 6.4516 \times 10^{-4}\ m^2$
Density $1\ lb_m/ft^3 = 16.0185\ kg/m^3$ $1\ lb_m/gal = 1.198264 \times 10^2\ kg/m^3$ Density of dry air at 0°C, 760 mm Hg = 1.2929 g/l 1 kg mol ideal gas at 0°C, 760 mm Hg = 22.414 m^3
Diffusivity $1\ ft^2/h = 2.581 \times 10^{-5}\ m^2/s$
Energy 1 Btu = 1055 J = 1.055 kJ 1 Btu = 252.16 cal 1 kcal = 4.184 kJ $1\ J = 1\ N\ m = 1\ kg\ m^2/s^2$ $1\ kW\ h = 3.6 \times 10^3\ kJ$
Enthalpy $1\ Btu/lb_m = 2.3258\ kJ/kg$
Force $1\ lb_f = 4.4482\ N$ $1\ N = 1\ kg\ m/s^2$ $1\ dyne = 1\ g\ cm/s^2 = 10^{-5}\ kg\ m/s^2$
Heat flow 1 Btu/h = 0.29307 W 1 Btu/min = 17.58 W $1\ kJ/h = 2.778 \times 10^{-4}\ kW$ 1 J/s = 1 W
Heat flux $1\ Btu/(h\ ft^2) = 3.1546\ W/m^2$
Heat transfer coefficient $1\ Btu/(h\ ft^2\ °F) = 5.6783\ W/(m^2\ K)$ $1\ Btu/(h\ ft^2\ °F) = 1.3571 \times 10^{-4}\ cal/(s\ cm^2\ °C)$
Length 1 ft = 0.3048 m $1\ micron = 10^{-6}\ m = 1\ \mu m$ $1\ Å = 10^{-10}\ m$ $1\ in = 2.54 \times 10^{-2}\ m$ $1\ mile = 1.609344 \times 10^3\ m$
Mass $1\ carat = 2 \times 10^{-4}\ kg$ $1\ lb_m = 0.45359\ kg$ $1\ lb_m = 16\ oz = 7000\ grains$ 1 ton (metric) = 1000 kg
Mass transfer coefficient $1\ lb\ mol/(h\ ft^2\ mol\ fraction) = 1.3562 \times 10^{-3}\ kg\ mol/(s\ m^2\ mol\ fraction)$

(Continued)

Table A.1.2 (*Continued*)

Power
1 hp = 0.7457 kW
1 W = 14.34 cal/min
1 hp = 550 ft lb_f/s
1 Btu/h = 0.29307 W
1 hp = 0.7068 Btu/s
1 J/s = 1 W
Pressure
1 psia = 6.895 kPa
1 psia = 6.895×10^3 N/m^2
1 bar = 1×10^5 Pa = 1×10^5 N/m^2
1 Pa = 1 N/m^2
1 mm Hg (0°C) = 1.333224×10^2 N/m^2
1 atm = 29.921 in. Hg at 0°C
1 atm = 33.90 ft H_2O at 4°C
1 atm = 14.696 psia = 1.01325×10^5 N/m^2
1 atm = 1.01325 bar
1 atm = 760 mm Hg at 0°C = 1.01325×10^5 Pa
1 lb_f/ft^2 = 4.788×10^2 dyne/cm^2 = 47.88 N/m^2
Specific heat
1 Btu/(lb_m °F) = 4.1865 J/(g K)
1 Btu/(lb_m °F) = 1 cal/(g °C)
Temperature
$T_{°F} = T_{°C} \times 1.8 + 32$
$T_{°C} = (T_{°F} - 32)/1.8$
Thermal conductivity
1 Btu/(h ft °F) = 1.731 W/(m K)
1 Btu in/(ft^2 h °F) = 1.442279×10^{-2} W/(m K)
Viscosity
1 lb_m/(ft h) = 0.4134 cp
1 lb_m/(ft s) = 1488.16 cp
1 cp = 10^{-2} g/(cm s) = 10^{-2} poise
1 cp = 10^{-3} Pa s = 10^{-3} kg/(m s) = 10^{-3} N s/m^2
1 lb_f s/ft^2 = 4.7879×10^4 cp
1 N s/m^2 = 1 Pa s
1 kg/(m s) = 1 Pa s
Volume
1 ft^3 = 0.02832 m^3
1 U.S. gal = 3.785×10^{-3} m^3
1 L = 1000 cm^3
1 m^3 = 1000 L
1 U.S. gal = 4 qt
1 ft^3 = 7.481 U.S. gal
1 British gal = 1.20094 U.S. gal
Work
1 hp h = 0.7457 kW h
1 hp h = 2544.5 Btu
1 ft lb_f = 1.35582 J

Table A.1.3 Conversion Factors for Pressure

	lb_f/in^2	kPa	kg_f/cm^2	in Hg (at 21°C)	mm Hg (at 21°C)	in H_2O (at 21°C)	atm
1 lb_f/in^2	= 1	689.473×10^{-2}	0.07031	2.036	51.715	27.71	0.06805
1 kPa	= 0.1450383	1	101.972×10^{-4}	0.2952997	7.5003	4.0188	986.923×10^{-5}
1kg_f/cm^2	= 14.2234	980.665×10^{-1}	1	28.959	735.550	394.0918	967.841×10^{-3}
1 in Hg (21°C)	= 0.4912	338.64×10^{-2}	0.03452	1	25.40	13.608	0.03342
1 mm Hg (21°C)	= 0.01934	0.1333273	1.359×10^{-3}	0.03937	1	0.5398	1.315×10^{-3}
1 in H_2O (21°C)	= 0.03609	24.883×10^{-2}	2.537×10^{-3}	0.0735	1.8665	1	2.458×10^{-3}
1 atm	= 14.6959	101.3251	1.03323	29.9212	760	406	1

A.2 PHYSICAL PROPERTIES OF FOODS

Table A.2.1 Specific Heat of Foods

Product	Composition (%)					Specific heat	
	Water	Protein	Carbohydrate	Fat	Ash	Eq. (4.4) (kJ/kg K)	Experimental[a] (kJ/[kg K])
Beef (hamburger)	68.3	20.7	0.0	10.0	1.0	3.35	3.52
Fish, canned	70.0	27.1	0.0	0.3	2.6	3.35	
Starch	12.0	0.5	87.0	0.2	0.3	1.754	
Orange juice	87.5	0.8	11.1	0.2	0.4	3.882	
Liver, raw beef	74.9	15.0	0.9	9.1	1.1	3.525	
Dry milk, nonfat	3.5	35.6	52.0	1.0	7.9	1.520	
Butter	15.5	0.6	0.4	81.0	2.5	2.043	2.051–2.135
Milk, whole pasteurized	87.0	3.5	4.9	3.9	0.7	3.831	3.852
Blueberries, syrup pack	73.0	0.4	23.6	0.4	2.6	3.445	
Cod, raw	82.6	15.0	0.0	0.4	2.0	3.697	
Skim milk	90.5	3.5	5.1	0.1	0.8	3.935	3.977–4.019
Tomato soup, concentrate	81.4	1.8	14.6	1.8	0.4	3.676	
Beef, lean	77.0	22.0	–	–	1.0	3.579	
Egg yolk	49.0	13.0	–	11.0	1.0	2.449	2.810
Fish, fresh	76.0	19.0	–	–	1.4	3.500	3.600
Beef, lean	71.7	21.6	0.0	5.7	1.0	3.437	3.433
Potato	79.8	2.1	17.1	0.1	0.9	3.634	3.517
Apple, raw	84.4	0.2	14.5	0.6	0.3	3.759	3.726–4.019
Bacon	49.9	27.6	0.3	17.5	4.7	2.851	2.01
Cucumber	96.1	0.5	1.9	0.1	1.4	4.061	4.103
Blackberry, syrup pack	76.0	0.7	22.9	0.2	0.2	3.521	
Potato	75.0	0.0	23.0	0.0	2.0	3.483	3.517
Veal	68.0	21.0	0.0	10.0	1.0	3.349	3.223
Fish	80.0	15.0	4.0	0.3	0.7	3.651	3.60
Cheese, cottage	65.0	25.0	1.0	2.0	7.0	3.215	3.265
Shrimp	66.2	26.8	0.0	1.4	0.0	3.404	3.014
Sardines	57.4	25.7	1.2	11.0	0.0	3.002	3.014
Beef, roast	60.0	25.0	0.0	13.0	0.0	3.115	3.056
Carrot, fresh	88.2	1.2	9.3	0.3	1.1	3.864	3.81–3.935

Source: Adapted from Heldman and Singh (1981).
[a]Experimental specific heat values from Reidy (1968).

Table A.2.2 Thermal Conductivity of Selected Food Products

Product	Moisture content (%)	Temperature (°C)	Thermal conductivity (W/[m °C])
Apple	85.6	2–36	0.393
Applesauce	78.8	2–36	0.516
Beef, freeze dried			
1000 mm Hg pressure	–	0	0.065
0.001 mm Hg pressure	–	0	0.037
Beef, lean			
Perpendicular to fibers	78.9	7	0.476
Perpendicular to fibers	78.9	62	0.485
Parallel to fibers	78.7	8	0.431
Parallel to fibers	78.7	61	0.447
Beef fat	–	24–38	0.19
Butter	15	46	0.197
Cod	83	2.8	0.544
Corn, yellow dust	0.91	8–52	0.141
	30.2	8–52	0.172
Egg, frozen whole	–	–10 to –6	0.97
Egg, white	–	36	0.577
Egg, yolk	–	33	0.338
Fish muscle	–	0–10	0.557
Grapefruit, whole	–	30	0.45
Honey	12.6	2	0.502
	80	2	0.344
	14.8	69	0.623
	80	69	0.415
Juice, apple	87.4	20	0.559
	87.4	80	0.632
	36.0	20	0.389
	36.0	80	0.436
Lamb			
Perpendicular to fiber	71.8	5	0.45
		61	0.478
Parallel to fiber	71.0	5	0.415
		61	0.422
Milk	–	37	0.530
Milk, condensed	90	24	0.571
	–	78	0.641
	50	26	0.329
	–	78	0.364

(*Continued*)

Table A.2.2 (*Continued*)

Product	Moisture content (%)	Temperature (°C)	Thermal conductivity (W/[m°C])
Milk, skimmed	–	1.5	0.538
	–	80	0.635
Milk, nonfat dry	4.2	39	0.419
Olive oil	–	15	0.189
	–	100	0.163
Oranges, combined	–	30	0.431
Peas, black-eyed	–	3–17	0.312
Pork			
Perpendicular to fibers	75.1	6	0.488
		60	0.54
Parallel to fibers	75.9	4	0.443
		61	0.489
Pork fat	–	25	0.152
Potato, raw flesh	81.5	1–32	0.554
Potato, starch gel	–	1–67	0.04
Poultry, broiler muscle	69.1–74.9	4–27	0.412
Salmon			
Perpendicular to fibers	73	4	0.502
Salt	–	87	0.247
Sausage mixture	65.72	24	0.407
Soybean oil meal	13.2	7–10	0.069
Strawberries	–	14–25	0.675
Sugars	–	29–62	0.087–0.22
Turkey, breast			
Perpendicular to fibers	74	3	0.502
Parallel to fibers	74	3	0.523
Veal			
Perpendicular to fibers	75	6	0.476
		62	0.489
Parallel to fibers	75	5	0.441
		60	0.452
Vegetable and animal oils	–	4–187	0.169
Wheat flour	8.8	43	0.45
		65.5	0.689
		1.7	0.542
Whey		80	0.641

Source: Reidy (1968).

Table A.2.3 Thermal Diffusivity of Some Foodstuffs

Product	Water content (% wt.)	Temperature[a] (°C)	Thermal diffusivity ($\times 10^{-7}$ m²/s)
Fruits, vegetables, and by-products			
Apples, whole, Red Delicious	85	0–30	1.37
Applesauce	37	5	1.05
	37	65	1.12
	80	5	1.22
	80	65	1.40
	–	26–129	1.67
Avocado, flesh	–	24, 0	1.24
Seed	–	24, 0	1.29
Whole	–	41, 0	1.54
Banana, flesh	76	5	1.18
	76	65	1.42
Beans, baked	–	4–122	1.68
Cherries, tart, flesh	–	30, 0	1.32
Grapefruit, Marsh, flesh	88.8	–	1.27
Grapefruit, Marsh, albedo	72.2	–	1.09
Lemon, whole	–	40, 0	1.07
Lima bean, pureed	–	26–122	1.80
Pea, pureed	–	26–128	1.82
Peach, whole	–	27, 4	1.39
Potato, flesh	–	25	1.70
Potato, mashed, cooked	78	5	1.23
	78	65	1.45
Rutabaga	–	48, 0	1.34
Squash, whole	–	47, 0	1.71
Strawberry, flesh	92	5	1.27
Sugarbeet	–	14, 60	1.26
Sweet potato, whole	–	35	1.06
	–	55	1.39
	–	70	1.91
Tomato, pulp	–	4, 26	1.48
Fish and meat products			
Codfish	81	5	1.22
	81	65	1.42
Corned beef	65	5	1.32
	65	65	1.18
Beef chuck[b]	66	40–65	1.23
Beef, round[b]	71	40–65	1.33
Beef, tongue[b]	68	40–65	1.32
Halibut	76	40–65	1.47
Ham, smoked	64	5	1.18
Ham, smoked	64	40–65	1.38
Water	–	30	1.48
	–	65	1.60
Ice	–	0	11.82

*Source: Singh (1982). Reprinted from Food Technology **36**(2), 87–91. Copyright © Institute of Food Technologists.*

[a] *Where two temperatures separated by a comma are given, the first is the initial temperature of the sample, and the second is that of the surroundings.*

[b] *Data are applicable only where the juices that exuded during the heating remain in the food samples.*

Table A.2.4 Viscosity of Liquid Foods

Product	Composition	Temperature (°C)	Viscosity (Pa s)
Cream	10% fat	40	0.00148
	10% fat	60	0.00107
	10% fat	80	0.00083
Cream	20% fat	60	0.00171
	30% fat	60	0.00289
	40% fat	60	0.00510
Homogenized milk	–	20	0.0020
	–	40	0.0015
	–	60	0.000775
	–	80	0.0006
Raw milk	–	0	0.00344
	–	10	0.00264
	–	20	0.00199
	–	30	0.00149
	–	40	0.00123
Corn oil	–	25	0.0565
	–	38	0.0317
Cottonseed oil	–	20	0.0704
	–	38	0.0306
Peanut oil	–	25	0.0656
	–	38	0.0251
Safflower oil	–	25	0.0522
	–	38	0.0286
Soybean oil	–	30	0.04
Honey, buckwheat	18.6% T.S.	24.8	3.86
sage	18.6% T.S.	25.9	8.88
white clover	18.2% T.S.	25.0	4.80
Apple juice	20° Brix	27	0.0021
	60° Brix	27	0.03
Grape juice	20° Brix	27	0.0025
	60° Brix	27	0.11
Corn syrup	48.4% T.S.	27	0.053

Source: Steffe (1983).

Table A.2.5 Properties of Ice as a Function of Temperature

Temperature (°C)	Thermal conductivity (W/m °C)	Specific heat (kJ/kg °C)	Density (kg/m^3)
−101	3.50	1.382	925.8
−73	3.08	1.587	924.2
−45.5	2.72	1.783	922.6
−23	2.41	1.922	919.4
−18	2.37	1.955	919.4
−12	2.32	1.989	919.4
−7	2.27	2.022	917.8
0	2.22	2.050	916.2

Source: Adapted from Dickerson (1969).

Table A.2.6 Approximate Heat Evolution Rates of Fresh Fruits and Vegetables When Stored at Temperatures Shown

		Watts per megagram (W/Mg)[a]		
Commodity	**0°C**	**5°C**	**10°C**	**15°C**
Apples	10–12	15–21	41–61	41–92
Apricots	15–17	19–27	33–56	63–101
Artichokes. globe	67–133	94–177	161–291	229–429
Asparagus	81–237	161–403	269–902	471–970
Avocados	–	59–89	–	183–464
Bananas, ripening	–	–	65–116	87–164
Beans, green or snap	–	101–103	161–172	251–276
Beans, lima (unshelled)	31–89	58–106	–	296–369
Beets, red (roots)	16–21	27–28	35–40	50–69
Blackberries	46–68	85–135	154–280	208–431
Blueberries	7–31	27–36	69–104	101–183
Broccoli, sprouting	55–63	102–474	–	514–1000
Brussels sprouts	46–71	95–143	186–250	282–316
Cabbage	12–40	28–63	36–86	66–169
Cantaloupes	15–17	26–30	46	100–114
Carrots, topped	46	58	93	117
Cauliflower	53–71	61–81	100–144	136–242
Celery	21	32	58–81	110
Cherries, sour	17–39	38–39	–	81–148
Corn, sweet	125	230	331	482
Cranberries	–	12–14	–	–
Cucumbers	–	–	68–86	71–98
Figs, Mission	–	32–39	65–68	145–187
Garlic	9–32	17–29	27–29	32–81

(*Continued*)

Table A.2.6 (*Continued*)

Commodity	0°C	Watts per megagram (W/Mg)[a]		
		5°C	10°C	15°C
Gooseberries	20–26	36–40	–	64–95
Grapefruit	–	–	20–27	35–38
Grapes, American	8	16	23	47
Grapes, European	4–7	9–17	24	30–35
Honeydew melons	–	9–15	24	35–47
Horseradish	24	32	78	97
Kohlrabi	30	48	93	145
Leeks	28–48	58–86	158–201	245–346
Lemons	9	15	33	47
Lettuce, head	27–50	39–59	64–118	114–121
Lettuce, leaf	68	87	116	186
Mushrooms	83–129	210	297	–
Nuts, kind not specified	2	5	10	10
Okra	–	163	258	431
Onions	7–9	10–20	21	33
Onions, green	31–66	51–201	107–174	195–288
Olives	–	–	–	64–115
Oranges	9	14–19	35–40	38–67
Peaches	11–19	19–27	46	98–125
Pears	8–20	15–46	23–63	45–159
Peas, green (in pod)	90–138	163–226	–	529–599
Peppers, sweet	–	–	43	68
Plums, Wickson	6–9	12–27	27–34	35–37
Potatoes, immature	–	35	42–62	42–92
Potatoes, mature	–	17–20	20–30	20–35
Radishes, with tops	43–51	57–62	92–108	207–230
Radishes, topped	16–17	23–24	45–47	82–97
Raspberries	52–74	92–114	82–164	243–300
Rhubarb, topped	24–39	32–54	–	92–134
Spinach	–	136	327	529
Squash, yellow	35–38	42–55	103–108	222–269
Strawberries	36–52	48–98	145–280	210–273
Sweet potatoes	–	–	39–95	47–85
Tomatoes, mature green	–	21	45	61
Tomatoes, ripening	–	–	42	79
Turnips, roots	26	28–30	–	63–71
Watermelons	–	9–12	22	–

[a] *Conversion factor: (watts per megagram) × (74.12898) = Btu per ton per 24 hr. From American Society of Heating, Refrigerating and Air-Conditioning Engineers; with permission of the American Society of Heating, Refrigerating, and Air-Conditioning Engineers, Atlanta, Georgia (1978).*

Table A.2.7 Enthalpy of Frozen Foods

Temperature (°C)	Beef (kJ/kg)	Lamb (kJ/kg)	Poultry (kJ/kg)	Fish (kJ/kg)	Beans (kJ/kg)	Broccoli (kJ/kg)	Peas (kJ/kg)	Mashed potatoes (kJ/kg)	Cooked rice (kJ/kg)
−28.9	14.7	19.3	11.2	9.1	4.4	4.2	11.2	9.1	18.1
−23.3	27.7	31.4	23.5	21.6	16.5	16.3	23.5	21.6	31.9
−17.8	42.6	45.4	37.7	35.6	29.3	28.8	37.7	35.6	47.7
−12.2	62.8	67.2	55.6	52.1	43.7	42.8	55.6	52.1	70.0
−9.4	77.7	84.2	68.1	63.9	52.1	51.2	68.1	63.9	87.5
−6.7	101.2	112.6	87.5	80.7	63.3	62.1	87.5	80.7	115.1
−5.6	115.8	130.9	99.1	91.2	69.8	67.9	99.1	91.2	133.0
−4.4	136.9	157.7	104.4	105.1	77.9	75.6	104.4	105.1	158.9
−3.9	151.6	176.8	126.8	115.1	83.0	80.7	126.8	115.1	176.9
−3.3	170.9	201.6	141.6	128.2	90.2	87.2	141.6	128.2	177.9
−2.8	197.2	228.2	142.3	145.1	99.1	95.6	142.3	145.1	233.5
−2.2	236.5	229.8	191.7	170.7	112.1	107.7	191.7	170.7	242.3
−1.7	278.2	231.2	240.9	212.1	132.8	126.9	240.9	212.1	243.9
−1.1	280.0	232.8	295.4	295.1	173.7	165.1	295.4	295.1	245.6
1.7	288.4	240.7	304.5	317.7	361.9	366.8	304.5	317.7	254.9
4.4	297.9	248.4	313.8	327.2	372.6	377.5	313.8	327.2	261.4
7.2	306.8	256.3	323.1	336.5	383.3	388.2	323.1	336.5	269.3
10.0	315.8	263.9	332.1	346.3	393.8	398.9	332.1	346.3	277.2
15.6	333.5	279.6	350.5	365.4	414.7	420.3	350.5	365.4	292.8

Source: Mott (1964), by permission of H.G. Goldstein, editor, Aust. Refrig. Air Cond. Heat.

Table A.2.8 Composition Values of Selected Foods

Food	Water (%)	Protein (%)	Fat (%)	Carbohydrate (%)	Ash (%)
Apples, fresh	84.4	0.2	0.6	14.5	0.3
Applesauce	88.5	0.2	0.2	10.8	0.6
Asparagus	91.7	2.5	0.2	5.0	0.6
Beans, lima	67.5	8.4	0.5	22.1	1.5
Beef, hamburger, raw	68.3	20.7	10.0	0.0	1.0
Bread, white	35.8	8.7	3.2	50.4	1.9
Butter	15.5	0.6	81.0	0.4	2.5
Cod	81.2	17.6	0.3	0.0	1.2
Corn, sweet, raw	72.7	3.5	1.0	22.1	0.7
Cream, half-and-half	79.7	3.2	11.7	4.6	0.6
Eggs	73.7	12.9	11.5	0.9	1.0
Garlic	61.3	6.2	0.2	30.8	1.5
Lettuce, Iceburg	95.5	0.9	0.1	2.9	0.6
Milk, whole	87.4	3.5	3.5	4.9	0.7
Orange juice	88.3	0.7	0.2	10.4	0.4
Peaches	89.1	0.6	0.1	9.7	0.5
Peanuts, raw	5.6	26.0	47.5	18.6	2.3
Peas, raw	78.0	6.3	0.4	14.4	0.9
Pineapple, raw	85.3	0.4	0.2	13.7	0.4
Potatoes, raw	79.8	2.1	0.1	17.1	0.9
Rice, white	12.0	6.7	0.4	80.4	0.5
Spinach	90.7	3.2	0.3	4.3	1.5
Tomatoes	93.5	1.1	0.2	4.7	0.5
Turkey	64.2	20.1	14.7	0.0	1.0
Turnips	91.5	1.0	0.2	6.6	0.7
Yogurt (whole milk)	88.0	3.0	3.4	4.9	0.7

Table A.2.9 Coefficients to Estimate Food Properties

Property	Component	Temperature function	Standard error	Standard % error
k (W/m°C)	Protein	$k = 1.7881 \times 10^{-1} + 1.1958 \times 10^{-3}T - 2.7178 \times 10^{-6}T^2$	0.012	5.91
	Fat	$k = 1.8071 \times 10^{-1} - 2.7604 \times 10^{-3}T - 1.7749 \times 10^{-7}T^2$	0.0032	1.95
	Carbohydrate	$k = 2.0141 \times 10^{-1} + 1.3874 \times 10^{-3}T - 4.3312 \times 10^{-6}T^2$	0.0134	5.42
	Fiber	$k = 1.8331 \times 10^{-1} + 1.2497 \times 10^{-3}T - 3.1683 \times 10^{-6}T^2$	0.0127	5.55
	Ash	$k = 3.2962 \times 10^{-1} + 1.4011 \times 10^{-3}T - 2.9069 \times 10^{-6}T^2$	0.0083	2.15
	Water	$k = 5.7109 \times 10^{-1} + 1.7625 \times 10^{-3}T - 6.7036 \times 10^{-6}T^2$	0.0028	0.45
	Ice	$k = 2.2196 - 6.2489 \times 10^{-3}T + 1.0154 \times 10^{-4}T^2$	0.0079	0.79
α (mm²/s)	Protein	$\alpha = 6.8714 \times 10^{-2} + 4.7578 \times 10^{-4}T - 1.4646 \times 10^{-6}T^2$	0.0038	4.50
	Fat	$\alpha = 9.8777 \times 10^{-2} - 1.2569 \times 10^{-4}T - 3.8286 \times 10^{-8}T^2$	0.0020	2.15
	Carbohydrate	$\alpha = 8.0842 \times 10^{-2} + 5.3052 \times 10^{-4}T - 2.3218 \times 10^{-6}T^2$	0.0058	5.84
	Fiber	$\alpha = 7.3976 \times 10^{-2} + 5.1902 \times 10^{-4}T - 2.2202 \times 10^{-6}T^2$	0.0026	3.14
	Ash	$\alpha = 1.2461 \times 10^{-1} + 3.7321 \times 10^{-4}T - 1.2244 \times 10^{-6}T^2$	0.0022	1.61
	Water	$\alpha = 1.3168 \times 10^{-1} + 6.2477 \times 10^{-4}T - 2.4022 \times 10^{-6}T^2$	0.0022×10^{-6}	1.44
	Ice	$\alpha = 1.1756 - 6.0833 \times 10^{-3}T + 9.5037 \times 10^{-5}T^2$	0.0044×10^{-6}	0.33
ρ (kg/m³)	Protein	$\rho = 1.3299 \times 10^{3} - 5.1840 \times 10^{-1}T$	39.9501	3.07
	Fat	$\rho = 9.2559 \times 10^{2} - 4.1757 \times 10^{-1}T$	4.2554	0.47
	Carbohydrate	$\rho = 1.5991 \times 10^{3} - 3.1046 \times 10^{-1}T$	93.1249	5.98
	Fiber	$\rho = 1.3115 \times 10^{3} - 3.6589 \times 10^{-1}T$	8.2687	0.64
	Ash	$\rho = 2.4238 \times 10^{3} - 2.8063 \times 10^{-1}T$	2.2315	0.09
	Water	$\rho = 9.9718 \times 10^{2} + 3.1439 \times 10^{-3}T - 3.7574 \times 10^{-3}T^2$	2.1044	0.22
	Ice	$\rho = 9.1689 \times 10^{2} - 1.3071 \times 10^{-1}T$	0.5382	0.06
c_p (kJ/kg°C)	Protein	$c_p = 2.0082 + 1.2089 \times 10^{-3}T - 1.3129 \times 10^{-6}T^2$	0.1147	5.57
	Fat	$c_p = 1.9842 + 1.4733 \times 10^{-3}T - 4.8008 \times 10^{-6}T^2$	0.0236	1.16
	Carbohydrate	$c_p = 1.5488 + 1.9625 \times 10^{-3}T - 5.9399 \times 10^{-6}T^2$	0.0986	5.96
	Fiber	$c_p = 1.8459 + 1.8306 \times 10^{-3}T - 4.6509 \times 10^{-6}T^2$	0.0293	1.66
	Ash	$c_p = 1.0926 + 1.8896 \times 10^{-3}T - 3.6817 \times 10^{-6}T^2$	0.0296	2.47
	Water[a]	$c_p = 4.0817 - 5.3062 \times 10^{-3}T + 9.9516 \times 10^{-4}T^2$	0.0988	2.15
	Water[b]	$c_p = 4.1762 - 9.0864 \times 10^{-5}T + 5.4731 \times 10^{-6}T^2$	0.0159	0.38
	Ice	$c_p = 2.0623 + 6.0769 \times 10^{-3}T$	0.0014	0.07

[a] *For the temperature range of –40 to 0°C.*
[b] *For the temperature range of 0 to 150°C.*

A.3 PHYSICAL PROPERTIES OF NONFOOD MATERIALS

Table A.3.1 Physical Properties of Metals

Metal	Properties at 20°C			
	ρ (kg/m^3)	c_p (kJ/kg °C)	k (W/m °C)	α ($\times 10^{-5}$ m^2/s)
Aluminum				
Pure	2707	0.896	204	8.418
Al-Cu (Duralumin, 94–96% Al, 3–5% Cu, trace Mg)	2787	0.883	164	6.676
Al-Si (Silumin, copper-bearing: 86.5% Al, 1% Cu)	2659	0.867	137	5.933
Al-Si (Alusil, 78–80% Al, 20–22% Si)	2627	0.854	161	7.172
Al-Mg-Si, 97% Al, 1%Mg, 1% Si, 1% Mn	2707	0.892	177	7.311
Lead	11,373	0.130	35	2.343
Iron				
Pure	7897	0.452	73	2.034
Steel				
(C max = 1.5%):				
Carbon steel				
C= 0.5%	7833	0.465	54	1.474
1.00%	7801	0.473	43	1.712
1.50%	7753	0.486	36	0.970
Nickel steel				
Ni= 0%	7897	0.452	73	2.026
20%	7933	0.46	19	0.526
40%	8169	0.46	10	0.279
80%	8618	0.46	35	0.872
Invar 36% Ni	8137	0.46	10.7	0.286
Chrome steel				
Cr= 0%	7897	0.452	73	2.026
1%	7865	0.46	61	1.665
5%	7833	0.46	40	1.110
20%	7689	0.46	22	0.635
Cr-Ni (chrome-nickel)				
15% Cr, 10% Ni	7865	0.46	19	0.526
18% Cr, 8% Ni (V2A)	7817	0.46	16.3	0.444
20% Cr, 15% Ni	7833	0.46	15.1	0.415
25% Cr, 20% Ni	7865	0.46	12.8	0.361
Tungsten steel				
W = 0%	7897	0.452	73	2.026
W = 1%	7913	0.448	66	1.858
W = 5%	8073	0.435	54	1.525
W = 10%	8314	0.419	48	1.391
Copper				
Pure	8954	0.3831	386	11.234
Aluminum bronze (95% Cu, 5% Al)	8666	0.410	83	2.330
Bronze (75%, 25% Sn)	8666	0.343	26	0.859

(*Continued*)

Table A.3.1 *(Continued)*

Metal	Properties at 20°C			
	ρ (kg/m^3)	c_p (kJ/kg °C)	k (W/m °C)	α ($\times 10^{-5}$ m^2/s)
Red brass (85% Cu, 9% Sn, 6% Zn)	8714	0.385	61	1.804
Brass (70% Cu, 30% Zn)	8522	0.385	111	3.412
German silver (62% Cu, 15% Ni, 22% Zn)	8618	0.394	24.9	0.733
Constantan (60% Cu, 40% Ni)	8922	0.410	22.7	0.612
Magnesium				
Pure	1746	1.013	171	9.708
Mg-Al (electrolytic), 6–8% Al, 1–2% Zn	1810	1.00	66	3.605
Molybdenum	10,220	0.251	123	4.790
Nickel				
Pure (99.9%)	8906	0.4459	90	2.266
Ni-Cr (90% Ni, 10% Cr)	8666	0.444	17	0.444
80% Ni, 20% Cr	8314	0.444	12.6	0.343
Silver				
Purest	10,524	0.2340	419	17.004
Pure (99.9%)	10,524	0.2340	407	16.563
Tin, pure	7304	0.2265	64	3.884
Tungsten	19,350	0.1344	163	6.271
Zinc, pure	7144	0.3843	112.2	4.106

Source: Adapted from Holman (2002). Reproduced with permission from the publisher.

Table A.3.2 Physical Properties of Nonmetals

Substance	Temperature (°C)	k (W/m °C)	ρ (kg/m^3)	c (kJ/kg °C)	α ($\times 10^{-7}$ m^2/s)
Asphalt	20–55	0.74–0.76			
Brick					
Building brick, common	20	0.69	1600	0.84	5.2
Fireclay brick, burnt					
133°C	500	1.04	2000	0.96	5.4
	800	1.07			
	1100	1.09			
Cement, Portland		0.29	1500		
Mortar	23	1.16			
Concrete, cinder	23	0.76			
Glass, window	20	0.78 (avg)	2700	0.84	3.4

(Continued)

Table A.3.2 (*Continued*)

Substance	Temperature (°C)	k (W/m °C)	ρ (kg/m^3)	c (kJ/kg °C)	α ($\times 10^{-7}$ m^2/s)
Plaster, gypsum	20	0.48	1440	0.84	4.0
Metal lath	20	0.47			
Wood lath	20	0.28			
Stone					
Granite		1.73–3.98	2640	0.82	8–18
Limestone	100–300	1.26–1.33	2500	0.90	5.6–5.9
Marble		2.07–2.94	2500–2700	0.80	10–13.6
Sandstone	40	1.83	2160–2300	0.71	11.2–11.9
Wood (across the grain)					
Cypress	30	0.097	460		
Fir	23	0.11	420	2.72	0.96
Maple or Oak	30	0.166	540	2.4	1.28
Yellow pine	23	0.147	640	2.8	0.82
White pine	30	0.112	430		
Asbestos					
Loosely packed	−45	0.149			
	0	0.154	470–570	0.816	3.3–4
	100	0.161			
Sheets	51	0.166			
Cardboard, corrugated	—	0.064			
Corkboard, 160 kg/m^3	30	0.043	160		
Cork, regranulated	32	0.045	45–120	1.88	2–5.3
Ground	32	0.043	150		
Diatomaceous earth (Sil-o-cel)	0	0.061	320		
Fiber, insulating board	20	0.048	240		
Glass wool, 24 kg/m^3	23	0.038	24	0.7	22.6
Magnesia, 85%	38	0.067	270		
	93	0.071			
	150	0.074			
	204	0.080			
Rock wool, 24 kg/m^3	32	0.040	160		
Loosely packed	150	0.067	64		
	260	0.087			
Sawdust	23	0.059			
Wood shavings	23	0.059			

Source: Adapted from Holman (2002). Reproduced with the permission of the publisher.

Table A.3.3 Emissivity of Various Surfaces

Material	Wavelength and average temperatures				
	9.3 μm 38°C	5.4 μm 260°C	3.6 μm 540°C	1.8 μm 1370°C	0.6 μm Solar
Metals					
Aluminum					
Polished	0.04	0.05	0.08	0.19	~0.3
Oxidized	0.11	0.12	0.18		
24-ST weathered	0.4	0.32	0.27		
Surface roofing	0.22				
Anodized (at 1000°F)	0.94	0.42	0.60	0.34	
Brass					
Polished	0.10	0.10			
Oxidized	0.61				
Chromium, polished	0.08	0.17	0.26	0.40	0.49
Copper					
Polished	0.04	0.05	0.18	0.17	
Oxidized	0.87	0.83	0.77		
Iron					
Polished	0.06	0.08	0.13	0.25	0.45
Cast, oxidized	0.63	0.66	0.76		
Galvanized, new	0.23	–	–	0.42	0.66
Galvanized, dirty	0.28	–	–	0.90	0.89
Steel plate, rough	0.94	0.97	0.98		
Oxide	0.96	–	0.85	–	0.74
Magnesium	0.07	0.13	0.18	0.24	0.30
Silver, polished	0.01	0.02	0.03	–	0.11
Stainless steel					
18-8, polished	0.15	0.18	0.22		
18-8, weathered	0.85	0.85	0.85		
Steel tube					
Oxidized	–	0.80			
Tungsten filament	0.03	–	–	~0.18	0.36[a]
Zinc					
Polished	0.02	0.03	0.04	0.06	0.46
Galvanized sheet	~0.25				
Building and insulating materials					
Asphalt	0.93	–	0.9	–	0.93
Brick					
Red	0.93	–	–	–	0.7
Fire clay	0.9	–	~0.7	~0.75	
Silica	0.9	–	~0.75	0.84	
Magnesite refractory	0.9	–	–	~0.4	
Enamel, white	0.9				
Paper, white	0.95	–	0.82	0.25	0.28
Plaster	0.91				
Roofing board	0.93				
Enameled steel, white	–	–	–	0.65	0.47

(Continued)

Table A.3.3 (*Continued*)

Material	Wavelength and average temperatures				
	9.3 μm 38°C	5.4 μm 260°C	3.6 μm 540°C	1.8 μm 1370°C	0.6 μm Solar
Paints					
Aluminized lacquer	0.65	0.65			
Lacquer, black	0.96	0.98			
Lampblack paint	0.96	0.97	–	0.97	0.97
Red paint	0.96	–	–	–	0.74
Yellow paint	0.95	–	0.5	–	0.30
Oil paints (all colors)	~0.94	~0.9			
White (ZnO)	0.95	–	0.91	–	0.18
Miscellaneous					
Ice	~0.97[b]				
Water	~0.96				
Carbon, T-carbon, 0.9% ash	0.82	0.80	0.79		
Wood	~0.93				
Glass	0.90	–	–	–	(Low)

Source: Adapted from Kreith (1973).

[a] *At 3315°C*

[b] *At 0°C*

A.4 [illegible]CAL PROPERTIES OF WATER AND AIR

Table A.4.1 Physical Properties of **Water** at the Saturation Pressure

Temperature T (°C)	Temperature T (K)	Density ρ (kg/m^3)	Coefficient of volumetric thermal expansion β ($\times 10^{-4}$ K^{-1})	Specific heat c_p (kJ/[kg °C])	Thermal conductivity k (W/[m °C])	Thermal diffusivity α ($\times 10^{-6}$ m^2/s)	Absolute viscosity μ ($\times 10^{-6}$ Pa s)	Kinematic viscosity ν ($\times 10^{-6}$ m^2/s)	Prandtl number N_{Pr}
0	273.15	999.9	−0.7	4.226	0.558	0.131	1793.636	1.789	13.7
5	278.15	1000.0	–	4.206	0.568	0.135	1534.741	1.535	11.4
10	283.15	999.7	0.95	4.195	0.577	0.137	1296.439	1.300	9.5
15	288.15	999.1	–	4.187	0.587	0.141	1135.610	1.146	8.1
20	293.15	998.2	2.1	4.182	0.597	0.143	993.414	1.006	7.0
25	298.15	997.1	–	4.178	0.606	0.146	880.637	0.884	6.1
30	303.15	995.7	3.0	4.176	0.615	0.149	792.377	0.805	5.4
35	308.15	994.1	–	4.175	0.624	0.150	719.808	0.725	4.8
40	313.15	992.2	3.9	4.175	0.633	0.151	658.026	0.658	4.3
45	318.15	990.2	–	4.176	0.640	0.155	605.070	0.611	3.9
50	323.15	988.1	4.6	4.178	0.647	0.157	555.056	0.556	3.55
55	328.15	985.7	–	4.179	0.652	0.158	509.946	0.517	3.27
60	333.15	983.2	5.3	4.181	0.658	0.159	471.650	0.478	3.00
65	338.15	980.6	–	4.184	0.663	0.161	435.415	0.444	2.76
70	343.15	977.8	5.8	4.187	0.668	0.163	404.034	0.415	2.55
75	348.15	974.9	–	4.190	0.671	0.164	376.575	0.366	2.23
80	353.15	971.8	6.3	4.194	0.673	0.165	352.059	0.364	2.25
85	358.15	968.7	–	4.198	0.676	0.166	328.523	0.339	2.04
90	363.15	965.3	7.0	4.202	0.678	0.167	308.909	0.326	1.95
95	368.15	961.9	–	4.206	0.680	0.168	292.238	0.310	1.84
100	373.15	958.4	7.5	4.211	0.682	0.169	277.528	0.294	1.75
110	383.15	951.0	8.0	4.224	0.684	0.170	254.973	0.268	1.57
120	393.15	943.5	8.5	4.232	0.684	0.171	235.360	0.244	1.43
130	403.15	934.8	9.1	4.250	0.685	0.172	211.824	0.226	1.32
140	413.15	926.3	9.7	4.257	0.686	0.172	201.036	0.212	1.23
150	423.15	916.9	10.3	4.270	0.684	0.173	185.346	0.201	1.17
160	433.15	907.6	10.8	4.285	0.680	0.173	171.616	0.191	1.10
170	443.15	897.3	11.5	4.396	0.679	0.172	162.290	0.181	1.05

(Continued)

Table A.4.1 (*Continued*)

Temperature		Density	Coefficient of volumetric thermal expansion	Specific heat	Thermal conductivity	Thermal diffusivity	Absolute viscosity	Kinematic viscosity	Prandtl number
T (°C)	T (K)	ρ (kg/m^3)	β ($\times 10^{-4}$ K^{-1})	c_p (kJ/[kg °C])	k (W/[m °C])	α ($\times 10^{-6}$ m^2/s)	μ ($\times 10^{-6}$ Pa s)	ν ($\times 10^{-6}$ m^2/s)	N_{Pr}
180	453.15	886.6	12.1	4.396	0.673	0.172	152.003	0.173	1.01
190	463.15	876.0	12.8	4.480	0.670	0.171	145.138	0.166	0.97
200	473.15	862.8	13.5	4.501	0.665	0.170	139.254	0.160	0.95
210	483.15	852.8	14.3	4.560	0.655	0.168	131.409	0.154	0.92
220	493.15	837.0	15.2	4.605	0.652	0.167	124.544	0.149	0.90
230	503.15	827.3	16.2	4.690	0.637	0.164	119.641	0.145	0.88
240	513.15	809.0	17.2	4.731	0.634	0.162	113.757	0.141	0.86
250	523.15	799.2	18.6	4.857	0.618	0.160	109.834	0.137	0.86

Source: Adapted from Raznjevic (1978)

Table A.4.2 Properties of Saturated Steam

Temperature (°C)	Vapor pressure (kPa)	Specific volume (m^3/kg)		Enthalpy (kJ/kg)		Entropy (kJ/[kg °C])	
		Liquid	Saturated vapor	Liquid (H_c)	Saturated vapor (H_v)	Liquid	Saturated vapor
0.01	0.6113	0.0010002	206.136	0.00	2501.4	0.0000	9.1562
3	0.7577	0.0010001	168.132	12.57	2506.9	0.0457	9.0773
6	0.9349	0.0010001	137.734	25.20	2512.4	0.0912	9.0003
9	1.1477	0.0010003	113.386	37.80	2517.9	0.1362	8.9253
12	1.4022	0.0010005	93.784	50.41	2523.4	0.1806	8.8524
15	1.7051	0.0010009	77.926	62.99	2528.9	0.2245	8.7814
18	2.0640	0.0010014	65.038	75.58	2534.4	0.2679	8.7123
21	2.487	0.0010020	54.514	88.14	2539.9	0.3109	8.6450
24	2.985	0.0010027	45.883	100.70	2545.4	0.3534	8.5794
27	3.567	0.0010035	38.774	113.25	2550.8	0.3954	8.5156
30	4.246	0.0010043	32.894	125.79	2556.3	0.4369	8.4533
33	5.034	0.0010053	28.011	138.33	2561.7	0.4781	8.3927
36	5.947	0.0010063	23.940	150.86	2567.1	0.5188	8.3336
40	7.384	0.0010078	19.523	167.57	2574.3	0.5725	8.2570
45	9.593	0.0010099	15.258	188.45	2583.2	0.6387	8.1648
50	12.349	0.0010121	12.032	209.33	2592.1	0.7038	8.0763

(*Continued*)

Table A.4.2 *(Continued)*

Temperature (°C)	Vapor pressure (kPa)	Specific volume (m^3/kg)		Enthalpy (kJ/kg)		Entropy (kJ/[kg °C])	
		Liquid	Saturated vapor	Liquid (H_c)	Saturated vapor (H_v)	Liquid	Saturated vapor
55	15.758	0.0010146	9.568	230.23	2600.9	0.7679	7.9913
60	19.940	0.0010172	7.671	251.13	2609.6	0.8312	7.9096
65	25.03	0.0010199	6.197	272.06	2618.3	0.8935	7.8310
70	31.19	0.0010228	5.042	292.98	2626.8	0.9549	7.7553
75	38.58	0.0010259	4.131	313.93	2635.3	1.0155	7.6824
80	47.39	0.0010291	3.407	334.91	2643.7	1.0753	7.6122
85	57.83	0.0010325	2.828	355.90	2651.9	1.1343	7.5445
90	70.14	0.0010360	2.361	376.92	2660.1	1.1925	7.4791
95	84.55	0.0010397	1.9819	397.96	2668.1	1.2500	7.4159
100	101.35	0.0010435	1.6729	419.04	2676.1	1.3069	7.3549
105	120.82	0.0010475	1.4194	440.15	2683.8	1.3630	7.2958
110	143.27	0.0010516	1.2102	461.30	2691.5	1.4185	7.2387
115	169.06	0.0010559	1.0366	482.48	2699.0	1.4734	7.1833
120	198.53	0.0010603	0.8919	503.71	2706.3	1.5276	7.1296
125	232.1	0.0010649	0.7706	524.99	2713.5	1.5813	7.0775
130	270.1	0.0010697	0.6685	546.31	2720.5	1.6344	7.0269
135	313.0	0.0010746	0.5822	567.69	2727.3	1.6870	6.9777
140	361.3	0.0010797	0.5089	589.13	2733.9	1.7391	6.9299
145	415.4	0.0010850	0.4463	610.63	2740.3	1.7907	6.8833
150	475.8	0.0010905	0.3928	632.20	2746.5	1.8418	6.8379
155	543.1	0.0010961	0.3468	653.84	2752.4	1.8925	6.7935
160	617.8	0.0011020	0.3071	675.55	2758.1	1.9427	6.7502
165	700.5	0.0011080	0.2727	697.34	2763.5	1.9925	6.7078
170	791.7	0.0011143	0.2428	719.21	2768.7	2.0419	6.6663
175	892.0	0.0011207	0.2168	741.17	2773.6	2.0909	6.6256
180	1002.1	0.0011274	0.19405	763.22	2778.2	2.1396	6.5857
190	1254.4	0.0011414	0.15654	807.62	2786.4	2.2359	6.5079
200	1553.8	0.0011565	0.12736	852.45	2793.2	2.3309	6.4323
225	2548	0.0011992	0.07849	966.78	2803.3	2.5639	6.2503
250	3973	0.0012512	0.05013	1085.36	2801.5	2.7927	6.0730
275	5942	0.0013168	0.03279	1210.07	2785.0	3.0208	5.8938
300	8581	0.0010436	0.02167	1344.0	2749.0	3.2534	5.7045

Source: Abridged from Keenan et al. (1969).

Table A.4.3 Properties of Superheated Steam

Absolute pressure (kPa, with sat. temperature, °C)[a]		Temperature (°C)							
		100	150	200	250	300	360	420	500
10	*V*	17.196	19.512	21.825	24.136	26.445	29.216	31.986	35.679
(45.81)	*H*	2687.5	2783.0	2879.5	2977.3	3076.5	3197.6	3320.9	3489.1
	s	8.4479	8.6882	8.9038	9.1002	9.2813	9.4821	9.6682	9.8978
50	*V*	3.418	3.889	4.356	4.820	5.284	5.839	6.394	7.134
(81.33)	*H*	2682.5	2780.1	2877.7	2976.0	3075.5	3196.8	3320.4	3488.7
	s	7.6947	7.9401	8.1580	8.3556	8.5373	8.7385	8.9249	9.1546
75	*V*	2.270	2.587	2.900	3.211	3.520	3.891	4.262	4.755
(91.78)	*H*	2679.4	2778.2	2876.5	2975.2	3074.9	3196.4	3320.0	3488.4
	s	7.5009	7.7496	7.9690	8.1673	8.3493	8.5508	8.7374	8.9672
100	*V*	1.6958	1.9364	2.172	2.406	2.639	2.917	3.195	3.565
(99.63)	*H*	2676.2	2776.4	2875.3	2974.3	3074.3	3195.9	3319.6	3488.1
	s	7.3614	7.6134	7.8343	8.0333	8.2158	8.4175	8.6042	8.8342
150	*V*		1.2853	1.4443	1.6012	1.7570	1.9432	2.129	2.376
(111.37)	*H*		2772.6	2872.9	2972.7	3073.1	3195.0	3318.9	3487.6
	s		7.4193	7.6433	7.8438	8.0720	8.2293	8.4163	8.6466
400	*V*		0.4708	0.5342	0.5951	0.6458	0.7257	0.7960	0.8893
(143.63)	*H*		2752.8	2860.5	2964.2	3066.8	3190.3	3315.3	3484.9
	s		6.9299	7.1706	7.3789	7.5662	7.7712	7.9598	8.1913
700	*V*			0.2999	0.3363	0.3714	0.4126	0.4533	0.5070
(164.97)	*H*			2844.8	2953.6	3059.1	3184.7	3310.9	3481.7
	s			6.8865	7.1053	7.2979	7.5063	7.6968	7.9299
1000	*V*			0.2060	0.2327	0.2579	0.2873	0.3162	0.3541
(179.91)	*H*			2827.9	2942.6	3051.2	3178.9	3306.5	3478.5
	s			6.6940	6.9247	7.1229	7.3349	7.5275	7.7622
1500	*V*			0.13248	0.15195	0.16966	0.18988	0.2095	0.2352
(198.32)	*H*			2796.8	2923.3	3037.6	3.1692	3299.1	3473.1
	s			6.4546	6.7090	6.9179	7.1363	7.3323	7.5698
2000	*V*				0.11144	0.12547	0.14113	0.15616	0.17568
(212.42)	*H*				2902.5	3023.5	3159.3	3291.6	3467.6
	s				6.5453	6.7664	6.9917	7.1915	7.4317
2500	*V*				0.08700	0.09890	0.11186	0.12414	0.13998
(223.99)	*H*				2880.1	3008.8	3149.1	3284.0	3462.1
	s				6.4085	6.6438	6.8767	7.0803	7.3234
3000	*V*				0.07058	0.08114	0.09233	0.10279	0.11619
(233.90)	*H*				2855.8	2993.5	3138.7	3276.3	3456.5
	s				6.2872	6.5390	6.7801	6.9878	7.2338

Source: Abridged from Keenan et al. (1969). Copyright © 1969 by John Wiley and Sons. Reprinted by permission of John Wiley and Sons, Inc.

[a] *V, specific volume, m^3/kg; H, enthalpy, kJ/kg; s, entropy, kJ/kg K.*

Table A.4.4 Physical Properties of Dry **Air** at Atmospheric Pressure

Temperature t (°C)	Temperature T (K)	Density (ρ) (kg/m^3)	Volumetric coefficient of expansion (β) ($\times 10^{-3}$ K^{-1})	Specific heat (c_p) (kJ/[kg K])	Thermal conductivity (k) (W/[m K])	Thermal diffusivity (α) ($\times 10^{-6}$ m^2/s)	Viscosity (μ) ($\times 10^{-6}$ N s/m^2)	Kinematic viscosity (ν) ($\times 10^{-6}$ m^2/s)	Prandtl number (N_{Pr})
−20	253.15	1.365	3.97	1.005	0.0226	16.8	16.279	12.0	0.71
0	273.15	1.252	3.65	1.011	0.0237	19.2	17.456	13.9	0.71
10	283.15	1.206	3.53	1.010	0.0244	20.7	17.848	14.66	0.71
20	293.15	1.164	3.41	1.012	0.0251	22.0	18.240	15.7	0.71
30	303.15	1.127	3.30	1.013	0.0258	23.4	18.682	16.58	0.71
40	313.15	1.092	3.20	1.014	0.0265	24.8	19.123	17.6	0.71
50	323.15	1.057	3.10	1.016	0.0272	26.2	19.515	18.58	0.71
60	333.15	1.025	3.00	1.017	0.0279	27.6	19.907	19.4	0.71
70	343.15	0.996	2.91	1.018	0.0286	29.2	20.398	20.65	0.71
80	353.15	0.968	2.83	1.019	0.0293	30.6	20.790	21.5	0.71
90	363.15	0.942	2.76	1.021	0.0300	32.2	21.231	22.82	0.71
100	373.15	0.916	2.69	1.022	0.0307	33.6	21.673	23.6	0.71
120	393.15	0.870	2.55	1.025	0.0320	37.0	22.555	25.9	0.71
140	413.15	0.827	2.43	1.027	0.0333	40.0	23.340	28.2	0.71
150	423.15	0.810	2.37	1.028	0.0336	41.2	23.732	29.4	0.71
160	433.15	0.789	2.31	1.030	0.0344	43.3	24.124	30.6	0.71
180	453.15	0.755	2.20	1.032	0.0357	47.0	24.909	33.0	0.71
200	473.15	0.723	2.11	1.035	0.0370	49.7	25.693	35.5	0.71
250	523.15	0.653	1.89	1.043	0.0400	60.0	27.557	42.2	0.71

Source: Adapted from Raznjevic (1978).

A.5 PSYCHROMETRIC CHARTS

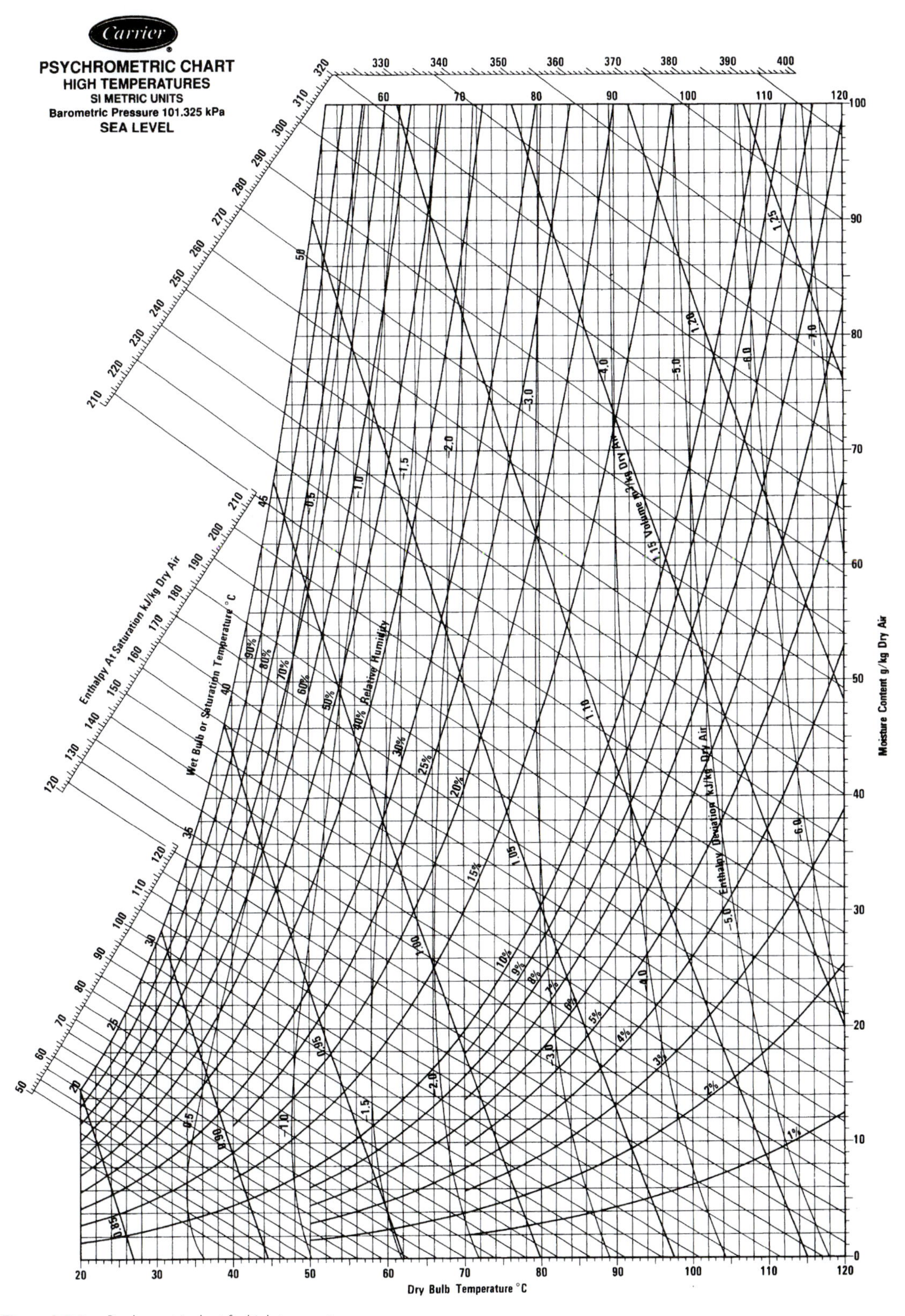

■ **Figure A.5.1** Psychrometric chart for high temperatures.

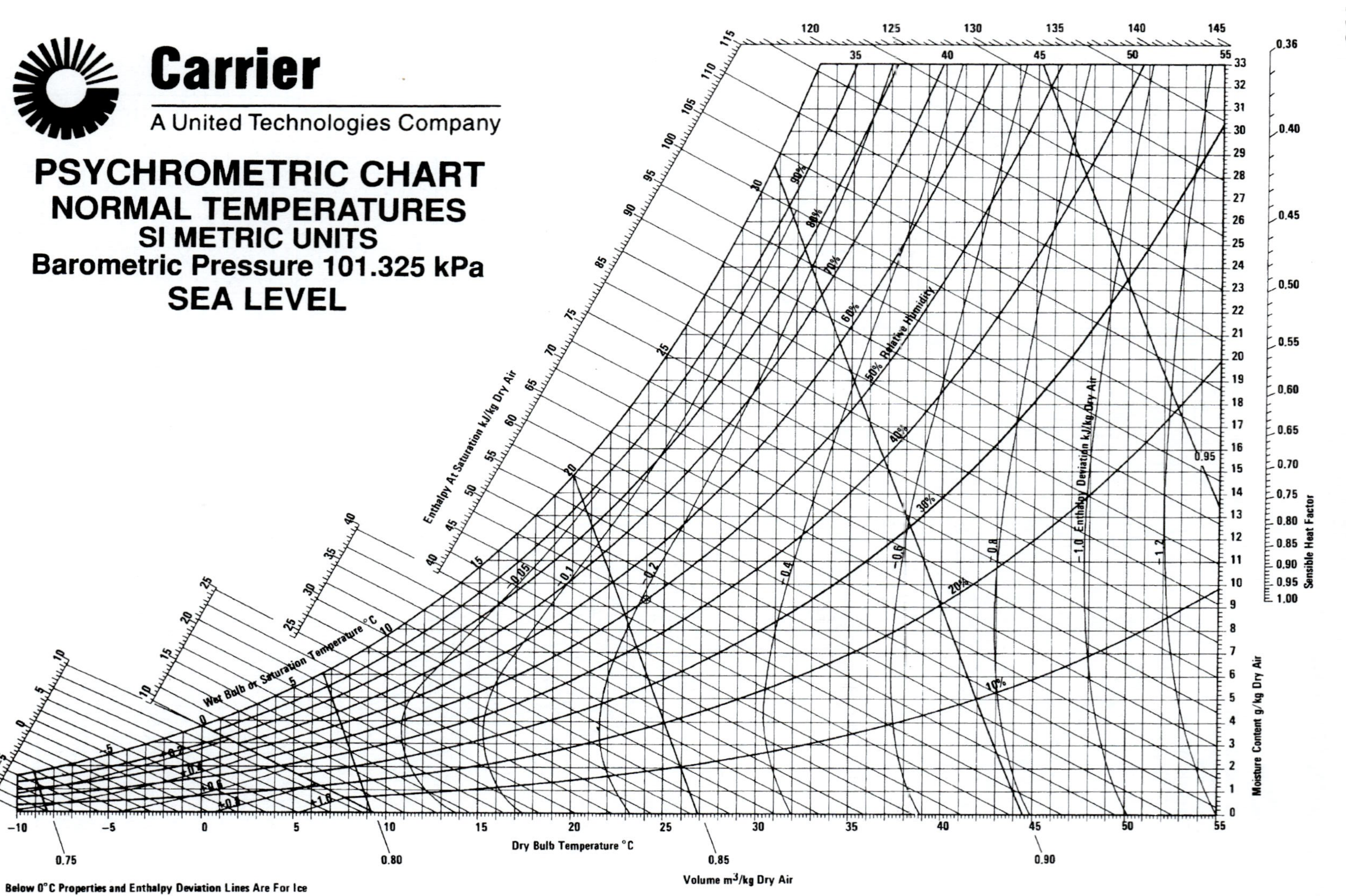

Figure A.5.2 Psychrometric chart for low temperatures.

A.6 PRESSURE–ENTHALPY DATA

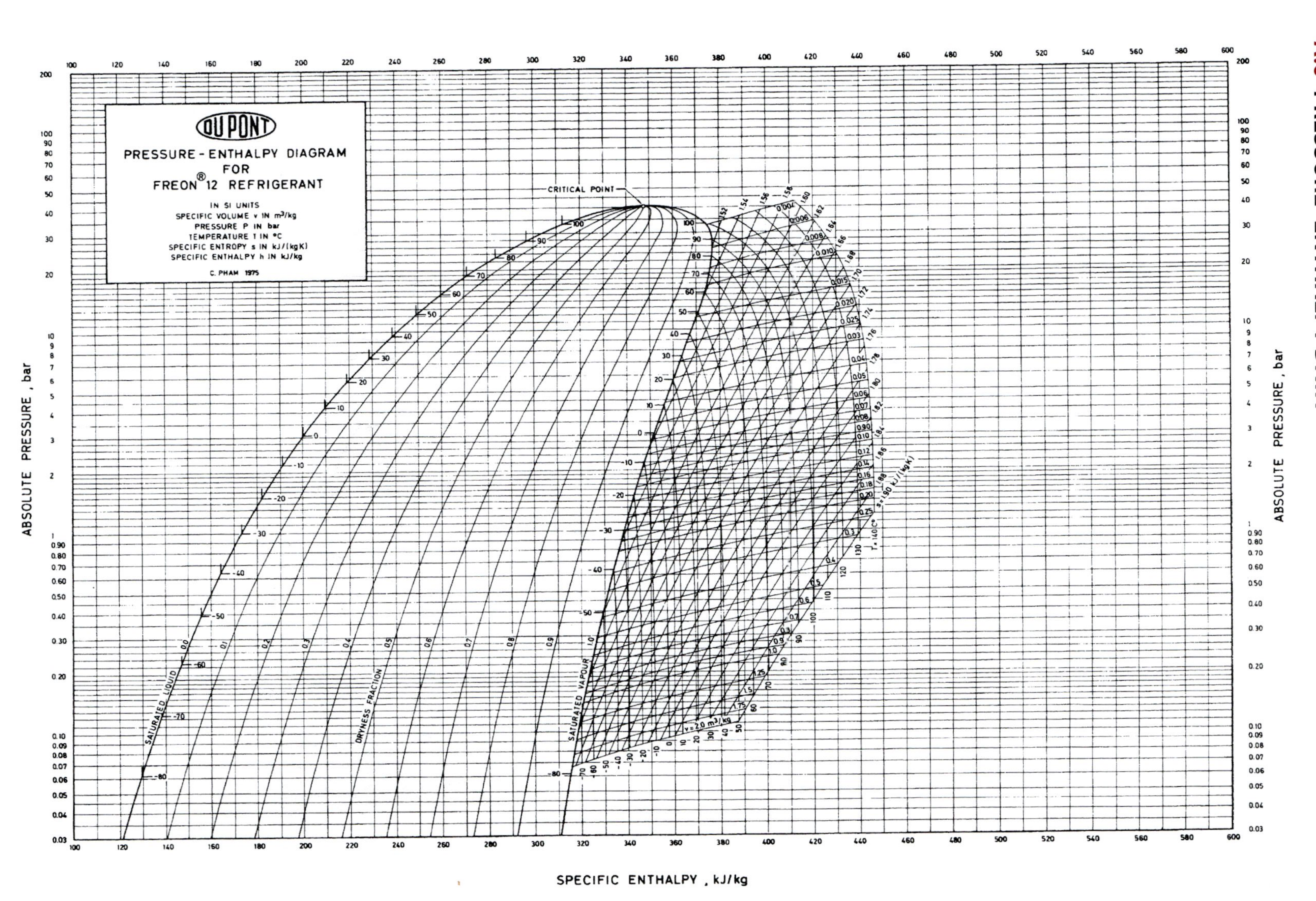

Figure A.6.1 Pressure–enthalpy diagram for Refrigerant 12. (Reproduced by permission of Du Pont de Nemours International S.A.)

Table A.6.1 Properties of Saturated Liquid and Vapor R-12[a]

Temp (°C)	Absolute pressure (kPa)	Enthalpy (kJ/kg)		Entropy (kJ/kg K)		Specific volume (L/kg)	
		h_f	h_g	s_f	s_g	v_f	v_g
−60	22.62	146.463	324.236	0.77977	1.61373	0.63689	637.911
−55	29.98	150.808	326.567	0.79990	1.60552	0.64226	491.000
−50	39.15	155.169	328.897	0.81964	1.59810	0.64782	383.105
−45	50.44	159.549	331.223	0.83901	1.59142	0.65355	302.683
−40	64.17	163.948	333.541	0.85805	1.58539	0.65949	241.910
−35	80.71	168.396	335.849	0.86776	1.57996	0.66563	195.398
−30	100.41	172.810	338.143	0.89516	1.57507	0.67200	159.375
−28	109.27	174.593	339.057	0.90244	1.57326	0.67461	147.275
−26	118.72	176.380	339.968	0.90967	1.57152	0.67726	136.284
−24	128.80	178.171	340.876	0.91686	1.56985	0.67996	126.282
−22	139.53	179.965	341.780	0.92400	1.56825	0.68269	117.167
−20	150.93	181.764	342.682	0.93110	1.56672	0.68547	108.847
−18	163.04	183.567	343.580	0.93816	1.56526	0.68829	101.242
−16	175.89	185.374	344.474	0.94518	1.56385	0.69115	94.2788
−14	189.50	187.185	345.365	0.95216	1.56250	0.69407	87.8951
−12	203.90	189.001	346.252	0.95910	1.56121	0.60703	82.0344
−10	219.12	190.822	347.134	0.96601	1.55997	0.70004	76.6464
−9	227.04	191.734	347.574	0.96945	1.55938	0.70157	74.1155
−8	235.19	192.647	348.012	0.97287	1.55897	0.70310	71.6864
−7	243.55	193.562	348.450	0.97629	1.55822	0.70465	69.3543
−6	252.14	194.477	348.886	0.97971	1.55765	0.70622	67.1146
−5	260.96	195.395	349.321	0.98311	1.55710	0.70780	64.9629
−4	270.01	196.313	349.755	0.98650	1.55657	0.70939	62.8952
−3	279.30	197.233	350.187	0.98989	1.55604	0.71099	60.9075
−2	288.82	198.154	350.619	0.99327	1.55552	0.71261	58.9963
−1	298.59	199.076	351.049	0.99664	1.55502	0.71425	57.1579
0	308.61	200.000	351.477	1.00000	1.55452	0.71590	55.3892
1	318.88	200.925	351.905	1.00355	1.55404	0.71756	53.6869
2	329.40	201.852	352.331	1.00670	1.55356	0.71324	52.0481
3	340.19	202.780	352.755	1.01004	1.55310	0.72094	50.4700
4	351.24	203.710	353.179	1.01337	1.55264	0.72265	48.9499
5	363.55	204.642	353.600	1.01670	1.55220	0.72438	47.4853
6	374.14	205.575	354.020	1.02001	1.55176	0.72612	46.0737

(Continued)

Table A.6.1 (Continued)

Temp (°C)	Absolute pressure (kPa)	Enthalpy (kJ/kg)		Entropy (kJ/kg K)		Specific volume (L/kg)	
		h_f	h_g	s_f	s_g	v_f	v_g
7	386.01	206.509	354.439	1.02333	1.55133	0.72788	44.7129
8	398.15	207.445	354.856	1.02663	1.55091	0.72966	43.4006
9	410.58	208.383	355.272	1.02993	1.55050	0.73146	42.1349
10	423.30	209.323	355.686	1.03322	1.55010	0.73326	40.9137
11	436.31	210.264	356.098	1.03650	1.54970	0.73510	39.7352
12	449.62	211.207	356.509	1.03978	1.54931	0.73695	38.5975
13	463.23	212.152	356.918	1.04305	1.54893	0.73882	37.4991
14	477.14	213.099	357.325	1.04632	1.54856	0.74071	36.4382
15	491.37	214.048	357.730	1.04958	1.54819	0.74262	35.4133
16	505.91	214.998	358.134	1.05284	1.54783	0.74455	34.4230
17	520.76	215.951	358.535	1.05609	1.54748	0.74649	33.4658
18	535.94	216.906	358.935	1.05933	1.54713	0.74846	32.5405
19	551.45	217.863	359.333	1.06258	1.54679	0.75045	31.6457
20	567.29	218.821	359.729	1.06581	1.54645	0.75246	30.7802
21	583.47	219.783	360.122	1.06904	1.54612	0.75449	29.9429
22	599.98	220.746	360.514	1.07227	1.54579	0.75655	29.1327
23	616.84	221.712	360.904	1.07549	1.54547	0.75863	28.3485
24	634.05	222.680	361.291	1.07871	1.54515	0.76073	27.5894
25	651.62	223.650	361.676	1.08193	1.54484	0.76286	26.8542
26	669.54	224.623	362.059	1.08514	1.54453	0.76501	26.1442
27	687.82	225.598	362.439	1.08835	1.54423	0.76718	25.4524
28	706.47	226.576	362.817	1.09155	1.54393	0.76938	24.7840
29	725.50	227.557	363.193	1.09475	1.54363	0.77161	24.1362
30	744.90	228.540	363.566	1.09795	1.54334	0.77386	23.5082
31	764.68	229.526	363.937	1.10115	1.54305	0.77614	22.8993
32	784.85	230.515	364.305	1.10434	1.54276	0.77845	22.3088
33	805.41	231.506	364.670	1.10753	1.54247	0.78079	21.7359
34	826.36	232.501	365.033	1.11072	1.54219	0.78316	21.1802
35	847.72	233.498	365.392	1.11391	1.54191	0.78556	20.6408
36	869.48	234.499	365.749	1.11710	1.54163	0.78799	20.1173
37	891.64	235.503	366.103	1.12028	1.54135	0.79045	19.6091
38	914.23	236.510	366.454	1.12347	1.54107	0.79294	19.1156
39	937.23	237.521	366.802	1.12665	1.54079	0.79546	18.6362

(Continued)

Table A.6.1 *(Continued)*

Temp (°C)	Absolute pressure (kPa)	Enthalpy (kJ/kg)		Entropy (kJ/kg K)		Specific volume (L/kg)	
		h_f	h_g	s_f	s_g	v_f	v_g
40	960.65	238.535	367.146	1.12984	1.54051	0.79802	18.1706
41	984.51	239.552	267.487	1.13302	1.54024	0.80062	17.7182
42	1008.8	240.574	367.825	1.13620	1.53996	0.80325	17.2785
43	1033.5	241.598	368.160	1.13938	1.53968	0.80592	16.8511
44	1058.7	242.627	368.491	1.14257	1.53941	0.80863	16.4356
45	1084.3	243.659	368.818	1.14575	1.53913	0.81137	16.0316
46	1110.4	244.696	369.141	1.14894	1.53885	0.81416	15.6386
47	1136.9	245.736	369.461	1.15213	1.53856	0.81698	15.2563
48	1163.9	246.781	369.777	1.15532	1.53828	0.81985	14.8844
49	1191.4	247.830	370.088	1.15851	1.53799	0.82277	14.5224
50	1219.3	248.884	370.396	1.16170	1.53770	0.82573	14.1701
52	1276.6	251.004	370.997	1.16810	1.53712	0.83179	13.4931
54	1335.9	253.144	371.581	1.17451	1.53651	0.83804	12.8509
56	1397.2	255.304	372.145	1.18093	1.53589	0.84451	12.2412
58	1460.5	257.486	372.688	1.18738	1.53524	0.85121	11.6620
60	1525.9	259.690	373.210	1.19384	1.53457	0.85814	11.1113
62	1593.5	261.918	373.707	1.20034	1.53387	0.86534	10.5872
64	1663.2	264.172	374.810	1.20686	1.53313	0.87282	10.0881
66	1735.1	266.452	374.625	1.21342	1.53235	0.88059	9.61234
68	1809.3	268.762	375.042	1.22001	1.53153	0.88870	9.15844
70	1885.8	271.102	375.427	1.22665	1.53066	0.89716	8.72502
75	2087.5	277.100	376.234	1.24347	1.52821	0.92009	7.72258
80	2304.6	283.341	376.777	1.26069	1.52526	0.94612	6.82143
85	2538.0	289.879	376.985	1.27845	1.52164	0.97621	6.00494
90	2788.5	296.788	376.748	1.29691	1.51708	1.01190	5.25759
95	3056.9	304.181	375.887	1.31637	1.51113	1.05581	4.56341
100	3344.1	312.261	374.070	1.33732	1.50296	1.11311	3.90280

Source: Stoecker (1988).

[a] *Subscripts: f = liquid, g = gas.*

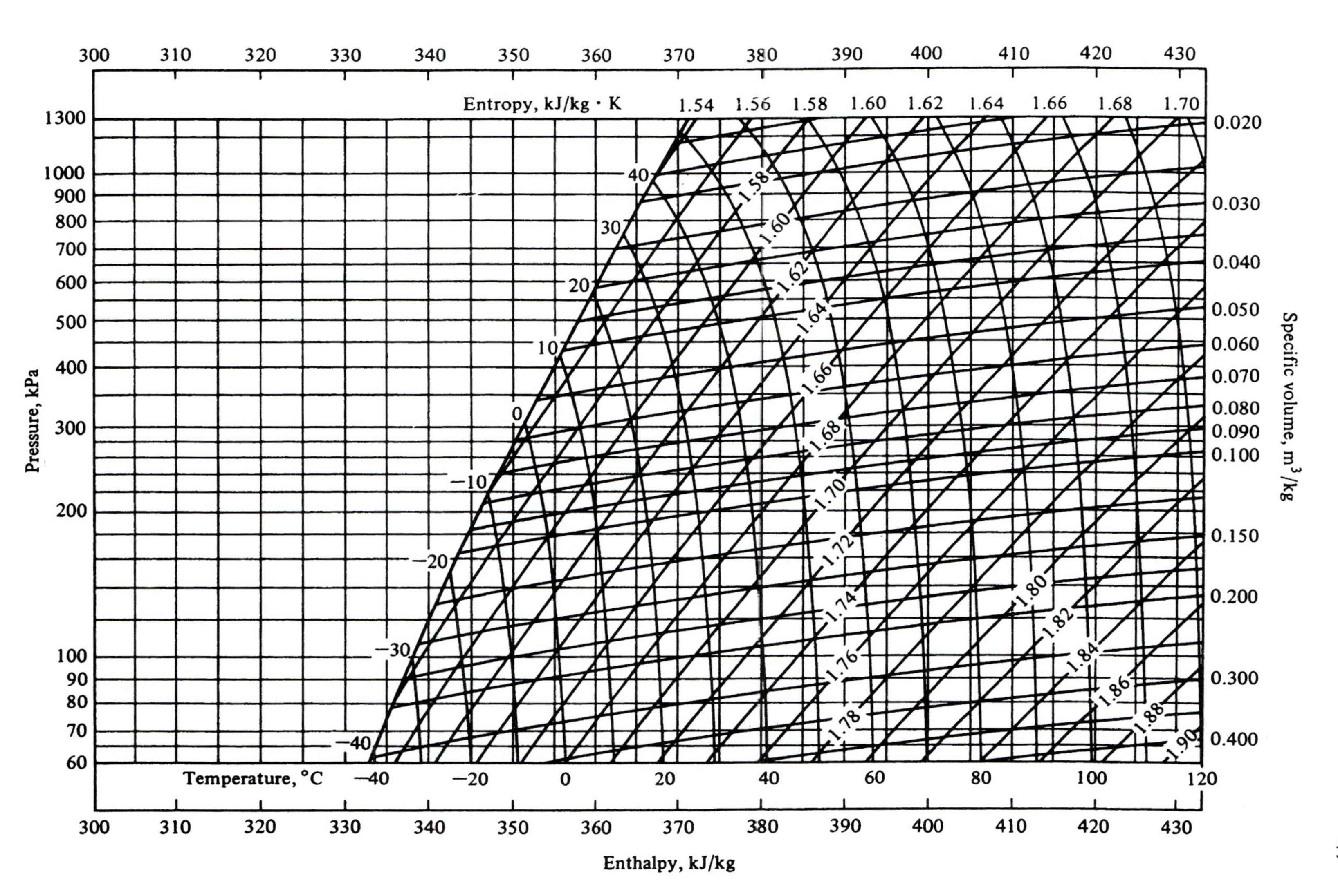

■ **Figure A.6.2** Pressure–enthalpy diagram of superheated R-12 vapor. (Courtesy, Technical University of Denmark.)

Table A.6.2 Properties of Saturated Liquid and Vapor R-717 (Ammonia)[a]

Temp (°C)	Absolute pressure (kPa)	Enthalpy (kJ/kg)		Entropy (kJ/[kg K])		Specific volume (L/kg)	
		h_f	h_g	s_f	s_g	v_f	v_g
−60	21.99	−69.5330	1373.19	−0.10909	6.6592	1.4010	4685.08
−55	30.29	−47.5062	1382.01	−0.00717	6.5454	1.4126	3474.22
−50	41.03	−25.4342	1390.64	−0.09264	6.4382	1.4245	2616.51
−45	54.74	−3.3020	1399.07	−0.19049	6.3369	1.4367	1998.91
−40	72.01	18.9024	1407.26	0.28651	6.2410	1.4493	1547.36
−35	93.49	41.1883	1415.20	0.38082	6.1501	1.4623	1212.49
−30	119.90	63.5629	1422.86	0.47351	6.0636	1.4757	960.867
−28	132.02	72.5387	1425.84	0.51015	6.0302	1.4811	878.100
−26	145.11	81.5300	1428.76	0.54655	5.9974	1.4867	803.761
−24	159.22	90.5370	1431.64	0.58272	5.9652	1.4923	736.868
−22	174.41	99.5600	1434.46	0.61865	5.9336	1.4980	676.570
−20	190.74	108.599	1437.23	0.65436	5.9025	1.5037	622.122
−18	208.26	117.656	1439.94	0.68984	5.8720	1.5096	572.875
−16	227.04	126.729	1442.60	0.72511	5.8420	1.5155	528.257
−14	247.14	135.820	1445.20	0.76016	5.8125	1.5215	487.769
−12	268.63	144.929	1447.74	0.79501	5.7835	1.5276	450.971
−10	291.57	154.056	1450.22	0.82965	5.7550	1.5338	417.477
−9	303.60	158.628	1451.44	0.84690	5.7409	1.5369	401.860
−8	316.02	163.204	1452.64	0.86410	5.7269	1.5400	386.944
−7	328.84	167.785	1453.83	0.88125	5.7131	1.5432	372.692
−6	342.07	172.371	1455.00	0.89835	5.6993	1.5464	359.071
−5	355.71	176.962	1456.15	0.91541	5.6856	1.5496	346.046
−4	369.77	181.559	1457.29	0.93242	5.6721	1.5528	333.589
−3	384.26	186.161	1458.42	0.94938	5.6586	1.5561	321.670
−2	399.20	190.768	1459.53	0.96630	5.6453	1.5594	310.263
−1	414.58	195.381	1460.62	0.98317	5.6320	1.5627	299.340
0	430.43	200.000	1461.70	1.00000	5.6189	1.5660	288.880
1	446.74	204.625	1462.76	1.01679	5.6058	1.5694	278.858
2	463.53	209.256	1463.80	1.03354	5.5929	1.5727	269.253
3	480.81	213.892	1464.83	1.05024	5.5800	1.5762	260.046
4	498.59	218.535	1465.84	1.06691	5.5672	1.5796	251.216
5	516.87	223.185	1466.84	1.08353	5.5545	1.5831	242.745
6	535.67	227.841	1467.82	1.10012	5.5419	1.5866	234.618

(Continued)

Table A.6.2 *(Continued)*

Temp (°C)	Absolute pressure (kPa)	Enthalpy (kJ/kg)		Entropy (kJ/[kg K])		Specific volume (L/kg)	
		h_f	h_g	s_f	s_g	v_f	v_g
7	555.00	232.503	1468.78	1.11667	5.5294	1.5901	226.817
8	574.87	237.172	1469.72	1.13317	5.5170	1.5936	219.326
9	595.28	241.848	1470.64	1.14964	5.5046	1.5972	212.132
10	616.25	246.531	1471.57	1.16607	5.4924	1.6008	205.221
11	637.78	251.221	1472.46	1.18246	5.4802	1.6045	198.580
12	659.89	255.918	1473.34	1.19882	5.4681	1.6081	192.196
13	682.59	260.622	1474.20	1.21515	5.4561	1.6118	186.058
14	705.88	265.334	1475.05	1.23144	5.4441	1.6156	180.154
15	729.29	270.053	1475.88	1.24769	5.4322	1.6193	174.475
16	754.31	274.779	1476.69	1.26391	5.4204	1.6231	169.009
17	779.46	279.513	1477.48	1.28010	5.4087	1.6269	163.748
18	805.25	284.255	1478.25	1.29626	5.3971	1.6308	158.683
19	831.69	289.005	1479.01	1.31238	5.3855	1.6347	153.804
20	858.79	293.762	1479.75	1.32847	5.3740	1.6386	149.106
21	886.57	298.527	1480.48	1.34452	5.3626	1.6426	144.578
22	915.03	303.300	1481.18	1.36055	5.3512	1.6466	140.214
23	944.18	308.081	1481.87	1.37654	5.3399	1.6507	136.006
24	974.03	312.870	1482.53	1.39250	5.3286	1.6547	131.950
25	1004.6	316.667	1483.18	1.40843	5.3175	1.6588	128.037
26	1035.9	322.471	1483.81	1.42433	4.3063	1.6630	124.261
27	1068.0	327.284	1484.42	1.44020	5.2953	1.6672	120.619
28	1100.7	332.104	1485.01	1.45064	5.2843	1.6714	117.103
29	1134.3	336.933	1485.59	1.47185	5.2733	1.6757	113.708
30	1168.6	341.769	1486.14	1.48762	5.2624	1.6800	110.430
31	1203.7	346.614	1486.67	1.50337	5.2516	1.6844	107.263
32	1239.6	351.466	1487.18	1.51908	5.2408	1.6888	104.205
33	1276.3	356.326	1487.66	1.53477	5.2300	1.6932	101.248
34	1313.9	361.195	1488.13	1.55042	5.2193	1.6977	98.3913
35	1352.2	366.072	1488.57	1.56605	5.2086	1.7023	95.6290
36	1391.5	370.957	1488.99	1.58165	5.1980	1.7069	92.9579
37	1431.5	375.851	1489.39	1.59722	5.1874	1.7115	90.3743
38	1472.4	380.754	1489.76	1.61276	5.1768	1.7162	87.8748
39	1514.3	385.666	1490.10	1.62828	5.1663	1.7209	85.4561

(Continued)

Table A.6.2 (*Continued*)

Temp (°C)	Absolute pressure (kPa)	Enthalpy (kJ/kg)		Entropy (kJ/[kg K])		Specific volume (L/kg)	
		h_f	h_g	s_f	s_g	v_f	v_g
40	1557.0	390.587	1490.42	1.64377	5.1558	1.7257	83.1150
41	1600.6	395.519	1490.71	1.65924	5.1453	1.7305	80.8484
42	1645.1	400.462	1490.98	1.67470	5.1349	1.7354	78.6536
43	1690.6	405.416	1491.21	1.69013	5.1244	1.7404	76.5276
44	1737.0	410.382	1491.41	1.70554	5.1140	1.7454	74.4678
45	1784.3	415.362	1491.58	1.72095	5.1036	1.7504	72.4716
46	1832.6	420.358	1491.72	1.73635	5.0932	1.7555	70.5365
47	1881.9	425.369	1491.83	1.75174	5.0827	1.7607	68.6602
48	1932.2	430.399	1491.88	1.76714	5.0723	1.7659	66.8403
49	1983.5	435.450	1491.91	1.78255	5.0618	1.7712	65.0746
50	2035.9	440.523	1491.89	1.79798	5.0514	1.7766	63.3608
51	2089.2	445.623	1491.83	1.81343	5.0409	1.7820	61.6971
52	2143.6	450.751	1491.73	1.82891	5.0303	1.7875	60.0813
53	2199.1	455.913	1491.58	1.84445	5.0198	1.7931	58.5114
54	2255.6	461.112	1491.38	1.86004	5.0092	1.7987	56.9855
55	2313.2	466.353	1491.12	1.87571	4.9985	1.8044	55.5019

Source: Stoecker (1988).
[a]*Subscripts: f = liquid, g = gas.*

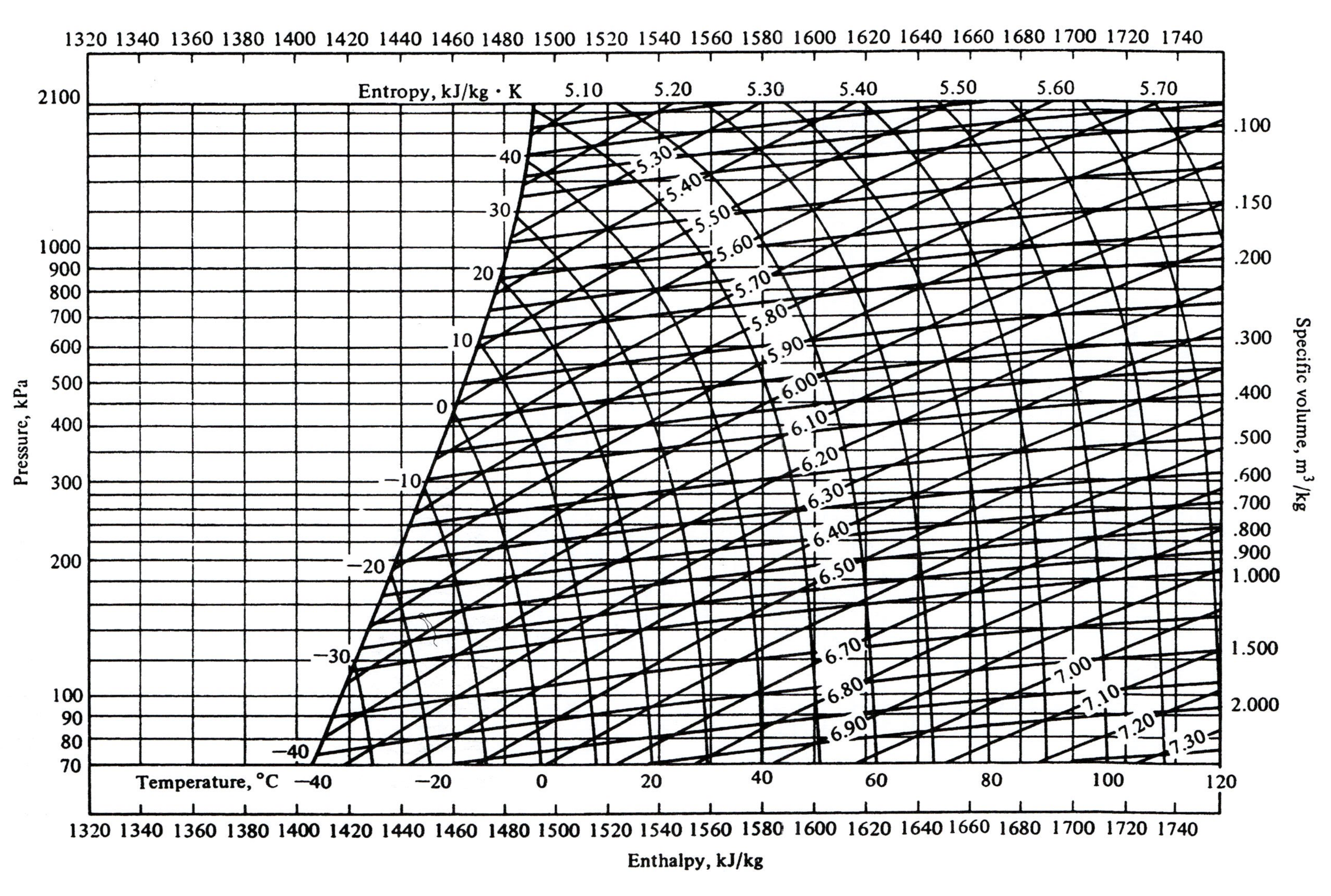

Figure A.6.3 Pressure–enthalpy diagram of superheated R-717 (ammonia) vapor. (Courtesy, Technical University of Denmark.)

Table A.6.3 Properties of Saturated Liquid and Vapor R-134a

Temp °C	Absolute pressure bar	Density		Enthalpy		Entropy	
		kg/m³ Liquid	kg/m³ Vapor	kJ/kg Liquid	kJ/kg Vapor	kJ/ (kg K) Liquid	kJ/ (kg K) Vapor
−60	0.15935	1472.0	0.9291	24.109	261.491	0.68772	1.8014
−55	0.21856	1458.5	1.2489	30.191	264.633	0.7159	1.79059
−50	0.29477	1444.9	1.6526	36.302	267.779	0.74358	1.7809
−45	0.39139	1431.0	2.1552	42.448	270.926	0.77078	1.77222
−40	0.51225	1417.0	2.7733	48.631	274.068	0.79756	1.76448
−35	0.66153	1402.7	3.5252	54.857	277.203	0.82393	1.75757
−30	0.84379	1388.2	4.4307	61.130	280.324	0.84995	1.75142
−28	0.92701	1382.3	4.8406	63.653	281.569	0.86026	1.74916
−26	1.01662	1376.4	5.2800	66.185	282.81	0.87051	1.74701
−24	1.11295	1370.5	5.7504	68.725	284.048	0.88072	1.74495
−22	1.21636	1364.4	6.2533	71.274	285.282	0.89088	1.743
−20	1.32719	1358.4	6.7903	73.833	286.513	0.901	1.74113
−18	1.44582	1352.3	7.3630	76.401	287.739	0.91107	1.73936
−16	1.57260	1346.2	7.9733	78.980	288.961	0.9211	1.73767
−14	1.70793	1340.0	8.6228	81.568	290.179	0.93109	1.73607
−12	1.85218	1333.7	9.3135	84.167	291.391	0.94104	1.73454
−10	2.00575	1327.4	10.047	86.777	292.598	0.95095	1.73309
−9	2.08615	1324.3	10.431	88.086	293.199	0.95589	1.73239
−8	2.16904	1321.1	10.826	89.398	293.798	0.96082	1.73171
−7	2.25446	1317.9	11.233	90.713	294.396	0.96575	1.73105
−6	2.34246	1314.7	11.652	92.031	294.993	0.97067	1.7304
−5	2.43310	1311.5	12.083	93.351	295.588	0.97557	1.72977
−4	2.52643	1308.2	12.526	94.675	296.181	0.98047	1.72915
−3	2.62250	1305.0	12.983	96.002	296.772	0.98537	1.72855
−2	2.72136	1301.7	13.453	97.331	297.362	0.99025	1.72796
−1	2.82307	1298.4	13.936	98.664	297.95	0.99513	1.72739
0	2.92769	1295.1	14.433	100.00	298.536	1	1.72684
1	3.03526	1291.8	14.944	101.339	299.12	1.00486	1.72629
2	3.14584	1288.5	15.469	102.681	299.701	1.00972	1.72577
3	3.25950	1285.1	16.009	104.027	300.281	1.01457	1.72525
4	3.37627	1281.8	16.564	105.376	300.859	1.01941	1.72474
5	3.49623	1278.4	17.134	106.728	301.434	1.02425	1.72425
6	3.61942	1275.0	17.719	108.083	302.008	1.02908	1.72377

(Continued)

Table A.6.3 (*Continued*)

Temp °C	Absolute pressure bar	Density		Enthalpy		Entropy	
		kg/m^3 Liquid	kg/m^3 Vapor	kJ/kg Liquid	kJ/kg Vapor	kJ/ (kg K) Liquid	kJ/ (kg K) Vapor
7	3.74591	1271.6	18.321	109.442	302.578	1.0339	1.7233
8	3.87575	1268.2	18.939	110.805	303.147	1.03872	1.72285
9	4.00900	1264.7	19.574	112.171	303.713	1.04353	1.7224
10	4.14571	1261.2	20.226	113.540	304.276	1.04834	1.72196
11	4.28595	1257.8	20.895	114.913	304.837	1.05314	1.72153
12	4.42978	1254.3	21.583	116.290	305.396	1.05793	1.72112
13	4.57725	1250.7	22.288	117.670	305.951	1.06273	1.72071
14	4.72842	1247.2	23.012	119.054	306.504	1.06751	1.72031
15	4.88336	1243.6	23.755	120.441	307.054	1.07229	1.71991
16	5.04212	1240.0	24.518	121.833	307.6	1.07707	1.71953
17	5.20477	1236.4	25.301	123.228	308.144	1.08184	1.71915
18	5.37137	1232.8	26.104	124.627	308.685	1.08661	1.71878
19	5.54197	1229.2	26.928	126.030	309.222	1.09137	1.71842
20	5.71665	1225.5	27.773	127.437	309.756	1.09613	1.71806
21	5.89546	1221.8	28.640	128.848	310.287	1.10089	1.71771
22	6.07846	1218.1	29.529	130.263	310.814	1.10564	1.71736
23	6.26573	1214.3	30.422	131.683	311.337	1.11039	1.71702
24	6.45732	1210.6	31.378	133.106	311.857	1.11513	1.71668
25	6.65330	1206.8	32.337	134.533	312.373	1.11987	1.71635
26	6.85374	1203.0	33.322	135.965	312.885	1.12461	1.71602
27	7.05869	1199.2	34.331	137.401	313.393	1.12935	1.71569
28	7.26823	1195.3	35.367	138.842	313.897	1.13408	1.71537
29	7.48241	1191.4	36.428	140.287	314.397	1.13881	1.71505
30	7.70132	1187.5	37.517	141.736	314.892	1.14354	1.71473
31	7.92501	1183.5	38.634	143.190	315.383	1.14826	1.71441
32	8.15355	1179.6	39.779	144.649	315.869	1.15299	1.71409
33	8.38701	1175.6	40.953	146.112	316.351	1.15771	1.71377
34	8.62545	1171.5	42.157	147.580	316.827	1.16243	1.71346
35	8.86896	1167.5	43.391	149.053	317.299	1.16715	1.71314
36	9.11759	1163.4	44.658	150.530	317.765	1.17187	1.71282
37	9.37142	1159.2	45.956	152.013	318.226	1.17659	1.7125
38	9.63052	1155.1	47.288	153.500	318.681	1.1813	1.71217
39	9.89496	1150.9	48.654	154.993	319.131	1.18602	1.71185

(*Continued*)

Table A.6.3 (*Continued*)

Temp °C	Absolute pressure bar	Density kg/m³ Liquid	Density kg/m³ Vapor	Enthalpy kJ/kg Liquid	Enthalpy kJ/kg Vapor	Entropy kJ/ (kg K) Liquid	Entropy kJ/ (kg K) Vapor
40	10.1648	1146.7	50.055	156.491	319.575	1.19073	1.71152
41	10.4401	1142.4	51.492	157.994	320.013	1.19545	1.71119
42	10.7210	1138.1	52.967	159.503	320.445	1.20017	1.71085
43	11.0076	1133.7	54.479	161.017	320.87	1.20488	1.71051
44	11.2998	1129.4	56.031	162.537	321.289	1.2096	1.71016
45	11.5978	1124.9	57.623	164.062	321.701	1.21432	1.70981
46	11.9017	1120.5	59.256	165.593	322.106	1.21904	1.70945
47	12.2115	1116.0	60.933	167.130	322.504	1.22376	1.70908
48	12.5273	1111.4	62.645	168.673	322.894	1.22848	1.7087
49	12.8492	1106.8	64.421	170.222	323.277	1.23321	1.70832
50	13.1773	1102.2	66.234	171.778	323.652	1.23794	1.70792
52	13.8523	1092.8	70.009	174.908	324.376	1.24741	1.7071
54	14.5529	1083.1	73.992	178.065	325.066	1.25689	1.70623
56	15.2799	1073.3	78.198	181.251	325.717	1.26639	1.7053
58	16.0339	1063.2	82.643	184.467	326.329	1.27592	1.70431
60	16.8156	1052.9	87.346	187.715	326.896	1.28548	1.70325
62	17.6258	1042.2	92.328	190.996	327.417	1.29507	1.70211
64	18.4653	1031.3	97.611	194.314	327.886	1.30469	1.70087
66	19.3347	1020.1	103.223	197.671	328.3	1.31437	1.69954
68	20.2349	1008.5	109.196	201.070	328.654	1.3241	1.69808
70	21.1668	996.49	115.564	204.515	328.941	1.3339	1.6965
75	23.6409	964.48	133.511	213.359	329.321	1.35876	1.69184
80	26.3336	928.78	155.130	222.616	329.095	1.38434	1.68585
85	29.2625	887.82	181.955	232.448	328.023	1.41108	1.67794
90	32.4489	838.51	216.936	243.168	325.655	1.43978	1.66692
95	35.9210	773.06	267.322	255.551	320.915	1.47246	1.65001
100	39.7254	649.71	367.064	273.641	309.037	1.5198	1.61466

Source: ICI Chemicals and Polymers Ltd. (KLEA 134a); Reference enthalpy 100 kJ/kg at 0ºC.

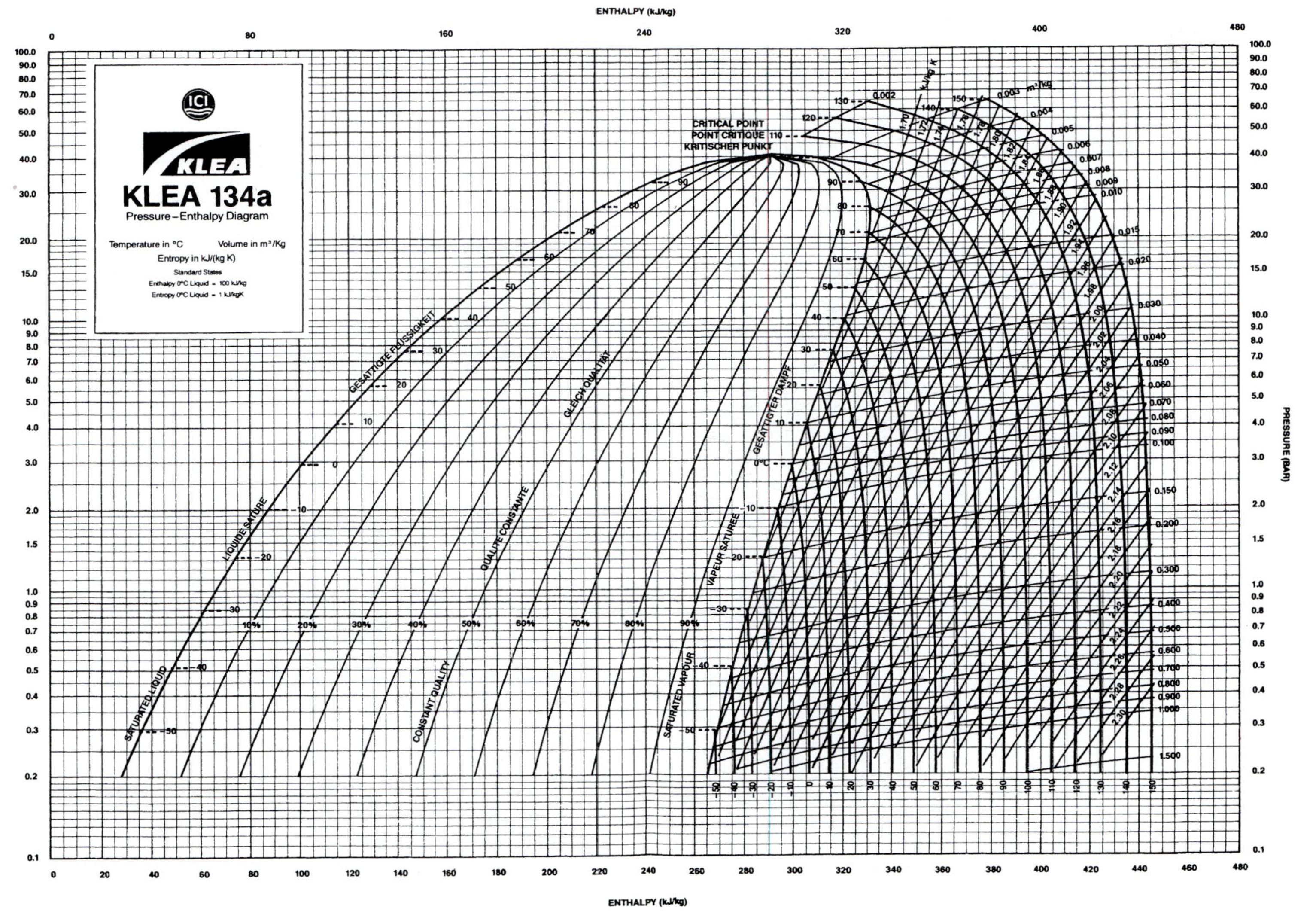

Figure A.6.4 Pressure–enthalpy diagram of R-134a. (Alternate P-H diagram with a datum of 200 kJ/kg AT 0°C is available from DuPont Fluorochemicals, Wilmington, Delaware, USA.)

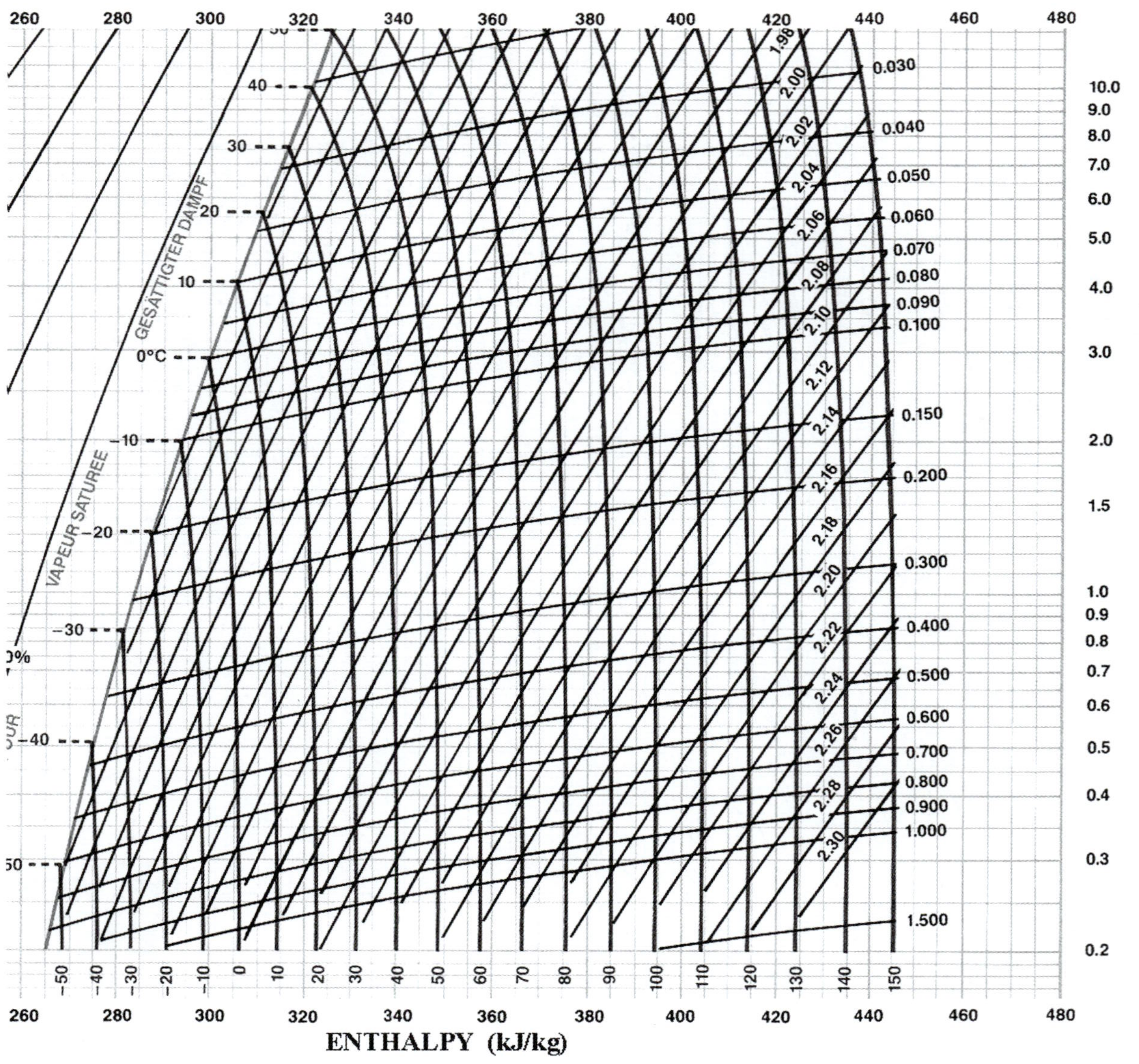

Figure A.6.5 Pressure–enthalpy diagram of R-134a (expanded scale). (Courtesy, ICI Co.)

A.7 SYMBOLS FOR USE IN DRAWING FOOD ENGINEERING PROCESS EQUIPMENT

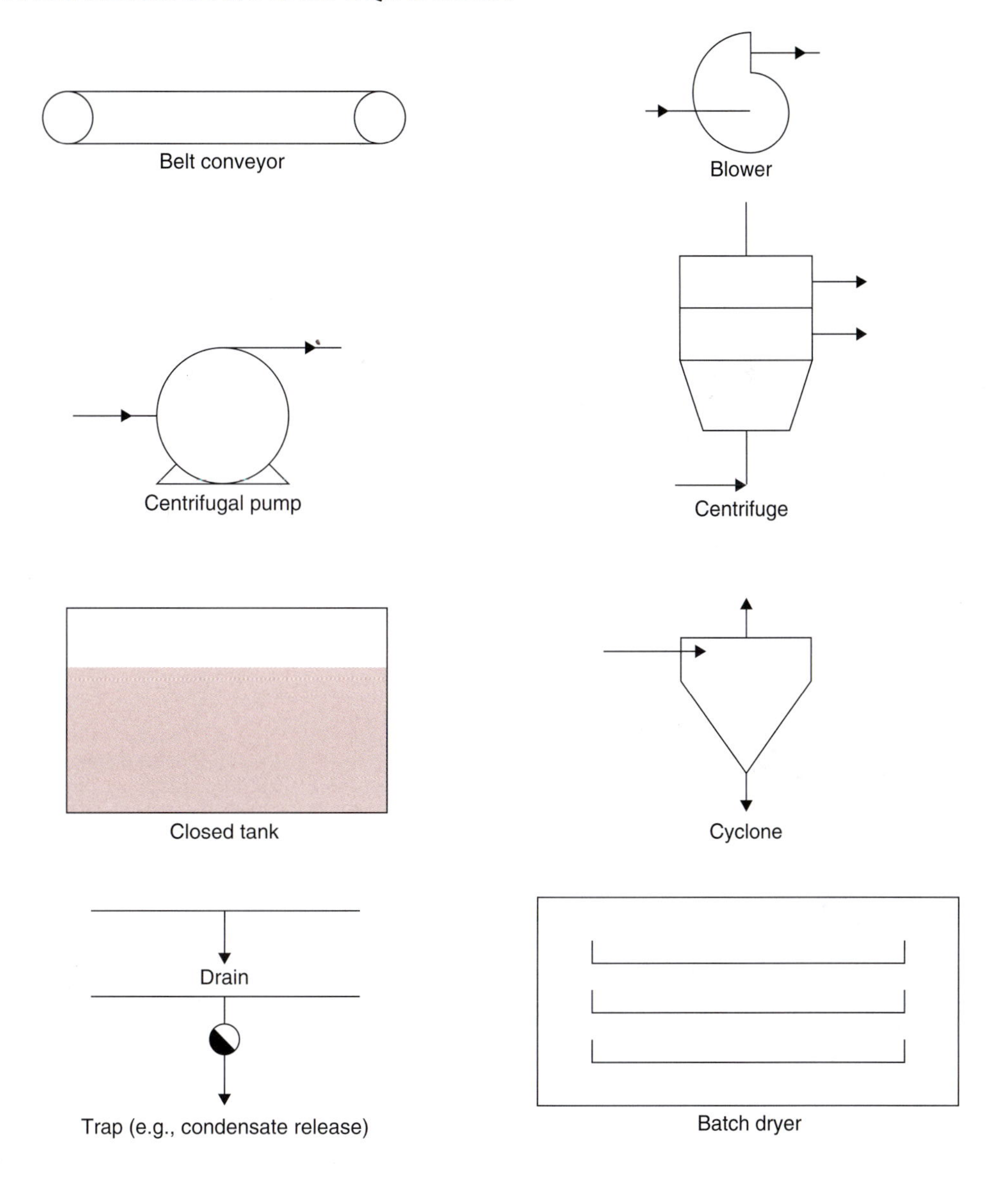

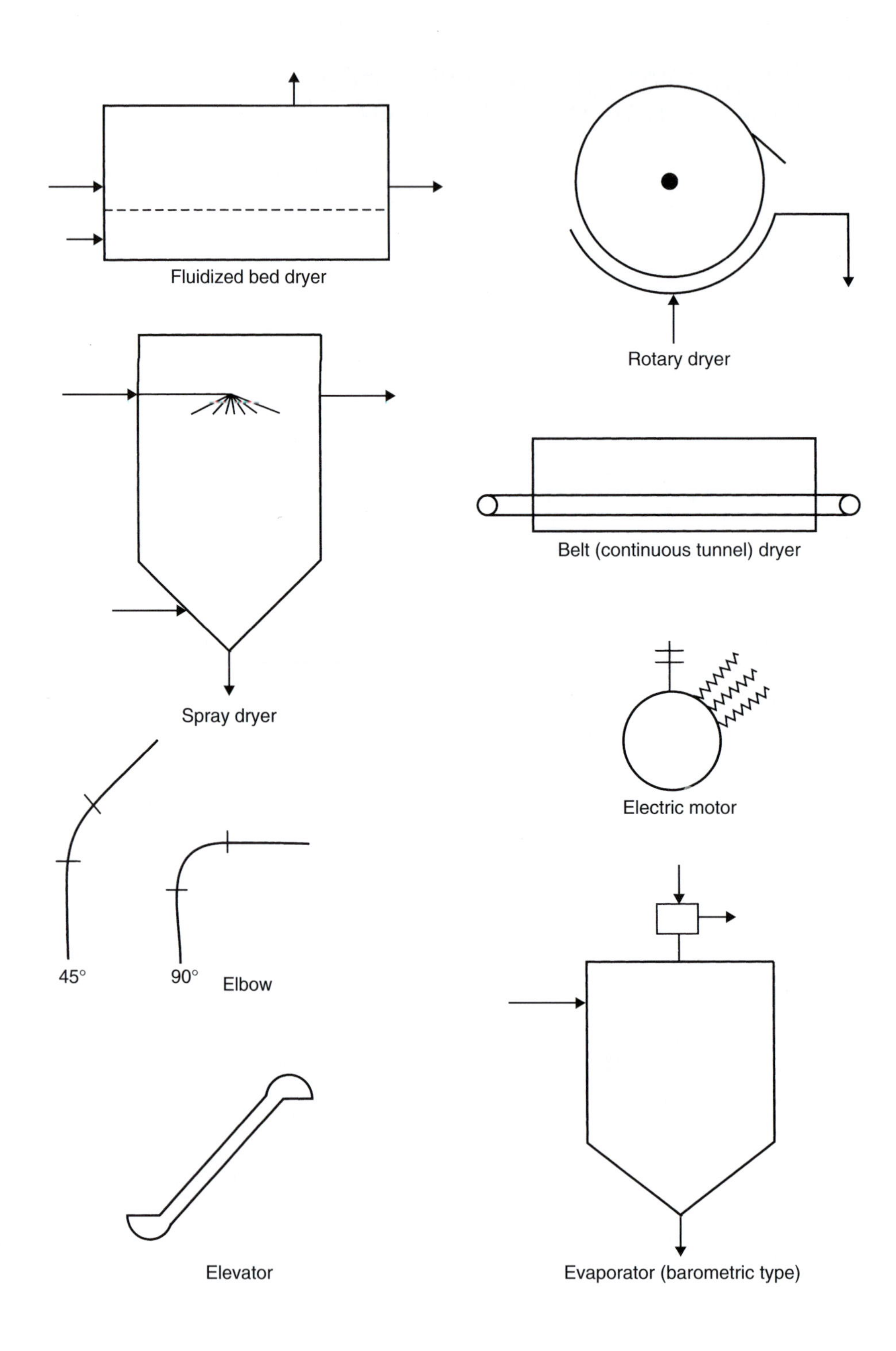
Fluidized bed dryer
Rotary dryer
Spray dryer
Belt (continuous tunnel) dryer
Electric motor
45°
90°
Elbow
Elevator
Evaporator (barometric type)

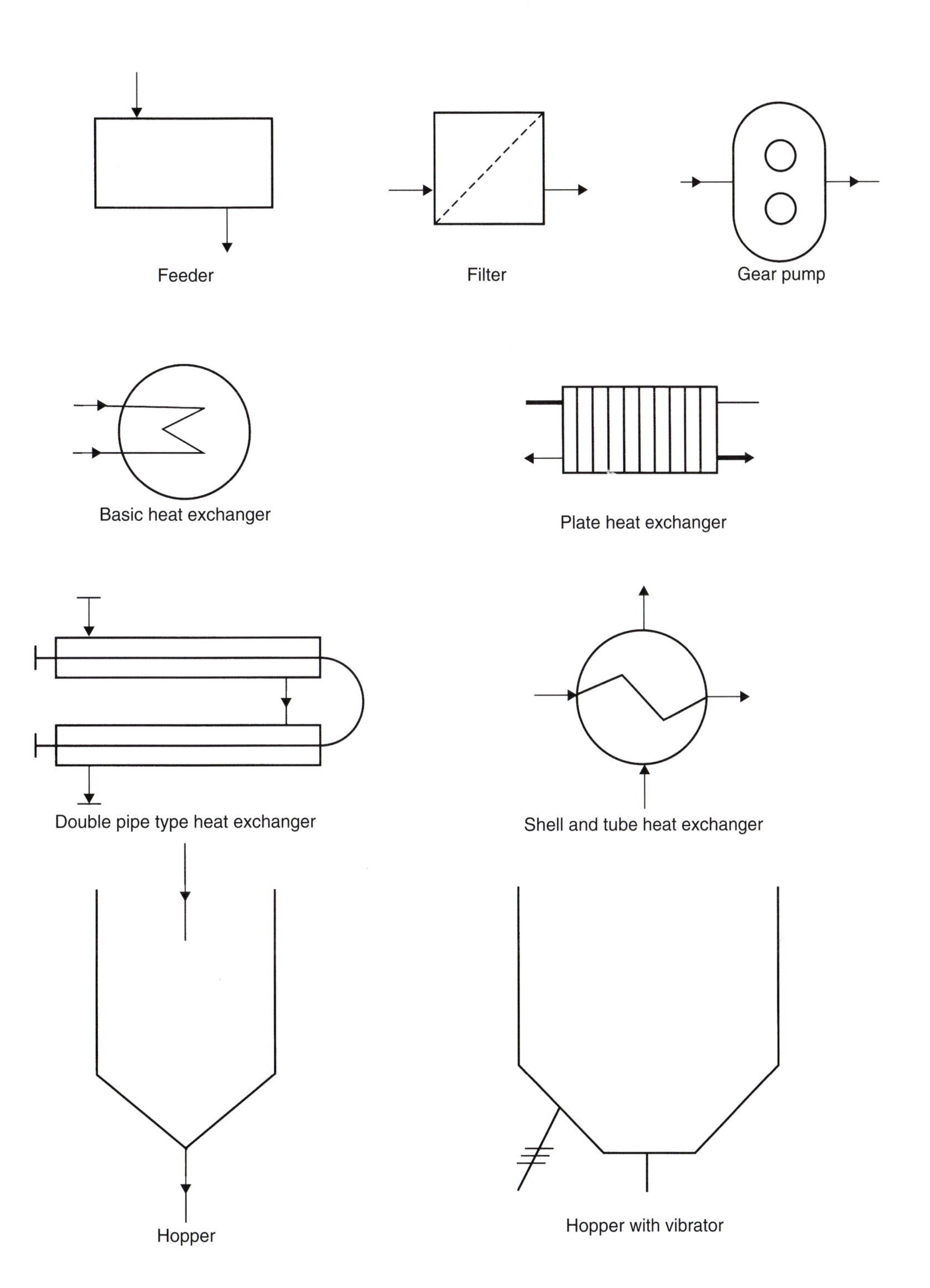
Feeder
Filter
Gear pump
Basic heat exchanger
Plate heat exchanger
Double pipe type heat exchanger
Shell and tube heat exchanger
Hopper
Hopper with vibrator

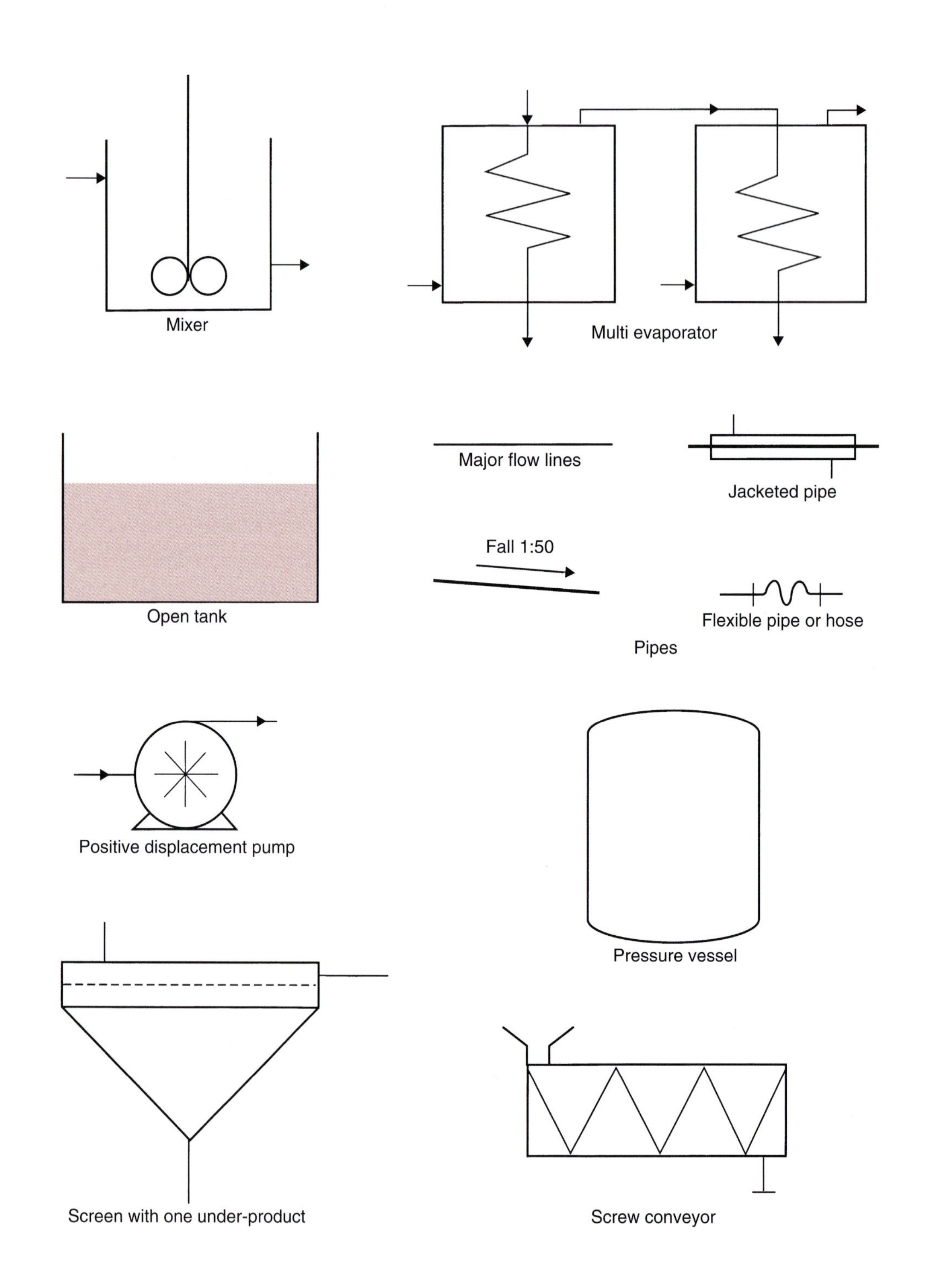
Mixer
Multi evaporator
Open tank
Major flow lines
Jacketed pipe
Fall 1:50
Flexible pipe or hose
Pipes
Positive displacement pump
Pressure vessel
Screen with one under-product
Screw conveyor

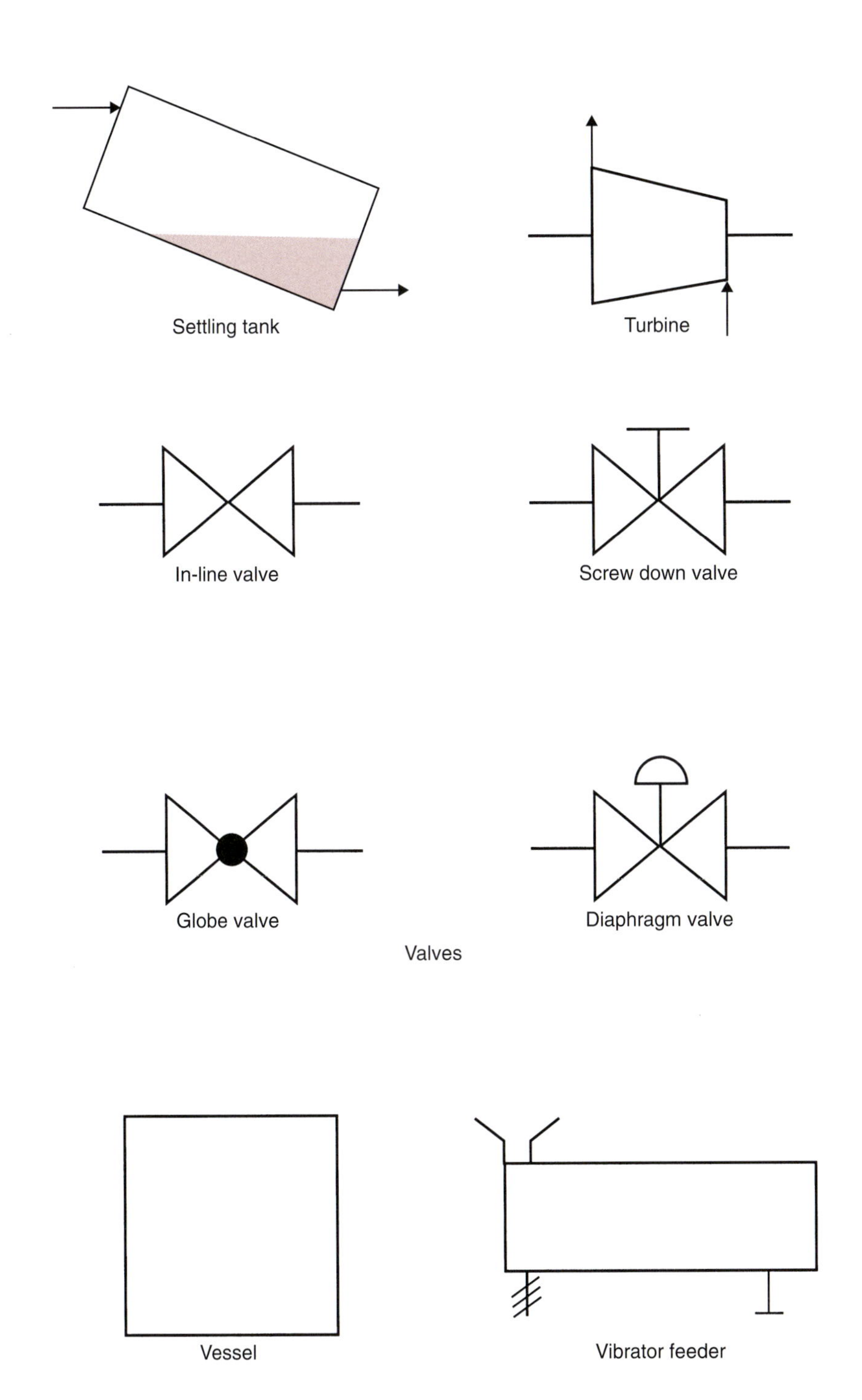
Settling tank
Turbine
In-line valve
Screw down valve
Globe valve
Diaphragm valve
Valves
Vessel
Vibrator feeder

A.8 MISCELLANEOUS

Table A.8.1 Numerical Data and Area/Volume of Objects

Numerical data

$\pi = 3.142$
$e = 2.718$
$\log_e 2 = 0.6931$
$\log_e 10 = 2.303$
$\log_{10} e = 0.4343$

Areas and Volumes

Object	Area/surface area	Volume
Circle, radius r 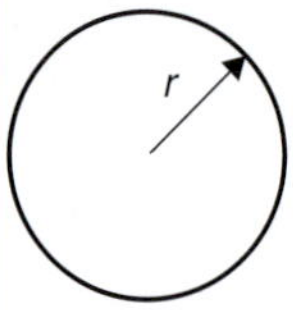	πr^2	– (Circumference = $2\pi r$)
Sphere, radius r 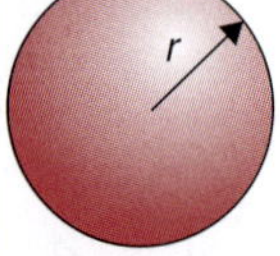	$4\pi r^2$	$\frac{4}{3}\pi r^3$
Cylinder, radius r, height h 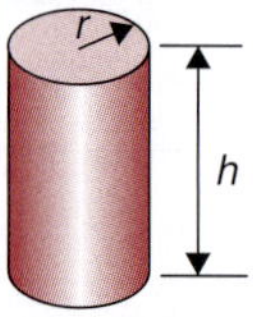	$2\pi r^2 + 2\pi rh$	$\pi r^2 h$
Brick 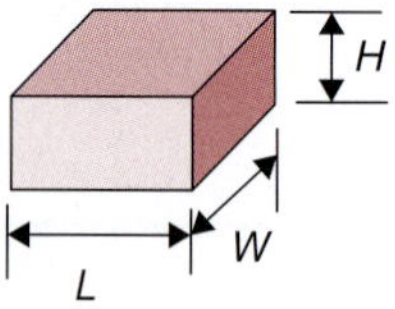	$2(L \times W + W \times H + L \times H)$	$L \times W \times H$

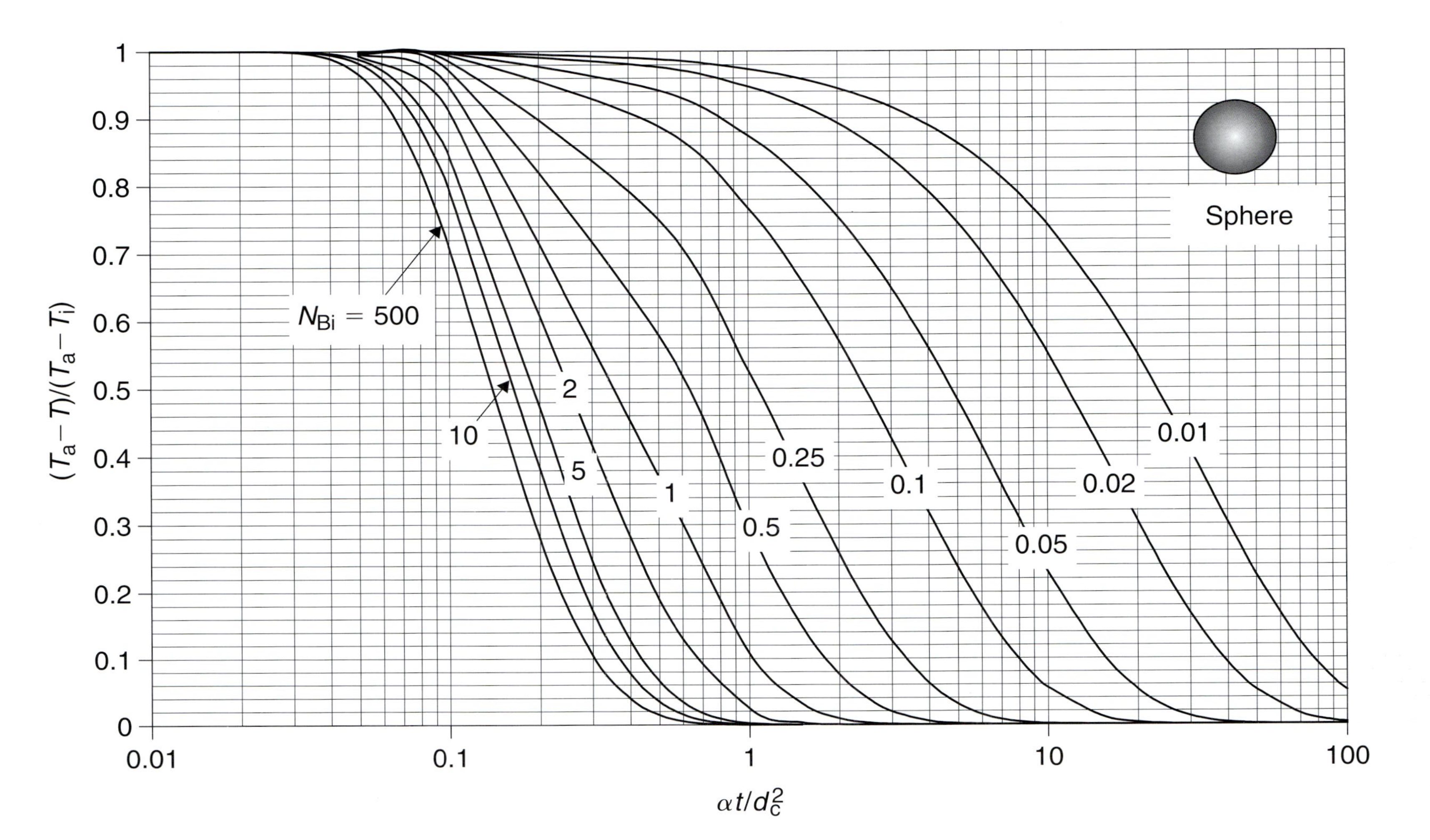

Figure A.8.1 Temperature at the geometric center of a sphere (expanded scale).

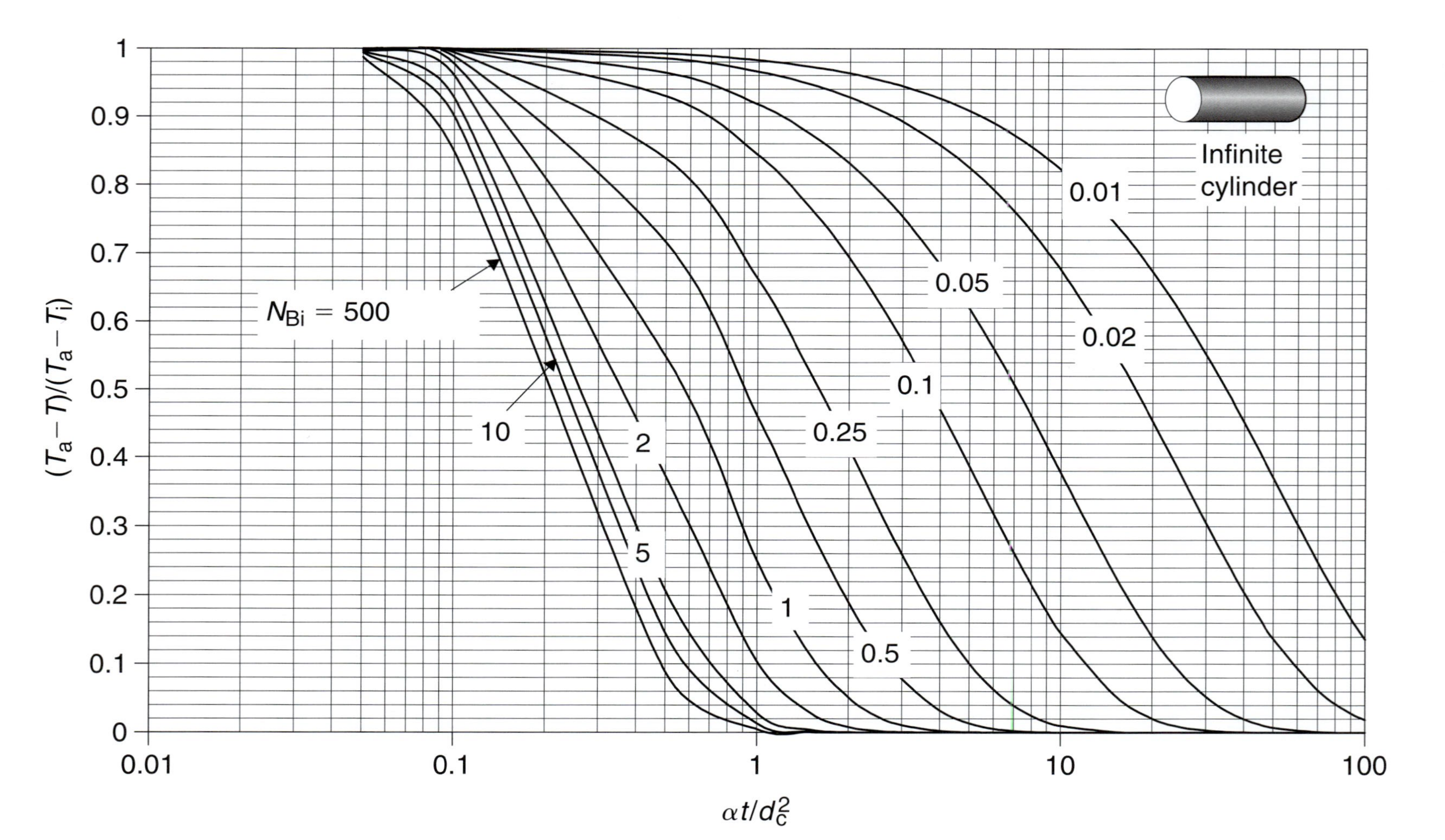

Figure A.8.2 Temperature at the axis of an infinitely long cylinder (expanded scale).

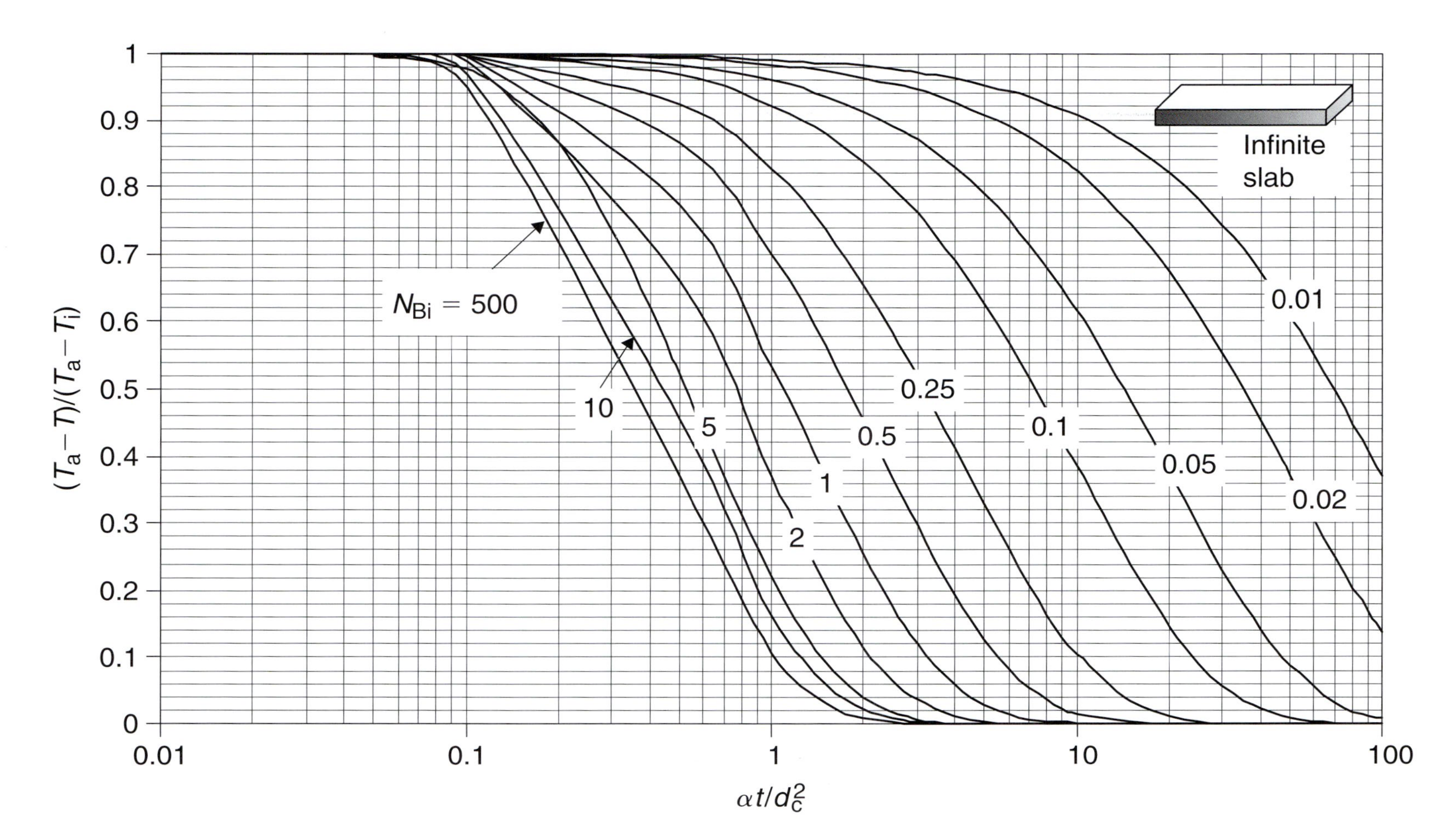

Figure A.8.3 Temperature at the midplane of an infinite slab (expanded scale).

A.9 DIMENSIONAL ANALYSIS

In engineering analysis, it is not uncommon to encounter problems that cannot be solved by using standard analytical procedures. For example, in the case of fluid flow, the highly complex nature of fluid flow in the immediate vicinity of a solid surface prevents us from developing a straightforward analytical solution. Similarly, heat transfer from a solid surface into a fluid is complicated, and analytical procedures to determine the convective heat transfer coefficient are possible only for much simplified situations. To solve these types of problems, we resort to experimental studies.

An experimental approach requires that we first clearly define the given physical system. Experiments are then are conducted to investigate the contribution of various factors that may influence a parameter of interest. It is quite possible that, once all the important factors are identified, the list of such factors may become quite large. The experimental procedure then becomes laborious. For example, consider a scenario where seven variables are identified to influence the convective heat transfer coefficient: fluid velocity, viscosity, density, thermal conductivity, specific heat, characteristic dimension, and axial distance. If we design an experiment that involves conducting trials holding all but one variable constant, and the trials are repeated at four levels of each variable, the number of trials required for this experimental plan will be 4^7 or 16,384 experiments! This will be a daunting task. To substantially reduce the number of experiments, and yet obtain all the necessary information, we use dimensional analysis. As we will see later in this section, with dimensional analysis, the same amount of information is obtained from a significantly reduced number of experiments, namely 64.

A.9.1 Buckingham π Theorem

A mathematically rigorous procedure used in dimensional analysis is the Buckingham π theorem. According to this theorem, the number of independent dimensional groups π_i, associated with a physical phenomenon, are equal to the total number of significant variables, A, minus the number of fundamental dimensions, J, required to define the dimensions of all variables. The relationship among various dimensionless groups is written as:

$$\pi_1 = \text{functions}(\pi_2, \pi_3 \ldots \pi_N) \tag{A.91}$$

Table A.9.1 Dimensions of Selected Experimental Variables

Factors	Symbol	Units (S.I.)	Dimensions
Convective heat transfer coefficient	h	$Js^{-1}m^{-2}K^{-1}$ $kg\,s^{-3}K^{-1}$	$[M/(t^3T)]$
Pipe diameter	D	m	$[L]$
Thermal conductivity	k	$Js^{-1}m^{-1}K^{-1}$ $kg\,m\,s^{-3}K^{-1}$	$[ML/(t^3T)]$
Density	ρ	$kg\,m^{-3}$	$[M/L^3]$
Viscosity	μ	$kg\,m^{-1}s^{-1}$	$[M/(Lt)]$
Velocity	u	$m\,s^{-1}$	$[L/t]$
Specific heat	c_p	$J\,kg^{-1}K^{-1}\,m^2\,s^{-2}K^{-1}$	$[L^2/(t^2T)]$
Entrance length	X	m	$[L]$

Let us use dimensional analysis to develop a correlation useful for estimating the convective heat transfer coefficient for a liquid flowing in a pipe. Assume that the liquid flow is turbulent. From previous knowledge of this system, we can list seven variables that are important in terms of their influence on the convective heat transfer coefficient.

From Table A.9.1, we note that there are eight variables, namely h, D, k, ρ, μ, u, c_p, and X. And, there are four fundamental dimensions, mass $[M]$, length $[L]$, temperature $[T]$, and time $[t]$. According to the Buckingham π theorem, $A = 8$ and $J = 4$, therefore,

$$N = A - J \tag{A.9.2}$$

$$N = 8 - 4 = 4$$

Thus, there are four dimensionless variables. We will call them π_1, π_2, π_3, and π_4.

Next, we will select groups of variables in such a way that in each case the selected group will contain all four fundamental dimensions. Variables D, k, ρ, μ are selected to be common for each group, and the remaining variables h, u, c_p, and X will be added as the last variable in each case. The dimensionless groups are selected as follows:

$$\pi_1 = D^a k^b \rho^c \mu^d h^e \tag{A.9.3}$$

$$\pi_2 = D^a k^b \rho^c \mu^d u^e \tag{A.9.4}$$

$$\pi_3 = D^a k^b \rho^c \mu^d c_p^e \tag{A.9.5}$$

$$\pi_4 = D^a k^b \rho^c \mu^d X^e \tag{A.9.6}$$

Next, we will substitute dimensions for each variable in Equations (A.9.3) through (A.9.6):

$$\pi_1 = [L]^a[M]^b[L]^b[t]^{-3b}[T]^{-b}[M]^c[L]^{-3c}[M]^d[L]^{-d}[t]^{-d}[M]^e[t]^{-3e}[T]^{-e} \tag{A.9.7}$$

For π_1 to be dimensionless, each fundamental dimension must have a power of zero, therefore

$$\text{For dimension } L: \quad a + b - 3c - d = 0 \tag{A.9.8}$$

$$\text{for } M: \quad b + c + d + e = 0 \tag{A.9.9}$$

$$\text{for } t: \quad -3b - d - 3e = 0 \tag{A.9.10}$$

$$\text{for } T: \quad -b - e = 0 \tag{A.9.11}$$

solving the preceding equations in terms of e

$$b = -e$$
$$d = 0$$
$$c = 0$$
$$a = e$$

Using the calculated values for a, b, c, and d in Equation (A.9.3),

$$\pi_1 = (D)^e \, (k)^{-e} \, (h)^e \tag{A.9.12}$$

$$\pi_1 = \frac{(h)^e \, (D)^e}{(k)^e} = (N_{Nu})^e \tag{A.9.13}$$

where π_1 is Nusselt number, $N_{Nu.}$

Equation (A.9.4) written in terms of dimensions

$$\pi_2 = [L]^a[M]^b[L]^b[t]^{-3b}[T]^{-b}[M]^c[L]^{-3c}[M]^d[L]^{-d}[t]^{-d}[L]^e[t]^{-e} \tag{A.9.14}$$

$$\text{for } L: \quad a + b - 3c - d + e = 0 \tag{A.9.15}$$

$$\text{for } M: \quad b + c + d = 0 \tag{A.9.16}$$

$$\text{for } t: \quad -3b - d - e = 0 \tag{A.9.17}$$

$$\text{for } T: \quad -b = 0 \tag{A.9.18}$$

Again solving the previous equations in terms of e

$$b = 0$$
$$d = -e$$
$$c = e$$
$$a = e$$

Therefore $$\pi_2 = (D)^e\,(\rho)^e(\mu)^{-e}(u)^e \qquad \text{(A.9.19)}$$

$$\pi_2 = \left(\frac{D\rho u}{\mu}\right)^e = \left(N_{\text{Re}}\right)^e \qquad \text{(A.9.20)}$$

Equation (A.9.5) written in terms of its dimensions:

$$\pi_3 = [L]^a[M]^b[L]^b[t]^{-3b}[T]^{-b}[M]^c[L]^{-3c}[M]^d[L]^{-d}[t]^{-d}[L]^{2e}[t]^{-2e}[T]^{-e} \qquad \text{(A.9.21)}$$

To make Equation (A.9.21) dimensionless:

$$L: \quad a + b - 3c - d + 2e = 0 \qquad \text{(A.9.22)}$$
$$\text{for } M: \quad b + c + d = 0 \qquad \text{(A.9.23)}$$
$$\text{for } t: \quad -3b - d - 2e = 0 \qquad \text{(A.9.24)}$$
$$\text{for } T: \quad -b - e = 0 \qquad \text{(A.9.25)}$$

Solving the previous equations in terms of e

$$b = -e$$
$$d = e$$
$$c = 0$$
$$a = 0$$

Therefore,

$$\pi_3 = (k)^{-e}\,(\mu)^e(c_p)^e \qquad \text{(A.9.26)}$$

or,

$$\pi_3 = \left(\frac{\mu c_{\text{p}}}{k}\right)^e = N_{\text{Pr}}^e \qquad \text{(A.9.27)}$$

Equation (A.9.6) written in terms of dimensions:

$$\pi_4 = [L]^a[M]^b[L]^b[t]^{-3b}[T]^{-b}[M]^c[L]^{-3c}[M]^d[L]^{-d}[t]^{-d}[L]^e \qquad \text{(A.9.28)}$$

To make Equation (A.9.28) dimensionless,

$$\text{for } L\text{:} \quad a + b - 3c - d + e = 0 \tag{A.9.29}$$

$$\text{for } M\text{:} \quad b + c + d = 0 \tag{A.9.30}$$

$$\text{for } t\text{:} \quad -3b - d = 0 \tag{A.9.31}$$

$$\text{for } T\text{:} \quad -b = 0 \tag{A.9.32}$$

Then solving the preceding equations in terms of e:

$$b = 0 \tag{A.9.33}$$

$$d = 0 \tag{A.9.34}$$

$$c = 0 \tag{A.9.35}$$

$$a = -e \tag{A.9.36}$$

Therefore,

$$\pi_4 = (D)^{-e}(k)^0(\rho)^0(\mu)^0(X)^e \tag{A.9.37}$$

or,

$$\pi_4 = \left(\frac{X}{D}\right)^e \tag{A.9.38}$$

From Equations (A.9.1), (A.9.13), (A.9.20), (A.9.27), and (A.9.38), we obtain

$$N_{Nu} = \text{function } (N_{Re}, N_{Pr}, X/D) \tag{A.9.39}$$

To obtain the functional relationship as suggested in Equation (A.9.39), experiments must be conducted varying each of the three dimensionless numbers. The number of experiments required to obtain the functional correlation, assuming that each number is varied four times, while holding others constant, is $4^3 = 64$; certainly more manageable than our original estimate of 16,400 measurements.

■ BIBLIOGRAPHY

American Society of Agricultural Engineers (1982). *Agricultural Engineers Yearbook*, ASAE, St. Joseph, Michigan.

American Society of Heating, Refrigerating and Air-Conditioning Engineers, Inc. (1978). *Handbook and Product Directory. 1978 Applications.* ASHRAE, Atlanta, Georgia.

Choi, Y. and Okos, M. R. (1986). Effects of temperature and composition on the thermal properties of foods. In *Food Engineering and Process Applications*, Volume 1: Transport Phenomena. M. Le Maguer and P. Jelen, eds., 93–101. Elsevier Applied Science Publishers, London.

Dickerson, R. W., Jr. (1969). Thermal properties of foods. In *The Freezing Preservation of Foods 4th ed*, D. K. Tressler, W. B. Van Arsdel, and M. J. Copley, eds., Volume. 2, 26–51. AVI Publ. Co., Westport, Connecticut.

Heldman, D. R. and Singh, R. P. (1981). *Food Process Engineering*, 2nd ed. AVI Publ. Co., Westport, Connecticut.

Holman, J. P. (2002). *Heat Transfer*, 9th ed. McGraw-Hill, New York.

Keenan, J. H., Keyes, F. G., Hill, P. G., and Moore, J. G. (1969). *Steam Tables—Metric Units*. Wiley, New York.

Raznjevic, K. (1978). *Handbook of Thermodynamic Tables and Charts*. Hemisphere Publ. Corp, Washington, D.C.

Reidy, G.A. (1968). Thermal properties of foods and methods of their determination. M.S. Thesis, Food Science Department, Michigan State University, East Lansing.

Singh, R. P. (1982). Thermal diffusivity in food processing. *Food Technol.* **36**(2): 87–91.

Steffe, J.F. (1983). Rheological properties of liquid foods. ASAE Paper No. 83-6512. ASAE, St. Joseph, Michigan.

Stoecker, W. F. (1988). *Industrial Refrigeration*. Business News Publishing Company, Troy, Michigan.

Index

A

B

C

Page numbers followed by "t" or "f" indicate Tables and Figures within the text.

K

L

M

N

O

R

Food Science and Technology
International Series

Maynard A. Amerine, Rose Marie Pangborn, and Edward B. Roessler, *Principles of Sensory Evaluation of Food*. 1965.

Martin Glicksman, *Gum Technology in the Food Industry*. 1970.

Maynard A. Joslyn, *Methods in Food Analysis*, second edition. 1970.

C. R. Stumbo, *Thermobacteriology in Food Processing*, second edition. 1973.

Aaron M. Altschul (ed.), *New Protein Foods*: Volume 1, *Technology, Part A*—1974. Volume 2, *Technology, Part B*—1976. Volume 3, *Animal Protein Supplies, Part A*—1978. Volume 4, *Animal Protein Supplies, Part B*—1981. Volume 5, *Seed Storage Proteins*—1985.

S. A. Goldblith, L. Rey, and W. W. Rothmayr, *Freeze Drying and Advanced Food Technology*. 1975.

R. B. Duckworth (ed.), *Water Relations of Food*. 1975.

John A. Troller and J. H. B. Christian, *Water Activity and Food*. 1978.

A. E. Bender, *Food Processing and Nutrition*. 1978.

D. R. Osborne and P. Voogt, *The Analysis of Nutrients in Foods*. 1978.

Marcel Loncin and R. L. Merson, *Food Engineering: Principles and Selected Applications*. 1979.

J. G. Vaughan (ed.), *Food Microscopy*. 1979.

J. R. A. Pollock (ed.), *Brewing Science*, Volume 1—1979. Volume 2—1980. Volume 3—1987.

J. Christopher Bauernfeind (ed.), *Carotenoids as Colorants and Vitamin A Precursors: Technological and Nutritional Applications*. 1981.

Pericles Markakis (ed.), *Anthocyanins as Food Colors*. 1982.

George F. Stewart and Maynard A. Amerine (eds), *Introduction to Food Science and Technology*, second edition. 1982.

Hector A. Iglesias and Jorge Chirife, *Handbook of Food Isotherms: Water Sorption Parameters for Food and Food Components*. 1982.

Colin Dennis (ed.), *Post-Harvest Pathology of Fruits and Vegetables*. 1983.

P. J. Barnes (ed.), *Lipids in Cereal Technology*. 1983.

David Pimentel and Carl W. Hall (eds), *Food and Energy Resources*. 1984.

Joe M. Regenstein and Carrie E. Regenstein, *Food Protein Chemistry: An Introduction for Food Scientists*. 1984.

Maximo C. Gacula, Jr. and Jagbir Singh, *Statistical Methods in Food and Consumer Research*. 1984.

Fergus M. Clydesdale and Kathryn L. Wiemer (eds), *Iron Fortification of Foods*. 1985.

Robert V. Decareau, *Microwaves in the Food Processing Industry*. 1985.

S. M. Herschdoerfer (ed.), *Quality Control in the Food Industry*, second edition. Volume 1—1985. Volume 2—1985. Volume 3—1986. Volume 4—1987.

F. E. Cunningham and N. A. Cox (eds), *Microbiology of Poultry Meat Products*. 1987.

Walter M. Urbain, *Food Irradiation*. 1986.

Peter J. Bechtel, *Muscle as Food*. 1986. H. W.-S. Chan, *Autoxidation of Unsaturated Lipids*. 1986.

Chester O. McCorkle, Jr., *Economics of Food Processing in the United States*. 1987.

Jethro Japtiani, Harvey T. Chan, Jr., and William S. Sakai, *Tropical Fruit Processing*. 1987.

J. Solms, D. A. Booth, R. M. Dangborn, and O. Raunhardt, *Food Acceptance and Nutrition*. 1987.

R. Macrae, *HPLC in Food Analysis*, second edition. 1988.

A. M. Pearson and R. B. Young, *Muscle and Meat Biochemistry*. 1989.

Marjorie P. Penfield and Ada Marie Campbell, *Experimental Food Science*, third edition. 1990.

Leroy C. Blankenship, *Colonization Control of Human Bacterial Enteropathogens in Poultry*. 1991.

Yeshajahu Pomeranz, *Functional Properties of Food Components*, second edition. 1991.

Reginald H. Walter, *The Chemistry and Technology of Pectin*. 1991.

Herbert Stone and Joel L. Sidel, *Sensory Evaluation Practices*, second edition. 1993.

Robert L. Shewfelt and Stanley E. Prussia, *Postharvest Handling: A Systems Approach*. 1993.

Tilak Nagodawithana and Gerald Reed, *Enzymes in Food Processing*, third edition. 1993.

Dallas G. Hoover and Larry R. Steenson, *Bacteriocins*. 1993.

Takayaki Shibamoto and Leonard Bjeldanes, *Introduction to Food Toxicology*. 1993.

John A. Troller, *Sanitation in Food Processing*, second edition. 1993.

Harold D. Hafs and Robert G. Zimbelman, *Low-fat Meats*. 1994.

Lance G. Phillips, Dana M. Whitehead, and John Kinsella, *Structure-Function Properties of Food Proteins*. 1994.

Robert G. Jensen, *Handbook of Milk Composition*. 1995.

Yrjö H. Roos, *Phase Transitions in Foods*. 1995.

Reginald H. Walter, *Polysaccharide Dispersions*. 1997.

Gustavo V. Barbosa-Cánovas, M. Marcela Góngora-Nieto, Usha R. Pothakamury, and Barry G. Swanson, *Preservation of Foods with Pulsed Electric Fields*. 1999.

Ronald S. Jackson, *Wine Tasting: A Professional Handbook*. 2002.

Malcolm C. Bourne, *Food Texture and Viscosity: Concept and Measurement*, second edition. 2002.

Benjamin Caballero and Barry M. Popkin (eds), *The Nutrition Transition: Diet and Disease in the Developing World*. 2002.

Dean O. Cliver and Hans P. Riemann (eds), *Foodborne Diseases*, second edition. 2002.

Martin Kohlmeier, *Nutrient Metabolism*, 2003.

Herbert Stone and Joel L. Sidel, *Sensory Evaluation Practices*, third edition. 2004.

Jung H. Han, *Innovations in Food Packaging*. 2005.

Da-Wen Sun, *Emerging Technologies for Food Processing*. 2005.

Hans Riemann and Dean Cliver (eds) *Foodborne Infections and Intoxications*, third edition. 2006.

Ioannis S. Arvanitoyannis, *Waste Management for the Food Industries*. 2008.

Ronald S. Jackson, *Wine Science: Principles and Applications*, third edition. 2008.

Da-Wen Sun, *Computer Vision Technology for Food Quality Evaluation*. 2008.

Kenneth David and Paul Thompson, *What Can Nanotechnology Learn From Biotechnology?* 2008.

Elke K. Arendt and Fabio Dal Bello, *Gluten-Free Cereal Products and Beverages*. 2008.

Debasis Bagchi, *Nutraceutical and Functional Food Regulations in the United States and Around the World*, 2008.

R. Paul Singh and Dennis R. Heldman, *Introduction to Food Engineering*, fourth edition. 2008.

Zeki Berk, *Food Process Engineering and Technology*. 2009.

Abby Thompson, Mike Boland and Harjinder Singh, *Milk Proteins: From Expression to Food*. 2009.

Wojciech J. Florkowski, Stanley E. Prussia, Robert L. Shewfelt and Bernhard Brueckner (eds) *Postharvest Handling*, second edition. 2009.